More information about this series at http://www.springer.com/series/7899

Communications in Computer and Information Science 744

Commenced Publication in 2007
Founding and Former Series Editors:
Alfredo Cuzzocrea, Orhun Kara, Dominik Ślęzak, and Xiaokang Yang

Giacomo Boracchi · Lazaros Iliadis
Chrisina Jayne · Aristidis Likas (Eds.)

Engineering Applications
of Neural Networks

18th International Conference, EANN 2017
Athens, Greece, August 25–27, 2017
Proceedings

 Springer

Editors

Giacomo Boracchi
Dipartimento di Elettronica
   Informazione e Bioingegneria
Politecnico di Milano
Milan
Italy

Lazaros Iliadis
School of Engineering
   Department of Civil Engineering
Democritus University of Thrace
   University Campus
Xanthi
Greece

Chrisina Jayne
School of Computing Science
   and Digital Media
Robert Gordon University
Aberdeen
UK

Aristidis Likas
Department of Computer Science
Univesity of Ioannina
Ioannina
Greece

ISSN 1865-0929   ISSN 1865-0937 (electronic)
Communications in Computer and Information Science
ISBN 978-3-319-65171-2   ISBN 978-3-319-65172-9   (eBook)
DOI 10.1007/978-3-319-65172-9

Library of Congress Control Number: 2017948185

Printed on acid-free paper

This Springer imprint is published by Springer Nature
The registered company is Springer International Publishing AG
The registered company address is: Gewerbestrasse 11, 6330 Cham, Switzerland

Preface

The impact of technology on all areas of science and industry in the first two decades of the 21st century has been enormous. The rise of deep learning has been a milestone for the evolution of artificial neural networks (ANN). Deep ANN using multiple hidden layers are employed offering high performance. In deep networks each layer categorizes some kind of information, then it refines it before passing it to the next one, achieving a hierarchical representation. This way computers can use this technology to teach themselves. "You essentially have software writing software" (Jen-Hsun-Huang, CEO of graphics processing leader Nvidia).

The consequences are enormous in the United States. Equity funding of AI-focused start-ups reached an all-time high of more than U.S. $1 billion [CB Insights research firm]. There were 121 funding rounds for this kind of start-up during the last 3 months of 2016, which is really impressive compared with the 21 for the corresponding quarter of 2011. Google is running several deep learning research projects.

EANN, a well-established event with a very long and successful history, is always following the evolution of AI and moreover it aims at spreading it. Twenty-two years have passed since the first event in Otaniemi, Finland, in 1995. For the following years, it has had a continuous and dynamic presence as a major European scientific event. An important milestone was reached in 2009, when its guidance by a Steering Committee of the INNS (EANN Special Interest Group) was initiated. Thus, from that moment the conference has been continuously supported technically by the International Neural Network Society (INNS).

This CCIS Springer volume contains papers that were accepted for oral presentation at the 18th EANN conference and its satellite workshops. The event was held (August 25–27) in the "ZAFOLIA Hotel," Athens, Greece, and was supported by the Aristotle University of Thessaloniki and the Democritus University of Thrace.

Two workshops on timely AI subjects were organized successfully in the 2017 event:

- The 6th Mining Humanistic Data Workshop (MHDW) supported by the Ionian University and the University of Patras. The 6th MHDW was organized by Professor Christos Makris (University of Patras, Greece), Dr. Andreas Kanavos (University of Patras, Greece), and Phivos Mylonas (Ionian University, Greece). The Steering Committee of the MHDW comprises Dr. Ioannis Karydis, (Ionian University, Greece), Professor Katia Lida Kermanidis (Ionian University, Greece), and Professor Spyros Sioutas (Ionian University, Greece). We wish to express our gratitude to all of these colleagues for their invaluable contribution.
- The second Workshop on 5G-Putting Intelligence to the Network Edge (5G-PINE), which was driven by Dr. Ioannis P. Chochliouros (Hellenic Telecommunications Organization, OTE, Greece), Dr. Leonardo Goratti (Fondazione Bruno Kessler FBK, Italy), Professors Oriol Sallent and Jordi Pérez (Romero Universitat Politècnica de Catalunya UPC, Spain), Dr. Ioannis Neokosmidis (INCITES

Consulting S.A.R.L, Luxembourg), Professor Fidel Liberal (Universidad del Pais Vasco/ Euskal Herriko Unibertsitatea EHU, Spain), Dr. Emmanouil Kafetzakis (ORION Innovations Company, Greece), and Mr. Athanassios Dardamanis (Smartnet S.A., Greece). We would like to thank all of these colleagues for their hard work.

The diverse nature of papers presented demonstrates the vitality of neural computing and related soft computing approaches and proves the very wide range of ANN applications as well.

The Organizing Committee was delighted by the overwhelming response to the call for papers. All papers have passed through a peer-review process by at least two independent academic referees. Where needed, a third referee was consulted to resolve any conflicts. In total, 83 papers were submitted to the main event and 40 of them, around 48% were accepted as full papers in contract to the 5 papers that were accepted as short papers each and to be included in the proceedings with 12 pages maximum. Owing to the high quality of the submissions, the Program Committee decided that it should accept additionally five short papers that will be given 15 minutes for oral presentation and 10 pages each for the proceedings. The workshops also followed the same rules. More specifically, 5G-PINE accepted seven full papers out of 14 submissions, whereas MHDW accepted seven full out of 16 submissions.

The accepted papers of the 18th EANN conference are related to the following thematic topics:

- Spiking ANN
- Ensemble ANN
- Neuro Fuzzy
- Deep ANN
- Theoretical Aspects of ANN
- Agents and Constraints
- Fuzzy Modeling
- Medical ANN
- Feature Selection

- Emotion Recognition
- Hybrid Intelligent models
- Filtering
- Robotics-Machine vision
- Classification-Pattern Recognition
- Cryptography Applications
- Optimization
- Games
- Unsupervised Machine Learning

The authors of submitted papers came from 30 different countries from all over the globe, namely: Europe (Bulgaria, Czech Republic, Denmark, France, Germany, Greece, Italy, Norway, Romania, Spain, Slovakia, Turkey, UK), America (Brazil, Canada, Chile, USA, Mexico), Asia (China, India, Japan, Kazakhstan, Pakistan, Thailand, Taiwan, United Arab Emirates, Vietnam), Africa (Tunisia, Algeria), and Oceania (New Zealand). The authors of the accepted papers came from 15 countries of Europe, Asia, Africa, and America.

Three keynote speakers were invited and they gave lectures on timely aspects of AI and ANN.

- Professor Plamen Angelov from Lancaster University, UK, delivered a talk on "Empirical Data Analytics: Learning Autonomously from Data Streams." He leads the Data Science groups at the School of Computing and Communications of

Lancaster University, which includes over 20 academics, researchers, and PhD students and is one of the eight groups of the school. He is a Fellow of IEEE for contributions to neuro-fuzzy and autonomous learning systems. He is also member of the Board of Governors of the International Neural Networks Society (INNS), and Chair of the Technical Committee on Evolving Intelligent Systems of the IEEE Systems, Man and Cybernetics Society. He has (co-)authored over 200 peer-reviewed publications in leading journals, peer-reviewed conference proceedings, five patents, two research monographs (by Wiley, 2012, and Springer, 2002) and over a dozen other books. These publications have been cited over 5,000 times (Google Scholar) with an h-index of 34. He received a number of IEEE best paper awards (2006, 2009, 2012, 2013) and one of his papers was nominated for outstanding IEEE Transactions paper (2010). He leads numerous projects (including several multimillion ones) funded by UK research councils, EU, industry, UK Ministry of Defense. His research was recognized by The Engineer Innovation and Technology 2008 Special Award and "For Outstanding Services" (2013) by IEEE and INNS. He is also the founding Co-Editor-in-Chief of Springer's journal on *Evolving Systems* and Associate Editor of the leading international scientific journals in this area, including *IEEE Transactions on Cybernetics, IEEE Transactions on Fuzzy Systems* and several other journals including *Applied Soft Computing, Fuzzy Sets and Systems, Soft Computing,* etc. He was general chair of prime conferences (IJCNN 2013; INNS inaugural Conference on Big Data) and PC Co-chair of prime conferences (FUZZ-IEEE 2014, IEEE Intelligent Systems 2014, IJCNN 2016). He has given over a dozen plenary and keynote talks at high-profile conferences. More information can be found on his website www.lancs.ac.uk/staff/angelov.

- Professor Stefanos Kollias from University of Lincoln, UK, delivered a talk on "Developing Performance-Aware Trustful Neural Architectures for Complex Data Analysis." Stefanos Kollias has been the founding professor of Machine Learning in the College of Science of the University of Lincoln since September 2016. He has been Professor with the School of Electrical and Computer Engineering of the National Technical University of Athens since 1997 and Director of the Intelligent Systems, Content and Interaction Laboratory. He is an IEEE Fellow (2015, as suggested by the IEEE Computational Intelligence Society). He has been member of the Executive Committee of the European Neural Network Society (2007–2016). He has world-leading research activity in the fields of machine learning, intelligent systems (with emphasis on artificial neural networks), semantic multimedia analysis, semantic metadata interoperability, and affective computing. He has published over 100 papers in international journals and 300 papers in proceedings of international conferences. His research has been highly referenced (about 8,000 citations with an h-index of 41 in Google Scholar). He has supervised more than 40 PhD students. He has led his group participation in more than 100 European R&D projects, in which his group funding has been more than 20 million euro. He has received the following awards: Fellow in Intelligent Systems by IEEE (2015), Best Learning Game Award for the SIREN system (Games and Learning Alliance Network of Excellence, 2013), Beta Sprint Award for the MINT system (Digital Public Library of America, 2011), as well as several best paper awards at international conferences.

- Professor Wlodzislaw Duch from the Nicolaus Copernicus University, Torun, Poland, delivered a talk on "From Understanding the Brain to Neurocognitive Technologies." He received his MSc degree in theoretical physics (1977), PhD in quantum chemistry (1980), postdoc at University of Southern California (1980–1982), DSc in applied mathematics (1987). He worked at the University of Florida, Max Planck Institute, Kyushu Institute of Technology, Meiji and Rikkyo University in Japan, and several other institutions. Currently, he heads the Neurocognitive Laboratory in the Center of Modern Interdisciplinary Technologies, and the Department of Informatics, both at Nicolaus Copernicus University. During 2014–2015 he served as a Deputy Minister for science and higher education in Poland, and during 2011–2014 as the Vice-President for Research and ICT Infrastructure at his university. Before that he worked as Visiting Professor (2010–2012) in the School of Computer Engineering, Nanyang Technological University, Singapore, where he also worked as a visiting professor during 2003–2007. He is/was on the editorial board of IEEE TNN, CPC, NIP-LR, *Journal of Mind and Behavior*, and 14 other journals. He was also co-founder and scientific editor of the *Polish Cognitive Science* journal. For two terms he served as the President of the European Neural Networks Society executive committee (2006–2011). The International Neural Network Society Board of Governors elected him to their most prestigious College of Fellows. He works as an expert of the European Union science programs; he published over 300 scientific and over 200 popular articles on diverse subjects; he has written or co-authored four books and co-edited 21 books, his DuchSoft company has made GhostMiner software package marketed by Fujitsu company. He is well known for development of computational intelligence (CI) methods that facilitate understanding of data, general CI theory based on similarity evaluation and composition of transformations, meta-learning schemes that automatically discover the best model for a given data.

We hope that these proceedings will help researchers worldwide to understand and to be aware of new ANN aspects. We believe that they will be of major interest for scientists all over the globe and that they will stimulate further research in the domain of ANN and AI in general.

August 2017

Giacomo Boracchi
Lazaros Iliadis
Chrisina Jayne
Aristidis Likas

Organization

General Chairs

Yannis Manolopoulos   Aristotle University of Thessaloniki, Greece
Vera Kurkova   Czech Academy of Sciences, Czech Republic

Organizing chairs

Lazaros Iliadis   Democritus University of Thrace, Greece
Ilias Maglogiannis   University of Piraeus, Greece

Honorary Committee

John MacIntyre   University of Sunderland, UK
Nikola Kasabov   Auckland University of Technology, New Zealand

Program Chairs

Aristidis Likas   University of Ioannina, Greece
Giacomo Boracchi   Politechnico, Milano, Italy
Chrisina Jayne   Robert Gordon University, UK

Advisory committee

Kostas Margaritis   University of Macedonia, Greece
Marley Vellasco   PUC-Rio, Brazil

Workshop Chairs

George Magoulas   University of London, Birkbeck College, UK
Christos Makris   University of Patras, Greece
Spyros Sioutas   Ionian University, Greece

Tutorial Chairs

Bernardette Ribeiro   Universidade de Coimbra, Portugal
Elias Pimenidis   University of West of England
Mario Malcangi   Politecnico di Milano, Italy
Apostolos Papadopoulos   Aristotle University of Thessaloniki, Greece

Publicity Chairs

Simone Scardapane	Sapienza University, Italy
Ioannis Karydis	Ionian University, Greece

Website Chair

Ioannis Karydis	Ionian University, Greece

Award Chair

Vassilis Plagianakos	University of Thessaly, Greece

Program Committee

Michel Aldanondo	University of Toulouse, France
Athanasios Alexiou	Ionian University, Greece
Ioannis Anagnostopoulos	University of Thessaly, Greece
George Anastassopoulos	Democritus University of Thrace, Greece
Costin Badica	University of Craiova, Romania
Rashid Bakirov	Bournemouth University, UK
Zbigniew Banaszak	Warsaw University of Technology, Poland
Ramazan Bayindir	Gazi University, Turkey
Bartlomiej Beliczynski	Warsaw University of Technology, Poland
Kostas Berberidis	University of Patras, Greece
Nik Bessis	Edge Hill University, UK
Monica Bianchini	University of Siena, Italy
Farah Bouakrif	University of Jijel, Algeria
Antônio Pádua Braga	Federal University of Minas Gerais, Brazil
Peter Brida	University of Zilina, Slovakia
Diego Carrera	Politecnico di Milano, Italy
Cristiano Cervellera	National Research Council of Italy, Italy
Ivo Bukovsky	Czech Technical University in Prague, Czech Republic
Anne Magaly de Paula Canuto	Universidade Federal Do Rio Grande Do Norte, Brazil
George Caridakis	National Technical University of Athens, Greece
Ioannis Chamodrakas	National and Kapodistrian University of Athens, Greece
Aristotelis Chatziioannou	National Hellenic Research Foundation, Greece
Jefferson Rodrigo De Souza	FACOM/UFU, Brazil
Kostas Demerztis	Democritus University of Thrace, Greece
Ioannis Dokas	Democritus University of Thrace, Greece
Ruggero Donida Labati	University of Milan, Italy
Javier Fernandez De Canete	University of Malaga, Spain
Maurizio Fiasché	Politecnico di Milano, Italy

Mauro Gaggero	National Research Council of Italy, Italy
Ignazio Gallo	Università dell'Insubria, Italy
Christos Georgiadis	University of Macedonia, Greece
Giorgio Gnecco	Institute for Advanced Studies, Italy
Avrilia Gogetsov	Democritus University of Thrace, Greece
Denise Gorse	University College London, UK
Foteini Grivokostopoulou	University of Patras, Greece
Hakan Haberdar	University of Houston, USA
Petr Hajek	University of Pardubice, Czech Republic
Ioannis Hatzilygeroudis	University of Patras, Greece
Martin Holena	Institute of Computer Science, Czech Republic
Jacek Kabziński	Technical University of Lodz, Poland
Antonios Kalampakas	Democritus University of Thrace, Greece
Achilles Kameas	Hellenic Open University, Greece
Ryotaro Kamimura	Tokai University, Japan
Stelios Kapetanakis	University of Brighton, UK
Ioannis Karydis	Ionian University, Greece
Petros Kefalas	The University of Sheffield International Faculty, CITY College, Greece
Katia Kermanidis	Ionian University, Greece
Muhammad Khurram Khan	King Saud University, Saudi Arabia
Kyriaki Kitikidou	Democritus University of Thrace, Greece
Yiannis Kokkinos	University of Macedonia, Thessaloniki, Greece
Mikko Kolehmainen	University of Eastern Finland, Finland
Petia Koprinkova-Hristova	Bulgarian Academy of Sciences, Bulgaria
Konstantinos Koutroumbas	National Observatory of Athens, Greece
Paul Krause	University of Surrey, UK
Ondrej Krejcar	University of Hradec Kralove, Czech Republic
Adam Krzyzak	Concordia University, Canada
Efthyvoulos Kyriacou	Frederick University, Cyprus
Florin Leon	Gheorghe Asachi Technical University of Iasi, Romania
Spyros Likothanasis	University of Patras, Greece
J.M. Luna	University of Cordoba, Spain
Danlio Macciò	National Research Council of Italy, Italy
Ilias Maglogiannis	University of Piraeus, Greece
George Magoulas	Birkbeck College, UK
Mario Natalino Malcangi	University of Milan, Italy
Francesco Marcelloni	University of Pisa, Italy
Konstantinos Margaritis	University of Macedonia, Thessaloniki, Greece
Nikolaos Mitianoudis	Democritus University of Thrace, Greece
Valeri Mladenov	Technical University Sofia, Bulgaria
Haralambos Mouratidis	University of Brighton, UK
Phivos Mylonas	Ionian University, Greece
Nicoletta Nicolaou	Imperial College London, UK
Stavros Ntalampiras	National Research Council of Italy, Italy

Contents

Deep Learning Convolutional ANN

Deep Learning Image Analysis

Fuzzy - Neuro Fuzzy

Learning Generalization

Recommendation Systems

Robotics and Machine Vision

MHDW2017

5GPINE2017

ANN in Engineering Applications

Motion-Specialized Deep Convolutional Descriptor for Plant Water Stress Estimation

Shun Shibata[1], Yukimasa Kaneda[1], and Hiroshi Mineno[2,3(✉)]

[1] Graduate School of Integrated Science and Technology, Shizuoka University,
3-5-1 Johoku, Naka-ku, Hamamatsu, Shizuoka 432-8011, Japan
shibata@minelab.jp
[2] College of Informatics, Academic Institute, Shizuoka University,
3-5-1 Johoku, Naka-ku, Hamamatsu, Shizuoka 432-8011, Japan
mineno@inf.shizuoka.ac.jp
[3] JST, PRESTO, 4-1-8 Honcho, Kawaguchi, Saitama 332-0012, Japan

Abstract. Mechanical water stress assessment is needed in agriculture to mechanically cultivate high-sugar-content crops. Although previous methods estimate water stress accurately, no method has been practically applied yet due to the high cost of equipment. Thus, the previous methods have a trade-off relationship between cost and estimation accuracy. In this paper, we propose a method for estimating water stress on the basis of plant images and sensor data collected from inexpensive equipment. Specifically, a motion-specialized deep convolutional descriptor (MDCD), which is a novel image descriptor that extracts motion features among multiple sequential images without considering appearance in each image, expresses plant wilt strongly related to water stress. Implicit exclusion of appearance enables extraction of general features of plant wilt, which is insulated from the effect of differences in shapes and colors of places and individual plants. We evaluated the performance of the proposed method using enormous agricultural data collected from a greenhouse. Accordingly, the proposed method reduced the error of mean absolute error (MAE) by approximately 25% compared with a naive convolutional neural network (CNN) using original images. The results show that the MDCD enhances temporal information, while reducing spatial information, and expresses the features of plant wilt appropriately.

Keywords: Water stress · Convolutional neural network · Preprocessing · Feature extraction

1 Introduction

Mechanical reproduction of cultivation techniques is important because cultivation techniques, which largely depend on the experience and intuition of farmers, are being lost as aging farmers retire without any successors in some developed countries. For example, skilled farmers produce high-sugar-content crops by restricting the water given to plants by exploiting the fact that water stress causes the sugar content of fruits to rise. However, advanced irrigation control is essential to provide sufficient water stress throughout the cultivation period because inefficient water stress does not cause

© Springer International Publishing AG 2017
G. Boracchi et al. (Eds.): EANN 2017, CCIS 744, pp. 3–14, 2017.
DOI: 10.1007/978-3-319-65172-9_1

the sugar content of plants to increase but excessive water stress causes plants to wither. Additionally, sufficient water stress changes depending on the climatic environment and plant condition. Thus, it is difficult for beginners who have insufficient experience and knowledge of agriculture to produce crops with high-sugar contents. Therefore, many water stress estimation methods have been developed to enable anyone to appropriately control irrigation [1–4]. In particular, a method based on plant wilt using image processing has attracted attention due to the low price of its equipment. However, the traditional algorithm in image processing tends to use only one image to estimate water stress without considering past images, resulting in insufficient expression of plant motion, which is more strongly related to plant wilt thana plant's appearance in an image.

To overcome this problem, and establish an inexpensive and highly accurate water stress estimation, we propose a method for estimating water stress on the basis of plant images and sensor data collected from inexpensive equipment. Specifically, the main contribution of this work is that we propose a novel image descriptor, motion-specialized deep convolutional descriptor (MDCD), to extract dynamic features of plant wilt appropriately from RGB image shot by general monocular camera.

2 Related Work

2.1 Water Stress Estimation

Water stress is caused by an imbalance between the water plants absorb and the water plants release. Previous methods estimate water stress on the basis of water content in soil, transpiration amount of plants, or water content in plants. The water content in soil, which is closely related to the plants' absorbed water, is measured using a soil moisture meter on the basis of capillary tension of soil water [1] or permittivity of soil [2]. However, many soil moisture meters are required for each soil area because the soil moisture tends to be uneven due to frequently douche. Moreover, high-precision soil meters are expensive for general consumers. Therefore, water stress estimation based on soil moisture content increases the introduction cost in proportion to measurement accuracy.

Meanwhile, the transpiration rate is traditionally estimated on the basis of weight loss measurements [3]. Weight loss accurately corresponds with transpiration when plants are grown sufficiently. However, when plant growth is not sufficient, the estimation error becomes larger because evaporation from the soil increases. Furthermore, since water stress is actually determined by the balance between the transpiration amount and soil moisture content, both the transpiration amount and soil water content or plant water content must be measured to evaluate water stress with higher accuracy.

The climatic environment (i.e., temperature, humidity, and sunlight) affects both plants' absorbed water and released water. For example, evaporation speed from the soil changes dynamically depending on temperature and humidity. In fact, climate data have been used by many applications such as environmental control systems to estimate water stress [5–7]. Moreover, almost all sensors for collecting climate data are inexpensive, and regular farmers can use them toestimate water stress easily. However, water stress estimation using only the climate data has a problem with accuracy because water stress is caused by not only climate but also other elements like plant condition.

Changes in the stem diameter have been shown to effectively indicate water potential because it strongly correlates with water potential [8, 9]. Therefore, methods have been developed to accurately estimate water stress on the basis of the stem diameter [10, 11]. Furthermore, the stem diameter is measured in real-time using sensors such as a laser displacement sensor. However, since an expensive measurement device is needed to measure minute changes of the stem diameter, it is too costly for general farmers to buy easily.

Meanwhile, an indirect water stress estimation method based on plant wilt using image processing has attracted attention because it can be easily used with a normal camera. Since water stress causes plant wilt, appropriately quantifying plant wilt enables water stress to be estimated. Traditional methods using image processing preferred to extract plant wilt features on the basis of plant contour from the image. For example, plant wilt detection based on projected plant area strongly correlates with water stress [4]. However, the traditional algorithm tends to use only one image to estimate water stress without considering past images, resulting in insufficient expression of plant motion, which is more strongly related to plant wilt thana plant's still appearance in an image. For example, if a slight, normal, and large plant motions are defined as "A," "B," and "C," the traditional algorithm does not distinguish between sequences "ABC" and "CCC" because it determines water stress from only a single image, in the current case "C".

2.2 Motion Feature Extraction Using Deep Learning

Convolutional neural networks (CNNs), which are a kind of deep neural network (DNN), are becoming more widespread and sophisticated in the areas of computer vision such as object detection, semantic segmentation, and image generation due to its great representation ability. In particular, CNNs outperform human in the area of action recognition by the extraction of spatiotemporal features from sequence of human images. We believe that the framework of action recognition using CNNs helps to sufficiently express the motion of plant wilt from sequence of plant images. A two-stream network architecture [12, 13] constructed from two different CNNs (temporal ConvNet and spatial ConvNet) classify human actions on the basis of motion features from a sequence of optical flow and appearance features from a still image because human actions are characterized by movement and the moving body region. In particular, through even only temporal ConvNet without spatial ConvNet sufficiently classified human actions, optical flow has been showed to be effective as motion features.

However, convolution operation is applied for each channels of input, and then all the channels are summed for each pixel in the convolution layer of CNN. Therefore, in the temporal ConvNet, which inputs sequence of optical flow as a channel, the temporal information included in each optical flow collapses after the first layer.

A 3D convolutional neural network (3D ConvNet) [14], which reserves temporal information in 3D convolution filters, enables spatiotemporal features of plant wilt to be expressed. Meanwhile, 3D ConvNet requires an enormous amount of data for fully learning because 3D convolution filters include not only spatial parameters but also temporal parameters. The amount of our plant images is much smaller than that of a large-scale image dataset like ImageNet. Furthermore, fine-tuning, which achieves high performance without a large-scale dataset, cannot be used because no existing model learning a similar task to ours has been published on the Internet. Therefore, a 3D ConvNet that learned our relatively small dataset from scratch may over-fitting our data to the training data resulting in a failure to learn.

Although almost all studies on action recognition focus onmovement classification, water stress is a continuous value, so its estimation is a regression problem. To deal with a different regression problem, a recurrent CNN (RCNN) estimates pain intensity from changes of a human face [15]. In particular, the RCNN adopts a frame vector as CNN's input to extract changes in facial expressions over time. The frame vector is a flattened image and the concatenation of the sequential frame vectors enables both temporal information among images and spatial information in a single image to be included in two-dimensional data. Therefore, by using a frame vector as input, a general CNN with a 2D convolutional filter can extract spatiotemporal information without collapsing temporal information while keeping the number of parameters smaller than other previous CNNs for action recognition. Whereas, the flattened operation partially collapses shapes in the image. Appearance in the image is important for many tasks in the computer vision. However, plant shape is not strongly related to water stress. The exclusion of unnecessary information due to flattened operation helps the convergence of neural network. Therefore, frame vector is effective for water stress estimation. Meanwhile, flattened operation keeps the colors original image. Appearance is mainly composed of shape and color. To extract general features of plant wilt, which is insulated from the effect of differences in appearance of places and individual plants, we should exclude both shape and color information constituting the appearance.

3 Proposed Method

3.1 Overview

In this paper, we propose a novel method for water stress estimation with low cost and high accuracy on the basis of plant images and sensor data. The proposed method is composed of three main elements: motion-specialized deep convolutional descriptor (MDCD) for extracting features of plant wilt on image, multi-layer perceptron (MLP) Memory (LSTM) for considering the time series of water stress. Specifically, the MDCD is a novel image descriptor that extracts motion features among multiple sequential images without considering the appearance in each image by using flow vectors, which are extensions of frame vectors. Flow vector implicitly excludes both colors for extracting features of climatic environment on sensor data, and Long Short-Term and shapes while emphasizing motion in original images. The exclusion of spatial information enables extraction of general features of plant wilt, which is insulated from the effect of differences in appearance of places and individual plants.

3.2   Motion-Specialized Deep Convolutional Descriptor

The overview of the MDCD is shown in Fig. 1(a). The MDCD extracts motion features from a temporal image, which is converted from multiple sequential images using special preprocessing. The preprocessing in MDCD consists of the following procedure: optical flow, pooled optical flow, and flow vector. The process is shown in Fig. 1(b). First, an optical flow, which is an image processing technology, is used to recognize the changes in appearance of plants. Optical flow represents momentary movement based on spatiotemporal changes between two images. MDCD adopts one algorithm, deep flow [16], as optical flow. Deep flow extracts dense optical flow and can determine the motion of non-rigid objects. Therefore, the changes of plants are easy to detect robustly in images taken in a greenhouse.

Meanwhile, optical flow tends to involve various noise vectors such as the change of wind conditions and displacement of camera position. If the shooting interval between two images used in optical flow is very short, the effect of the noise vector is negligibly small. However, in our case, since the two images are sequentially taken at few-minute intervals, the optical flow is susceptible to noise vectors. Therefore, we

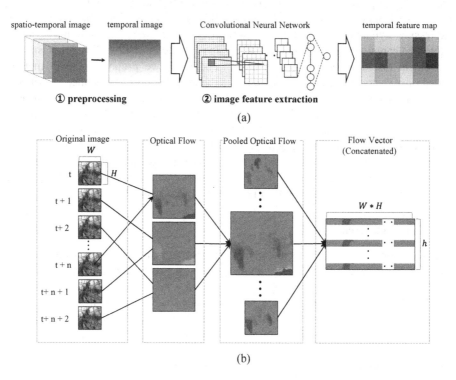

(a)

(b)

Fig. 1. (a) Overview of MDCD. (b) Preprocessing process in MDCD

propose a novel method to remove the noise using plural optical flows, pooled optical flow (POF). POF mainly pools optical flows the same way as CNN. Although a basic pooling processing is a non-linear down sampling method, in our pooling processing, the maximum value for each pixel from plural optical flows is calculated. The pooing processing in POF removes the noise vectors because the movement due to water stress on leaves is greater than the movement caused by wind and the displacement of camera position.

Although plant wilt is a sequence of leaf movements, optical flow is calculated from only two images without considering the movement before or after the images. Therefore, we express a sequence of movements using a flow vector, which is an extension of a frame vector [15]. A frame vector, which is the flattened image, is regarded as vector data containing pixel values of the original image in each dimension, and the adjacent dimension has the spatial locality of the original image. Therefore, the concatenation of sequential frame vectors whose row expresses the variation in value at a specific pixel contains spatiotemporal information. This approach enables motion tasks to be handled with fewer parameters for training the model due to including spatiotemporal information in two-dimensional data. Furthermore, in the case of water stress estimation, the flattened operation helps the convergence of neural network because unnecessary information like plant shapes, which is not strongly related to water stress, are excluded implicitly. However, a frame vector has two problems for water stress estimation. First, a frame vector contains much information irrelevant to plant wilt like plant colors. Learning unnecessary information leads to lower convergence speed and over-fitting to training data. Second, a frame vector has a limitation in expressing dynamic motion features because the concatenation of a frame vector expressly contains the changes of pixel value but not motion information such as movement distance and movement angle. Therefore, we propose a novel method, called flow vector, to express dynamic motion features without appearance information in two-dimensional data by constructing a frame vector using the optical flow instead of an original image. Since optical flow shows movement distance and movement angle in each pixel between adjacent images, each dimension of a flow vector also contains movement distance and movement angle between adjacent original images. Meanwhile, the number of concatenated flow vector is a parameter. Since our shooting interval is 5 min and plants can wither as quickly as about 30 min, we set the parameter to 5. Additionally, we adopt POF to remove the noise instead of optical flow.

3.3   Network Architecture

Figure 2 shows the network architecture for DNN that achieves compatibility between reducing the number of parameters and improving generalization performance. In the case of general tasks like object recognition and face detection, enormous amounts of training data and high-performance models, which are trained with enormous data using appropriate network architecture, are often obtained on the Internet. In our case, since no images of plant wilt are on the Internet, our training data must be collected from scratch. However, it is difficult to gather millions of images, and the training may be unsuccessful due to overfitting with a naive network architecture. Therefore, we

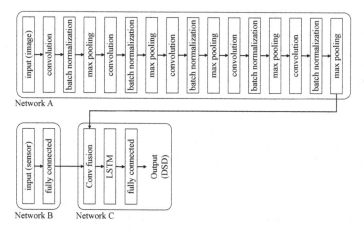

Fig. 2. Network architecture: network A is a 5 layer CNN with 5 convolution layers, network B is a MLP with a hidden layer, and network C has a LSTM layer and a fusion layer.

built a network architecture carefully while considering the correct kind of components and the number of learning parameters to prevent overfitting.

First, in network A, we adopt Pyramid Shape as a network architecture to eliminate redundancy and non-discrimination information while maximally maintaining the network representation ability [18]. Thus, we gradually increase the number of filters to acquire characteristic expressions while eliminating redundancy and non-discrimination information by down sampling. In particular, to reserve as much temporal information as possible until the last convolution layer, we take care of setting the parameters such as the kernel's vertical length and stride's vertical length in both the convolution operation and pooling operation. After the convolutional operation, the features are inputted to a batch normalization layer [19]. Batch normalization improves the generalization performance by adjusting activation distribution to vary it properly. Then, max-pooling is applied to shrink the activation map. The above procedure is repeated five times. Next, network B, which is MLP with a fully connected layer, increases the dimensions of environmental data using non-linear conversion to fuse the data with the image features extracted from network A. Finally, network C integrates image features and environmental features extracted from networks A and B, respectively, in a fusion layer. We adopt Conv fusion [13] as the fusion layer because it can integrate the features between different channels. After fusion, the integrated features are inputted to a LSTM and a fully connected layer to consider the time series of water stress. In networks A, B, and C, initial values of weight parameters are determined on the basis of He initialization [17] for acceleratory learning.

3.4 Definition of Water Stress

In our method, the stem-diameter is measured as the water stress of a dependent variable. Water stress is caused by a deficit of water in a plant, so water content in a plant is revealed by the stem-diameter thickness. Therefore, water stress can be measured indirectly on the basis of the stem-diameter [11]. To continuously measure the stem-diameter non-destructively and without contact, we used a laser displacement sensor that measures the approximate shape of objects on the basis of laser light injected by the sensor.

Although there is a strong relationship between the stem-diameter and water stress, the stem-diameter must not be used as a dependent variable. That is because a stem-diameter continues to increase as the plant itself grows, and the water stress is expressed as the relative change of the stem-diameter. Therefore, we defined the difference in stem-diameter (DSD) as a dependent variable. The DSD is the difference between the current stem-diameter and the greatest value of the observed stem-diameter:

$$DSD_i = \max(SD_0, \ldots, SD_i) - SD_i \tag{1}$$

The maximum is continuously updated as the plant grows. By calculating the amount of decrease from the stem-diameter that can be the thickest, change due to plant growth is ignored, and only the amount of water stress can be quantified from the stem-diameter.

4   Experiment

4.1   Experimental Environment

We developed a system to collect agricultural data in a greenhouse for low density planting of tomatoes at Shizuoka Prefectural Research Institute of Agriculture and Forestry. For a specific tomato seedling, we installed a small outdoor camera (GoPro HERO 4 Session by Woodman Labs) and a laser displacement sensor (HL-T1010A by Panasonic) for the stem-diameter measurement. Moreover, sensors for measuring temperature, humidity, and solar radiation are installed in the greenhouse. Our targeted places were four cultivation beds, where 24 tomatoes were cultivated. In our experiment, nursery trees cultivated in the center of each cultivation bed were targeted to estimate water stress. An overhead view of the experimental environment is shown in Fig. 3. All cameras were attached on iron pipes beside the cultivation beds and installed at places where the center of the taken image was the target plant. Temperature, humidity, and solar radiation sensors were also installed one by one in the upper iron pipe of the target plants. The all datawere collected at a frequency of once every five minutes from August 5 to 25, 2016.

4.2   Experimental Condition

We evaluated the performance of the proposed method using actual agricultural data. In the evaluation, we compared the estimation accuracy of water stress under nine kinds of conditions: three kinds of input data for CNN, with and without environmental sensor data, and with and without LSTM. The conditions are detailed in Table 1. To evaluate generalization performance of each kind of input data, each cultivation bed is different: areas 1 and 2 for training data, area 3 for validation data, and area 4 for testing

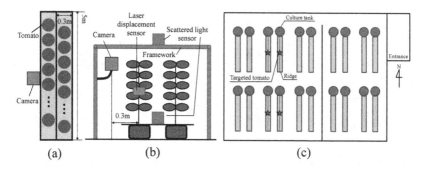

Fig. 3. Experimental environment. (a) Overhead view in one cultivation bed, (b) layout of measurement equipment for one targeted tomato, and (c) overhead view in entire protected

Table 1. Comparative features

Name	Image	Environmental data	LSTM
Org	Original	Unused	Unused
Org-S	Original	Used	Unused
Org-SL	Original	Used	Used
POF	POF	Unused	Unused
POF-S	POF	Used	Unused
POF-SL	POF	Used	Used
MDCD	Flow vector	Unused	Unused
MDCD-S	Flow vector	Used	Unused
MDCD-SL	Flow vector	Used	Used

data. Meanwhile, the important parameters are tuned by using random sampling: learning rate, batch size on mini-batch learning, drop-out rate, and sequence length for LSTM. Finally, used error indicators are mean absolute error (MAE), root mean squired error (RMSE), relative absolute error (RAE), relative squired error (RSE). In particular, when all models are tuned by using validation data, the models that have the lowest MAE were selected as tuned models on the basis of the best parameters.

4.3   Results and Discussion

Figure 4 shows the errors of each comparison for testing data. MDCD-SL obtained the lowest scores for all error indicators: 0.031 MAE, 0.0486 RMSE, 0.792 RSE, and 0.725 RAE. These values show that MDCD-SL effectively estimates water stress. In particular, MDCD-SL with the best features is able to reduce MAE by approximately 25% compared with Org. Furthermore, even MDCD without sensor data and LSTM performed better than Org and POF. Thus, the flow vector, which enhances temporal information while reducingspatial information, expresses the features of plant wilt appropriately. Meanwhile, adding sensor data or LSTM to the MDCD decreases each error sequentially. Climatic environment and time dependency also affect water stress estimation.

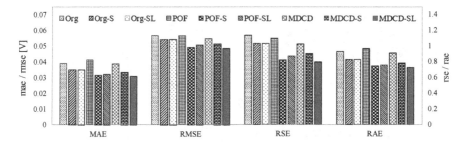

Fig. 4. Error indicators in test data when using each feature for input of DNN.

Figure 5 shows true and estimated values for Org, Org-SL, POF, POF-SL, MDCD, and MDCD-SL in test data. MDCD-SL (Fig. 5(f)) best tracks the characteristics of the true value. Specifically, only MDCD-SL correctly estimated the abrupt changes of the

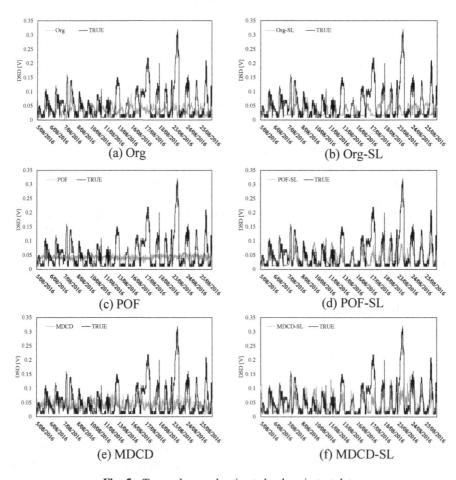

Fig. 5. True values and estimated values in test data

true value like the ascent on 07/08/2016. Meanwhile, when sensor data and LSTM are not used (Fig. 5(a), (c), and (e)), the direction of change from previous estimates is consistent with that of the true value, but the width of change is not. We concluded that the features in the POF and MDCD may have been related to the displacement but not the magnitude of water stress because both the POF and flow vector express relative motion of a plant from a certain point. Therefore, introducing sensor data related to the magnitude of water stress and LSTM to memorize past information contributes to improving estimation accuracy.

5 Conclusion

We proposed a novel method for water stress estimation with low cost and high accuracy on the basis of plant images and sensor data. Specifically, a motion-specialized deep convolutional descriptor (MDCD), which is a novel image descriptor that extracts motion features among multiple sequential images without considering appearance in each image, expresses plant wilt strongly related to water stress. Implicit exclusion of appearance enables extraction of general features of plant wilt, which is insulated from the effect of differences in shapes and colors of places and individual plants. We evaluated the performance of the proposed method using actual agricultural data. In the results, MDCD-SL reduced the error of mean absolute error (MAE) by approximately 25% compared with a naïve CNN using original images. Thus, the flow vector, which enhances temporal information while reducing spatial information, expresses the features of plant wilt appropriately.

In future work, through the evaluation of the general-purpose of the proposed method under various conditions with different fertilizer, water supply, and cultivation method, we consider the explanatory variables for building a model with high generalization ability. Meanwhile, we should evaluate MDCD using typical dataset as well as agricultural data because MDCD is an algorithm not limited to a specific application. Therefore, we build a classifier using features extracted by MDCD on the basis of the dataset for action recognition and compare it with typical action recognition algorithm.

Acknowledgements. This work was supported by JST PRESTO Grant Number JPMJPR15O5, Japan. And, we greatly appreciate Mr. Oishi, Mr. Imahara and Mr. Maejima, Shizuoka Prefectural Research Institute of Agriculture and Forestry.

References

1. Richards, L.A., Gardner, W.: Tensiometers for measuring the capillary tension of soil water. J. Am. Soc. Agron. **28**(1), 352–358 (1936)
2. Topp, G.C., Davis, J.L., Annan, A.P.: Electromagnetic determination of soil water content: measurements in coaxial transmission lines. Water Resour. Res. **16**(3), 574–582 (1980)
3. López-López, R., Ramírez, R.A., Sánchez-Cohen, I., Bustamante, W.O., González-Lauck, V.: Evapotranspiration and crop water stress index in mexican husk tomatoes (Physalis ixocarpa Brot). In: Evapotranspiration–From Measurements to Agricultural and Environmental Applications, p. 187. InTech (2011)

4. Takayama, K., Nishina, H., Iyoki, S., Arima, S., Hatou, K., Ueka, Y., Miyoshi, Y.: Early detection of drought stress in tomato plants with chlorophyll fluorescence imaging–practical application of the speaking plant approach in a greenhouse–. IFAC Proc. (IFAC-PapersOnline) **44**(1), 1785–1790 (2011)
5. Othman, M.F., Shazali, K.: Wireless sensor network applications: a study in environment monitoring system. Procedia Eng. **41**, 1204–1210 (2012)
6. Park, D.H., Park, J.W.: Wireless sensor network-based greenhouse environment monitoring and automatic control system for dew condensation prevention. Sensors **11**(4), 3640–3651 (2011)
7. Ibayashi, H., Kaneda, Y., Imahara, J., Oishi, N., Kuroda, M., Mineno, H.: A reliable wireless control system for tomato hydroponics. Sensors **16**(5), 644 (2016)
8. Huguet, J.G., Li, S.H., Lorendeau, J.Y., Pelloux, G.: Specific micromorphometric reactions of fruit trees to water stress and irrigation scheduling automation. J. Hortic. Sci. **67**(5), 631–640 (1992)
9. Goldhamer, D.A., Fereres, E.: Irrigation scheduling protocols using continuously recorded trunk diameter measurements. Irrig. Sci. **20**(3), 115–125 (2001)
10. Gallardo, M., Thompson, R.B., Valdez, L.C., Fernández, M.D.: Use of stem diameter variations to detect plant water stress in tomato. Irrig. Sci. **24**(4), 241–255 (2006)
11. Wang, X., Meng, Z., Chang, X., Deng, Z., Li, Y., Lv, M.: Determination of a suitable indicator of tomato water content based on stem diameter variation. Sci. Hortic. **215**, 142–148 (2017)
12. Simonyan, K., Zisserman, A.: Two-stream convolutional networks for action recognition in videos. In: NIPS, Montreal, pp. 568–576 (2014)
13. Feichtenhofer, C., Pinz, A., Zisserman, A.: Convolutional two-stream network fusion for video action recognition. In: CVPR, Las Vegas, pp. 1933–1941 (2016)
14. Tran, D., Bourdev, L., Fergus, R., Torresani, L., Paluri, M.: Learning spatiotemporal features with 3D convolutional networks. In: ICCV, Santiago, pp. 4489–4497 (2015)
15. Zhou, J., Hong, X., Su, F., Zhao, G.: Recurrent convolutional neural network regression for continuous pain intensity estimation in video. In: CVPR Workshops, Las Vegas, pp. 84–92 (2016)
16. Weinzaepfel, P., Revaud, J., Harchaoui, Z., Schmid, C.: DeepFlow: large displacement optical flow with deep matching. In: ICCV, Sydney, pp. 1385–1392 (2013)
17. He, K., Zhang, X., Ren, S., Sun, J.: Delving deep into rectifiers: surpassing human-level performance on imagenet classification. In: ICCV, Santiago, pp. 1026–1034 (2015)
18. Smith, L.N., Topin, N.: Deep convolutional neural network design patterns. arXiv preprint arXiv:1609.05672 (2016)
19. Ioffe, S., Szegedy, C.: Batch normalization: accelerating deep network training by reducing internal covariate shift. arXiv preprint arXiv:1502.03167 (2015)

Analysis of Parallel Process in HVAC Systems Using Deep Autoencoders

Antonio Morán[1]([✉]), Serafín Alonso[1], Miguel A. Prada[1], Juan J. Fuertes[1], Ignacio Díaz[2], and Manuel Domínguez[1]

[1] Grupo de investigación en Supervisión, Control y Automatización de Procesos Industriales (SUPPRESS), Esc. de Ing. Industrial e Informática, Universidad de León, Campus de Vegazana s/n, 24071 León, Spain
{a.moran,saloc,ma.prada,jj.fuertes,mdomg}@unileon.es
[2] Dept. de Ing. Elétrica, Electrónica, de Computadores y Sistemas, Universidad de Oviedo, Campus de Viesques s/n, Ed. Departamental 2, 33204 Gijón, Spain
idiaz@uniovi.es
http://suppress.unileon.es

Abstract. Heating, Ventilation, and Air Conditioning (HVAC) systems are generally built in a modular manner, comprising several identical subsystems in order to achieve their nominal capacity. These parallel subsystems and elements should have the same behavior and, therefore, differences between them can reveal failures and inefficiency in the system. The complexity in HVAC systems comes from the number of variables involved in these processes. For that reason, dimensionality reduction techniques can be a useful approach to reduce the complexity of the HVAC data and study their operation. However, for most of these techniques, it is not possible to project new data without retraining the projection and, as a result, it is not possible to easily compare several projections. In this paper, a method based on deep autoencoders is used to create a reference model with a HVAC system and new data is projected using this model to be able to compare them. The proposed approach is applied to real data from a chiller with 3 identical compressors at the Hospital of León.

Keywords: Dimensionality reduction · Information visualization · Data analysis · Deep autoencoder · HVAC systems

1 Introduction

Heating, ventilation and air conditioning (HVAC) systems represent about 50 % of the total consumption in the building sector, being the most energy-consuming

This work was supported in part by the Spanish *Ministerio de Ciencia e Innovación* (MICINN) and the European FEDER funds under project CICYT DPI2015-69891-C2-1-R/2-R.

© Springer International Publishing AG 2017
G. Boracchi et al. (Eds.): EANN 2017, CCIS 744, pp. 15–26, 2017.
DOI: 10.1007/978-3-319-65172-9_2

equipment. It is equivalent to 10–20 % of the final energy consumption in developed countries [1]. Due to that exponential growth of HVAC energy use, policies and regulations focus on promoting energy efficiency of those building systems.

In order to understand how to improve the energy efficiency in buildings, it is necessary to monitor the HVAC systems. The working state of these systems should be analyzed to check the operation and detect malfunctions in those systems [2]. The use of advanced visualization tools can help to improve the efficiency of the systems [3]. However, the main problem creating these visualizations is that HVAC systems may comprise several modules and a vast number of variables each. In addition, these modules are composed of identical machines and elements working in parallel (according to the HVAC stages), so it is expected to have variables with the same evolution, hindering data visualization and comparison. For these reasons, it is necessary to reduce the number of variables, so that the processes can be visualized in an easy way and the visualization is consistent, allowing the data to be compared among the parallel processes.

Tools for visualizing multivariate systems have already been tested previously in order to draw conclusions about the behavior of the process [4]. Nevertheless, one problem of these techniques arises when projecting new or out-of-sample data points from the high dimensional space onto the low dimensional space. In this case, it is required to use specific algorithm modifications or run the algorithm again. Since these algorithms generally use random initialization, if we train again to include the new data the projection output changes, we cannot compare the results between reruns. Furthermore, it is impossible to deduce process values in the projection areas that do not display projected points, since these techniques are not bijective. Thus, it is necessary to use an additional interpolation technique with the projection method, making the creation of maps more complex and obtaining less accurate results [5]. As an example, the tools proposed in [4] combine data projection by means of the dimension reduction techniques such as Isomap, MDS, CCA, etc. [6] and an interpolation technique.

This paper proposes the use of a dimensionality reduction technique (Deep autoencoder) to project data while overcoming the aforementioned issues. Using this technique, it would be possible to project new data without retraining the algorithm, making easier the comparison among projections of different process. Furthermore, the projection algorithm is simpler, because it is a bijective method. Real data from a HVAC system at the Hospital of León, a chiller of $1407\,kW$ comprising 3 identical subsystems (3 compressors), are used to test that dimensionality reduction technique.

This paper is structured as follows: The proposed approach is presented in Sect. 2. In Sect. 3, the testbed is described in detail. The experimental results are analyzed in Sect. 4. Finally, conclusions are drawn in Sect. 5.

2   Methodology

The main goal of our approach is to find a model that enables the projection of process data (composed by a large number of variables) onto a low dimensional space, where conclusions about the process behavior can be drawn in an intuitive manner. That model should allow the projection of new process data without modifying the distribution of the points already projected, in order to achieve a consistent comparison of the new points with the old ones, avoiding the need of model retraining. This kind of projection can be performed by means of *Deep Autoencoders* (DA) [7]. DAs do not only enable the projection of new points, but also define implicitly a *backward projection* from the low dimensional space to the high dimensional input space, allowing to infer input process data from areas of the output space where no previous projected data were available.

2.1 Deep Autoencoders

An autoencoder is a type of neural network in which the output target of the network is set to be equal to the input; i.e., it is an unsupervised learning method which uses a back propagation algorithm for training. The autoencoder has at least three layers: an input layer, a hidden (encoding) layer, and a decoding layer. Since it is trained to reconstruct its inputs, the hidden layer is forced to learn a good representation of the inputs, so if the hidden layer is limited to a fewer number of neurons than that of the input layer, a dimensionality reduction will be performed [8]. Autoencoders have been widely used not only for dimensionality reduction, but also for feature extraction and denoising applications [9].

Despite single-layer NNs have been proved to achieve universal approximation, the number of units required for that purpose might be unfeasibly large and generalization is not guaranteed [10]. It must be noted that the *complexity* of the relationships involved in HVAC systems is one of the elements under consideration, because the efficiency of a HVAC system depends on many factors (internal and external variables).

Deep learning (DL) *models* have turned out to be good at discovering intricate structures in high dimensional data [11]. A deep autoencoder increases the number of hidden layers creating always a symmetric network in which the first half of the net represents the encoding, and the second half represents the decoding [12]. Figure 1 shows the structure of a deep-autoencoder. Choosing the structure of the autoencoder involves selecting the number of hidden layers (depth) and the number of units for each layer (width) and it is not a trivial task.

Recent works [13,14] have showed that deep architectures allow computation of far more complex functions than shallow ones with a similar number of total units. The composition of layers allows identifying an exponentially growing number of input regions for large depths by means of successive space folding mechanisms, thereby enabling complex mappings by reusing pieces of computation. Moreover, the restriction imposed by the rigidity of the folding mechanism implicit in deep networks could be thought as a regularization mechanism that helps in achieving better generalization properties than shallow models [13].

Representation learning is another reason for choosing deep structures. The composition of successive nonlinear mappings that takes place in deep neural networks results in multiple levels of abstraction. The initial layers capture basic features of the input raw data, that are relevant for the problem, and subsequent

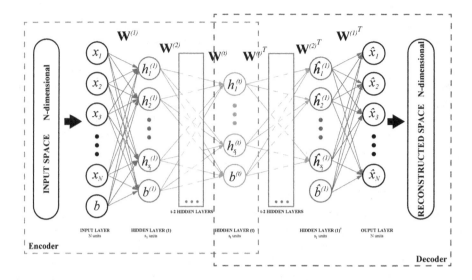

Fig. 1. Deep autoencoder structure

mappings result in more abstract and also problem-relevant features built upon the former ones. In other words, DL networks turn out to learn feature detectors. Moreover, introspection into the internal layers of DL networks has revealed that the extracted features were surprisingly intuitive or meaningful in most cases.

The ability of DL networks in learning complex functions and their capability for representation learning, as argued above, have thereby suggested the use of deep autoencoders in this work for finding meaningful visual representations of parallel processes from the same HVAC system.

2.2   Visual Analysis Using Autoencoders

We propose to project the data acquired from one of the process of a HVAC system, composed of several parallel units, in order to analyze its behavior. Using a set of variables, identical for all parallel units, allows us to obtain a *reference model*. Data from the reference unit are used to train a DA algorithm (see Fig. 2).

The proposed DA implements a "bottleneck" restriction of 2 output units in the *encoder* stage, because the aim is a mapping onto a 2D space. Such restriction forces the model to maximize the information flow in these two units in order to minimize the reconstruction error of the original process data carried out by the *decoder* stage.

- **Encoder:** once trained, this mapping of the high dimensional input space onto a 2D space allows to obtain projections for both *training* process data and new (*test*) process data.

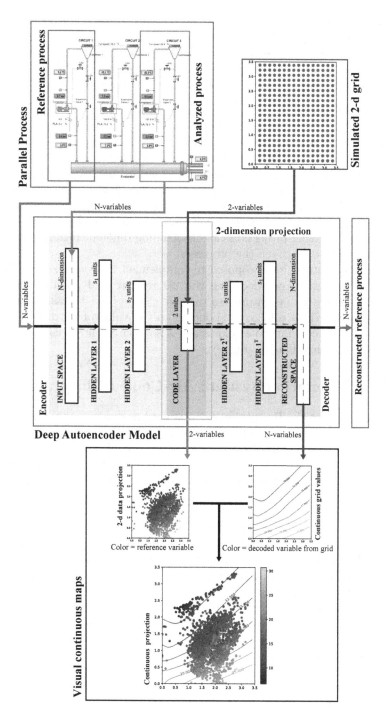

Fig. 2. Methodology

- **Decoder:** this stage maps back 2D projections to high dimensional process data, thereby allowing a reconstruction of process data vectors from the projections. The implicit interpolation that appears in this stage can be used to build visual component planes of the process variables, by applying it to a regular grid in the 2D space that covers the extent of the projections.

This model can be used to achieve an effective visual comparison between different parallel processes. This can be done by obtaining the projection of data from the process unit used for training the model and then obtaining the projection of the other process unit using the same model. Both projections can be compared in the same visualization, considering that two projections lying in a similar region of the visualization space will correspond to similar values of the process variables. Thus, if all projections of two or more parallel process units span the same region in the projection, it reveals that their whole behavior is similar, while having projection clouds spanning different regions shows some kind of dissimilarity between the processes.

To provide context, it is proposed to complement the visualization with the information obtained from decoding the 2D regular grid points. Component planes corresponding to each process variable can be built by assigning to each point of the 2D grid a color corresponding to the value of its reconstructed a process variable, according to a color scale. Also, contour levels can be added to improve the visualization (see Fig. 2).

3   Experimentation System: Air-Cooled Chillers

The chiller plant at the Hospital of León is used as experimental system. Basically, that plant consists of a chilled water production subsystem and a distribution subsystem. Air-cooled and water-cooled chillers can be found in the production subsystem, together with valves, sensors and pumps needed to complement the chiller operation. There are 5 identical air-cooled chillers, comprising 3 internal refrigeration circuits each one, whose data are used as testbed for the approach.

Each air-cooled chiller (model Petra APSa 400-3) has a maximum cooling capacity of 400 tons (approximately 1407 kW) and includes 3 identical and independent refrigeration circuits (see Fig. 3). Each one is composed of a screw compressor, an electronic expansion valve (EEV), and 3 individual condensers in V form. A common evaporator is used for the 3 circuits. The compressor, driven by a three-phase induction motor (400 V; 109 kW), has a maximum displacement of 791 m^3/h of R134a refrigeration gas. Its capacity can be regulated between 50–100 % of maximum value by means of two auxiliary load and unload valves. The condensers have 16 fans of 1.5 kW, driven by variable speed drives. Note that the compressor characterizes the electricity demand of the chiller (because it amounts to approximately 93 % of this demand). Each chiller requires external elements, such as a primary pump, which is driven by a variable speed drive to force water flow through the evaporator. Furthermore, an on/off valve is used to avoid water flow when the chiller is not running.

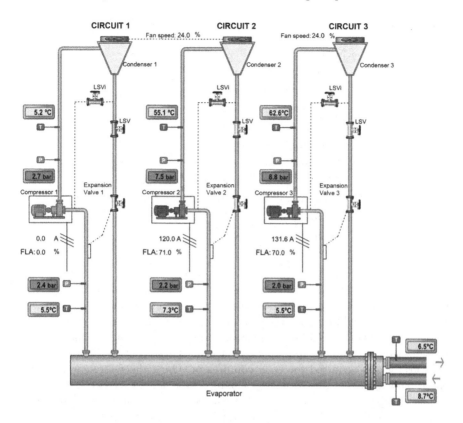

Fig. 3. Air-cooled chiller refrigeration circuits.

The control board acquires and controls several internal variables, being the most important ones listed in Table 1. It communicates with a central controller (Schneider Electric AS) which collects all chiller data using Modbus RTU protocol. Data is structured and stored in a SQLite database. Later, a Python service is used to preprocess raw data and build the training datasets.

4   Experimental Results

To perform the experiments, a *deep autoencoder* has been trained in *Python* using *TensorFlow* [15], which is an open source library capable of building and training neural networks, and *Keras* [16], which is a high-level API running on top of either TensorFlow or Theano. Both libraries together let us program and train an autoencoder in an easy way. In addition, they can run on the *GPU* of a graphics card so that models and the projection of new points can be calculated faster, allowing the use of these techniques in almost real-time.

The data used in the experiments are acquired from a chiller at the Hospital of León, as described in the previous section. A model is obtained from the one

Table 1. Internal variables to control each compressor circuit of the chiller.

Name	Unit
Evaporating temperature	°C
Evaporating pressure	Bar
Condensing temperature	°C
Condensing pressure	Bar
Compressor part load ratio	%
Compressor current	A
Chilled water leaving temperature	°C
Chilled water entering temperature	°C
Fan speed	%
Ambient temperature	°C

of the compressors, which will be known from this moment on as Compressor1. This compressor has been selected as the reference unit, taking into account the information obtained in tests which were made beforehand on the chiller. These tests determined that it was the best calibrated compressor and the most optimal one. The variables used as input space are the ones described in the Table 1. This set of variables is acquired for each compressor. Once the reference model has been trained with the data from the Compressor1, the other two compressors, noted as Compressor2 and Compressor3 will be projected using this model.

The parameters used to train the autoencoder model have been selected manually due to the difficulty of using a method that allows identifying the most satisfactory projection. Although there are indexes such as the continuity and dissimilarity that can be used to measure the effectiveness of a projection [17], the complexity of the autoencoders implies a high number of tests that do not guarantee to obtain an optimal solution. For this reason, several projections are performed with different parameterizations and the one that provides a more intuitive visualization and comparison is selected according to our experience [6], since the final purpose of the projection is to obtain information through the visual analysis.

The resulting autoencoder consists of a 10-dimensional input layer and three intermediate layers of 128, 64 and 2 dimensions respectively. The last two-dimensional intermediate layer is the encoding layer. As for the decoding part of the autoencoder, it is symmetric, as explained before.

Figure 4 shows the result of creating a reference model with the data from Compressor1 and then, applying this model to the other compressors. The point colors are related to the number of compressors that work simultaneously in the chiller, which provide information about the machine working load. The black line marks the density of the model projection (Compressor1), i.e., this line gives the visual information of the contour within which most of the points of the model are projected. If the compressor to be compared works similarly to the

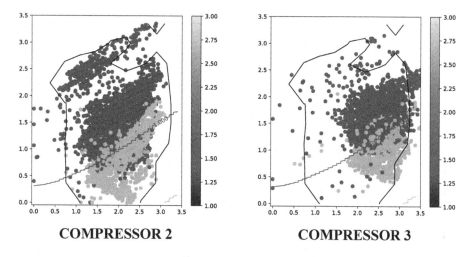

COMPRESSOR 2   **COMPRESSOR 3**

Fig. 4. Data projection of the compressors over the space created with the model of the first compressor. The black line shows the projection area of the reference model (Compressor1).

model, its points are projected within the contour line. Otherwise, it indicates that the compressor does not have the same behavior as the reference one.

Figure 4 shows that Compressor2 is almost entirely projected within the contour, so it can be deduced that its operation is basically the same as that of Compressor1 (the reference one). However, in the case of Compressor3 projection, part of the dots are outside the contour and are projected to the right. Thus, it reveals a clear difference between the operation of Compressor3 and the reference (Compressor1). A thorough study of the chiller configuration proved that both the structure and parameters of the Compressor3 circuit have slight variations with regard to the other compressors. Specifically, the condenser of Compressor3 circuit is 17 % larger and has two fans more than the other circuits.

Figure 5 shows the behavior of several variables of the three compressors. It also shows the projection of the reference model (Compressor1) so that the three circuits can be analyzed and compared in more detail. In these maps, the color is representative of the value of the variable that is visualized (some of the visualized variables as ΔT are calculated from the temperatures used during the projection) and the projection of the points is the result of applying the trained DA to each compressor circuit. The contour lines that appear on the projection come from applying the decoding model to the simulated grid that we used as a reference. The color of these curves has the same meaning as the color of the projected point, so they provide information about the distribution of the visualized variable values through the map.

Observing the projection of the points, we can check, using the maps, the similarity between the Compressor1 and 2, and the difference with the Compressor3. A priori, Compressor3 should take colder refrigeration gas (larger condenser),

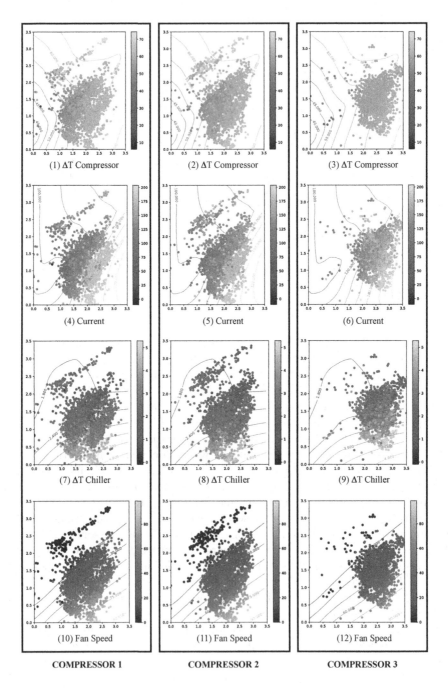

COMPRESSOR 1   **COMPRESSOR 2**   **COMPRESSOR 3**

Fig. 5. Different variables visualized for the three compressors of the chiller. The model is created with the first compressor and the lines shows the model reconstructed in the original space. (Color figure online)

causing lower power demand. However, it can be observed that ΔT and the demanded current in the Compressor3 are a bit higher. The revision of parameters of the fan speed drive allowed us to explain this behavior. Speed limits on that configuration were discovered to try to compensate its associated larger condenser. As seen on Compressor3 projection maps, fan speed is a little lower than in the other two compressors. It can be also seen how the values of the remaining variables are quite similar.

5   Conclusions

This paper presents a new approach to reduce the data from multivariate identical parallel processes so that they can be visualized. The resulting two-dimensional visualizations can be compared in order to check whether the processes are similar, since differences in the working process result in a different projection.

A deep autoencoder is used to reduce the dimension of the data because it provides a series of advantages with respect to other existing techniques. It facilitates the projection of new points and it is also possible to carry out the inverse process obtaining the values of the space input when a grid in the output space is provided. The decoder component of the autoencoder has been used to interpolate in the input space using a simulated grid. This data is used to create a continuous projection on the maps to improve the information visualization. Using this feature, only a single model is needed, instead of having to incorporate additional techniques for interpolate the data. In addition, it is possible to project new points on the created model of the process without retraining the model.

In this paper, the technique was applied to a chiller with three parallel refrigeration circuits which are assumed to be identical. Since these circuits involve several variables, we use a dimension reduction technique to project compressor data and analyze their operation. The internal variables influence on the power demand of the compressor, which typically represents more than 90% of total energy demand in a chiller. Thus, compressor data are important to verify energy efficiency of a chiller.

We showed that it is possible to use deep autoencoders to perform a dimensionality reduction for industrial-oriented data visualization. We also showed that it is possible to make comparisons using the projected data between the model created with one process and the data of the others, checking easily whether the processes behavior are the same or not. It was found, using these projections, that the Compressor3 circuit of the chiller is different from the other two since the coordinates of the projected points are not the same.

As future work, this approach will be applied to the remaining air-cooled chillers at the Hospital of León (5 chillers with 3 compressors each). In addition, this methodology will be improved by using algorithms which make more practical the parameter selection and the training phase of the reference model of the process.

References

1. Perez-Lombard, L., Ortiz, J., Maestre, I.R.: The map of energy flow in HVAC systems. Appl. Energy **88**(12), 5020–5031 (2011)
2. Wang, L., Greenberg, S., Fiegel, J., Rubalcava, A., Earni, S., Pang, X., Yin, R., Woodworth, S., Hernandez-Maldonado, J.: Monitoring-based HVAC commissioning of an existing office building for energy efficiency. Appl. Energy **102**, 1382–1390 (2013)
3. Meyers, S., Mills, E., Chen, A., Demsetz, L.: Building data visualization for diagnostics. ASHRAE J. **38**(6), 8 (1996)
4. Morán, A., Fuertes, J.J., Prada, M.A., Alonso, S., Barrientos, P., Díaz, I., Domínguez, M.: Analysis of electricity consumption profiles in public buildings with dimensionality reduction techniques. Eng. Appl. Artif. Intell. **26**(8), 1872–1880 (2003)
5. Van Der Maaten, L., Postma, E., Van den Herik, J.: Dimensionality reduction: a comparative. J. Mach. Learn. Res. **10**, 66–71 (2009)
6. Lee, J.A., Verleysen, M.: Nonlinear Dimensionality Reduction. Springer Publishing Company, Heidelberg (2007). doi:10.1007/978-0-387-39351-3
7. Hinton, G.E., Salakhutdinov, R.R.: Reducing the dimensionality of data with neural networks. Science **313**(5786), 504–507 (2006)
8. Wang, Y., Yao, H., Zhao, S.: Auto-encoder based dimensionality reduction. Neurocomputing **184**, 232–242 (2016)
9. Deng, L., Yu, D.: Deep learning: methods and applications. Found. Trends Sig. Process. **7**(3–4), 197–387 (2014)
10. Goodfellow, I., Bengio, Y., Courville, A.: Deep Learning. MIT Press, Cambridge (2016)
11. LeCun, Y., Bengio, Y., Hinton, G.: Deep learning. Nature **521**(7553), 436–444 (2015)
12. Baldi, P.: Autoencoders, unsupervised learning, and deep architectures. In: Proceedings of ICML Workshop on Unsupervised and Transfer Learning, vol. 27, pp. 17–49 (2012)
13. Montufar, G.F., Pascanu, R., Cho, K., Bengio, Y.: On the number of linear regions of deep neural networks. In: Advances in Neural Information Processing Systems, pp. 2924–2932 (2014)
14. Pascanu, R., Montúfar, G., Bengio, Y.: On the number of inference regions of deep feed forward networks with piece-wise linear activations. CoRR, abs/1312.6098 (2013)
15. Abadi, M., Agarwal, A., Barham, P., et al.: TensorFlow: Large-Scale Machine Learning on Heterogeneous Systems (2015). http://tensorflow.org/
16. Chollet, F.: Keras library. GitHub repository (2015). https://github.com/fchollet/keras
17. Venna, J., Kaski, S.: Comparison of visualization methods for an atlas of gene expression data sets. Inf. Vis. **6**, 139–154 (2007)

A Neural Network Approach for Predicting the Diameters of Electrospun Polyvinylacetate (PVAc) Nanofibers

Cosimo Ieracitano[1(⊠)], Fabiola Pantò[2], Patrizia Frontera[1], and Francesco Carlo Morabito[1]

[1] DICEAM, University Mediterranea of Reggio Calabria, Reggio Calabria, Italy
cosimo.ieracitano@unirc.it
[2] DIIES, University Mediterranea of Reggio Calabria, Reggio Calabria, Italy

Abstract. This study focuses on the design of a Neural Network (NN) model for the prediction of interpolated values of polyvinylacetate (PVAc) nanofiber diameters produced by the electrospinning process and it supposes to be a preliminary work for future and industrial applications. The experimental data gathered from the literature form the basis for generating a more consistent sample through standard interpolation. The inputs of the NN are the polymer concentration, the applied voltage, the nozzle-collector distance and the flow rate parameters of the process, whereas the average diameter acts as the unique output of the network. The generated model is able to approximate the mapping between process parameters and fiber morphology, which is of practical importance to help prepare homogeneous nano-fibers. The reliability of the model was tested by 7-fold cross validation as well as leave-one-out method, showing good performance in terms of both average RMSE (0.109, corresponding to 138.51 nm) and correlation coefficient (0.905) between the desired and the predicted diameters when a White Gaussian Noise with 2% power ($WGN_{2\%}$) is applied to the interpolations.

Keywords: Neural networks · Electrospinning · PVAc · Nanomaterials · Material informatics

1 Introduction

Electrospinning is the process of production of thin and continuous filaments (nano-fibers) of polymer solutions suitable in several applications [1–3]. It requires a great deal of effort on preparation of solution, expensive equipment, accurate working conditions and time consuming also for just one experiment. Recently, researchers have been applying Neural Networks (NNs) [4] for predicting the nanofiber average diameter produced by the electrospinning process, saving time and cost of the experiments. NNs are computational systems inspired by the human structure of the brain, able to learn from data. Mathematical models, called processing elements (PEs), are arranged in interconnected layers and synaptic weights are modified through the learning process to obtain the desired output. Carrera et al. [5, 6] employed a learning algorithm based on SEM images detection to predict diameter, porosity and

© Springer International Publishing AG 2017
G. Boracchi et al. (Eds.): EANN 2017, CCIS 744, pp. 27–38, 2017.
DOI: 10.1007/978-3-319-65172-9_3

defectiveness, and for the optimization of production parameters. Sakar et al. [7, 8] modelled aNN for predicting the diameter of electrospun polyethylene oxide (PEO) aqueous solution. They provided a nonlinear mapping between the four inputs (concentration, conductivity, flow rate, and electric field) and the nanofiber diameter, validated through a six-fold cross validation approach, estimating high prediction performance (93.19% accuracy) and high correlation value (r = 0.999). Faridi-Majidi et al. [9] trained a single hidden layer neural network for the prediction of the diameter of the electrospun nanofibers of Nylon 6,6, showing that the injection rate and concentration mostly affected the nanofiber diameter. Naghibzadeh et al. [10] used the polymer (gelatin) concentration, the solvent (acetic acid) concentration, the applied voltage and the temperature as input of a NN for predicting the diameter of gelatin nanofiber, obtaining acceptable performance: MSE of 0.1531 and coefficient regression of 0.9424. Vatankhah et al. [11] focused on the mechanical properties of scaffolds for tissue engineering (TE) [12] and modelled a single layer NN for investigating the influence of the composition, fiber diameter and orientation on the elastic modulus of electrospun polycaprolactone (PCL)/gelatin nanofiber. Karimi et al. [13] and Ketabchi et al. [14] modelled Multi-Layer Neural Networks to predict the mean diameter of polyvinylalcohol/chitosan (PVA/CS) and chitosan/polyethylene oxide (CS/PEO) nanofibers, respectively, proving the efficiency of the models, whereas, Brooks and Tucker [15] trained Artificial Neural Networks for prediction and classification purposes, basing on available literature data. In the past, NNs were compared with the response surface methodology (RSM) for predicting the average diameter of polyacrylonitrile (PAN) [16, 17], polymethyl methacrylate (PMMA) [18] and polyurethane (PU) [19] nanofibers concluding that the NN model outperformed the RSM technique. Nateri and Hasanzadeh [20] compared polynomial regression, neural network and fuzzy logic techniques to evaluate the diameter of polyacrylonitrile (PAN) nanofiber estimating a prediction error of 0.27, 0.13 and 0.22, respectively.

In the above discussed works, learning models have been trained only on experimental data, which is expensive. However, NN can be useful to generate interpolated models provided the database is suitably validated. The purpose of this work is accordingly to model a NN able to predict the nanofiber average diameter from both experimental and interpolated values. Several experimental dataset sexist in literature [21–24]: the experiments carried out by Park et al. [24] on polyvinylacetate (PVAc) polymer were chosen here as experimental dataset and were used to calculate the interpolated data. A white gaussian noise with 2% power (WGN$_{2\%}$) was added to the interpolations and the NN proposed was firstly tested on data without noise and then on noisy data, providing good performances in both cases.

The paper is organized as follow: Sect. 2 will introduce the electrospinning process and the material under analysis; Sect. 3 will describe the interpolation method; Sect. 4 will explain the ANN proposed; Sect. 5 will illustrate the results and Sect. 6 will address the main conclusion.

2 Electrospinning Process and Material

2.1 Electrospinning Process

The electrospinning process is schematized in Fig. 1. It mainly consists of a high voltage power supplier, a syringe pump and a metal screen (collector). The polymer solution is initially placed into a syringe and it is delivered to the tip of the metallic needle with a pump at a constant flow rate until a small droplet is formed. A high voltage is applied between the metal capillarity and the grounded collector. As the electric field increases, the hemispherical droplet deforms creating a conical shape (Taylor cone). When the electrostatic force overcomes the surface tension, a charged jet is ejected out from the Taylor cone [25]. Meanwhile, the jet moves towards the collector, the solvent evaporates and thin fibers are deposited on the collector plane.

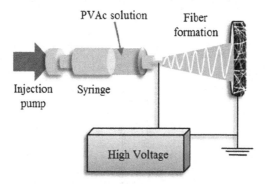

Fig. 1. Electrospinning setup

The morphology and the diameter of the nanofibers produced by the electrospinning process described above, depend on several factors such as polymer solution, processing and ambient conditions [26].

2.2 Material

The dataset used to design the learning model was derived from the literature, Park et al. [24]. They studied the morphology of polyvinylacetate (PVAc, Mn 140,000) dissolved in ethanol solvent studying the effects of concentration (X_1), applied voltage (X_2), tip-collector distance (TCD, X_3) and flow rate (X_4). The experiments were carried out varying one parameter at a time (e.g. X_1) under the same conditions of the remaining parameters (e.g. X_2, X_3, X_4), for a grand total of 19 experiments. Park and colleagues observed good electrospunPVAc diameters (Y) between 10–25 wt% polymer concentration (when $X_2 = 15$ kV, $X_3 = 10$ cm, $X_4 = 100\,\mu L/min$); 10–17.5 kV applied voltage (when $X_1 = 15$ wt%, $X_3 = 10$ cm, $X_4 = 100\,\mu L/min$ and when $X_1 = 20$ wt%, $X_3 = 10$ cm, $X_4 = 100\,\mu L/min$); 7.5–17.5 cm TCD (when $X_1 = 15$ wt%, $X_2 = 15$ kV, $X_4 = 100\,\mu L/min$); and 50–1000 $\mu L/min$ flow rate (when $X_1 = 15$ kV, $X_2 = 10$ cm, $X_3 = 10\,\mu L/min$).

3   Methodology

As the size of the dataset [24] was too small, it was actually not suitable for training a neural network, thus we need to augment the available data. A simple way of increasing the dataset dimension is applying an interpolation among data. In this study, a standard linear interpolation was used. Given two known measurements $\{x_k, y_k\}$, $\{x_{k+1}, y_{k+1}\}$, the unknown y* value corresponding to the new x* measurement ranged between x_k and x_{k+1} is obtained by the formula:

$$y^* = y_k + \frac{(x^* - x_k)(y_{k+1} - y_k)}{(x_{k+1} - x_k)} \tag{1}$$

In this study, x_k, x_{k+1} are two consecutive measurements of the same variable (e.g. X_1) under the same conditions of the remaining parameters (e.g. X_2, X_3, X_4); y_k, y_{k+1} are

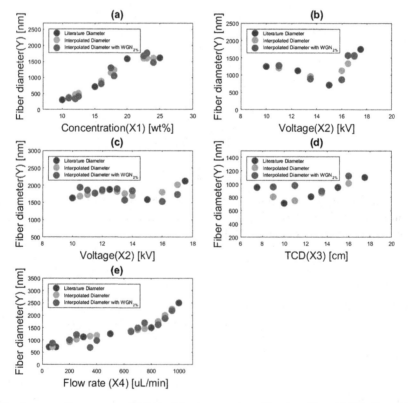

Fig. 2. Average diameters at different: (a) concentrations X_1 when $X_2 = 15$ kV, $X_3 = 10$ cm, $X_4 = 100$ µL/min; (b) voltages when $X_1 = 15$ wt% and $X_3 = 10$ cm, $X_4 = 100$ µL/min; (c) voltages when $X_1 = 20$ wt% and $X_3 = 10$ cm, $X_4 = 100$ µL/min (d) TCD when $X_1 = 15$ wt%, $X_2 = 15$ kV, $X_4 = 100$ µL/min; (e) flow rates when $X_1 = 15$ kV, $X_2 = 10$ cm, $X_3 = 10$ µL/min. Blue dots refer to the desired diameters, yellow dots to the interpolated diameters, red dots to the interpolated diameters with $WGN_{2\%}$. (Color figure online)

the average diameters corresponding to x_k, x_{k+1} respectively; x^* is the new working condition chosen arbitrarily (between x_k and x_{k+1}); y* is the interpolated value of the fiber diameter. In order to avoid linear trends that could be not realistic in the electrospinning process and to assess the stability of the simple model proposed, a white gaussian noise with 2% power ($WGN_{2\%}$) was added to the interpolated data:

$$\tilde{y} = y^* + WGN_{2\%} \tag{2}$$

Through interpolation, the original dataset was augmented from 19 to 57 measurements. The literature diameters (blue marker), the interpolated diameters (yellow marker) and the interpolated diameters with $WGN_{2\%}$ (red marker) vs. the input variables are showed in Fig. 2.

4   Neural Networks for Nano-Material

4.1   Neural Network Architecture

According to Kolmogorov's theorem [27], Neural Networks with a unique hidden layer are able to estimate any function to any degree of accuracy, so a feed-forward fully connected NN with a single hidden layer was modeled using NeuralWorks Professional II tool and executed on aHP xw4600 workstation with 12 GB RAM. The input variables included polymer concentration, voltage, nozzle-collector distance and flow rate. The output of the network was the average fiber diameter. The optimal architecture was chosen modifying the hidden layer size from 4 to 20 neurons and evaluating the RMSE, as shown in Fig. 3. The minimum RMSE was reached with 12 neurons (NN_{12}) but, as discussed in the Results, since the NN with 8 neurons (NN_8)

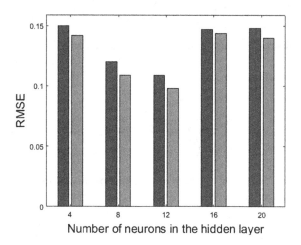

Fig. 3. RMSE value when the number of neurons in the hidden layer is: 4, 8, 12, 16, 20. Red bars refer to NNs performed with noisy data; whereas, yellow bars refer to NNs performed without noise. (Color figure online)

showed good performances and since having few hidden neurons reduce the complexity and number of parameters to evaluate, the NN_8 was chosen as optimal architecture. The hyperbolic tangent was used as transfer function for the hidden and output layers. A straight "in-out" link was added so that each input neuron was connected to the output directly, as showed in Fig. 4. Alternatively, a quasi-sigmoidal model has been used for the hidden layer, which is able to automatically separate the linear from the nonlinear part of the mapping [28]. In this study, the backpropagation (BP) was applied as learning algorithm and the k-fold cross validation and the leave-one-out methods were used to evaluate the validity of the model.

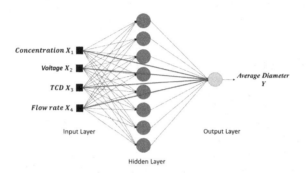

Fig. 4. Artificial neural network architecture

The Root Mean Square Error (RMSE) and the Pearson correlation coefficient (r) between the desired data (including literature [24] and interpolated values) and the predicted data were estimated. The standard RMSE is defined as follow:

$$RMSE = \sqrt{\frac{1}{n} \sum_{i=1}^{n} (y_i - \tilde{y}_l)^2} \tag{3}$$

where y is the desired average diameter; $\tilde{y}$ is the predicted diameter; n is the total number of data.

The Pearson coefficient measures the efficiency of the prediction model and is defined as follow:

$$r = \frac{n * \sum_{i=1}^{n} y_i * \tilde{y}_l - \sum_{i=1}^{n} y_i \sum_{i=1}^{n} \tilde{y}_l}{\sqrt{[n * (\sum_{i=1}^{n} y_i^2) - (\sum_{i=1}^{n} y_i)^2] * [n * (\sum_{i=1}^{n} \tilde{y}_l^2) - (\sum_{i=1}^{n} \tilde{y}_l)^2]}} \tag{4}$$

It is ranged between [−1, 1]: r = 1 indicates perfect positive correlation between y and $\tilde{y}$; r = −1 perfect negative correlation; r = 0 no correlation.

k-Fold Cross Validation Method
The dataset is randomly divided into k partitions, or folds, and k NNs are modelled. Each partition includes m samples and is used as testing set. The remaining k−1 partitions are used as training set [29]. In this study, the dataset was randomly splitted

into k = 7 subsets, so 7 ANNs with the architecture described above, were trained and tested. For each ANN, the normalized Root Mean Square Error (RMSE) and the correlation coefficient (r) were estimated on the testing set of both architectures (NN_8 and NN_{12}) as shown in Table 1.

Table 1. Normalized RMSE and r coefficient of NN_8 and NN_{12} evaluated on the test datasets used in the seven-fold cross validation method without noise and with the addition of $WGN_{2\%}$.

Testing set	NN$_8$				NN$_{12}$			
	RMSE	RMSE$_{WGN_{2\%}}$	r	r$_{WGN_{2\%}}$	RMSE	RMSE$_{WGN_{2\%}}$	r	r$_{WGN_{2\%}}$
{1}	0.0576	0.1593	0.993	0.9506	0.0426	0.1620	0.9958	0.9458
{2}	0.0476	0.1147	0.995	0.8835	0.1045	0.1425	0.9868	0.8556
{3}	0.0697	0.1217	0.984	0.9695	0.0588	0.1184	0.9910	0.9622
{4}	0.0951	0.1570	0.969	0.8143	0.0370	0.1806	0.9926	0.7701
{5}	0.1334	0.1562	0.919	0.9173	0.1161	0.1359	0.9432	0.9359
{6}	0.0946	0.1228	0.951	0.8226	0.0675	0.1144	0.9448	0.8450
{7}	0.0809	0.0874	0.989	0.9747	0.0727	0.0818	0.9848	0.9780
Mean	0.0827	0.1313	0.971	0.9046	0.0713	0.1336	0.977	0.8989

Leave-One-Out Method

The leave-one-out procedure consists in training as many NNs as the number of measurements (*n*) in the dataset. Each ANN is trained on the whole dataset, leaving out only one case. The single case left-out represents the testing set of the network [30]. In this study, because of the dimension of the dataset (57 measurements), only 10 different instances were chosen as left out cases, so 10 NNs were trained. Foreach ANN, the normalized Root Mean Square Error (RMSE) was estimated on the testing set of both architectures (NN_8 and NN_{12}) as shown in Table 2.

Table 2. Normalized RMSE of NN_8 and NN_{12} evaluated on the test datasets used in the leave one out method without noise and with the addition of $WGN_{2\%}$.

Testing set	NN$_8$		NN$_{12}$	
	RMSE	RMSE$_{WGN_{2\%}}$	RMSE	RMSE$_{WGN_{2\%}}$
{1}	0.0892	0.1180	0.0189	0.1342
{2}	0.1952	0.1860	0.1525	0.1739
{3}	0.0766	0.0087	0.0890	0.0543
{4}	0.0288	0.0958	0.0175	0.0707
{5}	0.0675	0.1132	0.0571	0.0087
{6}	0.0888	0.1208	0.0877	0.1130
{7}	0.1150	0.0453	0.1237	0.0760
{8}	0.0152	0.0854	0.0058	0.0814
{9}	0.1072	0.0471	0.0145	0.0701
{10}	0.0455	0.0662	0.1107	0.0456
Mean	0.0827	0.0886	0.0677	0.0827

5   Results

The prediction capabilities of the NN designed, were evaluated on the testing dataset, measuring the RMSE and the r correlation coefficient. Table 1 summarizes the results of the 7-fold cross validation procedure, comparing the performance of NN_8 and NN_{12} evaluated firstly on the data without noise and then with the addition of a White Gaussian Noise with 2% power ($WGN_{2\%}$) to the interpolated measurements. As regards NN_8, the average RMSE and r coefficient were 0.0827 and 0.971, respectively. After adding a $WGN_{2\%}$, good performances were still observed: average $RMSE_{WGN_{2\%}}$ and $r_{WGN_{2\%}}$ coefficient of 0.1313 and 0.9046, respectively. Figure 5 shows the performances of the 7 models and compares the desired diameter with the predicted diameter of NN_8. The results are quantified in the boxplot representation in Fig. 6: blue

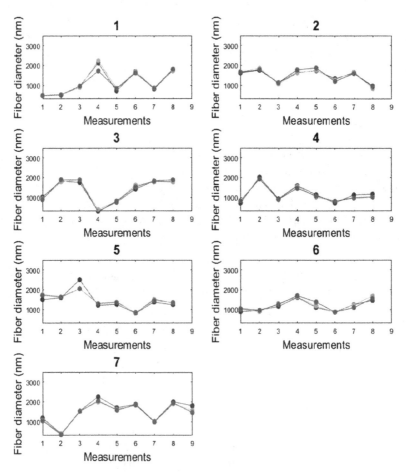

Fig. 5. Response of NN8 tested on the 7 partitions. Blue dots refer to the desired diameters, yellow dots to the predicted diameters, red dots to the predicted diameters with $WGN_{2\%}$ in the interpolated data. (Color figure online)

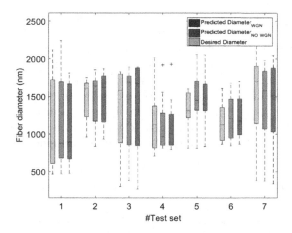

Fig. 6. Boxplot representation of the 7 partitions used as testing sets in the 7-fold cross validation method for NN_8. Blue boxes are associated to the desired diameters, yellow boxes to the predicted diameters, red boxes to the predicted diameters with $WGN_{2\%}$ in the interpolated data. The central mark of each boxplot is the median and the edges correspond to the 25th and 75th percentile. (Color figure online)

boxes represent the partitions used as test sets, yellow and red boxes represent the predicted values of the kth partition with no perturbation and with $WGN_{2\%}$, respectively. The median and the range of the noisy predicted diameters (red boxes) did not show substantial changes from the desired values especially for the first and second fold. A decreasing of the median of 11% was observed in the folds 4 and 7; whereas, the network overrated the average diameters in the 3, 5 and 6 partitions, observing an increasing of median of 5.5%, 7% and 5%, corresponding to 85 nm, 90 nm and 51 nm difference, respectively. Comparable results were obtained with NN_{12} as showed in Table 1.

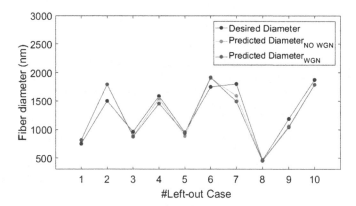

Fig. 7. Response of NN_8 tested on the 10 left out cases. Blue dots refer to the desired diameters, yellow dots to the predicted diameters, red dots to the predicted diameters with $WGN_{2\%}$ in the interpolated data. (Color figure online)

The reliability of the model was tested also with the leave one out method. Table 2 displays the RMSE of NN_8 and NN_{12}, measured for each net, evaluated firstly on the data without noise and then with the addition of the $WGN_{2\%}$ to the interpolated measurements. Figure 7 compares the response of NN_8 with the desired diameter tested on the 10 left out cases. The best prediction was measured for the left-out case number 8: desired diameter of 465.6 nm and predicted diameter of 458.5 nm also when $WGN_{2\%}$ is applied. As each test set included only one experiment, the Pearson coefficient could be not calculated. The average RMSE values were 0.0886 for NN_8 and 0.0827 for NN_{12} evaluated on data with $WGN_{2\%}$ in the interpolations.

6   Conclusions

In this paper, a single hidden layer Neural Network with a direct in-out link, or quasi-sigmoidal hidden layer transformation, was implemented for predicting interpolated diameters of electrospun nanofibers of polyvinylacetate (PVAc). The minimum RMSE value was obtained with 12 neurons in the single hidden layer (NN_{12}) but, the NN with 8 neurons (NN_8) was chosen as optimal network. NN_8 not only provided good performances in terms of RMSE and r coefficient as NN_{12} (Tables 1 and 2) but also allowed to reduce the complexity and number of parameters to evaluate. The experimental values of diameters measured at different working conditions were obtained from the literature [24] and a linear interpolation among data was applied, extending the cardinality of the database from 19 to 57 elements. A white Gaussian noise with 2% power ($WGN_{2\%}$) was added to the interpolated data to test the stability and the efficiency of the network designed, obtaining good performance in terms of RMES and r coefficient. The reliability of the model proposed was assessed through the k-fold cross validation (k = 7) and the leave-one-out methods (10 left out cases analyzed). As the results indicate acceptable performance (Tables 1 and 2), it was not adopted alternative and more complex networks, confirming the validity of the simple NN approach here proposed.

References

1. Huang, Z.-M., et al.: A review on polymer nanofibers by electrospinning and their applications in nanocomposites. Compos. Sci. Technol. **63**(15), 2223–2253 (2003)
2. Persano, L., et al.: Industrial upscaling of electrospinning and applications of polymer nanofibers: a review. Macromol. Mater. Eng. **298**(5), 504–520 (2013)
3. Pantò, F., Fan, Y., Frontera, P., Stelitano, S., Fazio, E., Patanè, S., Santangelo, S.: Are electrospun carbon/metal oxide composite fibers relevant electrodematerials for li-ion batteries? J. Electrochem. Soc. **163**(14), A2930–A2937 (2016)
4. Haykin, S.: Neural networks: a comprehensive foundation. Neural Netw. **2**(2004), 41 (2004)
5. Carrera, D., et al.: Defect detection in SEM images of nanofibrous materials. IEEE Trans. Ind. Inform. **13**(2), 551–561 (2017)
6. Borrotti, M., et al.: Defect minimization and feature control in electrospinning through design of experiments. J. Appl. Polym. Sci. **134**(17), 44740(1 of 10), 44740(2 of 10), .., 44740(10 of 10) (2017)

7. Sarkar, K., et al.: A neural network model for the numerical prediction of the diameter of electro-spun polyethylene oxide nanofibers. J. Mater. Process. Technol. **209**(7), 3156–3165 (2009)

8. Mirzaei, E., et al.: Artificial neural networks modeling of electrospinning of polyethylene oxide from aqueous acid acetic solution. J. Appl. Polym. Sci. **125**(3), 1910–1921 (2012)

9. Faridi-Majidi, R., et al.: Use of artificial neural networks to determine parameters controlling the nanofibers diameter in electrospinning of nylon-6, 6. J. Appl. Polym. Sci. **124**(2), 1589–1597 (2012)

10. Naghibzadeh, M., Adabi, M.: Evaluation of effective electrospinning parameters controlling gelatin nanofibers diameter via modelling artificial neural networks. Fibers Polym. **15**(4), 767–777 (2014)

11. Vatankhah, E., et al.: Artificial neural network for modeling the elastic modulus of electrospun polycaprolactone/gelatin scaffolds. Acta biomaterialia **10**(2), 709–721 (2014)

12. Pham, Q.P., Sharma, U., Mikos, A.G.: Electrospinning of polymeric nanofibers for tissue engineering applications: a review. Tissue Eng. **12**(5), 1197–1211 (2006)

13. Karimi, M.A., et al.: Using an artificial neural network for the evaluation of the parameters controlling PVA/chitosan electrospun nanofibers diameter. e-Polym. **15**(2), 127–138 (2015)

14. Ketabchi, N., et al.: Preparation and optimization of chitosan/polyethylene oxide nanofiber diameter using artificial neural networks. Neural Comput. Appl. 1–13 (2016). https://link.springer.com/article/10.1007/s00521-016-2212-0

15. Brooks, H., Tucker, N.: Electrospinning predictions using artificial neural networks. Polymer **58**, 22–29 (2015)

16. Nasouri, K., Shoushtari, A.M., Khamforoush, M.: Comparison between artificial neural network and response surface methodology in the prediction of the production rate of polyacrylonitrile electrospun nanofibers. Fibers Polym. **14**(11), 1849–1856 (2013)

17. Nasouri, K., et al.: Modeling and optimization of electrospun PAN nanofiber diameter using response surface methodology and artificial neural networks. J. Appl. Polym. Sci. **126**(1), 127–135 (2012)

18. Khanlou, H.M., et al.: Prediction and optimization of electrospinning parameters for polymethyl methacrylate nanofiber fabrication using response surface methodology and artificial neural networks. Neural Comput. Appl. **25**(3–4), 767–777 (2014)

19. Rabbi, A., et al.: RSM and ANN approaches for modeling and optimizing of electrospun polyurethane nanofibers morphology. Fibers Polym. **13**(8), 1007–1014 (2012)

20. Nateri, A.S., Hasanzadeh, M.: Using fuzzy-logic and neural network techniques to evaluating polyacrylonitrile nanofiber diameter. J. Comput. Theor. Nanosci. **6**(7), 1542–1545 (2009)

21. Son, W.K., et al.: The effects of solution properties and polyelectrolyte on electrospinning of ultrafine poly (ethylene oxide) fibers. Polymer **45**(9), 2959–2966 (2004)

22. Yördem, O.S., Papila, M., Menceloğlu, Y.Z.: Effects of electrospinning parameters on polyacrylonitrile nanofiber diameter: an investigation by response surface methodology. Mater. Des. **29**(1), 34–44 (2008)

23. Ojha, S.S., et al.: Morphology of electrospun nylon-6 nanofibers as a function of molecular weight and processing parameters. J. Appl. Polym. Sci. **108**(1), 308–319 (2008)

24. Park, J.Y., Lee, I.H., Bea, G.N.: Optimization of the electrospinning conditions for preparation of nanofibers from polyvinylacetate (PVAc) in ethanol solvent. J. Ind. Eng. Chem. **14**(6), 707–713 (2008)

25. Garg, K., Bowlin, G.L.: Electrospinning jets and nanofibrous structures. Biomicrofluidics **5**(1), 013403 (2011)

26. Ramakrishna, S.: An Introduction to Electrospinning and Nanofibers. World Scientific, Singapore (2005)

27. Chattopadhyay, R., Guha, A.: Artificialneural networks: applications to textiles. Textile Progress **35**(1), 1–46 (2004)
28. Morabito, F.C.: Independent component analysis and feature extraction techniques for NDT data. Mater. Eval. **58**(1), 85–92 (2000)
29. Refaeilzadeh, P., Tang, L., Liu, H.: Cross-validation. In: Liu, L., Özsu, M.T. (eds.) Encyclopedia of Database Systems, pp. 532–538. Springer, Heidelberg (2009). doi:10.1007/978-0-387-39940-9_565
30. Steyerberg, E.W., et al.: Internal validation of predictive models: efficiency of some procedures for logistic regression analysis. J. Clin. Epidemiol. **54**(8), 774–781 (2001)

Using Advanced Audio Generating Techniques to Model Electrical Energy Load

Michal Farkas and Peter Lacko$^{(\boxtimes)}$

Faculty of Informatics and Information Technologies,
Slovak University of Technology, Bratislava, Slovakia
michalfarkas1@gmail.com, peter.lacko@stuba.sk

Abstract. The prediction of electricity consumption has become an important part of managing the smart grid. Smart grid management involves energy production (from traditional and renewable sources), transportation and measurements (smart meters). Storing large amounts of electrical energy is not possible, therefore it is necessary to precisely predict energy consumption. Nowadays deep learning approaches are successfully used in different artificial intelligence areas. Deep neural network architecture called WaveNet was designed for text to speech task, improving speech quality over currently used approaches. In this paper, we present modification of the WaveNet architecture from speech (sound waves) generation to energy load prediction.

Keywords: Artificial neural networks · Deep learning · Time series prediction

1 Introduction

Due to the characteristics of production facilities, distribution networks and free market, accurate forecasts of various values is indeed very important for modern energy industry and market, if they have to operate at peak efficiency.

Electric energy production facilities can be dependent on external factors and may not be able to change their output on moments notice. For example various alternative sources of energy such as wind, solar energy have limited output determined by wind speed and sunlight exposure. Furthermore, they are often embedded within the local distribution network, thus they are seen from the outside only as a reduction of demand.

Distribution networks can be damaged by extremely high amount of energy in network, which is not consumed. There are risks of blackouts and brownouts, which are highly damaging and disrupting to economy and society and can even result in material damage. Because of the technological inability to efficiently store vast amount of electric energy, there is little to no balancing buffer. Without this ability, produced energy has to be consumed right away, otherwise it would be wasted. Naturally, limiting amounts of wasted energy leads to more efficiency, which in turn leads to competetive advantage and subsequently less

© Springer International Publishing AG 2017
G. Boracchi et al. (Eds.): EANN 2017, CCIS 744, pp. 39–48, 2017.
DOI: 10.1007/978-3-319-65172-9_4

ecological impact on environment, which has proved to be growing problem in recent years. With advent of smart-grids, deregularization of industry and other changes, energy load forecasting is even more important.

Due to how are energy markets, electric energy production and distribution industry constructed, there are various uses for load forecast. These are generally categorized by their prediction window, but they can also differ by time resolution, scope and specialization. In this paper we are considering multiple cases, which differ by time, area, resolution and scope. Artificial neural networks were successfully used for series analysis [9] and prediction. Nowadays deep learning approaches are massively used in different artificial intelligence areas providing state-of-the-art results.

2   State of the Art

There are great number of various methods proposed for problem of energy load forecasting, they fall under statistical, engineering and artificial intelligence categories.

Engineering methods can, if they are complicated enough, reach quite high accuracy. These tools are well developed and used, however they require high amount of information, which is unobtainable in practice on a larger scale. Coupled with their need for domain expert, their effectivness is severely limited. On the other hand, simplified engineering approaches don't require such vast amount of information, setting and expert work, however they are not as precise [7].

Statistical methods try to model data by mathematical functions, in which various factors act as input. Parameters of these models are then trained on historical data, thus using this regression, relationships between input factors and output value, in our case electric energy consumption, are found. In [5] authors discriminate between two different types of models, additive and multiplicative, former models load as a sum of various components (underlying trend, weather and day variations, holidays, etc.), latter is product of multiplication of various factors and informations. Statistical models are easy to understand and maintain and they achieve average results, but it has been concluded they lack flexibility.

Artificial intelligence approaches consists mainly of artificial neural networks, but various other solutions were used, for example support vector machines, expert systems, fuzzy logic and different possibly hybrid solutions [10–12]. In [4] wavelet decomposition coupled with PSO method, similar days approach and grey model method to predict electricity consumption on various case for two distinct datasets, which did consists of rather large area. On the other hand [1,2] predicted load on small areas, data of the former consisted of a single city and latter of a single buildings. Method for long term prediction both on single building and whole region, was proposed in [3], it was based on decomposition with various extensions of support vector regression. Incremental heterogeneous ensemble model for time series prediction was proposed in [8] and was successfuly used for electricity load prediction.

Parallel to domain of energy load forecasting is domain of energy prices forecasting. Naturally these two areas are closely related as in market economies, demand and price are closely related. Approaches used to predict price of energy are similar, if not same, as those used in energy load forecast [13].

3   Data Characteristics

Since method used in this article, was originally developed for audio generation purposes, it is quite approriate to describe characteristics of energy consumption time-series and how it differs from audio time-series.

In both cases there is strong periodicity of data, in case of energy it would naturally be daily, weekly and yearly periodicity. Audio data usually being composed of multiple components, which each had their own frequency, does naturally have strong periodicity.

One of the more obvious differences is sampling frequency. Energy time-series usually have sampling frequency of 15 min or one hour. In some special cases, dictated by data available or forecast specialization, its sampling frequency may differ. On the other hand, sampling frequency of audio data, while varying by standard, compression and recording quality, typically lie between 8 kHz–44.1 kHz. Even if we consider, that energy time-series cover vastly longer time-period than audio time-series, audio time-series will usually have much more information. To put it in perspective, energy time-series at 15 min sampling interval measured over 16 years, have same data length as 12.7 s of CD quality audio time-series. Direct consequences are smaller sizes of energy datasets, which can limit our prediction.

Audio data is mainly determined by its underlying information, whether it is music or speech. If we ignore content, which is far more complicated to predict in a meaningful way, there are still factors that need to be taken into consideration, such as speaker or emotion. Consumption of electric energy data lacks any meaningful content, instead it depends on socio-economical, environmental, cultural and legal factors. Most of these cannot be easily represented by a value, instead we can learn these from historical data. Factors that can be represented by a value, for example time, weather, workdays, can act as a additional input. However, even if these factors can be represented they must be available at the time of prediction. Unsuprisingly, accuracy of forecasts is highly dependent on factors taken into account.

Prime example of socio-economic or cultural factors are workdays and holidays, which have clear effect on electricity consumption. Naturally, national holidays are nation-specific and while most of the world shares the same workweeks, there are exceptions, such as Israel or some muslim countries. Weather factors, such as temperature, humidity, pressure and windspeed can affect electricity consumption in both directions. Most significant actors in this situation are HVAC systems and alternative power sources. As we mentioned in introduction chapter, some solar and wind power plants are embedded in network and their output is seen only as a reduction in local electricity consumption and since their output is dependent upon these weather factors, they have direct, inverted impact on electricity consumption. HVAC systems are mostly used in weather extremes to cool or to heat, thus generally in cold climates their impact is most pronounced in winter due to extensive heating and in tropical climates their impact is most noticeable in summer due to air conditioning.

Noticeable effect on varying scopes can have economic factors, such as economic performance, employment and others. However, impact of these factors can be marginal and they are rarely available at the time of prediction and thus are almost never used.

In data, there can be also anomalies, caused by one-time events, social unrest, conflicts, demonstrations various sporting events and competitions, list of potential disrupting events can be quite long. From their definition as anomalies these may not be predicted by model and their impact can be almost impossible to predict. In some cases, such disruptive events are announced with advance and thus human or automated system can feed some information into prediction system or different model altogether can be used.

4   Method Description

Neural network architecture used in this paper is called WaveNet, it was originally proposed by DeepMind as a audio generation model [6]. Authors tested it as a text-to-speech system on English and mandarin Chinese languages, as a music generation and as a speech recognition model. It operates on raw audio an its output is probability distribution. Main components of this model are 1×1 convolutions and causal dilated convolutions connected as a residual network. Due to its good results in audio generating domain, we decided it is worth to test it on energy load forecast domain.

Since we are predicting future values, we have to take into account only historical data, otherwise we would violate causality. This is ensured by shifting ordinary convolution by one timestep, by which we create causal convolution. Example of this type of convolution is shown in Fig. 1.

To enlarge receptive field of network, without changing number of layers or size of filter, convolution is modified so it takes inputs only at certain intervals.

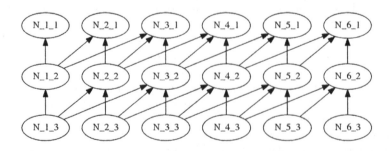

Fig. 1. Causal convolution.

In WaveNet model, these intervals are defined as $2^{(n \bmod 10)-1}$, where n stands for depth of a layer. This variant of convolution is called dilated or convolution with holes and is shown in Fig. 2.

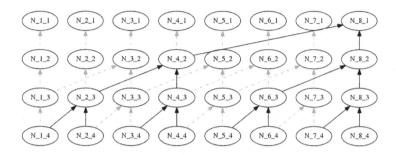

Fig. 2. Dilated causal convolution.

So called 1×1 convolutions have filter size of one in both dimensions, hence their name. They act as a fully connected layer on single collumn of input, thus they can only change number of channels. They are relatively cheap and dont interfere with causality.

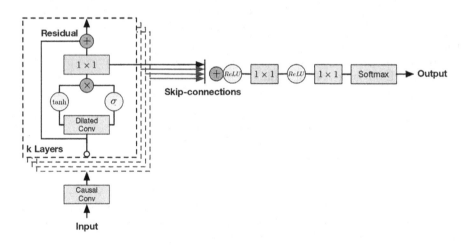

Fig. 3. WaveNet architecture; reproduced from [6].

These components are connected as shown in Fig. 3. Before dilatation layers, there is a single regular causal convolution layer, which changes channel size. Subsequent layers consists of two dilated convolution layers, their combination is depicted on following equation:

$$output = \tanh(W_f * x) \odot \sigma(W_g * x) \tag{1}$$

Dilatation layers are not only connected in serial fashion, but they also contain residual and skip connections, former are used to carry unchanged input to subsequent layer, latter are used to carry layer output to the end of neural network. There skip outputs are summed and processed to get final output.

Wavenet architecture implemented in this article does not have all features of original model, missing from our implementation are both local and global conditioning, context stacks and faster generation computation features.

however, since we do use weather and time data in prediction we need to feed it into neural network. We added this information as additional channels. By combining historical data, temperature, humidity, pressure, windspeed and time data, which consists of a sine and cosine components with weekly periods, our input has 7 channels. We tested different weather attributes and additional time channels with daily or yearly periods. We concluded that some weather attributes were not significant enough. Similarly daily time data did not help prediction, probably because network is able to infer time of day from historical data. We introduced yearly periodic data in hopes of identifying holidays such as Christmas, hovewer this was most likely hindered due to small data size.

Due to numerical stability of softmax function in Theano we had to add normalization function just before softmax to ensure proper results.

Output of neural network is probability distribution, size of it directly limit accuracy of prediction and more accurate prediction could be achieved by higher output resolution. To be able to predicted more than one sample ahead, we put most probable prediction at the end of input. To reduce error growth and encourage robustness, we added noise to the input during training, its size was dependent upon last evaluation error.

To measure accuracy in our experiments we used mean average percentage error - MAPE on value corresponding to most probable interval from neural network output converted to nonnormalized data. This error is defined by formula:

$$MAPE = \frac{100}{n} \sum |\frac{A_t - F_t}{A_t}| \tag{2}$$

5   Results

In first experiment, we tested this model on ELIA[1] dataset, which consists of whole energy distribution network belonging to a company in Belgium of the same name. Weather data was extracted from NASA Merra2 reanalysis. Sampling rate for both datasets was 15 min. Both weather and load data were normalized and any missing values were corrected by smoothing. We ran experiments on subset ranging from years 2007 to 2008. We tested the same model on different prediction scopes using MAPE (Table 2). On Fig. 4 are detailed examples of predictions on ELIA dataset.

In our second experiment, we used load data from city of Bratislava, between July of 2013 and February of 2015. This dataset did not consist of whole city

---

[1] http://www.elia.be/en/grid-data/data-download.

and it was produced mainly by industry customers. This was complemented by Merra2 dataset and had interval of 15 min too. Normalizing and filling missing values was done in the same way. Detail is shown in Fig. 5 and MAPE in Table 2.

In Table 1, comparisons with SVR and MLP are shown, both of these methods use time lagged variables. Predictions on longer periods were done in a similar fashion as with WaveNet. After inspecting results we concluded that SVR and MLP method rely mainly on repeating last value, which naturally leads to high errors when chained this way.

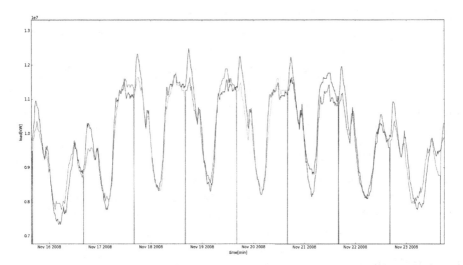

Fig. 4. Detail on one day ahead prediction on ELIA dataset, blue are real load values (Color figure online)

Table 1. Comparison of results

		ELIA	Bratislava
MLP	1-sample ahead	0.976%	0.81%
	24 h ahead	32.64%	25.125%
SVR	1-sample ahead	1.39%	2.85%
	24 h ahead	NaN	NaN
Wavenet	1-sample ahead	1.22%	1.09%
	24 h ahead	3.45%	1.934%

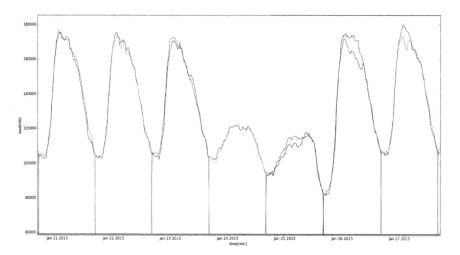

Fig. 5. Detail on day ahead prediction on Bratislava dataset, blue are real load values (Color figure online)

Table 2. Table of MAPE values for various datasets

Dataset	One sample ahead	1 day ahead	3 days ahead	7 days ahead	29 days ahead
ELIA dataset	1.22	3.45	3.52	4.20	5.02
Bratislava	1.09	1.934	2.14	2.29	3.53

6   Conlusion

In this paper, we presented a modification of the WaveNet architecture (originally used for sound generation) for electricity load forecasting. Two different datasets were used to evaluate this approach.

Our results in both cases show that this method can be precise enough, provided that enough data with small enough sampling interval is available. In line with our expectations, growth of error with longer prediction window was dependent upon initial error size. Thus, we can conclude that with error small enough we could predict greater time frames with acceptable precision.

Because this method seems to need small sampling interval and most datasets in use are hourly, we had problem of finding results with which we can compare ourselves. We decided to compare using simple SVR and MLP with time-lagged variables. In some cases we observed better results in these referential methods, but they failed when prediction was lengthened to 1 day. We strongly suspect that referential methods and Wavenet method could be significantly improved by better way of lengthening timeframe of prediction.

Forecast accuracy in this paper was severely limited by time and hardware constraints. We had to limit complexity of our models, which had only 8 layers, inner channel size of 128 and output size of 256. Higher values of these hyperparameters would almost surely result in even more accurate predictions.

Main limiting factor was memory avaible to gpu, however computation time of approximately 8 h was limiting as well. In some experiments with more complex models, which are not presented here, computation time could exceed 48 h. That being said, prediction on already trained model was fairly quick.

In our future work we would like to test this approach with more complex model with different hyperparameters, improve prediction over longer periods and perhaps further adapt it to better suit this domain.

Acknowledgments. This work was partially supported by the Scientific Grant Agency of the Slovak Republic, grant No. VG 1/0752/14 and with the support of the Research and Development Operational Programme for the project International centre of excellence for research of intelligent and secure information-communication technologies and systems, ITMS 26240120039, co-funded by the European Regional Development Fund.

References

1. Hernández, L., Baladrón, C., Aguiar, J., Carro, B., Sánchez-Esguevillas, A., Lloret, J.: Artificial neural networks for short-term load forecasting in microgrids environment. Energy **75**, 252–264 (2014)
2. Jurado, S., Nebot, Á., Mugica, F., Avellana, N.: Hybrid methodologies for electricity load forecasting: entropy-based feature selection with machine learning and soft computing techniques. Energy **86**, 276–291 (2015)
3. Ghelardoni, L., Ghio, A., Anguita, D.: Energy load forecasting using empirical mode decomposition and support vector regression. IEEE Trans. Smart Grid **4**, 549–556 (2013)
4. Bahrami, S., Hooshmand, R.-A., Parastegari, M.: Short term electric load forecasting by wavelet transform and grey model improved by PSO (particle swarm optimization) algorithm. Energy **72**, 434–442 (2014)
5. Feinberg, E.A., Genethliou, D.: Load forecasting. In: Chow, J.H., Wu, F.F., Momoh, J. (eds.) Applied Mathematics for Restructured Electric Power Systems. PEPS, pp. 269–285. Springer, Boston (2005). doi:10.1007/0-387-23471-3_12
6. van den Oord, A., et al.: WaveNet: A Generative Model for Raw Audio. CoRR abs/1609.03499 (2016)
7. Zhao, H., Magoulés, F.: A review on the prediction of building energy consumption. Renew. Sustain. Energy Rev. **16**, 3586–3592 (2012)
8. Grmanová, G., Laurinec, P., Rozinajová, V., Ezzeddine, A.B., Lucká, M., Lacko, P., Vrablecová, P., Návrat, P.: Incremental ensemble learning for electricity load forecasting. Acta Polytech. Hung. **13**(2) (2016)
9. Čerňanský, M., Makula, M., Trebatický, P., Lacko, P.: Text correction using approaches based on Markovian architectural bias. In: Proceedings of the 10th International Conference on Engineering Applications of Neural Networks, Publishing Centre Alexander Technological Educational Institute of Thessaloniki, EANN 2007, pp. 221–228 (2007)
10. Kavousi-Fard, A., Samet, H., Marzbani, F.: A new hybrid modified firefly algorithm and support vector regression model for accurate short term load forecasting. Expert Syst. Appl. **41**(13), 6047–6056 (2014)

11. Ko, C.N., Lee, C.M.: Short-term load forecasting using SVR (support vector regression)-based radial basis function neural network with dual extended Kalman filter. Energy **49**, 413–422 (2013)
12. Chaturvedi, D.K., Sinha, A.P., Malik, O.P.: Short term load forecast using fuzzy logic and wavelet transform integrated generalized neural network. Int. J. Electr. Power Energy Syst. **67**, 230–237 (2015)
13. Vardakas, J.S., Zenginis, I.: A survey on short-term electricity price prediction models for smart grid applications. In: Mumtaz, S., Rodriguez, J., Katz, M., Wang, C., Nascimento, A. (eds.) WICON 2014. LNICSSITE, vol. 146, pp. 60–69. Springer, Cham (2015). doi:10.1007/978-3-319-18802-7_9

Memristor Based Chaotic Neural Network with Application in Nonlinear Cryptosystem

N. Varsha Prasad$^{(\boxtimes)}$, Sriharini Tumu, and A. Ruhan Bevi

Department of Electronics and Communication Engineering, SRM University,
SRM Nagar, Kattankulathur, Chennai, India
{varshaprasad_rama,sriharini_tumu}@srmuniv.edu.in,
ruhan.b@ktr.srmuniv.ac.in

Abstract. The global shift towards digitization has resulted in intensive research on Cryptographic techniques. Chaotic neural networks, augment the process of cryptography by providing increased security. In this paper, a description of an algorithm for the generation of an initial value for encryption using neural network involving memristor and chaotic polynomials is provided. The chaotic series that is obtained is combined with nonlinear 1 Dimensional and 2 Dimensional chaotic equations for the encryption process. A detailed analysis is performed to find the fastest converging neural network, complemented by the chaotic equations to produce least correlated ciphertext and plaintext. The use of Memristor in Neural Network as a generator for chaotic initial value as the encryption key and the involvement of nonlinear equations for encryption, makes the communication more confidential. The network can further be used for secure multi receiver systems.

Keywords: Chaotic neural network · Memristor · Hermite polynomial · Chebyshev polynomials · 1 Dimensional chaotic maps · 2 Dimensional chaotic maps

1 Introduction

Chaos theory deals with the dynamical systems which are highly sensitive to the changes in the input. The integration of chaotic properties with artificial neural networks has been one of the most engaging tasks for researchers. The generation of Chaotic Neural Network (CNN) using memristor, as proposed by Wang et al. [1], has been an inspiration for the generation and analysis of the NN with the help of chaotic polynomials like Hermite, Chebyshev T and Chebyshev U. In this paper, inclusion of memristor for synaptic weight updation is an integral part of the network, as it helps in faster convergence.

Memristor was first introduced by Chua in 1971 [2], following which it was developed further by the team of HP Laboratories, led by Stanley Williams. The physical model was first realized using thin film of TiO_2 and platinum electrodes, in 2008 which was 37 years after Chua theoretically proposed it. Since then it

© Springer International Publishing AG 2017
G. Boracchi et al. (Eds.): EANN 2017, CCIS 744, pp. 49–60, 2017.
DOI: 10.1007/978-3-319-65172-9_5

has expanded to many applications. Due to its potential of having memory and drawing less power, it has been used in NVRAM (Non- volatile random access memory). Memristors can maintain a state without external bias and hence are used as switches. Additionally, as continuous memristive spectrum is attainable, they can be used as analogue memory elements [3]. Future developments can also lead to integration of memristor technology in computer systems to simulate brain like behaviour which remembers associate patterns as humans are capable of [4].

In the present digitalized world, information interchange with security, confidentiality and non repudiation has become essential. This calls forth the need of cryptographic techniques which are reliable and have least possibility of intruder decrypting the message with brute force attack or any other means. The sensitive dependence on initial condition, along with its randomness and broad band spectrum have led to the use of chaos theory in cryptography. References [5,6], are the initial works performed in the field of chaotic cryptography wherein they have proposed digital chaotic ciphers. Reference [7] has performed a review on the developments of chaos based cryptographic techniques.

In this paper, an architecture for chaotic neural network with synaptic weight updation using the conductance of the memristor is explained. Further chaotic polynomials are analysed and the network is evaluated to find the polynomial which has the fastest convergence rate and least computational complexity. The network mentioned above is used to generate a chaotic initial value, which is sent to the encryption process using 1 Dimensional or 2 Dimensional chaotic equations. Correlation coefficients between the output and the input are calculated and analyzed.

The paper is organized as follows. Section 2 consists of the non-linear properties of the memristor. Section 3 explains the chaotic polynomials such as Hermite, Chebyshev T and U. Section 4 describes the proposed algorithm used for CNN, along with the encryption/decryption processes which are performed with the help of one and two Dimensional chaotic maps, individually. Section 5 has the tabulation of the results, followed by the discussions and conclusion in Sects. 6 and 7.

2   Memristor

The non-linear, passive, two terminal electrical device, called memristor having non-volatile properties was first proposed by Leon Chua in 1971. The value of its electrical resistance is dependent on the previous current values.

The relation between the voltage $v(t)$ and current $i(t)$ in this device is given by the following equation:

$$v(t) = \left(R_{on} \frac{w(t)}{D} + R_{off} \left(1 - \frac{w(t)}{D} \right) \right) i(t) \tag{1}$$

where $D =$ Thickness of TiO_2 layer which is sandwiched, $w(t) =$ Thickness of doped area at time t, R_{on} is the low resistance and R_{off} is the high resistance value.

This is used to calculate the Memristance value which is the ratio between voltage and current. On substituting $x = w/D$ [0,1], we get:

$$M(t) = -(\Delta R)x(t) + R_{off} \tag{2}$$

$$M(t) = -(\Delta R)x_0 + R_{off} \ when \ t = 0 \tag{3}$$

$$\Delta R = R_{off} - R_{on} \tag{4}$$

The speed of the movement of the boundary between doped and undoped region.

$$\frac{dx}{dt} = ki(t)f(x) \ where \ k = (\mu_v R_{on})/D^2 \tag{5}$$

μ_v refers to average mobility which can be approximately given by $10^{-14}m^2s^{-1}V^{-1}$. The non-linearities in ionic transport can be produced by small voltages which produce huge electric fields. The non-linear dopant drift, which can be observed in the thin film edges, can be modelled by the window function $f(x)$. The paper choses Jokular window function as the one used by HP labs is not available. The non-linear behaviour of memristor is more evident when p = 1.

$$f(x) = 1 - \left(2\frac{w}{D} - 1\right)^{2p} \tag{6}$$

This function assures zero speed of the x-coordinate when approaching either boundary.

With the help of Jokular window function and Eq. (5), the expression for memristive conductance is given by:

$$G(t) = \frac{1}{M(t)} = \frac{1}{R_{on} + \Delta R \frac{1}{Ae^{4kq(t)}+1}} \tag{7}$$

$$where \ A = \frac{(R_{off} - R_{on})}{(R_0 - R_{on})}$$

Upon differentiating Eq. (7) with respect to t, we get

$$\frac{dG}{dt} = \frac{4kAe^{4kq(t)}\Delta R}{(R_{on}Ae^{(4kq(t))} + R_{off})^2} \tag{8}$$

The values of the parameters considered are, $R_{on} = 100\,\Omega$, $R_{off} = 20\,k\Omega$, $R_0 = 5\,k\Omega$ and D = 10 nm. We replace the change of weight in neural network Δw, by Δg. The non-volatile characteristic of the memristor helps in stabilizing the neural network and makes the process of weight updation convenient by accelerating towards convergence.

3   Chaotic Polynomials as Activation Functions

The uncertainty in the dynamical system is determined by Wiener chaos expansion or Polynomial chaos. It was first introduced by Norbert Wiener [8]. Hermite polynomials and Chebyshev polynomials [9] have the credit of exhibiting chaos and hence we use them as the activation functions in the algorithm and determine the convergence. These are given in Table 1.

Table 1. Analysis of chaotic polynomials

Polynomial	Generating function	Expansion
Hermite	$H_n(x) = (-1)^n e^{x^2} \frac{d^n}{dx^n} e^{-x^2}$	$H_n(x) = 2x H_{(n-1)}(x) - 2(n-1)H_{(n-2)}(x)$
Chebyshev T	$\frac{(1-tx)}{(1-2tx+t^2)} = \sum_{t=0}^{+\infty} T_n(x)t^n$	$T_{(n+1)}(x) = 2x T_n(x) - T_{(n-1)}(x)$
Chebyshev U	$\frac{1}{(1-2tx+t^2)} = \sum_{n=0}^{+\infty} U_n(x)t^n$	$U_{(n+1)}(x) = 2x U_n(x) - U_{(n-1)}(x)$

4   Cryptographic Process

The process of encryption is divided into two. First, we design a CNN with a memristor and consider the activation function to be the chaotic polynomials. This network acts as the generator for the given input key. The output of the CNN is given for proposed encryption process (inspired from Fiestel Network [10]), which involves mapping technique that use non-linear equations as key elements.

4.1   Structure of Neural Network

The network which we have considered consists of three layers, wherein there is one input layer having single neuron, a hidden layer having 5 neurons and a output layer having 1 neuron. The change in weight is given by the Eq. (11), wherein the calculated weight is added and is given to the input for the next epoch.

Further, the weights between the input and hidden layer are 1, and the weight between the hidden layer and output layer are calculated from the chaotic polynomials considered. In case of 2 Dimensional mapping, one of the input value is constant and known to the receiver.

4.2   Algorithm

1. Input layer $x = x_i$.
2. Input to the hidden neuron = x*v, where v is weight between input and hidden layer and '*' denotes matrix multiplication.
3. Output of the hidden neurons, considering t = 1:

$$y_i = \sum_{j=0}^{n-1} w_j C_{ip}(t) \tag{9}$$

where $p=0$ for Chebyshev U, $p=1$ for Chebyshev T and $p=2$ for Hermite polynomial. Also, j = number of neurons, i = number of epochs.

4. Calculation of error:

$$E = f(x) - y_i \tag{10}$$

5. Change in weight:

$$\Delta w_i = \frac{(4k A e^{4kq} \Delta r E \rho J_i(x))}{(R_{on} A e^{4kq} + R_{off})^2} \tag{11}$$

$$q = \int \left(\sum_{j=0}^{n-1} C_{jp}(x) \right) dt - f(x) \tag{12}$$

$$J_i(x) = Chaotic Polynomial$$

$$\rho = learning\, rate = 0.5$$

6. Updation of weight:

$$w_{i+1} = w_i + \Delta w_i \tag{13}$$

The error is calculated for the respective polynomials for fixed number of epochs. The output values using respective polynomials is given for the encryption process.

4.3   Encryption Process

1. Let the plain text sequence be $P = p_1 p_2 p_3 p_4 p_n$, where n is the length of the sequence.
2. With the help of the chaotic initial value, the non-linear sequence can be obtained as $a = [a_1(p_1), a_2(p_2), a_3(p_3), a_4(p_4), a_5(p_5).a_n(p_n)]$. This sequence is obtained by calculating subsequent values from the initial value in the chaotic map equations.
3. Perform the sorting process to retrieve the respective indices σ of the obtained values, which are in ascending order.
4. The text that is obtained by mapping is (i.e.) $g_i = p_\sigma$ where σ refers to the index as calculated above.
5. The obtained text $G = g_1 g_2 g_3 g_n$ is given as the input to the Network. Now it is divided into two halves: G_Left: Consisting of the first half of the set of characters in G, and G_Right: Consisting of the second half of the set of characters in G. These halves will be named G_Left(0) and G_Right(0) respectively.
6. For each iteration n:
   (a) $G\_Left(n) = Mapping(G\_Right(n - 1))$

(b) $G\_Right(n) = G\_Left(n-1)$

7. For computing the Mapping function, follow steps 2 to 4 with plaintext substituted with $G\_Right(n-1)$ and the result is stored in $G\_Left(n)$. A total of m iterations are performed. In case of 2 Dimensional Maps, any one of the two sequences generated separately can be considered. While evaluating, the x sequence was taken into account. A total of m iterations are performed.

8. The final $G\_Left(m)$ and $G\_Right(m)$ are concatenated to give the Ciphertext C. This C is sent across the communication channel, along with the key.

The following 1 Dimmensional [11] and 2 Dimmensional [12] chaotic maps are analyzed with the help of bifrucation diagram and the values of the control parameters are selected as follows and given in the respective Tables 2 and 3.

Table 2. Analysis of 1 dimensional chaotic equations

Name	Equation	Parameter value
Logistic	$a_{(n+1)} = r * a_n * (1 - a_n)$	$r = 4$
Cubic	$x_{(n+1)} = \lambda x_n (1 - x_n^2)$	$\lambda = 2.58$
Tent	$x_{n+1} = f_\mu(x_n) = \mu x_n \ for\ x_n < 1/2$ $= \mu(1 - x_n)\ for\ 1/2 \leqslant x_n$	$\mu = 2$
Sine	$x_{(n+1)} = f_\mu(x_n)$ $f_\mu(x) = \mu \sin(\pi x), \quad x[0,1], \mu > 0$	$\mu = 0.97$

Table 3. Analysis of 2 dimensional chaotic equations

Name	Equation	Parameter value
Henon	$x_{(n+1)} = 1 - a(x_n)^2 + by_n$ $y_{n+1} = x_n,$	$a = 1.4\ b = 0.3$
Ikeda	$x_{(n+1)} = 1 + u(x_n \cos(t_n) - y_n \sin(t_n))$ $y_{(n+1)} = u(x_n \sin(t_n) + y_n \cos(t_n)),$ $t_n = 0.4 - \frac{6}{1+(x_n)^2+(y_n)^2}$	$u = 0.918$
Duffing	$x_{(n+1)} = y_n$ $y_{n+1} = -bx_n + ay_n - (y_n)^3,$	$a = 2.75\ b = 0.2$
2D Logistic	$x_{(n+1)} = a(3y_n + 1)x_n(1 - x_n)$ $y_{(n+1)} = a(3x_{n+1} + 1)y_n(1 - y_n)$	$a = 1.19$

4.4 Decryption Process

1. Let the cipher text sequence be $C = c_1 c_2 c_3 c_4 c_n$, where n is the length of the sequence.

2. Divide the cipher text into two halves where $C\_Left(0)$: Consists of the first half of the set of characters in C, and $C\_Right(0)$: Consists of the second half of the set of characters in C.

3. These are given to the Network. Now a reverse mapping procedure is performed.

4. For each iteration n:

   (a) $C\_Right(n) = ReverseMapping(C\_Left(n-1))$

   (b) $C\_Left(n) = C\_Right(n-1)$

5. For Computing the Reverse Mapping function, follow steps 6 to 9 with input sequence as $C\_left(n)$

6. For a given input sequence $C = c_1c_2c_3c_4c_n$, with the help of the chaotic initial value, the non-linear sequence can be obtained as $a = [a_1(c_1), a_2(c_2)...a_n(c_n)]$. This sequence is obtained by calculating subsequent values from the initial value from the chaotic map equations.

7. Perform the sorting process to retrieve the respective indices σ of the obtained values, which are in ascending order.

8. Sort the indices that are obtained in the previous step and retrieve the respective indices. For the final obtained sequence. For 2D maps, the sequence to be considered is same as followed during encryption.

9. After m iterations, concatenate the obtained sequences to from a sequence C.

10. Then perform the steps 6 to 8 on the resulting concatenated sequence to obtain the indices σ.

11. The Plain text is obtained by mapping (i.e.) $P_i = C_\sigma$ where s refers to the index as calculated previously.

5   Results and Analysis

It is evident from Table 7 that the use of Chebyshev-U polynomial has resulted in a faster convergence to the output at approximately 100 cycles whereas the use of Chebyshev T result converged at approximately 500 cycles. For the proposed algorithm, the rate of convergence of the Hermite polynomial was found to be low. The computational time for the above results is given in Table 6, using Matlab2016a. It can be observed that Chebyshev U has lesser computational time among its counterparts. Intruder attack for the mentioned network is difficult pertaining to the fact that, the initial value used for encryption is highly sensitive to the output of the CNN. The encryption algorithm depends on the output of the CNN which is unique for every input sequence and cannot be directly estimated from the output function.

In this paper, we have analysed the effect of 1D chaotic maps in cryptographic communication using cross-correlation between the plain text and the cipher text. Table 4 shows that Cubic map has the least correlation followed by Sine map and Tent map for the given plain text. Similarly from Table 5, it can be observed that 2 D Logistic map gives the lowest correlated output.

Table 4. Correlation values for 1 dimensional maps

Initial value	Sine	Tent	Cubic	Logistic
0.1	0.0288	0.0665	0.0223	0.0336
0.2	0.0254	0.0301	0.0065	0.0177
0.3	0.0248	0.0248	0.0121	0.0413
0.4	0.0279	0.0279	0.0108	0.0485

Table 5. Correlation values for 2 dimensional maps

Initial value		Henon	2D Logistic	Duffing	Ikeda
x	y				
0.1	0.2	0.0199	0.0056	0.0643	0.0090
0.2	0.2	0.063	0.0083	0.0094	0.105
0.3	0.2	0.064	0.0085	0.1548	0.0124
0.4	0.2	0.0377	0.0062	0.0688	0.0362

6   Discussion

Shi et al. [13], proposed the use of memristor in a similar neural network architecture by the synaptic weight updation process and Hermite polynomial as the activation function. The use of memristor for chaos generation has been enhanced in the mentioned algorithm given in this paper, by using Chebyshev U polynomial. Moreover, the CNN described in the paper converges at lesser number of cycles, which is approximately 42.8% faster than the one described in by the former reference.

After obtaining the output from the chaotic neural network and sending it for the encryption process, which involves the use of 1 dimensional chaotic maps such as Logistic map, Cubic map, Tent map and Sine map or 2 Dimensional maps such as 2D Logistic, Henon, Ikeda and Duffing, we have analysed the results by calculating the correlation between the Cipher text and the Plain text. In general, all the correlation coefficients for the following maps using Hermite and Chebyshev (T and U) polynomials are very less indicating meagre dependence of the encrypted output with the input.

The proposed process is performed with a key of value 0.2, that is sent across the channel to the receivers CNN that uses the Chebyshev U polynomial to be

Table 6. Analysis of computational time

	Hermite	Chebyshev T	Chebyshev U
Time taken to compute polynomial	0.168 s	0.17 s	0.152 s
Overall computation time for 100 epochs	10.699 s	11.053 s	10.522 s

Table 7. Simulation results of chaotic neural network

Initial value	No of epochs	Hermite		Chebyshev T		Chebyshev U	
		Error	Output	Error	Output	Error	Output
0.2	100	0.5699	0.0701	0.2123	0.4277	1.44E−06	0.6400
	200	0.5069	0.1331	0.0696	0.5704	2.83E−12	0.6400
	500	0.3566	0.2834	0.0024	0.6375	2.22E−16	0.6400
0.3	100	0.8199	0.0201	0.1446	0.6954	3.37E−10	0.8400
	200	0.8000	0.0400	0.0244	0.8156	1.11E−16	0.8400
	500	0.7433	0.0967	1.18E−04	0.8399	1.10E−16	0.8400
0.4	100	0.9472	0.0128	0.1034	0.8566	7.84E−13	0.9600
	200	0.9344	0.0256	0.0109	0.9491	2.22E−16	0.9600
	500	0.8970	0.0630	1.27E−05	0.9600	2.22E−16	0.9600

converted into a chaotic initial value of 0.63999. This value is used in the cubic map (1D) to get the chaotic series and thus is used for encryption and decryption (16 iterations) (Figs. 1 and 2).

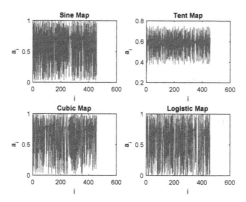

Fig. 1. Generated sequences of one dimensional maps for chaotic initial value 0.2. *i* on the *x axis* refers to the *indices of the plain text* and *a* on the *y axis* refers to the *calculated chaotic values.*

As an example, 2D encryption using two dimensional logistic map is also performed, keeping the value of y as 0.3 (set for intended receiver). The correlation between the input and output sequences for cubic and 2D Logistic is found to be is found to be 0.0065 and 0.0009 respectively. Also, Fig. 3 shows that 2D Logistic map which is two dimensional, is comparatively more efficient than 1 Dimensional Cubic map as the correlation values are relatively low.

Further, the proposed network can be used for a system containing n number of receivers, by providing secure and isolated communication in the same channel. The sender changes the bias according to the intended receiver. Each receiver is

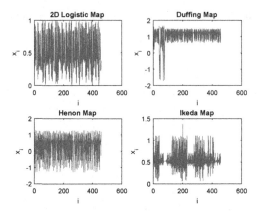

Fig. 2. Generated sequences of two dimensional maps for chaotic initial value 0.2 for x and 0.2 for y (set by receiver) sequences. i on the x *axis* refers to the *indices of the plain text* and a on the y *axis* refers to the *calculated chaotic values*.

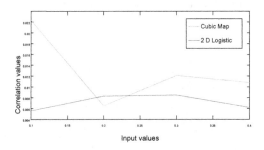

Fig. 3. Correlation values of *Cubic map* (1 Dimensional) and $2D$ *LogisticMap* (2 Dimensional) for various input values.

marked with the unique bias. Hence, the same key can produce different outputs and can be used for encryption.

Plain Text: *In the present digitized world, information exchange with security, confidentiality and non repudiation has become essential. This calls forth the need of cryptographic techniques which are reliable and have the least possibility of intruder decrypting the message with brute force attack or any other means. The sensitive dependence on initial condition, along with its randomness and broadband spectrum have led to the use of chaos theory in cryptography.*

Cipher Text - Cubic Map: *inhtaesdeunrnhsoeoylpi yteitispilouci nmtodhll tts h ycr gahltoe fwsaeer oe, miahe coe hnpt,b.ec ici drwnc.evb hfrc nsh si o ttoei-iphaos ttg gtbcohs eTlairyma ne ndtbcn tpoge efeev.r arhtoesnsoha y tyicieotieznx-denu di ehlit rip uionsntn ea, ee eeaehen nccs sid roentrrlt kiaipird radeasurte* *gdht c hoir ssanaaiihhnati pdwdodtwldaerfdofaihdenwe yetn tsohgueIhreithiri sbd-nrtyteaf riy matgaqcntpnadssreem nmlanabunTot seolft lcepnoonc nea avr ieoc.*

Cipher Text - 2D Logistic Map: *onnteo roeipoahoieowlatlra ania rzneect-noem dmyvuareawespsesann h. eefgta d,oc.h ithtn oinyyut gpfhdhcainsytaminrwr icoyesiad sa e sdsn prsy etgotd gola noir spofhtas mlnitt, deae tyr teira odh thao no ti ih heityifaieo cg nwehihi h egeariuecTfhoeeltttlmhrnrepevd ei ttdnnr nucy liasen ado stefth rimhd, pucailisraea etbntednabde osrgindncoehoos seihen-leh bbisnlt qindippteo ti c nbrescTe crcn lcr fhnu.ncoudkteds bchae eax i i sr wptct ieevrt lsIr e t.*

7 Conclusion

The chaotic neural network combined with the non linear chaotic equations generate an overall network which exhibits high dependence on initial value during both the processes. Further the memory property due to non-volatile nature and nano size of memristor for weight updation increases the convergence and speed of the algorithm.

The overall process does not require synchronization. The sender can send the encrypted message on a public channel and the initial value on a private channel. Further, the receiver has to use the initial value in CNN to generate a chaotic number which has to be used in the equation to obtain a non-linear series that can be used to decrypt and obtain the plain text. In case of 2 Dimensional cryptosystem, the receiver is tagged with the unique initial value for the y sequence of the generator, to establish secure communication. The overall process is secure and reliable with less complexity.

Acknowledgements. The authors would like to thank the faculty and staff of the Department of Electronics and Communication Engineering, SRM University for their support and guidance.

References

1. Wang, L., Duan, M., Duan, S.: Memristive chebyshev neural network and its applications in function approximation. Mathe. Probl. Eng. **2013** (2013). Article ID 429402. doi:10.1155/2013/429402
2. Chua, L.: Memristor-the missing circuit element. IEEE Trans. Circuit Theory **18**(5), 507–519 (1971)
3. Prodromakis, T., Toumazou, C.: A review on memristive devices and applications. In: 2010 17th IEEE International Conference on Electronics, Circuits, and Systems (ICECS). IEEE (2010)
4. Pershin, Y.V., Di Ventra, M.: Experimental demonstration of associative memory with memristive neural networks. Neural Netw. **23**(7), 881–886 (2010)
5. Matthews, R.: On the derivation of a encryption algorithm. Cryptologia **13**(1), 29–42 (1989)

6. Pecora, L.M., Carroll, T.L.: Synchronization in chaotic systems. Phys. Rev. Lett. **64**(8), 821 (1990)
7. Zhen, P., et al.: A survey of chaos-based cryptography. In: 2014 Ninth International Conference on P2P, Parallel, Grid, Cloud and Internet Computing (3PGCIC). IEEE (2014)
8. Kim, K.-K.K., et al.: Wiener's polynomial chaos for the analysis and control of nonlinear dynamical systems with probabilistic uncertainties [historical perspectives]. IEEE Control Syst. **33**(5), 58–67 (2013)
9. Wahab, H.B.A., Jaber, T.A.: Improve NTRU algorithm based on Chebyshev polynomial. In: 2015 World Congress on Information Technology and Computer Applications Congress (WCITCA). IEEE (2015)
10. Nyberg, K.: Generalized feistel networks. In: Kim, K., Matsumoto, T. (eds.) ASIACRYPT 1996. LNCS, vol. 1163, pp. 91–104. Springer, Heidelberg (1996). doi:10.1007/BFb0034838
11. Pareek, N.K., Patidar, V., Sud, K.K.: Cryptography using multiple one-dimensional chaotic maps. Commun. Nonlinear Sci. Numer. Simul. **10**(7), 715–723 (2005)
12. Fridrich, J.: Symmetric ciphers based on two-dimensional chaotic maps. Int. J. Bifurcat. Chaos **8**(06), 1259–1284 (1998)
13. Shi, X., et al.: A novel memristive electronic synapse-based Hermite chaotic neural network with application in cryptography. Neurocomputing **166**, 487–495 (2015)

Classification Pattern Recognition

DSS-PSP - A Decision Support Software for Evaluating Students' Performance

Ioannis E. Livieris[1(✉)], Konstantina Drakopoulou[2], Thodoris Kotsilieris[3], Vassilis Tampakas[1], and Panagiotis Pintelas[2]

[1] Department of Computer and Informatics Engineering (DISK Lab),
Technological Educational Institute of Western Greece, 263-34 Patras, Greece
`livieris@teiwest.gr, vtampakas@teimes.gr`
[2] Department of Mathematics, University of Patras, 265-00 Patras, Greece
`kdrak@math.upatras.gr, ppintelas@gmail.com`
[3] Department of Business Administration (LAIQDA Lab),
Technological Educational Institute of Peloponnese, 241-00 Kalamata, Greece
`tkots@teikal.gr`

Abstract. Prediction, utilizing machine learning and data mining techniques is a significant tool, offering a first step and a helping hand for educators to early recognize those students who are likely to exhibit poor performance. In this work, we introduce a new decision support software for predicting the students' performance at the final examinations. The proposed software is based on a novel 2-level classification technique which achieves better performance than any examined single learning algorithm. Furthermore, significant advantages of the presented tool are its simple and user-friendly interface and that it can be deployed in any platform under any operating system.

Keywords: Educational data mining · Machine learning · Student evaluation system · Decision support software

1 Introduction

Educational Data Mining (EDM) is an essential process where intelligent methods are applied to extract data patterns from students' databases in order to discover key characteristics and hidden knowledge. This new research field has grown exponentially and gained popularity in the modern educational era because of its potential to improve the quality of the educational institutions and system. The application of EDM is mainly concentrated on improving the learning process by the development of accurate models that predict students' characteristics and performance. The importance of EDM is founded on the fact that it allows educators and researchers to extract useful conclusions from sophisticated and complicated questions such as *"find the students who will exhibit poor performance"* in which traditional database queries cannot be applied [17].

G. Boracchi et al. (Eds.): EANN 2017, CCIS 744, pp. 63–74, 2017.
DOI: 10.1007/978-3-319-65172-9_6

Secondary education in Greece is a two-tied system in which the first three years cover general education followed by another three years of senior secondary education. Therefore, the three years of higher secondary education is an important and decisive factor in the life of any student since it acts like a bridge between school education and higher education, offered by universities and higher technological educational institutes [17]. Thus, the ability to monitor students' academic performance and progression is considered essential since the early identification of possible low performers could lead the academic staff to develop learning strategies (extra learning material, exercises, seminars, training tests) aiming to improve students' performance.

During the last decade, research focused on developing efficient and accurate Decision Support Systems (DSS) for predicting the students' future academic performance [5,7,10,17,21,22]. More analytically, an academic DSS is a knowledge-based information system to capture, handle and analyze information which affects or is intended to affect decision making performed by people in the scope of a professional task appointed by a user [4]. The development of an academic DSS is significant to students, educators and educational organizations and it will be more valuable if knowledge mined from the students' performance is available for educational managers in their decision making process.

In this paper, we present DSS-PSP (Decision Support Software for Predicting Students' Performance) which consists of an integrated software application that provides decision support for evaluating students' performance in the final examinations. The presented software incorporates a novel 2-level classifier which achieves better performance than any single learning algorithm. Furthermore, significant advantages of the presented tool are that it employs a simple and user-friendly interface, it is highly expandable due to its modular nature of design and implementation and it can be deployed in any platform under any operating system. Our primary goal is to support the academic task of successfully predicting the students' performance in the final examinations of the school year. Furthermore, decision-makers are able to evaluate various educational strategies and generate forecasts by means of simulating with the input data.

The remainder of this paper is organized as follows: The next section presents a survey of machine learning algorithms that have been used for predicting students' performance. Section 3 presents a description of the educational dataset utilized in our study and our proposed 2-level machine learning classifier. Finally, Sect. 4 presents the main features of our decision support software and Sect. 5 presents our conclusions.

2   Related Studies

During the last decade, the application of data mining for the development of accurate and efficient decision support systems for monitoring students' performance is becoming very popular in the modern educational era. Baker and Yacef [2], Romero and Ventura [26,27] and Dutt et al. [9] have provided excellent reviews of how EDM seeks to discover new insights into learning with new tools and techniques, so that those insights impact the activity of practitioners in all levels of education, as well as corporate learning. Furthermore, they described in detail the process of mining learning data, as well as how to apply the data mining techniques, such as statistics, visualization, classification, clustering and association rule mining.

Deniz and Ersan [7] demonstrated the usefulness of an academic decision-support system in evaluating huge amounts of student-course related data. Moreover, they presented the basic concepts used in the analysis and design of a new DSS software package and presented various ways in which student performance data can be analyzed and presented for academic decision making.

Kotsiantis et al. [13,14] studied the accuracy of six common machine learning algorithms in predicting students that tend to dropout from a distance learning course in Hellenic Open University. Based on previous works, Kotsiantis [12] introduced a prototype decision support system for predicting students' academic progress based on key demographic characteristics, attendance and their marks in written assignments.

Chau and Phung [5] proposed a knowledge-driven DSS for education with a semester credit system by taking advantage of educational data mining. Their proposed educational DSS is helpful for educational managers to make more appropriate and reasonable decisions about students' study and further give support to students for their graduation.

Romero et al. [25] studied how web usage mining can be applied in e-learning systems in order to predict the marks that university students will obtain in the final exam of a course. In addition, they developed a specific mining tool which takes into account the student's active involvement and daily usage in a Moodle forum.

Nagy et al. [21] proposed a "Student Advisory Framework" that integrates educational data mining and knowledge discovery to build an intelligent system. The system can be used to provide pieces of consultations to a first year university student to pursue a certain education track where he/she will likely succeed in, aiming to decrease the high rate of academic failure among these students. The framework acquires information from the datasets which stores the academic achievements of students before enrolling to higher education together with their first year grade after enrolling in a certain department. After acquiring all the relevant information, the intelligent system utilizes both classification and clustering techniques to provied recommendations for a certain department for a new student. Additionally, they presented a case study to prove the efficiency of the proposed framework. Students' data were collected from Cairo Higher Institute for Engineering, Computer Science and Management during the period from 2000 to 2012.

Mishra et al. [20] focused on the early identification of secondary school students who are at high risk of failure thereby helping the educator to take timely actions to improve the students' performance through extra coaching and counseling. Moreover, they classified the important attributes that influenced students' third semester performance and established the effects of emotional quotient parameters that influenced placement.

In more recent works, Livieris et al. [18] introduced a software tool for predicting the students' performance in the course of "Mathematics" of the first year of Lyceum. The proposed software is based on a neural network classifier which exhibits more consistent behavior and illustrates better accuracy than the other classifiers. Along this line, in [17] the authors presented a user-friendly decision support software for predicting students' performance, together with a case study concerning the final examinations in Mathematics. Their proposed tool is based on a hybrid predicting system which combines four learning algorithms utilizing a simple voting scheme. Their experimental results revealed that the application of data mining can gain significant insights in student progress and performance.

Márquez-Vera et al. [19] studied the serious problem of early prediction of high school dropout. They propose a methodology and a specific classification algorithm to discover comprehensible prediction models of student dropout as soon as possible. Additionally, they presented a case study using data from 419 first year high school Mexican students. They authors illustrated that their proposed method is possible of successfully predicting student dropout within the first 4–6 weeks of the course and trustworthy enough to be used in an early warning system.

3   Methodology

The aim of this study is to develop a decision support tool for predicting students' performance at the final examinations. For this purpose, we have adopted the following methodology that consists of three stages.

The first stage of the proposed methodology concerns data collection and data preparation and in the next stage, we introduce our proposed 2-level classification scheme. In the final stage, we evaluate the classification performance of our proposed 2-level classification algorithm with that the most popular and frequently used algorithms by conducting a series of tests.

3.1   Dataset

For the purpose of this study, we have utilized a dataset concerning the performance of 2206 students in courses of "Algebra" and "Geometry" of the first two years of Lyceum. The data have been collected by the *Microsoft showcase school "Avgoulea-Linardatou"* during the years 2007–2016.

Table 1 reports the set of attributes used in our study which concern information about the students' performance such as oral grades, tests grades, final examination grades and semester grades. The assessment of students during each semester consists of oral examination, two 15-minutes prewarned tests, a 1-hour exam and the overall semester performance of each student. The 15-minutes tests include multiple choice questions and short answer problems while the 1-hour exams include several theory questions and a variety of difficult mathematical problems requiring solving techniques and critical analysis. Finally, the overall semester grade of each student addresses the personal engagement of the student in the lesson and his progress. The students were classified utilizing a four-level classification scheme according to students' performance evaluation in the Greek schools: 0–9 (Fail), 10–14 (Good), 15–17 (Very good) and 18–20 (Excellent).

Table 1. List of features used in our study

Attributes	Type	Values
Oral grade of the 1st semester	Real	[0,20]
Grade of the 1st test of the 1st semester	Real	[0,20]
Grade of the 2nd test of the 1st semester	Real	[0,20]
Grade of the 1st semester's final examination	Real	[0,20]
Final grade of the 1st semester	Real	[0,20]
Oral grade of the 2nd semester	Real	[0,20]
Grade of the 1st test of the 2nd semester	Real	[0,20]
Grade of the 2nd test of the 2nd semester	Real	[0,20]
Grade of the 2nd semester's final examination	Real	[0,20]
Final grade of the 2nd semester	Real	[0,20]

Moreover, similar to [17,18], since it is of great importance for an educator to recognize weak students in the middle of the academic period, two datasets have been created based on the attributes presented in Table 1 and on the class distribution.

- $DATA_A$: It contains the attributes concerning the students' performance of the 1st semester.
- $DATA_{AB}$: It contains the attributes concerning the students' performance of the 1st and 2nd semesters.

3.2 2-Level Classifier

Our major challenge was to develop a new classification scheme which can achieve higher classification accuracy than individual classifiers. For this purpose, we introduce a two-level architecture classification scheme. Two-level classification schemes are heuristic pattern recognition tools that are supposed to yield better classification accuracy than single-level ones at the expense of a certain complication of the classification structure [3,15,29].

On the first level of our proposed classification scheme, we utilize a classifier to distinguish the students who are likely to "Pass" or "Fail" in the final examinations. More specifically, this classifier predicts if the student's performance is between 0 and 9 (Fail) or between 10 and 20 (Pass). In the rest of our work, we refer to this classifier as A-Level classifier. In case the verdict (or prediction) of the A-Level classifier is "Pass" in the final examinations, we utilize a second-level classifier in order to conduct a more specialized decision and distinguish between "Good", "Very good" and "Excellent". This classifier is titled as the B-Level classifier. An overview of our proposed 2-level classifier is depicted in Fig. 1.

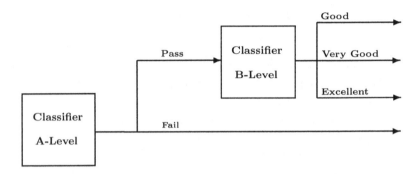

Fig. 1. An overview of the 2-level classifier

It worths noticing that, the decision of the A-Level classifier is more coarse and high level, while the decision of the B-Level is specialized and characterizes the performance of a successful student.

3.3   Experimental Results

In this section, we report a series of tests in order to evaluate the performance of our proposed 2-level classification scheme with that of the most popular and commonly used classification algorithms.

The Back-Propagation (BP) algorithm with momentum [28] was representative of the artificial neural networks which has been established as a well-known learning algorithm for building and training a neural network [16]. From the support vector machines, we have selected the Sequential Minimal Optimization (SMO) algorithm since it is one of the fastest training methods [23] while Naive Bayes (NB) algorithm was the representative of the Bayesian networks [8]. From the decision trees, C4.5 algorithm [24] was the representative in our study and RIPPER (JRip) algorithm [6] was selected as a typical rule-learning technique since it is one of the most usually used methods for producing classification rules. Finally, 3-NN algorithm was selected as instance-based learner [1] with Euclidean distance as distance metric. Moreover, in our experimental results Voting stands for simple voting scheme using RIPPER, 3-NN, BP and SMO as base classifiers presented in [17].

All algorithms have been implemented in WEKA Machine Learning Toolkit [11] and the classification accuracy was evaluated using the stratified 10-fold cross-validation i.e. the data was separated into folds so that each fold had the same distribution of grades as the entire data set.

Table 2 summarizes the accuracy of each individual classifier and the accuracy of the 2-level classification scheme, relative to both datasets. Clearly, our proposed 2-level scheme significantly improved the performance of each individual classifier from 3.8% to 9.5%.

Table 2. Individual classifier and 2-level classifier accuracy (%) for each dataset

| Classifier | DATA$_A$ | | DATA$_{AB}$ | |
	Individual	2-level	Individual	2-level
BP	81.1%	89.1%	76.2%	89.1%
SMO	84.1%	89.9%	82.7%	89.9%
NB	75.8%	81.2%	71.7%	79.0%
C4.5	84.2%	89.8%	84.1%	89.8%
JRip	84.7%	90.0%	86.1%	89.6%
3-NN	85.2%	88.1%	78.7%	85.3%
Voting	85.8%	88.2%	84.4%	87.0%

Hence, motivated by the previous results we evaluated the proposed 2-level scheme utilizing different classification algorithms at each level. Our aim is to find which of these classifiers is best suited for A-Level and B-Level for producing the highest performance.

Tables 3 and 4 summarize the performance of the proposed 2-level classifier utilizing various A-Level and B-Level classifiers for both datasets, respectively. The best performing technique for each dataset is illustrated in boldface.

Table 3. 2-level classifier classification accuracy (%) on DATA$_A$

| | | B-level classifier | | | | | |
		BP	SMO	NB	C4.5	JRip	3-NN
A-level classifier	BP	89.1%	89.7%	85.9%	89.7%	89.7%	88.0%
	SMO	89.8%	89.9%	85.8%	90.0%	89.9%	88.6%
	NB	84.4%	85.0%	81.2%	84.9%	85.0%	83.1%
	C4.5	89.7%	**90.1%**	86.2%	89.8%	89.8%	88.5%
	JRip	89.8%	90.0%	86.2%	89.9%	90.0%	88.2%
	3-NN	89.1%	89.7%	85.8%	89.7%	89.8%	88.1%

Firstly, we observe that the C4.5 reports the best classification performance as A-Level classifier, slightly outperforming JRip, relative to both datasets. In

Table 4. 2-level classifier classification accuracy (%) on $DATA_{AB}$

		B-level classifier					
		BP	SMO	NB	C4.5	JRip	3-NN
A-level classifier	BP	89.1%	89.6%	81.5%	89.7%	89.4%	86.3%
	SMO	89.7%	89.9%	81.8%	90.0%	89.8%	86.9%
	NB	78.5%	84.9%	79.0%	79.2%	78.9%	76.2%
	C4.5	89.6%	**90.3%**	81.8%	89.8%	89.8%	86.6%
	JRip	89.5%	90.0%	81.6%	89.9%	89.6%	86.7%
	3-NN	88.0%	89.7%	80.5%	88.6%	88.4%	85.3%

particular, it exhibits 86.2%–90.1% and 81.8%–90.3% classification performance, for $DATA_A$ and $DATA_{AB}$, respectively. Moreover, SMO is best B-Level classifier reporting the best overall performance, followed by C4.5. More specifically, SMO reported 84.9%–90.3% classification performance, regarding both datasets. Finally, it worths noticing that the best classification performance of the 2-level classifier was presented in case C4.5 was selected as A-Level classifier and SMO as a B-Level one.

4   DSS-PSP: Decision Support Software

For the purpose of this study, we developed a user-friendly decision support software, which is called DSS-PSP[1] for predicting the performance of an individual student at the final examinations based on its grades on the 1st and/or 2nd semester. The software is based on the WEKA Machine Learning Toolkit and has been developed in JAVA, making it platform independent and easily executed even by non-experienced users. Notice that DSS-PSP consists an updated version of the software presented in [17] with similar functionalities.

Figure 2 illustrates a screenshot of our proposed decision support software DSS-PSP illustrating its main features:

- **Student personal data:** This module is optionally used to import student's name, surname, father's name and remarks.
- **1st Semester's grades:** This module is used to import the student's grades of the first semester.
- **2nd Semester's grades:** This module is used to import the student's grades of the second semester.
- **Messages:** This module is used to print the messages, warnings and outputs of the tool.

---

[1] The tool is available at http://www.math.upatras.gr/~livieris/DSSPSP.zip. Notice that Java Virtual Machine (JVM) 1.2 or newer is needed for the execution of the program.

Subsequently, we demonstrate a use case in order to illustrate the functionalities of DSS-PSP. Firstly, the user/educator can use our data embedded in the software by clicking on the button "Import data" or he can load his/her data collected from his/her own past courses in XLSX file format.

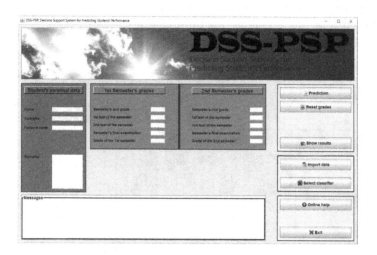

Fig. 2. DSS-PSP interface

Next, by clicking on the "Select classifier" button, DSS-PSP provides the ability to the user to choose between the old classifier based on a voting scheme [17] and the proposed 2-level classification algorithm (Fig. 3) which utilizes C4.5

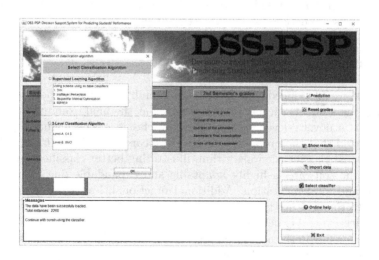

Fig. 3. DSS-PSP: select classifier

as a A-Level classifier and SMO as B-Level classifier. It worths noticing that the proposed 2-level classification algorithm is more accurate and it can be trained significantly faster than the voting scheme presented in [17].

After that, the user can import the new student's grades of the 1st and/or 2nd semester in the corresponding fields. Next, the DSS-PSP is able to predict the student's performance at the final examinations by simply clicking on the button "Prediction" as it is illustrated in Fig. 4.

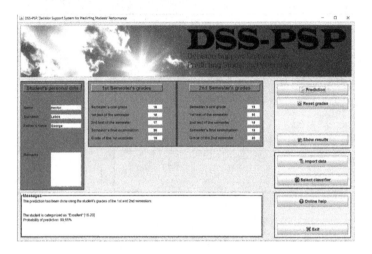

Fig. 4. DSS-PSP: prediction about the performance of a new student at the final examinations

Moreover, the tool provides on-line help, for novice users and the ability to see all previous predictions by clicking the button "Show results".

5   Conclusions

In this work, we introduced DSS-PSP, a user-friendly decision support system for predicting the students' academic performance at the final examinations which incorporates a novel 2-level machine learning classifier. Our numerical experiments revealed that the proposed scheme significantly improves the accuracy of each individual classification algorithm, illustrating better classification results. Moreover, the software is highly adaptable and expandable due to its modular design and implementation. Additional functionalities can be easily added according to the user needs.

Our objective and expectation is that this work could be used as a reference for decision making in the admission process and strengthen the service system in educational institutions by offering customized assistance according to students' predicted performance.

Acknowledgement. The authors are grateful to the Microsoft showcase school "Avgoulea-Linardatou" for the collection of the data used in our study, for the evaluation of the tool and for their kind and valuable comments, which essentially improved our work.

References

1. Aha, D.: Lazy Learning. Kluwer Academic Publishers, Dordrecht (1997)
2. Baker, R.S., Yacef, K.: The state of educational data mining in 2009: a review future visions. J. Educ. Data Min. **1**(1), 3–17 (2009)
3. Barabash, Y.L.: Collective Statistical Decisions in Recognition. Radio i Sviaz, Moscow (1983)
4. Bresfelean, V.P., Ghisoiu, N.: Higher education decision making and decision support systems. WSEAS Trans. Adv. Eng. Educ. **7**, 43–52 (2010)
5. Chau, V.T.N., Phung, N.H.: A knowledge driven education decision support system. In: 2012 IEEE RIVF International Conference on Computing Communication Technologies, Research, Innovation, and Vision for the Future, pp. 1–6 (2012)
6. Cohen, W.: Fast effective rule induction. In: International Conference on Machine Learning, pp. 115–123 (1995)
7. Deniz, D.Z., Ersan, I.: An academic decision support system based on academic performance evaluation for student and program assessment. Int. J. Eng. Educ. **18**(2), 236–244 (2002)
8. Domingos, P., Pazzani, M.: On the optimality of the simple Bayesian classifier under zero-one loss. Mach. Learn. **29**, 103–130 (1997)
9. Dutt, A., Ismail, M.A., Herawan, T.: A Systematic Review on Educational Data Mining. IEEE Access (2017)
10. Grivokostopoulou, F., Perikos, I., Hatzilygeroudis, I.: Utilizing semantic web technologies and data mining techniques to analyze students learning and predict final performance. In: 2014 International Conference on Teaching, Assessment and Learning (TALE), pp. 488–494. IEEE (2014)
11. Hall, M., Frank, E., Holmes, G., Pfahringer, B., Reutemann, P., Witten, I.: The WEKA data mining software: an update. SIGKDD Explor. Newsl. **11**, 10–18 (2009)
12. Kotsiantis, S.: Use of machine learning techniques for educational proposes: a decision support system for forecasting students' grades. Artif. Intell. Rev. **37**, 331–344 (2012)
13. Kotsiantis, S.B., Pierrakeas, C.J., Pintelas, P.E.: Preventing student dropout in distance learning using machine learning techniques. In: Palade, V., Howlett, R.J., Jain, L. (eds.) KES 2003 Part II. LNCS, vol. 2774, pp. 267–274. Springer, Heidelberg (2003). doi:10.1007/978-3-540-45226-3_37
14. Kotsiantis, S., Pierrakeas, C.J., Pintelas, P.: Predicting students performance in distance learning using machine learning techniques. Appl. Artif. Intell. **18**(5), 411–426 (2004)
15. Kuncheva, L.I.: "Change-glasses" approach in pattern recognition. Pattern Recognit. Lett. **14**, 619–623 (1993)
16. Lerner, B., Guterman, H., Aladjem, M., Dinstein, I.: A comparative study of neural network based feature extraction paradigms. Pattern Recognit. Lett. **20**(1), 7–14 (1999)
17. Livieris, I., Mikropoulos, T., Pintelas, P.: A decision support system for predicting students' performance. Themes Sci. Technol. Educ. **9**, 43–57 (2016)

18. Livieris, I.E., Drakopoulou, K., Pintelas, P.: Predicting students' performance using artificial neural networks. In: Information and Communication Technologies in Education (2012)
19. Márquez-Vera, C., Cano, A., Romero, C., Noaman, A.Y.M., Mousa-Fardoun, H., Ventura, S.: Early dropout prediction using data mining: a case study with high school students. Expert Syst. **33**(1), 107–124 (2016)
20. Mishra, T., Kumar, D., Gupta, S.: Mining students' data for prediction performance. In: 2014 Fourth International Conference on Advanced Computing Communication Technologies, pp. 255–262 (2014)
21. Nagy, H.M., Aly, W.M., Hegazy, O.F.: An educational data mining system for advising higher education students. World Acad. Sci. Eng. Technol. Int. J. Inf. Eng. **7**(10), 175–179 (2013)
22. Noaman, A.Y., Luna, J.M., Ragab, A.H.M., Ventura, S.: Recommending degree studies according to students attitudes in high school by means of subgroup discovery. Int. J. Comput. Intell. Syst. **9**(6), 1101–1117 (2016)
23. Platt, J.: Using sparseness and analytic QP to speed training of support vector machines. In: Kearns, M.S., Solla, S.A., Cohn, D.A. (eds.) Advances in Neural Information Processing Systems, pp. 557–563. MIT Press, Cambridge (1999)
24. Quinlan, J.R.: C4.5: Programs for Machine Learning. Morgan Kaufmann, San Francisco (1993)
25. Romero, C., Espejo, P.G., Zafra, A., Romero, J.R., Ventura, S.: Web usage mining for predicting final marks of students that use moodle courses. Comput. Appl. Eng. Educ. **21**(1), 135–146 (2013)
26. Romero, C., Ventura, S.: Educational data mining: a survey from 1995 to 2005. Expert Syst. Appl. **33**, 135–146 (2007)
27. Romero, C., Ventura, S.: Educational data mining: a review of the state of the art. IEEE on Trans. Syst. Man Cybern. - Part C: Appl. Rev. **40**(6), 601–618 (2010)
28. Rumelhart, D.E., Hinton, G.E., Williams, R.J.: Learning internal representations by error propagation. In: Rumelhart, D., McClelland, J. (eds.) Parallel Distributed Processing: Explorations in the Microstructure of Cognition, pp. 318–362. Massachusetts, Cambridge (1986)
29. Xu, L., Krzyzak, A., Suen, C.Y.: Methods of combining multiple classifiers and their applications to handwriting recognition. IEEE Trans. Syst. Man Cybern. **22**, 418–435 (1992)

Predicting Student Performance in Distance Higher Education Using Active Learning

Georgios Kostopoulos[1(✉)], Anastasia-Dimitra Lipitakis[3],
Sotiris Kotsiantis[1,4], and George Gravvanis[2,4]

[1] Educational Software Development Laboratory (ESDLab),
Department of Mathematics, University of Patras, Patras, Greece
kostg@sch.gr, sotos@math.upatras.gr
[2] Department of Electrical and Computer Engineering, School of Engineering,
Democritus University of Thrace, University Campus, Xanthi, Greece
ggravvan@ee.duth.gr
[3] Department of Informatics and Telematics,
Harokopio University of Athens, Kallithea, Greece
adlipita@hua.gr
[4] Hellenic Open University, Parodos Aristotelous 18, 26335 Patras, Greece

Abstract. Students' performance prediction in higher education has been identified as one of the most important research problems in machine learning. Educational data mining constitutes an important branch of machine learning trying to effectively analyze students' academic behavior and predict their performance. Over recent years, several machine learning methods have been effectively used in the educational field with remarkable results, and especially supervised classification methods. The early identification of in case fail students is of utmost importance for the academic staff and the universities. In this paper, we investigate the effectiveness of active learning methodologies in predicting students' performance in distance higher education. As far as we are aware of there exists no study dealing with the implementation of active learning methodologies in the educational field. Several experiments take place in our research comparing the accuracy measures of familiar active learners and demonstrating their efficiency by the exploitation of a small labeled dataset together with a large pool of unlabeled data.

Keywords: Distance higher education · Performance prediction · Unlabeled data · Pool-based active learning · Uncertainty sampling query

1 Introduction

In recent years, the prediction of students' progress, failure or drop out has become an important and challenging research problem in the educational field. A number of rewarding studies have been carried out leading to the development of several machine learning techniques, while investigating the factors influencing students' academic behavior and performance. The application of Data Mining (DM) techniques in the educational field have resulted to the formation of Educational Data Mining (EDM), a sub-field of DM. EDM constitutes an important branch of machine learning trying to

© Springer International Publishing AG 2017
G. Boracchi et al. (Eds.): EANN 2017, CCIS 744, pp. 75–86, 2017.
DOI: 10.1007/978-3-319-65172-9_7

effectively analyze students' academic behavior and predict their performance [23] using data usually originated from university databases, course management systems, online courses and educational software.

Nowadays, many universities offer innovative and high quality education via distance learning. Several undergraduate and postgraduate online courses are adapted to the personal needs and knowledge level of students. Students plan their own study schedule using a variety of learning materials under the continuous guidance and assistance of the academic staff. Unfortunately, successful completion of distance courses is influenced by even more factors, such as educational background, family and job obligations. The early identification of in case fail students in good time is of utmost importance for the academic staff and the universities, to improve the learning process and provide high quality education. In recent years, EDM has turned to a powerful tool for extracting and analyzing student data, as well as for decision making in educational institutions [30].

This study proposes the employment of several active learning algorithms to predict students' performance in higher education and more specifically in distance education. Numerous experiments are conducted measuring the accuracy of each algorithm based on several demographic, social, family and university variables. Moreover, we examine how early the prediction of low performance students could be done in order to offer them extra support and personalized guidance. Though active learning does reduce the amount of labeling needed [25], it is not clear how well the method applies in the educational field. To the best of our knowledge there exists no study dealing with the implementation of active learning methodologies for the prediction of students' academic progress, so this study is considered to be an initial step in this direction.

The rest of the paper is organized as follows: Sect. 2 review srecent studies concerning the implementation of machine learning techniques on detecting students' performance in higher education. Sect. 3 presents the central points of the active learning theory, while Sect. 4 provides a description of the dataset and the main study questions. In Sect. 5 we present the experiments that were conducted, the results obtained and a thorough analysis of these results. Finally, the paper concludes by summarizing the main aspects of the study and considering some thoughts for future research.

2 Educational Data Mining

As mentioned above, the need to predict the performance of high school and university students has resulted in the implementation of various machine learning techniques in the EDM field. A number of considerable studies examine the effectiveness of supervised methods, mainly to estimate the final output (pass or fail) of students, using predictive models based on several independent variables. On a recent work, Slater et al. [28] presented a variety of tools for conducting EDM and learning analytics research. Moreover, the lack of labeled data shows that the application of Semi-Supervised Learning (SSL) and Active Learning methods to identify students at risk in sufficient time is of major need.

Kotsiantis et al. [11] proposed an online ensemble of supervised algorithms to predict the performance (pass or fail) on the final examination test of students attending a distance university course. Naïve Bayes (NB) classifier, WINNOW (a linear online algorithm) and k-NN classifier constitute an online ensemble operating in incremental mode, while using the majority voting methodology for the output prediction. The proposed ensemble of classifiers outperforms well-known algorithms, such as the RBF, BP, C4.5, k-NN and SMO, and may be used as a predictive tool from tutors during the academic year to support and enhance low performers.

Romero et al. [22] proposed the use of different data mining approaches to gather information regarding first year students' knowledge on a computer science course. Using a subset of variables related to the subject of the course from the participation of students in on-line discussion forums, through a Moodle module, they seek to predict students' performance (pass or fail) in the final exam by applying classification and clustering algorithms.

Huang and Fung [6] developed four types of mathematical models to predict students' grade in the final exam of an engineering dynamics course. These models included multi-variable linear regression, multilayer perceptron network, radial basis function (RBF) network, and the support vector machine (SVM) model. Twenty four predictive models were developed based on six different combinations of predictors' variables.

Xing et al. [31] generated a student performance prediction model using data collected from a collaborative geometry problem solving environment granted byGeogebra. The model was based on GP technique and the online participation of students, and it takes precedence over traditional modeling algorithms.

Koprinska et al. [9] built a decision tree classifier for detecting students at risk of failing in a first year computer programming course. Several data related to students' assessment marks, activity in an automatic marking system and discussion forums were used for the early identification of low performers. Using only the assessment attributes, an accuracy of 83% is achieved, while using all attributes the accuracy reached to 87%.

Kostopoulos et al. [10] examined the possibility to predict the performance of students attending a distance course module in the Hellenic Open University using SSL methods. Several experiments were conducted using well-known SSL algorithms, such as Tri-training, Co-training and Self-training. The presented results were encouragingly positive showing the supremacy of SSL methods contrary to familiar supervised methods.

Santana et al. [24] examined the effectiveness of familiar supervised techniques (NB, SVM, J48 decision tree and multilayer neural networks) for early prediction of students that are possible to fail in two introductory programming courses in a Brazilian university. The first dataset refers to 262 students attending a 10 weeks distance course and the second one refers to 161 students attending a 16 weeks on-campus course. Both datasets contain several demographic, social and university attributes. The experimental results showed that EDM methods are effective in the early identification of low performers. Moreover, the effectiveness was improved after data preprocessing and fine-tuning of algorithms.

A thorough analysis of the studies concerning the educational field reveals that several DM techniques have been effectively applied for the detection of students' failure in higher education, but none of them employs active learning methods for the prediction of students' academic performance.

3 Active Learning

Machine learning has already been at the core of computer science and information technology. Data analysis is more crucial now than ever before, due to vast quantities of data which we are being overwhelmed with on a daily basis. There are mainly three basic and commonly used types of machine learning: supervised, unsupervised and SSL. Given a set of labeled data, supervised learning trains a function f to predict unseen future data. Classification and regression are the two main branches of supervised learning depending on the nature of f. As regards classification, it is one of the most commonly studied problems in machine learning [29]. On the other hand, unsupervised learning methods seek to discover hidden structure and interesting similarities on unlabeled data without human supervision. Clustering, novelty detection and dimensionality reduction are familiar unsupervised methods.

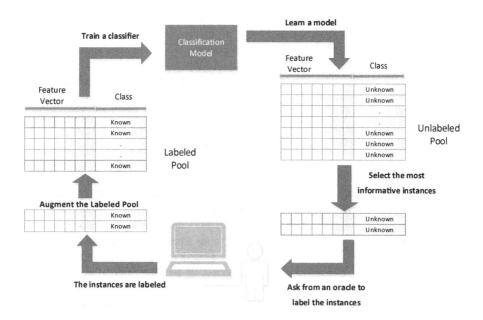

Fig. 1. The active learning framework

The wide availability of unlabeled data and, at the same time, the lack of labeled data has resulted in the development of SSL and Active or Query Learning which are regarded as the two primary paradigms for learning with labeled and unlabeled data [34]. In active learning the learning algorithm selects from unlabeled data the most informative instances and asks from an oracle to label them. These instances augment the training set and the procedure is repeated (Fig. 1). The key point in active learning is to build a high accuracy classifier (active learner) without making too many queries using a small labeled training set [2].

Various active learning scenarios and query strategies have been developed to maximize the classifier's performance by picking the "best" examples to label. Membership query synthesis, stream-based selective sampling and pool-based sampling are the three main scenarios in which active learners ask queries [25]. In this study we use the pool-based active learning scenario which is considered to be the most commonly used and well-studied active learning scenario. This scenario assumes that there is a small labeled dataset L and a large unlabeled dataset U (unlabeled data pool). Queries are selected from U according to specific evaluation criteria [12]. The selection of the most informative unlabeled instances can take place through several query strategies, such as uncertainty sampling and query-by-committee [20] and is usually based on two basic and widely used criteria: informativeness and representativeness [7]. Uncertainty sampling queries the label of the most uncertain to label instances. A shortcoming of uncertainty sampling is that it only considers information about the most probable label and fails to take account the remaining label distribution. Margin sampling aims to remedy this shortcoming by incorporating the posterior of the second most likely label [25]. For binary classification margin sampling query is the most popular strategy, while entropy sampling query is appropriate for multiclass problems using entropy as information, choice and uncertainty measure [26]. The pseudo-code description of active learning is shown in Algorithm 1:

Algorithm 1. Active Learning

Input: labeled dataset L, unlabeled dataset U.
1. Initially, apply base learner B to the training set L to obtain classifier C.
2. Apply C to unlabeled dataset U.
3. From U, select the most informative m instances to learn from (I).
4. Ask the teacher/expert for labeling the m instances (I).
5. Move I, with supplied classifications, from U to L ($U=U-I$).
6. Re-train using B on labeled set ($L=L+I$) to obtain a new classifier, C.
7. Repeat steps 2 to 6, until U is empty or until some stopping criterion is met
8. Output a classifier that is trained on L.

---

4   Study Questions and Dataset Description

Our study is focused on the following three questions:

1. How do active learning techniques perform for predicting students' performance in distance higher education?

2. Can we predict the students that are going to fail or pass the final examinations in a timely manner?
3. How accurate is the prediction and which are the influential variables?

The dataset used for this study was provided by the Hellenic Open University (HOU). A total of 344 instances part the dataset, and each instance is characterized by the values of 16 variables (Table 1). The instances correspond to students attending the "Introduction to Informatics" module of the "Computer Science" course. The completion of the course moduler equires the submission of four written assignments and a grade equal or higher than five in the final module examination. Students can undertake the final examination of the module only if they have successfully completed the written assignments with a total score of twenty or more (ten grade scale). Students have the opportunity of attending four optional four-hour contact sessions with their tutors during the academic year.

Table 1. Dataset variables

Variable	Type	Values	Description
Gender	Nominal	Male, female	Student's gender
Age	Integer	[24, 46]	Student's age
Domestic	Nominal	Single, married	Student's domestic
Children	Integer	0,1,2,3,4	Number of children
Work	Iinteger	0,1,2,3	Working time
Comp_Knowledge	Binary	Yes, no	Computer knowledge
Comp_Job	Integer	0,1,2	Job relation to computers usage
OCSi, i = 1,2,3,4	Binary	0, 1	Absence/presence in the i-thoptional contact session
TESTi, i = 1,2,3,4	Real	[−1, 10]	Grade of the i-th written assignment
Final	Ordinal	Fail, pass	Final performance

The first seven variables are related with student's demographic data and personal information such as gender, age, domestic, children, working time, computer knowledge and occupation, and are also being referred as pre-university or time-invariant variables. Especially, the variable related to student's knowledge of computers (Comp_Knowledge) is crucial for students attending online courses [27]. The next eight variables refer to the students' performance on the four written assignments (TESTi, i = 1,2,3,4) and their presence or absence in the four optional contact sessions (OCSi, i = 1,2,3,4). Grades in written assignments range from −1 to 10 (−1 corresponds to no submission of the written assignment). Presence or absence in contact sessions corresponds to values 1 and 0 respectively. The eight university variables are being added consecutively during the academic year and are influential for the timely prediction of students' performance, as demonstrated from the experimental results set out in the following section. The outcome binary variable "Final" is whether the student is going to pass or fail the final examination of the course module.

5   Experimental Setup

Initially, the dataset was partitioned into 10 folds of similar size using the 10-fold cross validation procedure. In each fold, 14 instances of the training set formed the labeled set and the rest 295 formed the unlabeled set. A Pool-based sampling scenario with the Margin Sampling Query strategy was used. We selected a batch size of two unlabeled examples for labeling at each of the iterations, and we have defined a maximum number of 15 iterations as a stopping criterion. On this basis, at the end of the learning process there will be 44 labeled instances.

A set of six familiar supervised algorithms from Weka were used as base classifiers to form six respective active learners. These algorithms are:

- J48 decision tree [19]
- JRip [1]
- Logistic Regression (LR) [16]
- Multilayer Perceptrons (MLPs), representative of Neural Networks [5]
- Naïve Bayes (NB) [8]
- Sequential Minimal Optimization (SMO), a very effective SVM algorithm [18]

SMO is an improved Support Vector Machines (SVMs) algorithm using an analytic quadratic programming step as an inner loop to optimize the objective function [18]. It is considered to be a simple, fast and easy to implement iterative algorithm, due to the fact that only two variables at each of the iterations are selected, while the others remain constant. The NB is a fast, effective and easy to implement algorithm for machine learning, and especially for classification [33]. It is originated on the Bayes theorem to estimate the probabilities of each class and is based on the assumption that all attributes are independent given the class attribute. It is the simplest form of the Bayesian Networks. J48 classifier is a Weka implementation of the frequently used C4.5 decision tree classification algorithm which was originally developed by Quinlan [17]. LR is a very powerful data analysis method trying to build a model to describe the relationship between a set of independent variables and a dependent variable which is considered to be binary [16]. JRip is an inference and rule learning algorithm which was originally introduced by Cohen [1]. It is a java optimized version of RIPPER, implemented in Weka. MLPs classifier is a form of neural networks consisting of a set of simple interconnected nodes connected by weights and output signals [3].

The experiments were carried out using the JCLALtool and were conducted in five successive steps. JCLAL is a computational tool implemented in java for performing active learning methods for both researchers and programmers. JCLAL provides a friendly and high-level environment facilitating the implementation of existing active learning methodologies or the development of new ones [21]. Since Weka is also implemented in java, JCLAL can be easy integrated. The 1st step includes all pre-university variables related with student's demographic data and personal information. In the 2nd step, variables OCS1, TEST1 and OCS2 are added to the previous set of variables. The 3rd step includes in addition the grade of the second written assignment (TEST2), while in the 4th step variables OCS3 and TEST3 are added. The final 5th step includes all the variables of the dataset.

In each step we measure the accuracy of the above mentioned active learners, which corresponds to the percentage of the correctly classified instances (Table 2). As Table 2 shows, the accuracy of all active learners is constantly increasing as new variables are added to the training set. Moreover, SMO, NB and MLPs appear to be superior in all steps of the experiments with an accuracy measure ranging from 58.18% to 81.09%. In the middle of the academic year (3rd step) SMO and NB accuracy exceed 75%, while in the final step the percentage of correctly classified instances exceeds 80% indicating that we can predict low performers in an efficient and timely manner.

Table 2. Correctly classified instances (%)

Algorithm	1st step	2nd step	3rd step	4th step	5th step
SMO	64.61	66.27	75.54	78.78	80.82
NB	59.86	66.58	75.30	77.60	80.60
J48	54.68	57.00	66.92	72.97	76.47
JRip	53.76	56.69	66.93	75.58	78.49
LR	54.66	66.61	72.09	73.29	79.34
MLPs	58.18	65.71	72.97	79.61	81.09

We evaluate the performance of the above active learners using the Friedman Aligned Ranks nonparametric test [4]. According to the test results (Table 3) the algorithms are ranking from the best performer to the lower one. So, SMO takes precedence over the rest active learners, followed by NB and MLPs. The accuracy measure of the active learner using SMO as base classifier ranges from 64.61% to 80.82%. We can make an initial prediction of students' performance at the beginning of the academic year based only on pre-university information with an accuracy of 64.61%, while the accuracy percentage rises to 78.78% at the middle of the second semester. Students at risk of failing can be quite accurately identified before the end of the first semester, since SMO accuracy is 75.74%, while NB accuracy is 75.30%.

Table 3. Friedman aligned ranks test (significance level of 0.05)

Algorithm	Rank
SMO	**6.2**
NB	9.4
MLPs	10.8
LR	17.0
JRIP	24.4
J48	25.2

In addition, a comparison is made between the active learner using SMO as base classifier and the SMO supervised classifier measuring the accuracy in each one of the experiments steps (Table 4). Using the Wilcoxon paired test [32] to test the null hypothesis Ho that the medians of the differences between the two algorithms are equal, it appears that Ho is accepted, since p-value is 0.0796 and the statistic is 1. So, both active learning and supervised learning achieve the same level of accuracy. However, supervised learning requires a large amount of labeled data to train the classifier (309 instances), while only a small amount of labeled data (44 instances) are needed for achieving similar accuracy using active learning.

Table 4 Active learning vs supervised learning

	Active learner (SMO)	Supervised SMO
1st step	64.61	64.2
2nd step	66.27	72.7
3rd step	75.54	77.6
4th step	78.78	79.4
5th step	80.82	81.4

Regarding the active learner using SMO, and for each one of the five steps of the experiments we present a graph (Fig. 2) illustrating the accuracy percentage rate related to the number of labeled instances during the learning process and measure the area (Table 5) bounded by the accuracy curve and the x-axis (AUC). AUC is another commonly used metric for evaluating the performance of a binary classifier [14].

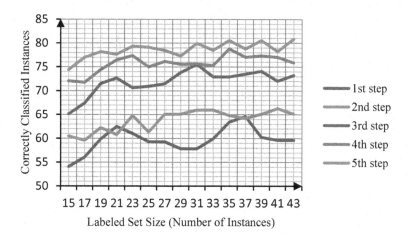

Fig. 2. AUC (SMO base classifier)

Table 5. AUC (SMO base classifier)

Step	1st step	2nd step	3rd step	4th step	5th step
Area	59.691	63.694	71.758	75.702	78.425

6 Conclusions

In this study we examine the effectiveness of active learning algorithms to predict students' performance (pass or fail) in the final exams of a distance learning undergraduate course module in HOU. The prediction of students' performance has been an interesting and highly important research topic for educational institutions in recent years. Identifying low performers as soon as possible could lead to the development of personalized learning strategies in accordance with students' needs and specificities enhancing their academic performance.

A number of experiments were conducted using several active learners and measuring the percentage of correctly classified instances in predicting students' performance in final exams. We have used several pre-university, as well as university variables showing their impact in students' success during the academic year. The experimental results indicate that a good predictive accuracy can be achieved using active learning algorithms. More specifically, SMO is the best performer scoring 64.61% accuracy based only on pre-university information at the beginning of the academic year. As new educational variables are added, the accuracy measure is continuously increased reaching 75.54% at the end of the first semester. Moreover, accuracy exceeds 80% before the final examinations. So, we can predict low performers in an efficient and timely manner. The grade of the second written assignment (TEST2) seems to be the influential variable, since there is substantial increase of accuracy measure (from 66.27% to 75.54%). Moreover, we compared the prediction accuracy of the active learner using SMO as base classifier to the corresponding supervised base classifier. It appears that both active learning and supervised learning achieve the same level of accuracy, however, the amount of labeled data used in active learning is clearly limited.

This study is a promising start in implementing active learning methodologies for the early identification of low performance students in higher education. A considerable advantage of active learning is the limited number of labeled examples needed to train a classifier, achieving a predictive accuracy as if all examples were labeled. On the other hand unlabeled examples are in great abundance. SSL methods have been effectively used in many scientific fields trying to take as much advantage of the unlabeled data as possible. An interesting aspect is to combine SSL and active learning [13].

Another interesting approach is the application of machine learning methodologies for knowledge discovery in educational data [15], as well as in the context of recommendation systems. An implemented decision support system as presented in [17] can provide personalized recommendations and that is an interesting topic to be studied in the future.

References

1. Cohen, W.W.: Fast effective rule induction. In: Proceedings of the Twelfth International Conference on Machine Learning, pp. 115–123 (1995)
2. Dasgupta, S.: Two faces of active learning. Theor. Comput. Sci. **412**(19), 1767–1781 (2011)

3. Gardner, M.W., Dorling, S.R.: Artificial neural networks (the multilayer perceptron)-a review of applications in the atmospheric sciences. Atmos. Environ. **32**(14), 2627–2636 (1998)
4. Hodges, J.L., Lehmann, E.L.: Rank methods for combination of independent experiments in analysis of variance. Ann. Math. Stat. **33**(2), 482–497 (1962)
5. Hornik, K., Stinchcombe, M., White, H.: Multilayer feedforward networks are universal approximators. Neural Netw. **2**(5), 359–366 (1989)
6. Huang, S., Fang, N.: Predicting student academic performance in an engineering dynamics course: a comparison of four types of predictive mathematical models. Comput. Educ. **61**, 133–145 (2013)
7. Huang, S.J., Jin, R., Zhou, Z.H.: Active learning by querying informative and representative examples. In: Advances in Neural Information Processing Systems, pp. 892–900 (2010)
8. John, G.H., Langley, P.: Estimating continuous distributions in Bayesian classifiers. In: Proceedings of the Eleventh Conference on Uncertainty in Artificial Intelligence, pp. 338–345. Morgan Kaufmann Publishers Inc. (1995)
9. Koprinska, I., Stretton, J., Yacef, K.: Students at risk: detection and remediation. In: Educational Data Mining (2015)
10. Kostopoulos, G., Kotsiantis, S., Pintelas, P.: Predicting student performance in distance higher education using semi-supervised techniques. In: Bellatreche, L., Manolopoulos, Y. (eds.) MEDI 2015. LNCS, vol. 9344, pp. 259–270. Springer, Cham (2015). doi:10.1007/978-3-319-23781-7_21
11. Kotsiantis, S., Patriarcheas, K., Xenos, M.: A combinational incremental ensemble of classifiers as a technique for predicting students' performance in distance education. Knowl. Syst. **23**(6), 529–535 (2010)
12. Kremer, J., Steenstrup Pedersen, K., Igel, C.: Active learning with support vector machines. Wiley Interdisc. Rev.: Data Min. Knowl. Discov. **4**(4), 313–326 (2014)
13. Leng, Y., Xu, X., Qi, G.: Combining active learning and semi-supervised learning to construct SVM classifier. Knowl. Syst. **44**, 121–131 (2013)
14. Ling, C.X., Huang, J., Zhang, H.: AUC: a statistically consistent and more discriminating measure than accuracy. IJCAI **3**, 519–524 (2003)
15. Luna, J.M., Castro, C., Romero, C.: MDM tool: a data mining framework integrated into Moodle. Comput. Appl. Eng. Educ. **25**(1), 90–102 (2017)
16. Ng, A.Y., Jordan, M.I.: On discriminative vs. generative classifiers: a comparison of logistic regression and naive bayes. Adv. Neural. Inf. Process. Syst. **2**, 841–848 (2002)
17. Noaman, A.Y., Luna, J.M., Ragab, A.H., Ventura, S.: Recommending degree studies according to students' attitudes in high school by means of subgroup discovery. Int. J. Comput. Intell. Syst. **9**(6), 1101–1117 (2016)
18. Platt, J.: Sequential minimal optimization: a fast algorithm for training support vector machines, Microsoft Research. Technical report MSR-TR-98-14 (1998)
19. Quinlan, J.R.: C4.5: Programs for Machine Learning. Elsevier, Amsterdam (1993)
20. Ramirez-Loaiza, M.E., Sharma, M., Kumar, G., Bilgic, M.: Active learning: an empirical study of common baselines. Data Min. Knowl. Discov. **31**, 1–27 (2016)
21. Reyes, O., Pérez, E., del Carmen Rodrıguez-Hernández, M., Fardoun, H.M., Ventura, S.: JCLAL: a Java framework for active learning. J. Mach. Learn. Res. **17**(95), 1–5 (2016)
22. Romero, C., López, M.I., Luna, J.M., Ventura, S.: Predicting students' final performance from participation in on-line discussion forums. Comput. Educ. **68**, 458–472 (2013)
23. Romero, C., Ventura, S.: Educational data mining a review of the state of the art. IEEE Trans. Syst. Man Cybern. Part C (Appl. Rev) **40**(6), 601–618 (2010)

24. Santana, M.A., Costa, E.B., Fonseca, B., Rego, J., de Araújo, F.F.: Evaluating the effectiveness of educational data mining techniques for early prediction of students' academic failure in introductory programming courses. Comput. Hum. Behav. **73**, 247–256 (2017)
25. Settles, B.: Active learning literature survey. University of Wisconsin, Madison, vol. 52, pp. 55–66 (2010) 11 p.
26. Shannon, C.E.: A mathematical theory of communication. ACM SIGMOBILE Mob. Comput. Commun. Rev. **5**(1), 3–55 (2001)
27. Simpson, O.: Predicting student success in open and distance learning. Open Learn. **21**(2), 125–138 (2006)
28. Slater, S., Joksimović, S., Kovanovic, V., Baker, R.S., Gasevic, D.: Tools for educational data mining a review. J. Educ. Behav. Stat. **42**, 85–106 (2016)
29. Smola, A., Vishwanathan, S.V.N.: Introduction to Machine Learning. Press syndicate of the University of Cambridge, Cambridge (2008)
30. Sullare, V.A., Thakur, R.S., Mishra, B.: Analysis of student performance using mining technique: a review. Artif. Intell. Syst. Mach. Learn. **8**(3), 94–97 (2016)
31. Xing, W., Guo, R., Petakovic, E., Goggins, S.: Participation-based student final performance prediction model through interpretable genetic programming: integrating learning analytics, educational data mining and theory. Comput. Hum. Behav. **47**, 168–181 (2015)
32. Wilcoxon, F.: Individual comparisons by ranking methods. Biom. Bull. **1**(6), 80–83 (1945)
33. Zhang, H.: The optimality of naive bayes. AA **1**(2), 3 (2004)
34. Zhou, Z.-H.: Learning with unlabeled data and its application to image retrieval. In: Yang, Q., Webb, G. (eds.) PRICAI 2006. LNCS, vol. 4099, pp. 5–10. Springer, Heidelberg (2006). doi:10.1007/978-3-540-36668-3_3

Heuristics-Based Detection to Improve Text/Graphics Segmentation in Complex Engineering Drawings

Carlos Francisco Moreno-García[(✉)], Eyad Elyan, and Chrisina Jayne

The Robert Gordon University, Garthdee Road, Aberdeen, UK
`c.moreno-garcia@rgu.ac.uk`

Abstract. The demand for digitisation of complex engineering drawings becomes increasingly important for the industry given the pressure to improve the efficiency and time effectiveness of operational processes. There have been numerous attempts to solve this problem, either by proposing a general form of document interpretation or by establishing an application dependant framework. Moreover, text/graphics segmentation has been presented as a particular form of addressing document digitisation problem, with the main aim of splitting text and graphics into different layers. Given the challenging characteristics of complex engineering drawings, this paper presents a novel sequential heuristics-based methodology which is aimed at localising and detecting the most representative symbols of the drawing. This implementation enables the subsequent application of a text/graphics segmentation method in a more effective form. The experimental framework is composed of two parts: first we show the performance of the symbol detection system and then we present an evaluation of three different state of the art text/graphic segmentation techniques to find text on the remaining image.

Keywords: Complex engineering drawing · Digitisation · Text/graphics segmentation · Connected component analysis

1 Introduction

We define a complex engineering drawing (CED) as any type of schematic diagram which aims at representing the flow or constitution of a circuit, a process, a plant or a device. Unlike the classical definition of an engineering drawing (ED) which includes standard logical gate circuits, mechanical or architectural drawings, through this new definition we intend to characterise a specific subset of EDs with a complexity that demands a more advanced series of methods for their automated digitisation. Some examples of CEDs are process and instrumentation diagrams (P&IDs), chemical process diagrams, complex circuit diagrams, telephone manholes and facility drawings. An example of a portion of a P&ID is shown in Fig. 1.

© Springer International Publishing AG 2017
G. Boracchi et al. (Eds.): EANN 2017, CCIS 744, pp. 87–98, 2017.
DOI: 10.1007/978-3-319-65172-9_8

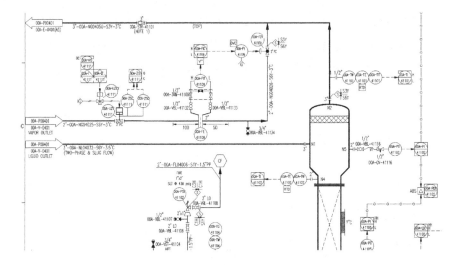

Fig. 1. Example of a portion of a process and instrumentation diagram

While ED digitisation has been largely reviewed and addressed in literature [1,2,7,8,15,20–22,27,34] the digitisation and contextualisation of CEDs still conveys several problems such as:

- *Size*: A single page of a CED contains about 100 different types of representations and around 150 symbols. Moreover, several pages (100 to 1000) may be required to represent a single process or structure.
- *Symbols*: Besides the conventional problem of variability in size, direction and position of symbols, CEDs present different symbol standards for different industries, and even when comparing two CEDs designed within a same company, standards may vary due to time. This leads to the constant employment of new symbols to describe incompatible elements. Hence, creating a symbol repository for training purposes [15] is sometimes not a viable solution.
- *Connections*: CEDs contain a dense and entangled structure of different types of connectors which represent physical and logical relations between symbols. Various type of connectors are usually depicted using different sizes and thickness, and thus methods based on thinning [27] become limited.
- *Text*: It is common that CEDs contain a large amount of codes and annotations (printed or handwritten). Moreover, connectors may also have corresponding text which contains important information such as the width of a connector. In general, CEDs are filled with a considerable amount of text which must be identified as well, since it is key for symbol recognition and drawing contextualisation.

Given the importance of text detection for CED digitisation and contextualisation, a particular kind of methodology called text/graphics segmentation can be considered for this task. For the past 35 years, it is possible to identify a vast amount of literature related to text/graphics segmentation methods for document images [9,12,13,17,24,30,31]. These methods may have a general purpose or be directed to a certain application domain, such as maps [5,29], book pages [6,32] and EDs [4,23]. While the characteristics of CEDs difficult the straightforward application of these methods, if a robust preprocessing and segmentation of symbols and connectors is applied to the CED in advance, then text/graphics segmentation becomes a viable option to locate the text remaining on the image.

The rest of the paper is structured as follows. In Sect. 2, we present the related work in text/graphics segmentation. Section 3 presents our proposed methodology to address the problem at hand. Section 4 contains the description of our experimental setting and the discussion of the results. Finally, Sect. 5 is reserved for conclusions and future work.

2   Related Work

In [1], Ablameyko and Uchida performed a review on ED digitisation methods, focusing on methods to detect symbols, connectors and text. Authors denote that most methods separate text from graphics either before or after ED vectorisation. Moreover, they noticed that text is commonly identified by using heuristics which help the system select either single characters or complete strings through certain constraints such as size, directional characteristics or complexity. Once text is isolated, the system either groups characters as strings within a certain space or erodes non-character shapes to keep text only. Most recently in [33], Wei et al. published a study on methodologies used for text detection in outdoor scene images. Authors found that text detection in most domains require two steps: character segmentation and string grouping. The first task is usually addressed through region-based methods, connected component (CC) analysis [28] or hybrid methods, while the second one is solved through a rule set approach, a clustering method or by learning algorithms. So far in literature, text/graphic segmentation methods mostly rely on CC analysis for character segmentation and rule-based string grouping.

In 1988, Fletcher and Kasturi [13] presented an algorithm to find text in printed drawings regardless of position, orientation or size of the text. The method consists on first applying CC analysis to the drawing in order to locate each character and graphic, discarding the ones longer than a size threshold and a height-to-width ratio threshold. To group characters into strings, authors introduce a methodology for linear analysis based on applying the Hough transform [18] to the centroids of the text CCs, which is a widely used method that has been applied to find lines [11,25], arbitrary shapes [3] and in more recent work, to locate partial images within their full counterparts [26]. This system has become a largely replicated solution due to its versatility and simplicity, however one of its greatest disadvantages is the incapability of the system to correctly identify individual characters and text overlapping lines or even other characters.

In 1998, Lu [23] presented a text/graphic separation method for characters in EDs. This method aims at erasing non-text and graphics from the ED to leave text only. Authors proposes a series of rule-based steps consisting on (1) erasing large line components, (2) erasing non-text shapes by analysing stroke density instead of size constraints and (3) grouping character into strings through a brush and opening operation to form new CCs, followed by a second parameter check on this newly formed CCs which restores miss-detected characters into their respective strings. The method deals better with the problem of text overlapping lines since most characters are left on the image and can be recovered on the last step. However, this method is very prone to identify false positives (such as small components or curved lines) and depends on text strings to be apart from each other so that the string grouping is executed correctly.

In 2002, Tombre et al. [31] presented an upgrade on [13] for document images rich in graphics. Authors increase the number of constraints on the first step of the original method so that the best enclosing rectangle of a shape identified as text is considered before analysing the CC. In addition, since the density and the elongation of the CCs are calculated and analysed for the text/graphic distinction, authors create a third layer where small elongated elements (i.e. "1", "—", "l", "-" or dot-dash connectors) are stored. At the second stage of the algorithm, authors propose alternative strategies to compute the string grouping in the Hough transform domain, which according to the characteristics of the document image, could lead to better or worse results. Finally authors add an extension of string step where shapes on the small elongated element space are restored into the text space into their respective strings according to an analysis of proximity. Other interesting papers that present improvements on CC analysis based text/graphics segmentation are He and Abe [17], where clustering is used to improve each step, Tan and Ng [30] where a pyramid version of the image is used to group strings, or Chowdhury et al. [6] that proposes a multi decision tree for a more specific segmentation.

Regarding work for text detection in other areas, the method for outdoor scenes proposed by Wei et al. [33] is based on an exhaustive segmentation approach, which means that multiple image binarisations are generated from a single image using the minimum and maximum gray pixel value as threshold range. Then, candidate character regions are determined for each binarisation based on CC analysis, and non-character regions are filtered out through a two-step strategy composed of a rule set and a Support Vector Machine (SVM) classifier working on a set of features i.e. area ratio, stroke-width variation, intensity, Euler number [16] and Hu moments [19]. After combining all true character regions through clustering, authors implement an edge cut approach for string grouping. This consists of first establishing a fully connected graph of all characters, and then calculating the true edges based on a second SVM classifier using a second set of features i.e. size, colour, intensity and stroke width. This method clearly results in a more complex and robust approach to the problem at hand, however it is difficult to implement on printed drawings. Nonetheless, methods of this nature lead us to realise that there are interesting alternatives to the classical text/graphics segmentation methods in literature.

3   Methodology

We propose a sequential heuristics-based methodology which is aimed at localising and removing the most representative symbols of the drawing, with the aim of preparing the image for the use of a text-graphics segmentation method which can detect text characters across the remaining image. In summary, the complete CED digitisation framework consists on the following steps:

1. Preprocessing.
2. Image resizing.
3. Detection of representative shapes.
   (a) Linear components.
   (b) Connectivity symbols.
   (c) Remaining geometrical symbols.
4. Text/Graphics Segmentation.

3.1   Heuristics-Based Symbol Detection

After applying preprocessing methods such as thresholding and noise removal, it has been noticed that several CEDs are surrounded by a blank frame which increase the size of the file and hence the time for digitisation. To discard this outer frame automatically, we apply a Canny edge detector on the image. Then, the resulting image is dilated using a cross structural element recursively, intending to connect all the schematics contained. Finally, a CC analysis is run, and only the portion of the original drawing located on the bounding box of the dilated image is considered as the input of the system.

While most text/graphics segmentation methods suggest to discard indistinctively all lines larger than a certain threshold (usually dependent on the average character height) either by analysing CCs of a large width or height [13] or by scanning the image for large sequences of pixels across different image inclinations [23], in CEDs large lines represent different aspects of the drawing based on their length and thickness. For instance, in P&IDs there are thick and long lines that represent pipelines or the outline of a vessel, thick and short lines that represent smaller symbols such as emergency shut down valves, and thin and long lines that represent the margin line, connectors, symbols. Therefore, a more thoughtful detection methodology has been implemented so that each long line is correctly localised within its context and thus the classification complexity can be reduced. To do so, the first aim is to detach symbols and text from connectors and large elements. Given h and w representing the height and the width of the input image respectively, the image is dilated two separate times using a rectangular structuring elements of size $(1 \times h/m)$ and $(w/m \times 1)$ respectively. Variable m must be set to a high value (i.e. one third of the size of the longer edge of the image) to allow the horizontal and vertical lines to be maintained to the most on each image after the dilation. Then, both images are combined to create a new image containing only large lines. The pixels of the input image which are not included in this image are considered either part of other symbols or text. Afterwards, a blur operation is applied to the resulting image so that the thicker lines can be distinguished from the thin ones. Thick line segments are analysed as follows; if one or more thick line segments conform a loop, then this is a representation of a thick line symbol/vessel, otherwise this line represents a connector. Searching for loop elements in an image can be addressed through several means, such as finding the Euler number [16], an enclosing chain code [14] or by contour detection. Regarding thin lines, these are classified according to their localisation. If the line is long and close to the image border, then it is a margin line; otherwise it is a connector line.

The next step is to locate symbols which are characteristic of the drawing and that have properties which allow their detection in more efficient manners. Such is the case of continuity labels, which are text-boxes in the end of thick connector lines which indicate the connection of the represented piping to another drawing. Since these labels are located on either side of the schematic, they can be located by scanning the new image either applying template matching or CC analysis. Given that these labels contain text, it is recommended to segment continuity labels along with the contained text for this to be used on later stages. For instance, a learning methodology could be applied to analyse this text in advance and deduce the average text size so that the subsequent text/graphics segmentation step can be automated and thus more effective.

Finally, geometrical symbols such as circles and polygons can be located. Circles may be found within dot-dash connectors or representing symbols, and can be segmented through the application of the Hough circles method [3] taking into consideration factors such as size and localisation to avoid false positives within text. On the other hand, polygons can be detected through contour detection and approximation, by means of methods such as the Douglas-Peucker algorithm [10]). If these instrumentation symbols contain text, this creates a second opportunity to read text or learn its properties in advance.

3.2   Text/Graphics Segmentation

Once the image without the aforementioned symbols and connectors has been generated, a text/graphics segmentation method can be applied. The main aim at this stage is to distinguish characters and group text strings considering the following limitations:

- *Symbols and connectors left*: Long or dashed lines representing connectors or measuring indicators, as well as symbols such as grid areas, irregular or disconnected polygons e.g. arrowheads, diamonds, trapezoids, or loop free symbols (a capacitor or a resistance) may still be present in the diagram. Examples are shown in Fig. 2a.
- *String size irregularity*: The characters to be grouped into strings present irregular shapes and sizes. While in some cases the string is vertical or horizontal, some others it is split into rows. Furthermore, in some cases a symbol splits the string either top-bottom or left-right in Fig. 2b, different string shapes and sizes are shown.

Fig. 2. Examples of limitations found in a P&ID once symbols found through the heuristics-based method have been removed. (a) Symbols and connectors left, (b) string size irregularity and (c) character overlapping

– *Punctuation signs*: It is of particular interest to avoid discarding punctuation sings (i.e. inches symbols, periods and commas) without wrongfully identifying them as noisy components
– *Character Overlapping*: Some of the text characters may overlap symbols and connectors, or even other characters, as shown in Fig. 2c.

To that aim, the most widely used text/graphics segmentation methods used on EDs are based on CC analysis. As described in Sect. 2, once CC analysis is done, characters and graphics are split into separate layers according to different variables according to the application and the system design, such as area, height-to-width ratio, stroke density, pixel density, elongation, number of loops, etc. Furthermore, some other methods create a third layer to store elongated elements (i.e. letters "l" and "i" or symbols "-" and "/").

4   Experiments and Results

After describing the drawings that compose the dataset used for experimentation, we briefly report the results of applying the heuristics-based methodology on the drawings to detect and segment as many representative symbols as possible. Then, we compare the character detection effectiveness of three state of the art text/graphics segmentation methods on either the original images of the dataset or on the image after the application of the heuristics-based symbol detection. This way, we aim to verify that detecting and segmenting the most representative symbols in advance leads to an improvement in the character detection.

4.1   Dataset Used

We have compiled from an industrial partner a collection of P&ID drawings with a large and dense quantity of symbols, connectors and text (an example of these drawings is shown in Fig. 1). Images have been scanned at a 300 dpi resolution resulting on average in 3508×2479 pixels size. In total, the drawings on the collection contain an average of 2.9 thick line symbols/vessels, 41 circular instrumentation symbols, 32.7 circles within dot-dash connectors, 6.6

continuity labels, 34.6 polygons (triangles, squares and hexagons) plus tenths of other irregular and unclassified symbols. Furthermore, drawings contain over one hundred text strings each, which range from 1 to 24 characters of length and may be grouped in different shapes and extensions according to the process or instrumentation described.

4.2 Heuristics-Based Method

After preprocessing, the image is reduced on average 85.28% from its original size. Afterwards, the image is inspected for line components. This system successfully distinguishes pipeline connectors (blue), margin lines (purple) and vessels (cyan) as shown in Fig. 3.

Successively, the method is capable of finding all continuity labels for all drawings easily, (red box in Fig. 3a). Furthermore, with the proper tuning of the radius we are capable of locating all large circles representing two types of instrumentations (red and yellow circles in Fig. 3b). Moreover, an average of 96.52% of small circles within dot-dash connectors are detected (light green circles in Fig. 3c) by using the circle detection method plus deducing the location of the missed small circles by analysing the dot-dash connector itself. To that aim, the image is scanned and all small linear segments adjacent to small circles (dark green) are identified as connectors. Using the small circles and line segments, the path of the dot-dash connector can be constituted and the missing circles are located. Also, circles within symbols such as valves can be found by using a similar approach. Finally, the polygon detection algorithm based on contour detection is capable of locating an average of 83.32% of the squares, diamonds and triangles on the datasets. Notice that this is the least accurate

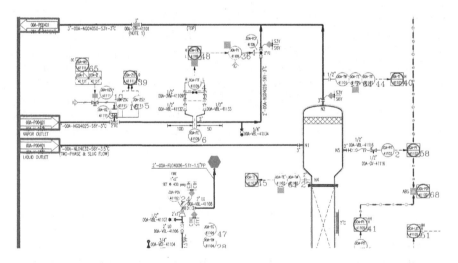

Fig. 3. Result of applying the sequential heuristics-based methodology on Fig. 1 (Color figure online)

of the detectors since this step depends on polygon approximation methods and thus, many of these symbols are not correctly approximated once their contours are detected.

4.3   Comparison of Character Detection in Text/Graphics Segmentation Methods

We have tested the character detection feature of three text/graphics segmentation methods in literature: Fletcher and Kasturi [13], Lu [23] and Tombre et al. [31]. These methods have been selected since they are the base of most existing methods and because they present the two step approach described in Sect. 2 (character detection and string grouping) and thus this enables a fair comparison.

In order to test whether the inclusion of the heuristics-based symbol detection method leads to an improvement and to compare the accuracy of the text/graphics segmentation methods, we have applied these methods both without and with the previous application of the heuristic-based symbol detection on our P&ID dataset. We present in Table 1 a comparison of precision, recall and runtime for the six possible scenarios.

Table 1. Comparison of accuracy (precision and recall) and average runtime between the character detection methods without or with the previous application of the heuristics-based symbol detection.

	No heuristics-based detection			With heuristics-based detection		
Character detection	Precision	Recall	Runtime (s)	Precision	Recall	Runtime (s)
Fletcher and Kasturi [13]	0.37	1	0.93	0.91	0.98	12.14
Lu [23]	0.39	1	14.88	0.93	0.98	18.35
Tombre et al. [31]	0.47	1	234.61	0.98	0.84	185.72

With respect to the accuracy, notice that the three methods present the highest possible recall given that they are capable of including all existing text; however since a large amount of false positives are included, precision is reduced. In contrast, the character detection methods after the application of the heuristics-based detection present high precision while mantaining a good recall, considering that in steps identify text. Notice that the precision using [31] is lower than in the other cases given that we are not considering the small elongated components that have yet to be classified as text or graphics during the string grouping phase.

Regarding the runtime to compute the full process, notice that the first two cases delay more when performing heuristics-based segmentation plus character detection than when applying character detection only. However, it can be appreciated that if [31] is used, it is less time consuming to apply both processes rather than applying character detection on the original image. This occurs because at the character detection stage there are less CCs to analyse after the symbol detection has taken place. Therefore, we infer that when applying more robust text/graphics segmentation methods using more complex filtering, applying a previous symbol detection could not only lead to an improvement in accuracy, but also in the runtime of the system overall. Tests where performed using a PC with Intel 3.4 GHz CPU and Windows 10 operating system.

5   Conclusions

In this paper, we present a symbol detection method aimed at improving the application of text/graphics segmentation on CEDs. This method uses an heuristic-based approach to detect and segment the most representative symbols of the drawing, using as example the case of a P&ID. We have tested our system on a collection of drawings with a large and dense quantity of symbols, connectors and text, and we have noticed that a high amount of symbols can be recognised if the algorithm is properly set and the characteristics of the drawing are understood in advance. Moreover, we have performed a comparison between different state of the art methods that perform character detection on engineering drawings. We apply three character detection methods in two cases: on the original image or on the image after the heuristics-based symbol detection has been applied. We have seen that the character detection after the symbol detection outperforms the application on the original image, since less false positives are detected and less strings have to be processed. Furthermore, the average runtime of applying each scenario has been calculated, noticing that for the most robust text/graphics segmentation method, an improvement in runtime can be achieved if the symbol detection is applied beforehand.

There is a clear room for further work in this area, given the large need for digitisation systems for CEDs. Firstly, we aim at completing the text/graphics segmentation process and test different grouping strings methodologies. Also, we aim at considering more advanced heuristics which allow to overcome usual problems in CEDs such as character overlapping. Finally, we intend to test our proposed methodology in more datasets containing a wider range of symbols and characteristics.

Acknowledgement. We would like to thank Dr. Brian Bain from DNV-GL Aberdeen for his feedback and collaboration in the project. This work is supported by a Scottish national project granted by the Data Lab Innovation Centre.

References

1. Ablameyko, S.V., Uchida, S.: Recognition of engineering drawing entities: review of approaches. Int. J. Image Graph. **07**(04), 709–733 (2007)
2. Arias, J.F., Lai, C.P., Chandran, S., Kasturi, R., Chhabra, A.: Interpretation of telephone system manhole drawings. Pattern Recognit. Lett. **16**(4), 365–368 (1995)
3. Ballard, D.H.: Generalizing the Hough transform to detect arbitrary shapes. Pattern Recognit. **13**(2), 111–122 (1981)

4. Bunke, H.: Automatic interpretation of lines and text in circuit diagrams. In: Kittler, J., Fu, K.S., Pau, L.F. (eds.) Pattern Recognition Theory and Applications, vol. 81, pp. 297–310. Springer, Dordrecht (1982)

5. Cao, R., Tan, C.L.: Text/graphics separation in maps. In: Blostein, D., Kwon, Y.-B. (eds.) GREC 2001. LNCS, vol. 2390, pp. 167–177. Springer, Heidelberg (2002). doi:10.1007/3-540-45868-9_14

6. Chowdhury, S.P., Mandal, S., Das, A.K., Chanda, B.: Segmentation of text and graphics from document images. In: Proceedings of the International Conference on Document Analysis and Recognition, ICDAR, vol, 2 (Sect. 4), pp. 619–623 (2007)

7. Cordella, L.P., Vento, M.: Symbol recognition in documents: a collection of techniques? Int. J. Doc. Anal. Recognit. **3**(2), 73–88 (2000)

8. De, P., Mandal, S., Bhowmick, P.: Identification of annotations for circuit symbols in electrical diagrams of document images. In: 2014 Fifth International Conference on Signal and Image Processing, pp. 297–302 (2014)

9. Dori, D., Wenyin, L.: Vector-based segmentation of text connected to graphics in engineering drawings. In: Perner, P., Wang, P., Rosenfeld, A. (eds.) SSPR 1996. LNCS, vol. 1121, pp. 322–331. Springer, Heidelberg (1996). doi:10.1007/3-540-61577-6_33

10. Douglas, D.H., Peucker, T.K.: Algorithms for the reduction of the number of points required to represent a digitized line or its caricature. Cartogr. Int. J. Geogr. Inf. Geovisualization **10**(2), 112–122 (1973)

11. Duda, R.O., Hart, P.E.: Use of the Hough transformation to detect lines and curves in pictures. Commun. ACM **15**, 11–15 (1971)

12. Fan, K.C., Liu, C.H., Wang, Y.K.: Segmentation and classification of mixed text/graphics/image documents. Pattern Recognit. Lett. **15**(12), 1201–1209 (1994)

13. Fletcher, L.A., Kasturi, R.: Robust algorithm for text string separation from mixed text/graphics images. IEEE Trans. Pattern Anal. Mach. Intell. **10**(6), 910–918 (1988)

14. Freeman, H.: On the encoding of arbitrary geometric configurations. IRE Trans. Electron. Comput. **EC−10**, 260–268 (1960)

15. Gellaboina, M.K., Venkoparao, V.G.: Graphic symbol recognition using auto associative neural network model. In: Proceedings of the 7th International Conference on Advances in Pattern Recognition, ICAPR 2009, pp. 297–301 (2009)

16. Gray, S.B.: Local properties of binary images in two dimensions. IEEE Trans. Comput. **20**(5), 551–561 (1971)

17. He, S., Abe, N.: A clustering-based approach to the separation of text strings from mixed text/graphics documents. Proc. - Int. Conf. Pattern Recognit. **3**, 706–710 (1996)

18. Hough, P.V.C.: Method and means for recognizing complex patterns. US Patent 3,069,654, 18 December 1962

19. Hu, M.K.: Visual pattern recognition by moment invariants. IRE Trans. Inf. Theory **8**, 179–187 (1962)

20. Kasturi, R., Bow, S.T., El-Masri, W., Shah, J., Gattiker, J.R.: A system for interpretation of line drawings. IEEE Trans. Pattern Anal. Mach. Intell. **12**(10), 978–992 (1990)

21. Kim, S.H., Suh, J.W., Kim, J.H.: Recognition of logic diagrams by identifying loops and rectilinear polylines. In Proceedings of the Second International Conference on Document Analysis and Recognition - ICDAR 1993, pp. 349–352 (1993)

22. Lladós, J., Valveny, E., Sánchez, G., Martí, E.: Symbol recognition: current advances and perspectives. In: Blostein, D., Kwon, Y.-B. (eds.) GREC 2001. LNCS, vol. 2390, pp. 104–128. Springer, Heidelberg (2002). doi:10.1007/3-540-45868-9_9

23. Lu, Z.: Detection of text regions from digital engineering drawings. IEEE Trans. Pattern Anal. Mach. Intell. **20**(4), 431–439 (1998)
24. Luo, H., Agam, G., Dinstein, I.: Directional mathematical morphology approach for line thinning and extraction of character strings from maps and line drawings. In: Proceedings of 3rd International Conference on Document Analysis and Recognition, vol. 1, pp. 257–260, 1 August 1995
25. Matas, J., Galambos, C., Kittler, J.: Robust detection of lines using the progressive probabilistic hough transform. Comput. Vis. Image Underst. **78**(1), 119–137 (2000)
26. Moreno-García, C.F., Cortés, X., Serratosa, F.: Partial to full image registration based on candidate positions and multiple correspondences. In: Bayro-Corrochano, E., Hancock, E. (eds.) CIARP 2014. LNCS, vol. 8827, pp. 745–753. Springer, Cham (2014). doi:10.1007/978-3-319-12568-8_90
27. Okazaki, A., Kondo, T., Mori, K., Tsunekawa, S., Kawamoto, E.: Automatic circuit diagram reader with loop-structure-based symbol recognition. IEEE Trans. Pattern Anal. Mach. Intell. **10**(3), 331–341 (1988)
28. Pratt, W.K.: Digital Image Processing, 4th edn. Wiley, Los Altos (2013)
29. Roy, P.P., Vazquez, E., Lladós, J., Baldrich, R., Pal, U.: A system to segment text and symbols from color maps. In: Liu, W., Lladós, J., Ogier, J.-M. (eds.) GREC 2007. LNCS, vol. 5046, pp. 245–256. Springer, Heidelberg (2008). doi:10.1007/978-3-540-88188-9_23
30. Tan, C., Ng, P.O.: Text extraction using pyramid. Pattern Recognit. **31**(1), 63–72 (1998)
31. Tombre, K., Tabbone, S., Pélissier, L., Lamiroy, B., Dosch, P.: Text/graphics separation revisited. In: Lopresti, D., Hu, J., Kashi, R. (eds.) DAS 2002. LNCS, vol. 2423, pp. 200–211. Springer, Heidelberg (2002). doi:10.1007/3-540-45869-7_24
32. Wahl, F.M., Wong, K.Y., Casey, R.G.: Block segmentation and text extraction in mixed text/image documents. Comput. Graph. Image Process. **20**(4), 375–390 (1982)
33. Wei, Y., Zhang, Z., Shen, W., Zeng, D., Fang, M., Zhou, S.: Text detection in scene images based on exhaustive segmentation. Signal Process. Image Commun. **50**, 1–8 (2017)
34. Yu, Y., Samal, A., Seth, S.C.: A system for recognizing a large class of engineering drawings. IEEE Trans. Pattern Anal. Mach. Intell. **19**(8), 868–890 (1997)

Intrinsic Plagiarism Detection with Feature-Rich Imbalanced Dataset Learning

Andrianna Polydouri[✉], Georgios Siolas, and Andreas Stafylopatis

Intelligent Systems, Content and Interaction Laboratory,
National Technical University of Athens, Athens, Greece
andriannapolyd@gmail.com, gsiolas@islab.ntua.gr, andreas@cs.ntua.gr

Abstract. In the context of intrinsic plagiarism detection, we are trying to discover plagiarised passages in a text, based on the stylistic changes and inconsistencies within the document itself. The main idea consists in profiling the style of the original author and marking as outliers the passages that seem to differ significantly. Besides some novel stylistic and semantic features, the present work proposes a new approach to the problem, where machine learning plays a significant role. Notably, we also consider, for the first time, the reality of unbalanced training dataset in intrinsic plagiarism detection as a major parameter of the problem. Our detection system is tested on the data corpora of PAN Webis intrinsic plagiarism detection's shared tasks of 2009 and 2011 and is compared to the results of the highest score participations.

Keywords: Intrinsic plagiarism detection · Stylometry · Supervised learning · Unbalanced training data · SMOTE · PAN Webis

1 Introduction

Plagiarism is the act of presenting part of someone else's work as one's own, without proper citation or acknowledgment. Current research proposes two major approaches for detecting plagiarism: extrinsic, where a corpus of reference documents is provided in order to detect the suspicious similarities, and intrinsic, where only source is the investigated document and the plagiarised passages are detected as outliers in comparison to the extracted stylistic signature of the original author.

Intrinsic plagiarism detection is based on the idea that every author has its own personal and unique writing style, which may be detected and quantified when analysing the document by stylistic and/or semantic means. If such an analysis is possible, then it is also possible to detect potential plagiarised passages, as those that seem not to fit that personal writing style. All these lie under one condition: the examined document must be mostly written by the same author, or else we lack an analysis object.

Intrinsic plagiarism detection is a problem closely related to other tasks, such as author identification and author verification. Author identification aims

© Springer International Publishing AG 2017
G. Boracchi et al. (Eds.): EANN 2017, CCIS 744, pp. 99–110, 2017.
DOI: 10.1007/978-3-319-65172-9_9

to determine the author of a book or passage given a set of documents and authors, while in author verification the task can be described as follows: given a document d written by an, already known, author A, is there a passage in d written by an author $B \neq A$?

The paper is structured as follows: in Sect. 2, the common methodology for intrinsic plagiarism detection is described, and the basic points of the most efficient and latest detection systems are presented. In Sect. 3, the PAN Webis 2009, 2011 & 2016 competitions and datasets are described in short. In Sect. 4, our detection system is described in detail. In Sect. 5, the results of our detection system are presented and compared to other systems' results. In Sect. 6, we make a conclusion and our proposals for future work.

2   Related Work

2.1   Intrinsic Plagiarism Detection Steps

Current research on intrinsic plagiarism detection follows a typical methodology consisting of three major steps: (1) text segmentation, (2) style analysis function extraction, (3) outliers detection.

Text Segmentation. The document is split into pieces, which will afterwards be the target units of the stylistic analysis. The most common splitting method is that of the *sliding window*, with parameters {*window length, window step*}. At each processing step the window moves by *window step* characters/sentences and it's contents are passed to the style analysis function for further processing.

Style Analysis Function. The style analysis function gives the stylistic imprint of each piece of the segmented document, which is later evaluated in order to determine whether that document piece contains plagiarised text or not. In addition, the stylistic signature of the original author may be extracted by applying the style analysis function to the whole document or by processing the pieces' imprints together. For the construction of the style analysis function stylometric and/or semantic features are used. The stylometric features may be lexical-, syntactical- or structural-oriented [1], while the semantic ones try to extract information about the vocabulary richness and the semantic context of the word or passage under analysis [2]. The output of the style analysis function is a real-valued vector containing the previously mentioned features.

Outlier Detection. The outliers detection step aims to extract the plagiarised passages. Two main methods are used in this purpose:

1. compare the stylistic imprint of the document's segments to the estimated signature of the original author (the stylistic analysis of the entire document or corpus attributed to him) and mark the outliers

2. compare the stylistic imprint of each document segment to each other and mark the outliers.

In most detection systems the comparison step is designed as follows: apply an appropriate distance function *(e.g. Cosine distance, Manhattan distance)* on the style feature vectors under comparison. In this way, a single scalar value is calculated, which is an expression of the stylistic distance between the two text segments. For the outliers detection step, a threshold distance value is defined, either in an ad-hoc manner, as in most current systems, by former knowledge and observation of sample-training data, or else by using machine learning methods on the training data. Text segments that exhibit distance values above this threshold are considered outliers.

2.2 Plagiarism Detection Approaches

In the present section, we will present short outlines of the most efficient and recent intrinsic plagiarism detection systems.

A participant in PAN 2009 competition, Stamatatos [3] uses character n-gram profiles for the style change function, which considers each text as a bag of characters, where *profile* of the text is a vector of normalised frequencies of all the character n-grams appearing in it. The dissimilarity level of each document piece against the whole document is evaluated using the proposed distance function.

The system of Oberreuter et al. [4], a participant in the 2011 PAN competition and winner of the competition, constructs a semantico-lexilogical style function, which is based on *term frequency*. The basic idea is the comparison of the words' term frequencies between each segment and the whole document. For the outliers detection part, the system uses an ad-hoc threshold.

The system of Kestemont et al. [5], that also participated in the 2011 PAN competition and ranked second, is based on character trigrams. They created a list with the most frequent character trigrams in the PAN 2009 dataset and used that list for their style function, calculating the occurrence frequency of these trigrams in each segment. For the outliers detection part, they compare the document segments to each other, instead of comparing them to the whole document.

The system of Kuznetsov *et al.* [6], winner of the PAN 2016 competition, segments each text document into sentences and constructs basic stylometric features for each sentence (character and word n-gram frequencies, punctuation and pronouns count), then trains a classifier using the PAN 2011 training data. The outliers detection method is threshold-based: all sentences with a classifier output higher than a certain threshold are marked as outliers.

Our contribution is twofold: In the style analysis function part, we introduce some promising stylistic and semantic features. In addition, we consider the outliers detection part as a pure machine learning task. In comparison to all the systems described above, the outliers detection step is not threshold based. A machine learning subsystem is responsible to classify the passages into plagiarised or non-plagiarised ones. Moreover, in contrast with the existing approaches, in our work we take into account the strongly unbalanced nature of the intrinsic plagiarism datasets (very few outliers), a decision that leads to important performance improvements.

3   PAN Webis and Intrinsic Plagiarism Detection

Until today, a total of three intrinsic plagiarism detection tasks have taken place in the PAN Webis[1] competition (years 2009, 2011 and 2016).

PAN 2009 Competition. The data corpus for the intrinsic plagiarism detection task consists of 3092 documents, which contain artificial plagiarism cases. The corpus of the competition is based on documents from the Project Gutenberg[2][7]. Our system is tested on this competition's data corpus and is compared to that of the winner system for this competition [3].

PAN 2011 Competition. The data corpus for this task consists of 4753 documents. This dataset has a weakness by design: the artificial plagiarism cases where randomly chosen and inserted into the documents, without taking into consideration the context of each document. As a result, the semantic content of the plagiarised cases is irrelevant to that of the original text, which is unrealistic as for real world plagiarism cases. The most efficient detection system was that of Oberreuter et al. [4], which had a style analysis function based on semantics. Despite of this system's good results, we consider that it took advantage of the dataset's weakness by using a strongly vocabulary oriented feature and hence it would not be able to generalise well, given a more neutral dataset [8,9]. Our system system is also tested on this completion's dataset and compared to Oberreuter's *et al.* system.

In both datasets the imbalance of the plagiarised and not plagiarised sentences is already present. As a sample, consider the sentence distribution in the documents 3803–4753 of the 2011 dataset:

	Plagiarised	Not plagiarised
Number of sentences	62,701	1,130,173

This means that approximately 6% of the total sentences in the dataset are plagiarised. Imbalanced datasets are an important issue to consider when using training data for machine learning [10].

---

[1] http://pan.webis.de/.

[2] http://www.gutenberg.org.

PAN 2016 Competition. In PAN 2016 intrinsic plagiarism detection was included as a subtask for *Author Diarisation* [11]. In Author Diarisation, portions of the text must be identified and clustered with respect to a given set of authors for a given document. In the case of one author against all, author diarisation coincides with intrinsic plagiarism detection. The results of the two pariticipants are rather poor and since there are no available results for the state of tha art approaches our system has not been tested on it, at least for the time being.

4   Intrinsic Plagiarism Detection System

We will now describe our detection system along with some comments on crucial modeling choices. The system consists of one pre-processing step, three major parts (text segmentation, style analysis - feature extraction, outliers detection) and a results' post-processing step.

The pre-processing steps are the following: (1) de-capitalisation, (2) sentence detection, (3) token detection, (4) removal of alphanumerics and special characters, (5) part-of-speech tagging, (6) stemming. For the steps *sentence detection, token detection, part-of-speech tagging* and *stemming*, the Apache OpenNLP library was used[3].

In the following subsections we describe in detail the three major parts of the system.

4.1   Text Segmentation

For the text segmentation we use the typical *sliding window* method. Experiments with various parameter sets lead us to *window length = 15 sentences* and *window step = 5 sentences*. We did not notice any performance gain when using a variable length window.

4.2   Style Analysis - Feature Extraction

For the style analysis part we extract 11 features, both stylistic and semantic. In the following list, features 1–4 are used by most existing systems [1, 2], 5–7 are novel stylistic features based on the idea of compression and 8–12 are novel semantic features:

1. average sentence length
2. average syllable count per token
3. Flesch-Kinkaid grade [12]
4. frequency of word "of"
5. verbs' compression rate
6. adverbs' compression rate
7. adjectives' compression rate

---

[3] https://opennlp.apache.org/.

8. mean value of positive subtraction of word frequency class between whole document - passage (wfc_1)
9. standard deviation of positive subtraction of word frequency class between whole document - passage (wfc_2)
10. percentage of the words having word frequency class subtraction value between whole document - passage greater than the mean value (wfc_3)
11. mean value of negative subtraction of word frequency class between whole document - passage, only for the frequent, in the whole document, words (wfc_4).

Novel Compression-Based Features. The novel stylistic features *(verbs', adverbs', adjectives' compression rate)* are based on an idea by Leanne and Matwin [13], who proposed stylistic metrics related to Kolmogorov Complexity. The basic idea is that in every document one may discover characteristic distribution for each word class. We constructed a simple algorithm that tries to express a potential distribution trough compression rate. For each one of the three word classes experimented with, a binary sequence was extracted, where the words belonging to the specific word class are represented with 1, while those who don't with 0. Afterwards, the binary sequence was compressed using the algorithm *run length encoding*. Finally, the compression rate is given from Eq. (1):

$$\text{Compression rate} = \frac{\text{Run-length encoding sequence length}}{\text{Binary sequence length}} \qquad (1)$$

Novel Semantic Features Based on Word Frequency Class. Our semantic features 8–11 are based on the concept of *word frequency class*, as described by zu Eissen and Stein [14].

Word Frequency Class. Let C a dataset and $f(w)$ the frequency of a word $w \in C$. The word frequency class $c(w)$ of the word w is calculated by Eq. (2):

$$c(w) = log_2 \left(\frac{f(w^*)}{f(w)} \right) \qquad (2)$$

where w^* the most common word in C.

More in detail, the most common word gets a zero word frequency class value, while the more rare a word is, the higher its word frequency class value. Note that word frequency class is dependent on the piece of document it is applied on, since it is relative to the most frequent word.

In our style analysis, we calculated the word frequency class of the words in the whole document, but also for each piece of document separately. The features we construct are related to the comparison of the word frequency class values of the words appearing in each document's segment towards their values in the context of the whole document. The idea is that rather rare words in the whole document but frequent in a passage, are a sign of potential plagiarism in that passage. This will be reflected by a high value in the subtraction between the document's - passage's word frequency classes.

Feature Normalisation. The last steps of the feature extraction for each piece of the segmented document are the comparison to the whole document and, finally, the normalisation of the values. Our semantic features are relative to the whole document by definition. For the stylistic features *(1–7)* we subtract the values of the segment from the corresponding value for the whole document.

Finally, we normalise the feature values applying Eq. (3):

$$\text{normalise}(x) = \frac{x}{1 + |x|} \qquad (3)$$

4.3 Outlier Detection Adapted to the Problem

For the outlier detection part we rely on machine learning. The intrinsic plagiarism detection task is by nature a task of unbalanced data: the plagiarised sections tend to be significantly less in number than the original ones, while, contrary to author identification, it is a pure binary classification problem.

Machine Learning. The input to the machine learning system is the feature vectors of document's segments. The machine learning system is trained and predicts (plagiarised - non plagiarised) on segment level *(15 sentences)*. We finally predict on sentence level: As described, every batch of 5 sentences is included in 3 segments. We evaluate a binary prediction (plagiarised sentences or not) according to the mean value of the machine learning system predictions for these 3 overlapping segments. Thanks to this post-processing step our system is able to predict with granularity equal to 1 (sentence level), by design. We experiment with two training algorithms: Decision Trees and Support Vector Machines. For the SVM we set the following parameters values: $C = 1.0$, Kernel = Radial Basis Function, $\gamma = \frac{1}{n_{features}}$. We use the Scikit-Learn python library[4].

Training Dataset Balancing. For the balancing of the training data we use the SMOTE borderline algorithm [15], which constructs synthetic examples of the minority class using the values of the existing minority class's examples and its neighbors, as long as they also belong to the minority class. We use Imbalanced-learn Python Toolkit [16] for this purpose.

5   Results

Mainly because of the weakness of the 2011 PAN competition dataset, we split our features (and resulting experiments) into two subcategories: stylistic features

---

[4] http://scikit-learn.org.

and semantic features. Hence, our results consist of experiments with the following 3 feature sets: (*i*) all of the features *(1–11)*, (*ii*) stylistic features only *(1–7)*, (*iii*) semantic features only *(8–11)*.

As for the machine learning part of the system, we split the datasets of the two PAN Webis competitions into training and test data applying *k-fold cross validation*. For the 2009 competition *(3092 documents)* we define $k = 4$ *(approximately 770 records per split)*, for 2011 *(4753 documents)* competition $k = 5$ *(approximately 1000 records per split)*.

We use Precision, Recall*(Sensitivity)* and F-Score for the evaluation of our system. As already mentioned, the result's granularity is always equal to 1. We experimented with two classifiers, Decision Trees and Support Vector Machines, as well as with balanced and unbalanced training datasets.

5.1 Decision Trees

In both Figs. 1 and 2 we can see that in all three feature sets, balancing the training datasets improves the system's efficiency. A $2\% - 3\%$ improvement in the prediction score is considered important in such a problem. In addition, considering the features sets experiment, the semantic features succeed higher score than the stylistic ones, while their combination outperforms both.

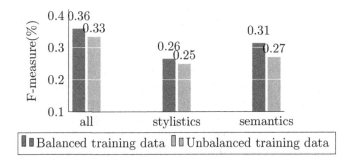

Fig. 1. Decision Trees, PAN 2009 data corpus

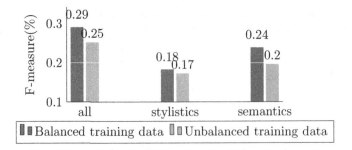

Fig. 2. Decision Trees, PAN 2011 dataset

5.2   Suppport Vector Machines

Concerning Support Vector Machines, the classifier was unable to predict correctly any plagiarised document while the dataset was unbalanced - all of the documents were predicted as *not plagiarised*. When balancing the training dataset using SMOTE borderline, though, the results are remarkable, as illustrated in Tables 1 and 2. Hence, we can argue that balancing the training dataset not only improves existing and well-working classification methods, like Decision Trees, but enables us to apply to the problem an important family of classifiers, that it would be impossible to use otherwise, like SVMs.

Table 1. SVM, PAN 2009 data corpus.

Feature set	F-score
Only stylistics	0.316
Only semantics	0.357
All features	0.419

Table 2. SVM, PAN 2011 data corpus.

Feature set	F-score
Only stylistics	0.232
Only semantics	0.277
All features	0.328

5.3   Stability of the Detection System

We, also, test our system's stability and reliability by examining the results on each individual split of the dataset, used in the k-fold cross validation method. Obviously, a stable system on the test dataset is more likely to give predictions of the expected precision, when applied on an unknown dataset.

As illustrated in Figs. 3 and 4, for the two data corpora, our system is rather stable. In no case do we get more than 4% deviation among the splits' results. In addition, it seems that when applying balancing of the training dataset the system tends to be more stable.

5.4   Feature Ranking

In Table 3, we present the results of each one of our system's features tested on PAN 2009 data corpus, for the best configuration of our system: balancing of the training dataset, using SMOTE borderline algortihm, and SVM classification.

It is obvious that our novel stylometric features *(verbs', adverbs' and adjectives' compression rate)* are highly competitive to the other well-known and widely used features. In addition, our semantic features give highly promising results, taking into consideration that, unlike PAN 2011 dataset, the PAN 2009 dataset has no semantic-related weakness.

However, the strength of our system lies in the variety of the applied features and their underlying association, and not in one powerful feature alone.

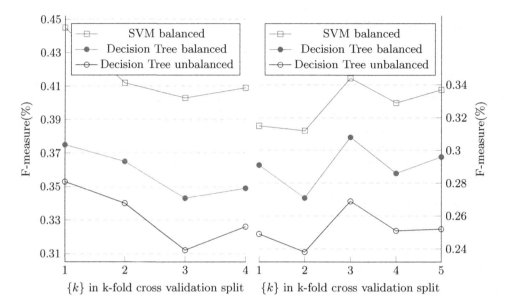

Fig. 3. PAN 2009 dataset - all features   **Fig. 4.** PAN 2011 dataset - all features

Table 3. F-score of each feature alone on PAN 2009 data corpus.

	Feature	F-score
Stylometric features	Average syllable count per token	0.280
	Verbs' compression rate	0.276
	Adjectives' compression rate	0.239
	Adverbs' compression rate	0.233
	Frequency of word *of*	0.226
	Flesch-Kinkaid score	0.199
	Average sentence length	0.198
Semantic features	wfc_4	0.359
	wfc_3	0.225
	wfc_1	0.210
	wfc_2	0.184

6   Comparison to Other Systems

In Tables 4 and 5 we compare our system to those who achieved the highest score predictions in the competitions: Stamatatos [3] for the 2009 competition and Oberreuter et al. for the 2011 competition. These two systems have been applied on the datasets of both the above PAN competitions and yielded the

Table 4. Comparative results: PAN 2009 data corpus.

	F-score	Precision	Recall	Granularity
Oberreuter et al.	0.346	**0.389**	0.311	1.001
Stamatatos	0.309	0.232	0.460	1.383
Our system	**0.419**	0.321	**0.600**	**1.000**

Table 5. Comparative results: PAN 2011 data corpus.

	F-Score	Precision	Recall	Granularity
Oberreuter et al.	0.325	**0.339**	0.312	1.000
Stamatatos	0.19	0.14	0.41	1.21
Our system	**0.328**	0.241	**0.513**	**1.000**

best results in comparison to all other participants. We consider our system with SVM classification and data balancing.

Overall, in both datasets our system achieves the best results in all metrics, except precision for which it is ranked second. In particular, in the 2011 corpus, our system is slightly better (F-Score) compared to Oberreuter's *et al.* system [4], while it is far superior to that of Stamatatos [3] achieving almost 13% higher total F1-score. As for the 2009 corpus, our system achieves more than 17% higher score than Stamatatos' system and 7% than Oberreuter's *et al.* system. At this point, we underline that the 2009 corpus is considered the "neutral" one, compared to that of 2011, where the plagiarised passages are of irrelevant semantic content, for which in Sect. 3 we postulated that Oberreuter's system was especially tweaked.

7   Conclusion and Future Work

Our approach proposes the implementation of a wide variety of features in combination to a strong machine learning subsystem, in order to discover not probable relations between the abnormalities that the plagiarised passages tend to present. The important role that machine learning plays in this approach leads us to consider the inherent, for all intrinsic plagiarism detection tasks, characteristic of unbalanced data as a problem's crucial parameter. The experiments proved that the balancing of the data played a key role for good classification results. In addition, the novel features we propose seem to be promising, yielding the best overall experimental results when used simultaneously.

The proposed model could also be tested on the data corpus of PAN 2016 competition. It can be generalised to fit other similar problems that concern current research, such as author identification and author profiling. We also believe that further research should be done in terms of machine learning and data balancing with respect to these tasks.

References

1. Stein, B., Lipka, N., Prettenhofer, P.: Intrinsic plagiarism analysis. Lang. Resour. Eval. **45**(1), 63–82 (2010)
2. Cheng, N., Chandramouli, R., Subbalakshmi, K.: Author gender identification from text. Digit. Invest. **8**, 78–88 (2011)
3. Stamatatos, E.: Intrinsic plagiarism detection using character n-gram profiles. In: Proceedings of the 3rd International Workshop on Uncovering Plagiarism, Authorship and Social Software Misuse (2009)
4. Oberreuter, G., L'Huillier, G., Rios, S.A., Velasquez, J.D.: Approaches for intrinsic and external plagiarism detection. In: Notebook for PAN at CLEF (2011)
5. Kestemont, M., Luyckx, K., Daelemans, W.: Intrinsic plagiarism detection using character trigrams distance scores. In: Notebook for PAN at CLEF (2011)
6. Kuznetsov, M., Motrenko, A., Kuznetsova, R., Strijov, V.: Methods for intrinsic plagiarism detection and author diarization. In: Notebook for PAN at CLEF (2016)
7. Potthast, M., Eiselt, A., Barrón-Cedeño, A., Stein, B., Rosso, P.: Overview of the 1st International Competition on Plagiarism Detection (2009)
8. Potthast, M., Eiselt, A., Barrón-Cedeño, A., Stein, B., Rosso, P.: Overview of the 3rd International Competition on Plagiarism Detection (2011)
9. Alzahrani, S.M., Salim, N., Abraham, A.: Understanding plagiarism linguistic patterns, textual features, and detection methods. Int. J. Educ. Integr. **9**(1), 55–71 (2013)
10. Tang, Y., Zhang, Y.Q., Chawla, N.V., Krasser, S.: SVMs modeling for highly imbalanced classification. In: Journal of LaTeX Class Files (2002)
11. Rosso, P., Rangel, F., Potthast, M., Stamatatos, E., Tschuggnall, M., Stein, B.: Overview of PAN 2016 new challenges for authorship analysis: cross-genre profiling, clustering, diarization, and obfuscation. In: 7th International Conference of the CLEF Initiative (2016)
12. DuBay, W.H.: The Principles of Readability, p. 21 (2004)
13. Leanne, S., Matwin, S.: Intrinsic plagiarism detection using complexity analysis (2009)
14. Meyer zu Eissen, S., Stein, B.: Intrinsic plagiarism detection. In: Lalmas, M., MacFarlane, A., Rüger, S., Tombros, A., Tsikrika, T., Yavlinsky, A. (eds.) ECIR 2006. LNCS, vol. 3936, pp. 565–569. Springer, Heidelberg (2006). doi:10.1007/11735106_66
15. Chawla, N., Bowyer, K., Hall, L., Kegelmeyer, W.: SMOTE: synthetic minority oversampling technique. J. Artif. Intell. Res. **16**(1), 321–357 (2002)
16. Lemaître, G., Nogueira, F., Aridas, C.K.: Imbalanced-learn: a python toolbox to tackle the curse of imbalanced datasets in machine learning. J. Mach. Learn. Res. **18**(1), 559–563 (2017)

Random Resampling in the One-Versus-All Strategy for Handling Multi-class Problems

Christos K. Aridas$^{(\boxtimes)}$, Stamatios-Aggelos N. Alexandropoulos,
Sotiris B. Kotsiantis, and Michael N. Vrahatis

Computational Intelligence Laboratory, Department of Mathematics,
University of Patras, GR-26110 Patras, Greece
char@upatras.gr, {alekst,sotos,vrahatis}@math.upatras.gr

Abstract. One of the most common approaches for handling the multi-class classification problem is to divise the original data set into binary subclasses and to use a set of binary classifiers in order to solve the binarization problem. A new method for solving multi-class classification problems is proposed, by incorporating random resampling techniques in the one-versus-all strategy. Specifically, the division used by the proposed method is based on the one-versus-all binarization technique using random resampling for handling the class-imbalance problem arising due to the one-versus-all binarization. The method has been tested extensively on several multiclass classification problems using Support Vector Machines with four different kernels. Experimental results show that the proposed method exhibits a better performance compared to the simple one-versus-all.

Keywords: Multi-class classification · One-versus-all · Random sampling

1 Introduction

Multi-class, also know as multinomial, classification refers to the problem of classifying patterns into three or more categories, whereas, binary classification is the task of classifying patterns into two distinct categories. Some classification algorithms like Decision Trees, Neural Networks and Bayesian Classifiers naturally handle multi-class problems. On the other hand, some other, like Support Vector Machines (SVMs) [4,16] are restricted to binary problems.

The most common approach for the generalization of binary classification to solve multi-class problems is to decompose the problem into several binary sub-problems [21]. Two of the most well-known approaches are: (a) the one-versus-one (OVO) strategy and (b) the one-versus-all (OVA) strategy [27]. The OVO strategy uses a binary classifier to discriminate piecewise the classes, while the OVA strategy uses a binary classifier to distinguish a single class from the rest classes.

© Springer International Publishing AG 2017
G. Boracchi et al. (Eds.): EANN 2017, CCIS 744, pp. 111–121, 2017.
DOI: 10.1007/978-3-319-65172-9_10

The OVO strategy, for a K class problem, trains $K \cdot (K-1)/2$ classifiers. The most straightforward combination is the majority voting rule where each classifier votes for the predicted class and the one with the largest amount of votes is predicted. On the other hand in OVA, K binary classifiers are trained and the decision is made by applying all binary classifiers to an unseen sample x and by predicting the class label for which the corresponding classifier reports the highest confidence score.

It is known that, the OVA strategy introduces class imbalance [26] during the binary reduction, which may lead classifiers towards the new generated majority class. In this research work a method that handles the problem of class imbalance is presented and its performance is measured in several well-known and widely used benchmark data sets.

The rest of the paper is organized as follows: In Sect. 2 similar works are briefly discussed. In Sect. 3 the proposed method is presented and analysed. In addition, experimental results obtained by using twenty multi-class benchmark data sets are exhibited. The paper ends in Sect. 4 with a synopsis and concluding remarks.

2   Related Work

The multi-class categorization problem [21] is one of the most known problems in Computer Science. Allwein *et al.* [1] have proposed a unifying framework to solve this problem by reducing it to various multiple binary problems. To achieve this, they used a margin-based learning algorithm. Specifically, they unified the most popular approaches: (a) each class is compared against all others, (b) all pairs are compared to each other and (c) codes with error-corresponding properties. In their paper, they have proposed a general method for combining the classifiers generated on the binary problem applying to the most well-known classification learning algorithms such as SVMs, AdaBoost, regression and others [17]. The experimental results with SVMs and AdaBoost have shown that this scheme provides an alternative solution to the mostly used multi-class algorithms.

Zadrozny and Elkan [30] have presented a method that solves the multi-class classification problem through class membership probability estimates using the probability estimates which are produced by binary classifiers. Their experimental results, using boosted naive Bayes, have shown that their method has similar classification accuracy to the loss-based decoding method.

One of the biggest difficulties that we have to deal with is the mapping of the multi-class problem onto a set of simpler binary classification problems, especially when we have to deal with hundreds of classes. Due to the fact that many of the statistical classification models do not have natural multi-class extensions, like SVMs, Rocha and Goldenstein [24] have introduced the correlation and joint probability of base binary learners. They have grouped the binary learners based on their independence and with Bayesian techniques they predict the class of new instances. They have also focused on the reduction of the number of the required learners and how to find new learners that complete the original set.

Despite the progress that has been made recently, the extension of SVMs as multi-class classification solvers is still ongoing. Most of the methods that have been proposed build a multi-class classifier by combing several binary classifiers or considering all classes at once. These kind of methods require to test on large-scale problems, a fact that makes them computationally expensive. Hsu and Lin [15], through their experimental work, have compared OVA, OVO and DAGSVM methods and they have shown that the last two methods are more suitable for practical use. In addition, they have indicated that SVM needs fewer vectors, in the case where all classes are considered at once.

Over the last decade many efforts have been made to construct methods with high classification efficiency for multi-class problems. Fei and Liu [9], have proposed Binary Tress of SVM (BTS), a method which decreases the number of binary classifiers without increasing the total complexity of the problem. The results of their work have shown that BTS maintains comparable accuracy and is much faster to be trained than DAGSVM or ECOC, especially in big problems (with a big number of classes).

Wu *et al.* [29], have proposed two multi-class classification methods for obtaining class probabilities. Both of them are easily implementable and more stable than the voting and Hastie-Tibshirani method [13].

The main goal of so called "binarization strategies" is to divide the original set into two classes, in order to address the multi-class classification problem as well as for each class to train a different binary algorithm. For this scope, two techniques are applied, namely: OVA and OVO. Hence, Galar *et al.* [11], have developed an experimental study on these strategies, to examine the potential of different classifiers, such as SVMs, Decision Trees, Instance Based Learning and Rule Based Systems, as well as the performance and robustness of these techniques, supported by several statistical tests. They have concluded that the best binarization technique highly depends on the base classifier and its confidence estimates.

In the attempt to exploit both the advantages of efficient computations and high classification accuracy, Cheong *et al.* [2], have proposed a binary Decision Tree architecture with SVMs classification. They have introduced a modified version of SOM, the K-SOM, which assists to the achievement of the conversion of multi-class problem into binary tree in order for the decision to be made by SVM. Their method overcomes the performance of trees and maintains comparable classification accuracy in comparison to OVA and OVO strategies.

Lorena *et al.* [21], have presented a survey of the main methods of binary classifiers that can be applied to multi-class classification problems. Another attempt to tackle the multi-class classification problem has been made by Crammer and Singer [5] who have focused on designing output codes and especially continuous codes that have not been viewed as a constrained optimization problem. More specifically, one of the aspects of their formulation was a scheme that built SVMs.

In [12], García-Pedrajas and Ortiz-Boyer have presented many capable binary classifier fusion methods for a multi-class classification problem. These methods require different assumptions, diverse influences and many aspects that need further study, in order to find out which of those techniques are better for a given multi-class classification problem. Following what we have mentioned above, Duan and Keerthi [8] have conducted an experimental study, trying to conclude on which multi-class SVM method is better.

In the literature, there are various other approaches for the above mentioned issues [7, 20, 25, 28]. Finally, we would like to dwell on the work of Chmielnicki and Stapor [3] who have used instance balancing to improve the performance of pairwise coupling, through the OVA strategy.

3   Proposed Method and Experimental Evaluation

Even in data sets where the patterns are equally distributed between the classes, the OVA approach could lead to high imbalanced binary data sets for each underlined classifier. For example, considering a ten class problem where the patterns are equally distributed, the OVA approach would train binary classifiers that would contain only 10% from the one class and 90% from the other. It is known that many supervised learning algorithms tend to prefer the more common classes using the prior knowledge of the training data set [26]. The proposed approach tackles the problem of class imbalance in data level, independently for each binary classifier, either by random over-sampling (ROS) the minority classes or by random under-sampling (RUS) the majority class. The proposed method is illustrated in Algorithm 1.

For the experiments twenty multi-class data sets have been chosen from the UCI Machine Learning Repository [19]. In Table 1 the name, the number of patterns, the number of input attributes, the number of different classes, as well as the percentage of the majority class for each data set are exhibited. All data sets have been preprocessed following the approach of [10]. Specifically, all discrete input attributes have been transformed to numeric by using a simple quantization. Each attribute has been scaled to have zero mean and standard deviation one. Also, all missing values have been treated as zero.

The classifiers' performance have been measured using the stratified 5-fold cross-validation procedure. The whole data set has been divided into five mutually exclusive folds and for each fold the classifier has been trained on the union of all of the other folds. The folds have been made by preserving the percentage of patterns for each class. Then, cross-validation has been run five times for each algorithm and the mean value of the five folds has been calculated. The performance metric that is reported is the F_1 score which is the weighted average of the precision and recall and has been calculated as:

$$F_1 = 2 \cdot \frac{Precision \cdot Recall}{Precision + Recall}$$

All experiments have been conducted with Python using the available implementations from the scikit-learn [23] and imbalanced-learn [18] libraries.

Algorithm 1

---

parameters
 Random Sampling Method M
 Base Classifier C
 Ratio R
procedure TRAINING(Data Set D)
 $f \leftarrow \emptyset$
 $m \leftarrow getNumberOfClasses(D)$
 for i to m **do**
  $onesDataset \leftarrow \emptyset$
  $allDataset \leftarrow \emptyset$
  for $\forall (x,y) \in D$ **do**
   if $y <> i$ **then**
    $y \leftarrow 0$
    $allDataset \leftarrow allDataset \cup \{(x,y)\}$
   else
    $y \leftarrow 1$
    $onesDataset \leftarrow onesDataset \cup \{(x,y)\}$
   end if
  end for
  if $M == \text{'}ros\text{'}$ **then**
   $onesDataset \leftarrow randomOverSample(onesDataset, R)$
  else
   $allDataset \leftarrow randomUnderSample(allDataset, R)$
  end if
  $binaryDataset \leftarrow onesDataset \cup allDataset$
  $f_i \leftarrow trainClassifier(C, binaryDataset)$
 end for
end procedure
procedure CLASSIFICATION(Data Set D)
 for $\forall (x,y) \in D$ **do**
  $f(x) \leftarrow \underset{i}{\operatorname{argmax}} f_i(x)$
 end for
end procedure

The experiments have been carried out using Support Vector Machines with four different kernels. The linear, the polynomial (with degree of 3), the RBF as well as the sigmoid kernel functions have been considered resulting to four different classifiers. The standard OVA approach has been compared to the proposed method using ROS and RUS with base classifier SVMs using the four different kernels mentioned.

In Tables 2, 3, 4 and 5 the obtained results for each kernel are exhibited. The best performer scheme for each data set is reported using boldface digits. It can be easily seen that the proposed method using ROS is the out-performer across all kernel functions, followed by the the RUS version.

The significance of the results have been examined using non-parametric statistical tests [6]. Particularly, the non-parametric Friedman Aligned Ranks [14]

Table 1. Collection of 20 multiclass data sets from the UCI Machine Learning Repository. The number of patterns (#patterns), the number of inputs (#inputs), the number of classes (#classes) as well as the percentage of the majority class (%majority) for each data set, are exhibited.

Data set	#patterns	#inputs	#classes	%majority
abalone	4177	8	3	34.6
arrhythmia	452	262	13	54.2
car	1728	6	4	70.0
contrac	1473	9	3	42.7
dermatology	366	34	6	30.6
ecoli	336	7	8	42.6
flags	194	28	8	30.9
glass	214	9	6	35.5
heart-cleveland	303	13	5	54.1
iris	150	4	3	33.3
lenses	24	4	3	62.5
libras	360	90	15	6.7
low-res-spect	531	100	9	51.9
nursery	12960	8	5	33.3
page-blocks	5473	10	5	89.8
seeds	210	7	3	33.3
steel-plates	1941	27	7	34.7
teaching	151	5	3	34.4
wine-quality-red	1599	11	6	42.6
wine-quality-white	4898	11	7	44.9

test has been performed because of the small number of the compared algorithms. The null-hypothesis states that the performance of all the compared methods are equivalent and therefore their ranks should be equal. In Table 6 the results obtained by using the Friedman test is presented. The p-value in all the cases, with the exception of the polynomial kernel, indicates that the null-hypotheses should be rejected. This means that there are methods whose performance difference was statistically significant to the others. Therefore, post-hoc tests using Nemenyi's [6] procedure have been employed and the obtained results are presented in Table 7. In the case of the linear kernel it can be seen that the ROS version outperforms both versions of the RUS and the standard approach, while in RBF and sigmoid kernels, the standard approach of the OVA scheme has been outperformed by both the ROS and RUS variation.

Table 2. Macro-averaged $F1$ scores using as base classifier SVM with linear kernel.

Data set	Standard-OVA	ROS-OVA	RUS-OVA
abalone	0.485 ± 0.02	$\mathbf{0.626 \pm 0.01}$	0.617 ± 0.02
arrhythmia	$\mathbf{0.422 \pm 0.08}$	0.407 ± 0.10	0.298 ± 0.05
car	0.441 ± 0.09	$\mathbf{0.541 \pm 0.09}$	0.475 ± 0.06
contrac	0.373 ± 0.03	0.490 ± 0.01	$\mathbf{0.487 \pm 0.02}$
dermatology	$\mathbf{0.972 \pm 0.03}$	0.967 ± 0.02	0.969 ± 0.02
ecoli	0.665 ± 0.15	0.609 ± 0.03	$\mathbf{0.668 \pm 0.12}$
flags	$\mathbf{0.294 \pm 0.06}$	0.289 ± 0.07	0.255 ± 0.09
glass	0.340 ± 0.13	$\mathbf{0.483 \pm 0.21}$	0.292 ± 0.11
heart-cleveland	0.283 ± 0.05	$\mathbf{0.383 \pm 0.08}$	0.379 ± 0.07
iris	0.831 ± 0.07	$\mathbf{0.885 \pm 0.04}$	$\mathbf{0.885 \pm 0.07}$
lenses	$\mathbf{0.849 \pm 0.17}$	0.773 ± 0.17	0.698 ± 0.21
libras	$\mathbf{0.568 \pm 0.10}$	0.564 ± 0.11	0.376 ± 0.07
low-res-spect	0.604 ± 0.13	$\mathbf{0.606 \pm 0.06}$	0.500 ± 0.08
nursery	0.518 ± 0.12	$\mathbf{0.537 \pm 0.12}$	0.429 ± 0.11
page-blocks	0.404 ± 0.11	$\mathbf{0.507 \pm 0.13}$	0.469 ± 0.10
seeds	0.909 ± 0.05	$\mathbf{0.928 \pm 0.04}$	0.919 ± 0.04
steel-plates	0.517 ± 0.08	$\mathbf{0.578 \pm 0.08}$	0.546 ± 0.09
teaching	0.405 ± 0.10	$\mathbf{0.529 \pm 0.08}$	0.513 ± 0.09
wine-quality-red	0.190 ± 0.03	0.271 ± 0.03	$\mathbf{0.273 \pm 0.02}$
wine-quality-white	0.175 ± 0.04	$\mathbf{0.262 \pm 0.07}$	0.248 ± 0.03

Table 3. Macro-averaged $F1$ scores using as base classifier SVM with polynomial kernel.

Data set	Standard-OVA	ROS-OVA	RUS-OVA
abalone	0.421 ± 0.03	$\mathbf{0.555 \pm 0.03}$	0.440 ± 0.05
arrhythmia	0.060 ± 0.01	$\mathbf{0.294 \pm 0.06}$	0.282 ± 0.08
car	0.448 ± 0.14	$\mathbf{0.516 \pm 0.12}$	0.495 ± 0.03
contrac	0.363 ± 0.01	$\mathbf{0.468 \pm 0.03}$	0.463 ± 0.02
dermatology	$\mathbf{0.801 \pm 0.09}$	0.616 ± 0.12	0.433 ± 0.12
ecoli	0.205 ± 0.03	$\mathbf{0.446 \pm 0.09}$	0.390 ± 0.09
flags	$\mathbf{0.278 \pm 0.10}$	0.126 ± 0.06	0.115 ± 0.04
glass	0.324 ± 0.07	$\mathbf{0.361 \pm 0.06}$	0.228 ± 0.03
heart-cleveland	0.155 ± 0.02	0.311 ± 0.04	$\mathbf{0.324 \pm 0.05}$
iris	0.532 ± 0.01	0.610 ± 0.14	$\mathbf{0.699 \pm 0.21}$
lenses	0.286 ± 0.08	0.621 ± 0.13	$\mathbf{0.667 \pm 0.24}$
libras	$\mathbf{0.598 \pm 0.12}$	0.338 ± 0.15	0.332 ± 0.15
low-res-spect	0.089 ± 0.01	$\mathbf{0.463 \pm 0.07}$	0.203 ± 0.07
nursery	$\mathbf{0.524 \pm 0.11}$	0.516 ± 0.11	0.428 ± 0.09
page-blocks	0.190 ± 0.00	$\mathbf{0.432 \pm 0.05}$	0.308 ± 0.06
seeds	0.540 ± 0.01	0.694 ± 0.16	$\mathbf{0.699 \pm 0.22}$
steel-plates	0.432 ± 0.09	$\mathbf{0.489 \pm 0.08}$	0.429 ± 0.05
teaching	0.468 ± 0.10	$\mathbf{0.488 \pm 0.07}$	0.487 ± 0.09
wine-quality-red	0.173 ± 0.02	0.081 ± 0.06	$\mathbf{0.174 \pm 0.04}$
wine-quality-white	$\mathbf{0.191 \pm 0.03}$	0.092 ± 0.03	0.109 ± 0.03

Table 4. Macro-averaged $F1$ scores using as base classifier SVM with RBF kernel.

Data set	Standard-OVA	ROS-OVA	RUS-OVA
abalone	0.462 ± 0.03	$\mathbf{0.611 \pm 0.03}$	0.603 ± 0.03
arrhythmia	0.060 ± 0.01	$\mathbf{0.361 \pm 0.08}$	0.309 ± 0.04
car	0.492 ± 0.18	$\mathbf{0.637 \pm 0.17}$	0.533 ± 0.06
contrac	0.391 ± 0.03	$\mathbf{0.472 \pm 0.02}$	0.470 ± 0.02
dermatology	0.966 ± 0.02	$\mathbf{0.970 \pm 0.03}$	0.948 ± 0.03
ecoli	0.549 ± 0.13	0.623 ± 0.04	$\mathbf{0.690 \pm 0.07}$
flags	0.258 ± 0.10	0.273 ± 0.05	$\mathbf{0.293 \pm 0.05}$
glass	0.261 ± 0.04	$\mathbf{0.404 \pm 0.17}$	0.262 ± 0.08
heart-cleveland	0.204 ± 0.03	0.345 ± 0.04	$\mathbf{0.349 \pm 0.06}$
iris	0.911 ± 0.08	$\mathbf{0.926 \pm 0.02}$	0.872 ± 0.04
lenses	0.663 ± 0.33	$\mathbf{0.849 \pm 0.17}$	$\mathbf{0.849 \pm 0.17}$
libras	$\mathbf{0.578 \pm 0.10}$	0.460 ± 0.08	0.456 ± 0.10
low-res-spect	0.373 ± 0.09	$\mathbf{0.454 \pm 0.06}$	0.343 ± 0.05
nursery	0.533 ± 0.10	$\mathbf{0.568 \pm 0.10}$	0.457 ± 0.08
page-blocks	0.311 ± 0.08	$\mathbf{0.506 \pm 0.11}$	0.493 ± 0.11
seeds	0.914 ± 0.06	0.914 ± 0.05	$\mathbf{0.918 \pm 0.05}$
steel-plates	0.484 ± 0.05	$\mathbf{0.582 \pm 0.08}$	0.519 ± 0.07
teaching	0.455 ± 0.11	$\mathbf{0.505 \pm 0.06}$	0.486 ± 0.09
wine-quality-red	0.177 ± 0.04	$\mathbf{0.266 \pm 0.02}$	0.259 ± 0.03
wine-quality-white	0.181 ± 0.04	$\mathbf{0.271 \pm 0.07}$	0.224 ± 0.01

Table 5. Macro-averaged $F1$ scores using as base classifier SVM with sigmoid kernel.

Data set	Standard-OVA	ROS-OVA	RUS-OVA
abalone	0.453 ± 0.03	$\mathbf{0.583 \pm 0.03}$	0.572 ± 0.04
arrhythmia	0.006 ± 0.01	$\mathbf{0.368 \pm 0.08}$	0.304 ± 0.04
car	0.298 ± 0.01	$\mathbf{0.411 \pm 0.06}$	0.389 ± 0.06
contrac	0.393 ± 0.02	$\mathbf{0.466 \pm 0.02}$	0.459 ± 0.02
dermatology	0.957 ± 0.02	$\mathbf{0.967 \pm 0.01}$	0.935 ± 0.04
ecoli	0.493 ± 0.12	0.590 ± 0.06	$\mathbf{0.611 \pm 0.14}$
flags	0.248 ± 0.07	0.263 ± 0.05	$\mathbf{0.269 \pm 0.08}$
glass	0.248 ± 0.05	$\mathbf{0.348 \pm 0.17}$	0.244 ± 0.04
heart-cleveland	0.231 ± 0.04	$\mathbf{0.392 \pm 0.07}$	0.289 ± 0.03
iris	0.573 ± 0.06	0.626 ± 0.08	$\mathbf{0.778 \pm 0.05}$
lenses	0.286 ± 0.08	0.760 ± 0.18	$\mathbf{0.849 \pm 0.17}$
libras	$\mathbf{0.441 \pm 0.09}$	0.405 ± 0.13	0.367 ± 0.15
low-res-spect	0.281 ± 0.10	$\mathbf{0.448 \pm 0.08}$	0.265 ± 0.06
nursery	0.402 ± 0.07	$\mathbf{0.465 \pm 0.10}$	0.409 ± 0.10
page-blocks	0.260 ± 0.06	0.469 ± 0.11	$\mathbf{0.481 \pm 0.09}$
seeds	0.790 ± 0.03	0.898 ± 0.05	$\mathbf{0.903 \pm 0.04}$
steel-plates	0.451 ± 0.04	$\mathbf{0.573 \pm 0.09}$	0.491 ± 0.05
teaching	$\mathbf{0.503 \pm 0.07}$	0.488 ± 0.13	0.378 ± 0.03
wine-quality-red	0.154 ± 0.02	$\mathbf{0.257 \pm 0.04}$	0.243 ± 0.02
wine-quality-white	0.164 ± 0.02	$\mathbf{0.245 \pm 0.04}$	0.211 ± 0.02

Table 6. Rankings of the algorithms using the Friedman Aligned Ranks test.

SVM-linear		SVM-poly		SVM-RBF		SVM-sigmoid	
ROS	42.42500	ROS	37.85000	ROS	42.95000	ROS	43.25000
RUS	25.32500	RUS	29.95000	RUS	31.82500	RUS	33.05000
Standard	23.75000	Standard	23.70000	Standard	16.72500	Standard	15.20000
Statistic	9.92085	Statistic	4.56181	Statistic	15.82051	Statistic	18.58499
p-value	0.00701	p-value	0.10219	p-value	0.00037	p-value	0.00009

Table 7. Post hoc comparisons using the Nemenyi's procedure.

Comparison	Statistic	Adjusted p-value	Result
SVM-linear			
ROS vs RUS	3.09632	0.00588	H_0 is rejected
ROS vs standard	3.38151	0.00216	H_0 is rejected
Standard vs RUS	0.28519	1.00000	H_0 is not rejected
SVM-RBF			
ROS vs RUS	2.01442	0.13190	H_0 is not rejected
ROS vs standard	4.74860	0.00001	H_0 is rejected
Standard vs RUS	2.73418	0.01876	H_0 is rejected
SVM-sigmoid			
ROS vs RUS	1.84693	0.19427	H_0 is not rejected
ROS vs standard	5.07906	0.00000	H_0 is rejected
Standard vs RUS	3.23213	0.00369	H_0 is rejected

4   Conclusions

An alternative scheme to the one-versus-all strategy for extending binary classifiers to multi-class cases is presented. The proposed scheme tackles the imbalanced problem that is introduced when the one-versus-all strategy decomposes a multi-class problem to several binary ones. Therefore, before training each binary classifier, the imbalanced problem is solved by the usage of a random resampling strategy. Experiments on several standard well-known and widely used benchmark data sets show that the application either of RUS or ROS to each binary classifier could enhance the performance compared to the standard one-versusall approach. Exploiting the adaptation of the proposed approach in multi-label classification tasks [22], could be an interesting area for further research.

Acknowledgements. Stamatios-Aggelos N. Alexandropoulos gratefully acknowledges the support of his work by the Hellenic State Scholarships Foundation (IKY), co-financed by the European Union (European Social Fund–ESF) and Greek national funds, "Reinforcement of the Human Research Potential through Doctoral Research" of the Operational Program "Development of Human Capital, Education and Lifelong Learning" of the National Strategic Reference Framework (NSRF 2014–2020).

References

1. Allwein, E.L., Schapire, R.E., Singer, Y.: Reducing multiclass to binary: a unifying approach for margin classifiers. J. Mach. Learn. Res. **1**, 113–141 (2000)
2. Cheong, S., Oh, S.H., Lee, S.Y.: Support vector machines with binary tree architecture for multi-class classification. Neural Inf. Process. Lett. Rev. **2**(3), 47–51 (2004)
3. Chmielnicki, W., Stąpor, K.: Using the one-versus-rest strategy with samples balancing to improve pairwise coupling classification. Int. J. Appl. Math. Comput. Sci. **26**(1), 191–201 (2016)
4. Christianini, N., Shawe-Taylor, J.: An Introduction to Support Vector Machines and Other Kernel-based Learning Methods. Cambridge University Press, Cambridge (2000)
5. Crammer, K., Singer, Y.: On the learnability and design of output codes for multiclass problems. Mach. Learn. **47**(2), 201–233 (2002)
6. Demšar, J.: Statistical comparisons of classifiers over multiple data sets. J. Mach. Learn. Res. **7**, 1–30 (2006)
7. Dogan, U., Glasmachers, T., Igel, C.: A unified view on multi-class support vector classification. J. Mach. Learn. Res. **17**(45), 1–32 (2016)
8. Duan, K.-B., Keerthi, S.S.: Which is the best multiclass SVM method? an empirical study. In: Oza, N.C., Polikar, R., Kittler, J., Roli, F. (eds.) MCS 2005. LNCS, vol. 3541, pp. 278–285. Springer, Heidelberg (2005). doi:10.1007/11494683_28
9. Fei, B., Liu, J.: Binary tree of SVM: a new fast multiclass training and classification algorithm. IEEE Trans. Netw. **17**(3), 696–704 (2006)
10. Fernández-Delgado, M., Cernadas, E., Barro, S., Amorim, D.: Do we need hundreds of classifiers to solve real world classification problems? J. Mach. Learn. Res. **15**, 3133–3181 (2014). bibtex: fernandez-delgado_we_2014. http://jmlr.org/papers/v15/delgado14a.html
11. Galar, M., Fernández, A., Barrenechea, E., Bustince, H., Herrera, F.: An overview of ensemble methods for binary classifiers in multi-class problems: experimental study on one-vs-one and one-vs-all schemes. Pattern Recogn. **44**(8), 1761–1776 (2011)
12. García-Pedrajas, N., Ortiz-Boyer, D.: An empirical study of binary classifier fusion methods for multiclass classification. Inf. Fusion **12**(2), 111–130 (2011)
13. Hastie, T., Tibshirani, R.: Classification by pairwise coupling. Ann. Stat. **26**(2), 451–471 (1998). http://dx.doi.org/10.1214/aos/1028144844
14. Hodges, J.L., Lehmann, E.L.: Rank methods for combination of independent experiments in analysis of variance. In: Rojo, J. (ed.) Selected Works of E.L. Lehmann, pp. 403–418. Springer, Heidelberg (2011). doi:10.1007/978-1-4614-1412-4_35
15. Hsu, C.W., Lin, C.J.: A comparison of methods for multiclass support vector machines. IEEE Trans. Neural Netw. **13**(2), 415–425 (2002)
16. Jian, L., Gao, C.: Binary coding SVMs for the multiclass problem of blast furnace system. IEEE Trans. Ind. Electro. **60**(9), 3846–3856 (2013)
17. Kotsiantis, S.B.: Bagging and boosting variants for handling classifications problems: a survey. Knowl. Eng. Rev. **29**(01), 78–100 (2014)

18. Lemaître, G., Nogueira, F., Aridas, C.K.: Imbalanced-learn: a python toolbox to tackle the curse of imbalanced datasets in machine learning. J. Mach. Learn. Res. **18**(17), 1–5 (2017). http://jmlr.org/papers/v18/16-365.html
19. Lichman, M.: UCI Machine Learning Repository (2013). http://archive.ics.uci.edu/ml
20. Liu, M., Zhang, D., Chen, S., Xue, H.: Joint binary classifier learning for ecoc-based multi-class classification. IEEE Trans. Pattern Anal. Mach. Intell. **38**(11), 2335–2341 (2016)
21. Lorena, A.C., De Carvalho, A.C., Gama, J.M.: A review on the combination of binary classifiers in multiclass problems. Artif. Intell. Rev. **30**(1), 19–37 (2008)
22. Madjarov, G., Kocev, D., Gjorgjevikj, D., Džeroski, S.: An extensive experimental comparison of methods for multi-label learning. Pattern Recogn. **45**(9), 3084–3104 (2012)
23. Pedregosa, F., Varoquaux, G., Gramfort, A., Michel, V., Thirion, B., Grisel, O., Blondel, M., Prettenhofer, P., Weiss, R., Dubourg, V., Vanderplas, J., Passos, A., Cournapeau, D., Brucher, M., Perrot, M., Duchesnay, E.: Scikit-learn: machine learning in python. J. Mach. Learn. Res. **12**, 2825–2830 (2011)
24. Rocha, A., Goldenstein, S.K.: Multiclass from binary: expanding one-versus-all, one-versus-one and ecoc-based approaches. IEEE Trans. Neural Netw. Learn. Syst. **25**(2), 289–302 (2014)
25. Santhanam, V., Morariu, V.I., Harwood, D., Davis, L.S.: A non-parametric approach to extending generic binary classifiers for multi-classification. Pattern Recogn. **58**, 149–158 (2016)
26. Kotsiantis, S., Kanellopoulos, D., Pintelas, P.: Handling imbalanced datasets: a review. Int. Trans. Comput. Sci. Eng. **30**, 25–36 (2006)
27. Tax, D.M., Duin, R.P.: Using two-class classifiers for multiclass classification. In: Proceedings of the 16th IEEE International Conference on Pattern Recognition, vol. 2, pp. 124–127. IEEE (2002)
28. Windeatt, T., Ghaderi, R.: Coding and decoding strategies for multi-class learning problems. Inf. Fusion **4**(1), 11–21 (2003)
29. Wu, T.F., Lin, C.J., Weng, R.C.: Probability estimates for multi-class classification by pairwise coupling. J. Mach. Learn. Res. **5**, 975–1005 (2004)
30. Zadrozny, B., Elkan, C.: Reducing multiclass to binary by coupling probability estimates. In: Advances in Neural Information Processing Systems, vol. 2, pp. 1041–1048 (2002)

A Spiking One-Class Anomaly Detection Framework for Cyber-Security on Industrial Control Systems

Konstantinos Demertzis[1(✉)], Lazaros Iliadis[1], and Stefanos Spartalis[2]

[1] Department of Civil Engineering, School of Engineering,
Faculty of Mathematics Programming and General Courses,
Democritus University of Thrace, Kimmeria, Xanthi, Greece
kdemertz@fmenr.duth.gr, liliadis@civil.duth.gr
[2] Department of Production and Management Engineering,
School of Engineering, Democritus University of Thrace,
Kimmeria, 67100 Xanthi, Greece
sspart@pme.duth.gr

Abstract. Developments and upgrades in the field of industrial information technology, particularly those relating to information systems' technologies for the collection and processing of real-time data, have introduced a large number of new threats. These threats are primarily related to the specific tasks these applications perform, such as their distinct design specifications, the specialized communication protocols they use and the heterogeneous devices they are required to interconnect. In particular, specialized attacks can undertake mechanical control, dynamic rearrangement of centrifugation or reprogramming of devices in order to accelerate or slow down their operations. This may result in total industrial equipment being destroyed or permanently damaged. Cyber-attacks against *Industrial Control Systems* which mainly use *Supervisory Control and Data Acquisition* (SCADA) combined with *Distributed Control Systems* are implemented with *Programmable Logic Controllers*. They are characterized as *Advanced Persistent Threats*. This paper presents an advanced *Spiking One-Class Anomaly Detection Framework* (SOCCADF) based on the *evolving Spiking Neural Network* algorithm. This algorithm implements an innovative application of the *One-class classification methodology* since it is trained exclusively with data that characterize the normal operation of ICS and it is able to detect divergent behaviors and abnormalities associated with APT attacks.

Keywords: Industrial Control Systems · SCADA · PLC · APT · evolving Spiking Neural Network · One-class classification · Anomaly detection

1 Introduction

1.1 Industrial Control System (ICS)

Automation and remote control are the most important methods used by critical infrastructure in order to improve productivity and quality of service. In this respect, the efficient management of industrial IT and the introduction of sophisticated

© Springer International Publishing AG 2017
G. Boracchi et al. (Eds.): EANN 2017, CCIS 744, pp. 122–134, 2017.
DOI: 10.1007/978-3-319-65172-9_11

automation systems, have contributed to the emergence of advanced Cyber-attacks against Industrial Control Systems (ICS) [1]. These systems are active devices of the infrastructure network whereas the successful completion of specialized activities requires all the devices used to be accurately controlled and totally reliable. Typical ICS automation devices are SCADA, Distributed Control Systems (DCS), Programmable Logic Controllers (PLC) together with the sensors used in control loops to collect the measurements [2]. The above systems are properly interconnected to allow remote monitoring and control of processes with high response rates, even in cases where the devices are distributed between different distant points. The most important categories of ICS applications concern water and sewage network infrastructure, natural gas, fuel and chemicals, building and building management systems in general, power generation and distribution, automation of road arteries - railways - airports - metro and telecom infrastructure management and Networks [3].

1.2 APT Against ICS

Integration into critical ICS infrastructures, especially where ICS includes features related to communications and internet technologies, introduces risks and new threats to the security and to the uninterrupted smooth operation of the critical infrastructure they include [4]. Exploiting the vulnerabilities of the wired and wireless communication networks used to interface these devices, as well as the vulnerabilities associated with their operating control, may cause total taking of critical devices and applications, or unavailability of necessary services even partial or total destruction of them [5]. The consequences may be severe. Critical infrastructures, however, are exposed not only to new risks due to the vulnerabilities of the communications and computer network (*malware, spyware, ransomware*) but also to the dangers inherent in the heterogeneity currently characterized by these systems [6]. Physical attacks interrupt service provision, while cyberattacks attempt to gain remote access for their benefit [7]. In any case, attacks against ICS are characterized as Advanced Persistent Threats (APT)s, as cybercriminals are fully familiar with specialized methods and tools for exploiting unknown vulnerabilities to the public (zero days). Most of the time, they are highly competent and organized, they are funded and they have significant incentives. The APT attack usually follows four steps [8].

Access: The attacker gathers as much information as possible and targets to specific ICSelements (SCADA systems) by using zero days' malware. In this way, he will exploit weaknesses that will provide access [6–8].

Discovery: After gaining access to the critical infrastructures network, discovery tactics are applied to the processes performed on it. For example, long term analysis and monitoring of the information flow in the location of the attacker. This is done to disclose information such as server mode, engineering workstation positions, architecture of local devices controlling individual components-units, connected Master Stations and so on [6–8].

Control: Once the network architecture and ICS mode have been understood, there are several ways to control the system. Typical potential targets that can control the network are the engineering workstation used to upgrade the software, database systems and the application server that hosts various applications used in the general system.

Hiding: In this step, attackers hide all the elements of the attacks by deleting specific folders that can betray their presence in automation systems [6–8]. Attacks designed to attack SCADA systems (e.g. the Stuxnet virus) have created many doubts about the level of critical infrastructure security and serious concerns about the consequences, as society is heavily dependent on the routine operations of these infrastructures. Intelligent anomaly detection is a process of high importance [6–8].

1.3 Anomaly Detection

The concept Anomaly Detection (AD) [9] refers to the recognition of standards from a set of data that exhibit a different behavior than expected. The goal is high level detection of possible anomalies combined with low false alert rates. The AD can be supervised, (performed on a training set containing normal versus anomalous classes) and semi-supervised where the training set is usually characterized as normal. The usual semi-supervised approach constructs a model to respond to normal behavior which is applied to determine the anomalies in the test data. In the unsupervised approach, no training is performed. It is based on the assumption that the normal incidents are more than the extreme ones in the testing data. If this reasoning does not apply, then the techniques have a large error rate [10]. Several abnormality detection techniques have been proposed in the literature, with more popular the *One-class classification (OCC)* methods, the *Distance Based* [11] ones, the *Replicator Neural Networks* [12] and the *Conditional Anomaly Detection* [13].

1.4 One-Class Classification

In machine learning (ML), the OCC method [14], tries to find objects of a particular class among all objects, by learning from a training set containing only objects of this class. Typically, these algorithms aim to implement classification models in which the negative class is absent, either because the missing class is not sampled, or due to the fact that it is difficult to do so. This mode of operation in which classifiers are required to determine effectively and reliably the boundaries of the class separation only based on the knowledge of the positive class, is a particularly complex problem of ML. When only data from the target class is available, the classifier is trained to receive target objects and to reject the ones that deviate significantly. Finally, it should be noted that the basic concept in OCC problem solving is the reverse of the generalization that is being pursued in other ML problems [15]. Particularly, it is intended that the parameter setting is fully defined, even if this exponentially increases the complexity of the classifier, provided it is able to correctly classify the target data.

1.5   The Proposed SOCCADF

Identifying anomalies that lead to scrapping or deprecation of ICS devices is an extremely complex matter, due to the fact that ART attacks are the most advanced and highly intelligent cyber-engineering techniques, operating under a chaotic architecture of industrial networks. Also, given the passive operation of traditional security systems which in most cases are unable to detect serious threats, alternative more active and more meaningful methods of locating ART attacks are necessary. Our research team has developed several innovative approaches of computational intelligence towards security threats identification. Herein we are proposing an intelligent system that significantly enhances the security level of critical infrastructures, by consuming the minimum level of resources. It is the *Spiking One-Class Anomaly Detection Framework (SOCCADF)*, which exploits for a first time a special operation form of the *evolving Spiking Neural Network (eSNN)* algorithm, in order to effectively classify the ICS anomalies resulting from APT attacks [16–30].

1.6   Innovation

An important innovation of SOCCADF is the use for first time of the *eSNN* algorithm (incorporated in Spiking Neural Networks) for the implementation of an OCC anomaly detection system. SNNs simulate the functioning of biological brain cells in a most realistic way and they rationally model data in a spatiotemporal mode. The produced signals are transmitted by discharges of temporal pulses, where duration and frequency of time pulses between neurons are the crucial factors. Also, innovation is attributed to the addition of artificial intelligence at the level of *real-time* analysis of industrial equipment, which greatly enhances the defensive mechanisms of critical infrastructures. It is much easier to locate ARTs Attacks by controlling the interdependencies of ICS at all times. Finally, there is innovation in the data selection process. This data has emerged after extensive research in the way ICS work and after comparisons and tests regarding the boundaries of their inherent behavior that determine their classification in normal or outliers.

2   Literature Review

Moya et al. [31] originated the term *One-Class Classification* in their research. Different researchers have used other terms such as *Single Class Classification* [32–34]. These terms originated as a result of different applications to which OCC has been applied. Juszczak [35] has defined *One-Class Classifiers* as class descriptors able to learn restricted domains in a multi-dimensional pattern space, using primarily just a positive set of examples. Luo et al. [36] have proposed a cost-sensitive *OCC-SVM* algorithm for intrusion detection problem. Their experiments have suggested that giving different cost or importance to system users than to processes results in higher performance in intrusion detection than other system calls. Shieh and Kamm [37] have introduced a *kernel* density estimation method to give weights to the training data objects, such that the outliers get the least weights and the positive class members get higher weights for creating bootstrap samples.

Soupionis et al. [38] proposed a combinatorial method for automatic detection and classification of faults and cyber-attacks occurring on the power grid system when there is limited data from the power grid nodes due to cyber implications. In addition, Tao et al. have described the network attack knowledge, based on the theory of the factor expression of knowledge, and studied the formal knowledge theory of SCADA network from the factor state space and equivalence partitioning. This approach utilizes the *factor neural network* (FNN) theory which contains high-level knowledge and quantitative reasoning described to establish a predictive model including analytic FNN and analogous FNN. This model abstracts and builds an equivalent and corresponding network attack and defense knowledge factors system.

Also, Qin et al. [39] have introduced an analytic factor neuron model which combines machine reasoning based on the cloud generator with the FNN theory. The FNN model is realized based on mobile intelligent agent and malicious behavior perception technology. The authors have acknowledged the potential of machine learning-based approaches in providing efficient and effective detection, but they have not provided a deeper insight on specific methods, neither the comparison of the approaches by detection performances and evaluation practices. Chen and Abdelwahed [40] have applied autonomic computing technology to monitor SCADA system performance, and proactively estimate upcoming attacks for a given system model of a physical infrastructure. Finally, Yasakethu and Jiang in [41] have introduced a new *European Framework-7* project "*CockpitCI (Critical Infrastructure)*" and roles of intelligent machine learning methods to prevent SCADA systems from cyber-attacks.

3   evolving Spiking Neural Network (eSNN)

The eSNNs based on the "Thorpe" neural model [42] are modular connectionist-based systems that evolve their structure and functionality in a continuous, self-organized, on-line, adaptive, interactive way from incoming information [43]. In order to classify real-valued data sets, each data sample is mapped into a sequence of spikes using the *Rank Order Population Encoding* (ROPE) technique [44, 45]. In this encoding method, neurons are organized into neuronal maps which share the same synaptic weights. Whenever a synaptic weight is modified, the same modification is applied to the entire population of neurons within the map. Inhibition is also present between each neuronal map. If a neuron spikes, it inhibits all the neurons in the other maps with neighboring positions. This prevents all the neurons from learning the same pattern. When propagating new information, neuronal activity is initially reset to zero. Then, as the propagation goes on, each time one of their inputs fire, neurons are progressively desensitized. This is making neuronal responses dependent upon the relative order of firing of the neuron's afferents [46, 47]. Also in this model, the neural plasticity is used to monitor the learning algorithm by using one-pass learning method. The aim of this learning scheme is to create a repository of trained output neurons during the presentation of training samples [48].

4   ICS Anomaly Datasets

In order to carry out the research and evaluate the proposed model, 3 suitable datasets were chosen to best match the ICS communication and transaction data [49]. These sets include data logs from a gas pipeline, a lab scale water tower, and a lab scale electric transmission system. The logs include flagged network transactions during the normal operation of specific ICSs, as well as transactions during 35 different cyber-attacks. In addition to the logs, measurements include normal behavior as well as abnormalities detected during attacks that were simulated in a virtual ICS environment, including Human Machine Interface (HMI), Virtual Physical Process, and Virtual Programmable Logic Controller (VPLC) and a Virtual Network (VN). Although the configured virtual systems do not have physical limits to the size of the modeling they simulate, the virtual platform on which the data was collected was implemented by escalating the ICS to represent their operating states in the most realistic way [49].

All three data sets contain network transaction data, preprocessed in a way to strip *lower layer transmission* data (*TCP, MAC*) [49]. The "*water_tower_dataset*" includes 23 independent parameters and 236,179 instances, from which 172,415 are normal and 63,764 outliers. In the case of the "*water_train_dataset*" the algorithm was trained by using 86,315 normal instances, whereas the rest 86,100 normal instances and the 63,764 outliers comprised the testing set "*water_test_dataset*". The "*gas_dataset*" contains 26 independent parameters and 97,019 instances, from which 61,156 are normal and 35,863 outliers. For the "*gas_train_dataset*" case 30,499 normal instances are used for the training process, whereas 30,657 normal instances and 35,863 outliers comprise the "*gas_test_dataset*". Finally, the "*electric_dataset*" includes 128 independent features and 146,519 instances, from which 90,856 are normal and 55,663 outliers. The "*electric_train_dataset*" has 45,402 normal instances. The rest 45,454 normal ones and the 55,663 outliers comprise the "*electric_test_dataset*". More details related to the dataset and to the data selection process can be found in [49].

5   Methodology

5.1   Description of the eSNN One-Class Classification Method

The proposed methodology uses an eSNN classification approach in order to detect and verify the anomalies on ICS. The topology of the developed eSNN is strictly feed-forward, organized in several layers and weight modification occurs on the connections between the neurons of the existing layers. The encoding is performed by ROPE technique with 20 Gaussian Receptive Fields (GRF) per variable [46]. The data are normalized to the interval $[-1, 1]$ and so the coverage of the Gaussians is determined by using i_min and i_max. Each input variable is encoded independently by a group of one-dimensional GRF. The GRF of neuron i is given by its center μ_i by Eq. (1) and width σ by Eq. (2) [46]

$$\mu_i = I_{min}^n + \frac{2i-3}{2} \frac{I_{max}^n - I_{min}^n}{M-2} \tag{1}$$

$$\sigma = \frac{1}{\beta} \frac{I_{max}^n - I_{min}^n}{M-2} \tag{2}$$

where $1 \leq \beta \leq 2$ and the parameter β directly controls the width of each Gaussian receptive field. When a neuron reaches its threshold, it spikes and inhibits neurons at equivalent positions in the other maps so that only one neuron will respond at any location [50]. Every spike triggers a time based Hebbian-like learning rule that adjusts the synaptic weights. For each training sample i with class label l, a new output neuron is created and fully connected to the previous layer of neurons, resulting in a real-valued weight vector $w^{(i)}$ with $w_j^{(i)} \in R$ denoting the connection between the pre-synaptic neuron j and the created neuron i. In the next step, the input spikes are propagated through the network and the value of weight $w_j^{(i)}$ is computed according to the order of spike transmission through a synapse [46]

$$j : w_j^{(i)} = (m_l)^{order(j)} \tag{3}$$

where j is the pre-synaptic neuron of i. Function $order(j)$ represents the rank of the spike emitted by neuron j. The firing threshold $\theta^{(i)}$ of the created neuron i is defined as the fraction $c_l \in R$, $0 < c_l < 1$, of the maximal possible potential [46]

$$u_{max}^{(i)} : \theta^{(i)} \leftarrow c_l u_{max}^{(i)} \tag{4}$$

$$u_{max}^{(i)} \leftarrow \sum_j w_j^{(i)} (m_l)^{order(j)} \tag{5}$$

The weight vector of the trained neuron is compared to the weights corresponding to neurons already stored in the repository. Two neurons are considered too "similar" if the minimal *Euclidean* distance between their weight vectors is smaller than a specified similarity threshold s_l (the eSNN object uses optimal similarity threshold s = 0.6) [46]. Both the firing thresholds and the weight vectors were merged according to Eqs. (6) and (7):

$$w_j^{(k)} \leftarrow \frac{w_j^{(i)} + N w_j^{(k)}}{1+N} \tag{6}$$

$$\theta^{(k)} \leftarrow \frac{\theta^{(i)} + N \theta^{(k)}}{1+N} \tag{7}$$

Integer N denotes the number of samples previously used to update neuron k. The merging is implemented as the average of the connection weights, and of the two firing thresholds. After merging, the trained neuron i is discarded and the next sample processed. If no other neuron in the repository is similar to the trained neuron i, the neuron i is added to the repository as a new output [46].

All parameters of eSNN included in this search space, are optimized according to the *Versatile Quantum-inspired Evolutionary* Algorithm (vQEA) [46].

5.2 Threshold Deciding Criteria for the Proposed Method

The choice of the threshold value used for class separation is the most important and critical factor for the success of the OCC approach. In order to determine the threshold, and given that the training set contains only positive samples, this paper proposes a reliable heuristic selection method based solely on criteria of merit. In particular, the proposed algorithm assumes that a distance function d between the objects and the target class is employed in the training phase. The determination of the *threshold* θ for the class separation (normal or outlier) is performed in a way that it discards a set of training samples, most of which diverge from the target class, in order to strengthen the classifier. Even when all samples are correctly labeled, the rejection of a small but representative rate of training samples helps the classifier to learn the most representative set of training samples. This approach significantly enhances active security of critical infrastructure. The following pseudocode presents the algorithmic approach for the determination of the class separation threshold θ.

Algorithm 1: Optimal Threshold

Optimal Threshold:
1: Calculate the error using Euclidean distance between actual and predicted on each training data;
2: Arrange the error in decreasing order;
3: Set the threshold at rejection of 10% most erroneous data (false negative rate at the rate of 10%);

---

6 Results and Comparative Analysis

The efficiency of the proposed OCC classifier is estimated by employing the following statistical indices. The numbers of misclassifications are related to the False Positive (FP) and False Negative (FN) indices A FP is the number of cases where you wrongfully receive a positive result and the FN is exactly the opposite. On the other hand, the True Positive (TP) is the number of records where you correctly receive a Positive result. The True Negative (TN) is defined respectively. The *True Positive rate* (TPR) also known as *Sensitivity*, the *True Negative rate* also known as *Specificity* (TNR) and the *Total Accuracy* (TA) are defined by using Eqs. 8, 9, 10 respectively [50]:

$$TPR = \frac{TP}{TP + FN} \tag{8}$$

$$TNR = \frac{TN}{TN + FP} \tag{9}$$

$$TA = \frac{TP + TN}{N} \tag{10}$$

The Precision (PRE) the Recall (REC) and the F-Score indices are defined as in Eqs. 11, 12 and 13 respectively:

$$PRE = \frac{TP}{TP + FP} \tag{11}$$

$$REC = \frac{TP}{TP + FN} \tag{12}$$

$$F - Score = 2 \times \frac{PRE \times REC}{PRE + REC} \tag{13}$$

The *ROC (Receiver Operating Characteristic)* is a standard technique for summarizing classifier performance over a range of trade-offs between TP and FP error rates. ROC curve is a plot of *Sensitivity (the ability of the model to predict an event correctly)* versus *1-Specificity* for the possible cut-off classification probability values π_0 [50]. The *Precision* measure shows what percentage of positive predictions where correct, whereas *Recall* measures the percentage of positive events that were correctly predicted. The *F-Score* can be interpreted as a weighted average of the precision and recall. Therefore, this score takes both false positives and false negatives into account. Intuitively it is not as easy to understand as accuracy, but F-Score is usually more useful than accuracy and it works best if false positives and false negatives have similar cost, in this case. Finally, the ROC curve is related in a direct and natural way to cost/benefit analysis of diagnostic decision making [50].

In the training process of the OCC, only the probability density of the positive class is known, which means that during training only the number of the positive class items that are not classified correctly (FN) can be minimized. Basically, this means that due to the fact that there are no examples of samples' distribution belonging to other classes (outliers) in the training phase, it is not possible to estimate the number of objects of other classes that were misclassified as Positive (FP) by the OCC classifier. So given the fact that *TP + FN = 1* the algorithm during training can provide estimations only for TP and FN. However during testing all four indices (TP, FN, TN, FP) can be obtained.

The proposed system manages to operate effectively in a particularly complex cyber security problem with high levels of accuracy. The performance of the SOC-CADF is evaluated by comparing it with OCC Support Vector Machines (OCC-SVM) and OCC *Combining Density and Class Probability Estimation* (OCC-CD/CPE) learning. Regarding the overall efficiency of the method, the results show that the proposed system significantly outperforms the other algorithms. The following Table 1, presents the analytical values of the predictive power of the SOCCADF and the corresponding results when competitive algorithms were used.

Table 1. Comparison between algorithms

water_tower_dataset						
Classifier	Classification accuracy & performance metrics					
	Total accuracy	RMSE	Precision	Recall	F-Score	ROC area
OCC-eSNN	**98.08%**	**0.1305**	**0.981**	**0.981**	**0.981**	**0.994**
OCC-SVM	98.01%	0.1312	0.980	0.980	0.980	0.995
OCC-CD/CPE	96.75%	0.1389	0.975	0.975	0.975	0.980
gas_dataset						
OCC-eSNN	**98.82%**	**0.0967**	**0.988**	**0.988**	**0.988**	**0.995**
OCC-SVM	97.98%	0.0981	0.980	0.980	0.980	0.990
OCC-CD/CPE	95.67%	0.1284	0.960	0.960	0.960	0.975
electric_dataset						
OCC-eSNN	**98.30%**	**0.1703**	**0.983**	**0.983**	**0.983**	**0.999**
OCC-SVM	97.63%	0.1840	0.978	0.978	0.978	0.990
OCC-CD/CPE	97.02%	0.1897	0.970	0.970	0.970	0.985

7 Discussion and Conclusions

This comparison generates expectations for the identification of the SOCCADF as a robust anomaly detection model suitable for difficult problems. According to this comparative analysis, it appears that SOCCADF is highly suitable method for applications with huge amounts of data such that traditional learning approaches that use the entire data set in aggregate are computationally infeasible. The eSNN algorithm successfully reduces the problem of entrapment in local minima in training process, with very fast convergence rates. These improvements are accompanied by high classification rates and low-test errors as well. The performance of proposed model was evaluated in a high complex dataset and the real-world sophisticated scenarios. The experimental results showed that the SOCCADF has better performance at a very fast learning speed and more accurate and reliable classification results. The final conclusion is that the proposed method has proven to be reliable and efficient and has outperformed the other approaches for the specific security problem.

Future research should include further optimization of the eSNN parameters, aiming to achieve an even more efficient and faster categorization process. Also, it would be important for the proposed framework to expand, based on *"metalearning"* methods to self-improve and redefine its parameters so that it can fully automate the process of locating APT attacks. Finally, an additional element that could be studied in the direction of future expansion is the creation of an additional cross-sectional anomaly analysis system. This could act counter-diametrically on the philosophy of the eSNN classifier with potential enhancement of the system's efficiency.

References

1. Falco, J., et al.: IT security for industrial control systems. NIST Internal Report (NISTIR) 6859 (2002). http://www.nist.gov/customcf/get_pdf.cfm?pub_id=821684
2. Bailey, D., Wright, E.: Practical SCADA for Industry. IDC Technologies, Vancouver (2003)
3. Boyer, S.: SCADA: Supervisory Control and Data Acquisition, 4th edn. International Society of Automation, Research Triangle Park, North Carolina (2010)
4. Weiss, J.: Current status of cybersecurity of control systems. In: Presentation to Georgia Tech Protective Relay Conference (2003)
5. Cárdenas, A.A., Amin, S., Sastry, S.: Research challenges for the security of control systems. In: 3rd USENIX Workshop on Hot Topics in Security (HotSec 2008), Associated with the 17th USENIX Security Symposium, San Jose, CA, USA (2008)
6. Raj, V.S., Chezhian, R.M., Mrithulashri, M.: Advanced persistent threats & recent high profile cyber threat encounters. Int. J. Innov. Res. Comput. Commun. Eng. 2(1) (2014). (An ISO 3297: 2007 Certified Organization)
7. Hutchins, E., Cloppert, M., Amin, R.: Intelligence-driven computer network defense informed by analysis of adversary campaigns and intrusion kill chains. In: The 6th International Conference on Information-Warfare & Security, pp. 113–125 (2010)
8. Sood, A.K., Enbody, R.J.: Targeted cyberattacks: a superset of advanced persistent threats. IEEE Secur. Priv. 11(1), 54–61 (2013). doi:10.1109/MSP.2012.90
9. Chandola, V., Banerjee, A., Kumar, V.: Anomaly detection: a survey. ACM Comput. Surv. 41(3), 1–58 (2009). doi:10.1145/1541880.1541882
10. Zimek, A., Schubert, E., Kriegel, H.-P.: A survey on unsupervised outlier detection in high-dimensional numerical data. Stat. Anal. Data Min. 5(5), 363–387 (2012). doi:10.1002/sam.11161
11. Knorr, E.M., Ng, R.T., Tucakov, V.: Distance-based outliers: algorithms and applications. VLDB J. Int. J. Very Large Data Bases 8(3–4), 237–253 (2000). doi:10.1007/s007780050006
12. Hawkins, S., He, H., Williams, G., Baxter, R.: Outlier detection using replicator neural networks. In: Kambayashi, Y., Winiwarter, W., Arikawa, M. (eds.) DaWaK 2002. LNCS, vol. 2454, pp. 170–180. Springer, Heidelberg (2002). doi:10.1007/3-540-46145-0_17
13. Valko, M., Cooper, G., Seybert, A., Visweswaran, S., Saul, M., Hauskrecht, M.: Conditional anomaly detection methods for patient-management alert systems. In: Workshop on Machine Learning in Health Care Applications in the 25th International Conference on Machine Learning (2008)
14. Skabar, A.: Single-class classifier learning using neural networks: an application to the prediction of mineral deposits. In: Proceedings of the Second International Conference on Machine Learning and Cybernetics, vol. 4, pp. 2127–2132 (2003)
15. Manevitz, L.M., Yousef, M.: One-class SVMS for document classification. J. Mach. Learn. Res. 2, 139–154 (2001)
16. Demertzis, K., Iliadis, L.: Intelligent bio-inspired detection of food borne pathogen by DNA barcodes: the case of invasive fish species Lagocephalus Sceleratus. In: Iliadis, L., Jayne, C. (eds.) EANN 2015. CCIS, vol. 517, pp. 89–99. Springer, Cham (2015). doi:10.1007/978-3-319-23983-5_9
17. Demertzis, K., Iliadis, L.: A hybrid network anomaly and intrusion detection approach based on evolving spiking neural network classification. In: Sideridis, A.B., Kardasiadou, Z., Yialouris, C.P., Zorkadis, V. (eds.) E-Democracy 2013. CCIS, vol. 441, pp. 11–23. Springer, Cham (2014). doi:10.1007/978-3-319-11710-2_2

18. Demertzis, K., Iliadis, L.: Evolving computational intelligence system for malware detection. In: Iliadis, L., Papazoglou, M., Pohl, K. (eds.) CAiSE 2014. LNBIP, vol. 178, pp. 322–334. Springer, Cham (2014). doi:10.1007/978-3-319-07869-4_30

19. Demertzis, K., Iliadis, L.: A bio-inspired hybrid artificial intelligence framework for cyber security. In: Daras, N.J., Rassias, M.T. (eds.) Computation, Cryptography, and Network Security, pp. 161–193. Springer, Cham (2015). doi:10.1007/978-3-319-18275-9_7

20. Demertzis, K., Iliadis, L.: Bio-inspired hybrid intelligent method for detecting android malware. In: Proceedings of the 9th KICSS 2014, Knowledge Information and Creative Support Systems, Cyprus, pp. 231–243, November 2014. ISBN 978-9963-700-84-4

21. Demertzis, K., Iliadis, L.: Evolving smart URL filter in a zone-based policy firewall for detecting algorithmically generated malicious domains. In: Gammerman, A., Vovk, V., Papadopoulos, H. (eds.) SLDS 2015. LNCS, vol. 9047, pp. 223–233. Springer, Cham (2015). doi:10.1007/978-3-319-17091-6_17

22. Demertzis, K., Iliadis, L.: SAME: an intelligent anti-malware extension for android ART virtual machine. In: Núñez, M., Nguyen, N.T., Camacho, D., Trawiński, B. (eds.) ICCCI 2015. LNCS, vol. 9330, pp. 235–245. Springer, Cham (2015). doi:10.1007/978-3-319-24306-1_23

23. Demertzis, K., Iliadis, L.: Computational intelligence anti-malware framework for android OS. Spec. Issue Vietnam J. Comput. Sci. (VJCS) **4**, 1–15 (2016). doi:10.1007/s40595-017-0095-3. Springer

24. Demertzis, K., Iliadis, L.: Detecting invasive species with a bio-inspired semi supervised neurocomputing approach: the case of Lagocephalus sceleratus. Spec. Issue Neural Comput. Appl. **28**, 1225–1234 (2016). doi:10.1007/s00521-016-2591-2. Springer

25. Demertzis, K., Iliadis, L.: SICASEG: a cyber threat bio-inspired intelligence management system. J. Appl. Math. Bioinform. **6**(3), 45–64 (2016). ISSN 1792-6602 (print), 1792-6939 (online). Scienpress Ltd.

26. Bougoudis, I., Demertzis, K., Iliadis, L.: Fast and low cost prediction of extreme air pollution values with hybrid unsupervised learning. Integr. Comput.-Aided Eng. **23**(2), 115–127 (2016). doi:10.3233/ICA-150505. IOS Press

27. Bougoudis, I., Demertzis, K., Iliadis, L.: HISYCOL a hybrid computational intelligence system for combined machine learning: the case of air pollution modeling in Athens. EANN Neural Comput. Appl. **27**, 1191–1206 (2016). doi:10.1007/s00521-015-1927-7

28. Anezakis, V.-D., Demertzis, K., Iliadis, L., Spartalis, S.: A hybrid soft computing approach producing robust forest fire risk indices. In: Iliadis, L., Maglogiannis, I. (eds.) AIAI 2016. IAICT, vol. 475, pp. 191–203. Springer, Cham (2016). doi:10.1007/978-3-319-44944-9_17

29. Anezakis, V.-D., Dermetzis, K., Iliadis, L., Spartalis, S.: Fuzzy cognitive maps for long-term prognosis of the evolution of atmospheric pollution, based on climate change scenarios: the case of Athens. In: Nguyen, N.-T., Manolopoulos, Y., Iliadis, L., Trawiński, B. (eds.) ICCCI 2016. LNCS, vol. 9875, pp. 175–186. Springer, Cham (2016). doi:10.1007/978-3-319-45243-2_16

30. Bougoudis, I., Demertzis, K., Iliadis, L., Anezakis, V.-D., Papaleonidas, A.: Semi-supervised hybrid modeling of atmospheric pollution in urban centers. In: Jayne, C., Iliadis, L. (eds.) EANN 2016. CCIS, vol. 629, pp. 51–63. Springer, Cham (2016). doi:10.1007/978-3-319-44188-7_4

31. Moya, M., Koch, M., Hostetler, L.: One-class classifier networks for target recognition applications. In: Proceedings World Congress on Neural Networks, pp. 797–801 (1993)

32. Munroe, D.T., Madden, M.G.: Multi-class and single-class classification approaches to vehicle model recognition from images. In: Proceedings of Irish Conference on Artificial Intelligence and Cognitive Science, Portstewart (2005)

33. Yu, H.: SVMC: single-class classification with support vector machines. In: Proceedings of International Joint Conference on Artificial Intelligence, pp. 567–572 (2003)
34. El-Yaniv, R., Nisenson, M.: Optimal single-class classification strategies. In: Proceedings of the 2006 NIPS Conference, vol. 19, pp. 377–384. MIT Press (2007)
35. Juszczak, P.: Learning to recognise. A study on one-class classification and active learning. Ph.D. thesis, Delft University of Technology (2006)
36. Luo, J., Ding, L., Pan, Z., Ni, G., Hu, G.: Research on cost-sensitive learning in one-class anomaly detection algorithms. In: Xiao, B., Yang, L.T., Ma, J., Muller-Schloer, C., Hua, Y. (eds.) ATC 2007. LNCS, vol. 4610, pp. 259–268. Springer, Heidelberg (2007). doi:10.1007/978-3-540-73547-2_27
37. Shieh, A.D., Kamm, D.F.: Ensembles of one class support vector machines. In: Benediktsson, J.A., Kittler, J., Roli, F. (eds.) MCS 2009. LNCS, vol. 5519, pp. 181–190. Springer, Heidelberg (2009). doi:10.1007/978-3-642-02326-2_19
38. Soupionis, Y., Ntalampiras, S., Giannopoulos, G.: Faults and cyber attacks detection in critical infrastructures. In: Panayiotou, C.G., Ellinas, G., Kyriakides, E., Polycarpou, M.M. (eds.) CRITIS 2014. LNCS, vol. 8985, pp. 283–289. Springer, Cham (2016). doi:10.1007/978-3-319-31664-2_29
39. Qin, Y., Cao, X., Liang, P., Hu, Q., Zhang, W.: Research on the analytic factor neuron model based on cloud generator and its application in oil&gas SCADA security defense. In: IEEE 3rd International Conference on Cloud Computing and Intelligence Systems (CCIS) (2014). doi:10.1109/CCIS.2014.7175721
40. Chen, Q., Abdelwahed, S.: A model-based approach to self-protection in computing system. In: Proceedings of the ACM Cloud and Autonomic Computing Conference, CAC 2013, Article No. 16 (2013)
41. Yasakethu, S.L.P., Jiang, J.: Intrusion detection via machine learning for SCADA system protection. In: Proceedings of the 1st International Symposium for ICS & SCADA Cyber Security Research, Learning and Development Ltd. (2013)
42. Thorpe, S.J., Delorme, A., Rullen, R.: Spike-based strategies for rapid processing. Neural Netw. 14(6–7), 715–725 (2001). Elsevier
43. Schliebs, S., Kasabov, N.: Evolving spiking neural network—a survey. Evol. Syst. 4, 87 (2013). doi:10.1007/s12530-013-9074-9. Springer
44. Delorme, A., Perrinet, L., Thorpe, S.J.: Networks of integrate-and-fire neurons using rank order coding. Neurocomputing 38–40(1–4), 539–545 (2000)
45. Thorpe, S., Gautrais, J.: Rank order coding. In: Bower, J.M. (ed.) CNS 1997, pp. 113–118. Springer, Boston (1998). doi:10.1007/978-1-4615-4831-7_19. Plenum Press
46. Kasabov, N.: Evolving Connectionist Systems: Methods and Applications in Bioinformatics, Brain Study and Intelligent Machines. Springer, Heidelberg (2002)
47. Wysoski, S.G., Benuskova, L., Kasabov, N.: Adaptive learning procedure for a network of spiking neurons and visual pattern recognition. In: Blanc-Talon, J., Philips, W., Popescu, D., Scheunders, P. (eds.) ACIVS 2006. LNCS, vol. 4179, pp. 1133–1142. Springer, Heidelberg (2006). doi:10.1007/11864349_103
48. Schliebs, S., Defoin-Platel, M., Kasabov, N.: Integrated feature and parameter optimization for an evolving spiking neural network. Neural Netw. 22(5–6), 623–632 (2009). 2009 International Joint Conference on Neural Networks
49. Morris, T.H., Thornton, Z., Turnipseed, I.: Industrial control system simulation and data logging for intrusion detection system research. Int. J. Netw. Secur. (IJNS) 17(2), 174–188 (2015)
50. Fawcett, T.: An introduction to ROC analysis. Pattern Recogn. Lett. 27(8), 861–874 (2006). doi:10.1016/j.patrec.2005.10.010. Elsevier Science Inc.

Deep Learning Convolutional ANN

Boosted Residual Networks

Alan Mosca$^{(\boxtimes)}$ and George D. Magoulas

Department of Computer Science and Information Systems,
Birkbeck, University of London, London, UK
{a.mosca,gmagoulas}@dcs.bbk.ac.uk

Abstract. In this paper we present a new ensemble method, called Boosted Residual Networks, which builds an ensemble of Residual Networks by growing the member network at each round of boosting. The proposed approach combines recent developements in Residual Networks - a method for creating very deep networks by including a shortcut layer between different groups of layers - with the Deep Incremental Boosting, which has been proposed as a methodology to train fast ensembles of networks of increasing depth through the use of boosting. We demonstrate that the synergy of Residual Networks and Deep Incremental Boosting has better potential than simply boosting a Residual Network of fixed structure or using the equivalent Deep Incremental Boosting without the shortcut layers.

1 Introduction

Residual Networks, a type of deep network recently introduced in [3], are characterized by the use of *shortcut* connections (sometimes also called *skip* connections), which connect the input of a layer of a deep network to the output of another layer positioned a number of levels "above" it. The result is that each one of these shortcuts shows that networks can be build in *blocks*, which rely on both the output of the previous layer and the previous block.

Residual Networks have been developed with many more layers than traditional Deep Networks, in some cases with over 1000 blocks, such as the networks in [5]. A recent study in [14] compares Residual Networks to an ensemble of smaller networks. This is done by unfolding the shortcut connections into the equivalent tree structure, which closely resembles an ensemble. An example of this can be shown in Fig. 1.

Dense Convolutional Neural Networks [6] are another type of network that makes use of shortcuts, with the difference that each layer is connected to all its ancestor layers directly by a shortcut. Similarly, these could be also unfolded into an equivalent ensemble.

True ensemble methods are often left as an *afterthought* in Deep Learning models: it is generally considered sufficient to treat the Deep Learning method

---

G.D. Magoulas—The authors gratefully acknowledge the support of NVIDIA Corporation with the donation of the Tesla Titan X Pascal GPUs used for this research.

© Springer International Publishing AG 2017
G. Boracchi et al. (Eds.): EANN 2017, CCIS 744, pp. 137–148, 2017.
DOI: 10.1007/978-3-319-65172-9_12

Fig. 1. A Residual Network of N blocks can be unfolded into an ensemble of $2^N - 1$ smaller networks.

as a "black-box" and use a well-known generic Ensemble method to obtain marginal improvements on the original results. Whilst this is an effective way of improving on existing results without much additional effort, we find that it can amount to a waste of computations. Instead, it would be much better to apply an Ensemble method that is aware, and makes use of, the underlying Deep Learning algorithm's architecture.

We define such methods as "white-box" Ensembles, which allow us to improve on the generalisation and training speed compared to traditional Ensembles, by making use of particular properties of the base classifier's learning algorithm and architecture. We propose a new such method, which we call Boosted Residual Networks (BRN), which makes use of developments in Deep Learning, previous other white-box Ensembles and combines several ideas to achieve improved results on benchmark datasets.

Using a white-box ensemble allows us to improve on the generalisation and training speed of other ensemble methods by making use of the knowledge of the base classifier's structure and architecture. Experimental results show that Boosted Residual Networks achieves improved results on benchmark datasets.

The next section presents the background on Deep Incremental Boosting. Then the proposed Boosted Residual Networks method is described. Experiments and results are discussed next, and the paper ends with conlusions.

2   Background

Deep Incremental Boosting, introduced in [9], is an example of such white-box ensemble method developed for building ensembles Convolutional Networks. The method makes use of principles from transfer of learning, like for example those used in [15], applying them to conventional AdaBoost [12]. Deep Incremental Boosting increases the size of the network at each round by adding new layers at the end of the network, allowing subsequent rounds of boosting to run much faster. In the original paper on Deep Incremental Boosting [9], this has been shown to be an effective way to learn the *corrections* introduced by the emphatisation of learning mistakes of the boosting process. The argument as to why this works effectively is based on the fact that the datasets at rounds t and $t + 1$ will be *mostly similar*, and therefore a classifier h_t that performs better than randomly on the resampled dataset X_t will also perform better than randomly on the resampled dataset X_{t+1}. This is under the assumption that both datasets are sampled from a common ancestor set $\boldsymbol{X}_a$. It is subsequently shown that such a classifier can be re-trained on the differences between $\boldsymbol{X}_t$ and $\boldsymbol{X}_{t+1}$.

This practically enables the ensemble algorithm to train the subsequent rounds for a considerably smaller number of epochs, consequently reducing the overall training time by a large factor. The original paper also provides a conjecture-based justification for why it makes sense to extend the previously trained network to learn the "corrections" taught by the boosting algorithm. A high level description of the method is shown in Algorithm 1, and the structure of the network at each round is illustrated in Fig. 2.

Algorithm 1. Deep Incremental Boosting

$D_{0,i} = 1/M$ for all i
$t = 0$
$\boldsymbol{W}_0 \leftarrow$ randomly initialised weights for first classifier
while $t < t_{end}$ **do**
    $\boldsymbol{X}_t \leftarrow$ sample from $\boldsymbol{X}_0$ with distribution $\boldsymbol{D}_t$
    $u_t \leftarrow$ create untrained classifier with additional layer of shape L_{new}
    copy weights from $\boldsymbol{W}_t$ into the bottom layers of u_t
    $h_t \leftarrow$ train u_t classifier on current subset
    $\boldsymbol{W}_{t+1} \leftarrow$ all weights from h_t
    $\epsilon_t = \frac{1}{2} \sum_{(i,y) \in B} D_{t,i}(1 - h_t(x_i, y_i) + h_t(x_i, y))$
    $\beta_t = \epsilon_t / (1 - \epsilon_t)$
    $D_{t+1,i} = \frac{D_{t,i}}{Z_t} \cdot \beta^{(1/2)(1 + h_t(x_i,y_i) - h_t(x_i,y))} |\forall i$
    where Z_t is a normalisation factor such that $\boldsymbol{D}_{t+1}$ is a distribution
    $\alpha_t = \frac{1}{\beta_t}$
    $t = t + 1$
end while
$H(x) = \text{argmax}_{y \in Y} \sum_{t=1}^{T} log\alpha_t h_t(x, y)$

---

3   Creating the Boosted Residual Network

In this section we propose a method for generating Boosted Residual Networks. This works by increasing the size of an original residual network by one residual block at each round of boosting. The method achieves this by selecting an *injection point* index p_i at which the new block is to be added, which is not necessarily the last block in the network, and by transferring the weights from the layers below p_i in the network trained at the previous round of boosting.

The boosting method performs an iterative re-weighting of the training set, which skews the resample at each round to *emphasize* the training examples that are harder to train. Therefore, it becomes necessary to utilise the entire ensemble at test time, rather than just use the network trained in the last round. This has the effect that the Boosted Residual Networks cannot be used as a way to train a single Residual Network incrementally. However, as we will discuss

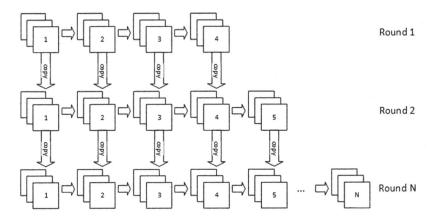

Fig. 2. Illusration of subsequent rounds of Deep Incremental Boosting

later, it is possible to alleviate this situation by deriving an approach that uses bagging instead of boosting; therefore removing the necessity to use the entire ensemble at test time. It is also possible to delete individual blocks from a Residual Network at training and/or testing time, as presented in [3], however this issue is considered out of the scope of this paper.

The iterative algorithm used in the paper is shown in Algorithm 2. At the first round, the entire training set is used to train a network of the original *base* architecture, for a number of epochs n_0. After the first round, the following steps are taken at each subsequent round t:

- The ensemble constructed so far is evaluated on the training set to obtain the set errors ϵ, so that a new training set can be sampled from the original training set. This is a step common to all boosting algorithms.
- A new network is created, with the addition of a new block of layers B_{new} immediately after position p_t, which is determined as an initial predetermined position p_0 plus an offset $t * \delta_p$ for all the blocks added at previous layers, where δ_p is generally chosen to be the size of the newly added layers. This puts the new block of layers immediately after the block of layers added at the previous round, so that all new blocks are effectively added sequentially.
- The weights from the layers below p_t are copied from the network trained at round $t - 1$ to the new network. This step allows to considerably shorten the training thanks to the transfer of learning shown in [15].
- The newly created network is subsequently trained for a reduced number of epochs $n_{t>0}$.
- The new network is added to the ensemble following the conventional rules and weight $\alpha_t = \frac{1}{\beta_t}$ used in AdaBoost. We did not see a need to modify the way β_t is calculated, as it has been performing well in both DIB and many AdaBoost variants [2,9,12,13].

Algorithm 2. Boosted Residual Networks

$D_{0,i} = 1/M$ for all i
$t = 0$
$\boldsymbol{W}_0 \leftarrow$ randomly initialised weights for first classifier
$p_0 \leftarrow$ initial injection position
while $t < T$ **do**
    $\boldsymbol{X}_t \leftarrow$ sample from $\boldsymbol{X}_0$ with distribution $\boldsymbol{D}_t$
    $u_t \leftarrow$ create untrained classifier with an additional block $\boldsymbol{B}_{new}$ of pre-determined
    shape $\boldsymbol{N}_{new}$
    determine block injection position $p_t = p_{t-1} + |\boldsymbol{B}_{new}|$
    connect the input of $\boldsymbol{B}_{new}$ to the output of layer $p_t - 1$
    connect the output of $\boldsymbol{B}_{new}$ and of layer $p_t - 1$ to a merge layer m_i
    connect the merge layer to the remainder of the network
    copy weights from $\boldsymbol{W}_t$ into the bottom layers $l < p_t$ of u_t
    $h_t \leftarrow$ train u_t classifier on current subset
    $\boldsymbol{W}_{t+1} \leftarrow$ all weights from h_t
    $\epsilon_t = \frac{1}{2} \sum_{(i,y) \in B} D_{t,i}(1 - h_t(x_i, y_i) + h_t(x_i, y))$
    $\beta_t = \epsilon_t/(1 - \epsilon_t)$
    $D_{t+1,i} = \frac{D_{t,i}}{Z_t} \cdot \beta^{(1/2)(1+h_t(x_i,y_i)-h_t(x_i,y))} |\forall i$
    where Z_t is a normalisation factor such that $\boldsymbol{D}_{t+1}$ is a distribution
    $\alpha_t = \frac{1}{\beta_t}$
    $t = t + 1$
end while
$H(x) = \text{argmax}_{y \in Y} \sum_{t=1}^{T} log\alpha_t h_t(x, y)$

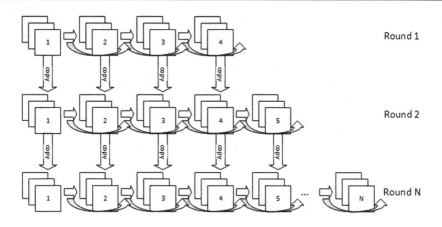

Fig. 3. Illusration of subsequent rounds of Boosted Residual Networks

Figure 3 shows a diagram of how the Ensemble is constructed by deriving the next network at each round of boosting from the network used in the previous round.

We identified a number of optional variations to the algorithm that may be implemented in practice, which we have empirically established as not having an impact on the overall performance of the network. We report them here for completeness.

- Freezing the layers that have been copied from the previous round.
- Only utilising the weights distribution for the examples in the training set instead of resampling, as an input to the training algorithm.
- Inserting the new block always at the same position, rather than after the previously-inserted block (we found this to affect performance negatively).

3.1 Comparison to Approximate Ensembles

While both Residual Networks and Densely Connected Convolutional Networks may be unfolded into an equivalent ensemble, we note that there is a differentiation between an actual ensemble method and an ensemble "approximation". During the creation of an ensemble, one of the principal factors is the creation of *diversity*: each base learner is trained independently, on variations (resamples in the case of boosting algorithms) of the training set, so that each classifier is guaranteed to learn a different function that represents an approximation of the training data. This is the enabling factor for the ensemble to perform better in aggregate.

In the case of Densely Connected Convolutional Networks (DCCN) specifically, one may argue that a partial unfolding of the network could be, from a schematic point of view, very similar to an ensemble of incrementally constructed Residual Networks. We make the observation that, although this would be correct, on top of the benefit of diversity, our method also provides a much faster training methodology: the only network that is trained for a full schedule is the network created at the first round, which is also the smallest one. All subsequent networks are trained for a much shorter schedule, saving a considerable amount of time. Additionally, while the schematic may seem identical, there is a subtle difference: each member network outputs a classification of its own, which is then aggregated by a weighted averaging determined by the errors on the test set, whilst in a DCCN the input of the final aggregation layer is the output of each underlying set of layers. We conjecture that this aggressive dimensionality reduction before the aggregation has a regularising effect on the ensemble.

4   Experiments and Discussion

In the experiments we used the MNIST, CIFAR-10 and CIFAR-100 datasets, and compared Boosted Residual Networks (BRN) with an equivalent Deep Incremental Boosting without the skip-connections (DIB), AdaBoost with the initial base network as the base classifier (AdaBoost) and the single Residual Network equivalent to the last round of Boosted Residual Networks (ResNet). It is to be noted that, because of the shortened training schedule and the differening architecture, the results on the augmented datasets are not the same as those reported in the original paper for Residual Networks.

Table 1. Test accuracy in the benchmark datasets for the methods compared.

	ResNet	AdaBoost	DIB	BRN
MNIST	99.41%	99.41%	99.47%	99.53%
CIFAR-10	89.12%	89.74%	90.83%	90.85%
CIFAR-10 (aug)	92.14%	91.66%	92.31%	92.94%
CIFAR-100	67.25%	68.18%	68.56%	69.04%
CIFAR-100 (aug)	69.72%	70.74%	71.55%	72.41%

MNIST [8] is a common computer vision dataset that associates 70000 pre-processed images of hand-written numerical digits with a class label representing that digit. The input features are the raw pixel values for the 28×28 images, in grayscale, and the outputs are the numerical value between 0 and 9. 50000 samples are used for training, 10000 for validation, and 10000 for testing.

CIFAR-10 is a dataset that contains 60000 small images of 10 categories of objects. It was first introduced in [7]. The images are 32×32 pixels, in RGB format. The output categories are *airplane, automobile, bird, cat, deer, dog, frog, horse, ship, truck*. The classes are completely mutually exclusive so that it is translatable to a *1-vs-all* multiclass classification. 50000 samples are used for training, and 10000 for testing. This dataset was originally constructed without a validation set.

CIFAR-100 is a dataset that contains 60000 small images of 100 categories of objects, grouped in 20 super-classes. It was first introduced in [7]. The image format is the same as CIFAR-10. Class labels are provided for the 100 classes as well as the 20 super-classes. A super-class is a category that includes 5 of the fine-grained class labels (e.g. "insects" contains *bee, beetle, butterfly, caterpillar, cockroach*). 50000 samples are used for training, and 10000 for testing. This dataset was originally constructed without a validation set.

In order to reduce noise in the results, we aligned the random initialisation of all networks across experiments, by fixing the seeds for the random number generators. All experiments were repeated 10 times and we report the mean performance values. We also report results with light data augmentation: we randomly rotated, flipped horizontally and scaled images, but did not use any heavy augmentation, including random crops. Results are reported in Table 1, while Fig. 4 shows a side-by-side comparison of accuracy levels at each round of boosting for both DIB and BRN on the MNIST and CIFAR-100 test sets. This figure illustrates how BRNs are able to consistently outperform DIB at each intermediate value of ensemble size, and although such differences would still fall within a Bernoulli confidence interval of 95%, we make the note that this does not take account of the fact that all the random initialisations were aligned, so both methods started with the exact same network. In fact, an additional Friedman Aligned Ranks test on the entire group of algorithms tested shows that there is a statistically significant difference in generalisation performance,

Table 2. Training times comparison. BRN and DIB are the fastest Ensemble methods compared. The time to train the base network and a ResNet of comparable performance is reported for comparison.

	ResNet	Base Net	AdaBoost	DIB	BRN
MNIST	217 min	62 min	442 min	202 min	199 min
CIFAR-10	1941 min	184 min	1212 min	461 min	449 min
CIFAR-10 (aug)	2228 min	213 min	2150 min	1031 min	911 min
CIFAR-100	2172 min	303 min	2873 min	607 min	648 min
CIFAR-100 (aug)	2421 min	328 min	3072 min	751 min	735 min

whilst a direct Wilcoxon test comparing only BRN and DIB shows that BRN is significantly better.

Table 2 shows that this is achieved without significant changes in the training time[1]. The main speed increase is due to the fact that the only network being trained with a full schedule is the first network, which is also the smallest, whilst all other derived networks are trained for a much shorter schedule (in this case only 10% of the original training schedule). If we excude the single network, which is clearly from a different distribution and only mentioned for reference, a Friedman Aligned Ranks test shows that there is a statistically significant difference in speed between the members of the group, but a Wilcoxon test between Deep Incremental Boosting and Boosted Residual Networks does not show a significant difference. This confirms what could be conjured from the algorithm itself for BRN, which is of the same complexity w.r.t. the number of Ensemble members as DIB.

The initial network architectures for the first round of boosting are shown in Table 3a for MNIST, and Table 3b for CIFAR-10 and CIFAR-100. The single networks currently used to reach state-of-the-art on these datasets are very cumbersome in terms of resources and training time. Instead, we used relatively simpler network architectures that were faster to train, which still perform well on the datasets at hand, with accuracy close to, and almost comparable to, the state-of-the-art. This enabled us to test larger Ensembles within an acceptable training time. Our intention is to demonstrate a methodology that makes it feasible to created ensembles of Residual Networks following a "white-box" approach to significantly improve the training times and accuracy levels achievable with current ensemble methods.

Training used the WAME method [11], which has been shown to be faster than Adam and RMSprop, whilst still achieving comparable generalisation. This is thanks to a specific weight-wise learning rate acceleration factor that is determined based only on the sign of the current and previous partial derivative $\frac{\partial E(x)}{\partial w_{ij}}$. For the single Residual Network, and for the networks in AdaBoost, we trained

---

[1] In some cases BRN is actually faster than DIB, but we believe this to be just noise due to external factors such as system load.

Table 3. Network structures used in experiments. The layers marked with "*" indicate the location after which we added the residual blocks.

2 × 96 conv, 3 × 3
96 conv, 3 × 3, 2 × 2 strides
96 conv, 3 × 3, 2 × 2 strides
96 conv, 3 × 3, 2 × 2 strides
2 × 2 max-pooling
2 × 192 conv, 3 × 3
192 conv, 3 × 3, 2 × 2 strides
192 conv, 3 × 3, 2 × 2 strides
192 conv, 3 × 3, 2 × 2 strides
2 × 2 max-pooling *
192 conv, 3 × 3
192 conv, 1 × 1
10 conv, 1 × 1
global average pooling
10-way softmax

64 conv, 5 × 5
2 × 2 max-pooling
128 conv, 5 × 5
2 × 2 max-pooling *
Dense, 1024 nodes
50% dropout

(a) MNIST

(b) CIFAR-10 and CIFAR-100

Table 4. Structure of blocks added at each round of DIB and BRN.

64 conv, 3 × 3
Batch Normalization
ReLu activation

(a) MNIST

192 conv, 3 × 3
Batch Normalization
ReLu activation
192 conv, 3 × 3
Batch Normalization
ReLu activation

(b) CIFAR-10 and CIFAR-100

each member for 100 epochs. For Deep Incremental Boosting and Boosted Residual Networks, we trained the first round for 50 epochs, and every subsequent round for 10 epochs, and ran all the algorithms for 10 rounds of boosting, except for the single network. The structure of each incremental block added to Deep Incremental Boosting and Boosted Residual Networks at each round is shown in Table 4a for MNIST, and in Table 4b for CIFAR-10 and CIFAR-100. All layers were initialised following the reccomendations in [4].

Distilled Boosted Residual Network: DBRN In another set of experiments we tested the performance of a Distilled Boosted Residual Network (DBRN). Distillation has been shown to be an effective process for regularising large Ensembles of Convolutional Networks in [10], and we have applied the same methodology to the proposed Boosted Residual Network. For the distilled network structure we used the same architecture as that of the Residual Network from the final round of boosting. Average accuracy results in testing over 10 runs are presented

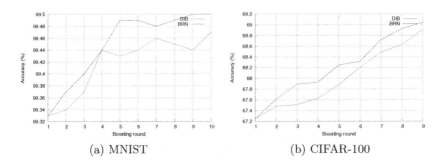

(a) MNIST   (b) CIFAR-100

Fig. 4. Round-by-round comparison of DIB vs BRN on the test set

Table 5. Comparative results in terms of testing accuracy.

	DBRN	DDIB
MNIST	99.49%	99.44%
CIFAR-10	91.11%	90.66%
CIFAR-10 (aug)	93.06%	92.71%
CIFAR-100	66.63%	65.91%
CIFAR-100 (aug)	70.24%	69.18%

in Table 5, and for completeness of comparison we also report the results for the distillation of DIB, following the same procedure, as DDIB. DBRN does appear to improve results only for CIFAR-10, but it consistently beats DDIB on all datasets. These differences are too small to be deemed statistically significant, confirming that the function learned by both BRN and DIB can be efficiently transferred to a single network.

Bagged Residual Networks: BARN We experimented with substituting the boosting algorithm with a simpler bagging algorithm [1] to evaluate whether it would be possible to only use the network from the final round of bagging as an approximation of the Ensemble. We called this the Bagged Approximate Residual Networks (BARN) method. We then also tested the performance of the Distilled version of the whole Bagged Ensemble for comparison. The results are reported as "DBARN". The results are reported in Table 6. It is not clear whether using the last round of bagging is significantly comparable to using the entire Bagging ensemble at test time, or deriving a new distilled network from it, and more experimentation would be required.

Table 6. Test accuracy for BARN and DBARN.

	BRN	Bagging	BARN	DBARN
MNIST	99.53%	99.55%	99.29%	99.36%
CIFAR-10	90.85%	91.43%	88.47%	90.63%
CIFAR-10 (aug)	92.94%	92.61%	92.73%	92.69%
CIFAR-100	69.04%	68.15%	69.42%	66.16%
CIFAR-100 (aug)	72.41%	71.90%	72.01%	70.44%

5   Conclusions and Future Work

In this paper we have derived a new ensemble algorithm specifically tailored to Convolutional Networks to generate Boosted Residual Networks. We have shown that this surpasses the performance of a single Residual Network equivalent to the one trained at the last round of boosting, of an ensemble of such networks trained with AdaBoost, and Deep Incremental Boosting on the MNIST and CIFAR datasets, with and without using augmentation techniques.

We then derived and looked at a distilled version of the method, and how this can serve as an effective way to reduce the test-time cost of running the Ensemble. We used Bagging as a proxy to test the generation of the approximate Residual Network, which, with the parameters tested, does not perform as well as the original Residual Network, BRN or DBRN.

Further experimentation of the Distilled methods presented in the paper, namely DBRN and DBARN, is necessary to fully investigate their behaviour. This is indeed part of our work in the near future. Additionally, the Residual Networks built in our experiments were comparatively smaller than those that achieve state-of-the-art performance. Nevertheless, it might be appealing in the future to evaluate the performance improvements obtained when creating ensembles of larger, state-of-the-art, networks. Additional further investigation could also be conducted on the creation of Boosted Densely Connected Convolutional Networks, by applying the same principle to DCCN instead of Residual Networks.

References

1. Breiman, L.: Bagging predictors. Mach. Learn. **24**(2), 123–140 (1996)
2. Freund, Y., Iyer, R., Schapire, R.E., Singer, Y.: An efficient boosting algorithm for combining preferences. J. Mach. Learn. Res. **4**, 933–969 (2003)
3. He, K., Zhang, X., Ren, S., Sun, J.: Deep residual learning for image recognition. arXiv preprint arXiv:1512.03385 (2015)
4. He, K., Zhang, X., Ren, S., Sun, J.: Delving deep into rectifiers: surpassing human-level performance on imagenet classification. In: Proceedings of the IEEE International Conference on Computer Vision, pp. 1026–1034 (2015)
5. He, K., Zhang, X., Ren, S., Sun, J.: Identity mappings in deep residual networks. arXiv preprint arXiv:1603.05027 (2016)

6. Huang, G., Liu, Z., Weinberger, K.Q.: Densely connected convolutional networks. arXiv preprint arXiv:1608.06993 (2016)
7. Krizhevsky, A., Hinton, G.: Learning multiple layers of features from tiny images (2009)
8. Lecun, Y., Cortes, C.: The MNIST database of handwritten digits. http://yann.lecun.com/exdb/mnist/
9. Mosca, A., Magoulas, G.: Deep incremental boosting. In: Benzmuller, C., Sutcliffe, G., Rojas, R. (eds.) GCAI 2016, 2nd Global Conference on Artificial Intelligence. EPiC Series in Computing, vol. 41, pp. 293–302. EasyChair (2016)
10. Mosca, A., Magoulas, G.D.: Regularizing deep learning ensembles by distillation. In: 6th International Workshop on Combinations of Intelligent Methods and Applications (CIMA 2016), p. 53 (2016)
11. Mosca, A., Magoulas, G.D.: Training convolutional networks with weight-wise adaptive learning rates. In: ESANN 2017 Proceedings, European Symposium on Artificial Neural Networks, Computational Intelligence and Machine Learning, Bruges, Belgium, 26–28 April 2017 (2017, in press). i6doc.com
12. Schapire, R.E.: The strength of weak learnability. Mach. Learn. 5, 197–227 (1990)
13. Schapire, R.E., Freund, Y.: Experiments with a new boosting algorithm. In: Machine Learning: Proceedings of the Thirteenth International Conference, pp. 148–156 (1996)
14. Veit, A., Wilber, M., Belongie, S.: Residual networks behave like ensembles of relatively shallow networks. arXiv e-prints, May 2016
15. Yosinski, J., Clune, J., Bengio, Y., Lipson, H.: How transferable are features in deep neural networks? In: Advances in Neural Information Processing Systems, pp. 3320–3328 (2014)

A Convolutional Approach to Multiword Expression Detection Based on Unsupervised Distributed Word Representations and Task-Driven Embedding of Lexical Features

Tiberiu Boros[✉] and Stefan Daniel Dumitrescu

Research Center for Artificial Intelligence, Romanian Academy,
Calea 13 Septembrie, 050711 Bucharest, Romania
{tibi,sdumitrescu}@racai.ro
http://www.racai.ro

Abstract. We introduce a convolutional network architecture aimed at performing token-level processing in natural language applications. We tune this architecture for a specific task - multiword expression detection - and we compare our results to state-of-the-art systems on the same datasets. The approach is multilingual and we rely on automatically extracted word embeddings from Wikipedia dumps. We also show that task-driven lexical features embeddings increase the speed and robustness of the system versus sparse encodings.

Keywords: Convolutional neural networks · Softmax · Unsupervised word embedding · Task-driven lexical features embedding · Natural language processing · Multiword expression detection · Language independent

1 Introduction

Convolutional neural networks have been previously used in natural language processing (NLP) tasks such as morphological analysis [4], sentiment analysis [5,16] and text categorization [8,9,21]. Though recurrent models should intuitively provide better results on NLP tasks, mainly because of their innate capacity of capturing long-range dependencies between words, convolutional networks are easier to train and provide up-to-par performance compared to their counterparts. We introduce a custom-designed convolutional neural network architecture aimed at tasks which use features from different categories (e.g.: lexical features such as words and lemmas with morphological features such as part-of-speech and specific attributes like gender, case, number etc.).

To test our proposed methodology we provide an evaluation on a well-known multilingual verbal multiword expression corpus (PARSEME [18]). The motivation behind our choice is that verbal multiword expressions depend on features that can be extracted by performing morphological analysis on the text and also

© Springer International Publishing AG 2017
G. Boracchi et al. (Eds.): EANN 2017, CCIS 744, pp. 149–159, 2017.
DOI: 10.1007/978-3-319-65172-9_13

on semantic features that can be derived either by employing standard NLP methods such as Latent Semantic Analysis (LSA) [7], Lexical Chains [1,6] or distributed word embeddings [12]. Additionally, this corpus was used in an open evaluation campaign and we are able to provide an easy comparison with state-of-the-art methods including one of our own, which is based on a similar feature template but uses Conditional Random Fields [11].

While semantic analysis produces values and vectors in a continuous domain, morphological features are multinomial and usually end up being used as sparse encodings (1-of-n). Clearly these features cannot be used uniformly in a convolutional network architecture, mainly because these models use filters to learn homogeneous local patterns inside their receptive field and it is highly undesirable to use the same kernel in handling both types of features.

2   Corpora and Task Description

By definition, the term "multiword expressions" (MWEs) denotes a group of words that act as a morphologic, syntactic and semantic unit in linguistic analysis. [17] explained the importance of identifying MWEs in general and we must state that Verbal MWEs (VMWEs) are especially useful in parsing because the verb is the central element in the syntactic organization of a sentence.

The PARSEME corpus contains multilingual annotated data for 18 languages and employs a custom annotation scheme and data-format called parsemetsv[1] (one-token per line, with tokenization and VMWEs information stored as tab-separated values). For 14 of the 18 languages, the annotated data is accompanied by morphological features in CONLL format[2]. The languages which have this type of annotations are: Romanian (RO), French (FR), Czech (CS), German (DE), Greek (EL), Spanish (ES), Hungarian (HU), Italian (IT), Maltese (MT), Slovene (SL), Swedish (SV), Turkish (TR), Farsi (FA) and Polish (PL).

Annotation is carried out at word level and contains the following categories (see Table 1 for distribution of categories for every language):

- **Light Verb Constructions (LVC)**: they are made up of a verb and a noun: the former has little if any semantic content, while the latter contributes to the semantics of the VMWE (e.g.: "to give a lecture")
- **Idioms (ID)**: these are expressions in which the verb can combine with any open-class word and their key-characteristic is the lack of compositional meaning (e.g.: "to go bananas", "fortune favors the bold")
- **Inherently reflexive verbs (IReflV)**: they are made up of a verb and a reflexive clitic and their meaning is different from those occurrences of the verb without the clitic (in case this is possible); the passive, reciprocal, possessive and impersonal constructions are excluded from annotation (e.g.: "to find oneself")

---

[1] http://typo.uni-konstanz.de/parseme/index.php/2-general/184-parseme-shared-task -format-of-the-final-annotation (last accessed 2017-02-15).

[2] http://universaldependencies.org/format.html (last accessed 2017-02-15).

- **Verb-Particle Constructions (VPC)**: they contain a verb and a particle and lack a compositional meaning (e.g.: "to do in")
- **Other (OTH)**: this is a residual class, including any VMWE that does not fit any of the above mentioned classes (e.g.: "to short-circuit", "to tumble dry")

Table 1. VMWE distribution in the training corpora

VMWE type	CS	DE	EL	ES	FA	FR	HU	IT	MT	PL	PT	RO	SL	SV	TR
IReflV	8851	111	0	336	0	1313	0	580	0	1548	515	2496	945	3	0
LVC	2580	178	955	214	0	1362	584	395	434	1284	2110	1019	186	13	2624
ID	1419	1005	515	196	0	1786	0	913	261	317	820	524	283	9	2911
VPC	0	1143	32	0	0	0	2415	62	0	0	0	0	371	31	0
OTH	2	10	16	2	2707	1	0	4	77	0	2	1	2	0	634

The morphological data comes in multicolumn, tab-separated format and contains relevant morphological information: (a) the word form, (b) lemma (dictionary form of the word), (c) coarse part-of-speech [15], (d) language-dependent part-of-speech (which defines a custom tag set of each individual language), (e) morphological attributes list (e.g. gender, case, number etc.) [20], (f) tokenization/detokenization information (SpaceAfter = yes/no) and (g) the parsing information composed of each word's head token (the index of the head word) and the relation type.

3   Network Architecture and Feature Encoding

First, we present the multiword expression detection task and motivate our chosen feature set. By their very definition the detection of LVC, IReflV and VPC expressions depends on the morphological attributes of the composing words, while the detection of ID expressions mostly depends on semantics.

Arguably, morphology is covered by the provided corpora, though the format imposed by the CONLL standard is violated for some languages: Romanian only provides a language dependent part-of-speech and parsing information is missing on several datasets.

Semantic information is not available in the training data. However, there are several ways in which this type of data can be added: WordNet for English [13] and EuroWordNet for European languages [19], Latent Semantic Indexing [7] and distributed word vectors [12]. The latter mentioned methodology is known to group contextually (and often-enough semantically) *related* words and has received significant interest from the research community.

As such, our chosen feature set is composed of: (a) semantic information in the form of word-embeddings computed using word2vec[3], (b) the coarse part-of-speech (whenever available), (c) morphological attributes (whenever available), (d) the language dependent part-of-speech (again, whenever available). The classifier is trained to produce an output label for each individual word based on

---

[3] https://github.com/dav/word2vec - accessed 2017-04-10.

the previously mentioned features extracted from a window of words centered on the current token. During our previous experiments (using a CRF classifier) [2] we established that a window size of 2 (2 previous words + current word + next 2 words) is a good choice and, in order to provide a fair comparison, we preserved this parameter in the current experiment set.

Before we proceed with the actual description of our network architecture we must re-state that convolutional networks work by learning "filters" which have different "responses" on various input patterns and are very efficient for image analysis. One can easily imagine that by converting each feature to a vector and by concatenating these vectors over the analysis window this would easily translate into a 2D matrix of values (or an image). However, the features we are using are heterogeneous (semantical representations vs morphological attributes) and it is likely that the same convolutional filter cannot be trained to work on both word embeddings and morphologic embeddings. In what follows we introduce our feature encoding mechanism (Subsect. 3.1) as well as our custom-designed network architecture which copes with heterogeneous input and can be applied to any NLP task without re-engineering.

3.1 Input Encoding

As previously mentioned our feature-set is composed of word forms and morphologic data such as coarse part of speech, morphological attributes and language dependent part of speech. In order to achieve best-results, each individual feature class must be carefully processed and converted into a numeric representation. So, we divided our methodology into 3 categories: (a) semantic features - which can be easily computed using external linguistic resources in the form of large-scale text corpora; (b) structured morphological features - which represent universal morphological attributes that require little linguistic knowledge for processing - this class contains coarse part-of-speech and morphological attributes; (c) unstructured morphological features - which are usually language-dependent compact representations of morphological attributes and require prior linguistic knowledge for decoding - this class represents the language-dependent part-of-speech.

Semantic Features: The distributed word representations are computed using word2vec on language specific Wikipedia dumps. Word2vec already produces n-dimensional dense vectors (in our case we set n=64) and no further processing is required.

Structured Features: The morphological attributes require special handling, depending on their role and meaning. Our proposed methodology treats the coarse part of speech along the morphological attributes as a single class, distinct from the language dependent tag set. The coarse part of speech is a label which takes values from a relatively small set (about 16–18 per language) and the morphological attributes are composed of a list of key-value pairs, in which the key represents the attribute name: for instance Gender = male, Gender = female,

Number = singular, Number = plural etc. As such, we created a joint representation for the combined feature composed of the coarse part of speech and morphological attributes using the following procedure:

1. Each coarse part of speech is sparely (1-of-n) encoded in a n-dimensional
   vector (where n is the total number of distinct tags);
2. We scan the training data and group the morphological attributes using their
   key, while counting the possible value-set for each attribute group
3. Each attribute group (g_i) is sparsely encoded (1-of-k_i) using a k_i dimensional
   vector. k_i is the total number of distinct values for the particular attribute
   group *key* plus 1. Whenever an attribute does not show up in the morphological attributes of an example-instance we use a special bit to encode the
   "missing value attribute" (hence the plus one). For instance, if one token
   does not contain the Gender attribute, the corresponding vector encodes an
   "invented" key-value pair that signifies Gender = NONE[4].
4. We concatenate the coarse part-of-speech vector with the corresponding vectors of each possible attribute group and use the newly obtained vector as
   the encoding for our combined feature-set.

Unstructured Features: Though the previous method works well when
applied to structured features, the language-dependent part-of-speech (XPOS)
does not share this property and requires a different approach. Creating a custom encoding scheme for these tags is impossible without prior linguistic knowledge and 1-hot encoding in undesirable since that would yield data sparsity (for
instance, the Romanian dataset has 606 unique XPOS labels). In order to obtain
a dense-vector representation for these features we used an approach inspired
after [3]. In our particular implementation we assigned each individual tag in
the tagset a "deconvolution" filter designed for projecting a "fixed-scalar" value
of 1 into an n-dimensional space [14]. The principle in our approach is fairly simple: each unique feature is associated with an n-dimensional weight vector; the
vector is tuned using standard backpropagation, in which the input is fixed at
value 1 (thus the gradient is only multiplied with the learning rate). The weights
of the deconvolution filters actually represent dense feature embeddings which
are also task-specific.

3.2 Proposed Network Architecture

As can easily be seen, the input data is composed of heterogeneous features
(semantic and morphological) and using the same convolutional input layer uniformly is likely to fail. In order to cope with feature set diversity we constructed
a custom convolutional network architecture which we refer to as "partitioned
CNN". To our knowledge, this is the first time this type of architecture is used
on NLP-related tasks.

---

[4] During our experiments we observed that doing so speeds up convergence of the
algorithm, with little impact over the computation time required by each training
iteration.

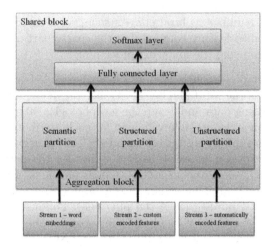

Fig. 1. The partitioned convolutional neural network architecture

The network architecture (see Fig. 1) is composed of two main blocks: (a) a feature aggregation block and (b) a shared block with an output softmax layer.

The feature aggregation block is actually a collection of separate convolutional neural networks, with different layer and filter configurations, which are separately trained for each individual data-stream (or feature class). For this particular task, the feature aggregation block is composed of 3 convolutional networks: one for word embeddings, one for structured features and one for unstructured features:

- **Word embeddings partition:** The word embeddings partition is a convolutional network with 64 convolutional filters as input, a max-pooling layer followed by a fully connected "code layer" of size 16.
- **Structured partition:** This is a convolutional network with 16 convolutional filters as input, a max-pooling layer and a fully connected layer, also of size 16.
- **Unstructured partition:** This is a convolutional network with 16 filters, a max-pooling layer and a code layer of size 16, but, in this case, the filters receive input from the deconvolution filters assigned to each unique input feature.

The shared block is composed of two layers: the first layer is a fully connected layer (size 120 - tanh activation) which connects to all the "code layers" of the input partitions; the second layer is a softmax output layer used to predict the probability of possible output labels.

All the values (64 filters for input, code layer of size 16, etc.) have been chosen heuristically - we attempted only a few runs with variations (e.g. we tested with 32, 64 and 128 filters for input) and we chose the configuration that showed best results. Please note that the exact parameters are not essential for the purpose of this paper as we attempt to prove that this method works; only after we obtain encouraging results would it be sensible to invest time into finding optimal parameters to try to squeeze max performance out of this architecture.

4   Experimental Validation

Corpus preparation. From the 18 PARSEME languages we selected a subset that had both the .parsemetsv and the .conllu annotations. An example word from the French .conllu train file (each word per line, new line for sentence boundaries):

```
9 a avoir VERB _ Mood=Ind|Number=Sing|Person=3|Tense=Pres|VerbForm
=Fin 0 root _ _
```

reads as follows: the 9-th word in the sentence is "a" (lemma: avoir) which is a Verb (detailed by the Mood, Number, etc. attributes) that is also a "root" of the parse tree; The same word in the .parsemetsv file is:

```
9 a \_ 1:LVC,
```

showing that the "a" verb is the head of a light verb construction (LVC). The first processing step was to create a development file by extracting 10% of the sentences from the train file. We extracted a balanced set of sentences by initially counting the frequency of sentences that had at least one VMWE, trying to create the same approximate distribution of IReflV, LVC, VPC, ID and OTH sentences, as well as sentences that had no VMWE. There were cases where this distribution could not be achieved; for example the Swedish train file had only 3 IReflV and 9 ID sentences - the dev file contains only one of the 9 ID sentences and no IReflV. The dev "file", at this point, is written in the same format as the train file, in the .parsemetsv and .conllu formats. Having the train, dev, and test files ready, we merged the parsemetsv and conllu formats into a single file ready for classifier training.

The classifier training format reads sentences iteratively (for each train, dev and test file) from both parsemetsv and conllu and extracts, for each word, three features and a relation type (label). The features are (1) the word itself, (2) structured feature - the part of speech concatenated with the attributes string, (3) unstructured feature - the second, language-dependent part of speech. We chose to use both parts pf speech because, while the column structure in the files is the same for all languages, there are differences in content: for some languages the first part of speech is usually a coarse POS (e.g. VERB) and the second is the fine-grained language-dependent description (e.g. Vaip3s), while in other cases the first POS is identical to the second; furthermore, as seen in the example above, for French, the second POS is missing altogether (noted as a "_"). The relation type is the word's label, either none or one of IReflV, LVC, VPC, ID or OTH, as well as any combination of them.

The train file has the following format: "W^{i-2} W^{i-1} W^i W^{i+1} W^{i+2} LABEL", where, for each word at position i in the sentence (one per line) we dump the context window of size 2 (2 words before and 2 after) and the label (or _) if word i has none. For each word we write its 3 features: the word itself and

Table 2. Detailed evaluation results per language

L	Rel	Sys	Prec	Rec	F-meas	L	Rel	Sys	Prec	Rec	F-meas
CS	IReflV	Best	0.7109	0.7554	0.7325	EL	IReflV	Best	-	-	-
		CRF	0.7109	0.7554	0.7325			CRF	-	-	-
		CNN	**0.7794**	**0.8796**	**0.8265**			CNN	-	-	-
	LVC	Best	0.7460	0.2741	0.4009		LVC	Best	0.4096	0.2798	0.3316
		CRF	0.7460	0.2741	0.4009			CRF	**0.4096**	**0.2798**	**0.3316**
		CNN	0.6396	0.1359	0.2237			CNN	0.3654	0.2746	0.3136
	ID	Best	0.5909	0.1354	0.2203		ID	Best	0.1392	**0.1732**	**0.1544**
		CRF	**0.5909**	**0.1354**	**0.2203**			CRF	**0.2321**	0.1024	0.1421
		CNN	0.7788	0.0369	0.0708			CNN	0.1844	0.1262	0.1503
	VPC	Best	-	-	-		VPC	Best	0.6667	0.2500	0.3636
		CRF	-	-	-			CRF	0.6667	**0.2500**	**0.3636**
		CNN	-	-	-			CNN	1.0000	0.1250	0.2225
	Overall	Best	**0.7897**	**0.6560**	**0.7167**		Overall	Best	0.3612	**0.4500**	**0.4007**
		CRF	0.7009	0.5918	0.6418			CRF	**0.4286**	0.2520	0.3174
		CNN	0.7776	0.6418	0.7037			CNN	0.3450	0.2300	0.2765
FA	IReflV	Best	-	-	-	FR	IReflV	Best	0.7000	0.6667	0.6829
		CRF	-	-	-			CRF	**0.7000**	**0.6667**	**0.6829**
		CNN	-	-	-			CNN	0.6672	**0.6864**	0.6763
	LVC	Best	-	-	-		LVC	Best	0.4486	**0.3542**	**0.3959**
		CRF	-	-	-			CRF	**0.7255**	0.1365	0.2298
		CNN	-	-	-			CNN	0.6145	0.1335	0.2195
	ID	Best	-	-	-		ID	Best	0.7294	0.5210	0.6078
		CRF	-	-	-			CRF	0.7294	**0.5210**	**0.6078**
		CNN	-	-	-			CNN	**0.7944**	0.4246	0.5525
	OTH	Best	0.8770	0.8560	0.8664		VPC	Best	-	-	-
		CRF	-	-	-			CRF	-	-	-
		CNN	**0.9290**	**0.8702**	**0.8991**			CNN	-	-	-
	Overall	Best	0.8770	0.8560	0.8664		Overall	Best	**0.7484**	**0.4700**	**0.5774**
		CRF	-	-	-			CRF	0.7415	0.3500	0.4755
		CNN	**0.9290**	**0.8702**	**0.8991**			CNN	0.7112	0.3272	0.4482
PT	IReflV	Best	**0.8125**	0.1605	0.2680	RO	IReflV	Best	**0.8197**	0.6897	**0.7491**
		CRF	-	-	-			CRF	0.8197	0.6897	0.7491
		CNN	0.5856	**0.2964**	**0.3933**			CNN	0.7172	**0.7448**	0.7304
	LVC	Best	**0.8523**	0.4407	0.5017		LVC	Best	0.9167	0.8184	0.8627
		CRF	-	-	-			CRF	**0.9167**	**0.8184**	**0.8627**
		CNN	0.7383	**0.4842**	**0.5853**			CNN	0.8566	0.7046	0.7726
	ID	Best	0.5172	**0.1667**	**0.2521**		ID	Best	0.8864	0.5200	0.6555
		CRF	-	-	-			CRF	**0.8864**	**0.5200**	**0.6555**
		CNN	**0.8571**	0.1332	0.2317			CNN	0.6368	0.1924	0.2956
	Overall	Best	**0.7543**	**0.6080**	**0.6733**		Overall	Best	0.8652	0.7060	0.7775
		CRF	-	-	-			CRF	**0.8652**	**0.7060**	**0.7775**
		CNN	0.7352	0.3854	0.5065			CNN	0.7643	0.6632	0.7102
SV	IReflV	Best	-	-	-	TR	IReflV	Best	-	-	-
		CRF	-	-	-			CRF	-	-	-
		CNN	-	-	-			CNN	-	-	-
	LVC	Best	0.4000	0.1429	0.2105		LVC	Best	0.6769	0.5226	0.5909
		CRF	**0.4000**	**0.1429**	**0.2105**			CRF	**0.6769**	**0.5226**	**0.5909**
		CNN	0.1111	**0.2143**	0.1463			CNN	0.6463	0.5132	0.5714
	ID	Best	**0.5000**	0.0196	0.0377		ID	Best	0.5921	0.3614	0.4489
		CRF	**0.5000**	0.0196	0.0377			CRF	0.5921	0.3614	0.4489
		CNN	0.1538	0.0392	0.0625			CNN	0.5354	0.3372	0.4145
	VPC	Best	0.5614	0.2065	0.3019		OTH	Best	0.5455	0.4528	0.4948
		CRF	**0.5614**	0.2065	0.3019			CRF	0.5455	**0.4528**	**0.4948**
		CNN	0.4237	**0.3226**	**0.3663**			CNN	**0.7732**	0.3212	0.4533
	Overall	Best	0.5100	0.2161	**0.3036**		Overall	Best	0.6106	**0.5070**	**0.5540**
		CRF	**0.5758**	0.1610	0.2517			CRF	**0.6304**	0.4391	0.5176
		CNN	0.3767	**0.2511**	0.3014			CNN	0.6172	0.4123	0.4944

coarse POS + attributes(POS-1) and language dependent POS (POS-2) as $W^i_{word} W^i_{POS-1} W^i_{POS-2}$.

For every language, the network was trained using Stochastic Gradient Descent (SGD) with ADAM Optimization [10], with a learning rate of 10^{-4} and the default values for β_1 and β_2, meaning 0.9 and respectively 0.999. During the validation procedure we kept the network configuration unchanged, regardless of the language and dataset.

Table 2 presents comparative per-MWE results of the best performing system in the PARSEME 2017 competition (noted as Best), of our CRF based system (also present in the competition, noted as CRF), and our current partitioned convolutional network (noted as CNN). Results are presented for each relation available for that language, as Precision, Recall and F-measure. The cases where the Best system has the same values as the CRF system indicate that, for that relation, our CRF system was ranked first in the PARSEME competition. The Best system was chosen by ranking F-measure first, meaning that the Precision and Recall belong to that particular system. Complete results on the PARSEME task are found here [18], and for space reasons we only list here 8 of the 15 languages we tested on.

The results show that, on average, the CNN system, while currently (without any tuning) having an overall slightly lower score, is on-par with the Best system (and with our previous CRF system). For several relation types the CNN system surpasses both our CRF and the Best system (e.g. for Czech, for IReflV relation, we have a 0.77/0.87/0.82 Precision/Recall/F-measure v.s. the CRF that has 0.71/0.75.0.73). For Farsi we have the best results so far: 0.89 F-measure versus the current Best of 0.86. We do note however, that while in several cases CNN precision is higher, we have a lower average recall which brings down the overall F-measure.

While results are mixed, analyzing the bigger picture we can say that: (1) The CNN has slightly better precision at the expense of slightly worse recall; (2) the CNN is on-par with both CRF and Best systems (where CRF is not Best itself), proving our methodology as valid and well-performing.

5   Conclusions

With this paper we introduced a novel convolutional-inspired architecture designed for token level labeling in natural language processing tasks. We applied this methodology to the specific task of multiword expression detection, mainly because it involved various types of features, from semantic word representation to custom encoded lexical features and task-driven feature embeddings. The topics covered here are: (a) task-specific details regarding the MWE task, (b) feature-sets and feature-window size, (c) dense feature embeddings, (d) a network architecture designed for multi-stream input and (e) a comparative validation of our methodology.

Though the proposed architecture is slightly adapted for the MWE task, the system can be directly used for any other NLP task such as part-of-speech tagging, tokenization, lemmatization, etc.

The results presented in Sect. 4, show that are system is on-par with current state-of-the art systems presented in the PARSEME 2017 competition, obtaining better results on Farsi and comparable results on the other languages tested on.

We must mention that we did not perform any language-specific hyper-parameter tunning for the network, keeping the same network configuration with the same inputs for all languages. The network parameters we chose (filter number for words, code layer and fully-connected hidden layer size, etc.) were selected heuristically after only a few runs on the development set, as the purpose of the paper is to prove our proposed methodology: if it works without parameter tuning on several languages (and on different types of POS tags - see discussion on input training data, especially language-dependent POS tags), then the approach is valid. This proposed methodology can obtain results beyond the state-of-the-art performance, given a proper parameter setup, something we wish to explore in upcoming NLP competitions.

Future work includes testing on other NLP tasks (simpler, like POS tagging, and more complex like parsing) as well as an in-depth theoretical and practical analysis of the architecture of the partitioned CNN where we will see how different partition setups affect final prediction results. Furthermore, we intend to develop our current network (written entirely in C++) into a freely downloadable library that contains not only CNNs but other types of neural nets.

Acknowledgements. This work was supported by UEFISCDI, under grant PN-II-PT-PCCA-2013-4-0789, project "Assistive Natural-language, Voice-controlled System for Intelligent Buildings" (2013–2017).

References

1. Barzilay, R., Elhadad, M.: Using lexical chains for text summarization. In: Advances in Automatic Text Summarization, pp. 111–121 (1999)
2. Boros, T., Pipa, S., Mititelu, V.B., Tufis, D.: A data-driven approach to verbal multiword expression detection. PARSEME shared task system description paper. In: MWE 2017, p. 121 (2017)
3. Chen, D., Manning, C.D.: A fast and accurate dependency parser using neural networks. In: EMNLP, pp. 740–750 (2014)
4. Collobert, R., Weston, J.: A unified architecture for natural language processing: deep neural networks with multitask learning. In: Proceedings of the 25th International Conference on Machine Learning, pp. 160–167. ACM (2008)
5. Dos Santos, C.N., Gatti, M.: Deep convolutional neural networks for sentiment analysis of short texts. In: COLING, pp. 69–78 (2014)
6. Hirst, G., St-Onge, D., et al.: Lexical chains as representations of context for the detection and correction of malapropisms. WordNet: Electron. Lex. Database **305**, 305–332 (1998)
7. Hofmann, T.: Probabilistic latent semantic indexing. In: Proceedings of the 22nd Annual International ACM SIGIR Conference on Research and Development in Information Retrieval, pp. 50–57. ACM (1999)
8. Johnson, R., Zhang, T.: Effective use of word order for text categorization with convolutional neural networks. arXiv preprint arXiv:14121058 (2014)

9. Kim, Y.: Convolutional neural networks for sentence classification. arXiv preprint arXiv:14085882 (2014)
10. Kingma, D., Ba, J.: A method for stochastic optimization. arXiv preprint arXiv:14126980 (2014)
11. Lafferty, J., McCallum, A., Pereira, F., et al.: Conditional random fields: probabilistic models for segmenting and labeling sequence data. In: Proceedings of the Eighteenth International Conference on Machine Learning, ICML, vol. 1, pp. 282–289 (2001)
12. Mikolov, T., Sutskever, I., Chen, K., Corrado, G.S., Dean, J.: Distributed representations of words and phrases and their compositionality. In: Advances in Neural Information Processing Systems, pp. 3111–3119 (2013)
13. Miller, G.A.: WordNet: a lexical database for english. Commun. ACM **38**(11), 39–41 (1995)
14. Noh, H., Hong, S., Han, B.: Learning deconvolution network for semantic segmentation. In: Proceedings of the IEEE International Conference on Computer Vision, pp. 1520–1528 (2015)
15. Petrov, S., Das, D., McDonald, R.: A universal part-of-speech tagset. arXiv preprint arXiv:11042086 (2011)
16. Poria, S., Cambria, E., Gelbukh, A.F.: Deep convolutional neural network textual features and multiple kernel learning for utterance-level multimodal sentiment analysis. In: EMNLP, pp. 2539–2544 (2015)
17. Sag, I.A., Baldwin, T., Bond, F., Copestake, A., Flickinger, D.: Multiword expressions: a pain in the neck for NLP. In: Gelbukh, A. (ed.) CICLing 2002. LNCS, vol. 2276, pp. 1–15. Springer, Heidelberg (2002). doi:10.1007/3-540-45715-1_1
18. Savary, A., Ramisch, C., Cordeiro, S., Sangati, F., Vincze, V., QasemiZadeh, B., Candito, M., Cap, F., Giouli, V., Stoyanova, I., Doucet, A.: The PARSEME shared task on automatic identification of verbal multiword expressions. In: Proceedings of the 13th Workshop on Multiword Expressions, Association for Computational Linguistics, Valencia, Spain (2017)
19. Vossen, P.: EuroWordNet: A Multilingual Database with Lexical Semantic Networks. Springer, Heidelberg (1998)
20. Zeman, D.: Reusable tagset conversion using tagset drivers. In: LREC (2008)
21. Zhang, X., Zhao, J., LeCun, Y.: Character-level convolutional networks for text classification. In: Advances in Neural Information Processing Systems, pp. 649–657 (2015)

Remarks on Tea Leaves Aroma Recognition Using Deep Neural Network

Kazuhiko Takahashi[1(✉)] and Iwao Sugimoto[2]

[1] Information Systems Design, Doshisha University,
Kyotanabe, Kyoto 610-0321, Japan
katakaha@mail.doshisha.ac.jp
[2] School of Computer Science, Tokyo University of Technology,
Hachioji, Tokyo 192-0982, Japan
sugimoto@stf.teu.ac.jp

Abstract. This study explored the application of a deep neural network to the task of recognising tea types from their aroma. The aroma was measured from tea leaves using an array of quartz crystal resonators coated with plasma organic polymer films. Frequency analysis based on continuous wavelet transform, with the Morlet function as the mother wavelet, was applied to the sensor signals to construct the input vectors of the deep neural network. Experiments were conducted using oolong, jasmine and pu'erh teas as the samples and dehumidified indoor air as the base gas. The deep neural network achieved a recognition accuracy of 96.3% for the three tea types and the base gas. The experimental results demonstrated the effectiveness of applying a deep neural network to this task.

Keywords: Deep neural network · Chemical sensor · Tea leaves aroma · Wavelet transform · Morlet function

1 Introduction

Demand is increasing for electronic noses using chemical sensor technology in fields such as environmental monitoring, food and beverage evaluation, flavour and fragrance testing, medical diagnosis, drug detection and robotics [1–6]. Research to support this demand has included studies of materials, sensor devices, sensor systems and signal processing algorithms [7].

Tea is a popular beverage that can be classified into two types, based on the manufacturing process: fermented and un–fermented. Each tea has its own unique aroma, reflecting the mix of chemicals produced by the enzymatic oxidation of tea polyphenols during the fermentation process. The use of aroma to categories the tea and evaluate its quality is an emerging application of electronic nose systems [8–10]. In previous studies, we developed a gas–sensing system that uses mass–sensitive chemical sensors based on an array of quartz crystal resonators (QCRs) coated with plasma organic polymer films (PPFs) [11,12].

© Springer International Publishing AG 2017
G. Boracchi et al. (Eds.): EANN 2017, CCIS 744, pp. 160–167, 2017.
DOI: 10.1007/978-3-319-65172-9_14

As an application of our gas–sensing system, we applied it to the identification of tea leaves from their aroma. We applied several standard machine–learning approaches, including neural networks (multi–layer perceptron), support vector machines and random forests and compared their performance [13]. The recognition rate of 96.7% was achieved for three tea types using the neural network, however, the number of sample obtained by sensor responses was small and the base gas was not considered as a class of recognition.

Deep neural networks [14,15] have been attracting increasing attention in a number of fields. In particular, they have been used to solve engineering problems such as image and speech recognition, because their representation capabilities make them suitable for use in perception–related tasks. The application of deep neural networks to the recognition of aroma may therefore be expected to improve accuracy and may allow recognition tasks to be addressed without requiring complex pre–processing of the sensor signal.

In this study, we investigated the recognition of tea leaves from their aroma, using a gas–sensing system comprising mass–sensitive chemical sensors with PPF–coated QCRs. The rest of the paper is organised as follows. Section 2 describes data acquisition from the gas–sensing system. Section 3 introduces the experiments we conducted on recognition of tea aroma using a deep neural network. Section 4 presents our conclusions.

2   Measurement of Aroma from Tea Leaves

Figure 1 shows the experimental setup for measuring aroma from tea leaves. The sensor module used was able to detect volatile organic compounds in dry air at the parts–per–billion level. This module comprised eight QCRs coated with PPFs of phenylalanine, tyrosine, glucose, histidine, adenine, polyethylene and polychlorotrifluoroethylene. Figure 2 shows a schematic of the gas–sensing system. The system had two gas–flow lines: a sample line and a reference line. Air at standard indoor conditions was dehumidified using silica gel before being

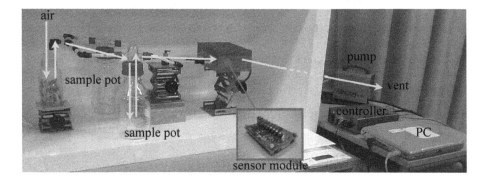

Fig. 1. Experimental setup.

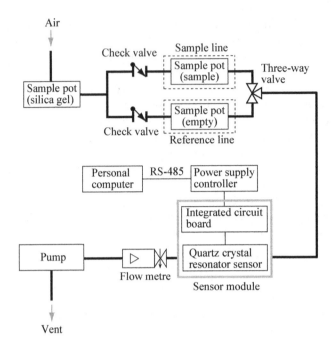

Fig. 2. Schematic of gas–sensing system.

introduced into the lines as the base gas. The flow rate was controlled by a mass flow metre and a dry vacuum pump.

Three kinds of tea leaf were used as the sample gas source: oolong (semi–fermented tea), jasmine (un–fermented, with the fragrance of jasmine blossoms added to the green leaves) and pu'erh (post–fermented tea). A 5 g sample of leaf was used to collect the sensor responses. The experiment was carried out as follows:

(1) To establish a baseline, the initial state of the sensors was recorded using the base gas.
(2) To measure the aroma, the flow was switched to the sample gas and measured for 60 s.
(3) The flow was then switched to the reference line for 180 s, to allow cleaning of the sensors.
(4) The cycle was repeated, starting at step 2.

Sensor responses were sampled 120 times, and the tea type identified was recorded.

3   Experiments Aroma Recognition

Features extracted from sensor response curves in the time domain, such as the peak amplitude and pulse width, are often used to construct feature vectors.

However, they are affected by variations in the temperature and humidity of the base gas. In a previous study [11], we applied continuous wavelet transform, using the Gabor function as a mother wavelet, to extract features from the sensor signal in the frequency domain. We demonstrated the effectiveness of this approach in composing the feature vectors for aroma recognition. Using the mother wavelet $g(t)$, the continuous wavelet transform of the i–th sensor signal $x_i(t)$ was derived as follows:

$$X_i(a,b) = \int_{-\infty}^{+\infty} x_i(t) g_{a,b}^*(t) dt,$$

where

$$g_{a,b}(t) = \frac{1}{\sqrt{a}} g\left(\frac{t-b}{a}\right),$$

where a is the scale factor, b is the shift factor and g^* denotes the conjugate of g. In the current study, the Morlet function that satisfied the admissibility condition for wavelets was used as the mother wavelet, to improve recognition accuracy. The number of frequencies in the wavelet transform was 64 and the number of scales was 4. To define the features, we calculated the mean of the power obtained by continuous wavelet transform through the measurement period of the sample from each sensor:

$$s_{i_j} = \frac{1}{K} \sum_{k=1}^{K} |X_i(j\Delta_a, k\Delta_b)|,$$

where each factor is discretised at the sampling rate Δ_i $(i = a, b)$, to construct the feature vector. Two types of feature vector $\boldsymbol{v}$ were then constructed, as follows:

$$\boldsymbol{v}_1 = \begin{bmatrix} \bar{s}_1 & \bar{s}_2 & \bar{s}_3 & \bar{s}_4 & \bar{s}_5 & \bar{s}_6 & \bar{s}_7 & \bar{s}_8 \end{bmatrix}^{\mathrm{T}},$$

$$\boldsymbol{v}_2 = \begin{bmatrix} s_{1_1} & \cdots & s_{1_{256}} & s_{2_1} & \cdots & s_{2_{256}} & s_{3_1} & \cdots\cdots & s_{7_{256}} & s_{8_1} & \cdots & s_{8_{256}} \end{bmatrix}^{\mathrm{T}},$$

where

$$\bar{s}_i = \frac{1}{N} \sum_{j=1}^{N} s_{i_j}.$$

For recognition of tea aroma, a deep neural network trained using stochastic gradient descent optimisation with an adaptive learning rate was applied. The deep neural network contained several hidden layers. For purposes of comparison, a hyperbolic tangent function, a rectified linear unit function and a maxout function were used as the activation function of the neuron in the hidden layers. In the output layer, a softmax function (log–linear model) was used as the activation function. Training and testing of the deep neural network was conducted using the leave–one–out cross–validation method.

Table 1. Confusion matrix of tea aroma recognition using a deep neural network in which the activation function in the hidden layer neurons is a hyperbolic tangent function.

		Predicted			
		Base gas	Jasmine	Oolong	Pu'erh
(a) 8–12–12–4 network topology					
Real	Base gas	75.0	12.5	6.25	6.25
	Jasmine	0	100	0	0
	Oolong	0	0	100	0
	Pu'erh	2.5	0	0	97.5
(b) 8–12–12–12–4 network topology					
Real	Base gas	85.41	4.17	6.25	4.17
	Jasmine	0	97.5	2.5	0
	Oolong	2.5	0	97.5	0
	Pu'erh	0	0	0	100

Table 2. Confusion matrix of tea aroma recognition using a deep neural network in which the activation function in the hidden layer neurons is a rectified linear unit function.

		Predicted			
		Base gas	Jasmine	Oolong	Pu'erh
(a) 8–12–12–4 network topology					
Real	Base gas	62.5	14.58	12.5	10.42
	Jasmine	2.5	97.5	0	0
	Oolong	2.5	0	97.5	0
	Pu'erh	2.5	0	0	97.5
(b) 8–12–12–12–4 network topology					
Real	Base gas	83.33	4.17	10.42	2.08
	Jasmine	0	97.5	0	2.5
	Oolong	2.5	0	97.5	0
	Pu'erh	0	0	0	100

First, the 8–dimensional input vector v_1 was used as the input to the deep neural network, and the 8–12–12–4 and 8–12–12–12–4 deep neural network topologies were compared. The recognition results are shown in Tables 1, 2 and 3. The recognition rate improved as the number of hidden layers was increased. When using the three–hidden–layer network topology with a hyperbolic tangent function, a high recognition rate of 95.1% was obtained for the four classes (three teas and the base gas). However, the base gas was sometimes misclassified as a sample gas.

Table 3. Confusion matrix of tea aroma recognition using a deep neural network in which the activation function in the hidden layer neurons is a maxout function.

		Predicted			
		Base gas	Jasmine	Oolong	Pu'erh
(a) 8–12–12–4 network topology					
Real	Base gas	77.08	12.5	8.34	2.08
	Jasmine	0	100	0	0
	Oolong	0	0	100	0
	Pu'erh	0	0	0	100
(b) 8–12–12–12–4 network topology					
Real	Base gas	83.33	6.25	6.25	4.17
	Jasmine	0	97.5	0	2.5
	Oolong	0	0	100	0
	Pu'erh	0	2.5	0	97.5

Table 4. Confusion matrix of tea aroma recognition using a deep neural network in which the activation function in the hidden layer neurons is a hyperbolic tangent function.

		Predicted			
		Base gas	Jasmine	Oolong	Pu'erh
(a) 2048–16–16–4					
Real	Base gas	100	0	0	0
	Jasmine	2.50	90.00	0	7.50
	Oolong	0	5.00	92.50	2.50
	Pu'erh	0	0	2.50	97.50
(b) 2048–16–16–16–4					
Real	Base gas	100	0	0	0
	Jasmine	0	85.00	7.50	7.50
	Oolong	0	2.50	95.0	2.50
	Pu'erh	0	2.50	0	97.50

Next, the 2048–dimensional input vector v_2 was used as the input to the deep neural network. The results are shown in Tables 4, 5 and 6. The recognition rate for the base gas was improved and a high recognition rate of 96.3% was obtained for the four classes when using the two–hidden layer network topology with the maxout function. The vector v_1 lost distribution information on the frequency component, because each element of the vector v_1 was defined by the mean of the power. However, the vector v_2 contained all the frequency components. As a result, the recognition accuracy for the base gas improved. These results suggest

Table 5. Confusion matrix of tea aroma recognition using a deep neural network in which the activation function in the hidden layer neurons is a rectified linear unit function.

		Predicted			
		Base gas	Jasmine	Oolong	Pu'erh
(a) 2048–16–16–4					
Real	Base gas	97.92	2.08	0	0
	Jasmine	2.50	87.50	0	10.00
	Oolong	0	2.50	97.50	0
	Pu'erh	0	0	0	100
(b) 2048–16–16–16–4					
Real	Base gas	100	0	0	0
	Jasmine	2.50	87.50	2.50	7.50
	Oolong	0	2.50	97.50	0
	Pu'erh	0	5.00	0	95.00

Table 6. Confusion matrix of tea aroma recognition using a deep neural network in which the activation function in the hidden layer neurons is a maxout function.

		Predicted			
		Base gas	Jasmine	Oolong	Pu'erh
(a) 2048–16–16–4					
Real	Base gas	97.92	0	2.08	0
	Jasmine	0	92.50	2.50	5.00
	Oolong	0	0	100	0
	Pu'erh	0	0	2.50	97.50
(b) 2048–16–16–16–4					
Real	Base gas	100	0	0	0
	Jasmine	2.50	90.00	0	7.50
	Oolong	0	0	100	0
	Pu'erh	0	2.50	2.50	95.00

that the use of the deep neural network improved the accuracy with which the tea types were identified from their aroma.

4   Conclusions

In this study, we investigated the capacity of a deep neural network to classify tea types from their aroma, measured using a mass–sensitive chemical sensor with PPF–coated QCRs. The input vector of the deep neural network was defined from the sensor signals via continuous wavelet transform, using the Morlet function as the mother wavelet. Recognition experiments were conducted using three tea types, with standard indoor air as the base gas. The highest recognition rate of 96.3% for all three tea leaves and the base gas was achieved when using a two–hidden–layer network topology with a maxout function in the hidden layer neurons. The experimental results confirmed the effectiveness of the proposed method for aroma classification.

References

1. Baldwin, E.A., Bai, J., Plotto, A., Dea, S.: Electronic noses and tongues: applications for the food and pharmaceutical industries. Sensors **11**, 4744–4766 (2011)
2. Wilson, A.D.: Review of electronic-nose technologies and algorithms to detect hazardous chemicals in the environment. Procedia Technol. **1**, 453–463 (2012)
3. Ishida, H., Wada, Y., Matsukura, H.: Chemical sensing in robotic applications: a review. IEEE Sens. J. **12**(11), 3163–3173 (2012)
4. Wilson, A.D.: Diverse applications of electronic-nose technologies in agriculture and forestry. Sensors **13**(2), 2295–2348 (2013)
5. Loutfi, A., Coradeschi, S., Mani, G.K., Shankar, P., Rayappan, J.B.B.: Electronic noses for food quality: a review. J. Food Eng. **144**, 103–111 (2015)
6. Kiani, S., Saeid, M., Mahdi, G.V.: Application of electronic nose systems for assessing quality of medicinal and aromatic plant products: a review. J. Appl. Res. Med. Aromat. Plants **3**(1), 1–9 (2016)
7. Boeker, P.: On 'Electronic Nose' methodology. Sens. Actuators B: Chem. **204**, 2–17 (2014)
8. Chen, Q., Liu, A., Zhao, J., Ouyang, Q.: Classification of tea category using a portable electronic nose based on an odor imaging sensor array. J. Pharm. Biomed. Anal. **84**, 77–83 (2013)
9. Huo, D., Wu, Y., Yang, M., Fa, H., Luo, X., Hou, C.: Discrimination of chinese green tea according to varieties and grade levels using artificial nose and tongue based on colorimetric sensor arrays. Food Chem. **145**, 639–645 (2014)
10. Torri, L., Massimiliano, R., Emma, C.: Electronic nose evaluation of volatile emission of chinese teas: from leaves to infusions. Int. J. Food Sci. Technol. **49**(5), 1315–1323 (2014)
11. Takahashi, K., Sugimoto, I.: Remarks on emotion recognition using breath gas sensing system. In: Mukhopadhyay, S.C., Gupta, G.S. (eds.) Smart Sensors and Sensing Technology. Lecture Notes Electrical Engineering, vol. 20, pp. 49–62. Springer, Heidelberg (2008). doi:10.1007/978-3-540-79590-2_4
12. Takahashi, K., Kawanobe, Y., Nishiwaki, N., Sugimoto, I.: Remarks on a computational model of a mass-sensitive chemical sensor with plasma-organic-polymer-film-coated quartz crystal resonators. In: Proceedings of 2014 IEEE/SICE International Symposium on System Integration, pp. 558–563 (2014)
13. Takahashi, K., Mune, M., Sugimoto, I.: Remarks on neural network-based tea aroma recognition with a mass-sensitive chemical sensor using plasma-organic-polymer-film-coated quartz crystal resonators. In: Proceedings of the 2nd International Conference on Frontiers of Signal Processing, pp. 38–42 (2016)
14. LeCun, Y., Bengio, Y., Hinton, G.: Deep learning. Nature **521**(7553), 436–444 (2015)
15. Schmidhuber, J.: Deep learning in neural networks: an overview. Neural Netw. **61**, 85–117 (2015)

Baby Cry Sound Detection: A Comparison of Hand Crafted Features and Deep Learning Approach

Rafael Torres[1], Daniele Battaglino[1,2(✉)], and Ludovick Lepauloux[1]

[1] NXP Semiconductors, Mougins, France
ludovick.lepauloux@nxp.com
[2] EURECOM, Biot, France
battagli@eurecom.fr

Abstract. Baby cry sound detection allows parents to be automatically alerted when their baby is crying. Current solutions in home environment ask for a client-server architecture where an end-node device streams the audio to a centralized server in charge of the detection. Even providing the best performances, these solutions raise power consumption and privacy issues. For these reasons, interest has recently grown in the community for methods which can run locally on battery-powered devices. This work presents a new set of features tailored to baby cry sound recognition, called hand crafted baby cry (HCBC) features. The proposed method is compared with a baseline using mel-frequency cepstrum coefficients (MFCCs) and a state-of-the-art convolutional neural network (CNN) system. HCBC features result to be on par with CNN, while requiring less computation effort and memory space at the cost of being application specific.

Keywords: Baby cry detection · Hand crafted baby cry features · Support vector data description · Convolutional neural networks

1 Introduction

Audio event detection (AED) has recently gained attention in the audio community [1–3]. AED is therefore pertinent to *smart-home* market where the presence of connected devices enables sounds to be detected through consumer microphones.

Thus, this work has focused on the detection of baby cry sounds specifically for home environment. This choice has been driven by the practical use case consisting of capturing the baby crying and automatically alerting his parents. AED based technologies are smarter than standard baby monitors or walkie talkies: in the former, monitoring is usually energy-based with the counterpart of being easily deceived by high energy sounds; in the latter, parents have to constantly listen to the receiver during their activities.

Existing approaches in baby cry detection literature consist of extracting meaningful features from audio signal frames. Most of them use spectral features

© Springer International Publishing AG 2017
G. Boracchi et al. (Eds.): EANN 2017, CCIS 744, pp. 168–179, 2017.
DOI: 10.1007/978-3-319-65172-9_15

such as mel-frequency cepstrum coefficients (MFCCs), combined with binary classifiers such as support vector machines (SVMs) [4]. Recent researches have explored the use of convolutional neural networks (CNNs) tailored to baby cry detection [5].

Whereas showing the most promising results, CNN computation and memory requirements render it more compliant with a client-server solution, where an end-node client (i.e. low-power device equipped with a microphone) streams the audio to a central server in charge of the entire process. Some works in AED have started to question this client-server approach by focusing on battery-powered devices [6,7] with a significant reduction of band-width and power. Moreover, moving the complexity towards the end-node has the advantage of respecting user privacy, since audio is analyzed locally in the device.

Thus, the need for an *always-active* baby cry detector calls for algorithm efficiency, essential to minimize battery consumption, and for classifier robustness, able to detect a baby cry sounds within a broad set of unknown conditions. As expressed in [8,9], the detection in real conditions requires new types of classifiers more robust to unknown classes. In that sense the support vector data description (SVDD) is a good candidate for modeling baby cry features without being influenced by the number and the type of classes in the training set.

We herein present three methods for the baby cry detection task: the first baseline employs MFCCs and SVDD as a classifier; the second is based on CNN applied on mel-spectrogram; the third proposes a novel set of features tailored to baby cry detection.

The contributions of this work can be summarized in: (i) hand crafted baby cry (HCBC) features; (ii) the adoption of SVDD classifier for both MFCC and HCBC features; (iii) improvements in terms of normalization and regularization of state-of-the-art CNN; (iv) the comparison between hand-crafted features and deep learning approaches.

The remainder of the paper is organized as follows: Sect. 2 describes in details the three methods; Sect. 3 presents the experimental set-up, database description and results; conclusions and future works are discussed in Sect. 4.

2   Methods

This section describes the aforementioned methods for baby cry sound detection.

2.1   One-Class Classifier - SVDD

Whereas baby cry sounds can be easily collected and modeled, non-baby cry samples are more difficult to identify and categorize. In domestic environment, many sounds may resemble a baby cry sound provoking false alarms during the detection. Standard binary classifiers may fail to learn both baby cry (*target*) and non-baby cry (*non-target*) samples when these latter represent a subset of those encountered during testing. This problem has been identified as *open-set* and specific classifiers have therefore been employed [8,9].

Instead of separating *target* from *non-target*, SVDD models only target samples with a hypersphere [10]. Once the radius R and the center of the hypersphere a have been found during training, the decision function f to determine if a new sample z belongs to the target class depends upon R and a:

$$f(z, R, a) = sign(R^2 - ||z - a||^2).\qquad(1)$$

Drawing upon the SVM theory, SVDD finds the support vectors (SVs) using Lagrangian procedure to optimize a and R. These SVs are training samples which are selected to represent the boundary between the two classes.

Differently from prior works on SVDD, the grid-search for finding the best classifier parameters has been modified. This routine points at the presence of non-target samples in the training set to automatically select the best pair of parameters by minimizing the following function:

$$\lambda = \sqrt{(\frac{\#SVs}{T})^2 + (1 - \text{AUC})^2}\qquad(2)$$

where $\#SVs$ corresponds to the number of SVs, T is the cardinality of the target class samples and AUC stands for the area under the receiver operating characteristic curve, used to evaluate the performance of SVDD on a validation set made of target and non-target samples. The function in Eq. 2 defines a trade-off between an estimation of the classifier complexity ($\frac{\#SVs}{T}$ term) and the global performance of the system expressed with AUC metric.

2.2 Baseline System - MFCC

A baby cry sound is generated by an excitation of the vocal cords producing a sequence of periodic impulses. In healthy babies the fundamental frequency ($F0$) reaches values between 250 Hz – 600 Hz, which has a higher range than that of adult females and males. Hence, a higher $F0$ characterizes most of the baby cry sounds, as depicted in Fig. 1.

Given these spectral properties, MFCCs have been proven in the literature [11] to be a good candidate for baby cry detection task since it represents each signal as rate of change in the frequency bands. Due to baby cry harmonics, MFCCs represent this information with higher MFCCs coefficients. This phenomenon is observable in Fig. 1. As a consequence of representing the entire spectral information, MFCCs are very dependent on overlapping sounds or additional noise which are mixed with the baby cry signal.

2.3 State-of-art System - CNN

Although primarily designed for image classification tasks, CNNs have proven to be successful in speech and music recognition [12]. Recently, these techniques have attracted interest also in AED [13] showing promising results. Inspired by

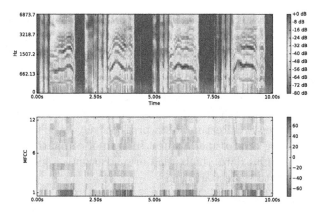

Fig. 1. The log-mel power spectrum on the top and the MFCC without the first coefficient C0 on the bottom for a sequence of baby cry sounds. The sound produced by the baby creates periodic harmonics in the log-mel power spectrogram that are captured in the MFCCs domain. In this example, each harmonic sequence is preceded by an unvoiced breath of the baby which produces noisy-like sounds in the lower frequency. This phenomenon is represented by a peak in the first MFCCs coefficients, while baby cry is captured by higher coefficients.

the architecture in [5], CNN performances are enhanced by introducing normalization and regularization layers. These modifications report a better generalization and they are even applicable on datasets of modest size.

A standard CNN is a deep architecture of successive layers, which are connected in different ways from the input data until the output layer. The global architecture is shown in Fig. 2. Differently from MFCCs which decorrelate the data with the discrete cosine transform (DCT), CNN takes as input the log mel-filtered spectrogram mimicking an image processing behavior. In the convolutional layer, each hidden unit is not connected to all the inputs from previous layer, but only to an area of the original input space, capturing local correlation. These small parts of the whole input space are connected to the hidden units through the weights. This operation is equivalent to a *convolutional filter* processing.

The pooling consists of merging close units according to some criteria (such as mean or max). This effectively performs a *downsampling* which smooths the resulting outputs of each convolutional layer, making the system more robust to small variations or translations. In the case of spectrograms as input, these local variations have to be attenuated in order to better recognize global patterns.

The main differences between the proposed CNN and the one in [5] are listed below:

1. convolutional filters of the first layer are 10×10 blocks, to capture both frequency and temporal resolution;
2. regularization techniques avoid overfitting of the network on a relatively small dataset. One of the most adopted is the so called *dropout*, which consists

of literally dropping out hidden units with a certain probability. At each training iteration, a random subset of hidden units is temporary disabled by multiplying the input to these units by 0. This forces the network to find robust features that do not depend on the presence of particular other neurons [14];

3. scaling inputs to zero mean and unit standard deviation is a common preprocessing step to uniform values across heterogeneous features. When they pass through a deep architecture, data progressively loose this normalization resulting in too big or too small values. Instead of computing the normalization only on the input data, the batch normalization is applied to the hidden layers so to avoid this effect [15].

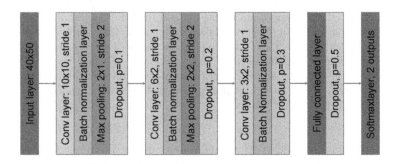

Fig. 2. The global architecture of the proposed CNN. More details of each layer are provided in Subsect. 3.3.

2.4 Proposed Approach - HCBC with SVDD Classifier

As explained in Subsect. 2.2 and Fig. 1, baby cry sounds are characterized by specific acoustic properties. Whereas the spectral content can be represented with MFCCs, there exists no standard features exploiting voiced-unvoiced recurrence in a baby cry signal. In the proposed algorithm, hand crafted features are specifically designed to characterize these temporal patterns.

HCBC features consist of frame-based descriptors which are then aggregated over longer audio-clips. For each frame, the fundamental frequency $\hat{F}0$ is estimated using an autocorrelation method. These features are composed of *voiced unvoiced counter*, *consecutive $F0$* and *harmonic ratio accumulation* which create a 3-D feature vector used by SVDD classifier to model the target baby-cry class. The explanation of each feature is described below:

Voiced unvoiced counter (VUVC) counts all frames having a significant periodic content. This is obtained by looking at the harmonic strength, called $R0$, defined in Boersma's work [16]. For each frame, the local maximum of the frame normalized autocorrelation $R0$ must be greater than a predefined threshold t_{vuvc}.

Consecutive $F0$ (CF0) acts as an accumulator, which tracks the temporal continuity of the estimated $\hat{F}0$. Let us define $Fref$ as the most occurring $\hat{F}0$ learned from the training set (see Fig. 4). First, the distance between $\hat{F}0$ of each frame and $Fref$ is calculated. As long as this distance is smaller than a tolerance parameter ϵ, a score is computed and accumulated in $CF0$ with a weight that follows a square law. The longer a sequence of consecutive $\hat{F}0$, the greater the weight is. The corresponding method is given in Algorithm 1.

input : An array of $\hat{F}0_{i=1\cdots M}$ for a given audio-clip
        Frequency sampling Fs
        Tolerance parameter ϵ
        Reference fundamental frequency $Fref$
output: $CF0$

begin
    $CF0 = 0$
    $counter = 1$

    for $i = 1$ **to** M **do**
        if $|\hat{F}0_i - Fref| < \epsilon$ **then**
            $score = \frac{Fs - |\hat{F}0_i - Fref|}{Fs}$
            $CF0 = CF0 + sqrt(counter) \times score$
            $counter = counter + 1$
        else
            $CF0$ not updated and $counter = 1$
        end
    end
end

Algorithm 1. Consecutive $F0$ ($CF0$) algorithm.

Harmonic ratio accumulation (HRA) is defined similarly to [11] as the ratio between the energy in harmonics and the overall frame energy. Let us define x as the microphone signal, X its discrete N points Fourier transform and n_i the closest bin of the ith harmonic with i from 1 to Ny the last harmonic before Nyquist frequency. Note that the first harmonic n_1 corresponds to $2 \times F0$. For a given frame, the harmonic ratio is defined as follows:

$$HR = \frac{\sum_{i=1}^{Ny} |X[n_i]|^2}{\sum_{j=1}^{N} |X[j]|^2}. \tag{3}$$

Thus, HRA of a given audio-clip is the sum of all its frames harmonic ratios. Considering M the number of frames in an audio-clip:

$$HRA = \sum_{1}^{M} HR. \tag{4}$$

The three presented features (VUVC, CF0 and HRA) capture correlated, mono-tonal and harmonic patterns. A baby cry sound has a specific pitch, duration and spectral distribution that requires fine tuning for optimal class differentiation as illustrated in Fig. 3. It must be also emphasized that HCBC features

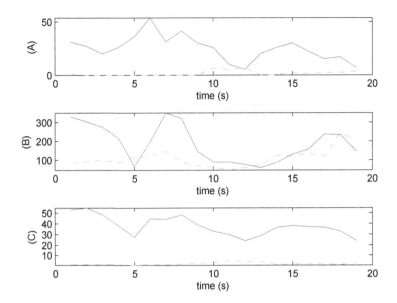

Fig. 3. Example of HCBC features behavior for a baby cry sample in solid blue line and a male cry sample in red dashed line with (A) voiced unvoiced counter, (B) consecutive F0 and (C) harmonic ratio accumulation. (Color figure online)

are energy independent. Differently from previous works, HCBC are self-content low-dimensional descriptors that are not concatenated with other features (such as MFCC) resulting in a better memory and computation efficiency.

3   Experimental Results

This section describes the datasets used for training together with testing protocols, metrics, implementation details and results.

3.1   Dataset

Due to its recent development in AED field, no public standardized dataset has yet been released for baby cry sound detection task.

However, collaborative databases may offer a good alternative. One major advantage is the diversity of the signals in terms of device audio path and signal to noise ratio (SNR) levels. This heterogeneity covers many possible scenarios with a more robust *on-field* evaluation.

The database employed in this work comes from a set of available on-line resources[1]. For training, it includes 102 baby cry sound events (1 h 07 m) and

---

[1] http://www.audiomicro.com.
   https://www.freesound.org.
   https://www.pond5.com.
   https://www.soundsnap.com.

93 non-baby cry (1 h 24 m) i.e. tv, toy, adult cry, baby talk/play, music, fan, vacuum cleaner which are used for modeling target and non-target class.

The testing set is composed of 10 files that are created by mixing 26 baby cry events (0 h 16 m) separated from each other by 30 s with 10 different 5 m looped background recordings at a SNR level of 18 dB. These background recordings are repeated to avoid a significant noise variation along the baby cry sequence that may unfairly affect the detection.

Hence, the testing set is composed of a 4 h recordings which sparsely contain target sounds and additional 2 h of whole non-target sounds. This non-target noise library is made up of home environment recordings from CHiMe-Home (more details are available in [17]). All signals are 16 kHz mono wave files.

3.2 Evaluation Protocols and Metrics

The three methods output a continuous score every second. For this type of application, it has been identified as a good trade-off between detection rate, computation cost and latency. According to that, the groundtruth of each file has been manually annotated.

Performance is evaluated by a receiver operating characteristic (ROC) curve and a precision-recall (PR) curve. These two curves provide complementary information: the ROC curve presents how the number of correctly classified samples varies with the number of negative incorrectly classified samples. Positive and negative samples are however separately counted and normalized.

When the absolute number of positive samples (i.e. baby cry) is significantly less than the possible number of negative samples (i.e. non-baby cry), ROC curve may give a too optimistic view of the algorithm performance. The precision of the PR curve, instead, directly compares absolute number of positive and negative samples. In the case of an highly unbalanced set, the precision will be affected by the number of false positive providing a view closer to real performances [18].

3.3 Implementation Details

Baseline MFCC features are extracted from a frame length of 32 ms overlapped by 16 ms. The filter-bank is built of 40 Mel-scaled filters up to 8 kHz, resulting in 13 MFCCs for each frame. The mean and standard deviation are then computed over the 3 s audio-clip overlapped by 2 s, resulting in a 24 dimensional feature vector (without C0). For the remaining parameters, the default ones of *rastamat* library[2] have been selected.

The SVDD classifier in the implementation of *libsvm* [19] uses the radial basis function (RBF) kernel while the best pair of parameters C, γ are selected by minimizing the function in Eq. 2 on the validation set. In our experiments, in order to adjust classifier and features parameters, a validation set is randomly selected from a 30% of the training set. Once the parameters have been estimated, the final model is trained using the entire training set.

---

[2] http://www.ee.columbia.edu/ln/rosa/matlab/rastamat/.

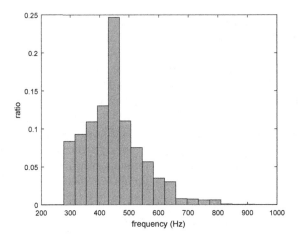

Fig. 4. Normalized histogram of the baby cry frame $\hat{F}0$ distribution of the training database with a bar width of 40 Hz. The maximum is reached at 430–468 Hz with a ratio of 24.7%. This is previously referenced as $Fref$.

The same strategy is then used to choose the CNN parameters and architecture, based on the lowest classification error on the validation set. Also in this case, the final model is then trained using the whole training set with a network in the order to 1 million trainable parameters. CNN structure is depicted in Fig. 2. The input layer takes a log mel-filtered spectrograms of 40 filters and 50 frames (corresponding to 1 s). The activation function is the standard rectifier (ReLU), except for the last layer where the *softmax* returns the probability for the two outputs (baby cry or non-baby cry). During the training phase, a stochastic gradient descent is evaluated over a mini-batch of 256 shuffled input spectrograms. The random shuffling of the inputs is important to represent the data not in their temporal order. The momentum is set to 0.9, over 50 training iterations with a learning rate of 0.001.

Details of filters are displayed in Fig. 2. The number of filters is set to 32. The stride term refers to the amount by which each convolutional filter shifts horizontally and vertically. There are 4 dropout layers in this architecture, with an increasing probability of dropping off units (from 0.1 to 0.5). This prevents to prune too many units in the first layers where features are being built. Thus an higher dropout probability is applied to the fully connected layer, particularly prone to overfit.

The library employed to implement the CNN is the *lasagne*[3] library, a wrapper of *Theano*[4]. Experiments have been run on a Nvidia Quadro M4000 GPU.

The frame length used in HCBC features is 32 ms overlapped by 16 ms. All frames are then aggregated over bigger audio-clips of 3 s with an overlap of 2 s. For each frame, $\hat{F}0$ is computed in a restricted range of 250 Hz–1000 Hz. VUVC

---

[3] http://lasagne.readthedocs.io/en/latest/.
[4] http://deeplearning.net/software/theano/.

threshold is set at tvuvc = 0.85. CF0 standard fundamental frequency $Fref$ is estimated at 449 Hz with a tolerance $\epsilon = 20$ Hz. These values are set during training phase based on the overall $\hat{F}0$ distribution (see Fig. 4) and validation results.

3.4   Results

Reported in Fig. 5 are experimental results for the three methods: MFCC in dotted black line, CNN in dashed red line and HCBC in solid blue line. The area under the curve (AUC) metric is also reported in the legend of the ROC curve.

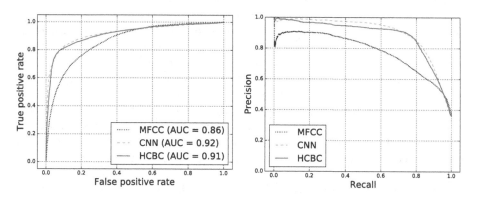

Fig. 5. The ROC on the left and the precision-recall curve on the right. (Color figure online)

Concerning the ROC curve, CNN and HCBC outperform the baseline MFCC system, passing from an AUC of 86% to more than 90%. CNN classifier is slightly better than HCBC, with 1% improvement in AUC. The precision-recall curve is more related to the type of application, where we can choose the trade-off between an high precision (therefore less false positive) or an higher recall (with less false negative).

Let us consider that a baby statistically cries 2 h per day. An acceptable metric for a baby cry detector would be a recall of 80% with a maximum of 5 m per week of false positives. These numbers result in a precision of 99,3% and a recall of 80%. CNN and HCBC are the closest algorithms to this requirement, while the baseline MFCC has a drop of 20% in precision compared to them.

Albeit showing similar performance, systems must be compared also in terms of computational and memory cost. From a feature computation point of view, not using the mel spectrogram as for MFCC or CNN is a clear advantage. Referring to [20], HCBC features employ only FFT and autocorrelation as basic processing, resulting in 20 times lower computational cost than standard MFCCs with no additional memory cost. From a classifier point of view, it has been demonstrated in [21] that CNNs may reach the highest performance at the expense of high computation complexity. This mainly prohibits their use on low-power devices. Finally, the proposed approach shows advantage over the 24 dimensional MFCC features used in the baseline system: knowing that complexity of SVDD is proportional to the number of SVs and feature vector dimensionality [22], HCBC outperforms the baseline system with only 3 dimensional feature vector.

4   Conclusions

In this work three methods were proposed for detecting baby crying in every day domestic environment: a baseline based on MFCC and SVDD classifier; a state-of-the-art CNN system and a new set of features specifically designed for this task. These 3 dimensional features capture repetition of voice-unvoiced pattern during time and therefore outperform the MFCC baseline, reaching the same level of performance of CNN. CNN is able to automatically extract meaningful patterns from log mel-filtered spectrograms, achieving the best results. Nevertheless, depending on the computational and memory resources available, the choice of CNN may not be compatible with low-power devices. The proposed method HCBC has the same level of performance and it is less computational and memory demanding. The drawback of this approach is to be suited uniquely for baby cry sound detection, with an high cost in designing specific baby cry features.

Further research should investigate ways of reducing complexity of CNN, by decreasing the number of filters and their size. Another track may encode the temporal patterns directly in the deep learning architecture.

References

1. Mesaros, A., Heittola, T., Virtanen, T.: TUT database for acoustic scene classification and sound event detection. In: 24th European Signal Processing Conference (EUSIPCO), pp. 1128–1132 (2016)
2. Barchiesi, D., Giannoulis, D., Stowell, D., Plumbley, M.: Acoustic scene classification: classifying environments from the sounds they produce. IEEE Sig. Process. Mag. **32**(3), 16–34 (2015)
3. Ntalampiras, S.: Audio pattern recognition of baby crying sound events. J. Audio Eng. Soc. **63**(5), 358–369 (2015)
4. Saraswathy, J., Hariharan, M., Yaacob, S., Khairunizam, W.: Automatic classification of infant cry: a review. In: International Conference on Biomedical Engineering (ICoBE), pp. 543–548, February 2012
5. Lavner, Y., Cohen, R., Ruinskiy, D., Ijzerman, H.: Baby cry detection in domestic environment using deep learning. In: IEEE International Conference on the Science of Electrical Engineering (ICSEE), pp. 1–5, November 2016
6. Saha, B., Purkait, P.K., Mukherjee, J., Majumdar, A.K., Majumdar, B., Singh, A.K.: An embedded system for automatic classification of neonatal cry. In: IEEE Point-of-Care Healthcare Technologies (PHT), pp. 248–251, January 2013

7. Bǎnicǎ, I.A., Cucu, H., Buzo, A., Burileanu, D., Burileanu, C.: Baby cry recognition in real-world conditions. In: 39th International Conference on Telecommunications and Signal Processing (TSP), pp. 315–318, June 2016

8. Battaglino, D., Lepauloux, L., Evans, N.: The open-set problem in acoustic scene classification. In: IEEE International Workshop on Acoustic Signal Enhancement (IWAENC), pp. 1–5, September 2016

9. Rabaoui, A., Davy, M., Rossignol, S., Lachiri, Z., Ellouze, N.: Improved one-class svm classifier for sounds classification. In: IEEE Conference on Advanced Video and Signal Based Surveillance (AVSS), pp. 117–122 (2007)

10. Tax, D.M.J., Duin, R.P.W.: Data domain description using support vectors. In: European Symposium on Artificial Neural Networks, pp. 251–256 (1999)

11. Cohen, R., Lavner, Y.: Infant cry analysis and detection. In: IEEE 27th Convention of Electrical and Electronics Engineers, pp. 1–5, November 2012

12. Deng, L., Yu, D.: Deep learning: methods and applications. Found. Trends Sig. Process. **7**(3–4), 197–387 (2014)

13. Piczak, K.J.: Environmental sound classification with convolutional neural networks. In: IEEE 25th International Workshop on Machine Learning for Signal Processing (MLSP), pp. 1–6, September 2015

14. Hinton, G.E., Srivastava, N., Krizhevsky, A., Sutskever, I., Salakhutdinov, R.R.: Improving neural networks by preventing co-adaptation of feature detectors. arXiv:1207.0580 (2012)

15. Ioffe, S., Szegedy, C.: Batch normalization: accelerating deep network training by reducing internal covariate shift. In: 32nd International Conference on Machine Learning, ICML, pp. 448–456 (2015)

16. Boersma, P.: Accurate short-term analysis of the fundamental frequency and the harmonics-to-noise ratio of a sampled sound. In: IFA Proceedings 17, pp. 97–110 (1993)

17. Foster, P., Sigtia, S., Krstulovic, S., Barker, J., Plumbley, M.D.: Chime-home: a dataset for sound source recognition in a domestic environment. In: IEEE Workshop on Applications of Signal Processing to Audio and Acoustics (WASPAA), pp. 1–5 (2015)

18. Saito, T., Rehmsmeier, M.: The precision-recall plot is more informative than the ROC plot when evaluating binary classifiers on imbalanced datasets. PloS One **10**(3), e0118432 (2015)

19. Chang, C.C., Lin, C.J.: LIBSVM: a library for support vector machines. ACM Trans. Intell. Syst. Technol. **2**, 27:1–27:27 (2011). http://www.csie.ntu.edu.tw/~cjlin/libsvm

20. Wang, J.C., Wang, J.F., Weng, Y.S.: Chip design of MFCC extraction for speech recognition. Integr. VLSI J. **32**(1–3), 111–131 (2002)

21. Wu, J., Leng, C., Wang, Y., Hu, Q., Cheng, J.: Quantized convolutional neural networks for mobile devices. In: 2016 IEEE Conference on Computer Vision and Pattern Recognition (CVPR), pp. 4820–4828, June 2016

22. Sigtia, S., Stark, A.M., Krstulovi, S., Plumbley, M.D.: Automatic environmental sound recognition: performance versus computational cost. IEEE/ACM Trans. Audio Speech Lang. Process. **24**(11), 2096–2107 (2016)

Deep Learning Image Analysis

Deep Convolutional Neural Networks for Fire Detection in Images

Jivitesh Sharma$^{(\boxtimes)}$, Ole-Christoffer Granmo, Morten Goodwin,
and Jahn Thomas Fidje

University of Agder (UiA), Kristiansand, Norway
jivitesh.sharma@uia.no

Abstract. Detecting fire in images using image processing and computer vision techniques has gained a lot of attention from researchers during the past few years. Indeed, with sufficient accuracy, such systems may outperform traditional fire detection equipment. One of the most promising techniques used in this area is Convolutional Neural Networks (CNNs). However, the previous research on fire detection with CNNs has only been evaluated on balanced datasets, which may give misleading information on real-world performance, where fire is a rare event. Actually, as demonstrated in this paper, it turns out that a traditional CNN performs relatively poorly when evaluated on the more realistically balanced benchmark dataset provided in this paper. We therefore propose to use even deeper Convolutional Neural Networks for fire detection in images, and enhancing these with fine tuning based on a fully connected layer. We use two pretrained state-of-the-art Deep CNNs, VGG16 and Resnet50, to develop our fire detection system. The Deep CNNs are tested on our imbalanced dataset, which we have assembled to replicate real world scenarios. It includes images that are particularly difficult to classify and that are deliberately unbalanced by including significantly more non-fire images than fire images. The dataset has been made available online. Our results show that adding fully connected layers for fine tuning indeed does increase accuracy, however, this also increases training time. Overall, we found that our deeper CNNs give good performance on a more challenging dataset, with Resnet50 slightly outperforming VGG16. These results may thus lead to more successful fire detection systems in practice.

Keywords: Fire detection · Deep Convolutional Neural Networks · VGG16 · Resnet50

1 Introduction

Emergency situations like floods, earthquakes and fires pose a big threat to public health and safety, property and environment. Fire related disasters are the most common type of Emergency situation which requires thorough analysis of the situation required for a quick and precise response. The first step involved

© Springer International Publishing AG 2017
G. Boracchi et al. (Eds.): EANN 2017, CCIS 744, pp. 183–193, 2017.
DOI: 10.1007/978-3-319-65172-9_16

in this process is to detect fire in the environment as quickly and accurately as possible.

Fire Detection in most places employs equipment like temperature detectors, smoke detectors, thermal cameras etc. which is expensive and not available to all [14]. But, after the advent of advanced image processing and computer vision techniques, detection of fire may not require any equipment other than cameras. Due to this expeditious development in vision-based fire detection models, there is a particular inclination towards replacing the traditional fire detection tools with vision-based models. These models have many advantages over their hardware based counterparts like accuracy, more detailed view of the situation, less prone to errors, robustness towards the environment, considerably lower cost and the ability to work on existing camera surveillance systems.

There have been many innovative techniques proposed in the past to build an accurate fire detection system which are broadly based on image processing and computer vision techniques. The state-of-the-art vision-based techniques for fire and smoke detection have been comprehensively evaluated and compared in [21]. The colour analysis technique has been widely used in the literature to detect and analyse fire in images and videos [2,13,16,20]. On top of colour analysis, many novel methods have been used to extract high level features from fire images like texture analysis [2], dynamic temporal analysis with pixel-level filtering and spatial analysis with envelope decomposition and object labelling [22], fire flicker and irregular fire shape detection with wavelet transform [20], etc. These techniques give adequate performance but are outperformed by Machine Learning techniques. A comparative analysis between colour-based models for extraction of rules and a Machine Learning algorithm is done for the fire detection problem in [19]. The machine learning technique used in [19] is Logistic Regression which is one of the simplest techniques in Machine Learning and still outperforms the colour-based algorithms in almost all scenarios. These scenarios consist of images containing different fire pixel colours of different intensities, with and without smoke.

Instead of using many different algorithms on top of each other to extract relevant features, we can use a network that learns relevant features on its own. Neural networks have been successfully used in many different areas such as Natural Language Processing, Speech Recognition, Text Analysis and especially Image Classification. Extracting relevant features from images is the key to accurate classification and analysis which is why the problem of fire detection is ideally suited for Deep Learning. Deep Neural Networks are used to automatically 'learn' hierarchy of pertinent features from data without human intervention and the type of neural network ideally suited for image classification is the Convolutional Neural Networks (CNN).

Therefore, our approach is to employ state-of-the-art CNNs to distinguish between images that containing fire and images that do not and build an accurate fire detection system. To make these models more robust, we use a custom-made image dataset containing images with numerous scenarios.

The rest of paper is organised in the following manner: Sect. 2 briefly describes the previous research that uses CNNs for detecting fire. In Sect. 3 give a description of our proposed work. Section 4 gives the experimental results along with an illustration of our dataset, which is available online for the research community. Finally, Sect. 5 concludes our paper.

2   Related Work

There have been many significant contributions from various researchers in developing a system that can accurately detect fire in the surrounding environment. But, the most notable research in this field involves Deep Convolutional Neural Networks (DCNN). DCNN models are currently among the most successful image classification models which makes them ideal for a task such as Fire detection in images. This has been demonstrated by previous research published in this area.

In [5], the authors use CNN for detection of fire and smoke in videos. A simple sequential CNN architecture, similar to LeNet-5 [11], is used for classification. The authors quote a testing accuracy of 97.9% with a satisfactory false positive rate.

Whereas in [23], a very innovative cascaded CNN technique is used to detect fire in an image, followed by fine-grained localisation of patches in the image that contain the fire pixels. The cascaded CNN consists of AlexNet CNN architecture [10] with pre-trained ImageNet weights [15] and another small network after the final pooling layer which extracts patch features and labels the patches which contain fire. Different patch classifiers are compared.

The AlexNet architecture is also used in [18] which is used to detect smoke in images. It is trained on a fairly large dataset containing smoke and non-smoke images for a considerably long time. The quoted accuracies for large and small datasets are 96.88% and 99.4% respectively with relatively low false positive rates.

Another paper that uses the AlexNet architecture is [12]. This paper builds its own fire image and video dataset by simulating fire in images and videos using Blender. It adds fire to frames by adding fire properties like shadow, fore-ground fire, mask etc. separately. The animated fire and video frames are composited using OpenCV [1]. The model is tested on real world images. The results show reasonable accuracy with high false positive rate.

As opposed to CNNs which extract features directly from raw images, in some methods image/video features are extracted using image processing techniques and then given as input to a neural network. Such an approach has been used in [4]. The fire regions from video frames are obtained by threshold values in the HSV colour space. The general characteristics of fire are computed using these values from five continuous frames and their mean and standard deviation is given as input to a neural network which is trained using back propagation to identify forest fire regions. This method performs segmentation of images very accurately and the results show high accuracy and low false positive rates.

In [8], a neural network is used to extract fire features based on the HSI colour model which gives the fire area in the image as output. The next step is fire area segmentation where the fire areas are roughly segmented and spurious fire areas like fire shadows and fire-like objects are removed by image difference. After this the change in shape of fire is estimated by taking contour image difference and white pixel ratio to estimate the burning degree of fire, i.e. no-fire, small, medium and large. The experimental results show that the method is able to detect different fire scenarios with relatively good accuracy.

All the research work done in this area has been exemplary. But, there are some issues associated with each of them that we try to alleviate in this paper. We use a dataset that consists of images that we have handpicked from the internet. The dataset contains images that are extremely hard to classify which results in poor generalization. The dataset also contains many different scenarios and is highly unbalanced to replicate real world behaviour. In this paper, we propose to use state-of-the-art pre-trained DCNN models. The reason behind using such complex models is explained in the next section. We also modify these models to improve accuracy at the cost of training time.

3   The Fire Detector

In this paper, we propose to employ Deep Convolutional Neural Networks instead of simple and shallow CNN models. The AlexNet has been used by researchers in the past for fire detection which has produced satisfactory results. We propose to use two Deep CNN architectures that have outperformed the AlexNet on the ImageNet dataset, namely VGG16 [17] and Resnet50 [7]. We use these models with pre-trained ImageNet weights. This helps greatly when there is lack of training data. So, we just have to fine-tune the fully-connected layers on our dataset.

3.1   Deep ConvNet Models

The Convolutional Neural Network was first introduced in 1980 by Fukushima [6]. The CNN is designed to take advantage of two dimensional structures like 2D Images and capture local spatial patterns. This is achieved with local connections and tied weights. It consists of one or more convolution layers with pooling layers between them, followed by one or more fully connected layers, as in a standard multilayer perceptron. CNNs are easier to train compared to Deep Neural Networks because they have fewer parameters and local receptive fields.

In CNNs, kernels/filters are used to see where particular features are present in an image by convolution with the image. The size of the filters gives rise to locally connected structure which are each convolved with the image to produce feature maps. The feature maps are usually subsampled using mean or max pooling. The reduction is parameters is due to the fact that convolution layers share weights. The reason behind parameter sharing is that we make an assumption, that the statistics of a patch of a natural image are the same as any other patch of the image, which suggests that features learned at a location can also be learned for other locations. So, we can apply this learned feature detector anywhere in the image. This makes CNNs ideal feature extractors for images.

The CNNs with many layers have been used for various applications especially image classification. In this paper, we use two state-of-the-art Deep CNNs that have achieved one of the lowest errors in image classification tasks.

VGG16: The VGG16 architecture was proposed by the Visual Geometry Group at the University of Oxford [17]. The main purpose of the paper was to investigate the effect of depth in CNN models. They developed a number of models with different depths ranging from 11 layers to 19 layers and tested them on different tasks. The results on these tasks show that increasing depth also increases performance and accuracy. The 19 layer architecture, VGG19 won the ImageNet challenge in 2014, but the 16 layer architecture, VGG16 achieved an accuracy which was very close to VGG19. Both the models are simple and sequential. The 3×3 convolution filters are used in the VGG models which is the smallest size and thus captures local features. The 1×1 convolutions can be viewed as linear transformations and can also be used for dimensionality reduction. We choose the VGG16 over the VGG19 because it takes less time to train and the classification task in hand is not as complex as ImageNet challenge. Both the models have the same number of fully connected layers, i.e. 3, but differ in the number of 3×3 filters.

VGG16 (Modified): In this work, we also test a modified version of VGG16 which consists of 4 fully connected layers, fine-tuned on the training data, which was able to increase the accuracy of classification. We also tested with more fully connected layers but the increase in accuracy was overshadowed by the increase in training time. The Figs. 1(a) and (b) show the original and modified VGG16 architectures respectively.

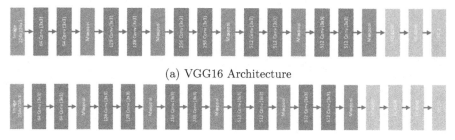

(a) VGG16 Architecture

(b) Modified VGG16 Architecture

Fig. 1. (a) VGG16 Architecture, (b) Modified VGG16 Architecture

Resnet50: After the success of the VGG architectures, it was established that deeper models outperform shallower networks. But, the problem with making models deeper was the difficulty in training them because model complexity increases as the number of layers increase. This issue was addressed by Microsoft Research, who proposed extremely deep architectures but with lower complexity [7]. They introduced a new framework of learning to ease training of such deep networks. This is called Residual learning and hence the models that employed this framework are called Residual Networks. Residual Learning involves learning residual functions. If a few stacked layers can approximate a complex function, $F(x)$ where, x is the input to the first layer, then they can also approximate the residual function $F(x) - x$. So, instead the stacked layers approximate the residual function $G(x) = F(x) - x$, where the original function becomes $G(x) + x$. Even though both can capable of approximating the desired function, the ease of training with residual functions is better. These residual functions are forwarded across layers in the network using identity mapping shortcut connections. The ImageNet 2015 results show that Resnet achieves the lowest error rates in image classification. The Resnet architectures consist of networks of various depths: 18-layers, 34-layers, 50-layers, 101-layers and 152-layers. We choose the architecture with intermediate depth, i.e. 50 layers. The Resnet consists of 3×3 and 1×1 filters, pooling layers and residual connections and a single softmax layer at the end.

Resnet50 (Modified): We also test a modified Resnet model by adding a fully connected layer fine-tuned on the training data, which increase accuracy further. We did not add any more fully connected layers since the model is already quite deep and takes a long time to train. The Figs. 2(a) and (b) show the original and modified Resnet50 architectures respectively.

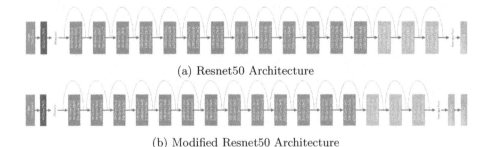

(a) Resnet50 Architecture

(b) Modified Resnet50 Architecture

Fig. 2. (a) Resnet50 Architecture, (b) Modified Resnet50 Architecture

4 Experiments

We conducted our experiments to compare training and testing accuracies and execution times of the VGG16 and Resnet50 models including modifications.

We also trained a simple CNN which is used in [5] and compare with much deeper models to show why deeper and more complex models are necessary for fire detection on our dataset. We also train the modified VGG16 and Resnet50 models and compare the performance. We used pre-trained Keras [3] models and fine-tuned the fully-connected layers on our dataset. The training of the models was done on the following hardware specifications: Intel i5 2.5 GHz, 8 GB RAM and Nvidia Geforce GTX 820 2 GB GPU. Each model was trained on the dataset for 10 training epochs with the ADAM optimizer [9] with default parameters $\alpha = 0.001$, $\beta_1 = 0.9$, $\beta_2 = 0.999$ and $\epsilon = 10^{-8}$. The details of the dataset are given in the next subsection.

4.1 The Dataset

Since there is no benchmark dataset for fire detection in images, we created our own dataset by handpicking images from the internet. [1]This dataset consists of 651 images which is quite small in size but it enables us to test the generalization capabilities and the effectiveness and efficiency of models to extract relevant features from images when training data is scarce. The dataset is divided into training and testing sets. The training set consists of 549 images: 59 fire images and 490 non-fire images. The imbalance is delibrate to replicate real world situations, as the probability of occurrence of fire hazards is quite small. The datasets used in previous papers have been balanced which does not imitate the real world environment. The testing set contains 102 images: 51 images each of fire and non-fire classes. As the training set is highly unbalanced and the testing set is exactly balanced, it makes a good test to see whether the models are able to generalize well or not. For a model with good accuracy, it must be able to extract the distinguishing features from the small amount of fire images. To extract such features from small amount of data the model must be deep enough. A poor model would just label all images as non-fire, which is the case shown in the results.

Apart from being unbalanced, there are a few images that are very hard to classify. The dataset contains images from all scenarios like fire in a house, room, office, forest fire, with different illumination intensity and different shades of red, yellow and orange, small and big fires, fire at night, fire in the morning; non-fire images contain a few images that are hard to distinguish from fire images like a bright red room with high illumination, sunset, red coloured houses and vehicles, bright lights with different shades of yellow and red etc.

The Figs. 3(a) to (f) show the fire images with different environments: indoor, outdoor, daytime, nighttime, forest fire, big and small fire. And the Figs. 4(a) to (f) show the non-fire images that are difficult to classify. Considering these characteristics of our dataset, detecting fire can be a difficult task. We have made the dataset available online so that it can be used for future research in this area.

---

[1] The dataset is available here: https://github.com/UIA-CAIR/Fire-Detection-Image-Dataset.

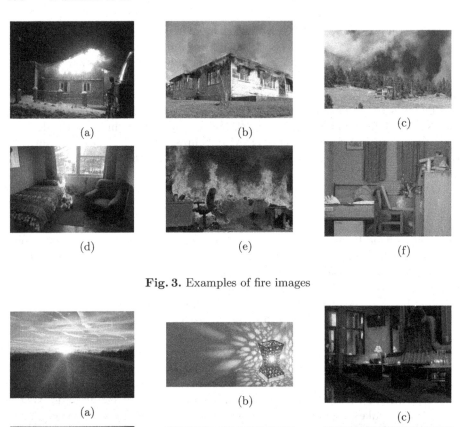

Fig. 3. Examples of fire images

Fig. 4. Examples of non-fire images that are difficult to classify

4.2   Results

Table 1 shows the results of our experiments. The simple CNN model labels all images as non-fire which means that it is unable to extract relevant features from the dataset and cannot handle unbalanced datasets, which we can see from the training accuracy which is exactly equal to the percentage of non-fire images in the training set. So, the simple CNN overfits on the majority class of the unbalanced training dataset. Since, the training and fine-tuning methods for all models used here are the same, at the end it comes down to the architecture

Table 1. Comparison between CNN models

Model	Training accuracy	Training time (in sec)	Testing accuracy	Testing time (in sec)
VGG16	100	7149	90.19	121
VGG16 (modified)	100	7320	91.18	122
Resnet50	100	15995	91.18	105
Resnet50 (modified)	100	16098	92.15	107
Simple CNN [5]	89.25	112	50.00	2

of the model. This justifies the use of deeper models like VGG16 and Resnet50. The simple CNN tested on our dataset is similar to the one used in [5]. The deep models achieve testing accuracy greater than 90%. And, the modified VGG16 and Resnet50 models outperform the base models by a small margin with slightly higher training time. It seems obvious that adding fully-connected layers to a network would increase accuracy. But on such a small dataset, the trade-off between accuracy and training time is quite poor, so we stop after adding just one fully connected layer. We also tested for more fully-connected layers(which is feasible since the model is pre-trained) but the increase in accuracy compared to increase in training time was too small.

Overall, the deep models perform well on this dataset. This shows that these models generalize well even when there is lack of training data. This means that if we want to slightly alter what the model does, we do not require large amount of data for retraining.

5   Conclusion

In this paper, we have proposed to use two state-of-the-art Deep Convolutional Neural Networks for fire detection in images, VGG16 and Resnet50. We test these models on our dataset which is made specifically to replicate real world environment. The dataset includes images that are difficult to classify and is highly unbalanced by including less fire images and more non-fire images since fire is a rare occurrence in the real world. We rationalize the use of such deep and complex models by showing that a simple CNN performs poorly on our dataset. To further increase accuracy, we added fully connected layers to both VGG16 and Resnet50. Results show that adding fully connected layers does improve the accuracy of the detector but also increases its training time. In practice, increasing the number of fully connected layers by more than one results in minute increase in accuracy compared to the large increase in training time, even if the models are pre-trained. To conclude, we found that deep CNNs provide good performance on a diverse and highly imbalanced dataset of small size, with Resnet50 slightly outperforming VGG16 and adding fully connected layers slightly improves accuracy but takes longer to train.

References

1. Bradski, G.: Opencv. Dr. Dobb's Journal of Software Tools (2000)
2. Chino, D.Y.T., Avalhais, L.P.S., Rodrigues Jr., J.F., Traina, A.J.M.: Bowfire: detection of fire in still images by integrating pixel color and texture analysis. CoRR, abs/1506.03495 (2015)
3. Chollet, F.: Keras (2015)
4. Zhao, J., Zhang, Z., Qu, C., Ke, Y., Zhang, D., Han, S., Chen, X.: Image based forest fire detection using dynamic characteristics with artificial neural networks. In: 2009 International Joint Conference on Artificial Intelligence, pp. 290–293, April 2009
5. Frizzi, S., Kaabi, R., Bouchouicha, M., Ginoux, J.M., Moreau, E., Fnaiech, F.: Convolutional neural network for video fire and smoke detection. In: IECON 2016–42nd Annual Conference of the IEEE Industrial Electronics Society, pp. 877–882, October 2016
6. Fukushima, K.: Neocognitron: a self-organizing neural network model for a mechanism of pattern recognition unaffected by shift in position. Biol. Cybern. **36**(4), 193–202 (1980)
7. He, K., Zhang, X., Ren, S., Sun, J.: Deep residual learning for image recognition. In: The IEEE Conference on Computer Vision and Pattern Recognition (CVPR), June 2016
8. Horng, W.-B., Peng, J.-W.: Image-based fire detection using neural networks. In: JCIS (2006)
9. Kingma, D.P., Ba, J.: Adam: a method for stochastic optimization. CoRR, abs/1412.6980 (2014)
10. Krizhevsky, A., Sutskever, I., Hinton, G.E.: ImageNet classification with deep convolutional neural networks. In: Pereira, F., Burges, C.J.C., Bottou, L., Weinberger, K.Q. (eds.) Advances in Neural Information Processing Systems, vol. 25, pp. 1097–1105. Curran Associates Inc. (2012)
11. Lecun, Y., Bottou, L., Bengio, Y., Haffner, P.: Gradient-based learning applied to document recognition. Proc. IEEE **86**(11), 2278–2324 (1998)
12. Tomas Polednik, Bc.: Detection of fire in images and video using cnn. Excel@FIT (2015)
13. Poobalan, K., Liew, S.C.: Fire detection algorithm using image processing techniques. In: 3rd International Conference on Artificial Intelligence and Computer Science (AICS2015), Ocotober 2015
14. Custer, R.B.R.: Fire detection: The state of the art. NBS Technical Note, US Department of Commerce (1974)
15. Russakovsky, O., Deng, J., Su, H., Krause, J., Satheesh, S., Ma, S., Huang, Z., Karpathy, A., Khosla, A., Bernstein, M., Berg, A.C., Fei-Fei, L.: ImageNet large scale visual recognition challenge. Int. J. Comput. Vis. (IJCV) **115**(3), 211–252 (2015)
16. Shao, J., Wang, G., Guo, W.: An image-based fire detection method using color analysis. In: 2012 International Conference on Computer Science and Information Processing (CSIP), pp. 1008–1011, August 2012
17. Simonyan, K., Zisserman, A.: Very deep convolutional networks for large-scale image recognition. CoRR, abs/1409.1556 (2014)
18. Tao, C., Zhang, J., Wang, P.: Smoke detection based on deep convolutional neural networks. In: 2016 International Conference on Industrial Informatics - Computing Technology, Intelligent Technology, Industrial Information Integration (ICIICII), pp. 150–153, December 2016

19. Toulouse, T., Rossi, L., Celik, T., Akhloufi, M.: Automatic fire pixel detection using image processing: a comparative analysis of rule-based and machine learning-based methods. Sig. Image Video Process. **10**(4), 647–654 (2016)
20. Treyin, B.U., Dedeoglu, Y., Gkbay, U., Enisetin, A.: Computer vision based method for real-time fire and flame detection. Pattern Recogn. Lett. **27**(1), 49–58 (2006)
21. Verstockt, S., Lambert, P., Van de Walle, R., Merci, B., Sette, B.: State of the art in vision-based fire and smoke dectection. In: Luck, H., Willms, I. (eds.) International Conference on Automatic Fire Detection, 14th, Proceedings, vol. 2, pp. 285–292. University of Duisburg-Essen. Department of Communication Systems (2009)
22. Vicente, J., Guillemant, P.: An image processing technique for automatically detecting forest fire. Int. J. Thermal Sci. **41**(12), 1113–1120 (2002)
23. Zhang, Q., Xu, J., Xu, L., Guo, H.: Deep convolutional neural networks for forest fire detection, February 2016

Improving Face Pose Estimation Using Long-Term Temporal Averaging for Stochastic Optimization

Nikolaos Passalis$^{(\boxtimes)}$ and Anastasios Tefas

Department of Informatics, Aristotle University of Thessaloniki, Thessaloniki, Greece
passalis@csd.auth.gr, tefas@aiia.csd.auth.gr

Abstract. Among the most crucial components of an intelligent system capable of assisting drone-based cinematography is estimating the pose of the main actors. However, training deep CNNs towards this task is not straightforward, mainly due to the noisy nature of the data and instabilities that occur during the learning process, significantly slowing down the development of such systems. In this work we propose a temporal averaging technique that is capable of stabilizing as well as speeding up the convergence of stochastic optimization techniques for neural network training. We use two face pose estimation datasets to experimentally verify that the proposed method can improve both the convergence of training algorithms and the accuracy of pose estimation. This also reduces the risk of stopping the training process when a bad descent step was taken and the learning rate was not appropriately set, ensuring that the network will perform well at any point of the training process.

Keywords: Pose estimation · Deep convolutional neural networks · Stochastic optimization

1 Introduction

Unmanned Aerial Vehicles (UAVs), or *drones*, are capable of capturing spectacular aerial shots of various events, e.g., athletic events, concerts, etc., that would be otherwise difficult and expensive to obtain. This has led many media producers to use drones to film outdoor events instead of resorting to other more established, yet expensive means, such as hiring a helicopter and a crew to film the aerial shots. However, flying a drone in a professional filming setting requires at least two operators. The first one is responsible for controlling the flight path of the drone and avoiding possible hazards, while the second one for controlling the main shooting camera. Assisting parts of the flying and the shooting process using machine learning techniques, such as deep neural networks [11], and reinforcement learning techniques [14], has the potential to reduce the load of the human operators, increasing the quality of the captured shots and reducing the cost of the film productions.

© Springer International Publishing AG 2017
G. Boracchi et al. (Eds.): EANN 2017, CCIS 744, pp. 194–204, 2017.
DOI: 10.1007/978-3-319-65172-9_17

Among the most crucial components of an intelligent system capable of assisting the filming process is estimating the pose of the main actors. For example, consider the problem of controlling the camera during a bicycle race event. After detecting a bicycle the drone must rotate its camera towards the bicyclist. To this end, the face pose, i.e., tilt and pan, of the bicyclist must be estimated in order to calculate the appropriate shooting angle according to the specifications of each shot type. Simple computer vision methods can be used to tackle this problem [24]. However, with the advent of deep convolutional neural networks (CNNs) it was established that it is possible to train CNNs capable of performing pose estimation significantly better than the conventional pose estimation methods [23].

Note that training deep CNNs is not always a straightforward task, slowing down the development of such systems. Several methods have been proposed to stabilize and smoothen the convergence of the training procedure [5–7,9,22]. During our experiments we also observed instabilities during the training process, even though a state-of-the-art stochastic optimization technique, the Adaptive Moment Estimation method (ADAM) [9], was used. If a slightly larger learning rate than the optimal was selected the training process was unstable. On the other hand, if the learning rate was too small, the optimization process slowed down significantly. Furthermore, the unstable behavior of CNNs used for pose estimation can be also partially attributed to the noisy nature of the data. After detecting a face using an object detector, such as the YOLO detector [19], or the SSD detector [13], the bounding box of the face is cropped, resized and then fed to the pose estimation CNN. However, the object detector is usually incapable of perfectly centering and determining the bounds of the face introducing a significant amount of noise into the aforementioned process.

The main contribution of this work is the proposal of a temporal averaging technique that is capable of stabilizing as well as speeding up the convergence of stochastic optimization for neural network training. The proposed technique uses an exponential running average on the parameters of the neural network to bias the current parameters towards a stabler state. As we show in Sect. 3, this is equivalent to first taking big descent steps to explore the solution space and then annealing towards stabler states. It was experimentally verified using two face pose estimation datasets that the proposed method can improve both the convergence of the utilized training algorithms and the accuracy of pose estimation. The more stable convergence of the algorithm also reduces the risk of stopping the training process when a bad descent step was taken and the learning rate was not appropriately set, ensuring that the network will perform well at any point of the training process (after a certain number of iterations have been performed).

The rest of the paper is structured as follows. First, the related work is briefly introduced and discussed in Sect. 2. Then, the proposed method is presented in detail in Sect. 3 and the experimental evaluation is provided in Sect. 4. Finally, conclusions are drawn and future work is discussed in Sect. 5.

2 Related Work

Several methods have been proposed for training deep neural networks as well as for improving the convergence of stochastic gradient descent. For example, using batch normalization [7], rectifier activation units [5], and residual connections [6,22], allows for effectively dealing with the problem of *vanishing gradients*. Furthermore, advanced optimization techniques, such as the ADAGRAD algorithm [1], and the ADAM algorithm [9], adjust the learning rate for each parameter separately effectively dealing with gradients of different magnitudes and allowing for improving the convergence speed. Each of these techniques deal with a specific problem that arises during the training of deep neural networks. The method proposed in this paper is complementary to these methods since it addresses a different problem, i.e., improves the stability of the training procedure regardless of the selected learning rate. This is also demonstrated in Sect. 4 where the proposed method is combined with some of the aforementioned methods to improve the convergence speed and the stability of the training process.

Parameter averaging techniques were also proposed in some works to deal with the noisy stochastic updates of gradient descent [18,20], where after completing the training of the neural network the weights of the network are replaced with the average weights, as calculated during the training process. A more deliberate technique was proposed in [9], where an exponential moving average over the parameters of the network is used to ensure that higher weight is given to the recent states of the network. A similar approach is also adopted for deep reinforcement learning tasks [12]. The method proposed in this paper is different from the method proposed in [9], since the weights of the network are not averaged after each iteration. Instead, a number of descent steps are taken, e.g., 10 optimization steps, and after them the parameters of the networks are updated. This allows for better exploring the solution space and increasing the convergence speed, while maintaining the stability that the averaging process offers, as demonstrated in Sect. 4.

3 Proposed Method

In this Section the used notation is introduced and the proposed Long-Term Temporal Averaging (LT-TA) algorithm is presented in detail. Then, we examine the behavior of the LT-TA algorithm by analyzing the proposed parameter update technique.

Let θ denote the parameters of the neural network that is to be optimized towards minimizing a loss function $\mathcal{L}$. The notation θ_t is used to denote the parameters after t optimization iterations. Also, let $f(\theta, x, \eta)$ be an optimization method that provides the updates for the parameters of the neural network, where x is the training data and η the hyper-parameters of the optimization method, e.g., the learning rate. Any optimization technique can be used ranging from the simple Stochastic Gradient Descent method [4], to more advanced techniques, such as the ADAM method [9].

Algorithm 1. Long-Term Temporal Averaging Algorithm

Input: A training set of data $\mathcal{X}$, the initial parameters of the network θ_0, and an optimization method $f(\cdot)$ along with its hyper-parameteres η
Parameters: the target update rate α_∞, the update rate decay m, the exploration window N_S, and the number of iterations N
Output: The optimized parameters θ

1: **procedure** PAST ALGORITHM
2:     $\theta_{past} \leftarrow \theta_0$
3:     **for** $t \leftarrow 1; t \leq N; t++$ **do**
4:         Sample a batch x from $\mathcal{X}$
5:         $\theta_t \leftarrow f(\theta_{t-1}, x, \eta)$
6:         **if** mod $(t, N_S) = 0$ **then**
7:             $\alpha \leftarrow \alpha_\infty + (1 - \alpha_\infty)\exp(-m(t/Ns))$
8:             $\theta_{past} \leftarrow \alpha\theta_t + (1 - \alpha)\theta_{past}$
9:             $\theta_t \leftarrow \theta_{past}$
        return θ_{past}

The proposed method is shown in Algorithm 1. The proposed algorithm keeps track of a exponentially averaged version of the parameters of the network denoted by θ_{past}. These parameters are initially set to the current state of the network (line 2) and represent the stable state of the network. During the optimization (lines 3–5) the proposed algorithm performs regular optimization updates (line 5). However, every N_S iterations the current weights of the network are annealed towards the previous (past) stable state θ_{past}. As we demonstrate later in this Section this is equivalent to performing large exploration descent steps during the N_S iterations and then slowing down the learning in order to update the stable version of the network. The rate a, used for the updating the stable parameters of the network (line 8), is determined using an exponential decay strategy (line 7). During the initial iterations less weight is given to the past state of the network, since usually we start with a randomly initialized neural network. However, as the optimization progress the past weights of the network converge towards their stable state. Therefore, the update rate is decayed towards a_∞. For all the conducted experiments, a_∞ is set to 0.5, while the decay rate m is set to 10^{-3}. Thus, for the initial iterations the update rate is close to 1, while as the training procedure converges it is slowly decayed to 0.5. After the performing N iterations the algorithm returns the stable version of the optimized parameters θ_{past}.

To better understand how the proposed algorithm works consider the first N_S iterations when the simple stochastic gradient descent algorithm with learning rate η is used to provide the optimization updates:

$$\theta_1 = \theta_0 - \eta \frac{\partial \mathcal{L}}{\partial \theta_0}$$

$$\theta_2 = \theta_1 - \eta \frac{\partial \mathcal{L}}{\partial \theta_1}$$

$$\vdots \tag{1}$$

$$\theta_{N_S-1} = \theta_{N_S-2} - \eta \frac{\partial \mathcal{L}}{\partial \theta_{N_S-2}}$$

$$\theta_{N_S} = \theta_{N_S-1} - \eta \frac{\partial \mathcal{L}}{\partial \theta_{N_S-1}}$$

It is easy to see that after N_S optimization steps the weights of the network can be expressed as a weighted sum over the descent steps:

$$\theta_{N_S} = \theta_{past} - \eta \sum_{i=0}^{N_S-1} \frac{\partial \mathcal{L}}{\partial \theta_i} \tag{2}$$

since $\theta_{past} = \theta_0$. After these iterations the exponentially averaged copy of the parameters of the network is updated (line 8 of Algorithm 1) as:

$$\theta_{past} = (1-\alpha)\theta_{past} + \alpha\theta_{N_S-1} = \theta_{past} - \alpha\eta \sum_{i=0}^{N_S} \frac{\partial \mathcal{L}}{\partial \theta_i} \tag{3}$$

Therefore, updating the exponentially averaged copy of the parameters θ_{past} is equivalent to lowering the learning rate of the previous updates to $\alpha\eta$ while updating the parameters θ_{past}. However this is *not equivalent* to performing optimization with the lowered learning rate. To understand this note that the intermediate states $\theta_1, \theta_2, \ldots, \theta_{N_S}$ are calculated using the original learning rate η instead of the lowered rate $\alpha\eta$:

$$\theta_i = \theta_{i-1} - \eta \frac{\partial \mathcal{L}}{\partial \theta_{i-1}} \tag{4}$$

That is, during the N_S steps the proposed algorithm explores the solution space by taking large steps towards the descent direction, while the stable state θ_{past} is updated using a lowered learning rate. This increases the convergence speed, while ensuring that relatively large descent steps that overshoot the local minima will not significantly affect the stability of the training procedure.

4   Experiments

Two datasets were used to evaluate the proposed method: the Annotated Facial Landmarks in the Wild dataset (AFLW) [10], and the Head Pose Image Dataset (HPID) [3]. The Annotated Facial Landmarks in the Wild (AFLW) dataset [10], is a large-scale dataset for facial landmark localization. The 75% of the images were used to train the models, while the rest 25% for evaluating the accuracy of the models. The face images were cropped according to the supplied annotations and then resized to 32 × 32 pixels. Face images smaller than 16 × 16 pixels were not used for training or evaluating the model. Some examples of cropped images are shown in Fig. 1. The Head Pose Image Dataset (HPID) [3] is a smaller dataset that contains 2790 face images of 15 subjects in various poses taken in a constrained environment. Some sample images are shown in Fig. 2. Again, all images were resized to 32 × 32 pixels before feeding them to the used CNN. For both datasets the horizontal pose (pan) is to be predicted. The AFLW dataset provides continuous pose targets, while the HPID dataset provides discrete targets (13 steps).

Fig. 1. Cropped face images from the ALFW dataset

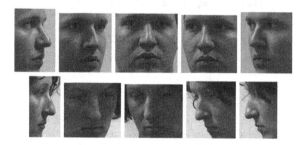

Fig. 2. Cropped face images from the HPID dataset

During the training the following data augmentation techniques were used:

1. random vertical flip with probability 0.5
2. random horizontal shift up to 5%
3. random vertical shift up to 5%
4. random zoom up to 5%
5. random rotation up to 10°

The vertical and horizontal shifts simulate the behavior of face detectors that are usually unable to perfectly align the face in the images, while the zooming and the rotation transformations increase the invariance of the network to various shooting specifications.

Deploying a deep learning model on a drone with limited processing and memory resources imposes significant restrictions on the complexity of the model. To this end, a lightweight CNN with less than 300 K parameters is used. The architecture of the used deep convolutional neural network is shown in Table 1. Local contrast normalization (LCN) is used after each of the two convolutional blocks [8], while the dropout technique is used to regularize the learning process [21]. Finally, the ADAM algorithm [9], with learning rate $\eta = 0.001$, and the default hyper-parameters ($\beta_1 = 0.9$ and $\beta_2 = 0.999$) is used for optimizing the network using mini-batches of 32 samples.

Table 1. Architecture of the used CNN

Layer type	Output shape
Input	$32 \times 32 \times 3$
Convolutional (3×3)	$30 \times 30 \times 16$
Convolutional (3×3)	$28 \times 28 \times 16$
Max pooling (2×2)	$14 \times 14 \times 16$
LCN + dropout $(p = 0.5)$	$14 \times 14 \times 16$
Convolutional (3×3)	$12 \times 12 \times 32$
Convolutional (3×3)	$10 \times 10 \times 32$
Max pooling (2×2)	$5 \times 5 \times 32$
LCN + dropout $(p = 0.5)$	$5 \times 5 \times 32$
Dense	256
Dense	1

Table 2. ALFW dataset evaluation: comparing the proposed Long-Term Temporal Averaging (LT-TA) method to the plain Temporal Averaging (TA) and the baseline learning methods. The mean angular error is reported. The train error deviation refers to the error deviation during the last 5,000 iterations.

Method	Train error	Train deviation	Test error
Baseline	7.41	0.32	8.16
TA	7.24	0.18	8.01
LT-TA	**6.98**	**0.17**	**7.72**

First, the ALFW dataset is used to evaluate the proposed method. The results are shown in Table 2. The proposed Long-Term Temporal Averaging method (abbreviated as LT-TA) outperforms both the plain Temporal Averaging (abbreviated as TA), as well as the baseline learning technique, i.e., using

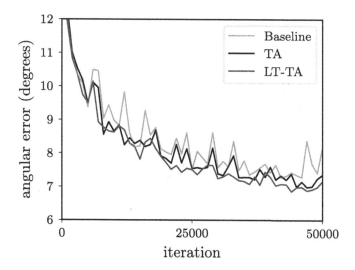

Fig. 3. ALFW dataset evaluation: comparing the mean angular error during the training process

only the ADAM algorithm without any temporal averaging. For both the TA and the LT-TA techniques the weight update parameter is exponentially decayed to 0.5 (Algorithm 1). The proposed LT-TA method reduces both the training and the testing mean angular error as well as it stabilizes the learning process by reducing the error deviation during the last training iterations (the train error deviation during the last 5,000 iterations is reported). These results are also confirmed by the learning curve depicted in Fig. 3, where the mean angular error during the training process is ploted. Note how the proposed LT-TA method stabilizes the convergence of the training process (the error spikes are significantly reduced). Using a smaller learning rate can also have similar effect, but it also slows down the convergence.

The evaluation results for the HPID dataset are reported in Table 3. Again, the proposed method reduces both the train and the test error, while reducing the instabilities of the algorithm during the training process (the deviation is reduced from 0.21 to 0.11). This is also confirmed by the learning curves shown in

Table 3. HPID dataset evaluation: comparing the proposed Long-Term Temporal Averaging (LT-TA) method to the plain Temporal Averaging (TA) and the baseline learning methods. The mean angular error is reported. The train error deviation refers to the error deviation during the last 5,000 iterations.

Method	Train error	Train deviation	Test error
Baseline	4.57	0.21	5.99
TA	4.02	0.21	5.89
LT-TA	**3.63**	**0.11**	**5.74**

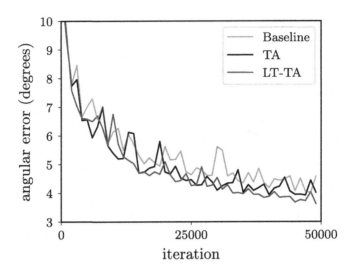

Fig. 4. HPID dataset evaluation: comparing the mean angular error during the training process

Fig. 4, where the LT-TA method reduces the fluctuations of the error (especially in the last iterations).

5   Conclusions

In this work we proposed a Long-Term Temporal Averaging technique that first takes big descent steps to explore the solution space and then uses an exponential running average on the parameters of the neural network to bias the current parameters towards stabler states. The more stable convergence of the algorithm also reduces the risk of stopping the training process when a bad descent step was taken and the learning rate was not appropriately set, ensuring that the network will perform well at any point of the training process (after a certain number of iterations have been performed). It is demonstrated, using two face image dataset for pose estimation, that the proposed technique is capable of stabilizing as well as speeding up the convergence of stochastic optimization techniques for neural network training.

There are several interesting future research directions. First, the proposed technique can be evaluated under a wider range of learning scenarios, e.g., different learning rates, network architectures, datasets, etc. This also includes several other tasks that are needed for intelligent drone-based cinematography, such as face recognition [2,16], and learning compact representations [15,17], that can be used to develop lightweight models for tasks running on-board. Furthermore, the update rate for the exponentially averaged parameters can be dynamically determined. For example, the relative change of the loss function can be used to adjust the update rate, or second order statistics can be also taken into account, similar to [9].

Acknowledgments. This project has received funding from the European Union's Horizon 2020 research and innovation programme under grant agreement No. 731667 (MULTIDRONE). This publication reflects the authors' views only. The European Commission is not responsible for any use that may be made of the information it contains.

References

1. Duchi, J., Hazan, E., Singer, Y.: Adaptive subgradient methods for online learning and stochastic optimization. J. Mach. Learn. Res. **12**(Jul), 2121–2159 (2011)
2. Goudelis, G., Tefas, A., Pitas, I.: Emerging biometric modalities: a survey. J. Multimodal User Interfaces **2**(3), 217–235 (2008)
3. Gourier, N., Hall, D., Crowley, J.L.: Estimating face orientation from robust detection of salient facial structures. In: FG NET Workshop on Visual Observation of Deictic Gestures (2004)
4. Haykin, S., Network, N.: A comprehensive foundation. Neural Netw. **2**(2004), 41 (2004)
5. He, K., Zhang, X., Ren, S., Sun, J.: Delving deep into rectifiers: surpassing human-level performance on ImageNet classification. In: Proceedings of the IEEE International Conference on Computer Vision, pp. 1026–1034 (2015)
6. He, K., Zhang, X., Ren, S., Sun, J.: Deep residual learning for image recognition. In: Proceedings of the IEEE Conference on Computer Vision and Pattern Recognition, pp. 770–778 (2016)
7. Ioffe, S., Szegedy, C.: Batch normalization: accelerating deep network training by reducing internal covariate shift. In: Proceedings of the 32nd International Conference on Machine Learning, pp. 448–456 (2015)
8. Jarrett, K., Kavukcuoglu, K., LeCun, Y., et al.: What is the best multi-stage architecture for object recognition? In: Proceedings of the IEEE International Conference on Computer Vision, pp. 2146–2153 (2009)
9. Kingma, D., Ba, J.: Adam: A method for stochastic optimization. arXiv preprint arXiv:1412.6980 (2014)
10. Koestinger, M., Wohlhart, P., Roth, P.M., Bischof, H.: Annotated facial landmarks in the wild: a large-scale, real-world database for facial landmark localization. In: First IEEE International Workshop on Benchmarking Facial Image Analysis Technologies (2011)
11. Krizhevsky, A., Sutskever, I., Hinton, G.E.: ImageNet classification with deep convolutional neural networks. In: Proceedings of the Advances in Neural Information Processing Systems, pp. 1097–1105 (2012)
12. Lillicrap, T.P., Hunt, J.J., Pritzel, A., Heess, N., Erez, T., Tassa, Y., Silver, D., Wierstra, D.: Continuous control with deep reinforcement learning. arXiv preprint arXiv:1509.02971 (2015)
13. Liu, W., Anguelov, D., Erhan, D., Szegedy, C., Reed, S., Fu, C.Y., Berg, A.C.: SSD: single shot MultiBox detector. In: Proceedings of the European Conference on Computer Vision, pp. 21–37 (2016)
14. Mnih, V., Kavukcuoglu, K., Silver, D., Rusu, A.A., Veness, J., Bellemare, M.G., Graves, A., Riedmiller, M., Fidjeland, A.K., Ostrovski, G., et al.: Human-level control through deep reinforcement learning. Nature **518**(7540), 529–533 (2015)
15. Nousi, P., Tefas, A.: Deep learning algorithms for discriminant autoencoding. Neurocomputing (2017)

16. Passalis, N., Tefas, A.: Learning neural bag-of-features for large-scale image retrieval. IEEE Trans. Syst. Man Cybern.: Syst. (2017)
17. Passalis, N., Tefas, A.: Neural bag-of-features learning. Pattern Recogn. **64**, 277–294 (2017)
18. Polyak, B.T., Juditsky, A.B.: Acceleration of stochastic approximation by averaging. SIAM J. Control Optim. **30**(4), 838–855 (1992)
19. Redmon, J., Divvala, S., Girshick, R., Farhadi, A.: You only look once: unified, real-time object detection. In: Proceedings of the IEEE Conference on Computer Vision and Pattern Recognition, pp. 779–788 (2016)
20. Ruppert, D.: Efficient estimations from a slowly convergent robbins-monro process. Cornell University Operations Research and Industrial Engineering, Technical report (1988)
21. Srivastava, N., Hinton, G.E., Krizhevsky, A., Sutskever, I., Salakhutdinov, R.: Dropout: a simple way to prevent neural networks from overfitting. J. Mach. Learn. Res. **15**(1), 1929–1958 (2014)
22. Srivastava, R.K., Greff, K., Schmidhuber, J.: Highway networks. arXiv preprint arXiv:1505.00387 (2015)
23. Toshev, A., Szegedy, C.: DeepPose: human pose estimation via deep neural networks. In: Proceedings of the IEEE Conference on Computer Vision and Pattern Recognition, pp. 1653–1660 (2014)
24. Zhu, X., Ramanan, D.: Face detection, pose estimation, and landmark localization in the wild. In: Proceedings of the IEEE Conference on Computer Vision and Pattern Recognition, pp. 2879–2886 (2012)

Discriminatively Trained Autoencoders for Fast and Accurate Face Recognition

Paraskevi Nousi$^{(\boxtimes)}$ and Anastasios Tefas

Department of Informatics, Aristotle University of Thessaloniki, Thessaloniki, Greece
`paranous@csd.auth.gr`, `tefas@aiia.csd.auth.gr`

Abstract. Accurate face recognition is vital in person identification tasks and may serve as an auxiliary tool to opportunistic video shooting using Unmanned Aerial Vehicles (UAVs). However, face recognition methods often require complex Machine Learning algorithms to be effective, making them inefficient for direct utilization in UAVs and other machines with low computational resources. In this paper, we propose a method of training Autoencoders (AEs) where the low-dimensional representation is learned in a way such that the various classes are more easily discriminated. Results on the ORL and Yale datasets indicate that the proposed AEs are capable of producing low-dimensional representations with enough discriminative ability such that the face recognition accuracy achieved by simple, lightweight classifiers surpasses even that achieved by more complex models.

Keywords: Autoencoders · Dimensionality reduction · Face recognition

1 Introduction

Face recognition is an important task in Machine Learning, and is vital in the problem of person identification as a person's facial image constitutes his/her most prominent biometric features [6]. Correctly identifying a person given a facial image may, for example, facilitate the task of automatic video capturing using Unmanned Aerial Vehicles. In such a scenario, after faces have been detected in the video stream provided by the UAV, the faces may be analyzed by a face recognition framework so as to identify persons of interest, e.g., important athletes in sports competitions. After correct identification has been established, the UAV may then use tracking algorithms to track a person, thus aiding the capture of opportunistic shots of this nature.

As face recognition is a very popular problem, many algorithms have been proposed for its solution over the past years. Autoencoders, in particular, have emerged as tools useful to this task and many variants have been proposed in literature, yielding impressive results. Typically, such methods on the one hand produce low-dimensional representations, lowering subsequent computational costs, but on the other hand, further analysis of the low-dimensional representations by computationally expensive classifiers is required, as lightweight

© Springer International Publishing AG 2017
G. Boracchi et al. (Eds.): EANN 2017, CCIS 744, pp. 205–215, 2017.
DOI: 10.1007/978-3-319-65172-9_18

classifiers may not possess enough discriminative ability to produce accurate predictions. However, deploying such methods on UAVs, which are afflicted by limited computational capabilities, is inefficient. Intuitively, if the data representation obtained by an AE contains enough discriminative ability itself, even a simple, lightweight classifier should yield significant recognition accuracy.

This is the main intuition behind this work: we seek to incorporate label information into the training process of an AE, for the purpose of dimensionality reduction in conjunction with producing discriminative features so as to facilitate even lightweight classifiers. The representations produced by such an AE should be discriminative enough, such that a very simple classifier should be able to differentiate well between samples of different classes and a more complex classifier may be trained faster and more effectively. Furthermore, the compression of the data dimensionality reduces the computational costs imposed on the classifiers.

The main contribution of this work is the proposal of a method to implicitly incorporate label information into the training process of autoencoders, by shifting the data samples in the input space in a way such that they become more easily classifiable. The proposed method is lightweight, as it doesn't require any specialized loss functions or tuples of samples for the training process, making it suitable for deployment in mobile applications, e.g., on UAVs for identification of persons of interest. Moreover, significantly improved accuracy results are achieved in various classifiers, including computationally inexpensive ones such as the Nearest Centroid classifier.

The rest of this paper is structured as follows. Section 2 provides insight into related work on the subject of discriminative dimensionality reduction approaches and highlights the advantages of the proposed method. In Sect. 3, after a brief summary of autoencoders, the proposed discriminatively trained autoencoders are introduced. Section 4 presents the experimental setup used for the evaluation of the proposed method as well as the performance of various classifiers when using the proposed methodology. Finally, our conclusions are drawn in Sect. 5.

2   Related Work

Autoencoders [12] have been widely deployed in the past to facilitate several classification tasks, including the task of face recognition [10]. As autoencoders are trained in an unsupervised fashion, research interest steered towards incorporating supervised information into their training process, so as to produce hidden representations better suited to specific tasks.

In facial expression recognition, for example, [13] proposed a method to discriminate between features that are relevant to the facial expression recognition and features that are irrelevant to this purpose. In [8] the use of pose variations at given degrees of yaw rotation of the same face was suggested for mapping the variations back to the neutral pose progressively, by manipulating the loss functions for the variations.

In [14], the label information is incorporated into an AE's training process by augmenting the loss function so as to include the classification error. In another approach, [16] employ discriminative criteria by forcing pairs of representations corresponding to the same face to be closer together in the latent Euclidean subspace than to other representations corresponding to different faces, using a triplet loss method. However, heuristically selecting such triplets is very computationally expensive. Similarly, in [4], gated autoencoders, which require pairs of samples as inputs, were deployed for the task of measuring similarity between parents and children.

Methods such as the aforementioned ones focus on either incorporating the classification error into the AE's objective, or by utilizing carefully selected tuples of samples. In contrast, the method proposed in this work does not require any complex loss functions, which may disrupt the convergence of the reconstruction error of the AE, or the selection of any specialized tuples, which imposes heavy computational costs during the training process of the AE.

3 Proposed Method

In the following sections, a summary of autoencoders is presented. Then, by exploiting supervised information, the proposed discriminative autoencoders are introduced and analyzed.

3.1 Autoencoders

Autoencoders (AEs) are neural networks which learn to map their input into a latent subspace of typically lower dimension so as to reconstruct their input through the latent (or hidden) representation [17,18]. As the input can be reconstructed given the hidden representation, the latter can be thought of as a low-dimensional representation of the input data.

The process through which the input is mapped to the latent representation is referred to as the encoding part of the AE and it may consist of several layers of neurons accompanied by non-linear activation functions. These non-linearities enable the AE to uncover more complex relations in the data, and separates autoencoders from linear dimensionality reduction algorithms.

Formally, an Autoencoder learns to map its input $\mathbf{x} \in \mathbb{R}^D$ into a hidden representation $\mathbf{y} \in \mathbb{R}^d$, using one or more layers of non-linearities:

$$\mathbf{y} = f(\mathbf{x}; \theta_{enc}) \tag{1}$$

where f denotes the encoding procedure and θ_{enc} is the set of parameters of the encoding part. The hidden representation $\mathbf{y}$ is then decoded through a similar procedure, i.e., one with a symmetrical architecture of layers, to produce the reconstruction $\hat{\mathbf{x}}$ of the input:

$$\hat{\mathbf{x}} = g(\mathbf{y}; \theta_{dec}) \tag{2}$$

where g denotes the decoding procedure and θ_{dec} is the set of parameters of the decoding part.

The parameters $\{\theta_{enc}, \theta_{dec}\}$ of the network are initialized either randomly or by using an improved initialization method [5], and updated through an error backpropagation algorithm, such as ADAM [9], so as to minimize the error between the produced reconstruction and the network's input, e.g., the mean squared error between the reconstruction and the input:

$$\ell = \|\hat{\mathbf{x}} - \mathbf{x}\|_2^2. \tag{3}$$

3.2   Discriminative Autoencoders

The latent representation produced by an AE is learned via minimizing the network's reconstruction error. Intuitively, if the target to be learned for each sample is a modified version of itself, such that it is closer to other samples of the same class, the network will learn to reconstruct samples which are more easily separable. This modification should intuitively be reflected by the intermediate representation, thus producing well-separated low-dimensional representations of the network's input data.

Let $\tilde{\mathbf{x}}_i^{(t)}$ be the target reconstruction of sample $\mathbf{x}_i$, then for $t = 0$, $\tilde{\mathbf{x}}_i^{(0)} \equiv \mathbf{x}_i$ corresponds to the standard AE targets. The *target shifting* process may be repeated multiple times, each time building on top of the previous iteration. The exponent t denotes the current iteration. The objective to be minimized for these *discriminative* autoencoders becomes:

$$\ell = \|\hat{\mathbf{x}} - \tilde{\mathbf{x}}^{(t)}\|_2^2 \tag{4}$$

The new targets may be shifted so that the samples are moved towards their class centroids, weighted by a small value α:

$$\tilde{\mathbf{x}}_i^{(t+1)} = (1 - \alpha)\tilde{\mathbf{x}}_i^{(t)} + \alpha\Big(\frac{1}{|\mathcal{C}_i|} \sum_{\tilde{\mathbf{x}}_j^{(t)} \in \mathcal{C}_i} \tilde{\mathbf{x}}_j^{(t)}\Big) \tag{5}$$

where $\mathcal{C}_i$ is set of samples belonging to the same class as the i-th sample.

Respectively, the distances between samples and centroids of rival classes could be enlarged by moving each sample away from the mean of all samples belonging to other classes:

$$\tilde{\mathbf{x}}_i^{(t+1)} = (1 + \alpha)\tilde{\mathbf{x}}_i^{(t)} - \alpha\Big(\frac{1}{N - |\mathcal{C}_i|} \sum_{\tilde{\mathbf{x}}_j^{(t)} \notin \mathcal{C}_i} \tilde{\mathbf{x}}_j^{(t)}\Big) \tag{6}$$

where N is the number of samples in the dataset. However, Eq. (6) can be modified to only include the centroids of rival classes within a given range of each sample, or only the top k nearest rivalling centroids to the i-th sample, instead of the mean of all rival class samples.

4   Experiments

The proposed methodology is evaluated on two popular face recognition datasets, through measuring the accuracy achieved by several classifiers using as input the latent representations obtained by the discriminative AEs.

4.1   Datasets

The proposed methodology is evaluated on the ORL faces dataset [15] as well as the Extended Yale B dataset [11]. For both datasets, the grayscale images depicting the faces to be recognized are resized to 32×32, meaning the original data dimension is 1024. The dimension is downscaled by a factor of 4, down to 256, by the AEs.

The ORL dataset consists of 400 pictures depicting 40 subjects under slight pose, expression and other variations. Five-fold cross-validation is commonly used for the conduction of experiments with this dataset, where five experiments are conducted using 80% of the images per person as training data and selecting a different portion of the dataset for each fold such that all images serve as training and testing data at different runs.

The cropped version of the Yale dataset is used in our experiments, which contains images depicting 38 individuals under severe lighting variations and slight pose variations. Typically, half of the images per person are selected as training data and evaluation is performed on the remaining half images. We follow the same dataset splitting methodology performed five times and average the results over all folds.

4.2   Classifiers

The performance of four different classifiers is compared for the representations obtained by a classical AE and the representations obtained by the proposed AEs as well as for the original 1024-dimensional feature vectors corresponding to the pixel intensities.

Multilayer Perceptron. A Multilayer Perceptron (MLP) [3], without hidden layers, maps its input to output neurons which correspond to the various classes describing the data. Thus the input layer has as many neurons as is the dimensionality of the input data, and the output layer has as many neurons as is the number of classes. The softmax function is typically used as the activation function in the output layer of neurons, in order to produce a probability distribution over the possible classes, which may then be used for the optimization of the network's parameters via the minimization of the categorical cross-entropy loss function.

Nearest Centroid. The Nearest Centroid (NC) classifier assigns samples to the class whose centroid (i.e., mean of samples belonging to that class) lies the closest to them in space. The dimensionality of the data heavily affects the performance of this classifier, as the distances between very high-dimensional data have been shown to be inefficient for determining neighboring samples [1].

k-Nearest Neighbors. Similarly to the NC classifier, the k-Nearest Neighbors (kNN) [2] classifier assigns samples to the class to which the majority of its k nearest neighbors belongs to. The dimensionality of the data affects the performance of this classifier as well, as it requires the computation of distances between all data samples.

Support Vector Machine. A Support Vector Machine (SVM) [7] aims to find the optimal hyperplane to separate samples belonging to different classes. The kernel method can be utilized by SVMs to map the input data into a higher-dimensional space which is more easily separable by linear hyperplanes. In our experiments, the Radial Basis Function (RBF) kernel was used for this classifier.

4.3   Experimental Results

The accuracy achieved by the above classifiers is evaluated and compared for six types of inputs (parentheses show the respective notation used in corresponding Tables and Figures):

1. the original 1024-dimensional feature vectors, corresponding to pixel intensities (No AE)
2. the 256-dimensional latent representations achieved by a standard AE (AE)
3. the 256-dimensional latent representations achieved by the proposed AEs where the targets were shifted:
   (a) towards their class centers (dAE-1)
   (b) towards their class center as well as away from the nearest rival-class center (dAE-2)

For fair comparison between the results obtained by the standard AE and the proposed AEs the same architecture, number of epochs and initialization was used. In total, the target shifting process is applied five times over the training process of the AE. As for the hyperparameter α, a value of 0.4 was used for the shift towards the class centers and a small value of 0.01 for the shift away from rival centers, to ensure the shift in the input space is smooth.

The performance achieved by the evaluated classifiers for all input representations for the ORL dataset is summarized in Table 1. The dAE-2 method yields the best improvement for all classifiers, even though using the dAE-1 method the results still surpass those achieved by using the pixel intensities and the low-dimensional representations obtained by the standard AE.

Although the performance achieved by the MLP using the pixel intensities representation is quite high, the high-dimensionality of the data imposes higher computational costs both in training and deployment. More importantly, the performance achieved by the less computationally intensive NC classifier is very close to the performance achieved by the MLP, and yields 10% and 15% improved accuracy results over the accuracy achieved when using the pixel intensities representation and the representation obtained by the standard AE respectively.

Table 1. ORL dataset accuracy results.

	MLP	NC	kNN	SVM
No AE	96.25 ± 1.58	85.25 ± 1.22	88.25 ± 5.51	90.75 ± 1.69
AE	92.75 ± 2.42	79.50 ± 1.87	82.25 ± 4.35	88.75 ± 2.50
dAE-1	96.00 ± 2.29	94.75 ± 2.29	95.25 ± 2.42	95.50 ± 1.69
dAE-2	$\mathbf{97.00 \pm 1.50}$	$\mathbf{95.50 \pm 1.87}$	$\mathbf{96.25 \pm 2.09}$	$\mathbf{96.50 \pm 1.87}$

Table 2 summarizes the accuracy achieved by all classifiers and input representations for the Yale dataset. The proposed methods outperform the baselines by a large margin. The disadvantage of the NC and kNN classifiers when data dimensionality is high becomes very clear in this dataset when the pixel intensities are used as the data representation, indicated by their extremely inaccurate predictions and low accuracy results.

Table 2. YALE dataset accuracy results.

	MLP	NC	kNN	SVM
No AE	93.52 ± 1.07	10.94 ± 1.61	54.91 ± 1.54	71.19 ± 1.40
AE	88.25 ± 1.48	63.24 ± 2.06	73.70 ± 1.10	85.07 ± 1.48
dAE-1	$\mathbf{94.40 \pm 0.78}$	$\mathbf{89.93 \pm 0.68}$	91.12 ± 1.33	94.51 ± 0.63
dAE-1	94.30 ± 0.79	89.64 ± 0.91	$\mathbf{91.43 \pm 0.88}$	$\mathbf{94.61 \pm 0.74}$

The results indicate that the proposed AEs are capable of generalizing well and implicitly applying the shifting process to unknown test samples. Figures 1 and 2 illustrate and ascertain this hypothesis. The left plot in both Figures is a 3-dimensional projection of the hidden representation learned by the standard AE, obtained via PCA. The middle plot is a 3-dimensional PCA projection of the representation learned by the dAE where the targets of the AE have been shifted five times in total towards their class centers. Finally, the plot on the right in both Figures is the 3-dimensional PCA projections of the hidden representation of the test samples, obtained by the same dAE as the middle projection. For both datasets, the 3D projections of the AE representation are difficult to unfold into separable manifolds. Through iterative repetitions of the target shifting process however, manifolds start to form and become obvious.

Furthermore, the test samples appear at positions close to their counterparts used in the training process in the 3D projection, meaning that the AE learns to map those samples closer to their manifold.

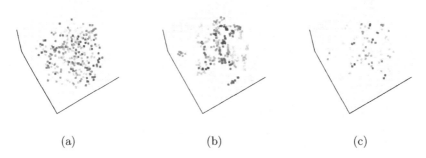

(a)   (b)   (c)

Fig. 1. ORL hidden representation 3-dimensional projection by PCA: (a) hidden representation of the training data obtained by standard AE, (b) hidden representation of the training data obtained by dAE-1, and (c) hidden representation of the test data also obtained by dAE-1.

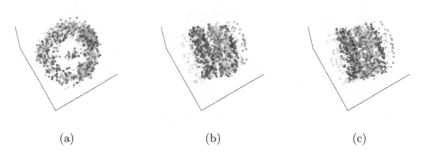

(a)   (b)   (c)

Fig. 2. YALE hidden representation 3-dimensional projection by PCA: (a) hidden representation of the training data obtained by standard AE, (b) hidden representation of the training data obtained by dAE-1, and (c) hidden representation of the test data also obtained by dAE-1.

The distribution shift that occurs to the training samples also affects the test samples belonging to the same classes. This is partly due to the fact that the testing samples follow more or less the distribution of the training samples in the input space as well. However, in the original input space as well as the subspace produced by the standard AE, the various class distributions are not well separated at all, which makes classification by lightweight classifiers very difficult. This is reflected by the extremely low accuracy results achieved by the NC and kNN classifiers especially in the Yale dataset (Table 2).

Figures 3 and 4 show samples (left) and their reconstructions (right) as given by the dAE-1 from the ORL and YALE dataset respectively. Figure 3a shows a sample from a training subset of the ORL dataset and its reconstruction. The pose variation of the input image is alleviated in the reconstruction, i.e. the face depicted is frontalized making it easier to recognize. Figure 3b shows a sample from a test subset of the ORL dataset and its reconstruction, where the facial expression of the depicted face is neutralized.

(a)   (b)

Fig. 3. Examples of ORL reconstructions: (a) input and reconstruction of a training sample, (b) input and reconstruction of a test sample.

Figure 4a shows a sample from a training subset of the YALE dataset and its reconstruction, as it is given by the dAE-1. As the sample is moved towards the centroid of its class, the illumination increases. Figure 4b shows a sample of a test subset of the YALE dataset and its reconstruction by the same AE. The network seems to have learned to generalize and is able to move the test sample towards its class centroid, removing the harsh obscurities imposed by the illumination imbalance.

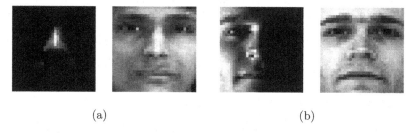

(a)   (b)

Fig. 4. Examples of YALE reconstructions: (a) input and reconstruction of a training sample, (b) input and reconstruction of a test sample.

The above reconstructions are consistent with the 3-dimensional projections shown in Figs. 1 and 2 as well as the results presented in Tables 1 and 2 for the

ORL and YALE datasets respectively: the discriminatively trained autoencoders are able to learn to map their input into a low-dimensional representation which is well-separated as well as to reconstruct a version of their input which is more informative about the depicted person's identity, by removing unrelated features such as pose, facial expression and illumination.

5   Conclusion

A method of training autoencoders, such that the low-dimensional representations of the data are more easily separable, has been proposed in this paper. The low-dimensional representation is learned in a way such that the reconstruction of the AE is a modified version of its input, which is shifted in space so that samples belonging to the same class will lie closer together and further from samples of rivalling classes. The proposed AE representations improve the performance of various classifiers, as illustrated by experimental results on two popular face recognition datasets. The classifiers' tolerance to pose, lighting and other variations is increased and they produce very accurate results while keeping the computational complexity low. Thus, the proposed AEs may be utilized in mobile environments, such as UAVs, for the task of fast and accurate person identification.

Acknowledgments. This project has received funding from the European Unions Horizon 2020 research and innovation programme under grant agreement No 731667 (MULTIDRONE). This publication reflects the authors' views only. The European Commission is not responsible for any use that may be made of the information it contains.

References

1. Aggarwal, C.C., Hinneburg, A., Keim, D.A.: On the surprising behavior of distance metrics in high dimensional space. In: Bussche, J., Vianu, V. (eds.) ICDT 2001. LNCS, vol. 1973, pp. 420–434. Springer, Heidelberg (2001). doi:10.1007/3-540-44503-X_27
2. Belhumeur, P.N., Hespanha, J.P., Kriegman, D.J.: Eigenfaces vs. fisherfaces: recognition using class specific linear projection. IEEE Trans. Pattern Anal. Mach. Intell. **19**(7), 711–720 (1997)
3. Bhuiyan, A.A., Liu, C.H.: On face recognition using gabor filters. World Acad. Sci. Eng. Technol. **28**, 51–56 (2007)
4. Dehghan, A., Ortiz, E.G., Villegas, R., Shah, M.: Who do i look like? Determining parent-offspring resemblance via gated autoencoders. In: Proceedings of the IEEE Conference on Computer Vision and Pattern Recognition, pp. 1757–1764 (2014)
5. Glorot, X., Bengio, Y.: Understanding the difficulty of training deep feedforward neural networks. In: AISTATS, vol. 9, pp. 249–256 (2010)
6. Goudelis, G., Tefas, A., Pitas, I.: Emerging biometric modalities: a survey. J. Multimodal User Interfaces **2**(3), 217–235 (2008)

7. Guo, G., Li, S.Z., Chan, K.: Face recognition by support vector machines. In: Proceedings of Fourth IEEE International Conference on Automatic Face and Gesture Recognition, pp. 196–201. IEEE (2000)

8. Kan, M., Shan, S., Chang, H., Chen, X.: Stacked progressive auto-encoders (SPAE) for face recognition across poses. In: Proceedings of the IEEE Conference on Computer Vision and Pattern Recognition, pp. 1883–1890 (2014)

9. Kingma, D., Ba, J.: Adam: a method for stochastic optimization. arXiv preprint arXiv:1412.6980 (2014)

10. Kotropoulos, C., Tefas, A., Pitas, I.: Frontal face authentication using variants of dynamic link matching based on mathematical morphology. In: Proceedings of 1998 International Conference on Image Processing, ICIP 98, vol. 1, pp. 122–126. IEEE (1998)

11. Lee, K.C., Ho, J., Kriegman, D.J.: Acquiring linear subspaces for face recognition under variable lighting. IEEE Trans. Pattern Anal. Mach. Intell. **27**(5), 684–698 (2005)

12. Nousi, P., Tefas, A.: Deep learning algorithms for discriminant autoencoding. Neurocomputing (2017)

13. Rifai, S., Bengio, Y., Courville, A., Vincent, P., Mirza, M.: Disentangling factors of variation for facial expression recognition. Comput. Vis.-ECCV **2012**, 808–822 (2012)

14. Rolfe, J.T., LeCun, Y.: Discriminative recurrent sparse auto-encoders. arXiv preprint arXiv:1301.3775 (2013)

15. Samaria, F.S., Harter, A.C.: Parameterisation of a stochastic model for human face identification. In: Proceedings of the Second IEEE Workshop on Applications of Computer Vision, pp. 138–142. IEEE (1994)

16. Schroff, F., Kalenichenko, D., Philbin, J.: Facenet: a unified embedding for face recognition and clustering. In: Proceedings of the IEEE Conference on Computer Vision and Pattern Recognition, pp. 815–823 (2015)

17. Vincent, P., Larochelle, H., Bengio, Y., Manzagol, P.A.: Extracting and composing robust features with denoising autoencoders. In: Proceedings of the 25th international conference on Machine learning, pp. 1096–1103. ACM (2008)

18. Vincent, P., Larochelle, H., Lajoie, I., Bengio, Y., Manzagol, P.A.: Stacked denoising autoencoders: learning useful representations in a deep network with a local denoising criterion. J. Mach. Learn. Res. **11**(Dec), 3371–3408 (2010)

Fish Classification in Context of Noisy Images

Adamu Ali-Gombe[(⊠)], Eyad Elyan, and Chrisina Jayne

Robert Gordon University, Aberdeen, Scotland
{a.ali-gombe,e.elyan,c.p.jayne}@rgu.ac.uk

Abstract. In this paper, we analysed the performance of deep convolutional neural networks on noisy images of fish species. Thorough experiments using four variants of noisy and challenging dataset was carried out. Different deep convolutional models were evaluated. Firstly, we trained models on noisy dataset of fishing boat images. Our second approach trained the models on a new dataset generated by annotating fish instances only from the initial set of images. Lastly, we trained the models by synthesizing more data through the application of affine transforms and random noise. Results indicate that deep convolutional network performance deteriorate in the absence of well annotated training set. This opens direction for future research in automatic image annotation.

1 Introduction

Fish detection and recognition is important for conservation agencies, marine live scientist, fishing industry and Governments to maintain fish supply and balance in the ecosystem. Increase in continental reef monitoring and deep sea surveillance has created the need for more imagery analysis. Images are generated by mounted cameras that capture continues data for marine biologist. The rate at which data is generated by underwater cameras, fishing boat cameras, automatic underwater vehicles (AUV) and conveyor belt cameras challenge human manual approach to count and sort dish species. Therefore, image based techniques are now more popular in this domain [1, 2, 20].

Because of its economic importance, a lot of approaches have been proposed in detection and classification of fishes. Researchers employ specialised software and hardware to monitor the marine eco-system. This has helped them in studying fish species behaviour [20], classifying fishes into different species [3, 12], count individual species and also track their movements [8]. To support growing needs of the research community, competitions such as Kaggle[1] and Seaclef/LifeClef[2] provides richly annotated datasets for researchers aiming at pushing the research boundaries.

However, challenges still exist in identifying fish species from these images and videos. In this domain, images obtained here are largely noisy and are affected by illumination. Furthermore, camouflage and presence of multiple

---

[1] https://www.kaggle.com.

[2] http://www.imageclef.org/lifeclef/2016/sea.

© Springer International Publishing AG 2017
G. Boracchi et al. (Eds.): EANN 2017, CCIS 744, pp. 216–226, 2017.
DOI: 10.1007/978-3-319-65172-9_19

objects in a frame affect segmentation and subsequent localization of object of interest. Hence, successful techniques rely heavily on preprocessing to achieve good results [1,2,8,10].

In this paper, we investigate the performance of state of the art convolutional neural network in context of noisy images. Our hypothesis is that deep learning based methods performances will deteriorate when lacking clean and well labelled set of images. To demonstrate this, we build an experimental framework to test using a challenging and complex set of images provided by kaggle[3]. The rest of the paper is organised as follows: Sect. 2 review related literatures in fish classification and related techniques. Section 3 outline methods employed with datasets and models used in this work. Section 4 discusses the results and evaluations in details. Section 5 contains the final remark.

2   Related Literatures

Fish classification is gradually becoming an interesting area in computer vision. Papp et al. research was motivated by the Seaclef of LifeClef competition to detect and track coral reefs from under water videos and recognize individual whales from images [15]. In recognizing individual whales, they applied segmentation after which SIFT-features (Scale Invariant Transform) and descriptors are generated. SIFT keys offer great resistance to deformation [11]. Image description was based on bag of words representation generated from Gaussian mixture model with a similarity measure calculated using RBF (Radial Basis Function). In 2010, Spampinato et al. applied texture, boundary and shape features in detecting and tracking fish species [20]. This is to assist marine biologist in sieving through massive videos from eco-grid feeds of reefs. Scientist are interested in studying fish behaviours in relation to aquatic movements. In their experiment, features were derived from grey level histograms using Gabor filters and grey level co-occurrence matrix (GLCM). Classification was carried out using discriminant analysis. To track clusters, group trajectories were build by clustering individual trajectories using I-kmeans. Li applied R-CNN (region convolutional neural network) [6] to provide real time detection of fish species [10]. He also demonstrated how segmentation could be achieved based on ROI (region of interest). Subsequently, they employed region proposal network (RPN) [16] which returns the best ROI scores at the ROI pooling layer [9]. This procedure was repeated on the same data set as in [10] and got a MAP of 82.7% which is slightly higher than the previous. But most importantly, this was at the expense of significantly larger training time. Closely related to this, is the work done by Zhang et al. in applying objectness to detect fishes from under water images [22]. Bridget et al. proposed a Haar classifier with a field programmable gate arrays framework for detecting fish species [1].

Established algorithms where also tested on noisy and real videos/images. Boudhane's experiment was based on images the authors acquired from the Baltic sea using AUV [2]. Their idea is to isolate fish from a turbid and noisy

---

[3] https://www.kaggle.com/c/the-nature-conservancy-fisheries-monitoring.

under water images. The authors made use of Poison-Gaussian theory in denoising and enhancing of image quality. They also applied posterior and log likely-hood probabilities in detecting objects in the images. The goal is to enable marine researchers to monitor underwater marine life through the AUV feed. SIFT, Viola and Jones and Kalman filters were used in detecting and tracking of fish by Ekaterina et al. [8]. Their approach was based on applying techniques on wild videos and those obtained in a controlled environment. Background subtraction technique with some algorithms they developed assist in isolating fish species. The authors reported a 73% accuracy on real videos. And they argued that not all established solutions are applicable to real images although they perform excellently on synthetic datasets. Noisy fish images were classified using SVM by Hossein et al. in [7]. They used Gaussian mix model for background subtraction and kalman filter in tracking fish species. However, they reported a significant drop in the detection accuracy of 40.1% in low quality images compared to 91.7% accuracy in high resolution images.

Similar studies were also conducted to investigate the quality of detection, classification and tracking algorithms on fish datasets. Comparison between PCA, SIFT and Viola and Jones performances in detecting and recognizing fish species from images was carried out by Matai et al. in [12]. Dataset used was privately collected and results suggested that more datasets used for training will increase the performance. Ogunlana et al. in [14] used an SVM in classifying fish species with 74.56% accuracy. However, the size of dataset used was small and some assumptions were made which might not hold in other cases. Rodriguez et al. in [17] suggested that artificial radius immune algorithm combined (ARIA) combined with PCA-features and a KNN classifier achieves better results in fish classification. Training was carried out using six species of fishes in formaldehyde. Nguyen et al. in [13] investigated a combination of GMM, kalman filters and frame-differencing in detecting and tracking fish species. Their approach shows robustness to different scenarios considered during experimentation such as speed, clarity of water and appearance than other approaches.

In this paper, we propose an experimental framework that study performance of convolutional neural networks in the context of noisy images. Results obtained shows that when images are well annotated, performances improves.

3   Dataset

In this paper we used a dataset of images provided by kaggle[4]. It contains 3777 images of fish. The fish categories include Albacore tuna, Bigeye tuna, Yellowfin tuna, Dolphin, Lampris guttatus, Sharks, other categories and images with no Fish, labelled as ALB, BET, YFT, LAG, DOL, SHARK, OTHER and NoF. It is worth pointing out that these images are extracted from video footage of fishing boat. Fish detection in these images is challenging even to the humans. Light variation in images, presence of multiple objects, pulse variation and partial

---

[4] https://www.kaggle.com/c/the-nature-conservancy-fisheries-monitoring.

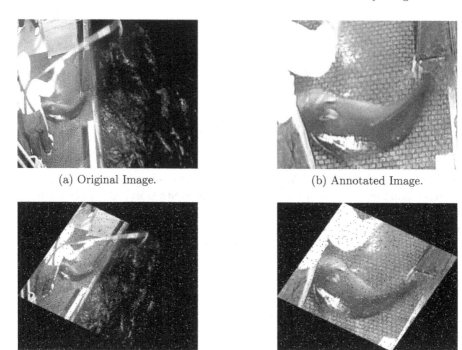

(a) Original Image.   (b) Annotated Image.

(c) Original + Noise Image.   (d) Annotated + Noise Image.

Fig. 1. Sample images

occlusion makes fish recognition very challenging. A sample image from this dataset is shown in Fig. 1a below.

A second dataset was generated from the original images by annotating all the images using an annotation tool[5]. It contains 3777 fish images. Annotation was done by isolating individual fish instances from an image using a bounding box. The bounding box was made big enough to incorporate other surrounding objects to maintain variability in the training set. Annotated instances contain a complete fish with head and tail visible or partially occluded head or tail but not both. It was observed that object view angles and light variation with shadows differs in similar images. This difference was considered visible enough to distinguish adjacent image frames as such no further cleaning was required. Figure 1b shows the result of this process.

A third dataset was created from the previous datasets. Images where generated from both the original and annotated images. The motive behind this is to address the biased nature of image distribution among classes. We also intend to achieve optimum model performance with more data. The new dataset contains 12,275 images across 8 categories. These images were synthesized by applying random noise and affine transform. Images were distorted using varying degree

---

[5] http://sloth.readthedocs.io/en/latest/.

Table 1. Dataset summary

Dataset	Number of images	Noise	Affine
Original	3777	-	-
Annotated	3777	-	-
Original+Noise	12275	X	X
Annotated+Noise	12275	X	X

of rotation angles (between 15 and 105°) and noise intensities. It is similar to the work done by Dostovistkiy et al. in [4] to generate training samples. This is to create enough distortion to generate distinct images from the originals. The result is shown in Fig. 1c and d. A summary of the datasets is shown in the following Table 1.

4   Methods

This section describes the techniques used in the study. Details of CNN architecture and model initialization are also discussed.

4.1   VGG Network

VGG networks were proposed by the Oxford visual geometry group (VGG) [19]. These networks where 11, 13, 16 and 19 layers deep also known as VGG-11, VGG-13, VGG-16 and VGG-19. These models ranked first and second place in the ImageNet classification challenge in 2014. Their models are one of the most widely used CNN models in image classification today. For the purpose of this experiment, we considered an untrained VGG-16 network. The model contains 5 blocks of 13 convolution layers and 3 fully connected layers. It makes use of a filter size of 3 for all convolution layers. It also employs a max-pooling layer between successive convolution blocks with a unit stride for down sampling. The 3 fully connected layers contains 4096, 4096 and 1000 ReLu activated units respectively (see [19] for details). The VGG-16 network architecture is illustrated below (Fig. 2).

Given that the dataset used has only 8 categories, the final layer was replaced with an 8 way soft-max classifier to suit experiment.

4.2   Transfer Learning

The second model used was proposed by applying transfer learning to the VGG-16 model from a pre-trained network on ImageNet dataset. Transfer learning attempts to reproduce similar results from experience on previous task. It enables a new model to inherit learned parameters from another trained model. This has proved to be effective where training images are scarce [5,21]. The intuition

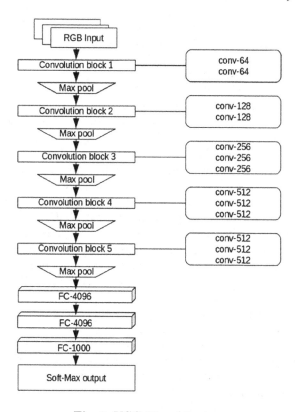

Fig. 2. VGG-16 architecture

behind this is to have a model that have already converged for comparison purposes. Model architecture is exactly the same as the one in Sect. 4.1 but its weights and biases were initialized from learned parameter from training on ImageNet dataset. The motivation behind using transfer learning is that given the size of the dataset, we try to fine tune the network as against learning new features from scratch with the hope that better results could be achieved.

5   Experiments and Results

Both Experiments where ran on NVIDIA DGX-1 machine[6]. Full advantage of the multiple GPU system was taken and this significantly reduced training and test time. Models were implemented using keras[7] with tensorflow[8] back end. Before training was initiated, all images were resized to 224 by 224. This is to accommodate them in the VGG-16 model. Each model was trained using all datasets

---

[6] http://www.nvidia.com/object/deep-learning-system.html.

[7] https://keras.io/.

[8] https://www.tensorflow.org/.

described above. At the beginning of training, images were shuffled, then split into test and train with 75% of data used for training and the remaining 25% of data for testing. Experiment on VGG-16 was carried out using a learning rate of 10^{-2} over 16 epochs and training was done using stochastic gradient descend with a batch size of 32. A weight decay was chosen as a ratio of learning rate to number of epochs and a momentum of 0.9 was maintained. The settings were to ensure faster convergence of models. Dataset normalization was applied by simply dividing each pixel by 255 for both training and test set where as the original VGG-16 experiment normalize by subtracting the mean pixel value from each pixel. This does not affect model accuracy but training time. We also differ in the choice of weight decay because subsequent experiments revealed that a dynamic weight decay works better than a statically chosen one. Training batch size was significantly lower than the one proposed in VGG-16 because the problem has significantly smaller dataset. Moreover, with smaller batch size shorter gradient updates can be realized. Apart from resizing, no further preprocessing was applied. During testing, we did not employ random crop or other methods as in [19], images are resized and the network is allowed to freely process the images.

Initial training settings for VGG-16 model were maintained for transfer learning as well. However, all the layers of the pre-trained model were fine tuned. No layer was fixed, hence the model was allowed the freedom to update parameter values for better performance similar to the methodology in [18].

Log loss and accuracy metrics were used to evaluate the models. Multi-class logarithmic loss is shown in Eq. (1) below;

$$logloss = -\frac{1}{N} \sum_{i=1}^{N} \sum_{j=1}^{M} y_{ij} log(p_{ij}) \tag{1}$$

Where N and M represents the size of the sample and categories respectively, y_{ij} is the correct prediction of sample i being in category j, and p_{ij} is the estimated probability that the sample i belongs to the class j. Logarithm loss penalizes the accuracy of the classifiers on false positives. Probabilities where obtained as predictions from the soft-max layer in the networks. Table 2 below shows the log loss summary of the experiments conducted.

The accuracy of a classifier is the ratio of number of correct prediction from sample to the total number of samples to be predicted. Accuracy is represented as follows;

$$accuracy = \frac{number \quad of \quad correct \quad predictions}{total \quad number \quad of \quad all \quad cases \quad to \quad be \quad predicted} * 100 \tag{2}$$

Table 2. Models log loss

Model	Original	Annotated	Original+Noise	Annotated+Noise
VGG-16	0.54	1.20	0.12	0.38
VGG-16 (transfer)	18.45	27.88	0.10	30.61

Table 3 shows test accuracy of models.

Table 3. Test accuracy of models

Model	Original	Annotated	Original+Noise	Annotated+Noise
VGG-16	97.20%	90.17%	99.38%	98.00%
VGG-16 (transfer)	86.60%	79.81%	99.54%	77.80%

High accuracy of models was observed during testing on original dataset. This could be attributed partly to the fact that images were obtained from fishing boat cameras. In a still camera with 24 frames per second set up, not much difference exists between adjacent frames. Although the object view angles and illumination may vary. Again, closely looking at the feature maps from the network layers revealed that prominent background objects also contributed to this. Learning was tuned towards these objects as against the fish instance. This can be seen clearly in the cross section of feature maps from the first and fourth convolutional layers in the figure below (Fig. 3).

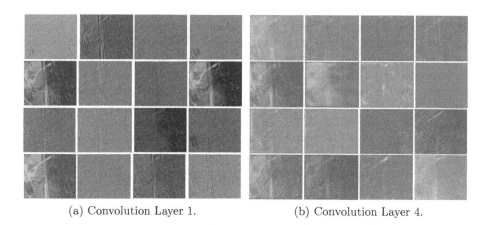

    (a) Convolution Layer 1.   (b) Convolution Layer 4.

Fig. 3. Feature maps from original image

This effect became more obvious as we go deeper into the network. When training is done on these noisy images, it over-fit on stationary objects that re-appear in images. This adds to the high accuracies recorded. However, these effects were minimal in the experiment with annotated dataset. Fish instances dominate images and this suggest that learning is based on object of interest. Fish parts are visible through the feature maps even as we go deeper into the network. A cross section of feature maps from the first and fourth convolutional layers is shown in the figure below (Fig. 4).

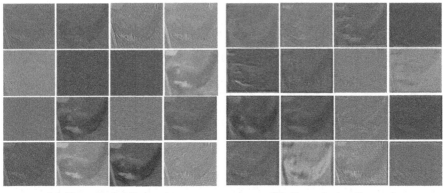

(a) Convolution Layer 1.   (b) Convolution Layer 4.

Fig. 4. Feature maps from annotated image

Unbalanced nature of image distribution among classes is also associated with weird model results. This issue was addressed when generating more data for the experiment. Large margin between classes was checked to reduce the variance. Test results shows higher recall when more training data is available. Initial experiments with annotated data performed poorly than the original dataset but we observed increase in performance when more training examples become available. A summary of sensitivity analysis of VGG-16 model (untrained) on the datasets is shown in Table 4. Experiments on the new dataset showed significant increase in accuracy by both models but did not reduce the effects observed. However, transfer learning model log loss was far worse than expected. This could be associated with its strong confidence in false classifications. Another reason could be the variation between images used and the ImageNet images. Transfer learning works best when the two datasets are closely similar.

Table 4. Summary of VGG-16 model performance

Dataset	Precision	Recall	F1-score
Original	0.82	0.82	0.81
Annotated	0.43	0.54	0.47
Original+Noise	0.92	0.91	0.91
Annotated+Noise	0.96	0.96	0.96

6   Conclusion

Deep convolutional neural network performances in context of noisy images was studied. Results shows that in noisy images, the network learns general features that are also common to all objects in the images. Features from these noisy prominent objects becomes more dominant as we go deeper into the network. As such, they prevent the network from learning specific fish features required for category classification. With well annotated images, the network learns deep features that are category specific and for correct classification. Transfer learning is an emerging area in CNN that has established its presence in recent literatures and has shown stringent results in recent times. But in this study, transfer learning from a pre-trained model on ImageNet was not effective. Learned features are transferable but a closely related dataset could have produced better results. These Results further solidifies that optimum performances are obtained when careful annotation of images is carried out. Manually annotating a massive dataset is challenging and automatic annotation require other techniques such as segmentation and objectness approaches to achieve good results. This opens new research direction in the area of image labelling and data annotation.

References

1. Benson, B., Cho, J., Goshorn, D., Kastner, R.: Field programmable gate array (FPGA) based fish detection using Haar classifiers. Am. Acad. Underwater Sci. (2009)
2. Boudhane, M., Nsiri, B.: Underwater image processing method for fish localization and detection in submarine environment. J. Vis. Commun. Image Represent. **39**, 226–238 (2016)
3. Demertzis, K., Iliadis, L.: Detecting invasive species with a bio-inspired semi-supervised neurocomputing approach: the case of lagocephalus sceleratus. Neural Comput. Appl. 1–10
4. Dosovitskiy, A., Springenberg, J.T., Riedmiller, M., Brox, T.: Discriminative unsupervised feature learning with convolutional neural networks. In: Advances in Neural Information Processing Systems, pp. 766–774 (2014)
5. Erhan, D., Szegedy, C., Toshev, A., Anguelov, D.: Scalable object detection using deep neural networks. In: Proceedings of the IEEE Conference on Computer Vision and Pattern Recognition, pp. 2147–2154 (2014)
6. Girshick, R.: Fast R-CNN. In: Proceedings of the IEEE International Conference on Computer Vision, pp. 1440–1448 (2015)
7. Hossain, E., Alam, S.S., Ali, A.A., Amin, M.A.: Fish activity tracking and species identification in underwater video. In: 2016 5th International Conference on Informatics, Electronics and Vision (ICIEV), pp. 62–66. IEEE (2016)
8. Lantsova, E., Voitiuk, T., Zudilova, T., Kaarna, A.: Using low-quality video sequences for fish detection and tracking. In: SAI Computing Conference (SAI), pp. 426–433. IEEE (2016)
9. Li, X., Shang, M., Hao, J., Yang, Z.: Accelerating fish detection and recognition by sharing CNNs with objectness learning. In: OCEANS 2016-Shanghai, pp. 1–5. IEEE (2016)
10. Li, X., Shang, M., Qin, H., Chen, L.: Fast accurate fish detection and recognition of underwater images with fast R-CNN. In: OCEANS 2015-MTS/IEEE Washington, pp. 1–5. IEEE (2015)
11. Lowe, D.G.: Object recognition from local scale-invariant features. In: The Proceedings of the Seventh IEEE International Conference on Computer vision, vol. 2, pp. 1150–1157. IEEE (1999)

12. Matai, J., Kastner, R., Cutter Jr., G.R., Demer, D.A.: Automated techniques for detection and recognition of fishes using computer vision algorithms. In: Williams K., Rooper C., Harms, J. (eds.) NOAA Technical Memorandum NMFS-F/SPO-121, Report of the National Marine Fisheries Service Automated Image Processing Workshop, Seattle, Washington, 4–7 September 2010 (2010)
13. Nguyen, N.D., Huynh, K.N., Vo, N.N., van Pham, T.: Fish detection and movement tracking. In: 2015 International Conference on Advanced Technologies for Communications (ATC), pp. 484–489. IEEE (2015)
14. Ogunlana, S.O., Olabode, O., Oluwadare, S.A.A., Iwasokun, G.B.: Fish classification using support vector machine. Afr. J. Comput. ICT **8**(2), 75–82 (2015)
15. Papp, D., Lovas, D., Szűcs, G.: Object detection, classification, tracking and individual recognition for sea images and videos
16. Ren, S., He, K., Girshick, R., Sun, J.: Faster R-CNN: towards real-time object detection with region proposal networks. In: Advances in Neural Information Processing Systems, pp. 91–99 (2015)
17. Rodrigues, M.T., Freitas, M.H., Pádua, F.L., Gomes, R.M., Carrano, E.G.: Evaluating cluster detection algorithms and feature extraction techniques in automatic classification of fish species. Pattern Anal. Appl. **18**(4), 783–797 (2015)
18. Shin, H.C., Roth, H.R., Gao, M., Lu, L., Xu, Z., Nogues, I., Yao, J., Mollura, D., Summers, R.M.: Deep convolutional neural networks for computer-aided detection: Cnn architectures, dataset characteristics and transfer learning. IEEE Trans. Med. Imaging **35**(5), 1285–1298 (2016)
19. Simonyan, K., Zisserman, A.: Very deep convolutional networks for large-scale image recognition. arXiv preprint arXiv:1409.1556 (2014)
20. Spampinato, C., Giordano, D., Di Salvo, R., Chen-Burger, Y.H.J., Fisher, R.B., Nadarajan, G.: Automatic fish classification for underwater species behavior understanding. In: Proceedings of the First ACM International Workshop on Analysis and Retrieval of Tracked Events and Motion in Imagery Streams, pp. 45–50. ACM (2010)
21. Yosinski, J., Clune, J., Bengio, Y., Lipson, H.: How transferable are features in deep neural networks? In: Advances in Neural Information Processing Systems, pp. 3320–3328 (2014)
22. Zhang, D., Kopanas, G., Desai, C., Chai, S., Piacentino, M.: Unsupervised underwater fish detection fusing flow and objectiveness. In: 2016 IEEE Winter Applications of Computer Vision Workshops (WACVW), pp. 1–7. IEEE (2016)

Fuzzy - Neuro Fuzzy

Neuro-Fuzzy Network for Modeling the Shoreline Realignment of the Kamari Beach, Santorini, Greece

George E. Tsekouras[1(✉)], Vasilis Trygonis[2], Anastasios Rigos[1],
Antonios Chatzipavlis[2], Dimitrios Tsolakis[1],
and Adonis F. Velegrakis[2]

[1] Department of Cultural Technology and Communication,
University of the Aegean, 81100 Mytilene, Lesbos Island, Greece
gtsek@ct.aegean.gr, a.rigos@aegean.gr
[2] Department of Marine Sciences, University of the Aegean, 81100 Mytilene,
Lesbos Island, Greece
{vtrygonis,a.chatzipavlis}@marine.aegean.gr,
beachtour@aegean.gr

Abstract. In this paper, a novel multiple-layer neuro-fuzzy network is proposed to model/predict shoreline realignment at a highly touristic island beach (Kamari beach, Santorini, Greece). A specialized experimental setup was deployed to generate a set of input-output data that comprise parameters describing the beach morphology and wave conditions and the cross-shore shoreline position at 30 cross-sections of the beach extracted from coastal video imagery, respectively. The proposed network consists of three distinct modules. The first module concerns the network representation of a fuzzy model equipped with a typical inference mechanism. The second module implements a novel competitive learning network to generate initial values for the rule base antecedent parameters. These parameters are, then, used to facilitate the third module that employs particle swarm optimization to perform a stochastic search for optimal parameter estimation. The network is compared favorably to two other neural networks: a radial basis function neural network and a feedforward neural network. Regarding the effectiveness of the proposed network to model shoreline re-alignment, the RMSE found (7.2–7.7 m, depending on the number of rules/nodes), reflects the high variability of the shoreline position of the Kamari beach during the period of observations: the RMSE is of a similar order to the standard deviation (up to 8 m) of the cross-shore shoreline position. The results are encouraging and the effectiveness of the proposed network could be further improved by changes (fine-tuning) of the input variables.

Keywords: Shoreline realignment · Beach morphodynamics · Neuro-fuzzy network · Competitive learning · Particle swarm optimization

© Springer International Publishing AG 2017
G. Boracchi et al. (Eds.): EANN 2017, CCIS 744, pp. 229–241, 2017.
DOI: 10.1007/978-3-319-65172-9_20

1   Introduction

Beaches, i.e. the low-lying coasts formed on unconsolidated sediments, are generally under erosion [1–3]. These significant on their own right ecosystems are critical components of the coastal zone, as they provide very important ecosystem services, such as flood protection to the valuable coastal assets/infrastructure they front [4] as well as substantial socio-economic benefits as they form the pillar of the ever-increasing coastal tourism [5]. Therefore, the long- and short-term beach morphological evolution (morphodynamics) and/or erosion and its controls have important implications for the sustainable development of the coastal zone.

Beach morphodynamics is controlled by complex forcing-response processes that operate at various spatio-temporal scales [6]. An important determinant of the current and future beach morphological evolution is the shoreline position. This is generally characterized by high spatio-temporal variability, mainly controlled by complex interactions between the beach morphology and sediments and the incident waves. Such interactions can lead to large localized shoreline position changesas well as an overall short- and long-term shoreline retreat (beach erosion) particularly under increasing sea levels and changing storminess [7, 8]. However, the study of these changes through traditional morphodynamic modeling (e.g. [7, 9, 10]) is subject to limitations arising from the: (a) high non-linearities involved in the shoreline re-alignment phenomenon due to complex nearshore hydrodynamical, sedimentological and morphological processes, and (b) high computational costs involved.

To address the above problem, we propose to use a novel neuro-fuzzy network (NFN), designed to model shoreline realignment on the basis of a small number of environmental variables that are based on high frequency observations of the shoreline position and the nearshore hydrodynamics. NFNs attempt to offset the approximate reasoning of fuzzy systems through the learning mechanisms and connectionist structures of neural networks [11–14]. The proposed network comprises three modules. The first module is a network representation of a standard fuzzy model. The second module is a competitive learning network structure, which generates a set of initial values for the fuzzy rule base antecedent parts. These values are then used to assist to a particle swarm optimization-based learning procedure that finally carries out the network's optimal parameter estimation. The model is trained and tested at Kamari beach (Santorini), one of the most touristic beaches of the Aegean Archipelago.

The material is organized as follows. Section 2 describes the experimental setup and the data acquisition process. Section 3 presents the analytical structure of the proposed neuro-fuzzy network. The simulation study is given in Sect. 4. Finally, the paper concludes in Sect. 5.

2   Experimental Setup and Raw Data Extraction

Kamari beach is a microtidal beach, located at the island of Santorini (Fig. 1). The southern section of beach (length of about 600 m) forms on coarse sediments, which along the shoreline have mean grain sizes of 2.2–9.5 mm and in the nearshore seabed mean sizes of 0.17–1.39 mm. The beach has a SE orientation and could be exposed to energetic wave conditions.

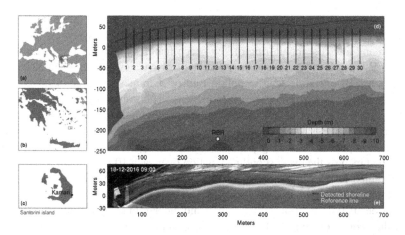

Fig. 1. (a)–(c) Location of Kamari beach in Santorini, Greece. (d) Bathymetric map of the studied beach part showing also the 30 cross-shore sections and the location of the deployed RBR pressure sensor; and (e) example of a detected shoreline plotted on the corresponding TIMEX image showing also the reference line (the blue line located above the shoreline) used to define the output parameter in the experiments. (Color figure online)

The experimental methodology consists of the collection/analysis of (a) beach topographic data and (b) time series of shoreline position and contemporaneous nearshore wave activity during a two-week period at the end of December 2016 (16-31/12/2016).

More specifically, nearshore bathymetric data were obtained through a single beam digital echo-sounder (Hi-Target HD 370) and a Differential GPS (Topcon Hipper RTK-DGPS) deployed from an inflatable boat, whereasland topographic data were obtained through a RTK-DGPS survey at the beginning of the examined period (16/12/2016). From these data, 30 cross-shore (perpendicular to the shoreline) profiles were extracted with a spacing of 20 m (Fig. 1(d)). For each of these cross-shore sections, the wet beach slope (W_S), a critical morphological variable of beach morphodynamics [7, 9], has been estimated forming the first input variable of the proposed network. A pressure sensor (RBR*virtuoso*) deployed offshore at about 9 m water depth of the beach (Fig. 1(d)) provided high frequency (4 Hz) nearshore wave data, during 10 min bursts each hour, from which hourly values of significant wave height (H_S), peak wave period (T_P), wave energy (E) and tidal elevations (T_S) were estimated. In addition, mean wave directions during these bursts were estimated from video records obtained from the coastal video system deployed in the area to monitor in high frequency the shoreline position (see below). Wave directions were expressed with regard to the North-South (denoted as W_{NS}) and East-West (denoted as W_{EW}) vector components. All the above hydrodynamic characteristics control shoreline position as they impose direct effects on beach morphodynamics [15], and form the remainder of the network's input variables. Based on the above nomenclature, the input variables taken into account are described as: $x_1 = W_S$, $x_2 = H_s$, $x_3 = T_p$, $x_4 = E$, $x_5 = T_s$, $x_6 = W_{NS}$, and $x_7 = W_{EW}$.

An autonomous coastal video monitoring system was installed at the beach (at the (0,0) point shown in Fig. 1(e)) to monitor shoreline position. The system, which allows automatic and low-cost monitoring of key beach features, comprised a station PC and two fixed Vivotek IP8362 video cameras (center of view elevation of about 22 m above mean sea level), operating at a sampling rate of 5 frames per secondand in hourly 10-minute bursts. Images were corrected for lens distortion and georectified and pro-jected on real-world coordinates using standard photogrammetric methods and ground control points (GCPs) collected during the dedicated RTK-DGPS survey. Shoreline detection was performed during post-processing, using an automated coastal feature detector that records the mean shoreline position over each 10-minute burst on the basis of the obtained TIMEX images (Fig. 1(e)); TIMEX images are the time-averages of the 3000 snapshots collected in each burst defined on the red–green–blue (RGB) colour model. A detailed description of the system and the automated procedure developed to extract the shoreline from the TIMEX imagery can be found in [10]. On the basis of this information, the network's output variable (y) (for each of the 30 cross-shore profiles and each burst) was defined as the distance between the video extracted shoreline position and a reference line (see Fig. 1(e)); this variable expresses the shoreline position change (re-alignment).

To summarize, the experimental setup generated $n = 3480$ input-output data symbolized as $\boldsymbol{x}_k = \begin{bmatrix} x_{k1} & x_{k2} & x_{k3} & x_{k4} & x_{k5} & x_{k6} & x_{k7} \end{bmatrix}^T$, and $y_k \in R$. These data are elaborated by the neuro-fuzzy network in order to model the shoreline realignment during the period of the experiment.

3   The Proposed Neuro-Fuzzy Network

The proposed network comprises three modules (see Fig. 2) namely: (a) the Takagi-Sugeno-Kang (TSK) fuzzy network, (b) the competitive learning network, and (c) the particle swarm optimization (PSO). In what follows, it is assumed that there are n input-output data pairs $\{\boldsymbol{x}_k; y_k\}|_{k=1}^{n}$ with $\boldsymbol{x}_k \in R^p$ and $y_k \in R$.

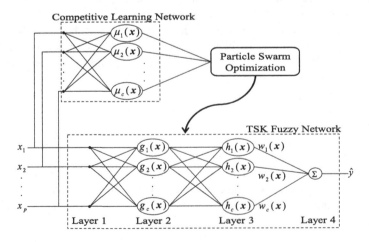

Fig. 2. The structure of the TSK neuro-fuzzy network.

3.1   TSK Fuzzy Network

The typical TSK fuzzy model establishes the input-output relations in terms of c fuzzy rules written as:

$$R_\ell : \quad \text{If } x_1 \text{ is } A_{\ell 1} \text{ and } .... \text{ and } x_p \text{ is } A_{\ell p} \text{ then } y \text{ is } w_\ell(\boldsymbol{x}), \quad (1 \le \ell \le c) \qquad (1)$$

where, $A_{\ell j}$ are Gaussian fuzzy sets:

$$A_{\ell j}(x_j) = \exp\left(-\left((x_j - v_{\ell j})/\sigma_{\ell j}\right)^2\right) \qquad (2)$$

and

$$w_\ell(\boldsymbol{x}) = a_{\ell 0} + a_{\ell 1} x_1 + \ldots + a_{\ell p} x_p \qquad (3)$$

with $v_{\ell j}$ and $\sigma_{\ell j}$ being the fuzzy set centers and widths, and $a_{\ell j} \in R$. The proposed network representation of a TSK fuzzy model is illustrated in Fig. 2.

There are four layers involved in the inference process. Layer 1 is the input layer. Layer 2 calculates the rule firing degrees as indicated next,

$$g_\ell(\boldsymbol{x}) = \prod_{j=1}^{p} A_{\ell j}(x_j) = \exp\left(-\sum_{j=1}^{p} \left((x_j - v_{\ell j})/(\sigma_{\ell j})\right)^2\right) \qquad (4)$$

In Layer 3, the fuzzy basis functions are calculated as follows:

$$h_\ell(\boldsymbol{x}) = g_\ell(\boldsymbol{x}) \Big/ \sum_{i=1}^{c} g_i(\boldsymbol{x}) \qquad (5)$$

Finally, Layer 4 estimates the output of the network:

$$\hat{y} = \sum_{\ell=1}^{c} h_\ell(\boldsymbol{x}) w_\ell(\boldsymbol{x}) \qquad (6)$$

3.2   Competitive Learning Network

The motivation of using the competitive learning network (CLN) is to facilitate the evolutionary process of the PSO algorithm. To do so, the CLN elaborates the input-output data and generates a set of parameter values for the rule antecedent parts. To provide an accurate initialization of the learning mechanism, the above values are then codified in the best position of the PSO algorithm (see Sect. 3.3). The structure of CLN (see Fig. 2) includes a hidden layer that comprises c nodes, the activation functions of which are given as,

$$\mu_\ell(\pmb{x}_k) = \left[\sum_{i=1}^c \left(\frac{\|\pmb{x}_k - \pmb{v}_\ell\|}{\|\pmb{x}_k - \pmb{v}_i\|} \right)^{\frac{2}{m-1}} \right]^{-1}, \qquad (1 \le \ell \le c; \ 1 \le k \le n) \qquad (7)$$

where $m > 1$, and $\pmb{v}_\ell = [v_{\ell 1}, v_{\ell 2}, \dots v_{\ell p}]^T \in R^p$ are the codewords obtained by the competitive learning process. In this sense the set $U = \{\pmb{v}_1, \pmb{v}_2, \dots, \pmb{v}_c\}$ is realized as the set of the center elements of the partition of the data set $X = \{\pmb{x}_1, \pmb{x}_2, \dots, \pmb{x}_n\}$ into c clusters. The CLN falls in the realm of batch learning vector quantization (BLVQ) introduced by Kohonen in [16]. The objective of the CLN is to provide an estimation of the $\pmb{v}_\ell \, (1 \le \ell \le c)$ by minimizing the following distortion function [16–18],

$$D = \frac{1}{n} \sum_{k=1}^n \min_{1 \le \ell \le c} \left\{ \|\pmb{x}_k - \pmb{v}_\ell\|^2 \right\} \qquad (8)$$

Assume that we are given a partition of the data set X into c clusters with centers (i.e. codewords) $U = \{\pmb{v}_1, \pmb{v}_2, \dots, \pmb{v}_c\}$. Then, the mean of the set U is evaluated as,

$$\tilde{v} = \frac{1}{c} \sum_{\ell=1}^c \pmb{v}_\ell \qquad (9)$$

In addition, we define the $p \times (c+1)$ matrix

$$\pmb{B} = [\pmb{b}_1 \quad \pmb{b}_2 \quad \dots \quad \pmb{b}_c \quad \pmb{b}_{c+1}] = [\pmb{v}_1 \quad \pmb{v}_2 \quad \dots \quad \pmb{v}_c \quad \tilde{v}] \qquad (10)$$

The codeword positions are directly affected by the distribution of the clusters across the feature space [17, 18]. An optimal cluster distribution should possess well separated and compact clusters. These two properties depend on the relative positions of the codewords [18, 19]. To quantify the effect the codeword relative positions impose on the quality of the partition, we view each codeword as the center of a multidimensional fuzzy set, the elements of which are the rest of the codewords. In this direction, the membership degree of the codeword $\pmb{b}_\ell = \pmb{v}_\ell$ in the cluster with center the codeword $\pmb{b}_i = \pmb{v}_i$ is defined as [18]

$$u_{i\ell} = \left[\sum_{\substack{l=1 \\ l \ne \ell}}^{c+1} \left(\frac{\|\pmb{b}_\ell - \pmb{b}_i\|}{\|\pmb{b}_\ell - \pmb{b}_l\|} \right)^{\frac{2}{m-1}} \right]^{-1}, \qquad (1 \le \ell \le c; \ 1 \le i \le c) \qquad (11)$$

where $m \in (1, \infty)$ is the fuzziness parameter. Note that the indices ℓ and i are not assigned the value $c + 1$, which corresponds to the position $\pmb{b}_{c+1} = \tilde{v}$. If the quantity $\tilde{v}$ was not taken into account, then in case there were only two codewords (i.e. $c = 2$) the function in $u_{i\ell}$ would not work properly because it would give one in all cases. Therefore the presence of $\pmb{b}_{c+1} = \tilde{v}$ is important. Next, for each $\pmb{x}_k$ the closest codeword $\pmb{v}_{i_k} = \pmb{b}_{i_k}$ is detected

$$\|x_k - v_{i_k}\| = \min_{1 \le \ell \le c} \{\|x_k - v_\ell\|\} \tag{12}$$

To this end, the codewords are updated by the subsequent learning rule,

$$v_\ell^{t+1} = v_\ell^t + \eta f_{i_k,\ell}^t \left(x_k - v_\ell^t\right) \tag{13}$$

where t is the iteration number, $\eta > 0$ is the learning rate, and $f_{i_k,\ell}$ is the neighborhood function:

$$f_{i_k,\ell} = \begin{cases} 1 \, , & if \, \ell = i_k \\ u_{i_k,\ell}, & otherwise \end{cases} \tag{14}$$

with $i_k \in \{1, 2, \ldots, c\}$, $1 \le \ell \le c$, and $u_{i_k,\ell}$ is calculated in (11). The function $f_{i_k,\ell}$ gives the relative excitation degree of each codeword having as reference point the winning codeword and the relative positions of the rest of the codewords. For a more detailed analysis of the properties of the function $f_{i_k,\ell}$ the interested reader is referred to [18]. As easily seen, the learning rule in Eq. (13) constitutes an on-line process. In a similar way to the BLVQ [16], we can produce a batch mechanism by employing the subsequent expectation measure:

$$E\left[f_{i_k,\ell}^t \left(x_k - v_\ell^t\right)\right] = 0 \quad as \, t \to \infty \tag{15}$$

The condition in (15) enables us to modify the codeword updating rule as,

$$v_\ell^{t+1} = \sum_{k=1}^n f_{i_k,\ell}^t x_k \Bigg/ \sum_{k=1}^n f_{i_k,\ell}^t \tag{16}$$

The above CLN appears to be less sensitive to initialization when compare to the BLVQ. This remark is justified by the fact that, based on the functions in (11) and (14), a specific codeword moves towards its new position considering the relative positions of the rest of codewords. Thus, before obtaining the new partition, the algorithm takes into account the overall current partition and forces all codewords to be more competitive. Therefore, in a single iteration all of the codewords are moving in an attempt to win as much as training vectors they can. This behavior enables the whole updating process to avoid undesired local minima.

Based on the nomenclature given in Eqs. (1)–(6), to extract initial values for the rule antecedents, the fuzzy set centers are determined by projecting the codewords v_ℓ $(1 \le \ell \le c)$ on each dimension:

$$v_{\ell j} = v_{\ell j}, \quad (1 \le \ell \le c; \, 1 \le j \le p) \tag{17}$$

while the corresponding fuzzy set widths by the next relation

$$\sigma_{\ell j} = (Diag(\boldsymbol{FC}_\ell))^{1/2} \tag{18}$$

where $\boldsymbol{FC}_\ell$ is the fuzzy covariance matrix [20], which based on Eq. (7) is

$$\boldsymbol{FC}_\ell = \sum_{k=1}^{n} (\mu_\ell(\boldsymbol{x}_k))^m (\boldsymbol{x}_k - \boldsymbol{v}_\ell)(\boldsymbol{x}_k - \boldsymbol{v}_\ell)^T \bigg/ \sum_{k=1}^{n} (\mu_\ell(\boldsymbol{x}_k))^m \tag{19}$$

3.3   Particle Swarm Optimization

The particle swarm optimization (PSO) elaborates on a population of N particles $\boldsymbol{p}_i \in R^q$ [21–23]. Each particle is assigned a velocity $\boldsymbol{h}_i \in R^q (1 \leq i \leq N)$. The positions with the best values of the objective function obtained so far by the ith particle and by all particles are respectively denoted as $\boldsymbol{p}_i^{best}$ and $\boldsymbol{p}_{best}$. The velocity is updated as

$$\begin{aligned} \boldsymbol{h}_i(t+1) = \ & \omega \boldsymbol{h}_i(t) + \varphi_1 \boldsymbol{U}(0, 1) \otimes \left(\boldsymbol{p}_i^{best}(t) - \boldsymbol{p}_i(t)\right) + \varphi_2 \boldsymbol{U}(0, 1) \\ & \otimes \left(\boldsymbol{p}_{best}(t) - \boldsymbol{p}_i(t)\right) \end{aligned} \tag{20}$$

where $\otimes$ is the point-wise vector multiplication; $\boldsymbol{U}(0, 1)$ is a vector with elements randomly generated in [0, 1]; ω, φ_1, and φ_2 are positive constant numbers called the inertia, cognitive and social parameter, respectively. The position of each particle is:

$$\boldsymbol{p}_i(t+1) = \boldsymbol{p}_i(t) + \boldsymbol{h}_i(t+1) \tag{21}$$

In our case, to implement the PSO, each particle codifies the antecedent parameters, i.e. the fuzzy set centers and widths. Therefore, the dimension of the particles' search space is equal to $q = 2cp$ (i.e. $\boldsymbol{p}_i \in R^q$, $i = 1, 2, \ldots, N$). All particles are randomly initialized. In addition, the values of the antecedent parameters that were calculated in Eqs. (17) and (18) are codified in the best overall position $\boldsymbol{p}_{best}$, which is expected to guide more efficiently the particles in their search for optimal parameter estimation. On the other hand, there are $c(p+1)$ consequent parameters, described as

$$\boldsymbol{a} = \left[a_{10}, a_{11}, \ldots, a_{1p}, a_{20}, a_{21}, \ldots, a_{2p}, \ldots, a_{c0}, a_{c1}, \ldots, a_{cp}\right]^T \tag{22}$$

The estimation of these parameters is carried out in terms of ridge regression [24]. To do so, we employ the following matrices:

$$\Lambda = [\Lambda_1 \quad \Lambda_2 \quad \ldots \quad \Lambda_n]^T \tag{23}$$

$$\Lambda_k = [\lambda_{k1} \quad \lambda_{k2} \quad \ldots \quad \lambda_{kc}] \quad (1 \leq k \leq n) \tag{24}$$

$$\lambda_{k\ell} = [h_\ell(\boldsymbol{x}_k) \quad h_\ell(\boldsymbol{x}_k) x_{k1} \quad \ldots \quad h_\ell(\boldsymbol{x}_k) x_{kp}] \quad (1 \leq \ell \leq c) \tag{25}$$

Note that the dimensionality of matrix Λ is $n \times c(p+1)$. The objective of ridge regression is to minimize the following error function [24]:

$$E = \|\Lambda a - Y\|^2 + \|\Gamma a\|^2 \tag{26}$$

where $Y = [y_1, y_2, \ldots, y_n]^T$, $\Gamma = \beta I$ with $\beta > 0$, and I the $c(p+1) \times c(p+1)$ identity matrix. The parameter β is called regularization parameter and, in this paper, its value is adjusted manually. The solution to the above problem reads as [24],

$$\tilde{a} = \left(\Lambda^T \Lambda + \Gamma^T \Gamma\right)^{-1} \Lambda^T Y \tag{27}$$

4 Simulation Study and Discussion

Based on the analysis described in Sect. 2, the data set includes $n = 3480$ input-output data pairs (corresponding to 30 beach cross-sections) of the form $\{x_k; y_k\}|_{k=1}^n$ with $x_k = [x_{k1} \quad x_{k2} \quad \ldots \quad x_{k7}]^T$ and $y_k \in \Re$. The data set was divided into a training set consisting of the 60% of the original data, and a testing set consisting of the remainder 40%. Parameter setting for the proposed methodology is as follows: $m = 2$, and $\beta = 20$. For the PSO we set $\varphi_1 = \varphi_2 = 2$, the parameter ω was randomly selected in $[0.5, 1]$, and the population size was $N = 20$.

For comparison, two more neural networks were designed and implemented. The first is a radial basis function neural network (RBFNN). The parameters of the basis functions were estimated in terms of the input-output fuzzy clustering algorithm developed in [20], while the connection weights were calculated by the orthogonal least squares method. The second is a feedforward neural network (FFNN), the activation functions of which read as:

$$f(x) = \tanh \frac{x}{2} \tag{28}$$

To train the FFNN we used the PSO algorithm (see previous Section) with the above parameter setting. All networks were implemented using the Matlab software.

The performance index used in the simulations was the root mean square error:

$$RMSE = \sqrt{\frac{1}{n} \sum_{k=1}^n |y_k - \hat{y}_k|^2} \tag{29}$$

For the three networks, we considered various numbers of rules/nodes, whereas for each number of rules/nodes we run 10 different initializations. The results are shown in Table 1. The proposed network appears to have the best performance compared with the other two networks. The best result for both the training and testing data sets is obtained by the proposed network for a number of rules equal to $c = 10$.

The number of rules plays an important role for the validity of the method since as it increases the RMSE decreases. The results reported in Table 1 are visualized also in Fig. 3. There are some interesting observations related to this figure. First, the superiority of the proposed method is clear, in comparison with the other two tested networks. Second, the behavior of the FFNN and the RBFNN appears to be similar, but with the latter achieving better performance in both the training and testing data. Third, in all networks, the general tendency is a decrease (as expected) of the RMSEs as the number of rules/nodes increases.

Table 1. RMSEs and the corresponding standard deviations for the three networks with respect to various numbers of rule/nodes

Method		$c = 4$	$c = 6$	$c = 8$	$c = 10$
Proposed	Train	7.540 ± 0.059	7.402 ± 0.044	7.327 ± 0.109	7.201 ± 0.099
	Test	7.626 ± 0.095	7.497 ± 0.079	7.351 ± 0.114	7.308 ± 0.115
FFNN	Train	8.003 ± 0.076	7.960 ± 0.055	7.899 ± 0.060	7.958 ± 0.108
	Test	8.097 ± 0.083	7.893 ± 0.116	7.846 ± 0.077	7.907 ± 0.102
RBFNN	Train	7.858 ± 0.004	7.751 ± 0.003	7.720 ± 0.005	7.710 ± 0.002
	Test	7.871 ± 0.005	7.727 ± 0.006	7.658 ± 0.004	7.734 ± 0.005

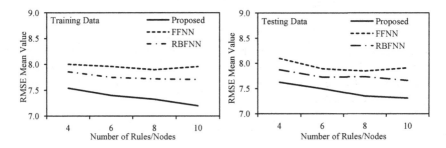

Fig. 3. Mean values of the RMSE as a function of the number of rules/nodes for the training and testing data.

With regard to the effectiveness of the proposed network to model shoreline re-alignment, the following should be noted. The RMSEs found, although smaller than those of the other tested networks is still considered high, being between 7.2–7.7 m, depending on the number of rules/nodes. However, it should be noted that the variability of the shoreline position of the Kamari beach was indeed high during the 15-day period of observations, showing, in some cross-sections, differences between the minimum and maximum cross-shore shoreline position in excess of 15 m and a standard deviation of up to 8 m. Therefore, it seems that the obtained high RMSEs reflect the high spatio-temporal variability of the Kamari shoreline position observed during the period of observations; thus, the proposed network could model/predict the shoreline re-alignment to accuracy similar to that of the actual variability of the beach.

The results suggest that in order to improve the proposed model effectiveness in the case of beaches the shorelines of which are characterised by high spatio-temporal variability, there should be some changes in the input variables. First, another input variable might be introduced that can define the 'antecedent' cross-shore position of the shoreline; a good guess should be the average shoreline position of the previous 24 h for each tested cross-section that could be easily defined from the video imagery. It is thought that the inclusion of such a parameter will improve the behavior of the proposed model. Secondly, as the tidal range at the Kamari beach was found to be rather small (<0.08 m), it may be that nearshore total water level during energetic events is controlled mostly by the storm-induced level increases. In this case, it makes sense for an input variable reflecting this process (i.e. the total nearshore water level) to be used instead of the tidal elevation (T_S). Generally, the encouraging results of the modelling exercise suggest that the proposed neural network, after a fine-tuning of the input variables, can be used to model the shoreline re-alignment of beaches characterised by highly variable morphodynamics, on the basis of a small number of environmental variables.

5   Summary and Conclusions

In the present contribution we present a systematic methodology that uses a sophisticated experimental setup and a novel neuro-fuzzy network to model and predict the phenomenon of shoreline realignment at a highly touristic beach (Kamari, Santorini). A set of morphological and wave variables were identified that can affect shoreline realignment, which together with records of hourly shoreline positions obtained using a coastal video imagery system, were used to generate the network's input-output training data. The proposed network consists of three modules. The main task of the first module is to provide the inference mechanism of a typical fuzzy system represented as network structure. The second module constitutes a competitive learning network able to provide an initial partition of the input feature space. This partition is used to extract a set of values for the fuzzy rule antecedent parameters that are codified in the overall best position of a PSO approach. Then, the PSO runs properly and optimizes the network's parameters. In each step of the iterative implementation of the PSO algorithm, the fuzzy rule consequent parameters are evaluated using ridge regression. Simulation experiments were carried and the results were compared with those by two other neural networks: a radial basis function neural network and a feedforward neural network. The comparison showed that the proposed network performs better in all cases. Regarding the effectiveness of the proposed network to model shoreline re-alignment, the RMSE found (7.2–7.7 m, depending on the number of rules/modes), reflects the high variability of the shoreline position of the Kamari beach during the period of observations: the RMSE is of a similar order to the standard deviation (up to 8 m) of the cross-shore shoreline position. The results are encouraging and the effectiveness of the proposed network could be further improved by changes (fine-tuning) of the input variables.

Acknowledgments. This research has been co-financed in 85% by the EEA GRANTS, 2009–2014, and 15% by the Public Investments Programme (PIP) of the Hellenic Republic. Project title ERABEACH: "Recording of and Technical Responses to Coastal Erosion of Touristic Aegean island beaches".

References

1. Eurosion: Living with coastal erosion in Europe: Sediment and Space for Sustainability. Part II. DG Environment, EC (2004). http://www.eurosion.org/reports-online/part2.pdf
2. Seneviratne, S.I., Nicholls, N., Easterling, D., et al.: Changes in climate extremes and their impacts on the natural physical environment. In: Field, C.B. et al. (eds.) Managing the Risks of Extreme Events and Disasters to Advance Climate Change Adaptation (Chap. 3). A Special Report of Working Groups I and II of the Intergovernmental Panel on Climate Change (IPCC), pp. 109–230. Cambridge University Press, Cambridge, UK, and New York, NY, USA (2012)
3. IPCC Climate Change: The Physical Science Basis. Contribution of Working Group I to the Fifth Assessment Report of the Intergovernmental Panel on Climate Change. In: Stocker, T. F., Qin, D., Plattner, G.-K., Tignor, M., Allen, S.K., Boschung, J., Nauels, A., Xia, Y., Bex, V., Midgley, P.M. (eds.) Cambridge University Press, Cambridge, United Kingdom and New York, NY, USA (2013)
4. Neumann, B., Vafeidis, A.T., Zimmerman, J., Nicholls, R.J.: Future coastal population growth and exposure to sea-level rise and coastal flooding - a global assessment. PLoS One (2015). doi:10.1371/journal.pone.0118571
5. Handmer, J., Honda, Y., Kundzewicz, Z.W., et al.: Changes in impacts of climate extremes: human systems and ecosystems. In: Field, C.B. et al. (eds.) Managing the Risks of Extreme Events and Disasters to Advance Climate Change Adaptation (Chap. 4). A Special Report of Working Groups I and II of the Intergovernmental Panel on Climate Change (IPCC), pp. 231–290. Cambridge University Press, Cambridge, UK, and New York, NY, USA (2012)
6. Short, A.D., Jackson, D.W.T.: Beach morphodynamics. Treatise Geomorphol. **10**, 106–129 (2013)
7. Monioudi, I.N., Velegrakis, A.F., Chatzipavlis, A.E., Rigos, A., Karambas, T., Vousdoukas, M.I., Hasiotis, T., Koukourouvli, N., Peduzzi, P., Manoutsoglou, E., Poulos, S.E., Collins, M.B.: Assessment of island beach erosion due to sea level rise: the case of the Aegean archipelago (Eastern Mediterranean). Nat. Hazards Earth Syst. Sci. **17**, 449–466 (2017)
8. Mentaschi, L., Vousdoukas, M.I., Voukouvalas, E., Dosio, A., Feyen, L.: Global changes of extreme coastal wave energy fluxes triggered by intensified teleconnection patterns. Geophys. Res. Lett. (2017). doi:10.1002/2016GL072488
9. Roelvink, D., Reniers, A., van Dongeren, A., van Thiel de Vries, J., Lescinsky, J., McCall, R.: Modeling storm impacts on beaches, dunes and barrier islands. Coastal Eng. **56**, 1133–1152 (2009)
10. Velegrakis, A.F., Trygonis, V., Chatzipavlis, A.E., Karambas, T., Vousdoukas, M.I., Ghionis, G., Monioudi, I.N., Hasiotis, T., Andreadis, O., Psarros, F.: Shoreline variability of an urban beach fronted by a beachrock reef from video imagery. Nat. Hazards **83**(1), 201–222 (2016)
11. Banakar, A., FazleAzeemb, M.: Parameter identification of TSK neuro-fuzzy models. Fuzzy Sets Syst. **179**, 62–82 (2011)

12. Chen, C., Wang, F.W.: A self-organizing neuro-fuzzy network based on first order effects sensitivity analysis. Neurocomputing **118**, 21–32 (2013)

13. Rigos, A., Tsekouras, G.E., Vousdoukas, M.I., Chatzipavlis, A., Velegrakis, A.F.: A Chebyshev polynomial radial basis function neural network for automated shoreline extraction from coastal imagery. Integr. Comput.-Aided Eng. **23**, 141–160 (2016)

14. Tsekouras, G.E., Rigos, A., Chatzipavlis, A., Velegrakis, A.: A neural-fuzzy network based on hermite polynomials to predict the coastal erosion. Commun. Comput. Inf. Sci. **517**, 195–205 (2015)

15. Thomas, T., Phillips, M.R., Williams, A.T.: A centurial record of beach rotation. J. Coastal Res. **65**, 594–599 (2013)

16. Kohonen, T.: The self-organizing map. Neurocomputing **21**, 1–6 (2003)

17. Tsolakis, D., Tsekouras, G.E., Niros, A.D., Rigos, A.: On the systematic development of fast fuzzy vector quantization for grayscale image compression. Neural Netw. **36**, 83–96 (2012)

18. Tsolakis, D.M., Tsekouras, G.E.: A fuzzy-soft competitive learning approach for grayscale image compression. In: Celebi, M., Aydin, K. (eds.) Unsupervised Learning Algorithms, pp. 385–404. Springer, Heidelberg (2016). doi:10.1007/978-3-319-24211-8_14

19. Tsolakis, D., Tsekouras, G.E., Tsimikas, J.: Fuzzy vector quantization for image compression based on competitive agglomeration and a novel codeword migration strategy. Eng. Appl. Artif. Intell. **25**, 1212–1225 (2012)

20. Tsekouras, G.E.: Fuzzy rule base simplification using multidimensional scaling and constrained optimization. Fuzzy Sets Syst. **297**, 46–72 (2016)

21. Clerc, M., Kennedy, J.: The particle swarm-explosion: stability, and convergence in a multidimensional complex space. IEEE Trans. Evol. Comput. **6**(1), 58–73 (2002)

22. Eberhart, R.C., Shi, Y.: Tracking and optimizing dynamic systems with particle swarms. In: Proceedings of the IEEE Congress on Evolutionary Computation, pp. 94–100, Seoul, Korea (2001)

23. Tsekouras, G.E.: A simple and effective algorithm for implementing particle swarm optimization in RBF network's design using input-output fuzzy clustering. Neurocomputing **108**, 36–44 (2013)

24. Tikhonov, A.N., Goncharsky A.V., Stepanov V.V., Yagola A.G.: Numerical methods for the solution of ill-posed problems. Kluwer Academic Publishers, Berlin (1995)

A Method for the Detection of the Most Suitable Fuzzy Implication for Data Applications

Panagiotis Pagouropoulos[1] , Christos D. Tzimopoulos[2],
and Basil K. Papadopoulos[1](✉)

[1] Democritus University of Thrace, Vas. Sofias 12, 67100 Xanthi, Greece
papadob@civil.duth.gr
[2] Aristotle University of Thessaloniki, 54124 Thessaloniki, Greece

Abstract. Fuzzy implications are widely used in applications where propositional logic is applicable. In cases where a variety of fuzzy implications can be used for a specific application, it is important that the optimal candidate be chosen in order valuable inference be drawn from a given set of data. This study introduces a method for detecting the most suitable fuzzy implication among others under consideration, which incorporates an algorithm for the separation of two extreme cases. According to the truth values of the corresponding fuzzy propositions the optimal implication is one of these two extremes. An example involving five such relations is used to illustrate the procedure of the method. The results obtained verify that the resulting implication is the optimal operator for inference making from the data.

Keywords: Fuzzy sets · Fuzzy implications · Fuzzy conditional propositions · Fuzzy set distance · Linguistic variables

1 Introduction

Logic is the study of reasoning. A fundamental area of logic, called propositional logic, deals with propositions, which involve logic variables and logic functions that assign a truth value to other logic variables or propositions. Logic variables are propositions of the form 's is P', where s is a subject and P designates a predicate that characterizes a property and takes a truth value. For example, the proposition '7 is a prime number' is true. Number '7' stands for a subject and 'a prime number' is a predicate that characterizes the property of being a natural number greater than one whose only positive integer factors are the number itself and number one. The most common logic functions used in propositional logic, as discussed in [20], are the negation (NOT), disjunction (OR) and conjunction (AND). Combinations of these operations form other logic functions. Such functions are the implications which are used in conditional propositions. Implications assign a truth value to two logic variables using the negation and disjunction operators. One of the two logic variables or propositions is the antecedent, while the other is the consequent. Once the truth values of an antecedent and a consequent are evaluated, the degree of truth of an implication indicates the truth or the

G. Boracchi et al. (Eds.): EANN 2017, CCIS 744, pp. 242–255, 2017.
DOI: 10.1007/978-3-319-65172-9_21

falsity that the hypothesis implies the consequent. Since implications are applicable to propositions of the form 'IF-THEN', one major issue for the structure of rule based systems is to select an implication for which valuable inference from these propositions can be made. In classical logic the truth value of a proposition is either true or false (i.e. the values 1 or 0 respectively). On the other hand, in fuzzy logic the truth value of a proposition is a matter of degree, and can be any number in the interval [0, 1]. This is what makes fuzzy concepts suitable for applications where the predicates are linguistic variables. In those cases, a predicate is considered to be a fuzzy set, the subject of the logic variable is an object of a universal crisp set and the degree of truth of the logic variable is the value of the membership function for each one of the objects in the crisp set. A fuzzy implication of a conditional fuzzy proposition maps a value in the close interval [0, 1] for the values for two linguistic variables and the degree of truth of the proposition is, in general, a transformation of the implication called qualifier or truth qualifier, which expresses the quality of truth of the proposition (i.e. True, Very true, Very-very true, False, Fairly false and so on), as discussed in [9, 23]. In this study the qualifier is considered to be the identity function (i.e. 'True'), so the degree of truth of the proposition is the value of the implication itself. The degree of truth of conditional fuzzy propositions is affected by the fuzzy implication that is involved. That said, the selection of a suitable fuzzy implication plays a pivotal role in the design of an inference system.

In the literature, fuzzy inference rules presented in [24] are used to assess the suitability of a fuzzy implication for approximate reasoning and rule based systems. Those are the generalized modus ponens (i.e. the mode that affirms), the generalized modus tollens (i.e. the mode that denies) and the generalized hypothetical syllogism. These theoretically supported guidelines suggest that a fuzzy implication is appropriate when the generalized tautologies coincide with their classical counterparts as discussed in [9, 15]. Yet, inference rules do not provide with a general solution, and can only be applied under a particular problem, as explained in [9]. A method that incorporates the inference rules, but with less strict theoretical limitations, is described in [17] and it is based on the modus ponens inference tautology. The most commonly used modus ponens formula is $(p \wedge (p \Rightarrow q)) \Rightarrow q$, where "$\wedge$" symbolizes a fuzzy disjunction and "$\Rightarrow$" a fuzzy implication IF-THEN. This inference tautology states that given two propositions (i.e. premises) p and $p \Rightarrow q$, the truth of the proposition (i.e. conclusion) q may be inferred. The resulting implication of that method is the one for which the distance between the fuzzy implication and the modus ponens is minimum. Although it can be implemented in data applications, it is still a set-theoretic alternative. A data-driven method, in which the suitability evaluation of fuzzy implicationsis based on similarity measures between fuzzy sets, is presented in [16]. The method introduced in this study is an improvement of the latter article, and it uses fuzzy distance metrics as a criterion for the detection instead. Due to the fact that the properties of the metric distance are weaker than those of the similarity measures, and since some degrees of similarity are distance-based relations that stem from distance metrics, this method can be applied in a wider variety of applications in comparison to the initial approach which incorporates the measures of similarity as a criterion for assessing the suitability of the implications.

In opposition to the above-mentioned methodologies, the method proposed in this article does not involve the restrictions of the fuzzy inference rules to evaluate the suitability of a fuzzy implication. In particular, the selection procedure relies on the dataset of an application and the expert's opinion, so the inference drawn from the resulting implication best reflects the data rather than the theoretical concept of the generalized tautologies.

The rest of the paper is structured as follows. In Sect. 2 some useful definitions and notations are presented. In Sect. 3 the methodology and procedure are described. In Sect. 4 an application of the method is provided through an illustrative example. Finally, Sect. 5 is devoted to the conclusions and future work.

2 Fuzzy Implications and Metrics

All the definitions and notations on fuzzy implications and fuzzy sets used in this paper can be found in [8, 9, 13, 18–20].

2.1 Implications

Definition 1. An implication in classical logic is a function: $l : \{0, 1\} \times \{0, 1\} \rightarrow \{0, 1\}$, which satisfies the boundary conditions:

1. $l(0, 0) = 1$
2. $l(0, 1) = 1$ (Falsity implies anything)
3. $l(1, 0) = 0$
4. $l(1, 1) = 1$

Definition 2. A fuzzy implication is a function: $R : [0, 1] \times [0, 1] \rightarrow [0, 1]$, which satisfies at least some of the following nine axioms that are not independent from one another. The axioms are listed from the weakest to the strongest axiom $\forall a, b, c \in [0, 1]$:

1. $a \leq b$ implies $R(a, x) \geq R(b, x)$
2. $a \leq b$ implies $R(x, a) \leq R(x, b)$
3. $R(0, a) = 1$ (Falsity implies anything)
4. $R(1, b) = b$
5. $R(a, a) = 1$
6. $R(a, R(b, x)) = R(b, R(a, x))$
7. $R(a, b) = 1$ iff $a \leq b$
8. $R(a, b) = R(c(b), c(a))$, where c is a fuzzy complement.
9. R is a continuous function.

If we let $a, b, c \in \{0, 1\}$, the first five axioms of fuzzy implications coincide with the boundary conditions of the implications in classical logic.

Similarly to classical logic, fuzzy implications are formed by a combination of fuzzy complements (i.e. negation), t-norms (i.e. conjunction) and t-conorms (i.e. disjunction) and it is required that this triplet satisfy the De Morgan laws.

The following implications used in this article are [4]:

1. $R_{Lukasiewicz}(a,b) = \min\{1, 1-a+b\}$
2. $R_{Reichenbach}(a,b) = 1-a+ab$
3. $R_{Early\,Zadeh}(a,b) = \max\{1-a, \min\{a,b\}\}$

In many applications, some operators often used are the fuzzy products. The most common are the Mamdani rule and the Larsen rule, as discussed in [2, 10, 12]. As mentioned in [14], when these operators are treated as implications they are called engineering implications. These implications are:

4. $R_{Mamdani}(a,b) = \min\{a,b\}$
5. $R_{Larsen}(a,b) = ab$

Since these implications are symmetric, they make no distinction between premise and conclusion, thus are suitable for applications where cause and effect are confused, as in [21]. For example, the variables "Customers' Overall Evaluation" of a product of a service and "Customers' Recommendatory Trust" that they have in that service can be considered either as an antecedent or as a consequent. More specifically, by using a symmetric implication, the degree of truth of the fuzzy proposition "p: IF Customers' Overall Evaluation is good, THEN Customers' Recommendatory Trust is high" is equal to the degree of truth of the proposition "p: IF Customers' Recommendatory Trust is high, THEN Customers' Overall Evaluation is good". In addition, these implications satisfy the boundary condition $R(0,a) = 0$ (i.e. falsity implies nothing), contrary to the third axiom of fuzzy implications (i.e. falsity implies anything). Fuzzy implications can also be treated as fuzzy sets, as explained in [9]. Let X and Y be crisp sets, $A = \{(x, A(x))|x \in X\}$ and $B = \{(y, B(y))|y \in Y\}$ two fuzzy sets on X and Y, with membership functions A and B respectively. An implication R is considered to be the fuzzy set $R = \{((x,y), R(x,y))|(x,y) \in X \times Y\}$ where $R(x,y) = R(A(x), B(y))$ is its membership function, $\forall x \in X$ and $\forall y \in Y$.

2.2   Metric Distance

Definition 3. Given a crisp set X, a metric distance d in X is a real function $d:$ $X \times X \to \mathrm{R}$ (where R is the set of real numbers) which satisfies the following conditions:

1. $d(x,y) = 0 \Leftrightarrow x = y$
2. $d(x,y) = d(y,x)$ (symmetric)
3. $d(x,y) + d(y,z) \geq d(x,z)$ (triangular inequality), where $x,y,z \in \mathrm{X}$

The pair (X, d) is called metric space.

A wide variety of metrics on fuzzy sets has been studied by [5], while in [11, 17, 22] similar metric distances between two fuzzy sets, like the Hamming, the Euclidean, and the Maximum distance are discussed. Let $X = \{x_1, x_2, \ldots, x_n\}$ be a finite crisp set and two fuzzy sets A and B on X. Some common metrics used to describe the distance between two fuzzy sets A and B are:

The Hamming distance: $d_H(A, B) = \sum_{i=1}^{n} |A(x_i) - B(x_i)|$

The Euclidean distance: $d_E(A, B) = \sqrt{\sum_{i=1}^{n} (A(x_i) - B(x_i))^2}$

The Maximum distance: $d_\infty(A, B) = \max_i |A(x_i) - B(x_i)|, \forall x_i \in X$

3 The Algorithm for the Selection of a Suitable Fuzzy Implication

In order to deduce which implication is the most appropriate one from an initial set of candidates under consideration a special fuzzy set is defined, which represents the ideal fuzzy implication for inference making from a given dataset. The membership function of this fuzzy set maps each one of the values of the data to 1. The construction of the ideal fuzzy set stems from the fact that data refer to real observations, therefore, regarding these values, the inference drawn from the conditional proposition that the premise implies the conclusion, ideally, has to be the absolute truth. As a result, the degree of truth of the fuzzy conditional proposition which corresponds to this implication for the measurement values is equal to 1 (i.e. True). In the rest of the paper this implication is symbolized with I.

Let X, Y be the crisp sets that contain the data, and A, B be fuzzy sets on X, Y. The fuzzy set I is defined as:

$$I = \{((x_i, y_i), I(x_i, y_i)) | (x_i, y_i) \in X \times Y\}, \qquad \text{where} \qquad I(x_i, y_i) = I(A(x_i), B(y_i)) = 1,$$
$\forall (x_i, y_i) \in X \times Y.$

The mechanism of the method introduced comprises five stages:

Stage 1: Fuzzification of the data (i.e. creation of the linguistic variables).
Stage 2: Decomposition of the dataset into a partition, where each subset contains the same number of elements and it is randomly populated so bias is avoided.
Stage 3: Evaluation of the membership functions of the linguistic variables created in Stage 1 with inputs the mean values of each subset of the partition.
Stage 4: Evaluation of the implications under consideration for the membership functions' values of Stage 3.
Stage 5: Calculation of the distance between each fuzzy implication and the fuzzy set I.

The most suitable fuzzy implication is the one which corresponds to the smallest or the largest distance from the fuzzy set I for each linguistic fuzzy set, according to the expert's opinion about the inference drawn from the degree of truth of the corresponding conditional fuzzy propositions of the application.

4 Illustrative Example

This section provides an example of applying the method for the identification of the most suitable fuzzy implication on a dataset of paired measurements given an initial set of implications. The dataset consists of thirty six yearly Rainfall measurements and the corresponding Overflow measurements that take place in Vogatsiko village, located in Northern Greece in the region of West Macedonia (Hellenic National Meteorological Service). The data are contained in Table 1. In this article the inference from the proposition 'p: Rainfall implies Overflow' is drawn by the evaluation of the degree of truth of

Table 1. The data set

Rainfall (mm)	Overflow (mm)
359.0000	3.5816
406.3000	13.8432
410.2000	14.8100
410.4000	14.8601
459.9000	28.7206
461.8000	29.3103
501.2000	42.4757
501.9000	42.7255
504.0000	43.4784
538.1000	56.3852
542.0000	57.9419
542.6000	58.1828
555.4000	63.4135
556.8000	63.9961
578.4000	73.2417
583.6000	75.5386
593.6000	80.0320
599.1000	82.5457
608.5000	86.8904
635.7000	100.0160
640.3000	102.3007
656.7000	110.6030
669.1000	117.0396
681.9000	123.8239
683.2000	124.5208
687.1000	126.6211
691.1000	128.7863
703.7000	135.6969
714.0000	141.4423
751.2000	162.8863
779.6000	179.9528
787.4000	184.7404
836.7000	215.9391
851.0000	225.2829
854.6000	227.6538
882.5000	246.2867

the fuzzy propositions 'Low: Low Rain implies Low Overflow', 'Medium: Medium Rain implies Medium Overflow' and 'High: High Rain implies High Overflow'.

Firstly, the three linguistic variables 'Medium Rainfall', 'Low Rainfall', and 'High Rainfall' for the variable Rainfall along with the corresponding variables 'Medium Overflow', 'Low Overflow', and 'High Overflow' for Overflow are created.

For the construction of the membership functions of the fuzzy sets 'Medium Rainfall' and 'Medium Overflow', the method of least-square curve fitting, as demonstrated in [9] has been applied, and the class of skew-normal distributions has been employed to fit the data. This parametrized class of functions is chosen as they best conform to the sample data according to the expert's opinion that the membership function values of the linguistic variables Medium Rainfall and Medium Overflow for the maximum and minimum values of Rainfall and Overflow should be the lowest and because there must be only one maximum membership value with grade 1 for these two fuzzy sets. The skew-normal distribution, as presented in [1, 3, 6] is defined by the formula:

$$f(x; a, b, c, d) = a \frac{2}{c} \varphi\left(\frac{x-b}{c}\right) \Phi\left(d\left(\frac{x-b}{c}\right)\right),$$

where b and c represent the usual location and scale parameters, d determines the skewness, a controls the height of the function, while φ and Φ denote the pdf and cdf of a standard Gaussian deviate. The two membership functions for the linguistic hedge 'Medium' are then defined by the formula:

$$m(x) = \max\{0, \min\{1, f(x; a, b, c, d)\}\},$$

So that their values belong to the close interval [0, 1].

The membership functions for the 'Medium Rainfall' and 'Medium Overflow' are:

$$m_{Medium\,Rainfall}(x) = 3.5297 \varphi\left(\frac{x-723.7342}{220.6916}\right) \cdot \Phi\left(-2.1556\left(\frac{x-723.7342}{220.6916}\right)\right) \quad (1)$$

$$m_{Medium\,Overflow}(x) = 3.2316 \varphi\left(\frac{x-3.7951}{109.3525}\right) \cdot \Phi\left(2.3429\left(\frac{x-3.7951}{109.3525}\right)\right) \quad (2)$$

The graphical representation of the membership functions (1) and (2) of Rainfall and Overflow for the hedge 'Medium' are shown in Figs. 1 and 2 respectively.

For the construction of the 'Low Rainfall' and 'Low Overflow' membership functions, the family of the sigmoidal membership functions is used, as explained in [7], under the formula:

$$f(x; a, c) = \frac{1}{1 + e^{-a(x-c)}},$$

where a controls the slope and c the center of the curve, which in this case is the mean value of Rainfall and that of Overflow. The decreasing sigmoidal membership functions are (see Figs. 3 and 4):

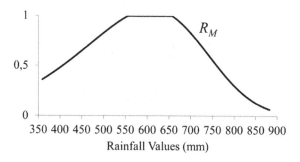

Fig. 1. Membership function of Medium Rainfall

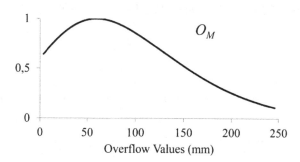

Fig. 2. Membership function of Medium Overflow

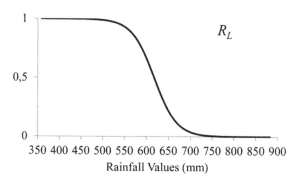

Fig. 3. Membership function of Low Rainfall

$$m_{LowRainfall}(x) = \frac{1}{1 + e^{-(-0.039)(x-617.1833)}} \tag{3}$$

$$m_{LowOverflow}(x) = \frac{1}{1 + e^{-(-0.11)(x-99.5990)}} \tag{4}$$

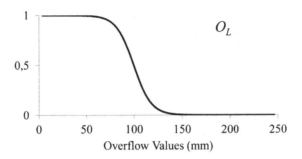

Fig. 4. Membership function of Low Overflow

Similarly, the 'High Rainfall' and 'High Overflow' membership functions follow the formula:

$$f(x; a, c) = \frac{1}{1 + e^{-a(-x+c)}}.$$

The increasing sigmoidal membership functions are (see Figs. 5 and 6):

$$m_{HighRainfall}(x) = \frac{1}{1 + e^{-(-0.038)(-x+617.1833)}} \qquad (5)$$

$$m_{HighOverflow}(x) = \frac{1}{1 + e^{-(-0.07)(-x+99.5990)}} \qquad (6)$$

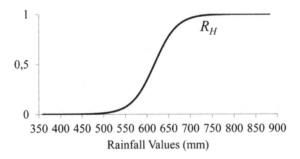

Fig. 5. Membership function of High Rainfall

Next, the partition of the data is created with nine subsets of four entries, where the objects of each subset of the partition are randomly chosen. Three of these clusters are used as the control subsets for testing the results. The arithmetic mean of each subset for Rainfall and Overflow is calculated. This measure of central tendency takes all the values of each subset into account. Then, the six membership functions are evaluated for the mean value of each subset of the partition. The next step is the evaluation of the

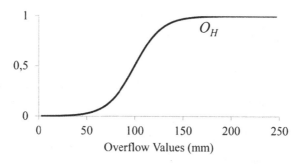

Fig. 6. Membership function of High Overflow

five implications, described in Sect. 2, using the membership functions created in the previous step. The implication values are included in Table 2. For the detection of the most suitable among the five implications of the example, the Euclidean distance between the implication and the fuzzy set I is employed as a criterion. The membership function for the fuzzy set I is:

$I(\bar{x}_i, \bar{y}_i) = I(A(\bar{x}_i), B(\bar{y}_i)) = 1, \forall i = 1, \ldots, 6$ where $\bar{x}_i, \bar{y}_i$ are the mean values of the $i - th$ subset, $A(x)$ represents one of the membership functions described by Eqs. (1), (3), and (5), and $B(y)$ one of those described by Eqs. (2), (4), and (6) respectively. The distance is:

Table 2. Values of the five implications for the fuzzy conditional propositions 'Low: Low Rainfall implies Low Overflow', 'Medium: Medium Rainfall implies Medium Overflow', 'High: High Rainfall implies High Overflow' of the subsets S1–S6 of the partition.

Subsets		R_{Larsen}	$R_{Mamdani}$	$R_{Lukasiewicz}$	R_{Zadeh}	$R_{Reichenbach}$
S1	Low	0.1814	0.3837	0.9108	0.5272	0.7086
	Med.	0.8343	0.8343	0.8343	0.8343	0.8343
	High	0.3027	0.5265	1.0000	0.5265	0.7762
S2	Low	0.6966	0.7871	1.0000	0.7871	0.9095
	Med.	0.9553	0.9553	0.9553	0.9553	0.9553
	High	0.0468	0.2144	0.9959	0.7815	0.8283
S3	Low	0.5237	0.6997	1.0000	0.6997	0.8240
	Med.	0.9159	0.9159	0.9159	0.9159	0.9159
	High	0.1016	0.3049	1.0000	0.6951	0.7967
S4	Low	0.0184	0.1241	0.9758	0.8517	0.8701
	Med.	0.7343	0.7461	0.7621	0.7462	0.7502
	High	0.6566	0.7762	0.9303	0.7762	0.8107
S5	Low	0.0899	0.2524	0.8962	0.6437	0.7337
	Med.	0.7987	0.7987	0.7987	0.7987	0.7987
	High	0.4265	0.6402	1.0000	0.6402	0.7862
S6	Low	0.0241	0.1360	0.9583	0.8224	0.8465
	Med.	0.7526	0.7526	0.7526	0.7526	0.7526
	High	0.6241	0.7644	0.9478	0.7644	0.8076

$$d(R, I) = \sqrt{\sum_{i=1}^{6} \left(R(A(\bar{x}_i), B(\bar{y}_i)) - I(A(\bar{x}_i), B(\bar{y}_i)) \right)^2} \tag{7}$$

Then, Eq. (7) is evaluated for each of the three hedges 'Low', 'Medium' and 'High' and for each one of the five implications. The distance values are included in Table 3, which shows that the Lukasiewicz implication corresponds to the smallest distance from I for all the corresponding propositions 'Low: Low Rainfall implies Low Overflow' with distance value $d(R_{Lukasiewicz}, I) = 0,1451$, 'Medium: Medium Rainfall implies Medium Overflow' having the distance value $d(R_{Lukasiewicz}, I) = 0,4414$, and 'High: High Rainfall implies High Overflow' taking the distance value $d(R_{Lukasiewicz}, I) = 0,0872$. This result means that the Lukasiewicz implication gives the highest values of all the rest implications, for every subset of the data and for every hedge 'Low', 'Medium', 'High', as presented in Table 2. For example, the first line of Table 2 means that for the values 0,4728 and 0,3837 of the linguistic variables Low Rainfall and Low Overflow respectively, the value of the Lukasiewicz implication is 0,9108 which is the highest in comparison to the remaining four implications. What is more, the corresponding conditional proposition 'Low: IF Low Rainfall is 0,4728, THEN Low Overflow is 0,3837' is true, with degree of truth 0,9108, which is the same as the value of the Lukasiewicz implication. The proximity of the degree of truth of the proposition to 1 (i.e. absolutely true) using the Lukasiewicz implication signifies that given the premise inference can be made for the conclusion.

Table 3. The distance between each implication and the fuzzy set I for the three hedges 'Low', 'Medium', 'High'.

$d(R_i, I)$	Low	Medium	High
R_{Larsen}	1.9321	0.4570	1.6704
$R_{Mamdani}$	1.6087	0.4502	1.2488
$R_{Lukasiewicz}$	0.1451	0.4414	0.0872
R_{Zadeh}	0.7345	0.4502	0.7746
$R_{Reichenbach}$	0.4852	0.4479	0.4894

On the other hand, the distance between the Larsen implication and the implication I is the largest for all the corresponding propositions as shown in Table 3. The above outcome indicates that the Larsen implication takes the smallest values of all the rest implications as shown in Table 2. As a consequence, since the degree of truth of the corresponding fuzzy propositions is equal to the implication values, the deduction is far from 1 (i.e. true), and therefore, no valuable inference can be made that the premise implies the conclusion for this dataset.

Lastly, the results are tested on the control subsets, which confirm the above outcome, as presented in Table 4.

Table 4. Values of the five implications for the control clusters (C1–C3).

Control subsets		R_{Larsen}	$R_{Mamdani}$	$R_{Lukasiewicz}$	R_{Zadeh}	$R_{Reichenbach}$
C1	Low	0.1106	0.3028	0.9378	0.6349	0.7455
	Med.	0.8135	0.8135	0.8135	0.8135	0.8135
	High	0.3977	0.6296	0.9980	0.6296	0.7661
C2	Low	0.0006	0.0167	0.9838	0.9668	0.9674
	Med.	0.4992	0.6160	0.8057	0.6160	0.6889
	High	0.8962	0.9297	0.9658	0.9297	0.9323
C3	Low	0.0006	0.0170	0.9838	0.9668	0.9674
	Med.	0.6672	0.7212	1.0000	0.7212	0.9459
	High	0.0001	0.0037	1.0000	0.9963	0.9964

5   Conclusions and Future Work

This article introduces an innovative method for detecting the most suitable fuzzy implication regarding a dataset of real observations. In rule based applications, given a set of fuzzy implications which satisfy the desired axioms, a suitable representative is the best-matching between the fuzzy inference rules and the observations. In the literature, the existing methods incorporate the rule induction involved rather than the inference drawn from the data entries, which is often not practical in applications. Given a set of potential fuzzy implications for a specific application, it is essential that an upper and a lower extreme of such a set of implications be found, in order to detect the most suitable one. This method distinguishes these two implications based on a distance criterion and the ideal fuzzy implication for the data of the application. Those implications are the one which corresponds to the smallest and the one with the largest distance from the implication I.

Due to the fuzzification process and the flexibility of fuzzy distance metrics incorporated, this method has no limitations in being applied to any data applications regarding different scientific areas for inference making. The algorithm of this method is such that the values of the implication which is closer to the ideal implication I are the highest in comparison to the rest of the implications examined, thus the corresponding fuzzy propositions in which this implication is involved take the highest degrees of truth. The opposing result is produced for the implication which is farther from the implication I. Then, according to the expert's opinion about the data, choosing between one of these two fuzzy implications results in the desired implication. The construction of the fuzzy sets using several linguistic hedges, in conjunction with the distance criterion makes the implication selection more data-driven rather than theoretically guided.

In this study, the method is applied to a dataset containing paired inputs of Rainfall and Overflow measurements. Given a set of five fuzzy implications the two extremes are found to be the Lukasiewicz implication, which corresponds to the smallest distance from I and the Larsen implication, which corresponds to the largest distance from the implication I. The results show that the Lukasiewicz implication takes the highest values in comparison to the four remaining implications while the Larsen takes the smallest values. Consecutively, the degree of truth of the conditional fuzzy propositions 'Low: Low Rainfall implies Low Overflow', 'Medium: Medium Rainfall implies Medium Overflow', and 'High: High Rainfall implies High Overflow' in which the Lukasiewicz implication is involved is the highest, for the examined application. This result indicates that the Lukasiewicz implication is the most suitable one for making the deduction that the Rainfall variable implies the Overflow variable for the data of the application. On the contrary, the Larsen implication is found to be the less suitable for inference making. Furthermore, in order to test the outcome, the method has been applied to all seven different partitions of the dataset and to different clustering for control, resulting in the same two fuzzy implications and verifying the previous conclusions.

In this study, the metric distance between fuzzy sets has been employed as a criterion for the implication selection. Future research will investigate other relations between the fuzzy implications and the implication I and a comparison analysis of all the different criteria obtained, along with other methods which address the same problem.

References

1. Azzalini, A.: A class of distributions which includes the normal ones. Scand. J. Stat. **12**, 171–178 (1985)
2. Balopoulos, V., Hatzimichailidis, A., Papadopoulos, B.: Distance and similarity measures for fuzzy operators. Inform. Sci. **177**, 2336–2348 (2007). doi:10.1016/j.ins.2007.01.005
3. Brito, P., Duarte Silva, A.P.: Modelling interval data with normal and skew-normal distributions. J. Appl. Stat. **39**, 3–20 (2012). doi:10.1080/02664763.2011.575125
4. Czogala, E., Leski, J.: On equivalence of approximate reasoning results using different interpretations of fuzzy if–then rules. Fuzzy Set. Syst. **117**, 279–296 (2001). doi:10.1016/S0165-0114(98)00412-6
5. Diamond, P., Kloeden, P.: Metric Spaces of Fuzzy Sets: Theory and Applications, 1st edn. Word Scientific, Singapore (1994)
6. Henze, N.: A probabilistic representation of the 'skew-normal' distribution. Scand. J. Stat. **13**, 271–275 (1986)
7. Jang, J.S.R., Sun, C.T., Mizutani, E.: Fuzzy Sets, in Neuro-Fuzzy and Soft Computing: A Computational Approach to Learning and Machine Intelligence, 1st edn. Prentice Hall, Upper Saddle River (1997)
8. Jayaram, B., Mesiar, R.: On special fuzzy implications. Fuzzy Set. Syst. **160**, 2063–2085 (2009). doi:10.1016/j.fss.2008.11.004
9. Klir, G.J., Yuan, B.: Fuzzy Sets and Fuzzy Logic: Theory and Applications, 1st edn. Prentice Hall, Upper Saddle River (1995)
10. Larsen, P.M.: Industrial applications of fuzzy logic control. Int. J. Man. Mach. Stud. **12**, 3–10 (1980)
11. Luo, M., Cheng, Ze.: The distance between fuzzy sets in fuzzy metric spaces. In: 12th International Conference on Fuzzy Systems and Knowledge Discovery (FSKD), pp. 190–194. IEEE Press, Zhangjiajie (2015). doi:10.1109/FSKD.2015.7381938
12. Mamdani, E.H., Assilian, S.: An experiment in linguistic synthesis with a fuzzy logic controller. Int. J. Man Mach. Stud. **7**, 1–13 (1975). doi:10.1016/s0020-7373(75)80002-2

13. Mas, M., Monserrat, M., Torrens, J., Trillas, E.: A survey on fuzzy implication functions. IEEE Trans. Fuzzy Syst. **15**, 1107–1121 (2007). doi:10.1109/tfuzz.2007.896304

14. Mendel, J.: Fuzzy logic systems for engineering: a tutorial. Proc. IEEE **83**, 345–377 (1995). doi:10.1109/5.364485

15. Mizumoto, M., Zimmermann, H.: Comparison of fuzzy reasoning methods. Fuzzy Set. Syst. **8**, 253–283 (1982). doi:10.1016/s0165-0114(82)80004-3

16. Pagouropoulos, P., Tzimopoulos, C., Papadopoulos, B.: Selecting the most appropriate fuzzy implication based on statistical data. Int. J. Fuzzy Syst. Adv. Appl. **3**, 32–42 (2016)

17. Papadopoulos, B.K., Trasanides, G., Hatzimichailidis, A.G.: Optimization method for the selection of the appropriate fuzzy implication. J. Optimiz. Theory Appl. **134**, 135–141 (2007). doi:10.1007/s10957-007-9246-5

18. Ross, T.J.: Fuzzy Logic with Engineering Applications, 2nd edn. Wiley, Chichester (2004)

19. Tick, J., Fodor, J.: Fuzzy implications and inference processes. Comput. Inform. **24**, 591–602 (2005)

20. Tzimopoulos, C., Papadopoulos, B.: Ασαφής ΛογικήΜεΕφαρμογές στις ΕπιστήμεςτουΜηχανικού [Fuzzy Logic with Application in Engineering] (in Greek). Ziti Publications, Thessaloniki (2013)

21. Wang, G., Chen, S., Zhou, Z., Liu, J.: Modelling and analyzing trust conformity in e-commerce based on fuzzy logic. Syst. **14**, 1–10 (2015)

22. Yager, R.R.: On the measure of fuzziness and negation part i: membership in the unit interval. Int. J. Gen. Syst. **5**, 221–229 (1979). doi:10.1080/03081077908547452

23. Zadeh, L.A.: A fuzzy-set-theoretic interpretation of linguistic hedges. J. Cybern. **2**, 4–34 (1972). doi:10.1080/01969727208542910

24. Zadeh, L.A.: The role of fuzzy logic in the management of uncertainty in expert systems. Fuzzy Set. Syst. **11**, 199–227 (1983). doi:10.1016/s0165-0114(83)80081-5

Applying the EFuNN Evolving Paradigm to the Recognition of Artefactual Beats in Continuous Seismocardiogram Recordings

Mario Malcangi[1(✉)], Hao Quan[1], Emanuele Vaini[2],
Prospero Lombardi[2], and Marco Di Rienzo[2]

[1] Computer Science Department, Università degli Studi di Milano,
Via Comelico 39, 20135 Milan, Italy
malcangi@di.unimi.it
[2] Fondazione Don Gnocchi, Milan, Italy
mdirienzo@dongnocchi.it

Abstract. Seismocardiogram (SCG) recording is a novel method for the prolonged monitoring of the cardiac mechanical performance during spontaneous behavior. The continuous monitoring results in a collection of thousands of beats recorded during a variety of physical activities so that the automatic analysis and processing of such data is a challenging task due to the presence of artefactual beats and morphological changes over time that currently request the human expertise. On this premise, we propose the use of the Evolving Fuzzy Neural Network (EFuNN) paradigm for the automatic artifact detection in the SCG signal. The fuzzy logic processing method can be applied to model the human expertise knowledge using the learning capabilities of an artificial neural network. The evolving capability of the EFuNN paradigm has been applied to solve the issue of the physiological variability of the SGC waveform. Preliminary tests have been carried out to validate this approach and the obtained results demonstrate the effectiveness of the method and its scalability.

Keywords: Seismocardiogram · Evolving Fuzzy Neural Network · Artfact identification

1 Introduction

The assessment of both the electrical and mechanical activity of the heart are essential for the full evaluation of the cardiac performance. It is worth noting that while the electrical activity of the heart is easily quantified by the electrocardiogram, the movement of the cardiac muscle (cardiac mechanics) is commonly checked by ultrasound techniques; this implies that the measure is often taken while the subject is at rest and in a clinical laboratory. This approach, although clinically efficient, leaves virtually out the possibility to explore the mechanical heart performance outside the laboratory setting.

A significant step forward with respect to such a traditional context may derive from the measure of the seismocardiogram (SCG). SCG is the quantification of the small thorax vibrations produced by the beating heart and by the blood ejection from the ventricles into the vascular tree. This signal may be simply detected by placing a

G. Boracchi et al. (Eds.): EANN 2017, CCIS 744, pp. 256–264, 2017.
DOI: 10.1007/978-3-319-65172-9_22

miniaturized accelerometer on the thorax of the subject [1]. For each heart beat, the SCG profile is characterized by a number of peaks and valleys. From the simultaneous assessment of SCG and ultrasound images, it was shown that each of these SCG displacements actually reflects a specific mechanical event of the heart cycle, including the opening and closure of the aortic and mitral valves [2], as illustrated Fig. 1. From the analysis of the SCG signal, when detected by wearable sensors, we may derive information on the mechanical performance of the heart even during the 24 h under the real challenges of the daily life [3].

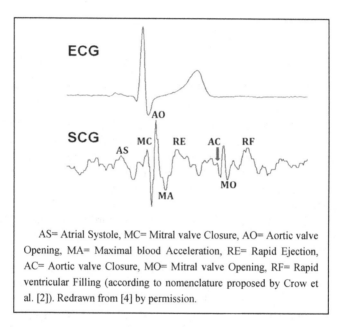

AS= Atrial Systole, MC= Mitral valve Closure, AO= Aortic valve Opening, MA= Maximal blood Acceleration, RE= Rapid Ejection, AC= Aortic valve Closure, MO= Mitral valve Opening, RF= Rapid ventricular Filling (according to nomenclature proposed by Crow et al. [2]). Redrawn from [4] by permission.

Fig. 1. Typical SCG waveform and its fiducial points associated with cardiac mechanical events *(lower panel)* as compared with the ECG complex *(upper panel).*

However, SCG is a low-amplitude signal, with a variability in the order of few mg (where 1 g is the terrestrial gravity acceleration, equal to 9.8 m/s^2). If we consider that the accelerations produced by the daily physical activity are in the order of 0.1–1 g, namely 10–100 times stronger, movement artifacts may be expected, and are actually present, in the SCG recordings carried out during the daily spontaneous behavior. As a consequence, the first action in the processing of SCG profiles, must be the identification and removal of artifacts. In case of short-term recordings this task is commonly achieved by a visual scrutiny of the signal, but when long term recordings should be handled, an automatic analysis is needed (Fig. 2).

At this moment no established procedure is available for the artifact rejection in the SCG signal and till now each research laboratory working in this area, included our lab, has its own deterministic rule-based algorithm for the artifact identification. However, this approach is not completely effective because the SCG morphology may vary from subject to subject, and changes may also occur over time in the same subject as a function

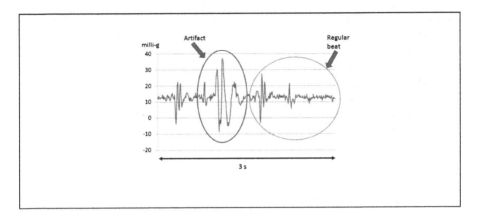

Fig. 2. Example of an artifactual beat and a good-quality SCG beat

of the respiratory phase, body position and heart rate. This means that it is difficult, if not impossible, to identify static deterministic rules which apply for all subjects and over long term recordings. The net result of this situation is that a tailoring of the algorithms is currently required when passing from the analysis of one subject to the other.

On this premise, the use of evolving machine learning techniques is expected to provide a significantly help in handling the SCG artifact identification. In addition, the same technique is deemed appropriate also for the subsequent phases of the SCG analysis, namely for the recognition of the specific fiducial points in the SCG profile associated to the opening and closure of the cardiac valves. So, our group decided to activate a long term project aimed to investigate the applicability of different paradigms of neural networks in all the steps of the SCG treatment. This paper refers to the very first step of the project, namely to the evaluation of the Evolving Fuzzy Neural Network (EFuNN) applicability in the artfact identification in SCG recordings.

It should be mentioned that we previously observed that artifactual distortions of the SCG signal may also be detected by the analysis of the SCG envelope, i.e. a derived signal containing much less details of the raw SCG signal. In this study the EFuNN analysis was carried out by considering both raw data and the envelope curve.

2   Data Set Preparation

One healthy volunteer (age: 38 years), was recruited for the data collection. In this subject a simultaneous ECG and SCG continuous recording was made during sleep by using a custom textile-based system, MagIC-SCG, developed in our laboratories. Briefly, this device is composed of a sensorized vest and an electronic unit (see Fig. 3). The vest is made of cotton and incorporates textile sensors for the ECG and respiratory detection. The electronic unit includes a tri-axial accelerometer and is positioned inside a pocket of the vest so to be in mechanical contact with the sternum and detect the SCG vibrations. All data, sampled at 200 Hz, were locally stored on a memory card. Details on the system may be found in [3]. As mentioned in the introduction, the artifact

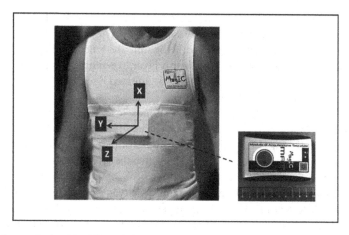

Fig. 3. *Left panel:* the MagIC-SCG garment with orientation of the accelerometric axes: x (longitudinal: foot-head), y (lateral: left-right), z (sagittal, back-front). *Right panel:* the electronic board, to be located into the vest pocket at the sternum level. Redrawn from [3] by permission.

identification was carried out by considering both raw data and the envelope of the SCG signal (see Fig. 4). The envelope curve was obtained by estimating the sample-by-sample absolute value of the SCG signal and then by filtering the output with a 31-sample FIR filter with triangular window.

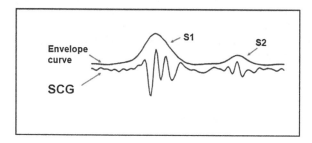

Fig. 4. SCG profile and the corresponding envelope curve

For this study, we selected a segment of 100 beats from the sleep recording and extracted the raw signal and the envelope data within this window to create the data sets to train and test the EFuNN.

3   The EFuNN Paradigm

The EFuNN [11] paradigm, as an implementation of the evolving [5–7] connectionist system (ECOS) paradigm [8], enables on-line adaptation and evolves in real-time. The evolving capability is incremental and adaptive making more effective the learning. EFuNN [9–12] is a connectionist paradigm based on fuzzy rules and a fuzzy inference engine.

EFuNN is a five layer architecture. Each layer deploys the full layers of the fuzzy logic framework. The first layer is the input layer. The second layer executes the fuzzification of the input data. The third layer runs the rules applied to the fuzzified inputs producing the fuzzy output. The fourth layer executes the defuzzification of the output data applying a weighted function and a saturated linear activation function. The fifth layer is the final output of the network.

The five layers fuzzy architecture corresponds to a five layers Artificial Neural Network (ANN) architecture so that the ANN's learning capabilities can be applied to set up the fuzzy logic engine's knowledge as nodes of the ANN. Such nodes evolve by learnig. As the rules are nodes of the ANN, after the training the ANN's nodes are feature's models of the input data.

EFuNN paradigm fuses both the fuzzy logic's advantages to infer by rules and the ANN's capability to learn by data, so the most challenging task of the fuzzy logic (the knowledge set up) is accomplished by a bio-inspired method to compile inferring rules and fuzzy representation of real (physical) world data.

4   Data Set and Training

As mentioned in Sect. 2, two data segments have been used to create the test and train data set, one from the envelope curve, and from the raw SCG signal. Each beat in the data segment was classified by an expert as good or artifactual. For each beat, two array constituted by the first 151 samples of the signal and its envelope were created.

In total, the data sets to train and test the EFuNN consisted of 100 signal arrays, 100 envelope arrays and the corresponding labels.

As to the analysis, first we trained the EFuNN with the envelope dataset. Figure 5 shows the sequence of the "good" ("1") and "artefactual" ("0") beats of the envelope dataset. Then we trained the EFuNN with the SGC raw data.

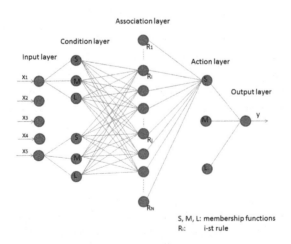

Fig. 5. EfuNN is a five layers artificial neural network where each layer corresponds to a layer of a fuzzy logic engine.

5 System Test and Validation

The training and the tests of the EFuNN have been executed in the simulation and modeling environment NeuCom [13] applying the following setup:

Sensitivity threshold: 0.9
Error threshold: 01
Number of membership functions: 3
Learning rate for W1: 0.1
Learning rate for W2: 0.1
Pruning: on
Node age: 60
Aggregation: on

The trained EFuNN has been tested with a new dataset to validate the EFuNN capability to recognize and classify each SCG beat period according to the expert knowledge.

The test results show that effective learning can be gained by the EFuNN at training-time. Some mismatches occurred on both envelope and row data (Figs. 6 and 7) after a single learning step. However, errors completely recovered after that some evolving training step was applied to the trained EFuNN (Figs. 8, 9 and 10).

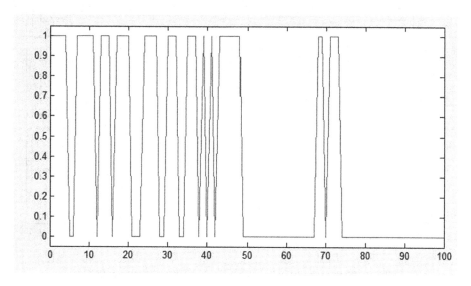

Fig. 6. Sequence of the "good" ("1") and "artefactual" ("0") beats of the envelope dataset.

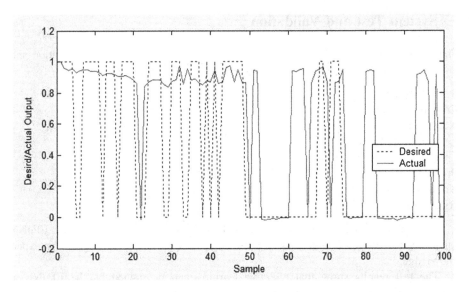

Fig. 7. Test of the EFuNN after a single learning step (envelope).

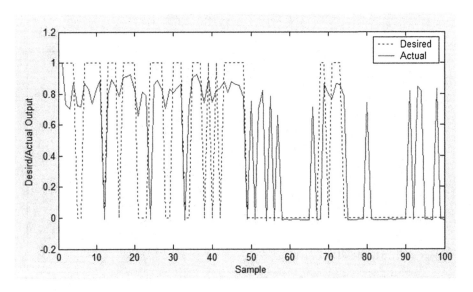

Fig. 8. Test of the EFuNN after a single learning step (raw SCG).

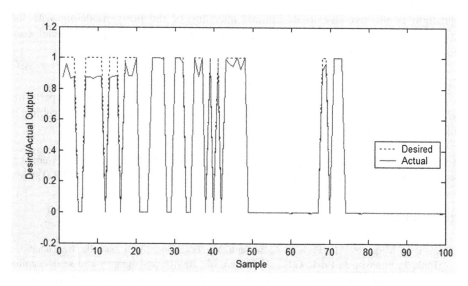

Fig. 9. Test of the EFuNN after one evolving step (envelope).

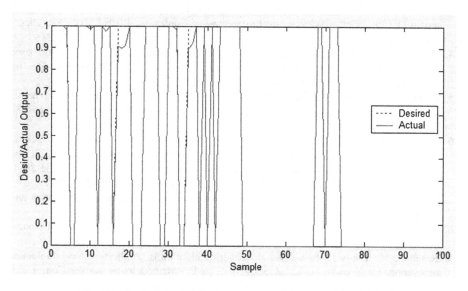

Fig. 10. Test of the EFuNN after one evolving step (raw SCG).

6 Results Evaluation and Future Developments

The first round of tests indicates that the use of EFuNN might solve the issue of the automatic rejection of the artefactual beats in continuous seismocardiogram recordings. This approach appears effective when evolving methods are applied. The EFuNN paradigm is effective due to its optimal matching of the fuzzy modeling with the knowledge, and because of the correct artificial neural network inference and connectionist capabilities.

Interestingly, the trained EFuNN correctly detect artifacts also in the raw SCG signal, which is much more complex and detailed than the envelope curve.

Future developments will include investigations on how this methodology performs when applied on longer recordings and if an EFuNN trained on data from one subject is effective to test data from a different subject.

Acknowledgements. The SCG data collection and the work of MDR, EV and PL were supported by the Italian Space Agency through the ASI 2013–061-I.0 and ASI 2013–079-R.0 grants.

References

1. Inan, O., Migeotte, P.F., Park, K.S., Etemadi, M., Tavakolian, K., Casanella, R., Zanetti, J., Tank, J., Funtova, I., Prisk, G.K., Di Rienzo, M.: Ballistocardiography and seismocardiography: a review of recent advances. J. Biomed. Health Inf. **19**(4), 1414–1427 (2015)
2. Crow, R.S., Hannan, P., Jacobs, D., Hadquist, L., Salerno, D.M.: Relationship between seismocardiogram and echocardiogram for events in cardiac cycle. Am. J. Noninvasive Cardiol. **8**, 39–46 (1994)
3. Di Rienzo, M., Meriggi, P., Vaini, E., Castiglioni, P., Rizzo, F.: 24 h seismocardiogram monitoring in ambulant subjects. In: Proceedings of Conference IEEE EMBS, San Diego, pp. 5050–5053 (2012)
4. Di Rienzo, M., Vaini, E., Castiglioni, P., Lombardi, P., Meriggi, P., Rizzo F.: A textile-based wearable system for the prolonged assessment of cardiac mechanics in daily life. In: Proceedings Conference IEEE EMBS, pp. 6896–6899 (2014)
5. Kasabov, N., Dhoble, K., Nuntalid, N., Indiveri, G.: Dynamic evolving spiking neural networks for on-line spatio- and spectro-temporal pattern recognition. Neural Netw. **41**, 188–201 (2013)
6. Ferrandez, M., Sanchez, J.R.A., de la Paz, F., Toledo, F.J. (eds.): New Challenges on Bioinspired Applications, Proceedings of 4th International Work-Conference on the Interplay Between Natural and Artificial Computation, IWINAC 2011, May–June 2011, Part 2. LNCS, vol. 6687. Springer, Heidelberg (2011). 10.1007/978-3-642-21326-7
7. Sutton, S., Barto, B.: Reinforzed learning: an introduction. In: Adaptive Computation and Machine Learning. MIT press (1998)
8. Kasabov, N.: Evolving Connectionist Systems: The Knowledge Engineering Approach. Springer, Heidelberg (2007)
9. Kasabov, N.: Evolving fuzzy neural networks. algorithms, applications and biological motivation. In: Yamakawa, T., Matsumoto, G. (eds.) Methodologies for the Conception, Design and Application of Soft Computing, World Scientific, pp. 271–274 (1998)
10. Kasabov, N.: DENFIS: dynamic evolving neural-fuzzy inference system and its applications to time-series prediction. IEEE Trans. Fuzzy Syst. **10**, 144–154 (2001)
11. Kasabov, N.: EFuNN. IEEE Tr SMC (2001)
12. Kasabov, N.: Evolving fuzzy neural networks – algorithms, applications and biological motivation. In: Yamakawa and Matsumoto (eds.) Methodologies for the Conception, Design and Application of the Soft Computing, World Computing, pp. 271–274 (1998)
13. http://www.kedri.aut.ac.nz/areas-of-expertise/data-mining-and-decision-support-systems/neuco

Learning Generalization

Application of Asymmetric Networks to Movement Detection and Generating Independent Subspaces

Naohiro Ishii[1]([⊠]), Toshinori Deguchi[2], Masashi Kawaguchi[3],
and Hiroshi Sasaki[4]

[1] Aichi Institute of Technology, Toyota, Japan
ishii@aitech.ac.jp
[2] National Institute of Technology, Gifu College, Gifu, Japan
deguchi@gifu-nct.ac.jp
[3] National Institute of Technology, Suzuka College, Suzuka, Mie, Japan
masashi@elec.suzuka-ct.ac.jp
[4] Fukui University of Technology, Fukui, Japan
h-sasaki@fukui-ut.ac.jp

Abstract. The prominent feature is the nonlinear characteristics as the squaring and rectification functions, which are observed in the retinal and visual cortex networks. Conventional model for motion processing in cortex, uses a symmetric quadratic functions with Gabor filters. This paper proposes a new motion processing model in the asymmetric networks. First, the asymmetric network is analyzed using Wiener kernels. It is shown that the asymmetric network with nonlinearities is effective and general for generating the directional movement compared with the conventional quadratic model. Second, independence maximization of data is an important issue in computational neural networks. To make clear the characteristics of the asymmetric network with Gabor functions, orthogonality is computed, which shows independent characteristics of the asymmetric network without maximizing optimization of independence in the quadratic model. The orthogonal analyses for the independence of the asymmetric networks are applied to the V1 and MT neural networks to generate independent subspaces by using selective Gabor functions.

Keywords: Asymmetric neural network · Gabor filters · Quadratic energy model · Symmetric network · Motion processing · Orthogonal relation · Independence

1 Introduction

The nonlinear characteristics as the squaring function and rectification function are observed in the retina [9, 10] and visual cortex networks [5–7]. Conventional model for cortical motion sensors is the use of symmetric quadratic functions with Gabor filters, which is called energy model [1, 2, 5]. These energy models have been studied to generate independent subspaces under the condition of some optimizations [2–4]. Recent study by Hess and Bair [5] discusses quadratic form is not necessary nor

© Springer International Publishing AG 2017
G. Boracchi et al. (Eds.): EANN 2017, CCIS 744, pp. 267–278, 2017.
DOI: 10.1007/978-3-319-65172-9_23

sufficient under certain stimulus condition. Then, minimal models for sensory processing are expected. This paper proposes a new motion sensing processing model in the biological asymmetric networks. The nonlinear function exists in the asymmetrical neural networks in the catfish retina. In this paper, first, to make clear the behavior of the asymmetric network with nonlinearity, Wieners nonlinear analysis is applied. It is shown that the asymmetric network with nonlinearities is effective and superior for generating the directional movement detection from the network computations. It is shown that the quadratic model works with Gabor functions [2–4], while the asymmetric network does not need their conditions. Next, independence maximization of data is an important issue in computational neural networks [2–4]. The proposed asymmetric networks with Gabor filters generate orthogonal relations. To make clear the asymmetric network with Gabor functions, orthogonality is computed, which shows independence characteristics of the asymmetric structure network without conventional maximizing independence [2–4]. The orthogonal analyses of the asymmetric networks are applied to the MT and V1 in the cortex to generate independent subspaces by using selective Gabor functions.

2 Basic Asymmetrical Neural Networks

Naka et al. [10] presented a simplified, but essential networks of catfish inner retina as shown in Fig. 1. The asymmetric structure network is composed of the linear pathway from the bipolar cell B1 to the amacrine cell N and the nonlinear pathway from the bipolar cell B2, via the amacrine cell C with squaring function to the amacrine cell N.

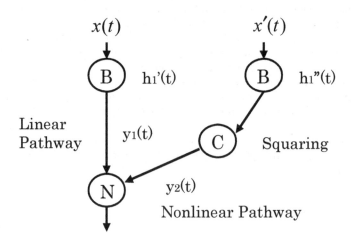

Fig. 1. Asymmetric network with linear and squaring nonlinear pathways

Here, the stimulus with Gaussian distribution is assumed to move from the left side stimulus $x(t)$ to the right side stimulus $x'(t)$ in front of the network in Fig. 1. The $h'_1(t)$ and $h''_1(t)$ show the impulse response functions of the B1 and B2 cellsrespectively.

By introducing a mixed ratio, α, of the different stimulus $x''(t)$, the input function of the right stimulus $x'(t)$, is described in the following equation, where $0 \leq \alpha \leq 1$ and $\beta = 1 - \alpha$ hold.

$$x'(t) = \alpha x(t) + \beta x''(t) \tag{1}$$

Let the power spectrums of $x(t)$ and $x''(t)$, be p and p'', respectively an equation $p'' = kp$ holds for the coefficient k. Figure 2 shows that the left slashed light is moving from the receptive field of B_1 cell to the right field of the B_2 cell in the schematic diagram.

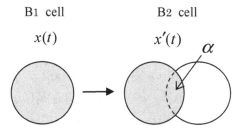

Fig. 2. Schematic diagram of the preferred stimulus direction

First, on the linear pathway in Fig. 1, the input function is $x(t)$ and the output function is $y(t)$ are given.

$$y(t) = \int h_1'''(\tau)(y_1(t - \tau) + y_2(t - \tau))d\tau + \varepsilon \tag{2}$$

where $y_1(t)$ shows the linear information on the linear pathway, while $y_2(t)$ shows the nonlinear information on the nonlinear pathway and ε shows error value. The $y_1(t)$ and $y_2(t)$ are given, respectively as follows,

$$y_1(t) = \int_0^\infty h_1'(\tau)x(t - \tau)d\tau \tag{3}$$

$$y_2(t) = \int_0^\infty \int_0^\infty h_1''(\tau_1)h_1''(\tau_2)x'(t - \tau_1)x'(t - \tau_2)d\tau_1 d\tau_2 \tag{4}$$

We assume that the impulse response function $h_1'''(t)$ is assumed to be value 1 without loss of generality.

2.1 Directional Equations from Optimized Conditions in the Asymmetric Networks

Under the assumption that the impulse response functions, $h_1'(t)$ of the cell B_1, $h_1''(t)$ of the cell B_2 and moving stimulus ratio α in the right to be unknown, the optimization of the network is carried out. By the minimization of the mean squared value ξ of ε in Eq. (2), the following necessary equations for the optimization of equations are derived,

$$\frac{\partial \xi}{\partial h_1'(t)} = 0, \frac{\partial \xi}{\partial h_2''(t)} = 0, \frac{\partial \xi}{\partial \alpha} = 0 \tag{5}$$

Then, the following three equations are derived for the optimization satisfying the Eq. (5).

$$E[y(t)x'(t-\lambda)] = \alpha p h_1'(\lambda)$$
$$E[(y(t)-C_0)x'(t-\lambda_1)x'(t-\lambda_2)] = 2\{(\alpha^2 + k\beta^2)p^2 h_1''(\lambda_1)h_1''(\lambda_2)\}$$
$$E[(y(t)-C_0)x(t-\lambda_1)x(t-\lambda_2)] = 2\alpha^2 p^2 h_1''(\lambda_1)h_1''(\lambda_2) \tag{6}$$
$$E[(y(t)-C_0)x''(t-\lambda_1)x''(t-\lambda_2)] = 2\beta^2 (kp)^2 h_1''(\lambda_1)h_1''(\lambda_2)$$

where C_0 is the mean value of, $y(t)$ which is shown in the following. Here, the Eq. (6) can be rewritten by applying Wiener kernels, which are computed from input and output correlations [11]. From the necessary optimization equation in (5), the following Wiener kernel equations are derived as shown in the following [8, 11–13]. We can compute the 0-th order Wiener kernel C_0, the 1-st order one and $C_{11}(\lambda)$ the 2-nd order one $C_{21}(\lambda_1, \lambda_2)$ as follows.

$$C_{11}(\lambda) = \frac{1}{p}E[y(t)x(t-\lambda)] = h_1'(\lambda) \tag{7}$$

$$C_{21}(\lambda_1, \lambda_2) = \frac{1}{2p^2}E[(y(t)-C_0)x(t-\lambda_1)x(t-\lambda_2)] = \alpha^2 h_1''(\lambda_1)h_1''(\lambda_2) \tag{8}$$

From Eqs. (1), (7) and (8), the ratio, α which is a mixed coefficient of $x(t)$ to, is $x'(t)$ shown by α^2 as the amplitude of the second order Wiener kernel. Second, on the nonlinear pathway, we can compute the 0-th order kernel, C_0 the 1-st order kernel $C_{12}(\lambda)$ and the 2-nd order kernel by the $C_{22}(\lambda_1, \lambda_2)$ cross-correlations between $x(t)$ and $y(t)$ as shown in the following, which are also derived from the optimization Eq. (6).

$$C_{12}(\lambda_1, \lambda_2) = \frac{1}{p(\alpha^2 + k\beta^2)}E[y(t)x'(t-\lambda)] = \frac{\alpha}{\alpha^2 + k(1-\alpha)^2}h_1'(\lambda) \tag{9}$$

and

$$C_{22}(\lambda_1, \lambda_2) = h_1''(\lambda_1)h_1''(\lambda_2) \tag{10}$$

The second order kernels C_{21} and C_{22} are abbreviated in the representation in Eqs. (8) and (10).

$$(C_{21}/C_{22}) = \alpha^2 \tag{11}$$

Then, from the Eq. (13) the ratio α is shown as follows

$$\alpha = \sqrt{\frac{C_{21}}{C_{22}}} \tag{12}$$

The Eq. (12) is called here α- equation, which implies the directional stimulus on the network. The directional equation from the left to the right, holds as shown in the following,

$$\frac{C_{12}}{C_{11}} = \frac{\sqrt{\frac{C_{21}}{C_{22}}}}{\frac{C_{21}}{C_{22}} + k(1 - \sqrt{\frac{C_{21}}{C_{22}}})^2} \tag{13}$$

The Eq. (13) shows the direction of th stimulus from the left to the right.

2.2 Algorithm for Movement Detection

When α increases to α' in the preferred direction, the movement will take place. Then, necessary conditions of the movement is to satisfy the directional Eq. (13) at α and that at α', respectively. The detection of the directional movement is carried out in the following steps.

① In the case of the directional movement from the left to the right, the Eq. (12) for α at time t, $\alpha(t)$, i.e., the root of kernel correlations ratio in the left side of the Eq. (13) is computed. Then, the Eq. (13) at time t is checked whether the equation holds.

② α' at time $(t + \Delta t)$, $\alpha'(t + \Delta t)$ is computed similarly. The Eq. (13) is checked whether the equation holds.

③ Assume here that the following holds,

$$\alpha(t) < \alpha'(t + \Delta t) \tag{14}$$

④ When ① and ② are satisfied, the directional movement is written as

$$\alpha \rightarrow \alpha' \tag{15}$$

3   Comparison with Conventional Quadratic Models

Motion detection of the conventional quadratic models under the same conditions in this paper is analyzed. The quadratic model in Fig. 3 is well known as the energy model for motion detection [1, 2, 5], which is a symmetric network model. The model with Gabor filters are used as the functions $h_1(t)$ and $h'_1(t)$ in the models [1, 2]. Under the same stimulus conditions in the asymmetric network in Figs. 1 and 2, the Wiener kernels are computed in the symmetric quadratic model in Fig. 3. In Fig. 3, the first order kernels disappear on the left and the right pathways. Only the second order kernels are computed in Fig. 3. On the left pathway in Fig. 3, the second order kernel $C_{21}(\lambda_1, \lambda_2)$ is computed as follows,

$$
\begin{aligned}
C_{21}(\lambda_1, \lambda_2) &= \frac{1}{2p^2} \iint h_1(\tau)h_1(\tau')E[x(t-\lambda)x(t-\lambda')x(t-\lambda_1)x(t-\lambda_2)]d\tau d\tau' \\
&+ \frac{1}{2p^2} \iint h'_1(\tau)h'_1(\tau')E[x'(t-\lambda)x'(t-\lambda')x(t-\lambda_1)x(t-\lambda_2)]d\tau d\tau' \quad (16) \\
&= h_1(\lambda_1)h_1(\lambda_2) + \alpha^2 h'_1(\lambda_1)h'_1(\lambda_2)
\end{aligned}
$$

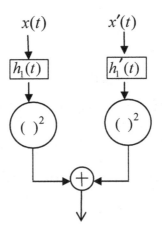

Fig. 3. Quadratic energy model with Gabor filters

On the right pathway in Fig. 5, the 2nd order kernel $C_{22}(\lambda_1, \lambda_2)$ is computed similarly,

$$
C_{22}(\lambda_1, \lambda_2) = \frac{\alpha^2}{(\alpha^2 + k\beta^2)^2} h_1(\lambda_1)h_1(\lambda_2) + h'_1(\lambda_1)h'_1(\lambda_2) \quad (17)
$$

In the conventional energy model of motion [1, 2, 5], the Gabor functions are given as

$$h_1(t) = \frac{1}{\sqrt{2\pi}\sigma}\exp(-\frac{t^2}{2\sigma^2})\sin(2\pi\omega t), \ h_1'(t) = \frac{1}{\sqrt{2\pi}\sigma}\exp(-\frac{t^2}{2\sigma^2})\cos(2\pi\omega t) \quad (18)$$

When Gabor functions in the quadratic model are given as the Eq. (18), the motion parameter α and the motion equation are computed as follows,

$$\alpha = \sqrt{\frac{C_{21}(\lambda_1, \lambda_2) - h_1(\lambda_1)h_1(\lambda_2)}{h_1'(\lambda_1)h_1'(\lambda_2)}} \quad (19)$$

$$C_{22}(\lambda_1, \lambda_2) = \frac{\alpha^2}{(\alpha^2 + k\beta^2)^2}h_1(\lambda_1)h_1(\lambda_2) + h_1'(\lambda_1)h_1'(\lambda_2) \quad (20)$$

Note that the conventional quadratic model generates the motion Eq. (20) under the condition of the given Gabor functions (20), while the asymmetric network in Fig. 1 generate the motion Eq. (13) without the condition of the Gabor functions. Thus, the asymmetric network has also general ability compared to the conventional quadratic model.

4 Orthogonality of Asymmetric Networks with Gabor Filters

Generating independence of data is an important issue in computational neural networks [2–4]. To make clear the asymmetric network with Gabor functions, orthogonality is computed, which shows independence characteristics of the asymmetric structure network without conventional maximizing independence in the quadratic model [2–4]. The orthogonality is checked in the asymmetric networks as shown in Fig. 4.

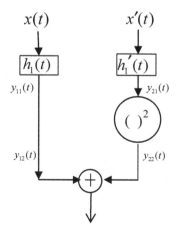

Fig. 4. Orthogonality of asymmetric network

The variable t in the Gabor filters is changed to t', where by setting $\xi \triangleq 2\pi\omega$ in the Eq. (18), $t' = 2\pi\omega t = \xi t$ and $dt = dt/\xi$ hold. Then, Gabor filters become to the following equation.

$$G_s(t') = \frac{1}{\sqrt{2\pi}\sigma} e^{-\frac{t^2}{2\sigma^2\xi^2}} \sin(t') \text{ and } G_c(t') = \frac{1}{\sqrt{2\pi}\sigma} e^{-\frac{t^2}{2\sigma^2\xi^2}} \cos(t').$$

The impulse response functions $h_1(t)$ and $h_1'(t)$ are replaced by $G_s(t')$ and $G_c(t')$ or vice versa.

The orthogonality of asymmetric networks are computed as follows. In Fig. 6,

$$
\begin{aligned}
\int_{-\infty}^{\infty} y_{11}(t)y_{21}(t)dt &= \frac{1}{2\pi\xi\sigma^2} \int_{-\infty}^{\infty} e^{-\frac{t^2}{\sigma^2\xi^2}} \sin(t')\cos(t')dt' \\
&= \frac{1}{2} \cdot \frac{1}{2\pi\xi\sigma^2} \int_{-\infty}^{\infty} e^{-\frac{t^2}{\sigma^2\xi^2}} \sin(2t')dt' = 0
\end{aligned}
\tag{21}
$$

The Eq. (21) shows the $y_{11}(t')$ and $y_{21}(t')$ in Fig. 6 are orthogonal. Next, the orthogonality between $y_{12}(t')(= y_{11}(t'))$ and $y_{22}(t')$ is computed as follows.

$$
\int_{-\infty}^{\infty} y_{12}(t)y_{22}(t)dt = \frac{1}{\sqrt{2\pi}\sigma} \cdot \left(\frac{1}{2\pi\xi\sigma^2}\right) \int_{-\infty}^{\infty} e^{-\frac{3t^2}{2\sigma^2\xi^2}} \cos^2(t')\sin(t')dt' = 0
\tag{22}
$$

The Eq. (22) shows the orthogonality between $y_{12}(t')(= y_{11}(t'))$ and $y_{22}(t')$. Thus, the independence holds between $y_{12}(t')(= y_{11}(t'))$ and $y_{22}(t')$ in the asymmetric network in Fig. 6.

5 Non-orthogonal Property of Conventional Energy Model

Conventional energy model has non-orthogonal properties for the Gabor function inputs.

The non-orthogonality of the energy model is computed as follows. The orthogonality in Fig. 5 is checked by the following Eq. (23).

$$
\begin{aligned}
\int_{-\infty}^{\infty} y_{11}(t)y_{21}(t)dt &= \frac{1}{2\pi\xi\sigma^2} \int_{-\infty}^{\infty} e^{-\frac{t^2}{\sigma^2\xi^2}} \sin(t')\cos(t')dt' \\
&= \frac{1}{2} \cdot \frac{1}{2\pi\xi\sigma^2} \int_{-\infty}^{\infty} e^{-\frac{t^2}{\sigma^2\xi^2}} \sin(2t')dt' = 0
\end{aligned}
\tag{23}
$$

The Eq. (23) shows to be orthogonal. Thus, $y_{11}(t)$ and $y_{21}(t)$ are independent. Next, the orthogonality of $y_{12}(t)$ and $y_{22}(t)$ are checked in the following.

$$\int_{-\infty}^{\infty} y_{12}(t)y_{22}(t)dt = (\frac{1}{2\pi\sigma^2})^2 \cdot \frac{1}{2\xi} \int_{-\infty}^{\infty} e^{-\frac{t'^2}{2\sigma^2\xi^2}}dt'$$

$$- (\frac{1}{2\pi\sigma^2})^2 \cdot \frac{1}{2\xi} \int_{-\infty}^{\infty} e^{-\frac{t'^2}{2\sigma^2\xi^2}}cos(2t')dt' - (\frac{1}{2\pi\sigma^2})^2 \cdot \frac{1}{2\xi} \int_{-\infty}^{\infty} e^{-\frac{t'^2}{2\sigma^2\xi^2}}sin^4(t')dt' \qquad (24)$$

$$= \frac{1}{16\sqrt{2\pi\sigma^2}}(1 - e^{-2\xi^2\sigma^2}) > 0$$

The Eq. (24) shows the non-orthogonality. Thus, this shows dependent between $y_{12}(t)$ and $y_{22}(t)$ in the conventional energy model in Fig. 5.

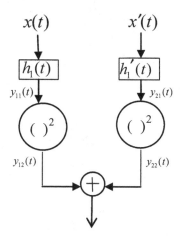

Fig. 5. Non-orthogonal properties of energy model

6 Application of Asymmetric Network to Bio-Inspired Neural Networks

We present an example of layered neural network in Fig. 6, which is developed from the neural network in the brain cortex [7]. Figure 6 is a connected network model of V1 followed by MT, where V1 is the front part of the total network, while MT is the rear part of it. The network model of V1 and MT was proposed by Simoncelli and Heager [7]. Neuron in V1 cell computes a weighted sum of its inputs followed by half-wave rectification, squaring and response normalization. In Fig. 6, the half-wave rectification followed by the normalization, is approximated as follows,

$$f(x) = \frac{1}{1 + e^{-\eta(x-\theta)}} \qquad (25)$$

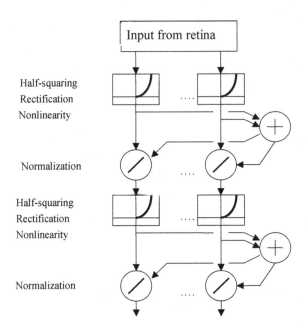

Fig. 6. Model of neural network of brain cortex V1 followed by MT [7]

By Taylor expansion of the Eq. (25) at $x = \theta$, the Eq. (26) is derived as follows,

$$
\begin{aligned}
f(x)_{x=\theta} &= f(\theta) + f'(\theta)(x - \theta) + \frac{1}{2!}f''(\theta)(x - \theta)^2 + \dots \\
&= \frac{1}{2} + \frac{\eta}{4}(x - \theta) + \frac{1}{2!}(-\frac{\eta^2}{4} + \frac{\eta^2 e^{-\eta\theta}}{2})(x - \theta)^2 + \dots
\end{aligned}
\tag{26}
$$

The Taylor expansion network is shown in Fig. 7. Figure 7 shows two layered networks, in which the upper layer is the V1 network, while the lower layer is MT network. There exist many combination of the asymmetric networks in the upper layer of V1. After the upper layer, by the operations of the linear, squaring, tripling, ..., many combinations of non linear pairs are made.

6.1 Selective Combinations of Orthogonal Pairs to Generate Independent Subspaces

In Fig. 7 the nonlinear terms, $x^2, x^3, x^4, \dots$ are generated in the Eq. (26). Thus, the selective combination of Gabor function pairs are generated in Fig. 7. Gabor functions are simply defined as follows.

$$
G_s(t') = Ae^{-\frac{t'^2}{2\sigma^2\xi^2}}sin(t') \text{ and } G_c(t') = Ae^{-\frac{t'^2}{2\sigma^2\xi^2}}cos(t'),
\tag{27}
$$

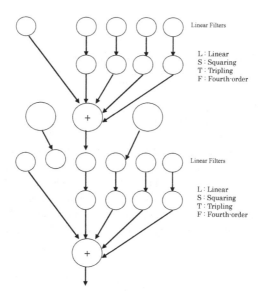

Fig. 7. A transformed network model for the layered network for one pathway in Fig. 6.

where $A \triangleq 1/\sqrt{2\pi}\sigma$ is used. Then, the example of selective combination pairs of the Gabor functions is shown in the following Eq. (28).

As shown in the Eq. (27), $Ae^{-\frac{t'^2}{2\sigma^2\xi^2}}sin(t')$ and $Ae^{-\frac{t'^2}{2\sigma^2\xi^2}}cos(t')$ are orthogonal, thus they have independent relation. This is shown in bold line in Fig. 8. Similarly, $Ae^{-\frac{t'^2}{2\sigma^2\xi^2}}sin(t')$ and $A^2(e^{-\frac{t'^2}{2\sigma^2\xi^2}})^2cos^2(t')$ are orthogonal, thus they have independent relation. Similar independent relation holds among $cos(t')$ terms in the Eq. (29).

$$Ae^{-\frac{t'^2}{2\sigma^2\xi^2}}sin(t') \quad Ae^{-\frac{t'^2}{2\sigma^2\xi^2}}cos(t') \quad A^2(e^{-\frac{t'^2}{2\sigma^2\xi^2}})^2cos^2(t') \quad A^3(e^{-\frac{t'^2}{2\sigma^2\xi^2}})^3cos^3(t')... \quad (28)$$

Fig. 8. Selective orthogonal combination pairs among Gabor sine and cosines functions

These selective orthogonal combination pairs in Figs. 8 and 9 generate independent subspaces.

$$Ae^{-\frac{t'^2}{2\sigma^2\xi^2}}sin(t') \quad Ae^{-\frac{t'^2}{2\sigma^2\xi^2}}cos(t') \quad A^2(e^{-\frac{t'^2}{2\sigma^2\xi^2}})^2cos^2(t') \quad A^3(e^{-\frac{t'^2}{2\sigma^2\xi^2}})^3cos^3(t')... \quad (29)$$

Fig. 9. Selective orthogonal combination pairs among Gabor cosines functions

7 Conclusion

The neural networks are analyzed to make clear functions of the biological asymmetric neural networks with nonlinearity. This kind of networks exits in the biological network as retina and brain cortex of V1 and MT areas. In this paper, it is shown that the asymmetric network with nonlinearities is effective and superior for generating the directional movement detection from the network computations. It is shown that the quadratic model works with Gabor functions, while the asymmetric network does not need their conditions for movement detection. Next, the asymmetric networks with Gabor filters generate orthogonal relations. Thus, independent subspaces are generated in the network, while the conventional symmetric model generates non-orthogonal relation. These results will suggest efficient functions of the detection behavior of the movement and generation of independent subspaces in the cortex, V1 and MT areas, in which efficient learning is expected by using independent subspaces.

References

1. Adelson, E.H., Bergen, J.R.: Spatiotemporal energy models for the perception of motion. J. Opt. Soc. Am. A **2**, 284–298 (1985)
2. Hyvarinen, A., Hoyer, P.: Emergence of phase-and shift-invariant features by decomposition of natural images into independent feature subspaces. Neural Comput. **12**, 1705–1720 (2000)
3. Olshausen, B.A., Field, D.A.: Emergence of simple-cell receptive field properties by learning a sparse code from natural images. Nature **381**, 607–609 (1996)
4. Hashimoto, W.: Qudratic forms in natural images. Netw.: Comput. Neural Syst. **14**, 765–788 (2003)
5. Heess, N., Bair, W.: Direction opponency, not quadrature, is key to the 1/4 cycle preference for apparent motion in the motion energy model. J. Neurosci. **30**(34), 11300–11304 (2010)
6. Taub, E., Victor, J.D., Conte, M.: Nonlinear preprocessing in short-range motion. Vis. Res. **37**, 1459–1477 (1997)
7. Simoncelli, E.P., Heeger, D.J.: A model of neuronal responses in visual area MT. Vis. Res. **38**, 743–761 (1996)
8. Marmarelis, P.Z., Marmarelis, V.Z.: Analysis of Physiological Systems – The White Noise Approach. Plenum Press, New York (1978)
9. Sakuranaga, M., Naka, K.-I.: Signal transmission in the catfish retina. III. Transmission to type-C cell. J. Neurophysiol. **53**(2), 411–428 (1985)
10. Naka, K.-I., Sakai, H.M., Ishii, N.: Generation of transformation of second order nonlinearity in catfish retina. Ann. Biomed. Eng. **16**, 53–64 (1988)
11. Lee, Y.W., Schetzen, M.: Measurements of the Wiener kernels of a nonlinear by cross-correlation. Int. J. Control **2**, 237–254 (1965)
12. Ishii, N., Ozaki, M., Sasaki, H.: Correlation computations for movement detection in neural networks. In: Negoita, M.G., Howlett, R.J., Jain, L.C. (eds.) KES 2004. LNCS, vol. 3214, pp. 124–130. Springer, Heidelberg (2004). doi:10.1007/978-3-540-30133-2_17
13. Ishii, N., Deguchi, T., Kawaguchi, M.: Neural computations by asymmetric networks with nonlinearities. In: Beliczynski, B., Dzielinski, A., Iwanowski, M., Ribeiro, B. (eds.) ICANNGA 2007. LNCS, vol. 4432, pp. 37–45. Springer, Heidelberg (2007). doi:10.1007/978-3-540-71629-7_5

Two Hidden Layers are Usually Better than One

Alan J. Thomas[1]([⊠]), Miltos Petridis[2], Simon D. Walters[1],
Saeed Malekshahi Gheytassi[1], and Robert E. Morgan[1]

[1] School of Computing Engineering and Mathematics,
University of Brighton, Brighton, UK
alan.j.thomas@gmail.com, {s.d.walters,m.s.malekshahi,
r.morgan2}@brighton.ac.uk
[2] Faculty of Science and Technology, Middlesex University, London, UK
m.petridis@mdx.ac.uk

Abstract. This study investigates whether feedforward neural networks with two hidden layers generalise better than those with one. In contrast to the existing literature, a method is proposed which allows these networks to be compared empirically on a hidden-node-by-hidden-node basis. This is applied to ten public domain function approximation datasets. Networks with two hidden layers were found to be better generalisers in nine of the ten cases, although the actual degree of improvement is case dependent. The proposed method can be used to rapidly determine whether it is worth considering two hidden layers for a given problem.

Keywords: Feedforward neural networks · How many hidden layers · Universal function approximation · Transformative optimisation · Optimal FNN topology · One or two hidden layers

1 Introduction

The most important aspect of the design of a neural network is its structure or topology, since this is crucial to its generalisation capability. In the case of a fully interconnected feedforward neural network (FNN), and given a fixed set of inputs and outputs, the topology is directly determined by the number of hidden nodes and layers. Whilst there is an extraordinary volume of literature on the subject of hidden node selection, there is scarcely any about hidden layer selection. This is almost certainly due in part to proofs that networks with a single hidden layer are sufficient for universal approximation [1–3]. Furthermore, the search space of candidate topologies is linear - so they are easier to find and train. Consequently, there is less interest in neural networks with two or more hidden layers and they are rarely used in practice [4].

However, it has been shown that two-hidden-layer feedforward networks (TLFNs) can outperform single-hidden-layer ones (SLFNs) in some cases. Indeed there is some evidence that certain problems can only be solved with a second hidden layer [5–7]. What is lacking in the literature is any indication about how SLFNs and TLFNs compare in practical situations. To redress this, SLFNs and TLFNs compete head to head on ten public domain datasets. In order to ensure a fair competition, all factors other than the number of hidden layers are kept constant.

© Springer International Publishing AG 2017
G. Boracchi et al. (Eds.): EANN 2017, CCIS 744, pp. 279–290, 2017.
DOI: 10.1007/978-3-319-65172-9_24

In Sect. 2, related work on the subject of the number of hidden layers is discussed. Section 3 describes "transformative optimisation", the approach used in this paper. This is a new name for a technique previously developed by the Authors in [8, 9]. It is ideally suited for the hidden-node-by-hidden-node comparisons used in the experiments. Sections 4 and 5 detail the experimental setup and methods. In Sect. 6, the results are summarised and discussed. It was found that TLFNs showed improvement in generalisation performance over SLFNs in most of the cases.

2   Related Work

The volume of literature comparing SLFNs and TLFNs is very scarce [10]. Funahashi [11] proved that any continuous function can be approximated with a single hidden layer. However it was subsequently shown by Chester that two hidden layers were better when dealing with pinnacle functions [5]. He states that "the problem with a single hidden layer is that the neurons interact with each other globally, making it difficult to improve an approximation at one point without worsening it elsewhere". He goes on to point out that in a TLFN, the first hidden layer can partition the input space into smaller regions, leaving the second hidden layer free to improve the approximation. Sontag [7] approached the question from a different angle. He showed that some nonlinear control systems require two hidden layers to achieve stabilization. Brightwell et al. [6] went on to show that certain classes of problems are not realizable with a single hidden layer: "XOR-situation, XOR-bow-tie, XOR-at-infinity, and critical cycle".

In 2011, Nakama conducted a "fair and systematic" comparison of multiple and single hidden layer networks [10]. These had the same number of inputs, outputs, nodes, and connections, and approximate the same target functions. He shows that the learning rate is more flexible for a single hidden layer (Fig. 1(C)), and that it also converges faster than a multiple hidden layer (Fig. 1(A)). However, the activation functions used are solely a linear function of the weights a-d. Thus these can be rearranged to reduce both networks to the linear perceptrons shown in Fig. 1(B) and (D). Therefore the results are possibly subject to misinterpretation, such as justification to exclude multiple hidden layers from investigations [12].

The current study is empirical in contrast to [5–7]. It is similar to [10] in its aim to compare the networks fairly, although the definition of fairness differs slightly. Here

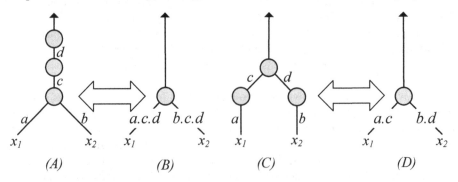

Fig. 1. Nakama networks – multiple (A) and single hidden layer (C), with equivalents (B&D)

"fairness" is defined as the networks having the same number of hidden nodes, the same activation functions, and trained with the same training algorithm with the same default training parameters. The Levenberg-Marquardt algorithm [13] is used because it is well suited and commonly used for function approximation problems [14]. The current study differs from [10] in several key areas:

- **Network Structure** – Only conventional, fully interconnected feedforward networks are used. These have non-linear activation functions in the hidden nodes, and a linear activation function in the output node. In contrast, the networks in [10] are not fully interconnected and use only linear activation functions. As has been shown, the consequence of this is that the networks can be reduced to linear perceptrons with no hidden layers at all.
- **Method of Comparison** – The generalisation capability of the networks is compared on a hidden-node-by-hidden-node basis over an adequate range of 1–64 nodes. This method of comparison was chosen as it is the main area of concern for designers of neural networks. The node-by-node comparison is facilitated by leveraging transformative optimisation, a technique developed by the Authors further described in Sect. 3. In contrast, [10] concerns itself with the learning rate and convergence properties in two specific cases with a total of 3 nodes.
- **Training Data** – Ten public domain datasets are used. These contain varying amounts of inputs, training samples and noise, as it is important to compare the networks over a wide range of different problems. In contrast [10] considers two simple mathematical functions ($y = x_1 + x_2$ and $y = x_1^2 + x_2^2$).

To the best of the Authors' knowledge, this is the first study of its kind to be undertaken.

3   Transformative Optimisation

Attempting to perform a hidden node for node comparison between SLFNs and TLFNs is a complicated affair for two reasons. Firstly, there are many different ways to allocate a given number of nodes between the first and second hidden layers of a TLFN; and secondly, the number of candidate topologies to be considered for a TLFN is quadratic. Thus it can take a prohibitive amount of time to test all candidates. Transformative optimisation is a new name given to a technique developed by the Authors which can be applied to solve both of these problems. It has been so named because it transforms the space of candidate TLFN topologies from quadratic to linear.

Consider the set of candidate TLFN topologies T, where the number of nodes in each hidden layer varies between 1 and k, i.e.

$$T = \left\{ \tau_{11}, \tau_{21}, \ldots, \tau_{ij}, \ldots, \tau_{kk} \right\}, \tag{1}$$

where

$$\tau_{ij} = N_0 : i : j : N_3 \tag{2}$$

The constants, N_0 and N_3 represent the number of inputs and outputs respectively for any given domain, and i and j are the number of nodes in the first and second hidden layers respectively. Transformative optimisation works by transforming this set into a different set $T' = \{\tau_4, \ldots, \tau_{n_h}, \ldots, \tau_{2k}\}$, where $T' \subset T$. This only has a single degree of freedom, n_h, which represents the total number of hidden nodes. Here

$$\tau_{n_h} = N_0 : [\varrho n_h + c] : [(1 - \varrho)n_h - c] : N_3, \tag{3}$$

where ϱ is a ratio determining the node allocation between the first and second hidden layers, and c is a constant. It has been shown in the Authors' previous studies [8, 9], that the optimal values are $\varrho = 0.5$ and $c = 1$. These had the highest probability of finding the best generalisers over the same ten datasets used in the current experiments. Substituting these values into (3), we have

$$\tau_{n_h} = N_0 : [0.5n_h + 1] : [0.5n_h - 1] : N_3. \tag{4}$$

It should be noted that the lowest value of n_h which yields a two hidden layer network is four, organised as three nodes in the first hidden layer and a single node in the second. Additionally, odd values of n_h yield fractional node values. This can be dealt with in two ways. The first rounds the nodes in the first hidden layer down and rounds those in the second up or vice versa. Alternatively, only even values of n_h are considered:

$$\tau_{n_h} = N_0 : [i + 1] : [i - 1] : N_3, \tag{5}$$

where $n_h = 2i$ and $2 \leq i \leq k$. In this case the set of candidate topologies T' is

$$T' = \{\tau_2, \ldots, \tau_i, \ldots, \tau_k\}, n_h = 2i. \tag{6}$$

Note that from (1), $|T| = k^2$, and from (6), $|T'| = k - 1$. This corresponds to the transformation of the candidate space from quadratic to linear. In this study, the rounding method is used to compare the performance of SLFNs and TLFNs node by node.

4 Experiments

4.1 Data Acquisition

A total of ten datasets were acquired for the experiments. These were selected on the basis of their availability in the public domain, and suitability for function approximation. One consequence of these selection criteria is that the datasets all have a single output. An exception to this was the engine dataset available in Matlab, which has two inputs and two outputs. However, the torque output was reassigned as an input, thus converting it to a 3 input, 1 output dataset. The datasets were sourced from the UCI Machine Learning Repository [15], Bilkent University Function Approximation Repository [16], University of Porto Regression Datasets [17], and Matlab. The datasets used are summarised in Table 1.

Following the data acquisition phase, Matlab R2014b was used to prepare the data, create and train the neural networks and generate the raw results for analysis.

Table 1. Dataset summary

Name	Samples	Inputs	Source
Abalone	4177	8	UCI Machine Learning Repository
Airfoil self-noise	1503	5	UCI Machine Learning Repository
Chemical	498	8	Matlab chemical_dataset
Concrete	1030	8	UCI Machine Learning Repository
Delta elevators	9517	6	University of Porto Regression Datasets
Engine	1199	3	Matlab engine_dataset
Kinematics	8292	8	BU Function Approx. Repository
Mortgage	1049	15	BU Function Approx. Repository
Simplefit	94	1	Matlab simplefit_dataset
White wine	4898	11	UCI Machine Learning Repository

4.2 Data Preparation

This one-off phase split each dataset into three sub-sets: training, validation and test. The validation set was used to stop the training early when the validation error began to rise, and the test set was used exclusively as an estimate of each network's generalisation error. Eighty percent of the original data was randomly allocated to the training set, and the remaining twenty percent was equally allocated between the validation and test sets. In order to ensure consistency, the same sub-sets were used to create and test all networks within the scope of a given dataset.

4.3 Network Creation, Training and Evaluation

All networks in the experiments were created, trained and evaluated in an identical fashion. They were created using the 'fitnet' function of the Neural Network Toolbox, with the 'mapminmax' function for input and output processing. The activation function of all hidden nodes was 'tansig' and that of the output node was linear. The training function used was Levenberg-Marquardt, 'trainlm', which often yields the best results for function approximation problems. The networks were all trained using Matlab's default training parameters: 1000 epochs, training goal of 0, minimum gradient of 10^{-7}, 6 validation failures, $\mu = 0.001$, $\mu_{dec} = 0.1$, $\mu_{inc} = 10$ and $\mu_{max} = 10^{-10}$.

The error function used during training was the mean squared error function 'mse', however the generalisation error was reported using the normalised root mean squared error (NRMSE). This is a function of the number of samples n, the target output $\hat{y}_i$, and the actual output y_i. It is normalised to the target output swing $\hat{y}_{max} - \hat{y}_{min}$, in order to faciltate comparison between different datasets. The NRMSE ε is given by:

$$\varepsilon = \frac{1}{\hat{y}_{max} - \hat{y}_{min}} \sqrt{\frac{\sum_{i=1}^{n} (y_i - \hat{y}_i)^2}{n}} \tag{7}$$

4.4   Experimental Method

A total of 20 experiments were carried out on each of the datasets. Half of these were using SLFNs, and the other half using TLFNs. Each experiment consisted of varying the number of hidden nodes n_h between 1 and 64. For SLFNs, the number of hidden nodes was simply n_h. For TLFNs, the number of nodes in each of the hidden layers was calculated according to Eq. (3), using the rounding method described in Sect. 2. Because the random weight initialisation can cause training to getting trapped in local minima [4], each topology is trained 30 times, and the network with the most favourable generalisation error is chosen as the "champion" network for that particular topology. For clarity, the pseudo-code for a single experiment is shown in Fig. 2.

5   Results and Discussion

For each experiment, the overall "champion" is the network with the lowest generalisation error. Its total number of hidden nodes n_h, and generalisation error ε are recorded. The results are summarised in Table 2, were $\mu(n_h)$ represents the mean number of hidden nodes and $\mu(\varepsilon)$ the mean generalisation error (as a percentage) over ten experiments. The relative improvement is calculated as $\delta\mu(\varepsilon) = \mu(\varepsilon_{SLFN}) - \mu(\varepsilon_{TLFN})$, and the improvement factor is calculated as $f = \delta\mu(\varepsilon)/\varepsilon_{SLFN}$. The winners are in Table 2 are emboldened.

As might be expected, the results were varied and dataset dependent. The generalisation error is known to be dependent on the complexity of the function to be approximated as well as the level of noise within it. For example, the Simplefit dataset

```
function e = singleExperiment(type, nh)
  for nh = 1 to 64 do
    if type is 'SLFN'          % calc nodes in each layer
      n1 = nh                  % First hidden layer nodes
      n2 = 0                   % SLFN has no hidden layer 2
    elseif type is 'TLFN'
      n1 = int(0.5nh + 1)      % round down for layer 1
      n2 = nh - n1             % Calculate nodes in layer 2
    end if
    for candidate = 1 to 30  do % process 30 networks
      net = createNetwork(n1,n2)
      nrmse[candidate] = trainNetwork(net)
    end do
    e[nh] = min(nrmse) % Calculate winner's error
  end do
  return e
end function
```

Fig. 2. Pseudo-code for a single experiment

is a very simple function which has no noise at all. It yielded almost zero generalisation error, and there is not much to be gained by using a TLFN. On the other end of the scale, the Airfoil dataset showed significant gains in generalisation error for a TLFN. Overall, 9 out of 10 datasets showed an improvement of some kind when using two hidden layers. Some showed more significant gains than others. The only exception to this is with the Chemical dataset, where a single hidden layer performed best. However this was only a 0.1% relative improvement $\delta\mu(\varepsilon)$. Since the generalisation error is so low in some cases, it was thought that a fairer method of comparison was using the improvement factor f in the final column of Table 2. The overall average improvement factor was 5.72% with a standard deviation of 8.5%. In over half of the cases, the TLFNs also achieved this improvement with fewer hidden nodes. It should be remembered, however, that node for node TLFNs are more complex as they have more weights. Since it is the weights which learn the problem, TLFNs have a larger storage capacity, which might account for this improvement.

Table 2. Results summary

Dataset	$\mu(n_h)$		$\mu(\varepsilon)$ (%)		$\delta\mu(\varepsilon)$ (%)	f (%)
	SLFN	TLFN	SLFN	TLFN		
Abalone	34.1	24.7	6.5280	**6.4597**	0.0683	1.05
Airfoil	40.9	38.7	3.8244	**2.8222**	1.0022	26.21
Chemical	31.6	36.4	**3.3738**	3.4762	−0.1024	−3.04
Concrete	44.5	41.2	4.0272	**3.6937**	0.3335	8.28
Delta Elevators	10.7	18.6	5.0334	**5.0315**	0.0019	0.04
Engine	56.9	46.9	0.8963	**0.8428**	0.0535	5.97
Kinematics	54.1	35.2	4.5220	**4.3730**	0.1490	3.30
Mortgage	51.7	50.1	0.3497	**0.3424**	0.0073	2.10
Simplefit	10.4	10.8	0.0014	**0.0012**	0.0002	12.63
White Wine	36.3	47.0	11.4743	**11.3983**	0.0761	0.66

So which is better? The evidence in these experiments point to TLFNs. However, the actual amount of improvement is case dependent. Furthermore, low complexity might be a design consideration or requirement and this might need to be balanced against the potential gains in generalisation error. So perhaps a better question might be: Is it *worth* using a TLFN? Fortunately, through transformative optimisation, this can easily be checked. Although the full 1–64 node scans used in these experiment (Figs. 4, 5 and 6 in Appendix A) each took several hours, this need not be the case. Binary sampling techniques can be used to reduce this process to a matter of minutes, whilst giving a broad idea of the likely gains (if any) and sometimes even an idea of where to look. This is illustrated in Fig. 3, which shows the result of binary sampling applied to the Airfoil dataset.

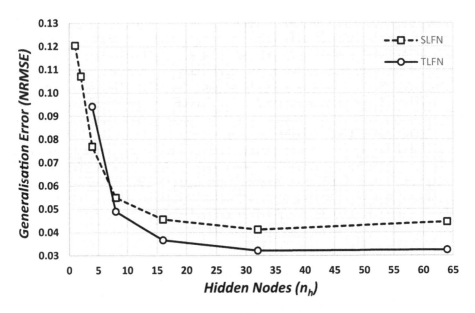

Fig. 3. Airfoil using 12 point binary sampling

6 Conclusion

This study set out to discover whether networks with two hidden layers generalise better than those with a single hidden layer in practical situations. In order to answer this question, a method called "transformative optimisation" was proposed and applied to perform a hidden-node-by-hidden-node comparison of SLFNs and TLFNs across ten separate datasets. It was found that in nine out of ten cases TLFNs outperformed SLFNs, but that the amount of improvement was very case dependent. However, the proposed method can be used in conjunction with binary sampling to rapidly determine whether it is worth using two hidden layers for any given problem. Although the results presented here indicate that TLFNs outperform SLFNs, further investigation with more complicated real-life datasets is necessary.

On a final note, this method could potentially be used for other training algorithms, although it would need to be verified whether the optimal values of $\varrho = 0.5$ and $c = 1$ still hold. Early indications are that this could well be the case for the 'trainscg' training algorithm [8], although more extensive testing is required. This verification, as well as equivalent comparisons of TLFNs and SLFNs for other training algorithms could be the subject of further work.

Acknowledgements. We thank Prof. Martin T. Hagan of Oklahoma State University for kindly donating the Engine dataset used in this paper to Matlab. Thanks also to Prof. I-Cheng Yeh for permission to use his Concrete Compressive Strength dataset [18], as well as the other donors of the various datasets used in this study.

Appendix A – Full Results: Average Node for Node Comparisons

See Figs. 4, 5 and 6.

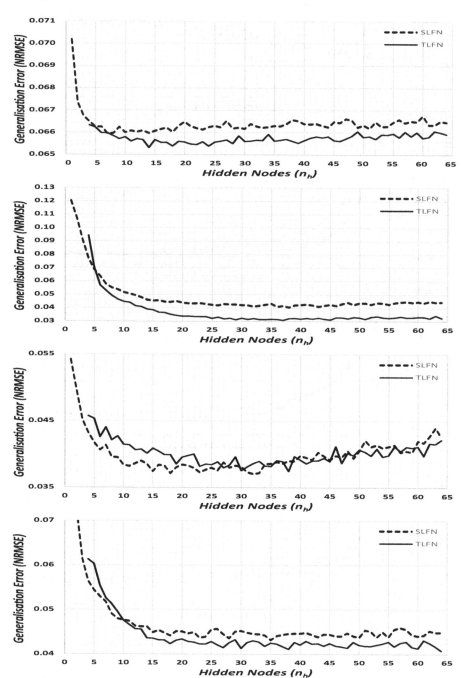

Fig. 4. Abalone (top), Airfoil, Chemical and Concrete (bottom)

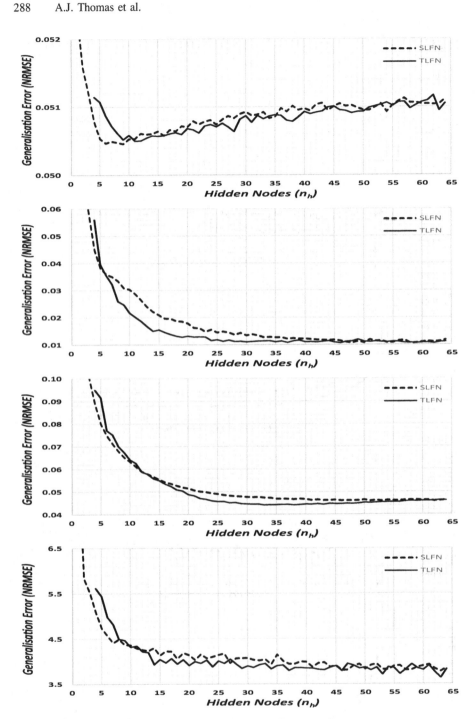

Fig. 5. Delta Elevators (top), Engine, Kinematics, and Mortgage (bottom)

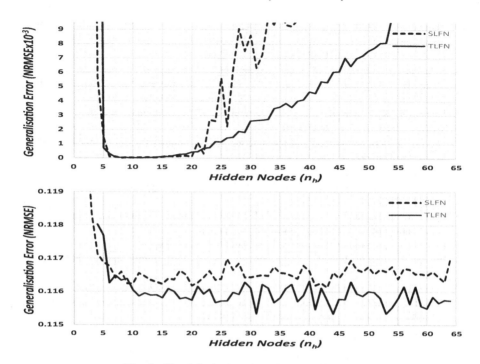

Fig. 6. Simplefit (top) and White Wine (bottom)

References

1. Hornik, K., Stinchcombe, M., White, H.: Multilayer feedforward networks are universal approximators. Neural Netw. **2**, 359–366 (1989)
2. Hornik, K., Stinchcombe, M., White, H.: Some new results on neural network approximation. Neural Netw. **6**, 1069–1072 (1993)
3. Huang, G.-B., Babri, H.A.: Upper bounds on the number of hidden neurons in feedforward networks with arbitrary bounded nonlinear activation functions. IEEE Trans. Neural Netw. **9**, 224–229 (1998)
4. Zhang, G.P.: Avoiding pitfalls in neural network research. IEEE Trans. Syst. Man Cybern. Part C Appl. Rev. **37**, 3–16 (2007)
5. Chester, D.L.: Why two hidden layers are better than one. In: Caudhill, M. (ed.) International Joint Conference on Neural Networks, vol. 1, pp. 265–268. Laurence Erlbaum, New Jersey (1990)
6. Brightwell, G., Kenyon, C., Paugam-Moisy, H.: Multilayer neural networks: one or two hidden layers? In: Mozer, M.C., Jordan, M.I., Petsche, T. (eds.) Advances in Neural Information Processing Systems, vol. 9, pp. 148–154. MIT Press, Cambridge (1997)
7. Sontag, E.D.: Feedback stabilization using two-hidden-layer nets. IEEE Trans. Neural Netw. **3**, 981–990 (1992)
8. Thomas, A.J., Walters, S.D., Petridis, M., Malekshahi Gheytassi, S., Morgan, R.E.: Accelerated optimal topology search for two-hidden-layer feedforward neural networks. In: Jayne, C., Iliadis, L. (eds.) EANN 2016. CCIS, vol. 629, pp. 253–266. Springer, Cham (2016). doi:10.1007/978-3-319-44188-7_19

9. Thomas, A.J., Walters, S.D., Malekshahi Gheytassi, S., Morgan, R.E., Petridis, M.: On the optimal node ratio between hidden layers: a probabilistic study. Int. J. Mach. Learn. Comput. **6**, 241–247 (2016). doi:10.18178/ijmlc.2016.6.5.605

10. Nakama, T.: Comparisons of single- and multiple-hidden-layer neural networks. In: Liu, D., Zhang, H., Polycarpou, M., Alippi, C., He, H. (eds.) Advances in Neural Networks – ISNN 2011 Part 1. LNCS, vol. 6675, pp. 270–279. Springer, Heidelberg (2011)

11. Funahashi, K.-I.: On the approximate realization of continuous mappings by neural networks. Neural Netw. **2**, 183–192 (1989)

12. Idler, C.: Pattern recognition and machine learning techniques for algorithmic trading. MA thesis, FernUniversität, Hagen, Germany (2014)

13. Moré, J.J.: The Levenberg-Marquardt algorithm: implementation and theory. In: Watson, G.A. (ed.) Numerical Analysis. LNM, vol. 630, pp. 105–116. Springer, Heidelberg (1978). doi:10.1007/BFb0067700

14. Beale, M.H., Hagan, M.T., Demuth, H.B.: Neural Network Toolbox User's guide. https://www.mathworks.com/help/pdf_doc/nnet/nnet_ug.pdf

15. UCI Machine Learning Repository. https://archive.ics.uci.edu/ml/

16. Bilkent University Function Approximation Repository. http://funapp.cs.bilkent.edu.tr/DataSets/

17. Regression Datasets. http://www.dcc.fc.up.pt/~ltorgo/Regression/DataSets.html

18. Yeh, I.-C.: Modeling of strength of high performance concrete using artificial neural networks. Cem. Concr. Res. **28**, 1797–1808 (1998)

Neural Networks as a Learning Component for Designing Board Games

Alexandros Nikolakakis and Dimitris Kalles$^{(\boxtimes)}$

School of Science and Technology, Hellenic Open University, Patras, Greece
alexandros_nkl@hotmail.com, kalles@eap.gr

Abstract. In this paper we present a new strategy game, with machine learning computer players, which have been developed using temporal difference reinforcement learning coupled with neural networks; the latter are used for value approximation and for storing the players' knowledge. We set out the game rules and then design and implement a comprehensive experimentation session to allow us to explore a large state space for investigating learning and playing behavior, without placing unreasonable demands on speed and accuracy. Our experiments demonstrate how computer players manage to adapt to their environment and improve their tactic over time, based on experience only, while still accommodating a variety of behaviors which are tuned via the conventional parameters of the reinforcement learning and neural network mechanisms.

Keywords: Neural networks · Machine learning · Reinforcement learning · Artificial intelligence · Strategy games

1 Introduction

Most people would likely believe that a computer can never become "better" than a human at playing strategy games, just because humans can deploy an imaginative way of thinking. That belief was shattered initially in 1997, when the then-reigning world chess champion Gary Kasparov was defeated by IBM's Deep Blue [1]. Strategy games are an ideal environment to apply machine learning because the set of defined rules which govern such games allows experimentation with potentially flawless (yet, expensive to implement) opponents. Besides chess, machine learning has been applied to backgammon [2, 3], go [4] and othello [5], to name the most widely known games, and to lesser known ones, like RL Game [6] and many more [7–9], many of which were initially conceived as research and/or education tools.

The main contribution of this paper is the evolution of a relatively novel and simple game, RLGame [6], into a more complex game, more akin to games like chess and stratego. We still attempt to couple such evolution with a fully-fledged capability to automatically learn how to play such a game using reinforcement learning, in a fairly attractive 3D environment, to facilitate the recording and the analysis of games between people, as well as the comparisons with machine learning based strategies. The application allows for experimentation with a variety of reinforcement learning parameters and makes extensive use of conventional feed-forward neural networks as a mechanism for approximating the value function of a TD(λ) based learning mechanism

© Springer International Publishing AG 2017
G. Boracchi et al. (Eds.): EANN 2017, CCIS 744, pp. 291–302, 2017.
DOI: 10.1007/978-3-319-65172-9_25

as well as a transfer learning mechanism for storing and re-using knowledge about how to play a game.

The rest of this paper is structured in four subsequent sections. Section 2 briefly reviews some related work and sets the path that led to the development of the new game. Details about design and rules appear in Sect. 3 while the results of the experimentation appear in Sect. 4. The last section summarizes the work and sets out key directions for further work.

2   A Brief Review of Related Work

Scientific research in artificial intelligence has a central component in intelligent agent research, where an agent is broadly conceptualized as an entity which can perceive its environment by using sensors (in the generic sense) and react using actuators [10]. An agent is usually implemented in software, whereas the environment could be another program (for example, a video game implementation), or a subset of the world (as in robotics).

Reinforcement learning, contrary to supervised learning, does not rely on examples provided by a knowledgeable external supervisor. As, during learning in games, the best action at a given game state might not be clearly defined, reliance on some reward (or penalty) signal is the closest means we have to interact with a supervisor. Such signals can be dispersed throughout a learning experience but, more often than not, are concentrated near final game states. An agent will then discover which actions will maximize its reward by trying some of them and observing the rewards it reaps [11] and, accordingly, by maybe modifying its approach to how it selects its actions [12–16]. Recent research has confirmed that an agent playing in a social environment learns better than a self-playing one [14, 15, 17].

An RL agent has to resolve the exploitation – exploration dilemma. To obtain maximum reward RL it should prefer an action that gave it a good reward in the past, but this prevents it from trying new actions which could give an even larger reward. An agent's policy is a balanced way to choose between exploitation and exploration.

Temporal-difference (TD) learning is an efficient combination of Monte Carlo and dynamic programming, as the estimated value of a game state can be modified by another estimation or an actual reward. A key component of TD learning is the concept of eligibility traces, which allow us to specify how many of an agent's previous actions are considered important for its current state, so that we can back-propagate any change in our estimate about the current state's worth. This is done through a lambda (λ) factor, where $\lambda = 1$ indicates that all previous game state values are affected by its current action, whereas $\lambda = 0$ indicates that only the previous one is affected.

Developing an accurate value function is of paramount importance to reinforcement learning. A conventional approach of state-value pairs is hardly a recommended approach for games beyond toy-size ones. Herein is where artificial neural networks have been deployed with profound success; not only can they deliver efficiency in terms of speed and space, by approximating the calculation of value functions, but they can also help capture deeper associations arising from game topology and be further used as mechanisms for transfer learning.

3   A New Game: Rules, Learning and Interaction

RL Strategy Game is a java application, using the libGDX graphics development framework. It implements a two-player, turn based, strategy game and allows the training of computer players. Computer players are controlled by intelligent agents and use reinforcement learning to adapt to their environment. Additionally, these agents use multilayer neural networks to approximate the evaluation of game state values during learning and playing.

3.1   Game Description

RL Strategy Game is a strategy game being played on a chessboard and having 3 types of pawns. At first sight RL Strategy Game looks like chess because of the chessboard, but it is not rectangular, and it does not have specific dimensions. Furthermore, there are positions on the board which create barriers and cannot be occupied by any pawn. Countless variants can be played by changing the chessboard dimensions, the barrier positions, the number and the type of each player's pawns, as well as the initial position of each pawn.

Each player has a base, marked by a flag, red or blue (see Fig. 1 for an indicative snapshot of a game situation). Bases are located at diagonally opposing corners of the board, and can be occupied by any pawn. Every pawn has a front side which is called "face". In order to describe a pawn's state, we need to specify its co-ordinates and the direction of its face (north, west, south, east).

Fig. 1. An indicative snapshot of RL Strategy Game in a 3D board

There are three types of pawns: Infantry, Spearman and Knight. Infantry and spearmen can perform one action per turn while a knight can perform two. Some games might contain just one or two of these types of pawns.

There are, also, three types of actions: transposition, rotation and attack. A rotation is the action that changes a pawn's face direction without changing its position. In a transposition, a pawn occupies one of the nearby board positions, in a succession of two steps: at first it rotates and then it makes a front step, (so, a rotation is not considered as an action when it occurs during a transposition, but when it happens individually it is considered as one). With an attack, a pawn eliminates one opponent pawn and occupies its position. Note that the transposition is a move that facilitates attacks whereas a defensive tactic incurs a larger number of moves.

Apart from these general rules, there are specific rules regarding each pawn type. Infantries can attack all opponent pawns which are located in the four nearby positions (diagonal moves are not allowed in infantry). Also, in case there are two opponent infantries, infantry A cannot attack infantry B, if B faces A. Spearmen can attack any opponent pawn located in the four nearby diagonal locations. When the opponent pawn is eliminated the spearman's face direction is west or east, depending on the eliminated pawn's location.

Knights can attack any pawn located in the four nearby positions as long as the opponent pawn is not facing the knight. It can be claimed that the knight is the most powerful pawn because of the two actions per turn, which is why two extra restrictions apply on the transpositions and rotations it is allowed to perform: (1) A knight cannot select a rotation as its first action, (2) In a succession of two transpositions, the second one cannot revert the first one.

There are three ways for a player to achieve a victory: (1) By eliminating all opponent pawns, (2) By occupying the opponent's base for two successive turns, (3) By eliminating all legal actions from all opponent pawns. A draw is declared when none of the two players emerges as winner after a pre-specified maximum number of turns have been played.

3.2 Designing and Implementing the Neural Network Mechanism for Learning and Playing

The RL agent uses a three layer neural network to estimate the value of each game state. Both the input layer and the hidden layer have the same number of neurons, depending on the size of the variant and the type of pawns used. The output layer has only one neuron which corresponds to the "value" (in the RL sense) of the board, as encoded by the input layers neurons.

A key issue is to decide how many neurons will be used to encode the game state for the input layer, as a large number of neurons can make the neural net more accurate but at the cost of slower learning. Below, we describe the eventually used network configuration, which our experimentation delivered as being most effective. We note that this was not an arbitrary choice and, for the sake of completeness, we also summarize, in Table 1, all the network variants which we tested and which proved either too slow (variants 1 and 2) or too ineffective (variants 2, 3 and 4); however, for space limitations we do not show individual performance/speed results which led us to rule out those architectures.

The specific numbers of input and hidden layer neurons (n) depend on the number of positions (cells) on the chessboard which can be occupied by pawns (p), and the number of pawn types which are used on the variant (a):

Table 1. Neural network input encodings to be avoided

#	Neurons (number, type)	Description
1	24 binary per board cell	Every board cell can have 24 occupied states (2 players $\times$ 3 pawn types $\times$ 4 faces) and 1 non-occupied - Non-occupied cell: all 24 neurons are set 0 - Occupied cell: one neuron is set to 1
2	9 binary per board cell	Semantics of a neuron being set to 1: - 1st: occupied by agent's pawn - 2nd: occupied by opponent's pawn - 3rd–5th: pawn type - 6th–9th: pawn face
3	$X + Y + 4$ binary per pawn (X, Y: board dimensions) All neurons except 3 are set to 0 according to the description	One of the $1^{st} - X^{th}$ neurons is set to 1, depending on cell X One of the $(X + 1)^{th} - (X + Y)^{th}$ neurons is set to 1, depending on cell Y One of the last 4 neurons is set to 1 depending on the pawn's face
4	4 non binary per board cell Neurons 1 and 2 depend on the agent's pawn while 3 and 4 depend on the opponent's pawn	Neuron 1 is set as follows: - 0: no occupation - 1: occupied by infantry - 2: occupied by spearman - 3: occupied by knight Neuron 2 is set as follows: - 0: no occupation - 1...4: depending on pawn face Neurons 3 and 4 are set correspondingly

$$n = p * (a + 1) + a \qquad (1)$$

Each chessboard position is allocated $a + 1$ input neurons. At most one of the first a neurons will assume a value depending on the type of the agent's pawn which sits on it. The $a + 1^{th}$ neuron will assume a value if an opponent pawn sits on the chessboard position. The rules for value assignment are as follows:

- If the position is occupied by the agent's pawn and:
  - it might be eliminated, the value is -1
  - it cannot be eliminated nor can it eliminate an opponent, the value is 2.5
  - it might eliminate an opponent but cannot be eliminated, the value is 3
- If the position is occupied by an opponent's pawn, the $a + 1^{th}$ neuron assumes a value of 1 and the a previous values are -3.
- If the position is not occupied, the value is -3 for all neurons.

The value of each one of the last a neurons of the input layer is set as follows (for each one of a types):

$$(RLpawns_a - OpponentPawns_a)/InitialNumberOfPawns_a \qquad (2)$$

3.3   Game Learning and Playing User Interactions

Training statistics are the primary means of conveying information regarding the quality of training. While the application is running, it always displays the performance statistics of the last $1/100^{th}$ of the total of games played during a particular experimental session. These statistics highlight the improvement of the RL agent's strategy over time and can be retrieved from log files for subsequent analysis and visualization.

A user can set-up an experiment by specifying key RL parameters which will guide the learning process; note that the neural network is essentially the transfer learning mechanism for a particular game configuration since it captures playing and learning behaviors which have been deployed. These parameters concern the ε-greedy policy of the agent for trading off exploitation with exploration, the λ parameter for controlling the eligibility traces and the γ discount rate. Additional options allow a user to create a variety of reward schemes which may used (for example, by extending a small reward each time an opponent pawn is eliminated or by setting minimum and maximum values for rewards).

Opponents are instrumental in learning how to play 2-player games. When an agent learns how to play a game, it really learns against a specific opponent (but, of course, may have learned from any number of opponents up to that point). To accommodate some diversity in playing and training against classes of opponents, we have also implemented some algorithmic opponents. CPU-Random, the simplest of all, just selects a legal move at random. CPU-Attack chooses an attack move at random, if attacks are possible, otherwise it reverts to CPU-Random. CPU-Defense evaluates the positions where it is subject to an opponent attack and attempts to minimize losses; if it is not threatened it reverts to CPU-Attack. Finally, CPU-Preset computes a composite measure which takes into account the benefit of attacking opponent pawns, the advance towards the opponent base and the need to avoid compromising its own pawns and base. CPU-x players do not use any (machine learning) technique to improve their play.

As a last note, we remind the reader that 3D graphics are usually attractive but slow down performance. An efficient training of an RL agent needs 20,000 to 500,000 games depending on the variant size and the number of pawns; these are simply impractical to visualize on the fly. We allow users to run a large series of games by toggling visualizations off and on, alternatively, for training the neural networks in batch mode and then for using them in an inspection mode.

4   Experimentation

We now present the results of the training sessions. We have experimented with 6 quite diverse variants which are graphically summarized in Fig. 2, while Table 2 sets out the number of games played per variant.

We have trained the agents using combinations of RL parameters (ε-greedy, γ, λ, min-reward, max-reward, elimination reward) and opponents drawn from the collection of opponent types specified in Sect. 3.3. The values of the parameters in Table 3 are indicative of distinct behaviors as observed over a large number of trials.

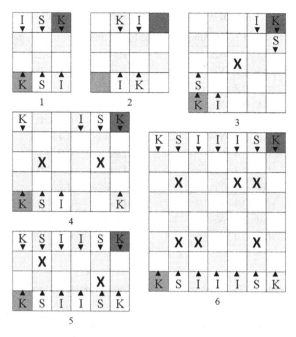

Fig. 2. The set up of all variants (in 2D); pawn faces are indicated by arrows

Table 2. Number of games per variant

Variant	Name	Number of games
1	3 × 4 Basic – 3 Pawns	200,000
2	4 × 4 Basic – 2 Pawns	200,000
3	5 × 5 Arena – 3 Pawns	500,000
4	6 × 5 Two Barriers – 4 Pawns	500,000
5	6 × 4 Two Barriers – 6 Pawns	500,000
6	7 × 8 Four Barriers – 7 Pawns	300,000

RL training is performed with each variant for all parameter sets. At the beginning of every training experiment a new neural network is created, with random weights; at the end of the experiment, the network can be stored and user at a subsequent time, either for resumed learning or for playing.

The quality of the training is evaluated by pitting the trained agent against an opponent in a playing-only mode (i.e. with full exploitation and its learning ability switched off, so the neural network is simply used for state value approximation).

A focused evaluation was first carried out by pitting each trained agent against its training opponent, for 1,000 successive games. Table 4 shows the results (for example, in the 7th row, under the 6 × 5 column, we see a figure of 90% which means that, we used the 4th variant – as shown in Table 2 - and the 7th parameter set – as shown in Table 3 - and during the evaluation session, the trained agent won 90% of the games against its opponent - a CPU-defense one).

Table 3. Configuration of learning parameters

Parameter set	ε	λ	Elimination reward	Opponent	γ	(min, max) reward
1	0.5	0	No	Random	0.9	−1,1
2	0.5	0.2	Yes	Random	0.9	0,1
3	0.6	0.2	No	Attack	0.9	−1,1
4	0.8	0.2	Yes	Attack	0.9	−1,1
5	0.6	0	No	Attack	0.9	0,1
6	0.8	0.2	Yes	Attack	0.9	0,1
7	0.9	0.4	No	Defense	0.9	−1,1
8	0.9	0.7	No	Defense	0.9	0,1
9	0.9	0.4	Yes	Defense	0.9	−1,1
10	0.9	0.4	Yes	Defense	0.9	0,1
11	0.9	0.4	No	Preset	0.9	−1,1
12	0.9	0.4	No	Preset	0.9	0,1

Table 4. Performance evaluation against training opponent

Parameter Set \ Variant	3x4	4x4	5x5	6x5	6x4	7x8
1	100%	99%	100%	100%	99%	99%
2	99%	99%	100%	100%	99%	99%
3	100%	100%	100%	100%	98%	97%
4	100%	100%	100%	100%	97%	99%
5	100%	100%	100%	100%	98%	95%
6	99%	100%	100%	100%	97%	97%
7	100%	100%	94%	90%	72%	67%
8	99%	90%	97%	89%	75%	69%
9	66%	100%	94%	79%	81%	69%
10	98%	100%	92%	82%	76%	73%
11	0%	0%	0%	0%	0%	0%
12	100%	0%	0%	0%	97%	0%

A further evaluation experiment was then carried out by pitting each trained agent (again, with full exploitation and no learning) against all possible opponents specified in Sect. 3.3 (for 1,000 games against each opponent). Table 5 shows the results.

A large percentage indicates a good strategy for the variant, however, a low percentage does not necessarily indicate a profound learning failure (for example, learning to play against a random opponent might allow an agent to beat that opponent but there is no experience to guide the agent into playing competitively against stronger opponents; hence, learning is limited not because of the learning mechanism but because of the data).

Table 5. Performance evaluation against all opponents

Variant / Parameter Set	3x4	4x4	5x5	6x5	6x4	7x8
1	74%	41%	46%	53%	34%	43%
2	46%	88%	38%	40%	40%	56%
3	75%	97%	71%	58%	63%	63%
4	75%	68%	72%	62%	77%	64%
5	75%	99%	79%	53%	58%	62%
6	77%	82%	75%	65%	56%	62%
7	73%	81%	77%	75%	58%	65%
8	94%	60%	79%	70%	51%	67%
9	59%	81%	77%	69%	53%	66%
10	74%	82%	76%	69%	53%	67%
11	58%	45%	32%	27%	37%	4%
12	95%	30%	19%	15%	71%	0,08%

Briefly reviewing the results, when the agent trains against CPU-Random (parameter sets 1 and 2), a relatively high adaptation to the environment is witnessed (performance is about 99–100%, as shown the first two rows in Table 4). As CPU-Random basically lacks any strategy, the agent develops a strategy which leads it directly to the opponent (which plays CPU-Random) base; however, this short-sighted approach severely limits its ability to face other opponents, as shown by the first two rows in Table 5.

Note that when the agent is trained against CPU Attack (parameter sets 3–6), it still basically learns to play near perfectly against a relatively short-sighted opponent (rows 3–6 in Table 4), though its ability to generalize is slightly improved, as witnessed by its performance against all opponents (rows 3–6 in Table 5).

Training against a CPU-Defense opponent raises performance quite consistently. As CPU-Defense is conservative, the trained agent still does manage to win a large fraction of the focused evaluation games (rows 7–10 in Table 4) but is also manages to do quite well against the mix of opponents, which contains the sophisticated CPU-Preset opponent (rows 7–10 in Table 5).

The two latter rows of Table 4 indicate a poor performance against the sophisticated CPU-Preset player. There are a couple of instances where the trained agent managed to learn how to address the opponent but these abilities did not readily extend to all other opponents, as they concerned games with a relatively large number of pawns relatively to board size, which compromised the quality of the CPU-Preset player. It all boils down to suggesting that the training session was limited to trying to respond to a stronger player and there was not enough room to develop an independent strategy that could be generalized to other opponents.

A brief note on eligibility traces is also due. An optimum value of λ is not easy to determine. We have observed that small values of λ, close to 0, result in slow, smooth learning whereas larger ones accelerate learning but also render it unstable. If λ is very high, close to 1, learning is close to failure.

In Fig. 3 we show the results of the same underlying experiment for 3 values of λ^1, namely 0, 0.4 and 0.7. The increase in λ leads to speeding up the training performance for the first games of each experiment. The increase from 0 to 0.4 has a positive effect, as higher winning percentages are observed throughout the experiment and also the final winning percentage rises from 75% to 80%. However, a further increase to 0.7 initially delivers (at the first 80,000 games) higher winning percentages than those at $\lambda = 0.4$ but then only compares well to the ones at $\lambda = 0$, with a final winning percentage decreasing back to 72% (which is the lowest overall). Similarly, the optimum value of λ for each variant depends on all the parameters and can be only determined by trial and error.

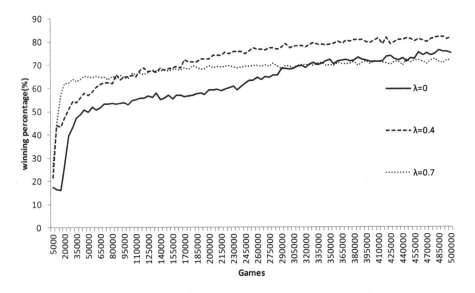

Fig. 3. Training performance in accordance with lambda

5   Conclusions and Directions for Further Work

We have elaborated on the development of a new strategy board game where, after specifying the rules, we have used neural networks as a learning approximation mechanism. By combining the neural networks with a reinforcement learning algorithm, we were able to experiment with a substantial variety of board variants, and semi-automatically gain substantial insight into the playability and learnability of the new game. The neural networks have been instrumental for speeding-up the

---

[1] The rest of the parameters are as follows: Variant = 6 × 5, ε-greedy = 0.9, gamma = 0.9, elimination reward = false, min-max reward = (−1, 1), opponent = CPU Defense.

experimentation and we believe that this is key evidence for using them in the life-cycle of game design. We do stress, however, that our implementation is modular enough to admit the possibility of embedding a variety of machine learning paradigms for exploring the performance-vs-speed relationship in more depth and in more scalable infrastructures to accommodate massive-scale experimentation [18].

Our work has already delivered results in terms of learning and playing but it is far from complete. We aim to further investigate the fine-tuning of the learning mechanisms so as to make it better adapt to a more elusive goal: not just learning to play a game as a cut-throat opponent but learning to play so as to also help a human opponent improve his/her game play.

References

1. Campbell, M., Hoane, A.J., Hsu, F.H.: Deep blue. Artif. Intell. **134**(1–2), 57–83 (2002)
2. Tesauro, G.: Programming backgammon using self-teaching neural nets. Artif. Intell. **134** (1–2), 181–199 (2002)
3. Papahristou, N., Refanidis, I.: Improving temporal difference learning performance in backgammon variants. In: Herik, H.J., Plaat, A. (eds.) ACG 2011. LNCS, vol. 7168, pp. 134–145. Springer, Heidelberg (2012). doi:10.1007/978-3-642-31866-5_12
4. Knight, W.: Google's AI masters the game of go a decade earlier than expected (2016), https://www.technologyreview.com/s/546066/googles-ai-masters-the-game-of-go-a-decade-earlier-than-expected/
5. Makris, V., Kalles, D.: Evolving multi-layer neural networks for Othello. In: 9th Hellenic Conference on Artificial Intelligence, Thessaloniki (2016)
6. Kalles, D., Kanellopoulos, P.: On verifying game design and playing strategies using reinforcement learning. In: Proceedings of ACM Symposium on Applied Computing, Special Track on Artificial Intelligence and Computation Logic, Las Vegas (2001)
7. Ram, A., Ontañón, S., Mehta, M.: Artificial intelligence for adaptive computer games. In: 20th International FLAIRS Conference on Artificial Intelligence (FLAIRS-2007). AAAI Press (2007)
8. Collection of Playable Experimental Games Created by Researchers in the Field of Artificial Intelligence (AI). http://www.aigameresearch.org/
9. Scott, J.P.: List of Online Game Learning Software (2001). http://satirist.org/learn-game/lists/software.html
10. Russell, S.J., Norving, P.: Artificial Intelligence: A Modern Approach, 3rd edn. Pearson Higher Education, Upper Saddle River (2010)
11. Sutton, R., Barto, A.: Reinforcement Learning: An Introduction, 2nd edn. MIT Press, London (2012)
12. Ferber, J.: Multi-agent Systems: An Introduction to Distributed Artificial Intelligence. Addison-Wesley, Boston (1999)
13. Shoham, Y., Leyton, K.B.: Multiagent Systems: Algorithmic, Game-Theoretic and Logical Foundations. Cambridge University Press, New York (2009)
14. Marivate, V.N.: Social learning methods in board game agents. In: IEEE Symposium Computational Intelligence and Games, Perth, Australia, pp. 323–328 (2008)
15. Kiourt, C., Kalles, D.: Social reinforcement learning in game playing. In: IEEE International Conference on Tools with Artificial Intelligence, Athens, Greece, pp. 322–326 (2012)

16. Lopes, M., Melo, F.S., Kenward, B., Santos-Victor, J.: A computational model of social-learning mechanisms. Adapt. Behav. **17**, 467–483 (2009)
17. Kiourt, C., Kalles, D.: Learning in multi agent social environments with opponent models. In: 13th European Conference on Multi-Agent Systems and 3rd International Conference on Agreement Technologies (EUMAS 2015 and AT 2015), pp. 137–144 (2016)
18. Kiourt, C., Kalles, D.: A platform for large-scale game-playing multi-agent systems on a high performance computing infrastructure. Int. J. Multiagent Grid Syst. **12**(1), 35–54 (2016)

Emotion Prediction of Sound Events Based on Transfer Learning

Stavros Ntalampiras[1(✉)] and Ilyas Potamitis[2]

[1] National Research Council of Italy, Milan, Italy
stavros.ntalampiras@ieiit.cnr.it
[2] Technological Educational Institute of Crete, Rethymno, Greece
potamitis@staff.teicrete.gr
https://sites.google.com/site/stavrosntalampiras/home

Abstract. Processing generalized sound events with the purpose of predicting the emotion they might evoke is a relatively young research field. Tools, datasets, and methodologies to address such a challenging task are still under development, far from any standardized format. This work aims to cover this gap by revealing and exploiting potential similarities existing during the perception of emotions evoked by sound events and music. o this end we propose (a) the usage of temporal modulation features and (b) a transfer learning module based on an Echo State Network assisting the prediction of valence and arousal measurements associated with generalized sound events. The effectiveness of the proposed transfer learning solution is demonstrated after a thoroughly designed experimental phase employing both sound and music data. The results demonstrate the importance of transfer learning in the specific field and encourage further research on approaches which manage the problem in a cooperative way.

Keywords: Audio emotion prediction · Transfer learning · Echo State Network · Regression

1 Introduction

Sound plays a fundamental role in out everyday lives carrying a great gamut of meanings and purposes, such as informative (i.e. door bell ringing), pleasant (e.g. a musical piece), alarming (e.g. a scream), relaxing (e.g. a sea wave splashing on the shore), etc. [1]. In this article we focus on the emotional meaning conveyed by sound events. Unlike speech signals, where the speaker is able to transmit certain emotional states by altering a range of his/her vocal parameters [3], such as fundamental frequency and loudness [4,5], we focus on the emotion conveyed to the listener. Such sounds may be a result of human activities (e.g. walking), natural phenomena (e.g. rock falling), animals (cow mooing), etc., carrying various types of information, such as movement, size of the source, etc. [6]. These may comprise the necessary stimuli for a receiver to perform various activities, for example one may decide to take the necessary precautions

G. Boracchi et al. (Eds.): EANN 2017, CCIS 744, pp. 303–313, 2017.
DOI: 10.1007/978-3-319-65172-9_26

in case a gunshot is heard. Such contexts demonstrate the close relationship existing between sound events and the emotions they evoke, i.e. sounds may cause emotional manifestations on the listener side, such as fear [7].

Affective computing has received a lot of attention in the last decades with a special focus on the analysis of emotional speech, where a great gamut of generative and discriminative classifiers have been employed [8–10], and music [11–13] where most of the literature is concentrated on regression methods. Even though generalized sound events play a major role in the emotion conveyed to the listener, they have received considerable less attention than the previously mentioned fields of research. One of the first attempts [14] considers emotions evoked by specific sounds such as a dental engine. A more generic approach [15] employed 1941 low level signal descriptors feeding a random subspace meta-learner for recognition. The authors used a dataset from FindSounds.com annotated by four labellers. A well organised methodology [16] defined the structure of a sound event from the prism of the associated emotion and aimed at its automatic prediction. The authors employed a wide range of well known acoustic features along with Support Vector Machine and Artificial Neural Network classifiers after mapping to 4 categories of emotions. However the results indicated there is no evident relationship between the waveform of a sound event and the evoked emotion(s). In the closest paper to this work [17], the authors investigate the relationships between musical, sound, and speech emotional spaces. In particular the authors use the characteristics of one space to predict the those of another one and vice versa, i.e. the emotion of a music piece is predicted using the speech emotional space, etc. The authors designed a cross-domain arousal, and valence regression model showing high correlations between their predictions and the observer annotations. Their methodology is based on the INTERSPEECH 2013 Computational Paralinguistics feature set [18] feeding a Support Vector Regression scheme. An organized methodology, which is actually a successor of the present work, for predicting the emotional content of generalized sound events is presented in [2] where the k-mediods regression scheme achieves encouraging performance.

This work is focused on the prediction of the emotional dimension of sound events, which is the area less studied in the related literature. We investigate whether the emotional space of music signals and generalized sounds is *common* since they both aim at capturing the emotions evoked to the listener(s).

We conducted thorough experimentations on a publicly available dataset, i.e. the International Affective Digital Sounds (IADS) emotionally annotated sound events database including a diverse range of sound events (usual events, like dog barking, to uncommon ones, like gunshots or vomiting) associated with different emotional states [19]. The specific dataset is quite useful as it has been annotated following the affective annotation protocols of music emotion recognition. The music dataset is the 1000 Songs Database, which is also publicly available [20]. An extensive experimental procedure was carried out, where the superiority of the transfer learning method over the existing ones is demonstrated.

This work is organised as follows: Sect. 2 details the design of the transfer learning framework including feature extraction. Section 3 presents the experiments starting from the specifics regarding datasets and parametrization of the considered solutions to the analysis of the obtained results. Finally, conclusions are drawn in Sect. 4.

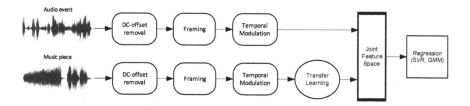

Fig. 1. The block diagram of the transfer learning framework.

2   The Transfer Learning Approach

We try to evaluate whether the emotional space of generalized sound events and musical pieces is shared, and if it is able to offer improved prediction of valence and arousal values. To this end the transfer learning framework, depicted in Fig. 1, includes the following three modules: (a) following the findings of our past work [21] we exploit the temporal modulation characteristics as they have provided a performance superior to the widely used Mel frequency cepstral coefficients when applied to a task of similar perceptual needs, i.e. predicting the unpleasantness level of a sound event, and (b) construction of the common feature space by means of transfer learning based on Echo State Networks (ESNs). The next two subsections provide the details regarding each module.

2.1   Temporal Modulation Features

The temporal modulation feature set is based on a modulation-frequency analysis via the Fourier transform and filtering theory [22–24]. Modulation filtering aims at retaining slow varying envelopes of spectral bands coming from non-stationary signals without affecting the signals phase and fine-structure. This feature set assigns high frequency values to the spectrum parts affecting the cochlea of the listener while emphasizing the temporal modulation.

Unlike the power spectrogram, the modulation one originates from human cochlea modelling. There, the existing inner-ear vibration is converted to electrically encoded signals. In general, sounds excite the basilar membrane while the associated response depends on the excitation frequency. Different components must be sufficiently distinct in frequency to stimulate unique areas of the membrane, which supports the hypothesis claiming that the output of the cochlea can be divided into frequency bands. The short-time excitation energy present in a specific channel is essentially the output of the associated band. It is important to note here that a harmonic sound event occupying many different auditory channels generates a similar modulation pattern across all bands. At this point lies the basic advantage of the modulation spectrogram since this redundancy does not exist in conventional spectral representations of harmonic sounds [25].

It should be mentioned that the implementation of the temporal modulation features is based on the one provided at [26].

2.2   Transfer Learning Based on Echo State Network

Feature space transformation is necessary for permitting the common handling of both feature sets by a regression methodology wishing to predict valence and arousal values of the sound events of interest. Such transformation is essential for addressing the diversities existing in the feature distributions. We overcome the particular obstacle by learning an ESN-based transformation [27,28]. It should be mentioned that this process could be performed in a vice-versa manner, i.e. exploiting sound event features for characterizing music genres, which is part of our future study.

A multiple-input multiple-output (MIMO) transformation is learnt using the training data of the music and sound signals. ESN modelling, and in particular Reservoir Network (RN), was employed at this stage as it is able to capture the non-linear relationships existing in the data. RNs represent a novel kind of echo-state networks providing good results in several demanding applications, such speech recognition [28], saving energy in wireless communication [29], etc.

An RN, the topology of which is depicted in Fig. 2, includes neurons with non-linear activation functions which are connected to the inputs (input connections) and to each other (recurrent connections). These two types of connections have randomly generated weights, which are kept fixed during both the training and operational phase. Finally, a linear function is associated with each output node.

RNs comprise a deep learning architecture as their main purpose is to capture the characteristics of high-level abstractions existing in the acquired data by designing multiple processing layers of complicated formations, i.e. non-linear functions. The associated depth is characterized by the amount of neurons included in the reservoir, which is usually called reservoir size (see Sect. 3.3 for the parametrization analysis). Deep learning is suitable for feature space transformation facilitating the extraction of useful information for the subsequent regression modeling phase.

Reservoir computing argues that since back-propagation is computationally complex but typically does not influence the internal layers severely, it may be totally excluded from the training process. On the contrary, the readout layer is a generalized linear classification/regression problem associated with low complexity. In addition any potential network instability is avoided by enforcing a simple constraint on the random parameters of the internal layers.

In the following we explain (a) how the transfer learning RN (in the following denoted as tRN) learns the transformation from the music feature space $\mathcal{M}$ to the sound event one $\mathcal{S}$, and (b) the exact way the transformation is employed.

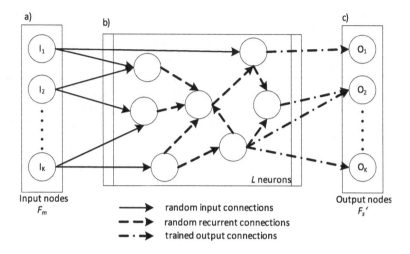

Fig. 2. The Echo State Network used for feature space transformation.

RN Learning. The tRN is used to learn the relationships existing in the features spaces of sound and music signals. We assume that an unknown system model is followed, which may be described as a transfer function f_{RN}.

f_{RN} comprises an RN with N inputs and N outputs. Its parameters are the weights of the output connections and are trained to achieve a specific result, i.e. a sound event feature vector. The output weights are learned by means of linear regression and are called read-outs since they "read" the reservoir state [30]. As a general formulation of the RNs, depicted in Fig. 2, we assume that the network has K inputs, L neurons (usually called reservoir size), K outputs, while the matrices $W_{in}(K \times L)$, $W_{res}(L \times L)$ and $W_{out}(L \times K)$ include the connection weights. The RN system equations are the following:

$$x(k) = f_{res}(W_{in}u(k) + W_{res}x(k)) \tag{1}$$
$$y(k) = f_{out}(W_{out})x(k), \tag{2}$$

where $u(k)$, $x(k)$ and $y(k)$ denote the values of the inputs, reservoir outputs and the read-out nodes at time k respectively. f_{res} and f_{out} are the activation functions of the reservoir and the output nodes, respectively. In this work we consider $f_{res}(x) = tanh(x)$ and $f_{out}(x) = x$.

Linear regression is used to determine the weights W_{out},

$$W_{out} = \underset{W}{argmin}(\frac{1}{N_{tr}}\|XW - D\|^2 + \epsilon\|W\|^2) \tag{3}$$
$$W_{out} = (X^T X + \epsilon I)^{-1}(X^T D), \tag{4}$$

where XW and D are the computed vectors, I a unity matrix, N_{tr} the number of the training samples while ϵ is a regularization term.

The recurrent weights are randomly generated by a zero-mean Gaussian distribution with variance v, which essentially controls the spectral radius SR of the reservoir. The largest absolute eigenvalue of W_{res} is proportional to v and is particularly important for the dynamical behavior of the reservoir [31]. W_{in} is randomly drawn from a uniform distribution $[-InputScalingFactor, +InputScalingFactor]$, which emphasises/deemphasises the inputs in the activation of the reservoir neurons. It is interesting to note that the significance of the specific parameter is decreased as the reservoir size increases.

Here, f_{RN} adopts the form explained in (1), (2) by substituting $y(k)$ with F_s and $u(k)$ with F_m, where F_s denotes an original sound event feature vector and F_m a feature vector associated with a music signal.

Application of f_{RN} After learning f_{RN}, it may be thought as a MIMO model of the form:

$$\begin{pmatrix} F_s^{1'}(t) \\ F_s^{2'}(t) \\ \vdots \\ F_s^{N'}(t) \end{pmatrix} = f_{RN} \begin{pmatrix} F_m^1(t) \\ F_m^2(t) \\ \vdots \\ F_m^N(t) \end{pmatrix}$$

where the music features $F_m^1 \ldots, F_m^N$ at time t are transformed using f_{RN} to observations belonging to the sound event features $F_s^{1'} \ldots, F_s^{N'}$, where N denotes the dimensionality of the feature vector shared by both domains. It should be noted that N depends on the feature set, i.e. temporal modulation, perceptual wavelet packets, and mel-scaled spectrum.

3   Experimental Set-Up and Results

This section explains (a) the experimental protocol that was followed towards revealing similarities between the emotions evoked by generalized sound events and music pieces, (b) the datasets including sound and music audio signals, (c) the parametrization of the modules included in the presented framework (see Fig. 1), and (d) the analysis of the obtained results.

3.1   Audio Databases

For the purposes of this work, two databases have been employed:

1. The International Affective Digital Sounds database (IADS-2) [19]: This dataset includes 167 emotionally-evocative sound stimuli that include contents across a wide range of semantic categories. Their annotations include two main dimensions, i.e. valence (ranging from pleasant to unpleasant) and arousal (ranging from calm to excited). Each stimuli was rated in 3 separate rating studies and the final values were averaged. The selected sounds cover a broad sample of contents across the entire affective space, while they communicate emotions relatively quickly.

2. The 1000 Songs Database [20]: This dataset includes 1000 songs has been selected from Free Music Archive[1]. Randomly (uniformly distributed) chosen excerpts with duration 45 s were subsequently isolated from each song. The songs were annotated by 100 subjects, 57 of which were males and 43 females. Their age average was 31.7 ± 10.1. A carefully designed data collection process was followed ensuring high level quality control.

The annotation values are normalized in the range [1,9] facilitating transfer learning from one dataset to the other. More specifically, rates are formed such that 9 represents a high rating on each dimension (i.e., high pleasure, high arousal), and 1 represents a low rating on each dimension (i.e., low pleasure, low arousal).

The stimulus existing in both datasets evoke reactions across the entire range of each emotional dimension, i.e. pleasure ratings for these sounds range from very unpleasant to very pleasant, and are distributed fairly evenly across the space. The observations are similar for the arousal levels as well.

3.2 Contrasted Approaches

We evaluated the transfer learning approach here on two levels, i.e. both feature extraction and regression. More specifically we used a Mel-scaled spectrogram and the Perceptual Wavelet Packets (PWPs) set [32,33] due to their ability to capture perceptual properties of audio signals. The PWP set analyses the audio signals across different spectral areas, while they are approximated by wavelet packets. They account for the fact that human perception is not affected in the same way by all parts of the spectrum [34] by employing a suitably-designed filterbank. The PWP feature set reflects upon the degree of variability exhibited by a specific wavelet coefficient within a critical band, thus they may capture useful information for characterizing emotional content. Moreover, as suggested by the related literature we employed support vector regression [17] and Gaussian mixture models clustering [21].

The contrasting experiments were designed such that all the approaches were compared on two feature spaces, i.e. the one constructed using the sound events alone and the joint one. Care was taken such that all approaches operated on identical train and test sets in order to achieve a fair comparison.

3.3 Parametrization of the Transfer Learning Framework

The audio files coming from both databases were sampled at 16 kHz with 16-bit quantization and preprocessed for eliminating any possible DC-offset. The feature extraction parametrization was kept constant with respect to every set facilitating comparison and fusion tasks. After early experimentations and in order to avoid possible misalignment(s), the low-level featureextraction window

---

[1] http://freemusicarchive.org/.

is 30 ms with 20 ms overlap. Furthermore to smooth any existing discontinuities the sampled data are hamming windowed while the FFT size, where applicable, is 512.

The parameters of the tRN were selected by means of exhaustive search based on the minimum reconstruction error criterion. The parameters were taken from the following sets: $SR \in \{0.8, 0.9, 0.95, 0.99\}$, $L \in \{0, 500, 1000, 5000, 10000\}$, and $InputScalingFactor \in \{0.1, 0.5, 0.7, 0.95, 0.99\}$. The combination of parameters providing the lowest reconstruction error on a validation set including both feature spaces. Its implementation was based on the Echo State Network Toolbox which is available at http://reservoir-computing.org/software.

The SVM's kernel function is a Gaussian radial basis ($k(x_i, x_j) = \exp(-\gamma||x_i - x_j||^2)$ for $\gamma > 0$) one while the soft margin parameter and γ where determined though a grid search guided by cross-validation on the training set.

3.4   Results and Analysis

In the first experiment we evaluated the emotion prediction approach without considering transfer learning. We employed different features sets and regression schemes for characterizing the emotional properties of generalized sound events. The results are tabulated in Table 1.

Table 1. The matrix tabulating the regression results with respect to various feature sets and regression schemes, while using the sound event feature space alone. MSE average values over 50 iterations are shown in the following format: Arousal/Valence, while the minimum errors, i.e. best performance, are emboldened.

Feature set	Regressor	
	SVR [17]	GMM clustering [21]
Mel-spectrum [35]	2.85/4.82	3.16/3.21
PWP [32]	3.05/4.01	3.10/3.36
Temporal modulation	2.85/4.85	**3.13/3.10**

As we can see in Table 1 the best prediction results with respect to both arousal and valence measurements are provided by the Temporal modulation feature set modelled by means of GMM clustering. The poor predictions of the SVR may be due to the limited amount of data included in the IADS-2 database. This burdens the flatness of the regression function which is of fundamental importance in SVR training [36]. Moving on, we observe that GMM clustering based on the Kullback-Leibler (KL) divergence captures better the relationships existing in the feature space. Generative modelling in the stochastic plane is able to provide the second best results in arousal and valence prediction.

In the next experimental phase we activated the ESN-based transfer learning component and included the music data to perform prediction of the emotional content of the sound events. The parameters of the tRN providing the lowest reconstruction error were $SR = 0.95$, $L = 5000$, and $InputScalingFactor = 0.99$. The results are tabulated in Table 2. As we can see most errors have decreased proving that (a) there exist similarities in the way song and generalized audio signals evoke emotions, and (b) transfer learning for automatic description of emotional content is beneficial. More specifically, the best performing method (Temporal modulation + GMM clustering) achieves MSE figures equal to 3.09 and 2.8 for arousal and valence prediction respectively.

Table 2. The matrix tabulating the regression results with respect to various feature sets and regression schemes, while using both feature spaces. MSE average values over 50 iterations are shown in the following format: Arousal/Valence, while the minimum errors, i.e. best performance, are emboldened.

Feature set	Regressor	
	SVR [17]	*GMM clustering* [21]
Mel-spectrum [35]	2.3/4.1	3.19/3.02
PWP [32]	3/4.13	3.24/3.20
Temporal modulation	1.75/4.45	**3.09/2.80**

While comparing Tables 1 and 2, the relevance of a transfer learning mechanism enabling feature space transformation becomes clear. The majority of MSEs have decreased confirming that such deep learning technique is able to transform the data successfully allowing the regressors to operate on a larger space where they are able to provide better performance.

4   Conclusions

This paper is an attempt towards the automatic assessment of the emotions evoked by generalized sound events by revealing perceptual similarities between music and sounds via transfer learning. In particular the presented approach proposes the usage of temporal modulation features and an ESN-based transfer learning module. The superiority of the proposed transfer learning approach was proven after a thorough evaluation employing sound and music datasets.

Our future work includes both development of transfer learning based solutions to deal with applications of the generalized sound recognition technology. For example, we intent to design a synergistic framework for transferring knowledge from the music information retrieval domain to address bioacoustic signal processing applications. Finally, in the next stage of this research, a much larger dataset of sound events will be employed.

Acknowledgment. The research leading to these results has received partial funding from European Union HORIZON 2020 fast track to innovation project no. 691131 REMOSIS.

References

1. Ntalampiras, S., Potamitis, I., Fakotakis, N.: Acoustic detection of human activities in natural environments. J. Audio Eng. Soc. **60**, 686–695 (2012)
2. Ntalampiras, S.: A transfer learning framework for predicting the emotional content of generalized sound events. J. Acoust. Soc. Am. **141**, 1694–1701 (2017)
3. Shigeno, S.: Effects of discrepancy between vocal emotion and the emotional meaning of speech on identifying the speakers emotions. J. Acoust. Soc. Am. **140**, 3399–3399 (2016)
4. Scherer, K.R.: Vocal communication of emotion: a review of research paradigms. Speech Commun. **40**, 227–256 (2003)
5. Hozjan, V., Kai, Z.: A rule-based emotion-dependent feature extraction method for emotion analysis from speech. J. Acoust. Soc. Am. **119**, 3109–3120 (2006)
6. Marcell, M., Malatanos, M., Leahy, C., Comeaux, C.: Identifying, rating, and remembering environmental sound events. Behav. Res. Methods **39**, 561–569 (2007)
7. Garner, T., Grimshaw, M.: A climate of fear: considerations for designing a virtual acoustic ecology of fear. In: Proceedings of 6th Audio Mostly Conference: A Conference on Interaction with Sound, pp. 31–38 (2011)
8. El Ayadi, M., Kamel, M.S., Karray, F.: Survey on speech emotion recognition: features, classification schemes, and databases. Pattern Recogn. **44**, 572–587 (2011)
9. Asadi, R., Fell, H.: Improving the accuracy of speech emotion recognition using acoustic landmarks and Teager energy operator features. J. Acoust. Soc. Am. **137**, 2303–2303 (2015)
10. Lee, C., Lui, S., So, C.: Visualization of time-varying joint development of pitch and dynamics for speech emotion recognition. J. Acoust. Soc. Am. **135**, 2422–2422 (2014)
11. Fukuyama, S., Goto, M.: Music emotion recognition with adaptive aggregation of Gaussian process regressors. In: IEEE International Conference on Acoustics, Speech and Signal Processing (ICASSP), pp. 71–75 (2016)
12. Markov, K., Matsui, T.: Music genre and emotion recognition using Gaussian processes. IEEE Access **2**, 688–697 (2014)
13. Yi-Hsuan, Y., Chen, H.: Machine recognition of music emotion: a review. ACM Trans. Intell. Syst. Technol. **3**, 40:1–40:30 (2012)
14. Gang, M.-J., Teft, L.: Individual differences in heart rate responses to affective sound. Psychophysiology **12**, 423–426 (1975)
15. Schuller, B., Hantke, S., Weninger, F., Han, W., Zhang, Z., Narayanan, S.: Automatic recognition of emotion evoked by general sound events. In: IEEE International Conference on Acoustics, Speech and Signal Processing (ICASSP), pp. 341–344 (2012)
16. Drossos, K., Floros, A., Kanellopoulos, N.-G.: Affective acoustic ecology: towards emotionally enhanced sound events. In: Proceedings of 7th Audio Mostly Conference: A Conference on Interaction with Sound, pp. 109–116 (2012)
17. Weninger, F., Eyben, F., Schuller, B., Mortillaro, M., Scherer, K.-R.: On the acoustics of emotion in audio: what speech, music and sound have in common. Front. Psychol. **292**, 1–12 (2013)
18. Schuller, B., Steidl, S., Batliner, A., Vinciarelli, A., Scherer, K.-R., Ringeval, F., Chetouani, M., Weninger, F., Eyben, F., Marchi, E., Mortillaro, M., Salamin, H., Polychroniou, A., Valente, F., Kim, S.: The INTERSPEECH 2013 computational paralinguistics challenge: social signals, conflict, emotion, autism. In: INTERSPEECH, pp. 148–152 (2013)

19. Bradley, M., Lang, P.-J.: The International Affective Digitized Sounds (2nd edn. IADS-2): Affective Ratings of Sounds and Instruction Manual. Technical report B-3, University of Florida, Gainesville, Fl (2004)
20. Soleymani, M., Caro, M.-N., Schmidt, E.-M., Sha, C.-Y., Yang, Y.H.: 1000 songs for emotional analysis of music. In: Proceedings of 2nd ACM International Workshop on Crowdsourcing for Multimedia, pp. 1–6 (2013)
21. Ntalampiras, S., Potamitis, I.: On predicting the unpleasantness level of a sound event. In: 15th Annual Conference of International Speech Communication Association (INTERSPEECH), pp. 1782–1785 (2014)
22. Clark, P., Atlas, L.: Time-frequency coherent modulation filtering of nonstationary signals. IEEE Trans. Signal Process. **57**, 4323–4332 (2009)
23. Schimmel, S.M., Atlas, L.E., Nie, K.: Feasibility of single channel speaker separation based on modulation frequency analysis. In: IEEE International Conference on Acoustics, Speech and Signal Processing, pp. 605–608 (2007)
24. Vinton, M.S., Atlas, L.E.: Scalable and progressive audio codec. In: 2001 IEEE International Conference on Acoustics, Speech, and Signal Processing, Proceedings (ICASSP 2001), pp. 3277–3280 (2001)
25. Klapuri, A.: Multipitch analysis of polyphonic music and speech signals using an auditory model. IEEE Trans. Audio Speech Lang. Process. **16**, 255–266 (2008)
26. Atlas, L., Clark, P., Schimmel, S.: Modulation Toolbox Version 2.1 for MATLAB. http://isdl.ee.washington.edu/projects/modulationtoolbox/. Accessed Sept 2010
27. Jalalvand, A., Triefenbach, F., Verstraeten, D., Martens, J.: Connected digit recognition by means of reservoir computing. In: Proceedings of 12th Annual Conference of the International Speech Communication Association, pp. 1725–1728 (2011)
28. Verstraeten, D., Schrauwen, B., Stroobandt, D.: Reservoir-based techniques for speech recognition. In: International Joint Conference on Neural Networks, IJCNN 2006, pp. 1050–1053 (2006)
29. Jaeger, H., Haas, H.: Harnessing nonlinearity: predicting chaotic systems and saving energy in wireless communication. Science **304**, 78–80 (2004)
30. Lukoševičius, M., Jaeger, H.: Survey: reservoir computing approaches to recurrent neural network training. Comput. Sci. Rev. **3**, 127–149 (2009)
31. Verstraeten, D., Schrauwen, B., d'Haene, M., Stroobandt, D.: An experimental unification of reservoir computing methods. Neural Netw. **20**, 391–403 (2007)
32. Ntalampiras, S., Potamitis, I., Fakotakis, N.: Exploiting temporal feature integration for generalized sound recognition. EURASIP J. Adv. Signal Process. **2009**, 1–12 (2009)
33. Ntalampiras, S.: Audio pattern recognition of baby crying sound events. J. Audio Eng. Soc **63**, 358–369 (2015)
34. Scharf, B.: Complex sounds and critical bands. Psychol. Bull. **58**, 205–217 (1961)
35. Yi-Lin, L., Gang, W.: Speech emotion recognition based on HMM and SVM. In: International Conference on Machine Learning and Cybernetics, vol. 8, pp. 4898–4901 (2005)
36. Smola, A.-J., Schölkopf, B.: A tutorial on support vector regression. Stat. Comput. **14**, 199–222 (2004)

Interval Analysis Based Neural Network Inversion: A Means for Evaluating Generalization

S.P. Adam[1,2(✉)], A.C. Likas[3], and M.N. Vrahatis[2]

[1] Department of Computer Engineering, Technological Education Institute of Epirus,
Arta, Greece
`adamsp@teiep.gr`
[2] Computational Intelligence Laboratory, Department of Mathematics,
University of Patras, Rion - Patras, Greece
`adamsp@upatras.gr`
[3] Department of Computer Science and Engineering,
University of Ioannina, Ioannina, Greece

Abstract. Inversion of a neural network trained on some classification problem has been an important issue related to the explanation of the neural classification function. Inversion based on Interval Analysis (IA) [1] showed that a reliable estimation of the neural network domain of validity is feasible and a number of quantitative issues arise from this inversion. This paper deals with the investigation of these quantitative issues and more precisely with those concerning the evaluation of the neural network classification function in terms of generalization, comparison of different network models and classification accuracy. Preliminary experimental results indicate that the IA-based inversion can offer a solid basis towards reliable evaluation of the neural classification function.

Keywords: Neural networks · Generalization · Interval Analysis · Reliable computation

1 Introduction

Inversion of a trained network has always been one of the objectives in neural network research as it permits to define the input space area covered by the network function, to delineate the decision boundaries learned by the network and to extract rules explaining the network operation. Hence, neural network classification is related to the so-called domain of validity of the network, which results from network inversion and it can be used either to provide a qualitative conclusion of the neural classification function [12,14,17] or to extract provably correct rules [5,20,22] explaining neural network operation. A number of approaches can be found in the literature which permit to define such a domain [3,16]. However, while an accurate definition of the domain of validity should be an obvious requirement of any approach used for this problem it seems that

© Springer International Publishing AG 2017
G. Boracchi et al. (Eds.): EANN 2017, CCIS 744, pp. 314–326, 2017.
DOI: 10.1007/978-3-319-65172-9_27

such a requirement had not effectively been tackled. As a result, to the best of our knowledge, there has been no research effort towards examining the domain of validity of a neural network in quantitative terms.

Recently, Adam et al. [1] proposed an IA-based approach for neural network inversion resulting in reliable definition of its domain of validity. Inversion of the network is carried out using an IA approach which, for any interval of the network output activity, defines a unique, consistent and guaranteed domain in the input space. The proposed method is termed to provide reliable estimation as it permits to define regions of validity in a guaranteed way. The results obtained in [1] are interesting in the sense that they provide quantitative information about the domain of validity of the neural network. This level of information seems to be inadequate for the classical explanation of neural network operation but as noted in [1] it may be used to provide useful insight to the neural classification task concerning the generalization ability of the trained network as well as the fitness of the neural network model adopted.

The aim of this paper is to advance on the hypotheses formulated in [1] by carrying out a number of experiments in order to evaluate the soundness of these statements. This paper deals with the following matters:

- Based on the volume of the domain of validity derive empirical metrics for the evaluation of the performance of a trained network in a classification task.
- Assess the generalization ability of a trained network using the previous metrics and compare the results with the classical cross-validation approach.
- Discuss open problems and future work.

The experimental results obtained provide concrete evidence that important aspects concerning the validity of the neural classification task can be evaluated using the empirical metrics defined. The reliability of the proposed metrics is supported by the IA-based inversion which provides verified results in a guaranteed way, as the interval computations permit to automatically verify the results obtained [13].

The paper is organized as following. Section 2 outlines the classification context along with the main assumptions and some theoretical results. Section 3 is dedicated to the description of the basic interval arithmetic concepts and the inversion procedure based on IA. In Sect. 4 we present the proposed approach along with the empirical metrics defined. Section 5 is devoted to the experimental evaluation and the discussion of the results obtained. Finally, Sect. 6 concludes the paper.

2   Problem Definition and Background

The analysis presented in this paper deals with, but is not limited to, multi-layer perceptrons (MLPs) which are considered to have one or more hidden layers, nodes with sigmoidal nonlinearities and being trained with some gradient descent procedure. The network is trained on a classification problem with M classes, $C_1, C_2, \ldots, C_M$, using a sample data set $\mathbf{D} = (\mathbf{X}, \mathbf{Y})$, of P examples defined by the N-dimensional patterns $\mathbf{X} = \{\mathbf{x}_1, \mathbf{x}_2, \ldots, \mathbf{x}_P\}$ instantiating the random variable $\mathcal{X}$, and the desired outputs $\mathbf{Y} = \{\mathbf{y}_1, \mathbf{y}_2, \ldots, \mathbf{y}_P\}$ for the random variable $\mathcal{Y}$. Output for each class is 1 of M, denoting that, there is one output unit corresponding to the correct class while all others are zero.

Classification decisions based on the *ad hoc* rule "the pattern $\mathbf{x}$ is assigned membership in class C_j if the corresponding output value is greater than some fixed threshold" are ambiguous as they may assign a pattern to multiple classes. Such decisions become unambiguous, if the following rule is used instead, "$\mathbf{x}$ is considered to belong to class C_j if the jth component of the network output is greater than all the other components" [9].

Important research focused on the operation of a neural network, in terms of defining decision rules governing their function and explaining how decision boundaries are formed by neural network outputs in classification problems [9]. A significant part of this research relates operation of an MLP classifier with Bayesian classification [2,4,6,15].

Hampshire and Pearlmutter [7] provided detailed proofs that, when dealing with asymptotically large sets of statistically independent training samples, MLP classifiers provide outputs which act as optimal Bayesian discriminant functions for these training samples. They also discussed necessary and sufficient conditions on the form of objective functions that yield Bayesian discriminant performance by engendering classifier outputs that are true estimates of the *a posteriori* probabilities $P(C_j|\mathbf{x})$. Richard and Lippmann [18], also, advanced on the previous statements, giving detailed proofs and analysis of some important network models such as MLPs, radial basis function (RBF) and high-order polynomial networks. They showed that the outputs of these networks provide good estimates of Bayesian probabilities. Estimation accuracy depends on network complexity the amount of training data, and the degree to which training data reflect true likelihood distributions and *a priori* class probabilities.

Here, let us assume that the network target outputs are considered to be binary, i.e. in the interval $[0,1]$. For an MLP classifier approximating the *a posteriori* class probabilities means that the conditional expectation of the desired output response vector, given the input data vector $\mathbf{x}$, that is $E(d_j = 1|\mathbf{x})$, equals the *posterior* class probability $P(\omega = \omega_j|\mathbf{x})$, where $\omega \in \Omega = \{\omega_1, \omega_2, \ldots, \omega_m\}$ and $\omega = \omega_j$ denotes membership of $\mathbf{x}$ to class C_j for $j = 1, 2, \ldots, M$.

As Hampshire and Pearlmutter [7] showed, in the case of perfect training, sufficiently large data set and complex network, the relation between the *a posteriori* class probabilities and the network output values is linear and so $P(\omega = \omega_j|\mathbf{x})$ is mapped to the interval $[0,1]$. However, due to various perturbations the target values are non binary and so $P(\omega = \omega_j|\mathbf{x})$, while still being linear, is mapped to the interval $[\epsilon, 1 - \epsilon]$ for some appropriate value of ϵ (e.g. $\epsilon = 0.2$). Similar conclusions were, also, provided by Richard and Lippmann in [18] regarding the degradation of the estimation accuracy of Bayesian posterior probabilities.

In many practical classification tasks when the jth output node is active assigning the input pattern to the jth class then its value is considered to be in an interval $[1 - \beta, 1]$, defined for some suitably chosen value of β (e.g. $\beta = 0.4$).

This constitutes an intuitive, yet practical and efficient way, to deal with output uncertainty due to imperfect training produced by various reasons such as an inexact error threshold, insufficient number of training epochs or when having a small sized training set. On the other hand, an output node is considered to be inactive if its activation is in the interval $[0, \beta]$. The greater the value of β the poorer the classification accuracy achieved by the trained network. Hence, defining the domain of validity generating output values in the interval $[1 - \beta, 1]$ seems to be a necessary tool to support the performance analysis of the network classification function. This can be achieved using IA concepts and SIVIA [10].

3   IA-Based Inversion and SIVIA

Interval arithmetic was introduced as a means to perform numerical computations with guaranteed accuracy and bounding the ranges of the quantities, used in the computations. An interval, or interval number, I is a closed interval $[a, b] \subset \mathbb{R}$ of all real numbers between (and including) the endpoints a and b, with $a \leqslant b$. In practical calculations interval arithmetic operation are reduced to operations between real numbers [10]. Hereafter, the bracketed notation $[x] = [\underline{x}, \overline{x}]$ denotes an interval object such as a number, variable, vector, matrix, etc. The set of n-dimensional vectors of real intervals is denoted by $\mathbb{IR}^n$. The definition of interval objects such as vectors, matrices, functions, etc. and their subsequent study resulted in the establishment of Interval Analysis.

If $[x] \subseteq D$ is an interval in the domain of a real function $f : D \subset \mathbb{R} \to \mathbb{R}$ then $f([x])$ is used to denote the range of values of f over $[x]$. Computing such a range, $f([x])$, using IA tools means to enclose it by an interval which is as narrow as possible. This constitutes an important matter in IA as it is used in various problems: localization and enclosure of global minimizers of f on $[x]$, verification of $f([x]) \subseteq [y]$ for given $[y]$, nonexistence of a zero of f in $[x]$ etc. In order to enclose $f([x])$ one needs to define a suitable interval function $[f] : \mathbb{IR} \to \mathbb{IR}$ such that $\forall [x] \in \mathbb{IR}, \ f([x]) \subset [f]([x])$, see Fig. 1.

SIVIA is an interval method introduced by Jaulin and Walter [11] in order to allow for the guaranteed estimation of nonlinear parameters from bounded error data. The method proceeds by defining a box or union of boxes enclosing a set of interest. Hence, given a function $f : X \to Y$, where $X \subset \mathbb{R}^n, Y \subset \mathbb{R}^m$ and an interval vector, i.e. a box, $[y] \subseteq Y$, the objective is to define the set of unknown vectors $x \in X$ such that $f(x) \in [y]$. This set can be defined as $S = \{x \in X \subseteq \mathbb{R}^n | f(x) \in [y]\} = f^{-1}([y]) \cap X$, where X is the search space containing the set of interest S; $[y]$ is known in advance to enclose the image of the set $f(S)$ and S denotes the unknown set of interest. Note that, here, f^{-1} denotes the reciprocal image of f, as f may not be invertible in the classical sense.

The solution proposed by SIVIA for this problem consists in computing boxes and unions of boxes S^- and $S^+ = S^- \cup \Delta S$ which form guaranteed outer and inner enclosures of S as they satisfy the relation $S^- \subseteq S \subseteq S^+$, [10]. SIVIA is a branch-and-bound approach which computes enclosures by recursively exploring the whole search space. During computation, a box $[x] \in \mathbb{R}^n$ is designated as feasible if $[x] \subseteq S$ and $f([x]) \subseteq [y]$, infeasible if $[f]([x]) \cap [y] = \emptyset$ and, in all other cases, $[x]$ is said to be indeterminate which means that $[x]$ may be feasible, unfeasible or ambiguous. The condition $[x] \subseteq S$ and $f([x]) \subseteq [y]$ is necessary and sufficient for $[x]$ to be feasible. Feasible boxes are added to S^- and infeasible become members of the complement of S^+. Finally, any indeterminate box is bisected and the method recursively examines the two resulting sub-boxes. Bisection is possible up to some limit, which is preset for the problem and defines its resolution. Boxes that are indeterminate and cannot be further bisected are added to the union ΔS. For a detailed description of the algorithm implementing SIVIA the reader should refer to [10]. Finally, note that SIVIA applies to any function f for which an inclusion function $[f]$ can be computed.

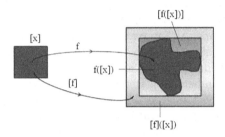

Fig. 1. A function f, an inclusion function $[f]$ and the images of $[x]$

4   Inversion-Based Generalization Metrics

The necessity of using an interval $[1 - \beta, 1]$ is illustrated by the 2-dimensional classification problem with two classes shown in Fig. 2a. A $2 - 10 - 2$ MLP, using the hyperbolic tangent activation for the hidden nodes and the logistic sigmoid one for the output nodes, was trained on this problem and produced the output shown in Fig. 2b. Patterns classified as class 1 form the white regions. Obviously, the gray level zone depicts the ambiguity of classification for patterns near the class 1 area which gives values of the MLP output in some interval $[1 - \beta, 1]$.

The value of β clearly extends or restricts the area in the input space classified by the MLP as class 1. One can verify this argument by simple reference to Figs. 3a and b where the red colored areas have been defined by inverting the MLP output using SIVIA. The striking difference between these two Figures concerns a significant part of the input space effectively belonging to class 1 which is present in Fig. 3b but not in Fig. 3a. This shows the importance of β which, here, needs to be given the value 0.1 if one wants to take into account a significant part of the input space. However, as shown in Figs. 4a and b with a higher number of training epochs and a smaller error threshold this deficiency seems to be remedied.

It is well known that the term generalization refers to the ability of a trained neural network to correctly classify previously unseen patterns. Achieving good generalization can be seen in two different ways; either have some fixed architecture of the network and determine the size of the training set, or start from a fixed size of the training set and define the best network architecture [9]. A number of studies carried out on these issues brought interesting research results, which however rely on theoretical assumptions such as: the a priori knowledge of the distribution of the network weight vectors, the confidence that the distribution of the training patterns is a good approximation of the distribution of the input space or even that the network architecture is well suited for the problem at hand, see [8] and references therein. In practice, none of these assumptions is verified and both researchers and engineers proceed with a fixed data set and an initial network model along with some complexity regularization technique in order to achieve the best network architecture. A well known approach for resolving the generalization issue is cross-validation [9] and especially the multifold cross-validation which is used in our experiments.

In the context of this paper classification of previously unseen patterns is considered under the following statement; the network will be able to correctly classify those patterns that fall into the area of the input space learned by the network during training. This area is, precisely, the domain of validity of the network as defined by the IA-based inversion of the network. The larger the domain of validity, the bigger its volume and so the higher the probability for some unknown pattern to be in this area and be classified. Hence, the first condition for some unknown pattern to be classified correctly is to be in the domain of validity of the network. Then, assigning the pattern to the right class requires the pattern to be in the area corresponding to the domain of validity of its respective class. These two conditions are the basis of the definition of two metrics.

In the case of an MLP classifier and a 1 of M encoding, in order to compute the total volume of the domain of validity of the network one needs to perform M inversions, one for each output node, compute the volume of each defined area and finally sum up the volumes of all the classes. Hence, if $V_1, V_2, \ldots, V_M$ are the volumes computed for the M classes, then the total volume of the domain of validity is $V_{net} = \sum_{i=1}^{M} V_i$.

In consequence, if V_{input} denotes the finite, non zero, volume of the input space then the ratio $\frac{V_{net}}{V_{input}}$, defines a measure of the probability for some new pattern to be effectively classified by the network. Moreover, this ratio is a metric for measuring over-training of the network. In order to cope with the second condition and evaluate the ability of the trained network to correctly classify a new pattern we need to rate the ability of the domain of validity of each class to contain the patterns effectively belonging to this class while excluding all the others. Actually for some class C_i some patterns may happen either to be misclassified to another class C_j or to be unclassified, which means that these patterns are outside the domain of validity of the network, i.e. the network

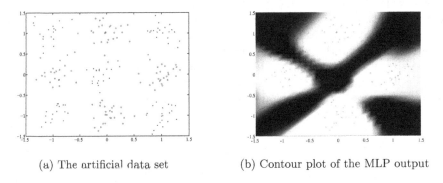

(a) The artificial data set   (b) Contour plot of the MLP output

Fig. 2. An artificial problem and the contour plot of an MLP trained on this data set

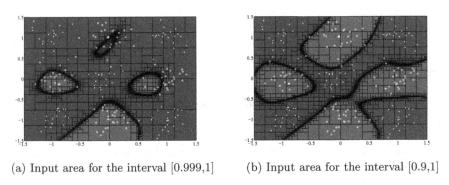

(a) Input area for the interval [0.999,1]   (b) Input area for the interval [0.9,1]

Fig. 3. How the interval $[1 - \beta, 1]$ affects the domain of validity of an MLP trained for 500 epochs and MSE $\leqslant 1e - 03$

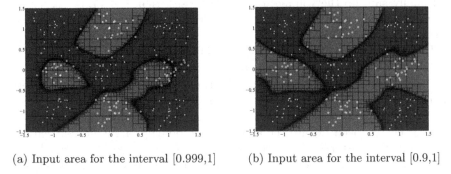

(a) Input area for the interval [0.999,1]   (b) Input area for the interval [0.9,1]

Fig. 4. How the interval $[1 - \beta, 1]$ affects the domain of validity of an MLP trained for 5000 epochs and MSE $\leqslant 1e - 05$

output signal for those patterns is not in the interval $[1 - \beta, 1]$. In order to tackle this problem two solutions are envisaged in this paper:

A. Penalize the domain of validity for the incorrect classifications by subtracting from its volume the volume corresponding to each misclassified or unclassified pattern.
B. For each box defined by SIVIA for any class C_j compute its volume by taking into account the number of patterns correctly classified in this box.

For the first approach we need to define the volume of the input space corresponding to each misclassified/unclassified pattern. A simple way to do this is to consider that to any input pattern corresponds an elementary volume V_{elem} of the input space which can be defined by simply taking $V_{elem} = \frac{V_{input}}{P}$, that is this elementary volume is proportional to the number of patterns available for the classification problem. This choice can be explained by the following argument; the more the input patterns the lower the probability of having a defective classification function due to the size of the training set. So, the penalty imposed on the domain of validity is proportional to the size of the training set. Hence, if l is the total number of misclassified/unclassified patterns then a generalization metric is given by the formula:

$$G_{net} = \frac{V_{net} - lV_{elem}}{V_{input}}. \tag{1}$$

This quantity G_{net} defines a metric measuring over-training while taking into account the classification errors of the trained network. Concerning the second approach let us consider the following:

– for each pattern $\mathbf{x}_n$ in the training set X let $C(\mathbf{x}_n)$ denote the class of this pattern,
– if C_k, $1 \leqslant k \leqslant M$, is the kth class then the following set of boxes is defined by IA-based inversion of the kth output node of the MLP: $\mathbb{B}^k = \{\boldsymbol{B}_1^k, \boldsymbol{B}_2^k, \ldots, \boldsymbol{B}_{N_k}^k\}$,
– let V_i^k denote the volume of $\boldsymbol{B}_i^k$,
– let X_i^k be the set of patterns $\mathbf{x}_n$ found to be inside the box $\boldsymbol{B}_i^k$.

Then the validity of the box $\boldsymbol{B}_i^k$ can be defined as

$$E_i^k = \left[\sum_{\mathbf{x}_n \in X_i^k} \left(\mathbb{1}_{\mathbb{C}} \left(C(\mathbf{x}_n) = C_k \right) - \mathbb{1}_{\mathbb{C}} \left(C(\mathbf{x}_n) \neq C_k \right) \right) \right] V_i^k, \tag{2}$$

where $\mathbb{1}_{\mathbb{C}}$ is a suitable indicator function. Finally, if one wants to obtain the total validity of the trained network according to these hypotheses then one has to compute the formula

$$E = \sum_{k=1}^{M} E_i^k. \tag{3}$$

The second approach for every box, identified by IA-based inversion, computes its part effectively belonging to some class by taking into account the patterns found in this box. In consequence, the metric provided by this approach has a twofold effect; first, for every box belonging to some class C_j, as determined by inversion, it accounts only the part of the box effectively containing patterns of the class C_j, and second it rejects those parts of the input space that do not contain any pattern at all. Doing so it rejects an important part of the interpolations and the extrapolations computed by the network during its training and hence it provides a validity index for the trained network.

5   Experimental Evaluation and Discussion

In this section we describe the setup for the experiments carried out in order to evaluate the statements of the previous section. Then, we discuss the results obtained pointing out some interesting characteristics and finally, we describe some potential issues for future work.

5.1   Experimental Evaluation

Experiments in this paper were executed using SCS Toolbox [23] which implements SIVIA using INTLAB, the MATLAB package of Rump [19] for interval computations.

Two data sets were used for our experiments. The first is the artificial data set for the classification of a pattern belonging to one of two classes originally defined in [21]. This data set is perfectly balanced as it accounts 100 patterns per class and the distribution of the classes has been modified in order to increase overlapping between them, Fig. 2a. For this data set we used 8 MLPs having an architecture $2 - H - 2$ with H taking on the values $4, 5, 6, 8, 10, 15, 20, 25$, and using the hyperbolic tangent for all nodes in the hidden layer and the logistic sigmoid for the output nodes. The data set was divided in twenty subsets in order to compare the results obtained by the proposed approach with the multifold (20-fold) cross-validation. Each MLP was trained until reaching a MSE of 0.001 or for a maximum number of 5000 epochs thus giving a total number of (20 trials × 8 models) = 160 training trials. For each trial the generalization was computed as the average of the correctly classified validation patterns for the cross-validation method.

For the proposed generalization metrics the network trained in each trial was first inverted using SIVIA. Hence, for each network model 20 trials resulted in 20 inversions and the volumes of the 20 domains of validity were used to compute the mean value for each metric and each network model. The interval of the output values inverted is $[0.8, 1]$. Table 1 outlines the results of this experiment.

The second data set used is the well known Fisher-Iris problem. Experiments were carried out for 6 MLPs having an architecture $4 - H - 3$ with H taking on the values $2, 3, 5, 8, 10, 15$, and using the hyperbolic tangent for all nodes in the hidden layer and the logistic sigmoid for the output nodes. The data set was

Table 1. Results for the artificial data set (higher values indicate better generalization)

Network model	Cross-validation mean (std)	Metric 1 mean (std)	Metric 2 mean (std)
2-4-2	90.50% (9.45%)	96.45% (2.81%)	16.92 (18.57)
2-5-2	91.50% (8.75%)	94.50% (2.76%)	10.52 (7.75)
2-6-2	91.00% (9.68%)	91.35% (1.72%)	9.98 (4.36)
2-8-2	93.00% (8.65%)	90.63% (2.64%)	11.29 (4.64)
2-10-2	93.50% (6.71%)	88.84% (1.54%)	9.51 (1.76)
2-15-2	92.00% (8.34%)	89.08% (2.10%)	9.87 (1.23)
2-20-2	93.00% (8.65%)	88.38% (1.35%)	8.45 (0.76)
2-25-2	93.00% (7.33%)	87.65% (1.07%)	7.49 (0.80)

divided in ten subsets in order to compare the results obtained by the proposed approach with the multifold (10-fold) cross-validation. Each MLP was trained until reaching a MSE of 0.001 or for a maximum number of 5000 epochs and for each trial the generalization was computed as the average of the correctly classified validation patterns for the cross-validation method. The proposed metrics were computed as in the previous experiment and the interval of the network output values inverted using SIVIA is again $[0.8, 1]$.

5.2 Discussion

Prior to a discussion of the results, we need to note the following; lower values of the Metric 1 indicate higher over-training (i.e. lower generalization ability), as the domain of validity of an over-trained network tends to concentrate around the input areas of higher density. On the other hand the values of the Metric 2, which "corrects" the domain of validity of the network, indicate the discrimination of the overlapping areas between different classes, thus contributing to a more accurate evaluation of the generalization. We consider that these two metrics should be taken as a set, and in future work we will be able to propose some aggregate form of them.

Comparing the results of the proposed metrics, as reported in Tables 1 and 2, against the results of the classical cross-validation one may easily notice that the proposed metrics detect and indicate both over-training and generalization of the network models tested, while the results of cross-validation do not confirm this essential theoretical issue.

The proposed Metric 2 evaluates the boxes as they are derived by the branch-and-bound searching performed by SIVIA which might result in rejecting areas that should be more carefully examined. This may be the cause of the high standard deviation observed for this metric in the case of the Fisher-Iris problem. Nevertheless, this metric merits to be further investigated in future work.

The proposed metrics are calculated based on the trained network itself, irrespectively of the data set used for training and validation. So, these metrics may

Table 2. Results for the Fisher-Iris data set (higher values indicate better generalization)

Network model	Cross-validation mean (std)	Metric 1 mean (std)	Metric 2 mean (std)
4-2-3	94.67% (6.13%)	80.42% (10.58%)	2.14 (4.09)
4-3-3	94.00% (6.63%)	79.66% (7.0%)	1.69 (1.28)
4-5-3	93.33% (7.70%)	73.59% (6.08%)	1.15 (1.05)
4-8-3	91.33% (8.92%)	67.67% (6.68%)	1.01 (0.74)
4-10-3	94.67% (10.33%)	66.98% (10.04%)	0.84 (0.75)
4-15-3	92.67% (7.98%)	63.91% (8.49%)	0.51 (0.51)

be used for a given data set in order to compare the performance of two different networks regardless of the way they were trained. Concerning the implementation of the approach, we foresee to take over a suitable implementation of SIVIA for neural network inversion which will perform inversion incrementally, i.e., for each class exclude from the search space the area found to belong to previously examined classes.

A hypothesis that seems to be strongly advocated by the experimental results, is that a suitable combination of the proposed metrics may constitute an effective means for comparing the performance of different network models on the same classification task. However, the proposed metrics should be examined from a mathematically defensible point of view and this constitutes another objective for future work. For the time being we need to note the strong similarity between the domain of validity derived as a level set and the density level sets that can be defined in the input space using clustering or other techniques.

6   Conclusion

In this paper we advanced the results obtained in [1] concerning the quantitative aspects of a neural network's domain of validity. The volume of the area defined by IA-based inversion constitutes a guaranteed quantity for computing the empirical metrics of the network classification performance, mainly, in terms of generalization and comparative evaluation of different network models. The results obtained provide concrete evidence of the potential suggested by the proposed metrics. However, we consider that these empirical metrics need to be validated from a mathematical point of view which will take into account the Bayesian aspects of the neural network classification function. This is one of the objectives of our current work.

Acknowledgments. The authors would like to thank the anonymous reviewers for their comments and suggestions on earlier draft of the manuscript.

References

1. Adam, S.P., Karras, D.A., Magoulas, G.D., Vrahatis, M.N.: Reliable estimation of a neural network's domain of validity through interval analysis based inversion. In: 2015 International Joint Conference on Neural Networks (IJCNN), pp. 1–8, July 2015
2. Bourlard, H., Wellekens, C.J.: Links between markov models and multilayer perceptrons. IEEE Trans. Pattern Anal. Mach. Intell. **12**(12), 1167–1178 (1990)
3. Courrieu, P.: Three algorithms for estimating the domain of validity of feedforward neural networks. Neural Netw. **7**(1), 169–174 (1994)
4. Duda, R.O., Hart, P.E.: Pattern Classification and Scene Analysis. Wiley, New York (1973)
5. Eberhart, R., Dobbins, R.: Designing neural network explanation facilities using genetic algorithms. In: 1991 IEEE International Joint Conference on Neural Networks, vol. 2, pp. 1758–1763, November 1991
6. Gish, H.: A probabilistic approach to the understanding and training of neural network classifiers. In: International Conference on Acoustics, Speech, and Signal Processing, vol. 3, pp. 1361–1364, April 1990
7. Hampshire II, J.B., Pearlmutter, B.A.: Equivalence proofs for multilayer perceptron classifiers and the Bayesian discriminant function. In: Proceedings of 1990 Connectionist Models Summer School, vol. 1, pp. 159–172, April 1991
8. Hassoun, M.H.: Fundamentals of Artificial Neural Networks. MIT Press, Cambridge (1995)
9. Haykin, S.: Neural Networks a Comprehensive Foundation, 2nd edn. Prentice-Hall, Upper Saddle River (1999)
10. Jaulin, L., Kieffer, M., Didrit, O., Walter, E.: Applied Interval Analysis. With Examples in Parameter and State Estimation, Robust Control and Robotics. Springer, London (2001)
11. Jaulin, L., Walter, E.: Set Inversion Via Interval Analysis for nonlinear bounded-error estimation. Automatica **29**(4), 1053–1064 (1993)
12. Jensen, C., Reed, R., Marks, R., El-Sharkawi, M., Jung, J.B., Miyamoto, R., Anderson, G., Eggen, C.: Inversion of feedforward neural networks: algorithms and applications. Proc. IEEE **87**(9), 1536–1549 (1999)
13. Kearfott, R.B.: Interval computations: introduction, uses, and resources. Euromath Bull. **2**(1), 95–112 (1996)
14. Kindermann, J., Linden, A.: Inversion of neural networks by gradient descent. Parallel Comput. **14**(3), 277–286 (1990)
15. Lippmann, R.P.: Pattern classification using neural networks. IEEE Commun. Mag. **27**(11), 47–50 (1989)
16. Lu, B.L., Kita, H., Nishikawa, Y.: Inverting feedforward neural networks using linear and nonlinear programming. IEEE Trans. Neural Netw. **10**(6), 1271–1290 (1999)
17. Reed, R., Marks, R.: An evolutionary algorithm for function inversion and boundary marking. In: IEEE International Conference on Evolutionary Computation, vol. 2, pp. 794–797, November 1995
18. Richard, M., Lippmann, R.: Neural network classifiers estimate Bayesian a posteriori probabilities. Neural Comput. **3**(4), 461–483 (1991)
19. Rump, S.M.: INTLAB - INTerval LABoratory. In: Csendes, T. (ed.) Developments in Reliable Computing, pp. 77–104. Kluwer Academic, Dordrecht (1999)

326   S.P. Adam et al.

20. Saad, E.W., Wunsch II, D.C.: Neural network explanation using inversion. Neural Netw. **20**(1), 78–93 (2007)
21. Theodoridis, S., Pikrakis, A., Koutroumbas, K., Kavouras, D.: Introduction to Pattern Recognition: A MATLAB Approach. Academic Press, Burlington (2010)
22. Thrun, S.B.: Extracting provably correct rules from artificial neural networks. Technical report IAI-TR-93-5, Institut fur Informatik III, Bonn, Germany (1993)
23. Tornil-Sin, S., Puig, V., Escobet, T.: Set computations with subpavings in MATLAB: the SCS toolbox. In: 2010 IEEE International Symposium on Computer-Aided Control System Design (CACSD), pp. 1403–1408, September 2010

A Novel Adaptive Learning Rate Algorithm for Convolutional Neural Network Training

S.V. Georgakopoulos[(✉)] and V.P. Plagianakos

Department of Computer Science and Biomedical Informatics, University of Thessaly,
Papassiopoulou 2–4, 35100 Lamia, Greece
{spirosgeorg,vpp}@uth.gr

Abstract. In this work an adaptive learning rate algorithm for Convolutional Neural Networks is presented. Harvesting already computed first order information of the gradient vectors of three consecutive iterations during the training phase, an adaptive learning rate is calculated. The learning rate is increasing proportionally to the similarity of the direction of the gradients in an attempt to accelerate the convergence and locate a good solution. The proposed algorithm is suitable for the time-consuming training of the Convolutional Neural Networks, alleviating the exhaustive and critical for the performance of trained network heuristic search for a suitable learning rate. The experimental results indicate that the proposed algorithm produces networks having good classification accuracy, regardless the initial learning rate value. Moreover, the training procedure is similar or better to the gradient descent algorithm with fixed heuristically chosen learning rate.

Keywords: Convolutional Neural Networks · Adaptive learning rate

1 Introduction

Convolutional Neural Networks (CNNs) are state-of-the-art classification algorithms with many recent successes, mainly in computer vision problems. CNNs have been suggested for pattern recognition [4], object localization [15], object classification in large scale database of real world image [10], abnormalities recognition on medical images [8], etc. Despite the increase of CNNs' usage in the recent years, the CNN model is rooted back to a model of 1980 [7], which had adopted layers of learnable filters and a supervised classifier for the output layer. This model is modified by simplifying the architecture and introducing the back-propagation algorithm to train it [11]. However, the high number of trainable weights that this type of networks require was an inhibitor for their wide use for many years. This problem was tackled with the advent of parallel architecture and the utilization of Graphic Processing Units (GPUs).

Because of the high volume of training data that the CNNs need for training, the on-line Stochastic Gradient Descent (SGD) learning algorithm is usually utilized. A crucial parameter for the SGD algorithm's convergence is the user

© Springer International Publishing AG 2017
G. Boracchi et al. (Eds.): EANN 2017, CCIS 744, pp. 327–336, 2017.
DOI: 10.1007/978-3-319-65172-9_28

chosen learning rate. The heuristic search of the proper learning rate value for each problem or for each specific dataset can be difficult and time consuming, since, the typical training time for a CNN varies from hours to days even when high-end GPUs are used. Additionally, it has been observed that a learning rate decay after a number of model training iterations tends to increase their performance. Thus, the human effort needed to find suitable parameter values, leads to the adoption of adaptive learning rate methods.

In this study, we present an adaptive learning rate algorithm for CNN training, which relies on the SGD algorithm and uses the last three gradients to adapt the learning rate in every training iteration.

The rest of the paper is structured as follows. In Sect. 2 related work of adaptive learning rate algorithms and preliminary materials are presented. In Sect. 3, we introduce the proposed algorithm for the adaptive learning rate and in Sect. 4 we present the experimental results and analysis. Finally, we conclude with pointers for future work.

2   Background Methods

In this section, we present related adaptive learning methods and review the basic tools used in the proposed methodology. In particular, we briefly present the CNN classification algorithm and the popular Stochastic Gradient Descent optimization method.

2.1   Convolutional Neural Networks

Convolutional Neural Networks are multistage trainable architectures used for classification tasks. Each of these stages consist of three types of layers [12]:

1. *Convolutional Layers*, which are the major components of the CNNs. A convolutional layer consists of a number of kernel matrices that perform convolution on their input and produce an output matrix (feature image) where a bias value is added. The learning procedures aim to train the kernel weights and biases as shared neuron connection weights.
2. *Pooling Layer*, which are also integral components of the CNNs. The purpose of a pooling layer is to perform dimensionality reduction of the input feature images. Pooling layers make a subsampling to the output of the convolutional layer matrices combing neighboring elements. The most common pooling function is the max-pooling function, which takes the maximum value of the local neighborhoods.
3. *Fully-Connected Layer*, is a classic Feed-Forward Neural Network (FNN) hidden layer. It can be interpreted as a special case of the convolutional layer with kernel size 1×1. This type of layer belongs to the class of trainable layer weights and it is used in the final stages of CNNs.

The training of CNN is relies on the BackPropagation (BP) training algorithm [12]. The requirements of the BP algorithm is a vector with input patterns x and a vector with targets y, respectively. The input x_i is associated with the output o_i. Each output is compared to its corresponding desirable target and their difference gives the training error. Our goal is to find weights that minimize the cost function

$$E(w) = \frac{1}{n} \sum_{p=1}^{P} \sum_{j=1}^{N_L} (o_{j,p}^L - y_{j,p})^2, \tag{1}$$

where P the number of patterns, $o_{j,p}^L$ the output of j neuron that belongs to L layer, N_L the number of neurons in output layer, $y_{j,p}$ the desirable target of j neuron of pattern p. To minimize the cost function $E(w)$, a pseudo-stochastic version of SGD algorithm, also called mini-batch Stochastic Gradient Descent (mSGD), is usually utilized [3].

2.2   Stochastic Gradient Descent

Stochastic Gradient Descent is an optimization algorithm with many successes in the Machine Learning field. Given a cost function $E(w)$ it aims to find the minimizer $w^* = (w_1^*, w_2^*, \cdots, w_v^*) \in \mathbb{R}^v$, such that:

$$w^* = \min_{w \in \mathbb{R}^v} E(w). \tag{2}$$

The SGD algorithm iteratively minimizes Eq. 1, updating the vector w after the presentation of each training example. To speedup the training process, the pseudo-stochastic variant of SGD uses a small number of training examples (mini batches) to update the weights. The mini-batches SGD has been proven very successful in CNN training [10].

To minimize the aforementioned function, the SGD calculates the gradient vector at every step and updates the network weights using a heuristically defined learning rate (n_0). Because of the CNN multi-layer architecture the gradient calculation is performed at each layer output with the chain rule. However, the learning rate parameter is very critical for the convergence of the SGD; usually large values accelerate convergence to local minima, while smaller values, result in larger training time, may locate "better" local minima.

2.3   Related Work

The problem of selecting a proper learning rate is crucial for the convergence of the SGD and in general to train high classification accuracy CNNs. Many adaptive schemes have proposed in the recent literature to deal with this problem. The AdaDelta [19] algorithm dynamically adapts the learning rate using information only from the gradient vector. The method does not require manual learning rate tuning, while it is robust to noisy gradient information. Another popular method AdaGrad [6] incorporates the knowledge of the geometry that the observed data provide on previous iterations to perform more informative gradient-based learning.

The most recently proposed adaptation method, which outperforms the previous ones, is the ADAM [2] algorithm. It is an optimization algorithm that utilizes the gradient information of the stochastic objective function, based on adaptive estimates of lower order moments. More specifically, the method calculates adaptive learning rates from estimates of first and second moments of the gradients. However, the selection of the initial learning rate is critical for the success of the adaptation.

3   The Proposed Method

The proposed algorithm is inspired by the methods presented in [1], where information of the gradient direction of the previous and the current step is used to modify the global learning rate of the SGD algorithm for FNNs. More specifically, when the gradient vector directions of two consecutive steps tend to be similar, convergence is accelerated by increasing the learning rate. On the contrary, the learning rate is decreased when the directions of the gradient vectors tend to significantly differ. This can be roughly approximated with the inner product $(\langle \cdot, \cdot \rangle)$ of the normalized gradient vectors. Note that the inner product of orthogonal gradient vectors is zero, while it is equal to one when the vectors are parallel. The update rule is following:

$$n_i = n_{i-1} + \gamma \langle \nabla E_{i-1}(w_{i-1}), \nabla E_i(w_i) \rangle, \tag{3}$$

where γ is a control parameter of the learning rate adaptation.

This work is based on an adaptive learning algorithm designed for FNN training [13,16]. The suggested Adaptive Learning Rate (AdLR) algorithm relies on the latest three gradient vector directions, exploiting already computed information. The learning rate is adjusted taking into consideration the current and the previous inner product of the gradient vectors. The update rule is the following:

$$\begin{aligned} n_i = n_{i-1} &+ \gamma_1 \langle \nabla E_{i-1}(w_{i-1}), \nabla E_i(w_i) \rangle + \\ &+ \gamma_2 \langle \nabla E_{i-2}(w_{i-2}), \nabla E_{i-1}(w_{i-1}) \rangle, \end{aligned} \tag{4}$$

where the γ_1 and γ_2 are control parameter called meta-learning rates. We suggest that $\gamma_1 > \gamma_2$, since the direction of the latest gradient vectors should contribute more to the adaptation of the learning rate. In addition, to prevent the divergence of the algorithm when the learning rate takes negative values, we apply the following reset rule:

$$n_i = \begin{cases} n_0 \cdot \gamma_3, & \text{if } n_i < 0 \\ n_i, & \text{if } n_i \geq 0 \end{cases} \tag{5}$$

where the n_0 is the initial learning rate value and γ_3 is a decrease factor of the original learning rate. The use of the decrease factor helps the stable convergence of the algorithm, especially in the case where the reset rule is activated after many epochs and the initial learning rate n_0 is rather large, the algorithm could overshoot the reached minimum and/or need many steps to properly readapt the learning rate.

The requirements of the proposed algorithm in computational resources is comparable to those of the SGD algorithm, since only the previous gradient vectors need to be stored and the inner products can be rapidly calculated. The AdLR[1] algorithm is implemented in C++ and CUDA using the CAFFE framework [9]. The pseudo-code of the proposed algorithm is presented below.

> **Input:** w_0, n_0, γ_1, γ_2, γ_3, *iter* ;
> $i = 0$;
> **while** $i < iter$ **do**
> > $i = i + 1$;
> > Calculate $E(w_i)$ and $\nabla E_i(w_i)$;
> > Update the weights
> > $w_{i+1} = w_i - n_i \nabla E_i(w_i)$;
> > Update learning rate
> > $n_i = n_{i-1} + \gamma_1 \langle \nabla E_{i-1}(w_{i-1}), \nabla E_i(w_i) \rangle +$
> > $\qquad + \gamma_2 \langle \nabla E_{i-2}(w_{i-2}), \nabla E_{i-1}(w_{i-1}) \rangle$
> > **if** $n_i < 0$ **then**
> > | $n_i = n_0 \cdot \gamma_3$
> > **end**
> **end**
> **Output:** w_{i+1}

Algorithm 1. The Stochastic Gradient Descent with Adaptive Learning Rate Algorithm (AdLR)

4 Experimental Results

To evaluate the proposed algorithm the CIFAR-10 object recognition dataset [17], was used. The dataset consists of 60,000 tiny color images with size 32×32 pixels, which contain ten classes with 6,000 images each. The dataset was split to 50,000 images for training, while the rest were kept for testing. The CNN architecture model for this problem is the one proposed in the CAFFE framework. More specifically, the CNN consists of two convolutional layers of 32 feature maps with 5×5 convolutional kernels, each followed by one 3×3 max pooling layer. Consecutively, another convolutional layer of 64 feature maps of 5×5 convolutional kernels followed by a 3×3 max pooling layer is utilized. Finally, a fully-connected layer with 10 neurons and a softmax logistic regression layer is integrated to the previous layers. After each convolutional layer a ReLu activation function [14] is applied, while after the first two pooling layer a local region normalization of their output is performed.

The proposed AdLR algorithm is compared to the SGD and the ADAM algorithms, using different initial learning rate values n_0, i.e. 10^{-2}, 10^{-3}, 10^{-4}

---

[1] The code is available on https://github.com/Georgakopoulos-Sp/.

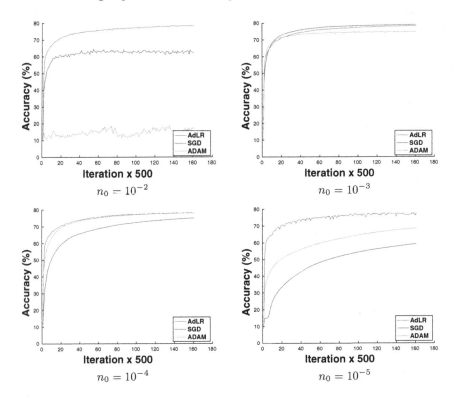

Fig. 1. The accuracy of the AdLR, the SGD and the ADAM algorithms with different initial learning rate values.

and 10^{-5}. The maximum number of iterations was set to $80,000$, the mini-batches of all algorithms contained 100 samples and the momentum term was set to 0.9. The parameters of the ADAM algorithm were set as proposed by its authors [2]. For the proposed AdLR algorithm the meta-learning parameters γ_1 and γ_2 had fixed values of 10^{-2} and 10^{-3}, respectively, while $\gamma_3 = 10^{-2}$. To validate the accuracy of the results, 100 independent executions for each initial learning rate n_0 for each algorithm were performed.

In Fig. 1 the mean accuracy is depicted, while Table 1 presents the average performance of the algorithms for 100 independent executions. The mean accuracy of the proposed algorithm is consistent and similar or better to the performance of the SGD and the ADAM algorithms, regardless the initial learning rate. As expected, the SGD algorithm is found to be very sensitive to the selection of the initial learning rate. On the other hand, although the ADAM algorithm adapts the direction of search, the user selected learning rate can be critical, as well. The proposed AdLR algorithm is suitable for CNN training, alleviating the exhaustive heuristic search for a suitable learning rate.

To investigate the results of the proposed AdLR algorithm against the other algorithms, a two-sided Wilcoxon non-parametric significant test [18] at the 5% significance level is conducted for every initial learning rate.

Table 1. The mean accuracy and standard deviation (Std) of the proposed Algorithm (AdLR), the SGD and the ADAM algorithm on CIFAR-10 dataset for different learning rate values.

	AdLR		SGD		ADAM	
	Mean (%)	Std	Mean (%)	Std	Mean (%)	Std
$n_0 = 10^{-2}$	78.4 (+/+)	0.7	62.3	1.9	17.2	8.3
$n_0 = 10^{-3}$	78.4 (=/+)	0.5	78.9	0.6	74.7	1.0
$n_0 = 10^{-4}$	78.3 (+/=)	0.7	75.0	0.3	78.1	0.5
$n_0 = 10^{-5}$	78.0 (+/+)	0.5	59.3	0.4	68.7	0.5

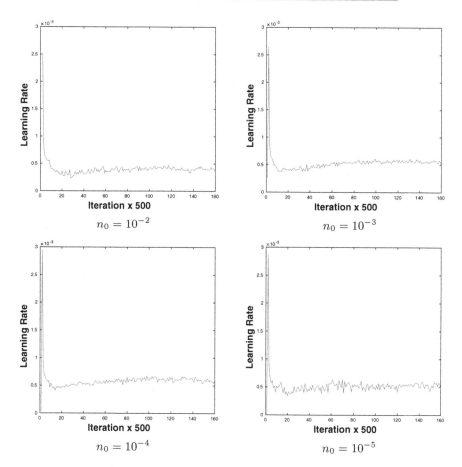

$$n_0 = 10^{-2} \qquad n_0 = 10^{-3}$$

$$n_0 = 10^{-4} \qquad n_0 = 10^{-5}$$

Fig. 2. The adaptation of learning rate of the proposed AdLR algorithm with different initial learning rate values.

The null hypothesis is that the samples of each comparison are independent and derived by identical continuous distributions with equal medians. In Table 1, we mark with the "+" sign the cases when the null hypothesis is rejected at the 5% significance level and the proposed algorithm exhibits superior performance, with the "−" sign when the null hypothesis is rejected at the same level of significance and the proposed algorithm exhibits inferior performance and with "=" when the performance difference is not statistically significant. The usage of the notation $(\cdot/\cdot)$ for the AdLR algorithm indicate the result of Wilcoxon test against the SGD and the ADAM algorithm, respectively.

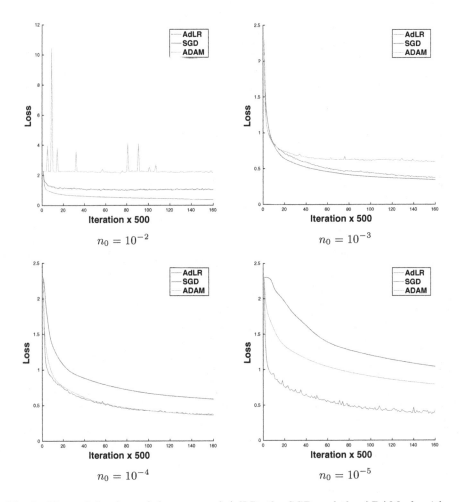

Fig. 3. The training loss of the proposed AdLR, the SGD and the ADAM algorithm with different initial learning rate values.

The proposed algorithm adjusts the learning rate on each iteration and the Fig. 2 presents the mean value of learning rate adaptations during the training for different initial values. Independently to the initial learning rate the learning rate adapts to similar values. In addition, in Fig. 3 the mean values of the loss function during training samples are depicted.

5   Conclusions

The heuristic search for the initial learning rate value for Convolutional Neural Networks training can be difficult and time consuming, since the training time is high even when powerful GPUs are used. Usually, a trial-and-error procedure is performed. Thus, the human effort needed to find suitable parameter values, leads to the adoption of adaptive learning rate methods.

In this work we present an adaptive learning rate algorithm for CNN training. Leveraging first order gradient information of three consecutive iterations during the training phase, the learning rate is adapted. The proposed AdLR algorithm is evaluated against the popular SGD optimization algorithm and the recently proposed ADAM adaptive algorithm on the CIFAR-10 dataset. Different initial learning rate values were used and the performance of the AdLR algorithm was found to be very promising.

In future work, we intent to establish a theoretical bound of the meta-learning values of the algorithm convergence and examine the accuracy of the algorithm with different meta-learning values. Additionally, we indent to evaluate the proposed algorithm on other datasets, such as the well known ImageNet [5] and perform a comparative analysis including more state-of-the-art adaptive learning rate algorithms.

Acknowledgments. We gratefully acknowledge the support of NVIDIA Corporation with the donation of the Titan X Pascal GPU used for this research. This work was partially supported by a Hellenic Artificial Intelligence Society (EETN) scholarship.

References

1. Almeidaa, L.B., Langloisa, T., Amaral, J.D., Plakhov, A.: Parameter adaptation in stochastic optimization. In: On-line Learning in Neural Networks, pp. 111–134. Cambridge University Press (1998)
2. Ba, J., Kingma, D.: Adam: a method for stochastic optimization. In: International Conference on Learning Representations (2015)
3. Bottou, L.: On-line learning and stochastic approximations. In: On-line Learning in Neural Networks, pp. 9–42. Cambridge University Press (1998)
4. Delibasis, K.K., Georgakopoulos, S.V., Kottari, K., Plagianakos, V.P., Maglogiannis, I.: Geodesically-corrected zernike descriptors for pose recognition in omni-directional images. Integr. Comput.-Aided Eng. **23**(2), 185–199 (2016)
5. Deng, J., Dong, W., Socher, R., Li, L.J., Li, K., Fei-Fei, L.: ImageNet: a large-scale hierarchical image database. In: CVPR 2009 (2009)
6. Duchi, J., Hazan, E., Singer, Y.: Adaptive subgradient methods for online learning and stochastic optimization. J. Mach. Learn. Res. **12**, 2121–2159 (2011)
7. Fukushima, K.: Neocognitron: a self-organizing neural network model for a mechanism of pattern recognition unaffected by shift in position. Biol. Cybern. **36**(4), 193–202 (1980)

8. Georgakopoulos, S.V., Iakovidis, D.K., Vasilakakis, M., Plagianakos, V.P., Koulaouzidis, A.: Weakly-supervised convolutional learning for detection of inflammatory gastrointestinal lesions. In: 2016 IEEE International Conference on Imaging Systems and Techniques (IST), pp. 510–514, October 2016

9. Jia, Y., Shelhamer, E., Donahue, J., Karayev, S., Long, J., Girshick, R., Guadarrama, S., Darrell, T.: Caffe: convolutional architecture for fast feature embedding. arXiv preprint arXiv:1408.5093 (2014)

10. Krizhevsky, A., Sutskever, I., Hinton, G.E.: Imagenet classification with deep convolutional neural networks. In: Pereira, F., Burges, C.J.C., Bottou, L., Weinberger, K.Q. (eds.) Advances in Neural Information Processing Systems, vol. 25, pp. 1097–1105. Curran Associates, Inc. (2012)

11. LeCun, Y., Boser, B., Denker, J.S., Henderson, D., Howard, R.E., Hubbard, W., Jackel, L.D.: Backpropagation applied to handwritten zip code recognition. Neural Comput. 1(4), 541–551 (1989)

12. Lecun, Y., Bottou, L., Bengio, Y., Haffner, P.: Gradient-based learning applied to document recognition. Proc. IEEE 86(11), 2278–2324 (1998)

13. Magoulas, G., Plagianakos, V., Vrahatis, M.: Adaptive stepsize algorithms for online training of neural networks. Nonlinear Anal.: Theory Methods Appl. 47(5), 3425–3430 (2001)

14. Nair, V., Hinton, G.E.: Rectified linear units improve restricted Boltzmann machines. In: Fnkranz, J., Joachims, T. (eds.) Proceedings of 27th International Conference on Machine Learning (ICML-2010), pp. 807–814. Omnipress (2010)

15. Oquab, M., Bottou, L., Laptev, I., Sivic, J.: Is object localization for free? - weakly-supervised learning with convolutional neural networks. In: 2015 IEEE Conference on Computer Vision and Pattern Recognition (CVPR), pp. 685–694, June 2015

16. Plagianakos, V.P., Magoulas, G.D., Vrahatis, M.N.: Global learning rate adaptation in on-line neural network training. In: Proceedings of 2nd International ICSC Symposium on Neural Computation (NC 2000), Berlin, Germany (2000)

17. Torralba, A., Fergus, R., Freeman, W.T.: 80 million tiny images: a large data set for nonparametric object and scene recognition. IEEE Trans. Pattern Anal. Mach. Intell. 30(11), 1958–1970 (2008). http://dx.doi.org/10.1109/TPAMI.2008.128

18. Wilcoxon, F.: Individual comparisons by ranking methods. Biom. Bull. 1(6), 80–83 (1945)

19. Zeiler, M.D.: ADADELTA: an adaptive learning rate method. CoRR abs/1212.5701 (2012)

Sparsity of Shallow Networks Representing Finite Mappings

Věra Kůrková[✉]

Institute of Computer Science, Czech Academy of Sciences,
Pod Vodárenskou věží 2, 18207 Prague, Czech Republic
vera@cs.cas.cz

Abstract. Limitations of capabilities of shallow networks to represent sparsely real-valued functions on finite domains is investigated. Influence of sizes of function domains and of sizes dictionaries of computational units on sparsity of networks computing finite mappings is explored. It is shown that when dictionary is not sufficiently large with respect to the size of the finite domain, then almost any uniformly randomly chosen function on the domain either cannot be sparsely represented or its computation is unstable.

Keywords: Shallow networks · Finite mappings · Sparsity · Model complexity · Concentration of measure · Signum perceptrons

1 Introduction

It is well-known that shallow networks with one hidden layer of many common types of computational units can approximate within any accuracy any reasonable function on a compact domain and also can exactly compute any function on a finite domain [1,2]. All these universality type results are proven assuming that numbers of network units are potentially infinite or, in the case of finite domains, are at least as large as sizes of the domains.

In practical applications, various constraints on numbers and sizes of network parameters often hold. In such cases, sparse solutions of computational tasks should be searched for. Sparsity, considered as a measure of network simplicity, has been promoted by various regularization techniques, such as the weight-decay [3]. In some cases, proper choice of a network architecture together with a type of computational units can decrease network complexity considerably.

Bengio and LeCun [4], who revived the interest in deep networks, conjectured that "most functions that can be represented compactly by deep architectures cannot be represented by a compact shallow architecture". On the other hand, a recent empirical study demonstrated that shallow networks can learn some functions previously learned by deep ones using the same numbers of parameters as the original deep networks [5].

Characterization of functions, which can be computed by deep networks more efficiently than by shallow ones can be derived by comparing lower bounds on

© Springer International Publishing AG 2017
G. Boracchi et al. (Eds.): EANN 2017, CCIS 744, pp. 337–348, 2017.
DOI: 10.1007/978-3-319-65172-9_29

complexity of shallow networks with upper bounds on complexity of deep ones. An important step towards this goal is exploration of functions which cannot be computed or approximated by shallow networks satisfying given sparsity constraints. Whereas many upper bounds on numbers of units in shallow networks sufficient for a given accuracy of function approximation are known (see, e.g., the survey article [6] and references therein), fewer lower bounds are available. Derivation of lower bounds is generally much more difficult. Some bounds hold merely for types of computational units that are not commonly used (see, e.g., [7,8].

Bengio et al. [9] suggested that a cause of large model complexities of shallow networks might be in the "amount of variations" of functions to be computed. They illustrated their suggestion by an example of classification of points in the d-dimensional Boolean cube according to their parities by Gaussian SVM algorithm. They proved that for this task, a shallow Gaussian kernel network requires at least 2^{d-1} units (support vectors). Bianchini and Scarselli [10] introduced a new approach to estimation of model complexities of networks with several hidden layers using the concept of the Betti Numbers from algebraic topology. In [11], we showed that complexity of networks computing the same functions strongly depends on types of their units. We proposed to use as a measure of sparsity norms tailored to dictionaries of computational units. Using a probabilistic approach based on Chernoff bound we derived non constructive lower bounds on these norms for functions representing binary classifiers. In [12,13] we described some concrete examples of such functions.

In this paper, we investigate sparsity of shallow networks computing real-valued functions on finite domains. Such domains can be 2-dimensional (e.g., pixels of photographs) or high-dimensional (e.g., digitized high-dimensional cubes), but often they are quite large. As minimization of numbers of non zero output weights is a difficult non convex task, searching for sparse solutions has been implemented by weigh-decay regularization methods with stabilizers in the form of l_1 and l_2-norms of output-weight vectors. Large l_1-norms of output-weight vectors imply that either the number of network units is large or some of output weights are large. Both are not desirable as networks with large numbers of units might require too large resources for an implementation, while large output weights might amplify small changes of inputs and thus lead to non stability of computation. We show that when domains are large, then most functions cannot be l_1-sparsely represented by networks with units from dictionaries which are not sufficiently large with respect to the sizes of the domains (more precisely, when the size of a dictionary is bounded by $e^{p(\ln m)}$, where p is a polynomial and m is the size of the domain). Our arguments are based on geometrical properties of high-dimensional spaces. We illustrate our probabilistic result by examples of some common dictionaries of computational units which satisfy this bound.

The paper is organized as follows. Section 2 contains basic concepts on shallow networks and dictionaries of computational units. In Sect. 3, sparsity is investigated in terms of l_1-norm and norms tailored to computational units. In Sect. 4 lower bounds on sparsity are derived in terms of sizes of dictionaries and sizes of finite domains of functions to be computed. In Sect. 5, general results are applied to perceptron and kernel networks. Section 6 contains a discussion.

2 Preliminaries

A *one-hidden-layer (shallow) network with a single linear output* computes input-output functions belonging to the set of the form

$$\text{span}_n\, G := \left\{ \sum_{i=1}^{n} w_i g_i \,\middle|\, w_i \in \mathbb{R},\ g_i \in G \right\},$$

where the coefficients w_i are called *output weights*, n denotes the number of network units, and G is a parameterized set of computational units called a *dictionary*.

In practical applications, domains $X \subset \mathbb{R}^d$ of functions to be computed are finite, but their sizes and/or input dimensions d can be quite large. Also dictionaries are finite. Thus in this paper, we focus only on finite mappings and finite dictionaries. We show that sizes of dictionaries play an important role in investigation of sparsity of representations of functions by shallow networks.

A common type of a computational unit is *perceptron*, which computes functions of the form $\sigma(v \cdot . + b) : X \to \mathbb{R}$, where $\sigma : \mathbb{R} \to \mathbb{R}$ is an *activation function*. It is called *sigmoid* when it is monotonic increasing and $\lim_{t\to-\infty} \sigma(t) = 0$ and $\lim_{t\to\infty} \sigma(t) = 1$. Important types of activation functions are the *Heaviside function* defined as

$$\vartheta(t) := 0 \text{ for } t < 0 \quad \text{and} \quad \vartheta(t) := 1 \text{ for } t \geq 0$$

and the *signum function* $\text{sgn} : \mathbb{R} \to \{-1, 1\}$, defined as

$$\text{sgn}(t) := -1 \text{ for } t < 0 \quad \text{and} \quad \text{sgn}(t) := 1 \quad \text{for } t \geq 0.$$

We denote by $P_d(X)$ the dictionary of functions on X computable by *signum perceptrons*, i.e.,

$$P_d(X) := \{\text{sgn}(v \cdot . + b) : X \to \{-1, 1\} \,|\, v \in \mathbb{R}^d, b \in \mathbb{R}\}. \tag{1}$$

Note that from the point of view of number of network units, there is only a minor difference between networks with signum and Heaviside perceptrons as

$$\text{sgn}(t) = 2\vartheta(t) - 1 \quad \text{and} \quad \vartheta(t) = \frac{\text{sgn}(t) + 1}{2}. \tag{2}$$

It is more convenient to consider the dictionary of signum perceptrons instead of Heaviside ones because all signum perceptrons have the same norms equal to $\sqrt{m}$, where m is the size of the domain X.

Another important class of dictionaries is formed by sets of kernel units. For $X, U \subseteq \mathbb{R}^d$ and a symmetric positive semidefinite kernel $K : \mathbb{R}^d \times \mathbb{R}^d \to \mathbb{R}$, we denote by

$$K(X, U) := \{K(., u) : X \to \mathbb{R} \mid u \in U\}$$

the *dictionary of kernel units on X with parameters (centers) in U*. When $X = U$, we write shortly $K(X)$. In the Support Vector Machine (SVM) algorithm, the set $U = \{u_i, \mid i = 1, \ldots, l\}$ is the set of points to be classified, among which some play the role of support vectors. The number of units in the trained network is equal to the number of support vectors.

For a domain $X \subset \mathbb{R}^d$ we denote by

$$\mathcal{F}(X) := \{f \mid f : X \to \mathbb{R}\}$$

the *set of all real-valued functions on X*.

It is easy to see that when card $X = m$ and $X = \{x_1, \ldots, x_m\}$ is a linear ordering of X, then the mapping $\iota : \mathcal{F}(X) \to \mathbb{R}^m$ defined as $\iota(f) := (f(x_1), \ldots, f(x_m))$ is an isomorphism. So, on $\mathcal{F}(X)$ we have the Euclidean inner product and the norm defined as

$$\langle f, g \rangle := \sum_{u \in X} f(u)g(u) \qquad \|f\| := \sqrt{\langle f, f \rangle}. \tag{3}$$

In contrast to the inner product $\langle ., . \rangle$ on $\mathcal{F}(X)$, we denote by $\cdot$ the inner product on $X \subset \mathbb{R}^d$, i.e., for $u, v \in X$,

$$u \cdot v := \sum_{i=1}^{d} u_i v_i.$$

3   Measures of Sparsity

It is desirable that architectures and computational units are chosen in such a way that networks computing given tasks are sufficiently sparse. The most natural measure of sparsity is the number of non-zero output weights. In some literature, the number of non-zero coefficients among w_i's in an input-output function

$$f = \sum_{i=1}^{n} w_i g_i \tag{4}$$

from span G is called an "l_0-pseudo-norm" in quotation marks and denoted $\|w\|_0$. However, it is neither a norm nor a pseudo-norm. The quantity $\|w\|_0$ is always an integer and thus $\| \cdot \|_0$ does not satisfy the homogeneity property of a norm ($\|\lambda x\| = |\lambda| \|x\|$ for all λ). Moreover, the "unit ball" $\{w \in \mathbb{R}^n \mid \|w\|_0 \leq 1\}$ is non convex and unbounded as it is equal to the set of all vectors having at most one non zero entry. For any $r > 0$, the ball of radius r is equal to $\text{span}_n G$, where $n = \lfloor r \rfloor$.

Minimization of "l_0-pseudo-norm" of the vector of output weights is a difficult non convex optimization task. Instead of "l_0", l_1-norm defined as $\|w\|_1 = \sum_{i=1}^{n} |w_i|$ and l_2-norm defined as $\|w\|_2 = \sum_{i=1}^{n} w_i^2$ of output weight vectors

$w = (w_1, \ldots, w_n)$ have been used in weight-decay regularization [3]. These norms can be implemented as stabilizers modifying error functionals to be minimized during learning [14]. A network with a large l_1 or l_2-norm of its output-weight vector must have either a large number of units or some output weights must be large. Both of these properties are not desirable as they imply either a large model complexity or non stability of the computation caused by ill-conditioning. When some of output weight of a network are large, then small errors in input data or small change of parameters of hidden units can lead to large differences in the network output.

Many dictionaries of computational units are not linearly independent and so they form over-complete spanning sets. Thus the representation (4) as a linear combination of units from the dictionary need not be unique. For finite dictionaries, the minimum of l_1-norms of output vectors of shallow networks with units from a dictionary G computing a function f is equal to a norm tailored to G. This norm, called *G-variation*, has been used as a tool for estimation of rates of approximation of functions by networks with increasing "l_0-pseudo-norms". G-variation is defined for a bounded subset G of a normed linear space $(\mathcal{X}, \|.\|)$ as

$$\|f\|_G := \inf \left\{ c \in \mathbb{R}_+ \ \middle| \ \frac{f}{c} \in \mathrm{cl}_\mathcal{X} \, \mathrm{conv} \, (G \cup -G) \right\},$$

where $-G := \{-g \mid g \in G\}$, $\mathrm{cl}_\mathcal{X}$ denotes the closure with respect to the topology induced by the norm $\|\cdot\|_\mathcal{X}$, and conv is the convex hull. Variation with respect to the dictionary of Heaviside perceptrons (called *variation with respect to half-spaces*) was introduced by Barron [15] and we extended it to general sets in [16].

As G-variation is a norm, it can be made arbitrarily large by multiplying a function by a scalar. Also in theoretical analysis of approximation capabilities of shallow networks, it has to be taken into account that the approximation error $\|f - \mathrm{span}_n \, G\|$ in any norm $\|.\|$ can be made arbitrarily large by multiplying f by a scalar. Indeed, for every $c > 0$, $\|cf - \mathrm{span}_n \, G\| = c\|f - \mathrm{span}_n G\|$. Thus, both G-variation and errors in approximation by $\mathrm{span}_n \, G$ have to be studied either for sets of normalized functions or for sets of functions of a given fixed norm.

G-variation is related to l_1-sparsity, it can be used for estimating its lower bounds. Proof of the next proposition follows easily from the definition.

Proposition 1. *Let G be a finite subset of $(\mathcal{X}, \|.\|)$ with $\mathrm{card} \, G = k$. Then, for every $f \in \mathcal{X}$*

$$\|f\|_G = \min \left\{ \sum_{i=1}^{k} |w_i| \ \middle| \ f = \sum_{i=1}^{k} w_i \, g_i \, , \, w_i \in \mathbb{R}, \, g_i \in G \right\}.$$

A small l_1-norm of an output-weight vector guarantees that an input-output function of a network can be well approximated by input-output functions computable by networks with small "l_0-pseudo-norms". This follows from the Maurey-Jones-Barron Theorem [17]. Here we state its reformulation in terms of

G-variation merely for finite dimensional Hilbert space $\mathcal{F}(X)$ with the Euclidean norm defined in (3) (see [16, 18]). By G^o is denoted the set of normalized elements of G, i.e., $G^o = \left\{ \frac{g}{\|g\|} \mid g \in G \right\}$.

Theorem 1. *Let* $X \subset \mathbb{R}^d$ *be finite,* G *be a finite subset of* $\mathcal{F}(X)$, $s_G = \max_{g \in G} \|g\|$, *and* $f \in \mathcal{F}(X)$. *Then for every* n,

$$\|f - \text{span}_n G\| \leq \frac{\|f\|_{G^o}}{\sqrt{n}} \leq \frac{s_G \|f\|_G}{\sqrt{n}}.$$

Theorem 1 implies that there exists an input-output function $f_n = \sum_{i=1}^n w_i g_i$, i.e., $\|w\|_0 \leq n$, such that $\|f - f_n\| \leq \frac{s_G \|f\|_G}{\sqrt{n}}$.

Lower bounds on variational norms can be obtained by geometrical arguments. The following theorem from [19] shows that functions which are "nearly orthogonal" to all elements of a dictionary G have large G-variations.

Theorem 2. *Let* $(\mathcal{X}, \|.\|_{\mathcal{X}})$ *be a Hilbert space and* G *its bounded subset. Then for every* $f \in \mathcal{X} \setminus G^{\perp}$,

$$\|f\|_G \geq \frac{\|f\|^2}{\sup_{g \in G} |g \cdot f|}.$$

4   Geometric Lower Bounds on Sparsity

In this section, we derive lower bounds on variational norms of functions on finite domains in terms of sizes of the domains and sizes of dictionaries. We show that almost any uniformly randomly chosen function on a domain of a large size m has variation at least $m^{1/4}$ with respect to any dictionary of size bounded by $e^{p(\ln m)}$, where p is a polynomial.

The main theorem of this section is based on rather counter-intuitive geometrical properties of high-dimensional Euclidean spaces. For large dimensions, most of areas of spheres S^{m-1} lie very close to their "equators". More precisely, let μ be a uniform probabilistic measure on the unit sphere $S^{m-1} := \{h \in \mathbb{R}^d \mid \|h\| = 1\}$, and for $g \in S^{m-1}$ and $\eta > 0$ let

$$C(g, \eta) := \{h \in S^{m-1} \mid |\langle h, g \rangle| \geq \eta\}$$

denote the *spherical cap* formed by all vectors within the angular distance $\alpha = \arccos \eta$ from g (see Fig. 1). Then

$$\mu(C(g, \eta)) \leq e^{-\frac{m\eta^2}{2}} = e^{-\frac{m(\cos \alpha)^2}{2}} \tag{5}$$

(see, e.g., [20, p. 11]). This is a special case of the phenomenon of *concentration of measure on high-dimensional spheres*, which states that for large dimensions

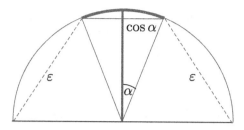

Fig. 1. Spherical cap

most of the values of Lipschitz continuous functions concentrate closely around their medians (see, e.g., [21, p. 337]).

The following theorem estimates uniform probability measures of sets of functions with large variations with respect to finite dictionaries. It is based on an argument comparing the measure of the whole sphere with its part formed by the union of "polar caps" centered at elements of a dictionary. When the dictionary is not sufficiently large, then this union covers only a small fraction of the sphere.

Theorem 3. *Let d be a positive integer, $X \subset \mathbb{R}^d$ with* $\operatorname{card} X = m$, $r > 0$, μ *be a uniform probabilistic measure on $S_r(X)$, $b > 0$, and $G(X)$ a finite subset of $\mathcal{F}(X)$ with* $\operatorname{card} G(X) = k$ *such that for all $g \in G(X)$, $\|g\| \geq r$. Then*

$$\mu(\{f \in S_r(X) \,|\, \|f\|_{G(X)} \geq b\}) \geq 1 - 2k\,e^{-\frac{m}{2b^2}}.$$

Proof. As $f \in S_r(X)$ and for all $g \in G$, $\|g\| \geq r$, we have by Theorem 2,

$$\|f\|_{G(X)} \geq \frac{\|f\|^2}{\max_{g \in G} |\langle f, g \rangle|} \geq \frac{1}{\max_{g \in G} |\langle f^o, g^o \rangle|}.$$

Hence

$$\{f \in S_r(X) \,|\, \|f\|_{G(X)} \geq b\} \supseteq S_r(X) \setminus \bigcup_{g \in G \cup -G} \bar{C}(g, 1/b),$$

where $\bar{C}(g, 1/b) = \{f \in S_r(X) \,|\, |\langle f^o, g^o \rangle| \geq \frac{1}{b}\}$. Setting $X = \{x_1, \ldots, x_m\}$, let $\iota : \mathcal{F}(X) \to \mathbb{R}^m$ be defined as $\iota(f) := (f(x_1), \ldots, f(x_m))$. As ι is an isometry between $\mathcal{F}(X)$ and $\mathbb{R}^m$, by the inequality (5) $\mu^o(\{h \in S_1(X) \,|\, |\langle h, g^o \rangle| \geq \frac{1}{b}\}) \leq e^{-\frac{m}{2b^2}}$, where μ^o is the uniform probability measure on $S_1(X)$ defined as $\mu^o(A^o) = \mu(A)$ for all $A \subset S_r(X)$. So the statement follows. $\qquad \square$

With a proper choice of b, for example b such that $b^4 = m = \operatorname{card} X$, we obtain from Theorem 3 large lower bounds on variational norms for functions on large domains. Theorem 3 implies existence of a function with G-variation at least $m^{1/4}$ for a dictionary of size smaller than $\frac{1}{2} e^{\frac{\sqrt{m}}{2}}$. When the size of the dictionary is bounded by $e^{p(\ln m)}$, this lower bound even holds for most functions on X with the fixed norm r.

Corollary 1. *Let d be a positive integer, $X \subset \mathbb{R}^d$ with $\operatorname{card} X = m$, $r > 0$, p be a polynomial, $G(X) \subset \mathcal{F}(X)$ with $\|g\| \geq r$ for all $g \in G$ and $\operatorname{card} G(X) \leq e^{p(\ln m)}$, μ be a uniform probabilistic measure on $S_r(X)$, and $b > 0$. Then*

$$\mu(\{f \in S_r(X) \mid \|f\|_{G(X)} \geq b) \geq 1 - 2e^{-(\frac{m}{2b^2} - p(\ln m))}.$$

Combining Corollary 1 with Proposition 1 we obtain a lower bound on l_1-sparsity of shallow networks with units from dictionaries of sizes which do not prevail the factor $e^{-\frac{\operatorname{card} X}{2b^2}}$.

Corollary 2. *Let d be a positive integer, $X \subset \mathbb{R}^d$ with $\operatorname{card} X = m$, $r > 0$, p be a polynomial, $G(X) \subset \mathcal{F}(X)$ with $\|g\| \geq r$ for all $g \in G$ and $\operatorname{card} G(X) \leq e^{p(\ln m)}$, μ be a uniform probabilistic measure on $S_r(X)$, and $b > 0$. Then the output-weight vector w of any shallow network with units from $G(X)$ computing a uniformly randomly chosen function $f \in S_r(X)$ satisfies $\|w\|_1 \geq b$ with probability at least $1 - 2e^{-(\frac{m}{2b^2} - p(\ln m))}$.*

For example, when the domain X is the d-dimensional cube, then Corollary 2 implies for almost all functions in $S_r(X)$ a lower bound $2^{d/4}$ on l_1-sparsity of all shallow networks with units from dictionaries of sizes bounded by $e^{\frac{1}{2}2^{d/2}}$ computing these functions. Such networks must have either at least $2^{d/4}$ hidden units or absolute values of some output weights have to be greater or equal to $2^{d/4}$, which might lead to non stability of computation.

5   Lower Bounds for SVM and Perceptron Networks

In this section, we apply lower bounds on l_1-sparsity to some dictionaries used in neurocomputing which are "small enough" to satisfy assumptions of Corollary 2.

Dictionaries used in SVM are indeed small, they contain kernel units parameterized by vectors which belong to the set of data to be classified with maximal margin or to data used for learning with generalization. Thus $\operatorname{card} X = \operatorname{card} G$. SVM algorithm assigns non zero output weights merely to those units which correspond to support vectors. A solution with a large "l_0-pseudo-norm" corresponds to a large number of support vectors. A solution with a large l_1-norm has a large number of support vectors or it is unstable because some output weights are large.

Theorem 3 implies that for a dictionary of kernel units with centers in the domain X

$$\mu(\{f \in S_r(X) \mid \|f\|_{G(X)} \geq b\}) \geq 1 - 2m\, e^{-\frac{m}{2b^2}}.$$

For the special case of the dictionary of Gaussian kernel units $K_a(x,y) = e^{-a\|x-y\|^2}$ with centers in $\{0,1\}^d$, in [11] we proved a lower bound on variation of a concrete function. Let $p_d : \{0,1\}^d \to \{-1,1\}$ denotes the parity function defined as $p_d(v) = -1^{v \cdot u}$, where $u = (1, \ldots, 1)$. Its variation with respect to the dictionary of Gaussian kernel units of any fixed width $a > 0$ is bounded from below by $2^{d/2}$.

Also the dictionary of signum perceptrons $P_d(X)$ on a finite domain X in $\mathbb{R}^d$ is small enough to satisfy assumptions of Corollary 2. Estimates of its size follow from bounds on numbers of linearly separable dichotomies to which finite subsets of $\mathbb{R}^d$ can be partitioned. Such dichotomies were studied already in 19th century by Schläfli [22]. Their sizes grow only polynomially with the sizes of the finite sets [23]. The degree of the polynomial is equal to the dimension d, so the sizes of these dictionaries grow with the dimension d exponentially, but for an application of Theorem 3 only dependence on the size of the domain is essential. The next theorem estimates probability distributions of functions with large variations with respect to signum perceptrons on finite domains.

Theorem 4. *Let d be a positive integer, $X \subset \mathbb{R}^d$ with $\operatorname{card} X = m$, μ a uniform probability measure on $S_{\sqrt{m}}(X)$, and $b > 0$. Then*

$$\mu(\{f \in S_{\sqrt{m}}(X) \mid \|f\|_{P_d(X)} \geq b\}) \geq 1 - 4\frac{m^d}{d!}e^{-\frac{m}{2b^2}}.$$

Proof. By [23, p. 330], for every d and every $X \subset \mathbb{R}^d$ such that $\operatorname{card} X = m$, $\operatorname{card} P_d(X) \leq 2\sum_{i=0}^{d}\binom{m-1}{i} \leq 2\frac{m^d}{d!}$. Combining this bound with an upper bound on partial sum of binomials we obtain an upper bound on $\operatorname{card} P_d(X)$. The statement then follows from Theorem 3.   □

Applying Theorem 4 to functions on the Boolean cube $\{0,1\}^d$, we obtain a lower bound on measures of sets of functions having variations with respect to signum perceptrons bounded from below by a given bound b. For example, for $b = 2^{d/4}$, we get a lower bound

$$1 - 4\frac{2^{d^2}}{d!}e^{-(2^{d/2}-1)}$$

on the probability that a uniformly randomly chosen function from $\mathcal{F}(\{0,1\}^d)$ with the norm $2^{d/2}$ has variation with respect to signum perceptrons greater or equal to $2^{d/4}$. Thus for computation of almost any uniformly randomly chosen function from the set of functions with norms equal to $2^{d/2}$ on the d-dimensional Boolean cube $\{0,1\}^d$ a shallow signum perceptron network needs either $2^{d/4}$ units or computation is unstable as absolute values of some output weights are at least $2^{d/4}$.

6   Discussion

Universality type results guarantee exact representations of all functions on finite domains by shallow networks having numbers of hidden units equal to sizes of the domains. As these numbers might be too large for practical applications, it is important to study classes of functions which can be computed or approximated by networks with considerably smaller numbers of units. Sparsity of shallow networks measured by the "l_0-pseudo-norm" counting the numbers of non zero output weights corresponding to the numbers of hidden units is difficult to investigate due to non convexity of "l_0-pseudo-norm". Thus we focused on sparsity measured by the l_1-norms of output-weight vectors. The concept of l_1-norm plays an important role in several fields. It has been used as a stabilizer in weight-decay regularization techniques to improve stability of solutions which leads to better generalization. Also, it was used in compressed sensing [24]. Classes of functions defined by bounds on their l_1-norms represent a similar type of a concept as classes of functions defined by bounds on both numbers of gates and sizes of output weights studied in theory of circuit complexity [25].

We derived lower bounds on sparsity measured by l_1-norms of output-weight vectors from lower bounds on variational norms tailored to dictionaries of computational units. Combining geometric properties of variational norms with concentration of measure phenomenon on high-dimensional spheres, we proved probabilistic lower bounds on l_1-sparsity.

Our results hold for almost any uniformly randomly chosen function on a large finite domain and can be applied to dictionaries of signum ad Heaviside perceptrons and dictionaries of kernel units used in SVM. Character of our results resemble the No Free Lunch Theorem which also assumes the uniform distribution. However in real applications, classes of functions of interest might be distributed non uniformly. They might belong to those small fractions of sets of all functions on given finite domains which can be computed by reasonably sparse shallow networks. This can explain capabilities of shallow networks to perform efficiently in many practical applications. Investigation of variational norms and l_1-sparsity of functions selected from non uniform distributions are subject of our future research.

Investigation of sparsity of artificial neural networks has also a biological motivation. Laughlin and Sejnowski [26] concluded from a number of studies that "the brain is organized to reduce wiring costs". They described both sparse activity (only a small fraction of neurons have a high rate of firing at any time) and sparse connectivity (each neuron is connected to only a limited number of other neurons). Our research was focused on investigation of sparse connectivity between hidden units and network outputs.

Acknowledgments. This work was partially supported by the Czech Grant Agency grant 15-18108S and institutional support of the Institute of Computer Science RVO 67985807.

References

1. Ito, Y.: Finite mapping by neural networks and truth functions. Math. Sci. **17**, 69–77 (1992)
2. Pinkus, A.: Approximation theory of the MLP model in neural networks. Acta Numerica **8**, 143–195 (1999)
3. Fine, T.L.: Feedforward Neural Network Methodology. Springer, Heidelberg (1999)

4. Bengio, Y., LeCun, Y.: Scaling learning algorithms towards AI. In: Bottou, L., Chapelle, O., DeCoste, D., Weston, J. (eds.) Large-Scale Kernel Machines. MIT Press (2007)
5. Ba, L.J., Caruana, R.: Do deep networks really need to be deep? In: Ghahrani, Z., et al. (eds.) Advances in Neural Information Processing Systems, vol. 27, pp. 1–9 (2014)
6. Kainen, P.C., Kůrková, V., Sanguineti, M.: Dependence of computational models on input dimension: tractability of approximation and optimization tasks. IEEE Trans. Inf. Theory **58**, 1203–1214 (2012)
7. Maiorov, V.E., Pinkus, A.: Lower bounds for approximation by MLP neural networks. Neurocomputing **25**, 81–91 (1999)
8. Maiorov, V.E., Meir, R.: On the near optimality of the stochastic approximation of smooth functions by neural networks. Adv. Comput. Math. **13**, 79–103 (2000)
9. Bengio, Y., Delalleau, O., Roux, N.L.: The curse of highly variable functions for local kernel machines. In: Advances in Neural Information Processing Systems, vol. 18, pp. 107–114. MIT Press (2006)
10. Bianchini, M., Scarselli, F.: On the complexity of neural network classifiers: a comparison between shallow and deep architectures. IEEE Trans. Neural Netw. Learn. Syst. **25**, 1553–1565 (2014)
11. Kůrková, V., Sanguineti, M.: Model complexities of shallow networks representing highly varying functions. Neurocomputing **171**, 598–604 (2016)
12. Kůrková, V.: Lower bounds on complexity of shallow perceptron networks. In: Jayne, C., Iliadis, L. (eds.) EANN 2016. CCIS, vol. 629, pp. 283–294. Springer, Heidelberg (2016)
13. Kůrková, V.: Constructive lower bounds on model complexity of shallow perceptron networks. Neural Comput. Appl. (2017). doi:10.1007/s00521-017-2965-0
14. Kůrková, V., Sanguineti, M.: Approximate minimization of the regularized expected error over kernel models. Math. Oper. Res. **33**, 747–756 (2008)
15. Barron, A.R.: Neural net approximation. In: Narendra, K.S. (ed.) Proceedings of the 7th Yale Workshop on Adaptive and Learning Systems, pp. 69–72. Yale University Press (1992)
16. Kůrková, V.: Dimension-independent rates of approximation by neural networks. In: Warwick, K., Kárný, M. (eds.) Computer-Intensive Methods in Control and Signal Processing, The Curse of Dimensionality, pp. 261–270. Birkhäuser, Boston (1997)
17. Barron, A.R.: Universal approximation bounds for superpositions of a sigmoidal function. IEEE Trans. Inf. Theory **39**, 930–945 (1993)
18. Kůrková, V.: Complexity estimates based on integral transforms induced by computational units. Neural Netw. **33**, 160–167 (2012)
19. Kůrková, V., Savický, P., Hlaváčková, K.: Representations and rates of approximation of real-valued Boolean functions by neural networks. Neural Netw. **11**, 651–659 (1998)
20. Ball, K.: An elementary introduction to modern convex geometry. In: Levy, S. (ed.) Flavors of Geometry, pp. 1–58. Cambridge University Press (1997)
21. Matoušek, J.: Lectures on Discrete Geometry. Springer, New York (2002)
22. Schläfli, L.: Theorie der Vielfachen Kontinuität. Zürcher & Furrer, Zürich (1901)
23. Cover, T.M.: Geometrical and statistical properties of systems of linear inequalities with applictions in pattern recognition. IEEE Trans. Electron. Comput. **14**, 326–334 (1965)
24. Candès, E.J.: The restricted isometry property and its implications for compressed sensing. C. R. Acad. Sci. Paris I **346**, 589–592 (2008)

25. Roychowdhury, V., Siu, K.Y., Orlitsky, A.: Neural models and spectral methods. In: Roychowdhury, V., Siu, K., Orlitsky, A. (eds.) Theoretical Advances in Neural Computation and Learning, pp. 3–36. Springer, New York (1994)
26. Laughlin, S.B., Sejnowski, T.J.: Communication in neural networks. Science **301**, 1870–1874 (2003)

Learning in Financial applications

Using Active Learning Methods for Predicting Fraudulent Financial Statements

Stamatis Karlos[1]([✉]), Georgios Kostopoulos[2], Sotiris Kotsiantis[2],
and Vassilis Tampakas[1]

[1] Department of Computer Engineering Informatics,
Technical Educational Institute of Western Greece, Antirrion, Greece
stkarlos@upatras.gr, vtampakas@teimes.gr
[2] Educational Software Development Laboratory (ESDLab),
Department of Mathematics, University of Patras, Patras, Greece
kostg@sch.gr, sotos@math.upatras.gr

Abstract. Detection of Fraudulent Financial Statements (FFS), or simpler fraud detection problem, refers to the falsification of financial statements with the aim either to demonstrate larger positive rates, such as assets and profit, or to conceal negative factors, such as expenses and losses. Since the expansion of contemporary markets and multinational trade are real phenomena, production of large volumes of data under which the operation of the current firms is facilitated constitutes a resulting consequence. Thus, analog upgrade of the antifraud mechanisms should be adopted, enabling the introduction of Machine Learning tools in the related field. However, because of the inability to collect trustworthy datasets that describe the corresponding ratios of a firm that has conducted fraud actions, strategies that exploit the existence of a few labeled instances for discovering useful patterns from a pool of unlabeled data could be proved really efficient. In this work, comparisons of algorithms that operate under Active Learning theory against their supervised variants are being conducted, using data extracted from Greek firms. To the best of our knowledge, this is the first study that uses Active Learning for predicting FFS. The obtained results prove the superior performance of the corresponding active learners.

Keywords: Active learning theory · Machine learning · Fraud detection · Financial ratios · Classification accuracy

1 Introduction

Nowadays, more and more scientific fields are getting affected by the innovations of technology. As a result, either some services are improved – towards more profitable or even more efficient directions – or their nature is totally transformed based on different ideas that merge either from the needs of the current society or from the new demands that have been posed by the problem that is encountered each time. One of the most interactive field with a great amount of services that people today come in contact with during their daily life is Machine Learning (ML).

Although many learning problems are tackled by ML and its subfields, the most well–known is classification. The objective target of this problem is the assignment of

G. Boracchi et al. (Eds.): EANN 2017, CCIS 744, pp. 351–362, 2017.
DOI: 10.1007/978-3-319-65172-9_30

each new instance with one of the pre-defined classes that have been provided during the construction of the problem based on the provided dataset. On the default case, only one value is assigned, while on multiclass problems more than one class values could be matched with an instance. Otherwise, if one or more feature values of any instance are not available, either the instances are discarded before the learning stage or are manipulated by appropriate methods that face the task of missing values [1].

The above scenario constitutes the supervised mode of classification problem. The other two modes are the semi-supervised and the unsupervised. They differentiate from the supervised mode because of the existence of unlabeled subset (U). These are instances whose the value of the output variable is unknown. The first mode requires a small amount of labeled instances (L) along with a larger pool, which plays the role of U subset, for mining useful information and formatting the learning model based on the assumptions of the selected classifier [2]. On the other hand, unsupervised learning does not demand any labeled instances, but tries to find possible similarities and/or same characteristics among all the unlabeled provided instances so as to separate them into any possible resulting class [3]. Another approach for tackling with such notable situations is Active Learning (AL) [4]. Even if some properties of Semi-supervised Learning (SSL) and AL are in common, such as the enlargement of the initial training set from selected instances that come from a foreign pool of U and their operation as a wrapper method that exploits one or more classifiers for building learning models per iteration, there are specific points that discern them, and would be discussed later. The most appealing asset of AL approach is the fact that is permitted to choose the number of the recordings that would be beneficial for the learning stage, though the optimization of some posed queries. This attribute allows the reduction of gathering vast amount of data and simultaneously favors the examination of the whole amount of the available instances under the selected restrictions, avoiding the extraction of decisions that were exported from randomly chosen subsets of instances or from conditions that were just verified and did not get optimized.

Detection of Fraudulent Financial Statements (FFS), or simpler fraud detection problem, constitutes a task that summarizes the peculiarities that have already been reported, as it concerns the collection of related with this topic data. To be more specific, FFS refers to the falsification of financial statements with the aim either to demonstrate larger positive rates, such as assets and profit, or to conceal negative factors, such as expenses and losses. It has to be cleared on this point that all the actions tagged with the term fraud exclude these made by mistake, even if the consequences may be the same for an external observer. Thus, only the actions that intend to deceive any antifraud mechanism are going to employ the auditors in charge, since they might appear under an iterative pattern through the official records with aim to deceive the body of the consumers, investors/creditors and the cooperating firms and/or industries by presenting a high–profile firm with profitable transactions. Besides, the majority of the counterfeits that are made by mistake could be detected easily enough even by the human factor, since they could be viewed either as "noisy" recordings produced by arithmetic carelessness or as manipulation of various values to not highly suspicious fields.

According to AICPA (The American Institute of Certified Public Accountants), the two distinct categories of fraud are: the intentional misstatement of financial information and the misappropriation of assets [5]. More recent works found in literature [6] separate these initial categories to deeper levels, analog to the size of the firm that fraud happens, or from whom these activities come from inside a business (e.g. employees, board of directors, CEO). Some of the most seen terms are: internal/ external, occupational or management/employee, small/large business, internet and investment fraud. It is evident that the more the different cases of fraud, the more difficult to find sufficient enough labeled recordings for building accurate and robust learning models. For these reasons, in this work examination of applying AL methods over fraud detection problem is presented. Our search is not oriented towards finding the optimum classifier for fraud data, since each dataset has its own specificities which may be determined by geographical parameters or time dependent conditions, but to compare the performance of several known supervised algorithms with their variants operating under specific queries that are set by AL theory.

The rest of this paper is organized as follows. Section 2 presents a short survey on other approaches of detecting FFS combined with a short description of their used methods. Section 3 highlights the most important aspects of AL theory, while in Sect. 4 a description of the exploited dataset is given. Section 5 contains the conducted experiments along with the produced results are placed. Finally, Sect. 6 provides conclusions and useful comments about the future work.

2 Related Work

Despite the fact that fraud detection has been appeared about 40 years ago as a phenomenon being recorded by several academic works and, at the same time, addressed by regulatory authorities during the past years – revealing corporate scandals such as these of ZZZZ Best, Waste Management, WorldCom, Enron, KPMG Malaysia and Xerox [6, 7] – the radical change of the nature of the operation of the firms has brought up the need of modernization of the strategies that the control bodies should adopt for ensuring their viability against other competitors by increasing their forecasting accuracy. The main reasons why such changes have been committed more drastically recently are the innovations over the communication networks and the current database systems. Due to the increased number of different tasks that auditors have to tackle with and of their inherent complexity, since more and more data are produced and many kinds of them (e.g. receipts, accounting records) are displayed only under on–line versions, the appropriate regulatory authorities have to be involved with FFS task under new views. Furthermore, the fact that the periodical publication of official reports on public networks is getting accelerated even in quarterly base, aggravates fast response of auditors or securities analysts for evaluating all these financial information combined with a satisfactory level of accuracy [9].

Moreover, thinking about the economic crisis that has stroke the worldwide market, both on international and on nation level, and the fact of the equivalence between dollar and euro is susceptible to facts like BREXIT, much more embezzlement activities could be conducted by several firms for preserving their wealthy profile. Thus, the role of audit committees is much more overcharged than previously, so as to retain the confidence over the financial reporting system of the public and reassuring the consolidation of the capital markets. In contrast with previous attempts to ensure better audit services, either by tightening the rules about the composition of the related audit committees (New York Stock Exchange – 1999, Sarbanes–Oxley Act and its provisions such as the creation of Public Company Accounting Oversight Board – 2002) [8] or by applying stricter corporate fraud policy inside each firm for restricting the individual opportunities of acting fraudulent activities, more recent approaches are oriented towards developing targeted antifraud tools. In order to achieve more automated strategies that offer improved forecasting accuracy than the older obsolete methods, incorporation of techniques related with the fields of data analytics, ML theory and Artificial Intelligence is taking place for encountering fraud detection (e.g. The Federal Court of Accounts of Brazil, 2015).

The beliefs of AICPA in 2001 about the future of audit services and their shift of role from audit to assurance have been confirmed [7, 8]. Thus, the role of current control bodies has shifted towards both collecting larger volumes of data and analyzing them for discovering any hidden information. According to more recent reports of ACFE (Association of Certified Fraud Experiments – 2014), it is estimated that more than 3$ trillion are getting lost through fraud loss by firms (about the 5% of the previously recorded Gross World Product). Moreover, the average duration of the period that is passed from the conduction of obstructive actions since the detection of them is 18 months [10], and the percentage of fraud detection through calls to hotlines or tips from whistle–blowers has been dramatically reduced. Summarizing these facts, it is easily conceivable that the adoption of techniques and methods which exploit data analysis and artificial intelligence theories for achieving better predictive performance constitutes an authoritative need, both for the candidate partners of any company and for the employee's morale [5].

Almost all the recorded works found in the literature that are related with the detection of FFS problem have made use of supervised learning methods [11]. Their main strategy is to collect the appropriate data, usually on national level, measuring either financial ratios or financial ratios coupled with non-financial ratios and apply some pre–processing stages for transforming each collected dataset to a more suitable version for boosting the learning accuracy of the selected classifier. During the initial appearance of such approaches, the use of Neural Networks (NNs) achieved the greater acceptance from the scientific community, either combined with other methods – linear discriminant analysis [12] or rule–based systems [13] – or operating standalone [14]. Their non-linear structure and their property not to demand any assumption about the distribution of the given input data were proven sufficient satisfactorily for establishing a new risk–based framework. According to [15], a top–down strategy is formatted for recognizing the factors that prevent a company from achieving its objective targets, examining at the final stage the corresponding financial statements. Several properties of NNs are discussed in this work, such as the sample size vs. the number of the input variables and the preferred since then architectures of NNs. Similar work has also been made by Koskivaara [9] two years later, reviewing plenty of works and tools related with financial topics and presenting a comprehensive survey of auditing with Artificial Neural Networks (ANNs). The drawback that was highlighted was the shortage of large amount of data for the majority of the searched articles. Although during the early stages of experimenting with ANNs this phenomenon might not be a problem, for implementing a robust tool that could generalize efficiently over any test data should be both numerically and representative enough.

Thus, several different approaches have also been demonstrated without exploiting black box models of NNs. Spathis, in 2002 [16] made use of Logistic regression analysis (Logit model) for discriminating fraud and non-fraud manufacturing companies in Greece. Their cardinality was equal to 76, equally separated, and were described with 17 financial ratios. Before the formatted dataset be screened by Logit model, correlation analysis and t–test were used for avoiding high correlated features and for preserving the most significant of them. At last, only 10 features were kept leading to an overall accuracy of 84.21% and some useful conclusions, such that the high values of total debt to total assets (TD/TA) should suspect the auditors in charge for fraud issues. The same year, an alternative method was presented by Spathis et al. [17]. They applied Multicriteria Decision Aid (MCDA) approach – this is a non-parametric framework that allows the mining of both numeric and nominal features – coupled with UTADIS classification method over the previously referred dataset and by using a more exhaustive evaluation method they recorded lower error rates. More specifically, there were examined two different feature sets, the original (the same 10 features that were examined on [16] after the pre-processing stages) and the reduced (4 only features that came from appropriate factor analysis oriented towards reducing the multicollinearity phenomenon), with the second one to be proven more reliable achieving 85.56% overall accuracy. This result facilitates any interested auditor, since by examining only a few variables a robust enough model may be designed and get mined for their aims. A more recent work based on ANNs that outperforms older related approaches is stated in [18], where Log–sigmoid transfer function, Gradient descent back propagation training function with 5 hidden layers were used along with 10 numerical variables.

A tool that is based on hybrid scheme that is designed using stacking theory is presented in [19]. Having exported data from 164 Greek companies, and measuring 21 financial ratios that were later reduced to 8 through an appropriate algorithm for selecting the most informative features, 7 heterogeneous classifiers were combined at the zero–level of stacking's structure – coming from decision trees, ANNs, Bayesian networks, Instance based learners, rule–learners, Support Vector Machines (SVM) and Logistic Regression – providing their predictions to a model tree classifier at the meta–level. The comparisons with other hybrid schemes certified the better learning quality of the proposed scheme. Cecchini et al. developed in 2010 a kernel specific to domain of finance, named as Financial Kernel (FK). Taking advantage of the kernel theory, which lets non–linear combinations of input features to take place, they expressed the produced features – whose cardinality is equal to $3n(n-1)$, where n equals to the number of the provided features – and forwarded them to SVM classifier. Theirs dataset included 205 fraudulent and 6427 non–fraudulent firms, giving a ratio of 1:31 (about 3%), which was closer to the real percentage of fraudulent occurrence (about 1% [20]) in comparison with techniques or previously reported works that adopted one–to–one match. Various metrics (AUC, ROC, accuracy) were tested for validating the power of SVM–FK method compared with 3 well–known methods, while the most informative attributes for this method were also highlighted for serving auditor committee's actions. Similar comparisons have been stated in [21], where 6 distinct algorithms – Multilayer Feed Forward Neural Network (MLFF), SVM, Genetic Programming (GP), Group Method of Data Handling (GMDH), Logit model and Probabilistic Neural Network (PNN) – were selected for classifying 202 Chinese companies that were listed in various Chinese stock exchanges (ratio of fraud and non-fraud was 1:1). Using 35 financial variables, two different scenarios were examined, using initially the top 18 features and secondly the top 10, judged by the t–statistic measure on both cases. The experiments showed that PNN and GP algorithms performed the best learning behaviors assessed by 10–cross–validation technique, compared also with previous works that manipulated the same dataset.

Exploitation of financial ratios exclusively has been demonstrated in [22] for predicting FFS phenomena in the case of small and medium scaled industries in Malaysia, which according to ACFE (2008) also suffer from obstructive actions, since their size – often less than 100 employers – do not, at least ostensibly, play the cardinal role on local or national markets. Use of three statistical methods (Beneish model, Altman Z–Score and Financial Ratio), proved that useful conclusions can be drawn for inform any antifraud agency. Lastly, researchers integrate both financial and non-financial ratios for extending the forecasting accuracy of their learning models. However, seldom any non-financial ratio plays cardinal role to the final learning stage, weighted with low significance values. Collecting data from the years 1998 to 2012 by companies of the Taiwan Economic Journal Data Bank (TEJ) and exploiting both kind of referred factors through stepwise regression for evaluating them, type I and type II classification errors of Logit model, C5.0 algorithm and SVM were compared over fraud detection [23]. Although 29 factors were initially recorded (24 financial, 5 non–financial), only 8 of them (7 financial, 1 non–financial) did finally selected from stepwise regression stage. Rough Set Theory (RST) coupled with SVM (RST+SVM) was applied to 100 recordings (50 fraud, 50 non-fraud) that come from Taiwan during the period of 1996–2007 [24]. The matching of the contained firms was made according to: (a) industry, (b) products, (c) capitalization and (d) values of assets. Comparisons against similar coupled approaches that make use of Instance Based learners and stepwise regression were conducted and proved the superiority of the proposed method, since its performance over type I and type II classification error was by far the best from the contained methods.

The only recorded work that scrutinizes the fraud problem under Semi–supervised classification (SSC) view is the recently published work [25]. There, 8 SSL schemes, all of them included in KEEL [26], were tested against their supervised variants under two different scenarios about the cardinality of the original data: 10% and 20%. This parameter (Labeled Ratio) lets the user to adjust the size of the initially available data (L) on contrast with the unknown instances (U) for simulating various scenarios related with real–life situations. Five out of the 8 SSL schemes support the choice of their base classifier among C4.5, KNN, SMO and NB. Bases on experimental procedure, schemes that exploited decision trees (C4.5 as base learner and CoForest) presented better improvement, besides the fact of the small cardinality of the unlabeled data subset, which may induce a poor source of information.

3   Active Learning Theory

In many real world applications, there is a lack of labeled data while large amounts of unlabeled data can easily be obtained. Labeling unlabeled data is difficult, expensive and time consuming, since it requires human effort. The need of exploiting U subsets to build efficient predictive models has resulted to the development of considerable ML methodologies. Semi-supervised learning, transductive learning and active learning are the three main approaches for learning with labeled and unlabeled data to improve learning performance [27].

Active or Query Learning (AL) aims to train a classifier by querying the labels of specific instances from an oracle. The primary objective of AL is to build high efficiency classifiers at the lowest possible cost and risk [28] by minimizing the number of queries posed [4]. The selection of the most helpful examples for labeling, in most of the cases the most informative examples, is the main issue in the active learning task [29]. Several scenarios and query strategies have been proposed to maximize the predictive performance of the active learner. There are three main active learning scenarios, i.e. pool-based, stream-based and membership query [29]. In this study we adopt the pool-based scenario which is considered as the most frequently used AL scenario. Given a small L subset and a large U subset, m examples are selected from U and asked for assigned with labels from an oracle [30]. These examples are added to U and the procedure is repeated until some stopping criterion is met (Fig. 1).

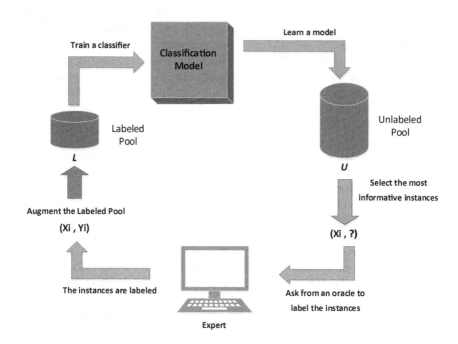

Fig. 1. Pool-based active learning

The two major queries strategies are uncertainty sampling and query by committee sampling [31]. Uncertainty sampling is a simple and commonly used strategy querying the label of the most uncertain to label instances, while query by committee sampling queries the label of an instance for which a committee of classifiers disagree the most. Margin and entropy sampling query are representative forms of the uncertainty sampling strategy. Entropy query is appropriate for multiclass problems, using entropy as information, choice and uncertainty measure [32], while margin query is considered to be quite effective for binary classification.

4 Dataset Description

The current dataset has been created using data from two consecutive years (2001–2002) from Greek businesses listed on the Athens Stock Exchange (ASE). Operating similarly with other related works, none of these firms could be characterized as a financial one. On the contrary, they belong to manufacturing field. This was firstly introduced by Kotsiantis et al. in [18, 32] and has been examined also in [25]. The most important properties are the following: it consists of 41 fraudulent and 123 non-fraudulent instances that are described through 25 predictive variables, leading to a dataset with cardinality equal to 164, described through nominal variables.

The certification of embezzlement activities inside each fraudulent firm stemmed from indications such as: existence of serious doubts on the control bodies' reports about the accuracy of the accounts, observations by the tax authorities regarding serious taxation intransigencies which drastically affected the annual balance sheet and income statement of each examined firm, involvement on court proceedings because of taxation contraventions and application of Greek legislation regarding negative worth [33]. Having detected the recordings of the fraudulent class, a matching phase of each contained fraudulent firm with a number of non–fraudulent firms was followed. The criteria under which this choice was conducted, determined through examining formal financial statements. The selected ratio was 1:3, regarding the size of the first vs the second class size.

As it concerns the financial ratios that were measured for recording each instance, similar strategies that were respected by other related works was followed also here [15, 17]. However, some innovations were also applied for obtaining some new predictive factors [19]. The most noticeable factors that are described by them is profitability, leverage, liquidity, efficiency, cash flow and financial distress. A feature reduction phase was implemented, maintaining the sequence of other experimental studies, not only for gaining computational resources during the building of classification models, since association of included predictive variables may offer informative enough patterns to auditors in charge and also to other fraud researchers. The statistical measure of ReliefF [34] was preferred among others, evaluating the ability of each variable to disambiguate similar samples. Principle of similarity is defined by proximity in the space that the selected variables form. The final outcome was the exportation of the 8 higher ranked variables, all of which are placed in Table 1 along with a short description and the corresponding ReliefF score, applied now only in the new set of features.

Table 1. Description of top–8 ranked predictive variables.

Abbreviation of variables	Description of variables	ReliefF score
NFA/TA	Net fixed assets/total assets	0.19695
WC/TA02	Working capital/total assets of 2002	0.17988
NDAP02	Number of days accounts payable of 2002	0.11098
TL/TA02	Total liabilities/total assets of 2002	0.10488
AR/TA02	Accounts receivable/total assets of 2002	0.10305
AR/TA01	Accounts receivable/total assets of 2001	0.08476
LTD/TCR02	Long term debt/total capital reserves	0.06768
DC/CA02	Deposits and cash/current assets of 2002	0.00183

5 Experimental Procedure

At first, we partitioned the dataset into 10 folds of similar size using the 10–fold cross validation procedure. One fold was kept to evaluate the predictive effectiveness of the model (its size is equal to 17 instances), while the rest 9 folds were used for the training process. In each fold, 6 instances of the training set formed the L subset and the rest 141 formed the U subset. A pool–based sampling scenario with the margin sampling query strategy was used. We have defined a maximum number of 30 iterations as a stopping criterion, while we selected a single example for labeling at each of the iterations. On this basis, at the end of the learning process 36 labeled instances formed the labeled pool.

A set of five familiar supervised algorithms from Weka tool [35] were used as base classifiers to form five respective active learners. These algorithms are:

- Bayes Net,
- Multilayer Perceptrons (MLPs),
- Naive Bayes (NB),
- Sequential Minimal Optimization (SMO),
- Random Forest (RF).

The experiments were carried out using the Java Class Library for Active Learning (JCLAL) tool and conducted in five successive steps. JCLAL is an open source software appropriate for implementing familiar AL strategies and integrating new developed ones [36]. Our experiments are presented in Table 2, where two distinct AL query strategies have been examined (Margin and Random), along with results of the

Table 2. Correctly classified instances (%)

Learner	AL (Margin query)	AL (Random selection)	Supervised
SMO	77.24	75.22	73.80
NB	77.87	69.63	71.30
Bayes Net	75.63	69.67	71.30
MLPs	71.36	74.56	68.90
RF	74.26	75.81	74.40

supervised version of the corresponding algorithms. It has to be mentioned that the whole training set has been used in supervised mode with the default procedure of 10–cross–validation, as it was described previously.

Although fluctuations of the learning behavior, especially during the initial iterations, were observed for all the selected classifiers, improvements of classification accuracy are detected. Except than NB (Random Selection) and Bayes Net (Random Selection) whose performance deteriorated and RF (Margin Query) whose learning performance was not hightly favored but reached almost the same accuracy, the AL theory proved to be harmonized with the detection of FFS problem, since all the other 7 cases achieved an accuracy improvement ranged from 1.9% to 9%. Moreover, the obtained results seem encouraging, since the majority of the mentioned classifiers managed to outperform their corresponding supervised rivals quite faster than the 30th iteration despite the restricted cardinality of the provided dataset. Due to space limitation we cannot present all such graphs (classification accuracy vs iterations), but on average, during the 10th and the 15th iteration the most of them reached and overpassed the referred limit.

In the sequel, we performed also a comparison of the three different learning schemes, assessing the classification accuracies of all the 5 selected learners. The preferred statistical test is Friedman test [37], a non-parametric statistical test that is used for one–way repeated measure analysis of variance by ranks. The recorded in Table 3 results, prove the supremacy of learning schemes supported by AL theory, even for a really small initial size of labeled instances. Furthermore, the Margin query offers a more confident strategy than Random query, despite the restricted cardinality of the instances of the examined fraud dataset, letting us to expect similar or improved learning behavior over more enriched datasets.

Table 3. Average ranking of the three different learning schemes.

Learning scheme	Friedman ranking
AL (margin query)	1.6
AL (random selection)	2
Supervised	2.4

6   Conclusions

In this work, examination of efficacy of AL theory over detection of FFS problem compared with supervised learning methodology has been performed, for first time, according to our research in literature. Although the cardinality of the gathered dataset is small (164 examples) and describes instances only from Greek firms for a two years period, the inherent inability to collect such kind of data easily due to both their sensitive financial nature and the fact that antagonism governs the current markets, is compensated by the learning strategy of AL schemes. Thus, only a small amount of labeled data are used, exploiting under some strategy the available unlabeled data.

To sum up, our main findings could be synthesized as follows: AL strategies outperformed supervised even for small initial L set, Margin query has been proved a more reliable solution than Random query, classifiers based on SMO and NB performed higher improvement of their learning behaviors when matched with AL scheme and financial ratios that express profitability and liquidity proved the most influential.

As future work, more in–depth experiments should be conducted testing several scenarios related with both the number of initial labeled recordings and these of iterations, and probably incorporating AL theory inside SSL schemes. As it concerns the choice of predictive variables, integration of financial and non–financial factors reviewing recent works [10] could be proved even more beneficial for classification problems.

References

1. Pigott, T.D.: A Review of Methods for Missing Data, vol. 7, no. 4, pp. 353–383 (2001)
2. Zhu, X., Goldberg, A.B.: Introduction to Semi-Supervised Learning, vol. 3, no. 1. Morgan & Claypool, San Rafael (2009)
3. Theodoridis, S., Koutroumbas, K.: Pattern recognition. Academic Press, Cambridge (2009)
4. Dasgupta, S.: Two faces of active learning. Theor. Comput. Sci. **412**(19), 1767–1781 (2011)
5. Coderre, D.: Computer-Aided Fraud Prevention & Detection. Wiley, Hoboken (2009)
6. Youngblood, J.: Fraud Identification and Prevention. CRC Press, Boca Raton (2015)
7. Rezaee, Z.: Financial Statement Fraud: Prevention and Detection. Wiley, Hoboken (2002)
8. Rezaee, Z., Riley, R.: Financial Statement Fraud Prevention and Detection. Wiley, Hoboken (2009)
9. Koskivaara, E.: Artificial Neural Networks in Auditing: State of the Art. ICFAI J. Audit Pract. **1**(4), 12–33 (2004)
10. Banarescu, A.: Detecting and preventing fraud with data analytics. Procedia Econ. Finan. **32**, 1827–1836 (2015)
11. Bao, Y., Ke, B., Li, B., Yu, J., Zhang, J.: Detecting accounting frauds in publicly traded U.S. firms: new perspective and new method, vol. 45, pp. 173–188 (2015)
12. Altman, E.I., Marco, G., Varetto, F.: Corporate distress diagnosis: Comparisons using linear discriminant analysis and neural networks (the Italian experience). J. Bank. Financ. **18**(3), 505–529 (1994)
13. Yoon, Y., Guimaraes, T., Swales, G.: Integrating artificial neural networks with rule-based expert systems. Decis. Support Syst. **11**(5), 497–507 (1994)
14. Green, B.P., Choi, J.H.: Assessing the risk of management fraud through neural network technology. Audit. A J. Pract. Theory **16**(1), 14–28 (1997)
15. Calderon, T.G., Cheh, J.J.: A roadmap for future neural networks research in auditing and risk assessment. Int. J. Account. Inf. Syst. **3**(4), 203–236 (2002)
16. Spathis, C.T.: Detecting false financial statements using published data: some evidence from Greece. Manag. Audit. J. **17**(4), 179–191 (2002)
17. Spathis, C., Doumpos, M., Zopounidis, C.: Detecting falsified financial statements: a comparative study using multicriteria analysis and multivariate statistical techniques. Eur. Account. Rev. **11**(3), 509–535 (2002)
18. Omar, N., Amirah Johari, Z., Smith, M.: Predicting fraudulent financial reporting using artificial neural network. J. Financ. Crime Iss. **24**(2), 362–387 (2017)

362   S. Karlos et al.

19. Kotsiantis, S., Koumanakos, E., Tzelepis, D., Tampakas, V.: Predicting Fraudulent Financial Statements with Machine Learning Techniques, pp. 538–542. Springer, Heidelberg (2006)
20. Beneish, M.D.: The detection of earnings manipulation. Financ. Anal. J. **55**(5), 24–36 (1999)
21. Ravisankar, P., Ravi, V., Raghava Rao, G., Bose, I.: Detection of financial statement fraud and feature selection using data mining techniques. Decis. Support Syst. **50**(2), 491–500 (2011)
22. Aris, N.A., Arif, S.M.M., Othman, R., Zain, M.M.: Fraudulent financial statement detection using statistical techniques: the case of small medium automotive enterprise. J. Appl. Bus. Res. **31**(4), 1469–1478 (2015)
23. Chen, S., Goo, Y.J., Shen, Z.: A hybrid approach of stepwise regression, logistic regression, support vector machine, and decision tree for forecasting fraudulent financial statements. Sci. World J. **2014**, 9 (2014)
24. Yeh, C.-C., Chi, D.-J., Lin, T.-Y., Chiu, S.-H.: A hybrid detecting fraudulent financial statements model using rough set theory and support vector machines. Cybern. Syst. **47**(4), 261–276 (2016)
25. Karlos, S., Fazakis, N., Kotsiantis, S., Sgarbas, K.: Semi-supervised forecasting of fraudulent financial statements. In: Proceedings of the 20th Pan-Hellenic Conference on Informatics, Article No. 34, pp. 1–6 (2016)
26. Alcalá-Fdez, J., Fernández, A., Luengo, J., Derrac, J., García, S., Sánchez, L., Herrera, F.: KEEL data-mining software tool: data set repository, integration of algorithms and experimental analysis framework. J. Mult. Log. Soft Comput. **17**(2–3), 255–287 (2011)
27. Zhou, Z.-H.: Learning with Unlabeled Data and Its Application to Image Retrieval, pp. 5–10. Springer, Heidelberg (2006)
28. Kremer, J., Steenstrup Pedersen, K., Igel, C.: Active learning with support vector machines. Wiley Interdiscip. Rev Data Min. Knowl. Discov. **4**(4), 313–326 (2014)
29. Settles, B.: Active learning literature survey. Univ. Wis. Madison **52**(55–66), 11 (2010)
30. Dwyer, K., Holte, R.: Decision tree instability and active learning. In: Kok, Joost N., Koronacki, J., Mantaras, RLd, Matwin, S., Mladenič, D., Skowron, A. (eds.) ECML 2007. LNCS, vol. 4701, pp. 128–139. Springer, Heidelberg (2007). doi:10.1007/978-3-540-74958-5_15
31. Ramirez-Loaiza, M.E., Sharma, M., Kumar, G., Bilgic, M.: Active learning: an empirical study of common baselines. Data Min. Knowl. Discov. **31**(2), 287–313 (2017)
32. Shannon, C.E.: A mathematical theory of communication. ACM SIGMOBILE Mob. Comput. Commun. Rev. **5**(1), 3 (2001)
33. Kotsianits, S., Koumanakos, E., Tzelepis, D., Tampakas, V.: Forecasting fraudulent financial statements using data mining. IT Prof. 1(12) (2007)
34. Sikonja, M.R., Kononenko, I.: An adaptation of Relief for attribute estimation in regression. In: Proceedings of 14th International Conference on Machine Learning, pp. 296–304 (1997)
35. Hall, M., Frank, E., Holmes, G., Pfahringer, B., Reutemann, P., Witten, I.H.: The WEKA data mining software. ACM SIGKDD Explor. Newsl. **11**(1), 10 (2009)
36. Reyes, O., Pérez, E., Del, M., Rodríguez-Hernández, C., Fardoun, H.M., Ventura, S.: JCLAL: a Java framework for active learning. J. Mach. Learn. Res. **17**, 1–5 (2016)
37. Friedman, J., Hastie, T., Tibshirani, R.: Additive logistic regression: a statistical view of boosting. Ann. Stat. **38**(2), 337–374 (1998)

Comparing Neural Networks for Predicting Stock Markets

Torkil Aamodt and Jim Torresen[✉]

Department of Informatics, University of Oslo, Oslo, Norway
torkil.aamodt@gmail.com, jimtoer@ifi.uio.no

Abstract. In this paper we compare a selection of artificial neural networks when applied for short-term stock market price prediction. The networks are selected due to their expected relevance to the problem. Further, the work aims at covering recent advances in the field of artificial neural networks. The networks considered include: Feed forward neural networks, echo state networks, conditional restricted Boltzmann machines, time-delay neural networks and convolutional neural networks. These models are also compared to another type of machine learning algorithm, support vector machine. The models are trained on daily stock exchange data, to make short-term predictions for one day and two days ahead, respectively. Performance is evaluated by following the models directly in a simple financial strategy; trade every prediction they make once during each day.

Possibly due to the noisy nature of stock data, the results are slightly inconsistent between different data sets. If performance is averaged across all the data sets, the feed forward network generates most profit during the three year test period: 23.13% and 30.43% for single-step and double-step prediction, respectively. Convolutional networks get close to the feed forward network in terms of profitability, but are found unreliable due to their unreasonable bias towards predicting positive price changes. The support vector machine delivered average profits of 17.28% for single-step and 11.30% for double-step, respectively. Low profits or large deviations were observed for the other models.

1 Introduction

Predicting the development of financial instruments like stocks carries obvious economical benefits. Short-term prediction carries a larger potential, as it allows one to follow market changes more closely. One is also more susceptible to noise in small, seemingly random price movements. The challenges and potential rewards associated with short-term prediction makes it an excellent problem for benchmarking.

Lately, an effort has been put into using machine learning techniques to model stock prices. The models have shown mixed success, and it can be difficult to compare the techniques when they are applied for different stocks and evaluated using different metrics. Considering the recent renaissance for artificial neural networks, a comparative evaluation of these models is timely.

© Springer International Publishing AG 2017
G. Boracchi et al. (Eds.): EANN 2017, CCIS 744, pp. 363–375, 2017.
DOI: 10.1007/978-3-319-65172-9_31

Stock markets are hard to predict. Driven by supply and demand, macroeconomic changes such as inflation or political instability can affect whole markets, while local events like company financial announcements or product releases impact individual stocks. Traders also look for signals in the price development that may indicate future prices. Because some trading is based on these indicators, price movement by itself can also trigger trading activity, causing further price adjustments. Investors and traders should ideally consider all these factors when evaluating a stock, however market participants are often partitioned into two groups: Traders that make decisions based on news, facts and numbers are known as *fundamental* analysts, while those who look for signals in price history are using *technical* analysis.

Both camps of traders are exploring how to utilize computers to efficiently find trading opportunities. Formulating the problem for a computer using the fundamental approach is challenging, because of difficult input like news written in natural language. Technical analysis on the other hand, uses only the trading history of a given stock, otherwise known as a time series. The methodology followed in this paper resembles technical analysis, in the sense that predictions are based solely on past trade information. Technical analysis traditionally relies on analyzing trades using a range of mathematical functions and visual inspection of the data. In this study however, such analysis is implicitly handled by a computer using machine learning.

1.1 Related Work

When it comes to modeling financial markets in particular, some research revolves around evaluating a single model for prediction. [9] presents a hybrid model using an Autoregressive Integrated Moving Average (ARIMA) and an Artificial Neural Network (ANN). The model was evaluated using several data sets, including one for exchange rates. Lower error rates were achieved for the hybrid solution, compared to either model alone. [3] finds that forecasting exchange rates using Deep Belief Network (DBNs) outperforms regular Feed Forward Neural Network (FFNNs).

Rather than researching one model, comparative studies are devoted to evaluating multiple types of models. [13] considers Time-Delay Neural Network (TDNNs), Recurrent Neural Network (RNNs) and probabilistic neural networks, and finds that they all deliver reasonable performance when predicting stock trends. [10] regards a range of single, independent studies in order to theoretically evaluate ANNs in general as a tool for financial forecasting.

1.2 Problem Formulation

The novelty in this work is to compare a selection of different neural networks, when applied for short-term, daily stock price prediction. Each model is evaluated in the context of following a financial strategy that trades every prediction of the model. Five neural networks are compared to each other and a baseline represented by an Support Vector Machine (SVM). The neural networks are selected such that they either were created explicitly to model time series, or have shown good results for other applications.

Models are trained independently across four stocks (BAC, MSFT, PBR and XOM) and two indices (S&P 500 and NASDAQ Composite) to make the results statistically relevant. Indices are aggregations of multiple stocks, and make a better representation of the market as a whole. They are commonly used for benchmarking financial models, and are therefore also included in this study.

The next section introduces the stock market data, followed by how the data is used for training the prediction models in Sect. 3. Experimental results are included in Sect. 4 and followed by conclusion in Sect. 5.

2   Stock Market Data

Data selection lay the foundation for the experiments and is outlined in this section. Stock data from both rising and declining markets have been selected. Investors often refer to markets that are expected to, or currently are rising as *bull* markets. *Bear* markets on the other hand, are declining markets. A model that has been fitted solely to bull market data, will probably struggle under bear market conditions.

2.1   Data Selection

There is a plethora of stocks to choose from, all with different attributes. Table 1 lists the stocks and indices, along with their ticker symbols, that was be used for learning and testing the models. The chosen instruments are well-known, highly liquid stocks and indices. The stocks are listed at the New York Stock Exchange (NYSE) or the National Association of Securities Dealers Automated Quotations (NASDAQ), and together span across multiple industries.

Table 1. The stock selection.

Symbol	Name	Type	Exchange
GSPC	S&P 500	Index	N/A
IXIC	NASDAQ Composite	Index	N/A
BAC	Bank of America Corporation	Stock	NYSE
MSFT	Microsoft Corporation	Stock	NASDAQ
PBR	Petrleo Brasileiro S.A. - Petrobras	Stock	NYSE
XOM	Exxon Mobile Corporation	Stock	NYSE

Indices: Two indices each consisting of multiple stocks are included in the data selection. **S&P 500**: Five hundred US leading companies constitute S&P 500 (GSPC), providing a diversified selection of stocks. GSPC price development is

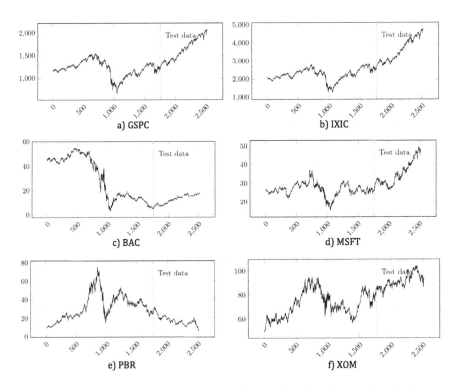

Fig. 1. The daily closing price development of the selected indices and stocks between 1-1-2005 and 1-1-2015.

shown in Fig. 1(a). Bull conditions are observed throughout the test data (to the right of the vertical line), while the training set (on the left of the vertical line) has an approximately neutral total development. **NASDAQ Composite**: Containing over 3000 instruments listed on the NASDAQ stock exchange. NASDAQ Composite (IXIC) consists mainly of stocks within the technology sector. Visually IXIC carries a strong resemblance to GSPC, as seen in Fig. 1(b); technology companies constitute a considerable part of the market as a whole, which would explain some similarities.

Stocks: Four liquid companies are included in the experiments, representing diverse combinations across sectors, stock exchanges and market conditions. **Bank of America Corporation**: Listed on the NYSE, Bank of America Corporation (BAC) is currently serving 49 million consumers and small businesses, according to [2]. Judging by the price plot in Fig. 1(c), the share has suffered a massive drop during the past ten years. Looking at the intersection between training and test samples however, we observe that it changes to less volatile bull conditions. Different market conditions between training and test data makes this a particularly interesting stock. **Microsoft Corporation**: Founded in 1975, Microsoft Corporation (MSFT) quickly grew to become a major name within the technology sector. In Fig. 1(d) we can observe how the training samples contain both bull and bear periods, ultimately resulting in a neutral trend. Test data on the other hand, is decidedly bullish. **Petrleo Brasileiro S.A. - Petrobras**: With upward of 135 production platforms and 31000 kilometers of pipeline [12], Petrleo Brasileiro S.A. - Petrobras (PBR) is a significant actor in the oil industry. As seen in Fig. 1(e), the training set is largely volatile, containing both bull and bear segments. The data has been adjusted for two stock splits in the training samples, both of which doubled the number of shares. The adjustment procedure is detailed in [1]. **Exxon Mobile Corporation**: According to its website [7], Exxon Mobile Corporation (XOM) is the largest publicly traded international oil and gas company in the world. Listed on the NYSE and categorized in the basic minerals sector, the company has grown about twice its original value since 2005, as seen in Fig. 1(f). On the other hand, the development has been volatile, including a solid period of bearish conditions halfway through the training set. Test samples trend in favor of the company, although with some volatility.

3   The Prediction Challenges

ANNs can be thought of as function approximators. We may view a stock as a function of time: $f(t) = v$, where t is the time and v the share value. Equation (1) describes a more realistic problem formulation:

$$f(v_{t-1}, v_{t-2}, ..., v_{t-n}) = v_t \tag{1}$$

where v_t is the share value at time t. By contrast to the previous function, we here use the n most recent share values as input rather than the time. This allows us to search for temporal patterns that lead to the same conclusion. The models are trained in this work using *ten* time steps, i.e. *ten* days, of data to make predictions for one day and two days ahead, respectively. This number was found following some preliminary tests, which showed that greater values tended to overfit models.

   To improve the prediction, we found it helpful to also take into account other variables including *open, high, low, close, volume* and *day of the week*. Open and close is the share price at the beginning and end of a trading day, while high and low refers to the highest and lowest point the value reached throughout a day. Lastly, volume is the number of shares that were traded in total. Day of the week is relevant to incorporate since some market participants are likely to enter the market in a weekly, cyclical pattern.

3.1   Implementation Notes

The goal of the experiments in this work is to train neural networks for daily stock data prediction and compare their performance.

Data Preprocessing. Data should not be partitioned directly into training samples, but rather *preprocessed* first in order to optimize learning. Preprocessing should be reversible in such a way that the predictions can be converted back to stock prices. Several approaches were tried, including simply normalizing the variables within a range, or calculating the percentage each variable changes between time steps.

The share value of a stock changes over time, and even the largest companies started off small. By training a model on raw data directly, the model may not be able to yield reliable predictions later on when the general prices are at another level. In an effort to make new data recognizable, data are considered as a *percentage of change*, rather than actual price.

Looking at the Change. Figure 2 illustrates how the data are initially prepared. O, H, L, C and V stand for open, high, low, close and volume respectively, while $D1...D5$ are the binary variables representing each working day. The value an arrow points to can be calculated by measuring its change in percent with respect to the value where the arrow begins. Assuming we have two values a and b, and an arrow connecting them from a to b, we can calculate the change p from a to b as follows: $p = \frac{b-a}{a}$.

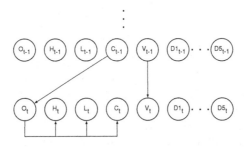

Fig. 2. Some values are seen as a percentage of change with respect to another variable. The figure shows calculation of a supervised input vector.

The data could have been preprocessed in a straight forward temporal manner, where a variable at time t is calculated with respect to itself at time $t-1$. However, as the opening price is a continuation from the closing price the day before, the temporal connection from C_{t-1} to O_t in Fig. 2 reflects that relationship. Similarly, high, low and close values are local within a day, and are therefore captured by spatial connections from O_t. Volume uses another metric than the prices do, and should be calculated with respect to itself in the previous time step. Note that $D1...D5$ are not touched at all, leaving them in their raw binary states.

The closing price is an appropriate *target* for the supervised models, and was chosen to be the value to be predicted by the models. That is, models predict the expected closing price for the following day.

Normalization. The second and final step of preprocessing is ensuring that the data has a mean of zero and a standard deviation of one. This is especially important for ANN models, to prevent large input values from slowing down the learning process; large values imply large weights, which require a long time to fit. Assuming a matrix p that consists of multiple variables i through time j, the normalized value x_{ij} is calculated from p_{ij} as follows:

$$x_{ij} = \frac{p_{ij} - \overline{p_i}}{\sigma_{p_i}} \tag{2}$$

where $\overline{p_i}$ and σ_{p_i} is the mean and standard deviation of variable i through time, denoted p_i. Input vectors and target values are normalized independently.

Libraries. Libraries are relied on as much as possible to reduce the development time, refer to Table 2 for an overview. Necessary new code was written in MATLAB® and used in combination with the libraries. MATLAB® was chosen for its scripting features, built-in matrix operators and active community. A number of parameters had to be set for each machine learning method. They had to be mostly left out of this paper due to limited space but are fully specified in [1].

Table 2. Libraries leveraged for experiments.

Library	Method	URL
Support Vector Regression	SVR	[4]
Deep Neural Network	FFNN	[14]
Neural Network Toolbox	TDNN	[11]
ESN Toolbox	ESN	[8]
Sample code, [16]	CRBM	[15]
ConvNet	CNN	[6]
Datafeed Toolbox	Data	[5]

3.2 Tuning Hyper-Parameters

Each kind of model in this experiment requires individual hyper-parameters to be set. Although these values sometimes can be set directly by reasoning, verifying and tuning through empirical testing is often required. Carefully adjusting hyper-parameters to each stock is undesirable, as we wish to avoid overfitting them to particular stocks. Instead, one of the data sets is selected to optimize hyper-parameters. Models are trained and evaluated on this time series, using the approach explained in Sects. 3.3 and 3.4. Care was taken to optimize the hyper-parameters of the various models as fairly as possible.

Because each data set is unique, the one used for tuning hyper-parameters should not be arbitrarily chosen. Indices have fewer peculiarities as they are composed of multiple stocks and closely represent the market as a whole. Hyper-parameters tuned to an index are therefore likely to suit other stocks as well. GSPC represents a wide selection of US companies from multiple industries, and was therefore used for the tuning process.

3.3   Training and Test Set

Each time series contains ten years of daily stock or index data, starting from 1-1-2005. The first 70% percent of each series is used for training, leaving the last 30% for testing. It can be argued that patterns found during training are more likely to disappear during the course of the rather large test set. For real-world applications, one could try to keep adapting the model to new data as time progresses. Models are trained and tested separately on each time series, which makes 36 models total (six models for every time series).

Neural networks are initialized from a random state, making learning non-deterministic. Conditional Restricted Boltzmann Machine (CRBMs) additionally rely on sampling random variables, while the Echo State Network (ESN) does not even optimize its random internal parameters. To help smooth out the randomness, non-deterministic models are trained identically *ten* times, each time with different initial states. The learning efficiency vary greatly between the models and each model was allowed to train as long time as needed but with a focus on avoiding overtraining. Results are then *averaged* and presented along with *standard deviation*. Support Vector Regression (SVR) learning is deterministic, and is therefore exempted from this rule. Taking the additional neural networks into consideration, the total number of trained models in this study is 306 (resulting from five models running 10 times each and SVR running only once for each stock).

Neural networks learn by being exposed to one training sample at a time. In order to prevent the network to be biased towards certain samples that are clustered towards the end of the training set, training samples are always *randomly shuffled*. Again, this is only relevant for neural networks, as the SVR is invariant to the sample order.

A common practice when evaluating machine learning models is to create the test set from random samples of the full data set, leaving the rest for training. Imagine for instance, a data set where the first 70% is bull, while the last 30% is bear. By randomly sampling the test set, we ensure that the model is tested and trained on both bull and bear samples. Although this will likely lead to better results, the technique conflicts with how stock market prediction works. Specifically, it does not make sense to train on data more recent than the samples we test on; it is the *future* prices we want to predict, not the *past*. To keep the experiments realistic, test data are defined as the most recent 30% of a given data set.

3.4 Evaluating

In evaluating a model, multiple aspects are considered. Equation (1) describes a *single-step* prediction problem: Given recent development, predict the next immediate value. Considering the noisy nature of stock data however, some values may be overshadowed by noise. Learning to predict price points further into the future might give more coherent patterns, as these values have had more time to develop. Models are therefore also evaluated for *double-step* prediction, where time step $t + 1$ is predicted given steps $t - 1$ through $t - n$. Thus, double-step training samples are identical to single-step vectors, except for the last time step of the price we are predicting. Following the theme of short-term modeling, multi-step prediction is not considered.

There are many ways prediction models can be incorporated into a financial system. The main evaluation metric used in this paper is the *profit* made by each model. We estimate profits by adopting a simple trading strategy: Given a certain amount of starting capital, we take a position in the market at the start of each trading day. If the model predicted a positive price development, we buy shares in the stock. This is known as entering a *long* position. If the predicted price movement is negative, we enter a *short* position. Shorting a stock is achieved by borrowing shares and selling them. Eventually the shares are bought back at a different price and returned to the lender. The trade can only be profitable if the shares are bought back at a lower price point. Every time we enter the market, we do so with all our assets. A position is closed the same time the next trading day. The difference between start and end capital expressed as a percentage gives an indication of performance in a trading context, and represents our evaluation metric. Profits are purely theoretical, and no effort has been put into simulating market slippage, broker commissions or other fees. A profit of 0.1 translates to 10% income during the whole test period, which spans approximately three years.

4 Results

In this section the test set results are presented, discussed and compared across the models. Figure 3 visualizes average single-step profits for each model, in addition to the natural development of each test set. The natural test set development is also known as the *buy and hold* strategy, since it represents the profits generated by buying shares at the onset of the test period, and selling at the end. Non-deterministic models have their *standard deviations* marked by thin lines. The performance of the double-step strategy is similar to the simple-step strategy and therefore omitted but included when average profit is compared in Sect. 4.3.

4.1 Index Performance

S&P 500. As detailed in Sect. 3.2, the GSPC index is used for tuning the hyperparameters. Looking at Fig. 3 it does not appear to result in any considerable advantage, with all models generating a profit *below* the buy and hold.

Single-step Profit (%)

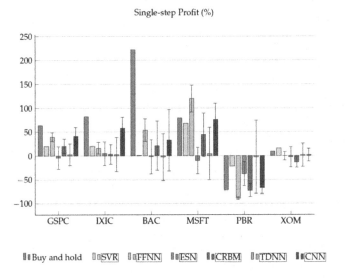

Fig. 3. Profits for single-step strategies and buy and hold.

FFNN and Convolutional Neural Network (CNN) stand out with a profit comparable to, but lower than GSPC itself. CNNs show some deviations. The SVR baseline underperforms in comparison, approximately returning a third of buy and hold. The CRBM has an average profit similar to the SVR, although is considered worse because of the significant deviations. ESN and TDNN models both meander around zero profit, with relatively large deviations.

NASDAQ Composite. IXIC is visually similar to GPSC as seen in Fig. 1, but differences between the two stocks are observed in the profits. The buy and hold strategy outperforms the average of any model, with only the CNN getting close. Its good performance is explained once again by the unlikely high bull ratio, and it is questionable whether the model has actually found any patterns beyond the general bull trend. SVR and FFNN make slight profits, although with considerable deviations for the FFNN. The ESN, CRBM and TDNN generate unreliable profits, staying barely positive on average with high standard deviations.

4.2   Stock Performance

Performance with respect to the four selected stocks is compared and discussed below. Dissimilar market conditions during training and testing make profits for BAC especially interesting (see Fig. 1c). 221.7% does the stock increase, which is more than any of the other test sets. The change in market conditions seems to confuse most of the models, as seen by their large standard deviations. SVR, ESn and TDNN profits are around zero on average. CRBM and CNN generate some profits, but are outweighed by large deviations. The FFNN is the only model that delivers considerable profits with reasonably low variance. With a bull test set development, buy and hold on MSFT is at the same level as IXIC. Unlike the index however, most models generate considerable profits on MSFT. The SVR and CNN profit similarly to buy and hold, and the FFNN surpasses it on average with significant margin. Once again, the ESN and TDNN fall behind.

Because of the bear development in PBR test samples, buy and hold for this stock is negative. However, a positive profit may still be achieved through *short market* positions. None of the models generate positive profit for single-step prediction on PBR, however some of them achieve a reduced loss compared to buy and hold. Because of large deviations in the TDNN, its small average loss is not considered. The SVR on the other hand, considerably reduces its losses from PBR, outperforming any of the neural networks. ESN performance approaches baseline, although with some deviations. Notably, the FFNN, CRBM and CNN all perform about as badly as buy and hold, failing to reduce the loss.

Price development for XOM is volatile with a slight positive change in the test set, as seen in Fig. 1(f). Referring to Fig. 3, neither models nor the buy and hold strategy generate considerable profits. The SVR outperforms the neural networks, also surpassing the stock itself.

4.3   Average Profits

In order to make a clear distinction between results, the profits for each model can be averaged across all time series. Reducing performance down to a single number does remove a lot of information, but since profit is the main performance measure, it is also relevant to consider.

Figure 4 plots the average profits. Despite their considerable losses on PBR, the FFNN and CNN deliver strong profits on the other time series which average greater than the SVR. Predicting double-step improves average profits for most of the neural networks, creating a greater gap between them and the SVR, which additionally generates less profit compared to single-step prediction. As expected the buy and hold strategy strongly outperforms the models in both cases.

Based on Fig. 4 alone, the FFNN and CNN perform similarly for single-step, while the FFNN outperforms all the models in double-step prediction. FFNN deliver best average profits overall, with the CNN following closely.

Single-Step vs. Double-Step. Double-step prediction was included in an attempt to reduce noise in the data, see Sect. 3.4. By comparing results however, models seem to gain from it about as many times as they degrade by it. Having said that, Fig. 4 shows improvement on average for the neural networks, when applied for double-step prediction. Given how the SVR performs worse for predicting two days ahead, drawing an absolute conclusion regarding the effect of double-step prediction is hard. Double-step prediction was intended to smooth out some of the noise, thereby increasing prediction quality and profitability. The neural networks seemingly adhere to this intuition, while the SVR performs better for single-step prediction.

Average Profit (%)

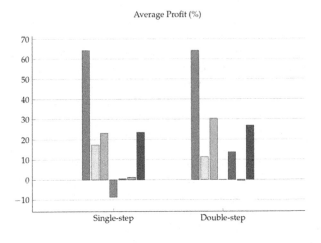

Fig. 4. Average profits for single-step and double-step strategies and buy and hold.

It can be argued that the double-step strategy described in Sect. 3.4 is over-simplified. As mentioned in Sect. 3.4 however, there are many ways machine learning models can be incorporated in a trading strategy. Since the goal of the study is to compare models, rather than trading strategies, it makes sense to use a strategy that fully follows every prediction the model makes.

5   Conclusion

This paper has aimed at comparing modern Artificial Neural Network (ANNs), when applied for short-term stock price prediction using daily data. Models are evaluated in the context of following a financial strategy that trades every prediction a given model makes on the closing price each day. The goal has been to compare the performance of the different models, and to see if any of them are viable to be used for trading.

The results varied, between both time series and models. The Support Vector Regression (SVR), Feed Forward Neural Network (FFNN) and Convolutional Neural Network (CNN) showed comparable results, mostly producing positive profits. Three models stand out with better performance: The SVR, FFNN and CNN. Due to large deviations and high bull ratios however, the CNN is regarded as potentially unreliable despite good profits. Both the SVR and FFNN show potential, and to some extent actually complement each other: The SVR is able to generate a profit on Exxon Mobile Corporation (XOM) while also significantly reducing losses on Petrleo Brasileiro S.A. - Petrobras (PBR). For Bank of America Corporation (BAC) and Microsoft Corporation (MSFT) on the other hand, the FFNN sees significant profits that outperform the other models. This encourages further research, for instance on using models in an optimized trading strategy, or together in a hybrid solution.

When comparing profits averaged across all time series, the FFNN outperforms the other models: 23.13% and 30.43% for single-step and double-step prediction, respectively. The CNN follows closely behind, with 23.50% single-step and 26.92% double-step profits. Average SVR profits were observed at 17.28% for single-step and 11.30% for double-step, thereby opposing the expectation that predicting two days ahead results in greater profits. Based on the average profits, the three abovementioned models all seem viable to trade according to. Given the extreme bull ratio of the CNN and the slightly lower profits for SVR, the FFNN is deemed most viable for trading.

Acknowledgment. This work is partially supported by The Research Council of Norway as a part of the *Engineering Predictability with Embodied Cognition (EPEC)* project, under grant agreement 240862.

References

1. Aamodt, T.: Predicting stock markets with neural networks - a comparative study. M.A. thesis, University of Oslo, Norway (2015). https://www.duo.uio.no/handle/10852/44765
2. Bank of America: Bank of America's history, heritage & timeline. http://tinyurl.com/y8lkycan
3. Chao, J., Shen, F., Zhao, J.: Forecasting exchange rate with deep belief networks. In: The 2011 International Joint Conference on Neural Networks (IJCNN), pp. 1259–1266, July 2011
4. Clark, R.: Support Vector Regression. http://www.mathworks.com/matlabcentral/fileexchange/43429-support-vector-regression
5. MATLAB®: Datafeed Toolbox. http://mathworks.com/products/datafeed/
6. Demyanov, S.: ConvNet. http://github.com/sdemyanov/ConvNet
7. ExxonMobil: About us. http://tinyurl.com/yca4pmd9
8. Jaeger, H.: ESN Toolbox. http://tinyurl.com/y6vthcfh
9. Khashei, M., Bijari, M.: An artificial neural network (p, d, q) model for timeseries forecasting. Expert Syst. **37**(1), 479–489 (2010)
10. Li, Y., Ma, W.: Applications of artificial neural networks in financial economics: a survey. In: 2010 International Symposium on Computational Intelligence and Design (ISCID), vol. 1, pp. 211–214 (2010)
11. MATLAB®: Neural Network Toolbox. http://tinyurl.com/yajrbqdl
12. Petrobras: About us - Petrobras. http://tinyurl.com/y7fs75an
13. Saad, E.W., Prokhorov, D.V., Wunsch, D.C.: Comparative study of stock trend prediction using time delay, recurrent and probabilistic neural networks. IEEE Trans. Neural Netw. **9**(6), 1456–1470 (1998)
14. Tanaka, M.: Deep Neural Network. http://www.mathworks.com/matlabcentral/fileexchange/42853-deep-neural-network
15. Taylor, G.: Modeling Human Motion Using Binary Latent Variables. http://www.uoguelph.ca/gwtaylor/publications/nips2006mhmublv/code.html
16. Taylor, G.W., et al.: Modeling human motion using binary latent variables. Adv. Neural Inf. Process. Syst. **19**, 1345 (2007)

Medical AI Applications

Beyond Lesion Detection: Towards Semantic Interpretation of Endoscopy Videos

Michael D. Vasilakakis[1], Dimitris K. Iakovidis[1(✉)],
Evaggelos Spyrou[1], Dimitris Chatzis[1], and Anastasios Koulaouzidis[2]

[1] Department of Computer Science and Biomedical Informatics,
University of Thessaly, Lamia, Greece
{vasilaka,diakovidis,espyrou,dchatzis}@uth.gr
[2] Endoscopy Unit, The Royal Infirmary of Edinburgh, Edinburgh, UK
akoulaouzidis@hotmail.com

Abstract. Several computer-based medical systems have been proposed for automatic detection of abnormalities in a variety of medical imaging domains. The majority of these systems are based on binary supervised classification algorithms capable of discriminating abnormal from normal image patterns. However, this approach usually does not take into account that the normal content of images is diverse, including various kinds of tissues and artifacts. In the context of gastrointestinal video-endoscopy, which is addressed in this study, the semantics of the normal content include mucosal tissues, the hole of the lumen, bubbles, and debris. In this paper we investigate such a semantic interpretation of the endoscopy video content as an approach to improve lesion detection in a weakly supervised framework. This framework is based on a novel salient point detection algorithm, the bag-of-words image representation technique and multi-label classification. Advantages of the proposed method include: (a) It does not require detailed, pixel-level annotation of training images, instead image-level annotations are sufficient; (b) It enables a richer description of image content, which is beneficial for the discrimination of lesions. The annotation of the multi-labeled training images was performed using a novel annotation tool called RATStream. The results of the experiments performed in a wireless capsule endoscopy dataset with inflammatory lesions promises an improved performance for future generation diagnostic systems.

Keywords: Endoscopy · Video analysis · Lesion detection · Weakly supervised learning · Multi-label classification · Bag-of-words

1 Introduction

The understanding of visual content from images and videos is still one of the most exciting and continuously growing research areas in the field of computer vision. Many research efforts have focused on the automatic extraction of image annotations in a way that imitates a human description based on their perception. This problem has often been referred to as "bridging the semantic gap" [1] i.e., automatically extracting high-level semantics given the low-level features captured from content with a systematic or standardized approach.

© Springer International Publishing AG 2017
G. Boracchi et al. (Eds.): EANN 2017, CCIS 744, pp. 379–390, 2017.
DOI: 10.1007/978-3-319-65172-9_32

Semantic interpretation of images can become even more challenging by extending its application to video scale. State-of-the-art works on semantic video interpretation are mainly based on supervised machine learning algorithms capable of classifying the contents of the video frames into semantically relevant categories [2]. Normal images could include various kinds of normal content that is not semantically differentiated. For example, in endoscopy video frames normal images may contain bubbles, debris and other content besides normal mucosa [3].

Conventionally, the ground truth description of the image content is provided by experts using annotations provided graphically. In order to minimize the requirements for annotation effort, which can prove particularly costly for large image collections, we have recently proposed two weakly supervised classification methods; a method based on Convolutional Neural Networks (CNNs) [4], and a method based on the Bag-of-visual-Words (BoW) technique [5]. These approaches do not require a detailed, pixel-level annotation of the ground truth images. Instead, they require image-level semantic annotations indicating only the classes in which their contents belong to. For example, in the case of an abnormality detection problem, the images can be annotated with only a label indicating whether they contain an abnormality or only normal tissue.

In this paper we present a novel extension of our BoW-based weakly supervised method [5] for semantic interpretation of endoscopy videos using multiple semantic categories. This approach is aided by a novel algorithm for the detection of salient points within the video frames. It acknowledges the fact that a normal or an abnormal video frame may include other normal contents beyond normal or abnormal mucosa, such as bubbles, debris etc. Considering that the image features of these contents are usually different (e.g., bubbles include white reflections, debris has green/yellow hues) the proposed approach identifies them as members of separate classes aiming to simplify lesion detection. Thus, for each video frame a more complete description is provided using multiple semantic identifiers (labels). To the best of our knowledge, semantic interpretation of endoscopy video using multi-label classification techniques has not been previously proposed.

The rest of this paper consists of 4 sections. In Sect. 2 we provide a brief medical background on gastrointestinal (GI) video endoscopy, focusing on wireless capsule endoscopy (WCE). Then, in Sect. 3 we describe the proposed methodology towards the multi-label semantic representation of video frames. Experimental results are presented and discussed in Sect. 4. Finally, conclusions are drawn in Sect. 5, where plans for future work are also presented.

2   Gastrointestinal Video Endoscopy

GI endoscopy is typically a minimally invasive procedure that allows physicians to examine and inspect the interior of the gastrointestinal tract. Traditional approach is based on a flexible (conventional) endoscope which is equipped with a Charged-Couple Device (CCD) camera. However, certain areas of the GI tract are not easily reached either by conventional endoscopes due to instrument length. To overcome such limitations, WCE has been proposed. It uses a swallowable capsule endoscope (CE) the size of a large vitamin pill, which is equipped with a color video camera and a light source. Current CEs are passive devices, in the sense that they move "naturally", i.e., by exploiting gravity as well as the peristaltic motion of the GI tract. A typical CE, during its 12-h journey within the GI tract continuously captures thousands of color images, which are wirelessly transmitted to an external video recorder [6].

Although WCE provides rich information regarding the inner GI tract, since it may capture approx. 100K (on average) color frames, this is considered to be its major limitation [3]. This set of frames needs to be manually reviewed by WCE readers, using specialized software, a process that may range from 45 min up to a couple of hours, while their attention needs to be undistracted. The lengthy sessions required for WCE reading/reporting mean that such tasks are frequently redirected to other, non-medical clinical personnel including nurses and clinical scientists [7, 11]. Nonetheless WCE reading is a time-consuming procedure, often tiring and certainly prone to human errors. For example, it has been shown that the detection rate of clinically significant findings by experts is limited to approx. 40% [8]. Thus, fully-automated lesion detection approaches are highly desirable [9]. Such approaches have been applied on the problem of the diagnosis of GI lesions in known or suspected inflammatory bowel disease (IBD) [10] with main aim to recognize abnormalities such as ulcers, aphthae etc.

Towards automatic lesion detection, several computer vision-based approaches have been proposed [3] and are often based on supervised approaches. Experts annotate abnormal areas within the video frames, while the rest are considered to be normal. Then, learning algorithms are applied on each category of abnormalities. Previously, we have presented a public, open access database, namely KID which provides high quality video frames annotated by medical experts [12], and has been used for the evaluation of some of our previous works [13].

3 Methodology

3.1 Weakly Labeled Image Representation

In order to overcome the aforementioned problem of pixel-wise level annotation, one possible solution is the weakly-supervised learning [14]. Contrary to supervised learning, it does not require explicit and detailed annotation. Instead, image-level annotation of the semantics of the image is needed. Thus, a given video frame may be annotated, e.g., with the semantic concept "abnormal", if it contains an abnormality, or with the semantic concept "normal", if it does not contain an abnormality. At a next level, further semantic concepts may be also added to the annotation process. In the context of lesion detection in WCE, different "normal" concepts can be associated with different normal intestinal content, e.g., "debris", "bubbles", whereas different "abnormal" concepts may include GI lesions, e.g., "inflammatory lesions", "vascular lesions", etc. Weakly-supervised methods are often criticized because they do not always provide localization information regarding the detected classes. However, in the context of manual reviewing of large WCE videos, such an approach could significantly reduce the amount of time required by the reviewer, if it could accurately and robustly detect frames that contain possible lesions. Since such frames are usually a rather small subset of the entire WCE video, the reviewers' task may be limited to localization of abnormalities within this subset, which should consist a less tiring and tedious task.

The Bag-of-visual-Words (BoW), when applied to images, is a popular weakly-supervised model. A given image (or a video frame) is described as a set of visual "words" that originate from a visual "vocabulary." The latter comprises of a set of representative quantized low-level feature descriptions of a training image set. These descriptions come from parts of images such as overlaid grids, regions (e.g., resulting upon a segmentation process) or patches (e.g., surrounding salient points). To this end, typical approaches adopt a clustering algorithm and the vocabulary consists of either the centroids or the medoids resulting from this process. These comprise the set of the visual words and accordingly an image is described (coded) using a histogram calculated on these words. Put differently, each feature vector captures the frequencies of all words of the visual dictionary within the corresponding image. Another advantage of the BoW approach is that it results to a fixed-size image description, i.e., appropriate to be used with typical machine learning approaches.

In this work, the adopted BoW approach is based on a set of extracted patches surrounding points that result from a salient point detection process. For the low-level description of the image patches, we choose to adopt a set of color-based features which we have previously [13] applied to the problem of lesion detection and yielded superior results compared to the state-of-the-art approaches. More specifically, a given image is first transformed to the CIE-Lab color space and then, from each patch we extract the Lab values of the interest point, and the max and min values of all three components within the entire patch. This way, a 9-value color feature vector is extracted from each patch.

3.2   Salient Point Detection

The BoW approach can be based on a set of extracted patches surrounding dense points that result from a sampling process using a regular grid (i.e., one with equal horizontal and vertical inter-pixel distances). Such dense approaches are often criticized as "naïve" when compared to more sophisticated approaches, such as the SIFT [15] or SURF [16] interest points. However, it has been shown that they carry valuable information regarding semantic interpretation of visual content [17]. In the context of endoscopic image analysis the application of SURF on channel a of the CIE-*Lab* color space (a-SURF) resulted in salient points on all the abnormalities included in that study. However, in [5] we showed that dense sampling may be more time consuming but it can result in higher abnormality detection rates.

The dense sampling process using regular grid, extracts a large number of feature vectors. These feature vectors are not easily separable by a clustering algorithm. For this reason there is a need to select some image points to extract fewer feature vectors without significant loss of information. A way to reduce these points in an image represented in CIE-*Lab* color space is to get points only from the image regions where a significant color change is observed. The purpose behind this idea is to discriminate and sample image regions, where a discontinuity in their color description appears. The discontinuity in color of channel a, indicates the region as a region of interest. In that sense these regions can be considered as "salient" points. In order to detect such salient points, the difference between two maximum values in a-channel around the densely sampled points is considered. The proposed Difference of Maxima (DoM) algorithm (Fig. 1) for salient point detection proceeds as follows: (a) Dense sampling of the image using window of size X. (b) For each point of the dense sampling grid calculate the maximum a_{max} and minimum a_{min} values of channel a from two windows of different sizes centered at each of these points; an outer window of size X, and an inner window of size $X/2$; (c) The central point of these windows (point 3 in Fig. 1c is included into the set of salient points, if the Euclidean distance between the two vectors $(a_{max}, a_{min})_X$ and $(a_{max}, a_{min})_{X/2}$ is larger than the mean Euclidean distances estimated from all windows in the same image; otherwise, it is rejected.

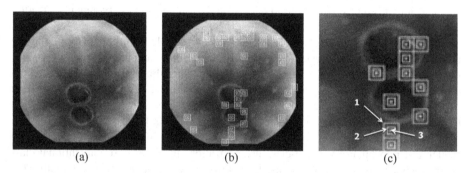

Fig. 1. DoM salient point detection. (a) Original image. (b) The remaining points after dense sampling, around which the Euclidean distances are estimated. (c) Magnified Fig. 1(b), clearly indicating the outer window (1) and the inner window (2) where the maxima are calculated, and the central point (3) of these windows.

3.3 Multi-label Classification

Following the BoW feature extraction process the content of the WCE images needs to be classified into semantic categories. Usually the classification of the endoscopic image content is performed into two categories, corresponding to normal and abnormal tissues. In such cases binary classifiers such as the Support Vector Machines (SVMs) [18] are used. However this approach provides only an abstract categorization of the GI image content. This happens due to the initial assumption that every image belongs in exactly one of the aforementioned categories. This assumption does not consider the multiple semantics within an image. For example, the semantics of a normal image besides mucosal tissues may include normal intestinal content such as bubbles and debris, and the lumen hole.

Let $Q = R^d$ be the feature space derived from a set of images used to train the supervised classification system, and L denotes the label space. In the binary case the classification system aims to learn a function $f : Q \rightarrow L$ using the feature vectors $q_i \in Q$, $i = 1, 2, \ldots, N$, extracted from N images labeled as l_j, $\in L$, $j = 1, 2$, from a training set $\{(q_i, l_i) | 1 \leq i \leq N, 1 \leq j \leq 2)\}$. In weakly supervised learning each label l_j refers to semantic content of the whole image, because the feature vector q_i describes the whole image and not a specific region.

The classification of the image content into multiple categories is usually performed via cascading binary or multi-class classifiers applied on image regions [4]. In the case of multi-class classification each whole image is labeled with a single label $l_j \in L$, $j = 1, 2, \ldots, m$, and m is the number of classes used to describe image content. However, this does not take into account that a single image may include various contents that can be described with different labels. To cope with this issue in this paper we consider multi-label classification [19, 20].

Let v be a vector of multiple labels m for every $q_i \in Q$, $i = 1, 2, \ldots, N$, where $v_j \in L, v_j = (l_1, l_2, \ldots l_m)$, $j = 1, 2, \ldots, z$. Each label is a binary flag denoting the presence of different kinds of image contents. In this paper a total of 5 labels are considered, indicating the presence of normal (l_1), abnormal (l_2), debris (l_3), bubbles (l_4), and lumen hole (l_5). The purpose of training a multi-label classifier is to learn a function $h: Q \rightarrow 2^L$.

There are two main learning strategies, namely the algorithm adaptation, and the problem transformation strategies [19]. The algorithm adaptation strategy tackles multiple labels by adapting existing learning algorithms from single label to multi-label. Examples of algorithms implementing this strategy include an adaptation of the k-Nearest Neighbor (k-NN) [18] classifier for Multi-Label classification (MLkNN) [21], kernel methods such as multi-label SVMs [22].

The problem transformation strategy deals with multi-label learning problem by reducing it into binary or multi-class categorization. In this way a conventional classifier, such as an SVM can be used. In this study we investigate the multi-label classification in the context of endoscopic image analysis using various problem transformation methods [19]: the binary relevance [19], the ranking and thresholding [19, 23], the pairwise classification [23, 24], and the label combination [25] methods.

The binary relevance method [19] trains different binary classifiers, each of which classifies the images according to one label. In the context of the endoscopic image classification investigated in this study, 5 binary classifiers are used to determine the existence of each one of the five categories of content considered, e.g., the existence of abnormalities or not, the existence of debris or not, etc. However, this methodology does not consider possible dependencies between labels.

The label combination method [25] transforms the task of multi-label learning into a standard, single-label, multi-class classification. It considers each different set of labels that exist in the multi-label data set as a single label. In this way treats every label combination in the training data as a unique class label in a binary label problem. There are classes derive from their combinations, except of the five classes that were referred earlier. For instance, one new class may be the images that have both the labels of normal and debris. The constructed class is the normal-debris class.

Ranking and thresholding methods [19, 23] transform the task of multi-label learning into a multi-class problem. In ranking the task is to order the set of labels. A threshold function is constructed from multi-label data, so that the topmost labels are more related with the new instance. A ranking of labels requires post-processing in order to give a set of labels, which is the proper output of a multi-label classifier.

The pairwise classification [23, 24] adopts the "one-*vs*-one" approach, where one classifier is associated with each pair of labels. This is contrary to binary relevance approach of "one-*vs*-all" where one classifier is associated with the relevance of each label. Hence, instead of five binary problems, ten binary problems are formed, because there are ten different pairs of labels. Typically, each pairwise problem is constructed from examples with which either labels (but not both) are associated, thus forming a decision boundary for these two labels.

3.4 Multi-label Annotation

The weak annotation of the training sets using multiple labels is more efficient than the detailed pixel-level annotation. The latter can be performed using state-of-the-art software tools, such as Ratsnake [26]. However, image annotation using solely semantic labels, although easier than the pixel-wise annotation process, can also become very time consuming. In order to speed up the multi-label annotation task we developed a novel software, called RATStream (Rapid Annotation Tool for image and video Streams)[1], which enables time-efficient annotation of image and video streams. It supports both weak and detailed multi-label annotation options (Fig. 2) for a variety of well-known image and video formats. Annotation is performed quickly as it provides access to full image or video streams, where the user can easily navigate back and forth, and add semantic tags and/or graphic, pixel-wise annotations with only a few clicks per image or video frame.

RATStream has been developed in C++ using JUCE library[2] for the development of the Graphical User Interface (GUI) and OpenCV library[3] for the implementation of computer vision algorithms, such as superpixel segmentation [27], which is exploited for faster pixel-level annotation (by clicking on superpixels as illustrated in Fig. 2b).

4   Experimental Evaluation

Experiments were conducted to evaluate the proposed methodology using different methods of multi-label and binary label classification. The experiments were performed on a subset of the publicly available KID dataset [12]. This subset consisted of 100 randomly selected normal images from the small bowel and 70 randomly selected images of the most common inflammatory lesions, vascular and polypoids. The 170 images were weakly annotated with the labels that indicate the presence of normal (l_1), abnormal (l_2), debris (l_3), bubbles (l_4), and lumen hole (l_5). For each image features extracted using the proposed DoM salient point detection method and the "naive" approach of dense feature extraction. The BoW model was constructed with a range of

---

[1] Ratstream is available upon request to the authors.

[2] https://sourceforge.net/projects/juce/.

[3] http://opencv.org/.

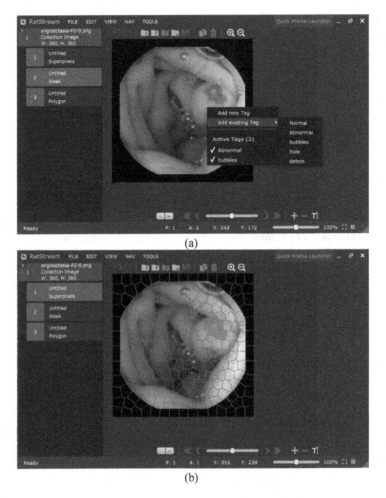

Fig. 2. The GUI of RATStream annotation tool developed for time-efficient weak multi-labeled annotation of image and video streams. (a) Weak annotation using semantic tags. (b) Pixel-level annotation using superpixels (other detailed annotations options are also available).

visual vocabulary sizes in the range from 500 to 4500 words. The classification performance was thoroughly investigated using Receiver Operating Characteristic (ROC) curves [28]. These curves illustrate the tradeoff between sensitivity and specificity for various decision thresholds. In order to enable comparisons between the ROC curves, the area under the ROC (AUC) was used as a classification performance measure, which unlike accuracy, it is relatively robust for datasets with imbalanced class distributions [29]. Experiments were performed using the 5-fold cross validation evaluation scheme, using SVMs as a binary classifier. Multi-label classification was implemented using a derivative of WEKA library [30] called MEKA [31]. The results with regards to lesion detection are presented in Table 1. It can be noticed that the use of the proposed DoM algorithm increases the binary classification performance to an

Table 1. Comparative lesion detection results.

Method	AUC
Binary classification	
DoM BoW	0.80
a-SURF BoW [5]	0.75
SIFT+LBP BoW [32]	0.74
SIFT+CLBP BoW [32]	0.70
Dense BoW [5]	0.75
DoM Multi-label classification	
Binary relevance	0.80
Pairwise classification	0.88
Label combination	0.78
Ranking and thresholding	0.81

AUC of 0.80. Using DoM for multi-label classification, results in an even higher classification performance than the conventional binary classification scheme. Best results were obtained using the pairwise multi-label classification method, with an AUC reaching up to 0.88 using a vocabulary of 2500 visual words. In addition, we performed experiments for comparison with a state of the art methodology, which used SIFT algorithm [15] for the detection of interest points and a concatenated feature vector of SIFT and LBP or SIFT and CLBP [32] for the description of image regions. In all experiments the proposed method achieved better results.

Table 2 summarizes the classification results of the pairwise multi-label classification method per class, for the maximum lesion detection performance. It can be noticed that bubbles are classified with an even higher AUC reaching up to 0.92. The lowest performance was observed for the detection of the lumen hole. However, this can be justified by the fact that it was clearly visible to fewer training images.

Table 2. Multi-label pairwise classification results using DoM salient point detection.

Category	AUC
Lesions	0.88
Bubbles	0.92
Debris	0.74
Lumen hole	0.67

The lower AUC obtained for debris and lumen hole can be explained by the smaller number of images containing debris and lumen hole; thus, the images used for training may not have included a sufficient amount of debris or lumen hole in order to train the classifier. The lowest AUC of lumen hole can be explained by the insufficient salient points detected in these regions, as the values of channel *a* are lower and there was no variation between them.

5 Conclusions

This work presented a novel extension of our BoW-based weakly supervised method [5] for semantic interpretation of endoscopy videos using DoM salient point detection and classification of the video frames into multiple semantic categories. Our approach takes under consideration the fact that a normal or an abnormal video frame possibly includes other valuable information beyond normal or abnormal mucosa, e.g., bubbles, as separate classes. Multi-label annotation was performed efficiently using Ratstream, a novel annotation tool for image and video streams.

The experimental results, lead us to the following conclusions:

- The semantic interpretation of whole endoscopy images is feasible using weakly supervised multi-label classification techniques;
- Lesion detection can be improved by the use of DoM in a BoW framework;
- Lesion detection can be improved by multi-label classification;
- The pairwise classification of problem transformation methods results in better classification of the endoscopic images than the use of the simple binary SVM classifiers.

From the above it can be derived that the use of multiple semantic identifiers to describe and interpret the whole content of endoscopy images is a promising direction for enhanced computer-aided analysis of endoscopic video streams. The performance of conventional tools for abnormality detection based on binary classifiers is usually affected by the presence of intestinal content such as debris and bubbles [3]. Currently, several supervised methods exist for detection and removal of uninformative frames due to the presence of intestinal [3]. Drawbacks of such methods include: (a) they increase the computational complexity of the overall video analysis task, because the video has to be processed in two steps (one for the removal of the uninformative frames and another one for the detection of the abnormalities; (b) by totally removing the video frames with the intestinal content there is a chance to also miss frames with abnormalities that are present with the intestinal content. On the other hand, the proposed approach, with one processing step provides information about both the presence of abnormalities and intestinal content. Furthermore, its comparative advantage is that it is weakly supervised.

Future research directions include investigation of scalability of multi-label classification, when larger dataset is available for the learning task. Moreover, the effect in classification performance may have, when we choose different feature extraction methods during the stage of clustering in BoW procedure. Last but not least, additional experiments need to be done for the investigation of classification performance, when algorithm adaptation methods are used.

Acknowledgements. The research presented in this paper was financially supported by the project "**Klearchos Koulaouzidis**" Grant No. 5151 and the Special Account of Research Grants of the University of Thessaly, Greece.

References

1. Smeulders, A., Worring, M., Santini, S., et al.: Content-based image retrieval at the end of the early years. IEEE Trans. Pattern Anal. Mach. Intell. **22**, 1349–1380 (2000). doi:10.1109/34.895972

2. Li, H., Liu, L., Sun, F., et al.: Multi-level feature representations for video semantic concept detection. Neurocomputing **172**, 64–70 (2016). doi:10.1016/j.neucom.2014.09.096

3. Iakovidis, D., Koulaouzidis, A.: Software for enhanced video capsule endoscopy: challenges for essential progress. Nat. Rev. Gastroenterol. Hepatol. **12**, 172–186 (2015). doi:10.1038/nrgastro.2015.13

4. Georgakopoulos, S., Iakovidis, D., Vasilakakis, M., et al.: Weakly-supervised convolutional learning for detection of inflammatory gastrointestinal lesions. In: IEEE International Conference on Imaging Systems and Techniques (IST), pp. 510–514. IEEE (2016)

5. Vasilakakis, M., Iakovidis, D., Spyrou, V., Koulaouzidis, A.: Weakly-supervised lesion detection in video capsule endoscopy based on a bag-of-colour features model. In: International Workshop on Computer-Assisted Robotic Endoscopy (CARE) at International Conference on Medical Image Computing and Computer Assisted Intervention (MICCAI) (2016)

6. Koulaouzidis, A.: Wireless endoscopy in 2020: will it still be a capsule? World J. Gastroenterol. **21**, 5119 (2015). doi:10.3748/wjg.v21.i17.5119

7. Yung, D., Fernandez-Urien, I., Douglas, S., Plevris, J., Sidhu, R., McAlindon, M., Panter, S., Koulaouzidis, A.: Systematic review and meta-analysis of the performance of nurses in small bowel capsule endoscopy reading. United Eur. Gastroenterol. J., 205064061668723. (2017) doi:10.1177/2050640616687232

8. Zheng, Y., Hawkins, L., Wolff, J., Goloubeva, O., Goldberg, E.: Detection of lesions during capsule endoscopy: physician performance is disappointing. Am. J. Gastroenterol. **107**, 554–560 (2012). doi:10.1038/ajg.2011.461

9. Iakovidis, D., Sarmiento, R., Silva, J., Histace, A., Romain, O., Koulaouzidis, A., Dehollain, C., Pinna, A., Granado, B., Dray, X.: Towards intelligent capsules for robust wireless endoscopic imaging of the gut. In: IEEE International Conference on Imaging Systems and Techniques, pp. 95–100. IEEE (2014)

10. Koulaouzidis, A.: Small-bowel capsule endoscopy: a ten-point contemporary review. World J. Gastroenterol. **19**, 3726 (2013). doi:10.3748/wjg.v19.i24.3726

11. Riphaus, A., Richter, S., Vonderach, M., Wehrmann, T.: Capsule Endoscopy Interpretation by an Endoscopy Nurse – a Comparative Trial. Zeitschrift für Gastroenterologie **47**, 273–276 (2009). doi:10.1055/s-2008-1027822

12. Koulaouzidis, A.: KID: Koulaouzidis-iakovidis database for capsule endoscopy (2015). http://is-innovation.eu/kid

13. Iakovidis, D., Koulaouzidis, A.: Automatic lesion detection in capsule endoscopy based on color saliency: closer to an essential adjunct for reviewing software. Gastrointest. Endosc. **80**, 877–883 (2014). doi:10.1016/j.gie.2014.06.026

14. Hoai, M., Torresani, L., De la Torre, F., Rother, C.: Learning discriminative localization from weakly labeled data. Pattern Recogn. **47**, 1523–1534 (2014). doi:10.1016/j.patcog.2013.09.028

15. Lowe, D.: Distinctive image features from scale-invariant keypoints. Int. J. Comput. Vis. **60**, 91–110 (2004). doi:10.1023/b:visi.0000029664.99615.94

16. Bay, H., Ess, A., Tuytelaars, T., Van Gool, L.: Speeded-up robust features (SURF). Comput. Vis. Image Underst. **110**, 346–359 (2008). doi:10.1016/j.cviu.2007.09.014

17. Tuytelaars, T.: Dense interest points. In: IEEE Conference on Computer Vision and Pattern Recognition (CVPR), pp. 2281–2288 (2010)
18. Theodoridis, S., Koutroumbas, K.: Pattern Recognition. Elsevier/Academic Press, Amsterdam (2008)
19. Tsoumakas, G., Katakis, I.: Multi-label classification. Int. J. Data Warehouse. Min. **3**, 1–13 (2007). doi:10.4018/jdwm.2007070101
20. Zhang, M., Zhou, Z.: A review on multi-label learning algorithms. IEEE Trans. Knowl. Data Eng. **26**, 1819–1837 (2014). doi:10.1109/tkde.2013.39
21. Zhang, M., Zhou, Z.: ML-KNN: a lazy learning approach to multi-label learning. Pattern Recogn. **40**, 2038–2048 (2007). doi:10.1016/j.patcog.2006.12.019
22. Elisseeff, A., Weston, J.: A kernel method for multi-labeled classification. In: NIPS, pp. 681–687 (2001)
23. Fürnkranz, J., Hüllermeier, E., Loza Mencía, E., Brinker, K.: Multilabel classification via calibrated label ranking. Mach. Learn. **73**, 133–153 (2008). doi:10.1007/s10994-008-5064-8
24. Mencia, E., Furnkranz, J.: Pairwise learning of multilabel classifications with perceptrons. In: 2008 IEEE International Joint Conference on Neural Networks, IJCNN 2008, (IEEE World Congress on Computational Intelligence), pp. 2899–2906. IEEE (2008)
25. Read, J., Pfahringer, B., Holmes, G.: Multi-label classification using ensembles of pruned sets. In: 2008 Eighth IEEE International Conference Data Mining, ICDM 2008 (2008)
26. Iakovidis, D., Goudas, T., Smailis, C., Maglogiannis, I.: Ratsnake: a versatile image annotation tool with application to computer-aided diagnosis. Sci. World J. **2014**, 1–12 (2014). doi:10.1155/2014/286856
27. Achanta, R., Shaji, A., Smith, K., Lucchi, A., Fua, P., Süsstrunk, S.: SLIC superpixels compared to state-of-the-art superpixel methods. IEEE Trans. Pattern Anal. Mach. Intell. **34**, 2274–2282 (2012). doi:10.1109/tpami.2012.120
28. Fawcett, T.: An introduction to ROC analysis. Pattern Recogn. Lett. **27**, 861–874 (2006). doi:10.1016/j.patrec.2005.10.010
29. Provost, F., Fawcett, T.: Analysis and visualization of classifier performance: comparison under imprecise class and cost distributions. In: KDD, pp. 43–48 (1997)
30. Witten, I., Frank, E., Hall, M., Pal, C.: Data Mining, 1st edn. Morgan Kaufmann, Amsterdam (2017)
31. Read, J., Reutemann, P., Pfahringer, B., Holmes, G.: MEKA: a multi-label/multi-target extension to WEKA. J. Mach. Learn. Res. **17**, 1–5 (2017)
32. Yuan, Y., Li, B., Meng, M.: Improved bag of feature for automatic polyp detection in wireless capsule endoscopy images. IEEE Trans. Autom. Sci. Eng. **13**, 529–535 (2016). doi:10.1109/tase.2015.2395429

Assessment of Parkinson's Disease Based on Deep Neural Networks

Athanasios Tagaris[✉], Dimitrios Kollias, and Andreas Stafylopatis

School of Electrical and Computer Engineering,
National Technical University of Athens,
Zografou Campus, 15780 Athens, Greece
thanos@islab.ntua.gr

Abstract. A novel system based on deep neural networks is presented, that performs analysis of medical imaging data. The aim is to study structural and functional alterations of the human brain in patients with Parkinson's Disease and to correlate them with epidemiological and clinical data. A new medical database, which is presently under development, is used for training the system and testing its performance. Preliminary experimental results are provided which illustrate the capability of the proposed system to analyze and provide an accurate estimation of the status of the disease.

Keywords: Deep neural networks · Parkinson's disease · Medical data analysis

1 Introduction

Parkinson's, is a neurodegenerative disease which develops progressively due to the lack of dopamine. It is one of the most common neurodegenerative disorders which usually starts between the ages of 50 to 70 years. In countries with ageing population, as in most EU countries and in the US, the number of patients is expected to triple during the next 50 years [1]. Although there is no definite treatment for Parkinson's Disease (PD), the early detection and appropriate management may highly improve the quality of patients' lives [2].

The presented approach is based on a large database related to Parkinson's disease, which is currently populated and includes the following:

- Magnetic Resonance Images (MRI) of the brain.
- Images obtained through scintigraphy with 123-ioflupane.
- Epidemiological data, such as the patient's current age, sex and disease duration.
- Treatment data, such as duration of dopaminergic treatment, dose.
- Clinical data, relating the patient status to several scales (UPDRS, PDQ-39).

This work was financed by the State Scholarships Foundation (IKY) through the "Research Projects for Excellence IKY/Siemens" Programme in the framework of the Hellenic Republic – Siemens Settlement Agreement.

G. Boracchi et al. (Eds.): EANN 2017, CCIS 744, pp. 391–403, 2017.
DOI: 10.1007/978-3-319-65172-9_33

These scales assess each patient's mobility, everyday activities, therapy complications and quality of life.

- Timed tests, on repetitive movements of upper limbs, gait.
- Mental activity tests, including the Mini Mental State Examination.
- Special Scales, such as the Unified Dyskinesia Rating Scale (UDysRS) and the Freezing of Gait Questionnaire.

All data are anonymous and, when completed, the database will refer to about 100 patients with Parkinson's disease and 40 people with other neurological diseases This database is being constructed, based on collaboration of the Intelligent Systems research group, National Technical University of Athens, with the Department of Neurology, Georgios Gennimatas General Hospital, Athens, Greece. It will be made fully available this summer. To the best of our knowledge, this is the first publicly available database of this type. Consequently, researchers will have the possibility of using it for the development of systems which learn to make predictions and assist medical doctors and patients in detection, analysis and assessment of Parkinson's disease.

The work presented here concerns the development of a novel system that is able of making predictions and assessments, using the above database and the state-of-the-art in machine learning. Our main approach is based on deep learning methodologies, which constitute the state-of-the-art in image analysis and computer vision. Deep neural systems are created, based on both deep convolutional (CNN) and recurrent neural networks (RNN), which prove to be able to process all types of available data and detect a Parkinson's or non-Parkinson's case according to the above-mentioned rating scales.

It is known that, in the great majority of cases, Parkinson's disease is idiopathic, in the sense that there is no specific explanation for the cause that the dopaminergic cells die. For this reason, we created several deep neural systems using transfer learning techniques. Deep CNN structures, have been proven able to extract rich features and internal representations from multiple image categories, over millions of images. For these reasons, this will be the starting structure for our analysis. The training procedure starts with the transfer of these representations to our novel systems and concludes with fine tuning the latter with training data from our database. Then, we apply the trained system on different subjects' test data from the database and evaluate the performance of the generated deep neural system.

Section 2 presents the dataset created for Parkinson's Disease. Section 3 describes the use of the neural system for the analysis of the MRI and DaT scans and for detection of Parkinson's. Experimental results are provided in Sect. 4, while Conclusions and Further Work are presented in Sect. 5.

2   Database Creation

We have been creating a database composed of about 100 patients with Parkinson's and 40 subjects with non-Parkinson's disease, which includes MRIs, DaT Scans and clinical data for each subject. This dataset will, therefore, be able to provide a training and test dataset of about 140 subjects appropriately classified in categories according to the epidemiological and clinical data. Up to now we have populated around one sixth of the database. In general, this size of dataset is not sufficient to train complex deep neural architectures. However, each of the MRI and DaT Scan sets includes sequences or multiple scans of each subject. Through different combinations of these images we have been able to create many thousands of training data, which form a sufficient data set, where deep neural networks with transfer learning can be effectively applied.

2.1   Magnetic Resonance Images (MRI)

The rapid evolution of non-invasive medical imaging techniques, over the past decades, has opened new possibilities for the analysis of the brain. The basic imaging technique is the Magnetic Resonance Imaging (MRI) which can yield from hundreds to even thousands of images per scan.

The assessment of this extremely large set of images per patient can be complicated and time-consuming for medical doctors. In Parkinson's Disease, the MRI can show the extent to which the different structures of the brain have been degenerated. Figure 1 shows an example of a MRI. Our main concern regarding PD, is the Lentiform Nucleus (green line in Fig. 2) and the capita of the Caudate Nucleus (red line in Fig. 2). Since we focus on volume estimation, we process the image sequences in batches, each composed of 3 consecutive frames.

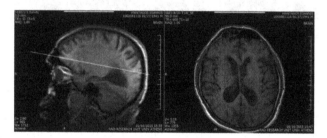

Fig. 1. A frame of an axial T1 sequence from a brain MRI (right). Location of the previous slice is placed with regard to a sagittal view of the brain (left).

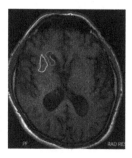

Fig. 2. An image from an axial T1 sequence. The Lentiform Nucleus is depicted with a green line while the capita of the caudate nucleus with a red line. (Color figure online)

2.2   Dopamine Transporters Scan (DaT Scan)

The second brain imaging technique included in the database is the Dopamine Tran-sporters (DaT) scan. This examination is a form of Single-Photon Emission Computer Tomography (SPECT) with Ioflupane Iodide-123 as it's contrast agent. In this exam-ination, we can see if the Striatum receives dopaminergic endings from the Substantia Nigra or not.

A series of images is produced in this way, as shown in Fig. 3.

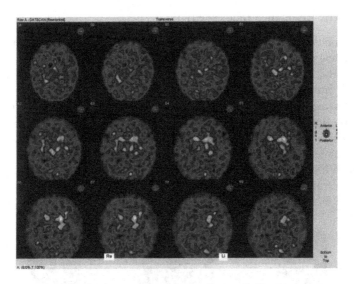

Fig. 3. A sequence of frames from a DaT scan.

The doctor selects the most representative of these (the 8th in the presented case) and marks the areas corresponding to the head of the caudate nucleus. An automated system then compares these with a neutral one (e.g. the cerebellum) and produces the ratios shown in Fig. 4. By comparing these ratios with normal ones, the doctor can make the diagnosis.

2.3   Clinical Data

As already mentioned these define the patient's clinical status. The scales we focus are the UPDRS [3], the patient stage [4], UDysRS [5], PDQ-39 [6], FOG [7], MMSE [8] and the two, timed tests.

The Unified Parkinson's Disease Rating Scale (UPDRS) is a metric that examines the patient's whole clinical performance in 4 parts: motor/non-motor experiences of daily living, motor examination and complications. These contain 13, 13, 18 and 6 elements respectively, with each ranging from 0-4 for a max score of 234.

The patient's stage represents the evolution of the disease and ranges from asymptomatic (0) to bedridden (5).

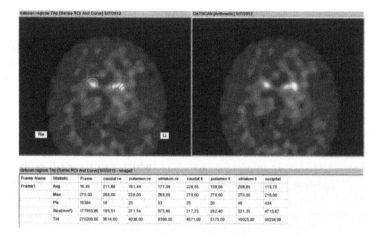

Frame Name	Statistic	Frame	caudat re	putamen re	striatum re	caudat li	putamen li	striatum li	occipital
Frame1	Avg	16,49	211,89	161,44	177,09	228,55	199,04	208,85	115,75
	Max	270,00	268,00	228,00	268,00	270,00	270,00	270,00	216,00
	Pix	16384	18	25	53	20	26	48	434
	Size(mm²)	177953,86	195,51	271,54	575,66	217,23	282,40	521,35	4713,87
	Tot	270205,00	3814,00	4036,00	9386,00	4571,00	5175,00	10025,00	50234,00

Fig. 4. DaT scan with expert selection (left). Same image without the markings (right). Ratios representing the dopamine deficiency, that are used for the diagnosis (bottom).

The Unified Dyskinesia Rating Scale (UDysRS) was created for evaluating the involuntary movements associated with PD; it has two parts measuring the dyskinesia and dystonia appearing "on" and "off" phases respectively. The first part has 11 while the second 15 elements, all ranging from 0 (asymptomatic) to 4 (severe symptoms), for a total of 150.

The Parkinson's Disease Questionnaire (PDQ) consists of 39 questions assessing patient's the functionality and quality of life. It can be separated in 8 different categories, while each question represents the frequency of a specific incident, ranging from 0 (never occurring) to 5 (always occurring), for a total of 156.

The "Freezing of Gait" (FOG) is one of the most characteristic PD symptoms. The quantification of this symptom is achieved through the homonymous questionnaire which contains 16 elements for a max rating of 24.

The Mini Mental State Examination (MMSE) is an 11-question questionnaire meant to measure the cognitive impairment associated with PD, with a max rating of 30.

3 Deep Neural Network Architectures

We next provide a brief account of the component architectures involved in our design and describe the resulting deep learning system.

3.1 Deep Convolutional Neural Networks

Deep CNNs are architectures that try to exploit the spatial structure of input information [9]. They have been used with great success in various applications, including image analysis, vision, object and emotion recognition [10]. The first successful CNN, the AlexNet CNN [11] has been used for classification of millions of images in 1000 categories. The network accepts color images of size 224 × 224 pixels and processes them through 253440–186624–64896–64896 – 43264–4096 – 4096 neurons in its respective layers, from input to output, with an output layer of 1000 neurons.

One of the most successful CNN architectures, is the Residual Network or ResNet [12], which was developed by Microsoft Research Laboratories and achieved 3.57% error rate winning the 1st place on the ISLVRC2015 classification task.

3.2 Transfer Learning in DCNN Adaptation

Transfer learning [13] is an important procedure, which can be used to avoid failure due to overfitting, when training complex CNN models with small amounts of training data. Transfer learning is achieved by initializing the weights of the model from a pre-trained CNN (previously trained with large datasets of irrelevant images - usually generic objects). Afterwards, a supervised fine-tuning procedure is performed, where the latter parts of the network are retrained from a smaller relevant dataset.

3.3 Recurrent Neural Networks

RNNs are powerful models which are naturally suited to processing sequential data [14]. RNNs can maintain a hidden state which encodes information about previous elements in the sequence. At every time step, the hidden state is updated as a function of both the input and the current hidden state, as shown in Fig. 5. One very successful approach, known as Long Short-Term Memory (LSTM) [15], modifies the architecture of the hidden units by introducing gates which explicitly control the flow of information as a function of both the state and the input. This allows information to be

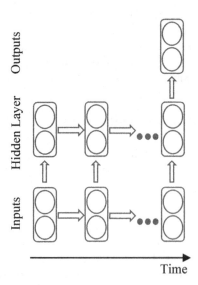

Fig. 5. Recurrent neural network

carried over long periods of time. By combining forward and backward processing of input data through time, a Bidirectional LSTM model is obtained.

3.4   The Deep Neural System

We target to develop a deep neural system that is designed (trained and tested) with raw input data and learns to produce the desired outputs. We consider various combinations of single, or ensembles of, networks, either of CNN, or CNN-RNN type. The former (CNN) are able to derive rich internal representations and features from the input data, while the latter are able to correlate and analyze time evolutions of the input data, finally predicting the desired outcome. The theoretical methodologies and the approaches we follow for this goal include many state-of-the-art techniques, especially transfer learning, with the RNN part being based on the above-mentioned Bidirectional Long Short-Term Memory (B-LSTM) neuron model.

The CNN system we consider follows the basic structure of the so called Deep Residual Net (ResNet), which contains 50 layers [12]. This network has won the 1st places on the tasks of ImageNet detection, ImageNet localization, COCO (object) detection, and COCO (object) segmentation. Following the convolutional and pooling layers we use up to 3 fully connected (FC) layers, with the so-called ReLU neuron models, i.e., neurons with a linear activation function, for positive input values, and a zeroing function elsewhere. Other networks (e.g. [16]) could also be used, but they have been mainly designed for human face analysis applications.

Figure 6 shows the proposed CNN architecture, in which the last layer is a linear one, so as to provide continuous estimates of the clinical data, in each considered scale.

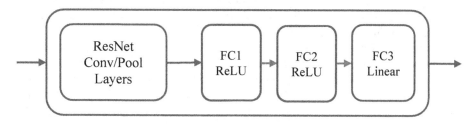

Fig. 6. ResNet structure

Figure 7 shows the CNN-RNN system: The CNN part of this network (upper part of Fig. 7) provides the outputs of the neurons of its second fully connected hidden layer to the RNN part. The RNN part accepts F1, F2, F3, …, FN and provides the predicted values, O(1),…,O(N), through time, at its output layer (lower part of Fig. 7).

First, we perform transfer learning for the weights of the convolutional and pooling parts of this network to our generated system. These parts are then fixed during the training phase, where we only train the fully connected layers of the system. The pre-trained convolutional networks have already learnt to generate rich image

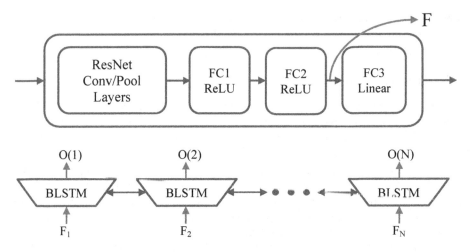

Fig. 7. ResNet and BLTSM

representations that have proven adequate for image classification and segmentation. These representations are enough abstract to help with specialized tasks, such as the analysis of MRIs and DaT Scans.

This leaves the fully connected part of the network, which is the only part of the network that we actually train. Many variants of this approach have been designed and tested. We selected to freeze the weights of some of the fully connected layers, particularly those belonging to the first FC layer. We have also considered some more weights of the network as free parameters, by applying fine tuning of the weights of (some of) the convolutional layers of the ResNet network with a smaller learning rate value in the updating of the weights of this part of the network, while using a normal learning rate value for the FC part of it.

We use the TensorFlow Platform as the main tool for generating the software implementation of the presented architecture. TensorFlow is a toolkit which got published by Google, under Apache License 2.0. It is mainly implemented using C++, with a significant bit of Python. Its architecture provides the ability to deploy computation to one or more CPUs or GPUs in a desktop, server, or mobile device with a single API.

4   Experimental Results

The current size of the generated fully annotated database is 21 subjects (about 15% of final size), with a ratio of 2:1 between Parkinson's patients and non-patients. We generated a dataset of about 20.000 combinations of color DaT scans with triplets of consecutive MRI gray scale images, for the patient category, and 6.000 such combinations for the non-patient category. To obtain a balance dataset, we applied various augmentation techniques, such as duplicating the latter category, or partitioning the former category. These were then used as a training dataset for the deep neural networks.

Each input consisted of three MRI images and an RBG DaT scan image. Moreover, we kept about 500 input combinations from each category for the validation/test data set. It should be emphasized that our target has been to test the ability of the networks to learn from a number of patients and generalize their performance to other cases, which have not been met in the training set. For this reason, the test data consisted of 3 new subjects, 2 with Parkinson's and one without. Currently, the networks have two linear outputs, (1,0) and (0,1), respectively, for the two categories.

Due to the incompleteness of the dataset, clinical data were not included in this experiment.

As a reference, 10 consecutive frames from an axial T1 brain MRI are presented in Fig. 8 for a patient without Parkinson's, and 10 more in Fig. 9 for a patient with the disease. Almost no differences can be seen with the naked eye, between these two classes.

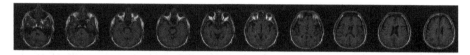

Fig. 8. MRI scan of a patient without Parkinson's disease. Axial orientation - T1 sequence.

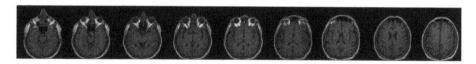

Fig. 9. MRI scan of a patient with Parkinson's disease. Axial orientation - T1 sequence.

Figure 10 shows two DaT scans of patients without and with Parkinson's Disease. The dopamine deficiency can be seen in these two images, as the relevant brain structures are depicted less brightly.

The dataset sizes were still small to train the CNN and CNN-RNN deep neural networks from scratch; starting from random initial weights in the convolutional and fully connected (FC) parts of the CNNs, or the convolutional and hidden layers of the CNN-RNNs. For this reason, we used transfer learning, i.e. transfer of the weights of the convolutional and pooling layers of a pretrained CNN, to our generated networks and designed and trained the 'upper' FC part of the targeted CNN network, as well as the RNN hidden layers of the CNN-RNN with the above dataset. For the initialization of these weights, we used the ResNet-50 CNN, which has been pre-trained with millions of general-type RGB images for this purpose. A separate system was used for each of the image types in our inputs, i.e., one focusing on the MRI triplets and another focusing on the DaT scan. We concatenated the outputs of these two ResNet sub-structures at the input of the first FC layer of the CNN network, or at the input of the first hidden layer of the RNN network.

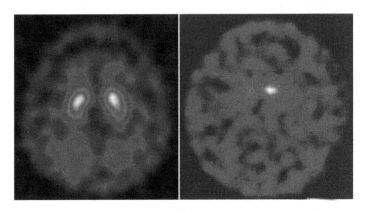

Fig. 10. DaT scan from a patient without Parkinson's disease (left). Same image from a patient with Parkinson's (right)

Based on this procedure, we trained first the resulting deep CNN network for Parkinson's and then the deep CNN-RNN network.

Table 1 summarizes the results obtained through different configurations of the CNN network, i.e. ones with different numbers of hidden layers and hidden units per layer. An accuracy of 70% was obtained in this experiment, which is quite satisfactory, for this, still small, database size.

Table 1. Accuracy results for CNN architectures

CNN architectures	No of fully connected (FC) layers	No of hidden units in each FC layer	Accuracy
1	3	4096-2622-2	0.64
2	3	2622-1000-2	0.70
3	3	1000-500-2	0.67
4	4	4096-4096-2622-2	0.68

Table 2 summarizes the accuracy obtained by the best CNN and the best CNN-RNN networks, with their respective structures. The addition of the RNN part provides the deep network with the ability to follow better the time varying correlations provided in the MRI sequence of triplets of frames, as far as the volume of the image regions of interest is concerned, increasing the accuracy of the Parkinson's prediction to 74%.

Table 2. Accuracy results for best CNN and CNN-RNN configurations

Best architectures	No of fully connected/hidden layers	No of hidden units in each FC/hidden layer	Accuracy
CNN	3	2622-1000-2	0.70
CNN-RNN	3	128-128-2	0.74

Figures 11 and 12 show, respectively, the CNN and CNN-RNN obtained accuracy on the validation/test data set, during training. It can be shown that the best accuracies are obtained early in the learning phase, afterwards reaching overfitting condition.

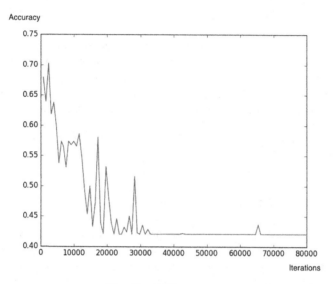

Fig. 11. CNN accuracy.

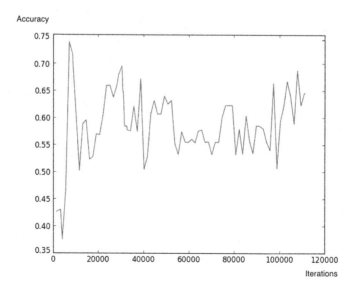

Fig. 12. CNN-RNN accuracy.

For the training of the network we used the Adam optimizer algorithm, in mini batches. The cost function was the Mean Squared Error (MSE).

For the CNN architecture, the hyper-parameters we elected to use are: a batch size of 30 (15 examples from each category), a constant learning rate of 0.001; 2622 and 1000 hidden units respectively in each fully connected layer and dropout after each fully connected layer with a value of 0.5. We also used biases in the fully connected layers.

For the CNN-RNN architectures the hyper-parameters were selected to match the previous, apart from the batch size which was 40 (20 examples from each category) and the number of hidden units in the LSTM layers, both of which were 128.

The weights of the fully connected layers were initialized from a Truncated Normal distribution with a zero mean and a variance equal to 0.1 and the biases were initialized to 1.

Training was performed on a single GeForce GTX TITAN X GPU and the training time was about 2–3 days.

5   Conclusions and Further Work

The presented Database is in its generation phase, so the experimental results reported here constitute preliminary results, which, nevertheless, show the potential of using the proposed approach for detection of Parkinson's based on medical imaging data. Our research is ongoing and will be extended to cover all cases met in the Parkinson's database. It is expected that the generated systems will provide significant support to clinicians and medical doctors for early detection of Parkinson's.

References

1. Goldman, S.M., Tanner, C.: Etiology of Parkinson's disease. In: Jankovic, J., Tolosa, E. (eds.) Parkinson's Disease and Movement Disorders, 3rd edn, pp. 133–158. Williams and Wilkins, Baltimore (1998)
2. DeMaagd, G., Philip, P.: Parkinson's disease and its management: Part 1: disease entity risk factors pathophysiology clinical presentation and diagnosis. Pharm. Ther. 40(8), 504 (2015)
3. Goetz, C., Tilley, B., Shaftman, S., et al.: Movement disorder society-sponsored revision of the unified Parkinson's disease rating scale (MDS-UPDRS): scale presentation and clinimetric testing results. Mov. Disord. 23(15), 2129–2170 (2008)
4. Hoehn, M., Yahr, M.: Parkinsonism: onset, progression, and mortality. Neurology 50(2), 318–318 (1998)
5. Goetz, C., Nutt, J., Stebbins, G.: The unified dyskinesia rating scale: presentation and clinimetric profile. Mov. Disord. 23(16), 2398–2403 (2008)
6. Jenkinson, C., Fitzpatrick, R., Peto, V., Greenhall, R., Hyman, N.: The Parkinson's disease questionnaire (PDQ-39): development and validation of a Parkinson's disease summary index score. Age Ageing 26(5), 353–357 (1997)
7. Giladi, N., Shabtai, H., et al.: Construction of freezing of gait questionnaire for patients with Parkinsonism. Parkinsonism Relat. Disord. 6(3), 165–170 (2000)

8. Tombaugh, T., McIntyre, N.: The mini-mental state examination: a comprehensive review. J. Am. Geriatr. Soc. **40**(9), 922–935 (1992)
9. Goodfellow, I., Bengio, Y., Courville, A.: Deep Learning. The MIT Press, Cambridge (2016)
10. LeCun, Y., Kavukcuoglu, K., Farabet, C.: Convolutional networks and applications in vision. In: Proceedings of ISCAS, pp. 253–256, Paris, France (2010)
11. Krizhevsky, A., Sutskever, I., Hinton, G.: Imagenet classification with deep convolutional NNs. Adv. Neural. Inf. Process. Syst. **25**, 1106–1114 (2012)
12. He, K., Zhang, X., Ren, S., Sun, J.: Deep Residual Learning for Image Recognition. https://arxiv.org/abs/1512.03385v1
13. Ng, H.W., Nguyen, V.D., et al.: Deep learning for emotion recognition on small datasets using transfer learning. In: Proceedings of ACM International Conference on Multimodal Interaction, pp. 443–449 (2015)
14. Ebrahimi Kahou, S., Michalski, V., et al.: Recurrent neural networks for emotion recognition in video. In: Proceedings of ACM International Conference on Multimodal Interaction, pp. 467–474 (2015)
15. Wollmer, M., Kaiser, M., et al.: Lstm-modeling of continuous emotions in an audiovisual affect recognition framework. Image Vis. Comput. **31**(2), 153–163 (2013)
16. Simonyan, K., Zisserman, A.: Very deep convolutional networks for large-scale image recognition. arXiv preprint arXiv:1409.1556 (2014)

Detection of Malignant Melanomas in Dermoscopic Images Using Convolutional Neural Network with Transfer Learning

S.V. Georgakopoulos[1], K. Kottari[1], K. Delibasis[1],
V.P. Plagianakos[1,2], and I. Maglogiannis[2(✉)]

[1] Department of Computer Science and Biomedical Informatics,
University of Thessaly, Papassiopoulou 2-4, Lamia, Greece
{spirosgeorg, vpp}@uth.gr,
kottarikonstantina@gmail.com, kdelibasis@yahoo.com
[2] Department of Digital Systems, University of Piraeus, Piraeus, Greece
imaglo@unipi.gr

Abstract. In this work, we report the use of convolutional neural networks for the detection of malignant melanomas against nevus skin lesions in a dataset of dermoscopic images of the same magnification. The technique of transfer learning is utilized to compensate for the limited size of the available image dataset. Results show that including transfer learning in training CNN architectures improves significantly the achieved classification results.

Keywords: Skin lesion · Melanoma detection · Computer Vision-based diagnostic systems · Convolutional neural networks - CNN · CNN architectures · Transfer learning

1 Introduction

Skin cancer is considered among the most frequent types of cancer and one of the most malignant ones. According to a recent study, its incidence has increased faster than that of almost all other cancers among young adults [1]. Melanoma is one of the most common and aggressive forms of skin cancer. Early-stage melanoma detection is particularly important for the survival of patients, according to a number of studies [2–5]. Moreover, a significant number of published works have shown that image-based quantification of tissue lesion features into measurable quantities may be of essential importance in clinical practice, due to correlation of these quantities with skin lesion malignancy [6–11]. Dermatoscopy (or Dermoscopy) is the examination of skin lesions using a dermatoscope, which magnifies the examined area of the skin. The acquired dermoscopic image can reveal many coloring and structural details of the skin lesion, impossible to observe with the unaided eye. During the last years, computer vision-based diagnostic systems have been used for the early detection of malignant melanoma tumour, by analysing the dermoscopy images, in terms of its morphological and color characteristics, as well as other measurable features.

G. Boracchi et al. (Eds.): EANN 2017, CCIS 744, pp. 404–414, 2017.
DOI: 10.1007/978-3-319-65172-9_34

Our team has previously dealt with the melanoma detection through dermatological differential structures called "streaks" [12]. Streaks are brownish-black linear structures of variable thickness, not clearly combined with pigment network lines, mostly at the periphery of the lesion. Irregular streaks are very likely to represent melanoma, especially when the streaks are distributed unevenly. The detection of streaks in dermoscopy images has been recently reported in [13, 14]. Other types of differential structures include dark dots – globules that have also been used to provide descriptors for melanoma detection [8]. Dots/globules may occur in both benign and malignant skin lesions, but with different spatial distribution. In typical melanocytic nevus, dots/globules with regular size and shape may be observed evenly distributed throughout the lesion. In melanomas, irregular dots/globules, however, occur predominantly at the periphery and vary in size and shape, and, moreover, are unevenly distributed.

In this paper, we propose the use of convolutional neural networks (CNN) for detecting malignant melanoma in dermoscopic images. Characteristic dermoscopic images that exhibit melanoma or non-melanoma lesions are presented in Figs. 1(d) and (a–c), respectively. We compare two different CNN architectures with and without pre-training in images from different domains. In the case of pre-trained architectures, we apply the transfer learning technique, instead of random initialization of the weights of CNN, in order to utilize the pre-learned weights as initial ones. This approach leads to a successful CNN training, which otherwise could not be achieved, due to the limited size of the available dataset. Preliminary results are presented using 69 images of melanoma and 972 images of non-melanoma lesions.

2 Methodology

The proposed methodology for melanoma and nevus dermoscopy images classification is based on Convolutional Neural Network (CNNs). In this section, we briefly present the CNNs classification algorithm and the popular transfer learning technique applied on the available image data.

2.1 Convolutional Neural Network

CNNs are multistage trainable architectures used for classification tasks. Each of these stages consists of three types of layers [15].

- *Convolutional Layer*, which is the major component of the CNNs. A convolutional layer consists of a number of kernel matrices that perform convolution on their input and produce an output matrix (feature image) where a bias value is added. The learning procedures aim to train the kernel weights and biases as shared neuron connection weights.
- *Pooling Layer*, which is also an integral component of the CNNs. The purpose of a pooling layer is to perform dimensionality reduction of the input feature images. Pooling layers make a subsampling to the output of the convolutional layer matrices combining neighboring elements. The most common pooling function is the max-pooling function, which takes the maximum value of the local neighborhoods.

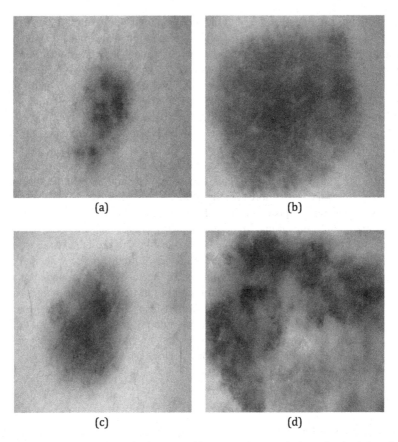

Fig. 1. Characteristic dermoscopic images with non-melanoma lesions (a–c) and melanoma lesions (d).

- *Fully-Connected Layer*, which is a classic Feed-Forward Neural Network (FNN) hidden layer. It can be interpreted as a special case of the convolutional layer with kernel size 1×1. This type of layer belongs to the class of trainable layer weights and it is used in the final stages of CNNs.

The training of CNN relies on the BackPropagation (BP) training algorithm [15], since the architecture relies on an acyclic directed graph and the gradient of the loss function propagates using the chain rule. The requirements of the BP algorithm is a vector with input patterns x and a vector with desirable targets y, respectively. The input x_i generates output o_i. Each output is compared to its corresponding desirable target y_i and their difference gives the training error. The goal of training is to find weights that minimize the cost function

$$E(w) = \frac{1}{n} \sum_{p=1}^{P} \sum_{j=1}^{N_L} \left(o_{j,p}^L - y_{j,p} \right)^2 \qquad (1)$$

where P the number of patterns, $o_{j,p}^L$ the output of j neuron that belongs to the output layer L, N_L the number of neurons in L, $y_{j,p}$ the desirable output of j neuron for pattern p. To minimize the cost function $E(w)$, a pseudo-stochastic version of SGD algorithm, also called mini-batch Stochastic Gradient Descent (mSGD), is usually utilized [16].

CNN Architectures. Two CNN architectures have been used in this work. The first architecture ("arch₁"), known as AlexNet [17], consists of 5 convolutional layers, each followed by a max pooling layer, and two fully-connected layers. AlexNet has been used on the ImageNet [18] dataset, in order to classify different objects from real world images. The second architecture (arch₂) consists of 4 convolutional layers, followed by a max pooling one, as well as a fifth one, followed by 2 fully-connected layers. Arch₂ is presented in [19], where it is utilized to identify images containing an abnormality, from a dataset of endoscopic images [20]. The details of both architectures are shown in Fig. 2.

As it will be further explained in the next subsection, each of the CNN architectures are utilized without, as well as with pre-training. The pre-training took place using the image-net [17] for arch₁. The second CNN architecture ("arch₂") was pre-trained with a set of endoscopic images of lower gastrointestinal (GI) tract [20].

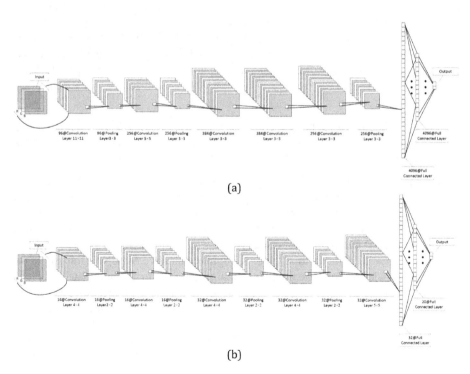

Fig. 2. CNN architectures used in this work: (a) arch₁ and (b) arch₂.

2.2   Data Handling

In general, the training set required for convolutional neural networks should be of sufficient size, if overfitting is to be avoided. In this work, we used 69 images of melanoma and 972 images of non-melanoma lesions of same magnification. However, the number of melanoma images in our dataset is significantly lower than the number of nevus images. In order to be able to use CNNs, the easiest and most common method is to artificially enlarge the dataset using label-preserving transformations. To tackle this problem, we adopt the following scheme: Dataset Expansion and Augmentation and Transfer Learning.

Dataset Expansion and Augmentation. In this study, we utilize two distinct forms of data augmentation [17], both of which allow transformed images to be produced from the original images with very little computational cost. This process is applied only on melanoma images.

- First, the small number of melanoma images is *expanded* using image geometric transformations, such as rotations and mirroring. More specifically, the step of data augmentation consists of rotations by multiples of 90° on each melanoma image and on each rotated image a mirror transformation. Thus, the number of melanoma images is expanded by eight times.
- Secondly, for $arch_1$, all images are resized to 256×256, while continuing to exhibit identical magnification. The CNN extracts *augmented* image patches of 224×224 at random locations of the resized image which are used as input. Similarly, for CNN $arch_2$, the images are resized to 340×340. The CNN extracts image patches of 320×320 at random locations of the resized image which are used as input to it.

Although this increases the size of the training set by a factor of 2048, the resulting training samples are highly interdependent. Finally, it must be noted that the transformed images are used during training and do not need to be stored.

Transfer Learning. The CNN is based on a pre-trained network and subsequently is being fine-tuned, using the available dermoscopy images. Even after the application of the above steps, the resulting number of melanoma images (552) is still not adequate for CNNs training. It is known that CNNs training requires a very high number of training samples, while for many problem domains, such as digital dermoscopy image classification, samples are difficult to acquire. A popular technique for CNNs training is the transfer of knowledge from one problem domain, namely A to another related problem, namely B. This methodology is also known as Transfer Learning (TL). Due to the internal structure of CNNs, the first layer tends to learn features that resemble either Gabor filters or color blobs [21], while the latest layers consists of domain's specific features. This observation leads to using the first n number of layers, of a network pre-trained to task A, to initialize a CNN for fine-tuning the solution of task B. Although, other approaches which have been held, maintain the total number of layers of the pre-trained network of task *A*, fine-tuning it for the domain *B*. TL is very useful in cases that the dataset is not adequate to train a full CNN network (small dataset, missing values, difficulties on annotations, etc.). Arch$_1$ is a very widely used CNN architecture, that, among others, has been pre-trained on the ImageNet dataset [18], which contains a number of 1000 different objects of real world images. This pre-trained network produces a large number of generic filters in the first layers, while in the last layers the filters tend to be more fine-tuned for the specific problem at hand. In this work we deal with a medical image type problem, so the use of non-medical images for CNN training seems to be inadequate. However, pre-trained networks with dermoscopy images are not publicly distributed. Thus, we examine a pre-trained network with a different type of medical images: a set of endoscopic images of lower GI tract [20]. The proposed methodology is presented in the following Figure (Fig. 3).

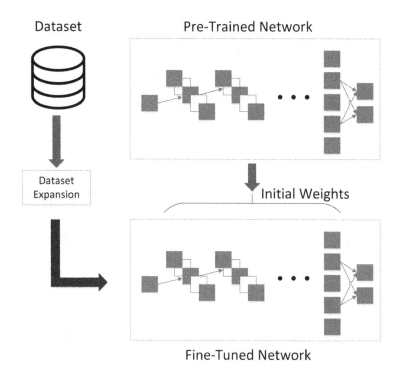

Fig. 3. The proposed methodology.

3   Experimental Results

3.1   Description of Image Data Set

The available data set consists of 1041 images, 972 of them are displaying nevus and the rest 69 images are containing malignant melanoma cases. All images were acquired with the same magnification using the ELM Molemax device at Hospital of Wien and at the department of Plastic Surgery and Dermatology in the General Hospital of Athens G. Gennimatas. The size of the melanomas image subset is substantial considering the fact that malignant melanoma cases in primordial state are very rare. It is very common that many patients arrive at specialized hospitals with partially removed lesions; lesion removal is a simple operation that may be performed in small health centres. A standard protocol was used for the acquisition of the skin lesion images ensuring the reliability and reproducibility of the collected images. Reproducibility is considered quite essential for the image characterization and recognition attempted in this study, since only standardized images may produce comparable results. The original size for the dermoscopy images is 632 × 387 and all images are resized as described for each CNN architecture. We apply the geometric transformations presented in subsection "*Dataset Expansion and Augmentation*", to expand the number of melanoma images by 8-fold, from 69 to 552, thus producing of more balanced dataset. For the needs of evaluation of the methods, we perform a 5-fold Cross Validation. The testing fold consists of only the original 69 melanoma images (excluding the melanoma images produced by geometric transformations).

3.2   Quantitative Results

As a first step, we study the performance of these two networks by training them without initialization, despite the very small size of the CNN training image dataset. Both arch$_1$ and arch$_2$ methods exhibit high accuracy results, but low sensitivity for the melanoma images, due to the small size of the melanoma training subset. To improve the achieved sensitivity for melanoma images, we integrated a priori knowledge by applying the transfer learning technique, as following. Both pre-trained architectures were fine-tuned (FT) using the dermoscopy dataset. The CNN based on arch1 was pre-trained with the ImageNet dataset and fine-tuned with dermoscopy dataset – denoted as FT-arch1. The arch2 CNN was pre-trained with endoscopic images and fine-tuned with dermoscopy dataset – denoted as FT-arch2. Experimental results are summarized in Table 1.

Finally, we also explore an alternative approach. Instead of just using the trained CNN as a classifier, we also utilise it to create features that will be used with another classifier [22]. More specifically, we feed the dermoscopy images to the pre-trained AlexNet [17] architecture and extract features from the first three convolutional layers (384 features). Subsequently, those features are used to train a Support Vector Machine

(SVM). The experimental results of this approach (SVM$_{CNN}$) are also presented in Table 1. The fine-tuned FT-arch$_1$ exhibits the highest classification results, followed by FT-arch$_2$. The corresponding ROC curves are illustrated in Fig. 4.

Table 1. The achieved classification results with 5-fold Cross Validation.

	Accuracy %	Sensitivity %	Specificity %
arch$_1$	92.3	100	0
arch$_2$	92.3	100	0
FT-arch$_1$	91.5	92.5	83
FT-arch$_2$	85.7	88.7	62
SVM$_{CNN}$	72.1	77.2	17.2

The 11×11 filters associated with the first convolutional layer of pre-trained arch$_1$, before TL are shown in Fig. 5(a). Some of the filters are easily recognisable due to their similarity with typical textbook image processing kernels. The difference induced by TL is shown in Fig. 5(b) (note the difference in grayscale colour-bar levels).

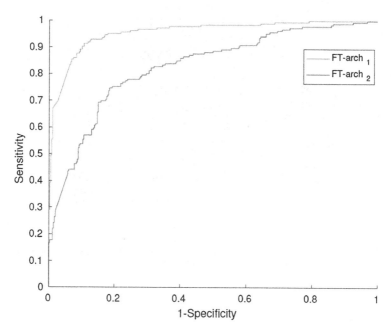

Fig. 4. The ROC curves obtained by TL for the two architectures.

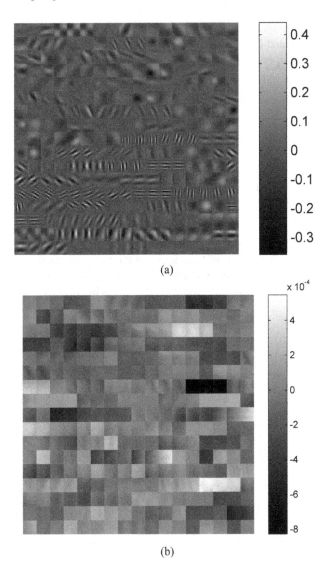

Fig. 5. The 11×11 filters associated with the first convolutional layer of pre-trained $arch_1$, (a) before TL and (b) the difference caused by TL. (Color figure online)

4   Conclusions

A methodology has been proposed for melanoma and nevus dermoscopy images classification, based on Convolutional Neural Network (CNNs). The novelty of the proposed method is the use of (a) data augmentation to increase the size of the melanoma subset in a meaningful way, as well as (b) transfer learning (TL): fine tuning the previously trained CNNs using the augmented dataset.

Results show that the above technique is able to compensate for the limited size of the available image dataset by utilizing two previously trained CNN architectures on images from different domains. The proposed approach of transfer learning achieves significantly better classification results compared to randomly initialized (without pre-training) CNNs. More specifically, the overall classification accuracy is not affected by the introduction of TL, while the sensitivity of melanoma detection achieved by the CNNs with TL increased to 92.5% and 88.7%, respectively for the two employed CNN architectures. It is clear that the application of CNNs without pre-training (without TL) has not been able to identify the melanoma images, thus it exhibits no clinical usefulness.

Initial examination of the image filters produced by CNNs gives an insight for the effect of transfer learning, which can be further exploited in future work, for faster CNN convergence of CNN training, using even smaller datasets.

Acknowledgments. We gratefully acknowledge the support of NVDIA Corporation for the donation of the Titan X Pascal GPU used for this research.

References

1. Reed, K.B., Brewer, J.D., Lohse, C.M., Bringe, K.E., Pruit, C.N., Gibson, L.E.: increasing incidence of melanoma among young adults: an epidemiological study in Olmsted County, Minnesota. Mayo Clin. Proc. **87**(4), 328–334 (2012)
2. Stern, R.S.: Prevalence of a history of skin cancer in 2007: results of an incidence-based model. Arch. Dermatol. **146**(3), 279–282 (2010)
3. Rogers, H.W., Weinstock, M.A., Harris, A.R., et al.: Incidence estimate of nonmelanoma skin cancer in the United States, 2006. Arch. Dermatol. **146**(3), 283–287 (2010)
4. American Cancer Society. Cancer Facts & Figures (2015). http://www.cancer.org/research/cancerfactsstatistics/cancerfactsfigures2015/ Accessed 12 May 2015
5. Maglogiannis, I., Doukas, C.N.: Overview of advanced computer vision systems for skin lesions characterization. IEEE Trans. Inf Technol. Biomed. **13**(5), 721–733 (2009)
6. Menzies, S.W.: Cutaneous melanoma: making a clinical diagnosis, present and future. Dermatol. Ther. **19**(1), 32–39 (2006)
7. Korotkov, K., Garcia, R.: Computerized analysis of pigmented skin lesions a review. Artif. Intell. Med. **56**(2), 69–90 (2012)
8. Maglogiannis, I., Delibasis, K.: Enhancing classification accuracy utilizing globules and dots features in digital dermoscopy. Comput. Methods Progr. Biomed. **118**(2), 124–133 (2015). ISSN 0169-2607
9. Dreiseitl, S., Ohno-Machado, L., Kittler, H., Vinterbo, S., Billhardt, H., Binder, M.: A comparison of machine learning methods for the diagnosis of pigmented skin lesions. J. Biomed. Inf. **34**(1), 28–36 (2001)
10. Maglogiannis, I., Zafiropoulos, E.: Utilizing support vector machines for the characterization of digital medical images. BMC Med. Inform. Decis. Mak. **4**(4) (2004). http://www.biomedcentral.com/content/pdf/1472-6947-4-4.pdf
11. Maragoudakis, M., Maglogiannis, I.: Skin lesion diagnosis from images using novel ensemble classification techniques. In: 2010 10th IEEE International Conference on Information Technology and Applications in Biomedicine (ITAB), pp. 1–5). IEEE, November 2010

12. Delibasis, K., Kottari, K., Maglogiannis, I.: Automated detection of streaks in dermoscopy images. In: Chbeir, R., Manolopoulos, Y., Maglogiannis, I., Alhajj, R. (eds.) AIAI 2015. IAICT, vol. 458, pp. 45–60. Springer, Cham (2015). doi:10.1007/978-3-319-23868-5_4

13. Sadeghi, M., Lee, T.K., McLean, D., Lui, H., Atkins, M.S.: Oriented pattern analysis for streak detection in dermoscopy images. In: Ayache, N., Delingette, H., Golland, P., Mori, K. (eds.) MICCAI 2012 Part I. LNCS, vol. 7510, pp. 298–306. Springer, Heidelberg (2012). doi:10.1007/978-3-642-33415-3_37

14. Sadeghi, M., Lee, T.K., McLean, D., Lui, H., Atkins, M.S.: Detection and analysis of irregular streaks in dermoscopic images of skin lesions. IEEE Trans. Med. Imaging 32(5), 849–861 (2013)

15. LeCun, Y., Bottou, L., Bengio, Y., Haffner, P.: Gradient-based learning applied to document recognition. Proc. IEEE 86(11), 2278–2324 (1998)

16. Bottou, L.: On-line learning and stochastic approximations. In: On-line Learning in Neural Networks, pp. 9–42. Cambridge University Press (1998)

17. Krizhevsky, A., Sutskever, I., Hinton, G.E.: Imagenet classification with deep convolutional neural networks. In: Pereira, F., Burges, C.J.C., Bottou, L., Weinberger, K.Q. (eds.) Advances in Neural Information Processing Systems 25, pp. 1097–1105. Curran Associates, Inc., New York (2012)

18. Deng, J., Dong, W., Socher, R., Li, L.J., Li, K., Fei-Fei, L.: ImageNet: a large-scale hierarchical image database. In: CVPR09 (2009)

19. Georgakopoulos, S.V., Iakovidis, D.K., Vasilakakis, M., Plagianakos, V.P., Koulaouzidis, A.: Weakly-supervised convolutional learning for detection of inflammatory gastrointestinal lesions. In: 2016 IEEE International Conference on Imaging Systems and Techniques (IST), pp. 510–514. October 2016

20. Iakovidis, D.K., Koulaouzidis, A.: Software for enhanced video capsule endoscopy: challenges for essential progress. Nat. Rev. Gastroenterol. Hepatol. 12(3), 172–186 (2015)

21. Yosinski, J., Clune, J., Bengio, Y., Lipson, H.: How transferable are features in deep neural networks? In: Advances in Neural Information Processing Systems, vol. 27, pp. 3320–3328 (2014)

22. Zhang, R., Zheng, Y., Mak, T.W.C., Yu, R., Wong, S.H., Lau, J.Y.W., Poon, C.C.Y.: Automatic detection and classification of colorectal polyps by transferring low-level CNN features from nonmedical domain. IEEE J. Biomed. Health Inform. 21(1), 41–47 (2017)

Optimization Data Mining

A New Metaheuristic Method
for Optimization: Sonar Inspired Optimization

Alexandros Tzanetos[(✉)] and Georgios Dounias

Management and Decision Engineering Laboratory,
Department of Financial and Management Engineering,
University of the Aegean, 41 Kountouriotou Str, 82100 Chios, Greece
{atzanetos,g.dounias}@aegean.gr

Abstract. In this study, we introduce a new population based optimization
algorithm named Sonar Inspired Optimization (SIO). This algorithm is based on
the underwater acoustics that war ships use for reckoning targets and obstacles.
The advantage of the proposed method is the ability for performing wider range
of search during the iterations of the algorithm. The proposed algorithm is tested
in known benchmarks and results are encouraging.

Keywords: Sonar Inspired Optimization · Nature-inspired intelligent
(NII) algorithm · Population based

1 Introduction

In the last twenty (20) years, a growth on Nature Inspired Intelligent (NII) methods
[1–3] is observed. Applications [4] and new challenges [5] are presented, underlying
the major contribution of these algorithms on the field of optimization. Except for
swarm based techniques [6], there are many others that are inspired by physical phenomena [7] and laws of science [8]. Recently the authors have extensively searched and
collected all the algorithms that are based in the above mentioned categories and
extracted some useful conclusions. The overwhelming majority is population based
schemes. A detail that highlights the need of multiple agents to achieve high exploration, while many of these algorithms are based also on attraction between their agents
through equations that model the main idea inspired from nature.

The most used schemes are based on the gravitational law (Gravitational Search
Algorithm [9]) or in attraction-based laws, e.g. Charged System Search [10], Electromagnetism-like optimization [11]. Based on these phenomena, the best solution attracts
all the others towards it. On the proposed scheme, introduced in this paper, each agent
doesn't interact with the others and thus, performs its independent search. The only
information shared between all agents is the best-so-far fitness achieved. That's a very
useful feature, because all better solutions are contributing to find the best one and the
algorithm cannot be trapped in local optima. So, a good balance between exploration
and exploitation is achieved.

What is more, in recent works, a major point of interest is the need for parameter
tuning of the metaheuristic for different kinds of problems [12–14]. Our goal here is to

© Springer International Publishing AG 2017
G. Boracchi et al. (Eds.): EANN 2017, CCIS 744, pp. 417–428, 2017.
DOI: 10.1007/978-3-319-65172-9_35

provide a new self-tuning algorithm, which overcomes the problem of setting the exact number of agents to solve a problem. Keeping the number of agents constant in the proposed algorithm, does not limit the area searched. Also, the self-tuning mechanism is based on the value of the solution; the worse that a solution is, the bigger will be the step for the current agent.

Similar concepts like the one of the proposed algorithm are used in Dolphin Echolocation algorithm [15] and Bat-inspired algorithm [16]. In Dolphin Echolocation algorithm, the authors use also a probability generator for alternative solutions. And they state that their algorithm performs no movement to the best answer, but works only with possibilities. On the other hand, Yang [16] built his algorithm on the logic of swarm models. So movement is done with vectors in the solution space area, whereas in our proposed scheme there is only replacement of solutions that each agent dis covers. Our goal was to build an equally effective tool, aiming at increasing the exploration grade without decreasing the corresponding exploitation. Bat-inspired algorithm is a swarm method, in which each agent performs one step in each iteration of the algorithmic process. On the other hand, the proposed algorithm is based on the search of multiple points around each agent during each iteration.

Furthermore, a significant detail is that our algorithm needs less parameterization. However, experimentation around parameter tuning of the proposed algorithm will be performed in future research on various real world problems. The values of the parameters presented in this work were carefully set by the authors and are in tune the literature findings on similar optimization algorithms. The concept we propose is based on the auto-tuning of the intensity parameter that determines how big search steps in the solution space the algorithm performs.

Finally, recent reviews of the nature inspired algorithms [17, 18] show that even more schemes are presented every year. The importance of a new algorithm can be shown by its effectiveness in a specific application or the usage as a hybrid component. The authors are working towards this direction, by applying the proposed algorithm in various optimization problems and also hybridizing with other schemes.

The rest of the paper is organized as follows; in Sect. 2 the actual sonar mechanism is briefly presented, in Sect. 3 the algorithm is explained analytically, in Sect. 4 the experimental results are presented and explained and in Sect. 5 there are further research recommendations and conclusion.

2   The Actual Sonar Mechanism

The mechanism that provides inspiration in the proposed algorithm is the sonar that the Navy uses for war ships' exploration for submarines. The basic idea behind the sonar application was to send an ultrasound and based on the sound level that the radio receives the size of an object or an obstacle can be estimated. So the ship can identify the position of possible targets (Fig. 1).

A characteristic feature of SONAR is the cyclic scan of the area around the ship. To model this phenomenon, the concept of intensity of sound is implemented [19]. Initially, the Acoustic Power Output or Sound Power (P) has to be calculated:

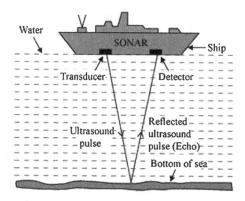

Fig. 1. Ship's SONAR (source: http://brightmags.com/how-does-sonar-work/)

$$P = \eta \cdot Pe \tag{2.1}$$

where, Pe is the Power Input and η is the Transducer efficiency, which is defined as the percentage of power output to power input. Then, the Intensity is calculated as the ratio between Sound Power (P) and the area scanned, as it is shown in Fig. 2:

$$I = \frac{P}{area} \tag{2.2}$$

where the area is calculated as:

$$area = 4 \cdot \pi \cdot r^2 \tag{2.3}$$

And r is the radius of the imaginary sphere around the ship that is scanned.

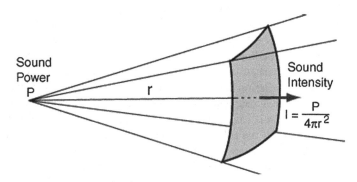

Fig. 2. Sound intensity depended upon sound power and area (source: http://hyperphysics.phy-astr.gsu.edu/hbase/Forces/isq.html)

As a result, someone can observe that the decrease of intensity I causes an increase of the effective radius r and thus, the area that is scanned. This relation is used also in the proposed scheme.

3 The Proposed SIO Algorithm

We consider each agent $X_i = \{x_1, x_2, x_3, \ldots, x_n\}$ to be a ship, where $i \in 1, 2, \ldots, N$ and N being the maximum number of agents, while n is the maximum number of dimensions of the problem. We decide at the beginning the number of ships for saving computational power. Although generally the more agents there are, the higher is the probability of finding the optimal solution, in the proposed algorithm this is not the case. As it can be seen in the next subsections, the multitude of generated points around each agent's work provides a wider search of the solution space, while the number of agents can remain the same. The strongest feature of the proposed scheme is the wider range of the solution area that is being searched, keeping the number of agents constant.

At start, we initialize the position of the agents somewhere in the solution space; the easiest way to do that is with random way via the normal distribution function, but this can be altered based on the values that every decision variable can take.

Using Eqs. (2.2) and (2.3), the initial radius and intensity for every agent is calculated. We set the Power Input as the fitness of each agent, and so we get:

$$Pe = fit_i, i \in \{1, 2, \ldots, N\} \tag{3.1}$$

On the other hand, Eq. (2.1) is reformed as:

$$P = e^{Pe} \tag{3.2}$$

in order to transform the fitness' value in positive numbers. This has to be done, because of the usage of logarithm for the rescale of intensity values. Logarithmic equations cannot take negative values, while fitness could be negative in some problems. Thus, we solve this difficulty with a transformation inspired by the physical analogue (i.e. the corresponding idea inspired from nature for the algorithm).

The next steps are repeated until the stopping criteria are met. In the experiments conducted, the stopping criterion is the maximum number of iterations, named "number of scans". For each ship we calculate the fitness function in order to find out the best solution. The best solution is saved and all agents change their intensity based on the solution they have found; if the solution is better from the previous best of the current agent, then the intensity increases and if the opposite exists, then the intensity decreases. That affects also the alteration of the effective radius.

Finally, one more useful mechanism is applied in our scheme. In reality, when a war ship doesn't detect anything in an area, it changes place. An easy way to relocate an agent is to take into consideration the position of the best solution found so far:

$$x_i^d = best^d + r_0^i \cdot rand \tag{3.3}$$

where x_i^d is the position of i-th agent in the d-th dimension, $best^d$ is the best position found in the current iteration, r_0^i is the effective radius of the i-th agent and $rand$ a random uniformly distributed number. However, this equation can be reformed to relocate an agent in areas that have not been checked before. Inspired by the similar concept of mutation rate [20]:

$$\mu_{opt} = \frac{1}{\tau} \tag{3.4}$$

where τ is the number of generations between environmental changes, we set the limit of time without improvement (or without environmental change) as the 1% of the number of iterations (scans). So, we get:

$$checkpoint = scans \cdot 0.01 \tag{3.5}$$

Sonar Inspired Optimization
_____

Initialization of ships' position
Initialize effective radius and intensity for every agent
While *stopping criteria not met*
    Update radius for every agent
    Calculate Intensity for every agent
    While full_scan = false
        Update the rotation angle in every dimension
        Calculate fitness of possible position
        Save the best so far for each agent in the current scan
    End
    Update best position and fitness
    Update intensity and acoustic power output for every agent
End
_____

Fig. 3. Pseudocode of the proposed Sonar Inspired Algorithm (SIO)

3.1 Intensity Parameter

The most important parameter in our algorithm is the intensity parameter. Intensity affects the change of effective radius and thus, the maximum size of area that each agent searches. Intensity is redefined at the end of each iteration based on the solution found by the corresponding agent. Using the exponential function's attribute:

$$I_i = I_i \cdot e^{magnitude} \tag{3.1.1}$$

Magnitude is a way to define the importance of the target found by an agent/ship and is calculated as:[1]

$$magnitude = scan\_best_i - best + s \qquad (3.1.2)$$

where $scan\_best_i$ is the fitness of the best solution found by the i-th agent in the current scan and $best$ is the globally best solution at the time. To avoid the zero value of magnitude that the agent with the global best solution will return, we add a very small value s.

The Eq. (3.1.1) is formed based on the graph of e^x. As it can be shown in Fig. 4, when the value of x for the e^x is below zero, the value of y is lower than one. Thus, if the magnitude is negative (meaning that the agent found better solution), then the intensity will be decreased because is multiplied with a number lower than the unity. On the other hand, if magnitude is positive, the intensity will increase because is multiplied with a number bigger than unity. And so, as further of the optimum an agent is so much bigger the increase of the intensity will be, resulting into bigger steps to find a better solution.

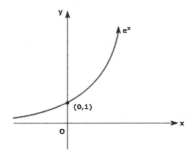

Fig. 4. Graph of y = e^x function

Although, to transform the high value of I_i to a more useful one, implementing the physical analogue:

$$I_i = 10 \cdot \log \frac{I_i}{I_0} \qquad (3.1.3)$$

where I_i is the Intensity of the i-th agent and I_0 is the Threshold of Hearing [19], which is defined as:

$$I_0 = 10^{-12} watts/m^2 = 10^{-16} watts/cm^2 \qquad (3.1.4)$$

In our experiments, the value of Threshold of Hearing I_0 is set to 10^{-12}, but can be changed based on the problem.

---

[1] Equation (3.1.2) is formed for minimization problems. In maximization problems, $scan\_best_i$ and $best$ reverse signs.

3.2   Effective Radius r_0

The initial value of r_0 should be considered, based on the solution space. A small value of the radius will drive the algorithm to perform smaller steps. On the other hand, the choice of a bigger radius will lead to longer jumps of the algorithm towards better optima, but with the risk of overlooking (bypassing) other solutions of desired quality.

By reversing Eq. (2.3), the effective radius r_0 is calculated as:

$$r_0 = \sqrt{\frac{area_i^k}{4 \cdot \pi}} \qquad (3.2.1)$$

where $area_i^k$ is the area scanned by the i-th agent in the k-th iteration.

As we see, this model represents the real relation between these measures; in higher intensity the area scanned is bigger than in lower intensity. Thus, the effective radius r_0 is smaller too. The aim here is to increase the radius, if no better solution is found, so that each agent searches further than its current position.

3.3   Full Scan Loop

In order to search wider areas of the solution space, in each iteration every agent checks the space around it that is limited by its effective radius r_0. This process is called full scan loop, because three steps are repeated until a full cyclic search has been done. Beginning from the angle of $0°$, random rotations in each dimension are executed. Each rotation covers a maximum of $a°$ and is calculated as follows:

$$angle^d = angle^d + rand \times a° \qquad (3.3.1)$$

where $rand$ is a random number produced from the uniform distribution function and $angle^d$ is the rotation angle in dimension d. When any of $angle^d$ exceeds $360°$, the loop is stopped. The vector of angles is converted in vector of movements in every dimension so that:

$$x_1^2 + x_2^2 + \ldots + x_n^d = r^2 \qquad (3.3.2)$$

where r is the random radius inside the cycle that is defined by the effective radius r_0 and n is the number of dimensions of the problem. In Fig. 5 below, an example of 36 points generated in one dimension for the Ackley's function is presented. A decrease of the maximum rotation angle $°$, leads into smaller rotations and thus, more generated points in every dimension. With this mechanism, each agent searches more points around its position, while other algorithms' agents check one point per iteration.

The new position is calculated as:

$$x_i^d = movement^d + x_i^d \qquad (3.3.3)$$

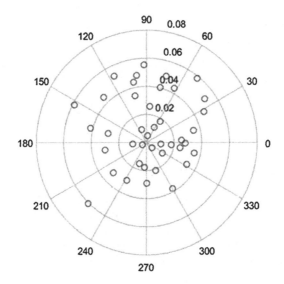

Fig. 5. Example of generated points checked in a single full scan loop by one agent for one dimension of Ackley's function

where, x_i^d is the position of the i-th agent in the d-th dimension and *movement*^{d} is the d-th element of the Eq. (3.3.2). In each one of the rotation phases, the fitness of the new position is calculated and if it is better than the best found by the current agent, the best position and its fitness are updated.

In order to avoid exceeding the bounds of the solution space, a correction mechanism has also been implemented. If an x_i^d is violating the bound constraints, it is relocated as:

$$x_i^d = lower\_bound^d + \left(upper\_bound^d - lower\_bound^d\right) \cdot \cos\left(x_i^d\right) \qquad (3.3.4)$$

in order to fulfil the relation: $lower\_bound^d < x_i^d < upper\_bound^d$.

4   Experimental Results

All experiments were conducted using Matlab on a 4 GB, 3.6 GHz Intel Core i7 Windows 10 Pro. In our experiments, we used the "hypersphere function", provided by Gianluca Dorini in Mathworks [21], to convert the angles vector into Cartesian coordinates vector. The maximum rotation angle a° was set to 20, the small value s was set to 0.0008 and the value of Threshold of Hearing I_0 was 10^{-12} for all runs.

In Table 1 is shown the performance of the algorithm on 12 known test functions with discrete solution space. The reason for selecting these functions as benchmarks is that the overwhelming majority of real world problems are described by decision variables that are defined within a discrete solution space. The parameters used for all test functions were:

- Number of iterations: 100
- Number of population: 100
- Number of independent runs: 100

Table 1. Results of SIO in benchmark test functions

Test function	Known minimum	Min	Average	Std
Ackley's function	$f_{min} = 0$	0.0003	0.0046	0.0023
Beale's function	$f_{min} = 0$	2.86e−11	1.34e−08	1.33e−08
Goldstein–Price function	$f_{min} = 3$	3.0000	3.0004	0.0004
Booth's function	$f_{min} = 0$	1.81e−10	1.69e−08	1.79e−08
Bukin function N. 6	$f_{min} = 0$	0.0237	0.1107	0.0558
Matyas function	$f_{min} = 0$	3.95e−12	3.55e−10	3.45e−10
Lévi function N. 13	$f_{min} = 0$	1.44e−11	3.31e−08	2.96e−08
Three-hump camel function	$f_{min} = 0$	2.35e−11	5.28e−09	5.53e−09
Easom function	$f_{min} = -1$	−1.0000	−0.0400	0.1969
Cross-in-tray function	$f_{min} = -2.06261$	−2.0626	−2.0609	0.0173
Schaffer function N. 2	$f_{min} = 0$	5.29e−12	0.0627	0.0417
Schaffer function N. 4	$f_{min} = 0.292579$	0.2926	0.3134	0.0152

In the rightmost columns of Table 1, the minimum, the average and standard deviation of all algorithm's runs are provided, while in the second column the known minimum of each function is provided.

As it can be seen in Table 1, the proposed algorithm is able to find a solution very close to the global minimum and in some cases succeeds in finding the minimum. In the majority of these functions, SIO has a small standard deviation of results. Also, in many cases it provides highly accurate solutions.

In Table 2 there is a comparison of the proposed algorithm with the Bat Algorithm (BA) [8], with which SIO shares a similar concept. For the experiments, the published BA code from Mathworks [22] was used and the alterations needed for each benchmark function and the parameters of loudness and pulse were made. Many optimization algorithms are presented and compared with other well-known schemes, but it is not clear if these schemes are altered in a way that affects their results, leading to the outperformance of the proposed algorithm. This is the reason why the version of the BA algorithm used for comparison, is mentioned. This version is tuned to provide better performance, so the parameters are not altered. What is more, the correction mechanism that is presented in Eq. (3.3.4) was added, in order to help BA remain in the solution space.

Finally, in Table 3 a convergence analysis on the previous results is presented. As it can be seen, the proposed algorithm is fast in finding optimum. On the other hand, the high value of standard deviation shows that this optimum may updated even in the last iterations (scans) due to the fact that SIO searches a large area of the solution space. The results of previous tables prove that statement.

Table 2. Results of SIO in test functions compared with BA

Test function	SIO			BA		
	Min	Average	Std	Min	Average	Std
Ackley's function	0.0003	0.0046	0.0023	0.0333	0.3785	0.2284
Beale's function	2.86e−11	1.34e−08	1.33e−08	3.94e−06	0.0005	0.0005
Goldstein–Price funct.	3.0000	3.0004	0.0004	3.0006	3.0719	0.0777
Booth's function	1.81e−10	1.69e−08	1.79e−08	1.31e−05	0.0007	0.0007
Bukin function N. 6	0.0237	0.1107	0.0558	0.4454	20.3861	9.4734
Matyas function	3.95e−12	3.55e−10	3.45e−10	1.24e−07	2.31e−05	2.18e−05
Lévi function N.13	1.44e−11	3.31e−08	2.96e 08	2.70e−05	0.0025	0.0026
Three-hump camel fun.	2.35e−11	5.28e−09	5.53e−09	6.45e−07	0.0003	0.0003
Easom function	−1.0000	−0.0400	0.1969	−1.0000	−0.8898	0.3144
Cross-in-tray function	−2.0626	−2.0609	0.0173	−2.0626	−2.0626	3.16e−05
Schaffer function N. 2	5.29e−12	0.0627	0.0417	1.20e−09	3.19e−05	0.0003
Schaffer function N. 4	0.2926	0.3134	0.0152	0.2926	0.2926	1.26e−05

Table 3. Convergence analysis of SIO in benchmark test functions compared with BA

Test function	Min iteration		Average iteration		Standard deviation	
	SIO	BA	SIO	BA	SIO	BA
Ackley's function	16	1	70.78	55.56	21.83	31.35
Beale's function	7	3	55.44	50.68	29.05	28.07
Goldstein–Price func.	13	4	65.41	53.30	22.58	27.61
Booth's function	11	2	55.98	56.40	23.91	27.2
Bukin function N. 6	12	2	69.38	2.32	21.69	0.62
Matyas function	5	6	51.39	59.84	28.57	27.26
Lévi function N. 13	5	3	57.75	51.44	26.54	28.96
Three-hump camel func.	4	2	51.74	50.74	26.23	29.25
Easom function	22	8	74.32	56.34	24.31	24.28
Cross-in-tray function	7	2	63.28	54.2	24.39	28.6
Schaffer function N. 2	19	14	88.03	62.97	19.60	22.39
Schaffer function N. 4	18	9	83.61	61.63	20.53	22.55
Average Performance	11.58	4.67	65.59	51.29	24.10	24.85

As mentioned above, the parameters of BA are chosen by its author. Increasing the population and the iterations would lead to new tuning of this algorithm to perform correctly. This will work in favor of the SIO, because the attraction-based concept of

BA will trap the algorithm in optima, while the scattering method of the proposed scheme avoids this problem.

The Bat Algorithm seems to be faster in locating optimum, but taking into consideration the above results as well as the fact that an already well-tuned algorithm has been used for comparison regardless of the problem coping with, SIO is a bit slower but performs a wider search and thus its performance is encouraging.

5 Conclusions and Future Research

In this paper, a new meta-heuristic algorithm named SIO (Sonar Inspired Optimization) has been proposed, based on the concept of SONAR mechanism used by war ships to detect targets and mines. The proposed scheme is easy to implement and the limited parameterization makes this algorithm useful for a wide range of problems. Another significant feature of SIO is the balance between exploration and exploitation, as the only parameter affecting the strictness of the algorithm is the intensity that self-adapts during the algorithmic process.

SIO was tested in known benchmark functions and was also compared with the Bat Algorithm, an approach which is similar to the proposed algorithm, where was found statistically comparable or superior in most of the cases. Also, a convergence analysis of the proposed algorithm has been provided. It can be seen that SIO follows the concept of evolutionary algorithms that keep evolving solutions, due to its good exploration and exploitation balance.

Furthermore, the main advantages should be highlighted; the minimal parameterization and the higher exploration of the solution space. Especially the second feature, SIO's agents search many possible positions around their current location in each iteration, while in other algorithms agents check only one new point. Resulting also in the main idea behind the introduction of Sonar Inspired Optimization, which is the increase of exploration but also ease of implementation. That will attract researchers to use this algorithm in opposition with another that is hard to implement or needs more parameterization.

Currently, work is underway on the application of Sonar Inspired Optimization in real world problems. The hybridization of this algorithm with other known algorithms in order to build effective hybrid schemes for specific problem areas is also a priority of research. Limited experiments have already taken place in this direction, in financial and industrial engineering problems. These experiments seem promising as the algorithm converges fast in the best (optimum or sub-optimum) solution found, while the concept of approaching optimum in steps, avoids the premature convergence that other competitive schemes face.

In addition, further improvement of the characteristics of SIO will be attempted, as well as proper parameter tuning of the rotation angles and the checkpoint value of the approach. Finally, the effect of Threshold of Hearing I_0 will be measured and studied.

Acknowledgments. The authors would like to thank Dr Jan Jantzen (Samso Energy Academy, Denmark and University of the Aegean, Chios, Greece), for his valuable comments and suggestions.

References

1. Yang, X.S.: Nature-Inspired Metaheuristic Algorithms. Luniver press, Bristol (2010)
2. Chiong, R. (ed.): Nature-Inspired Algorithms for Optimisation, vol. 193. Springer, Heidelberg (2009)
3. Liu, J., Tsui, K.C.: Toward nature-inspired computing. Commun. ACM **49**(10), 59–64 (2006)
4. Marrow, P.: Nature-inspired computing technology and applications. BT Technol. J. **18**(4), 13–23 (2000)
5. Yang, X.S.: Nature-inspired metaheuristic algorithms: success and new challenges. arXiv preprint arXiv:1211.6658 (2012)
6. Kennedy, J.F., Kennedy, J., Eberhart, R.C., Shi, Y.: Swarm Intelligence. Morgan Kaufmann, Burlington (2001)
7. Shah-Hosseini, H.: The intelligent water drops algorithm: a nature-inspired swarm-based optimization algorithm. Int. J. Bio-Inspired Comput. **1**(1–2), 71–79 (2009)
8. Nasir, A.N.K., Tokhi, M.O., Ghani, N.M.A., Raja Ismail, R.M.T.: Novel adaptive spiral dynamics algorithms for global optimization. In: 11th IEEE International Conference on Cybernetic Intelligent Systems (CIS), pp. 99–104. IEEE Press, Ireland, August 2012
9. Rashedi, E., Nezamabadi-Pour, H., Saryazdi, S.: GSA: a gravitational search algorithm. Inf. Sci. **179**(13), 2232–2248 (2009)
10. Kaveh, A., Talatahari, S.: A novel heuristic optimization method: charged system search. Acta Mech. **213**(3), 267–289 (2010)
11. Birbil, Şİ., Fang, S.C.: An electromagnetism-like mechanism for global optimization. J. Glob. Optim. **25**(3), 263–282 (2003)
12. Yang, X.S., Deb, S., Loomes, M., Karamanoglu, M.: A framework for self-tuning optimization algorithm. Neural Comput. Appl. **23**(7–8), 2051–2057 (2013)
13. Crawford, B., Valenzuela, C., Soto, R., Monfroy, E., Paredes, F.: Parameter tuning of metaheuristics using metaheuristics. Adv. Sci. Lett. **19**(12), 3556–3559 (2013)
14. Fallahi, M., Amiri, S., Yaghini, M.: A parameter tuning methodology for metaheuristics based on design of experiments. Int. J. Eng. Technol. Sci. **2**(6), 497–521 (2014)
15. Kaveh, A., Farhoudi, N.: A new optimization method: dolphin echolocation. Adv. Eng. Soft. **59**, 53–70 (2013)
16. Yang, X.S.: A new metaheuristic bat-inspired algorithm. In: González, J.R., Pelta, D.V., Cruz, C., Terrazas, G., Krasnogor, N. (eds.) Nature Inspired Cooperative Strategies for Optimization (NICSO 2010), pp. 65–74. Springer, Heidelberg (2010)
17. Vassiliadis, V., Dounias, G.: Nature-inspired intelligence: a review of selected methods and applications. Int. J. Artif. Intell. Tools **18**(04), 487–516 (2009)
18. Fister Jr., I., Yang, X.S., Fister, I., Brest, J., Fister, D.: A brief review of nature-inspired algorithms for optimization. arXiv preprint arXiv:1307.4186 (2013)
19. Lurton, X.: An Introduction to Underwater Acoustics: Principles and Applications. Springer Science & Business Media, Heidelberg (2002)
20. Nilsson, M., Snoad, N.: Optimal mutation rates in dynamic environments. Bull. Math. Biol. **64**(6), 1033–1043 (2002)
21. Mathworks File Exchange. https://www.mathworks.com/matlabcentral/fileexchange/5397-hypersphere
22. Mathworks File Exchange. https://www.mathworks.com/matlabcentral/fileexchange/37582-bat-algorithm–demo-/content/bat_algorithm.m

Data Preprocessing to Enhance Flow Forecasting in a Tropical River Basin

Jose Simmonds[1]([⊠]), Juan A. Gómez[2], and Agapito Ledezma[1]

[1] Departamento de Informática, Universidad Carlos III de Madrid,
Avenida de la Universidad 30, 28911 Leganes, Spain
jose.simmonds@alumnos.uc3m.es, ledezma@inf.uc3m.es
[2] Departamento de Biología Marina, Universidad de Panamá,
Estafeta Universitaria, Apartado 3366, Panama 4, Panama
juanay05@hotmail.com

Abstract. Missing hydrometric data is a critical issue for water resources management projects and problems related to flow damage and risk assessment. Though numerous ways can be found in the literature to impute them (i.e. Box-Jenkins models, Linear regression models, case deletion, listwise and pairwise deletion, etc.), not all will render effective on a given dataset. in tropical river basin, it's still needed to develop proven and simplified methods to deal with hydrometric data missingness and scarcity. This paper presents the analyses including an assessment of the condition of the existing hydrometric data and works related to the way in which the record was treated for flow forecasting purposes and the construction of the artificial neural network (ANN) models used for predicting the flows. The study was led based on 15-min rainfall, water surface elevation and discharge data, derived from the continuous real-time monitoring station located in the del Medio River Basin from the years 2012 to 2016. As a result, the proposed modeling approach followed two modeling methods, one employing the missing data record and the other was used a multiple imputation (MI) technique to impute the missing data and forecast flow for 1, 2 and 4 h ahead under each approach. The statistical metrics results for the two-modeling approaches, suggest the non-imputed data scenario to rule out the imputed data. This means it is recommended to further optimize the MI technique if to be used effectively to fill in the missing required days of measurements for estimating H3 gaps and afterwards to forecast the flow employing multilayer perceptron (MLP), artificial neural networks (ANNs) with 10-fold cross-validation.

Keywords: Artificial neural network · Computational intelligence · Data imputation · Discharge · River basin

1 Introduction

Hydrology is the study of the movement and storage of water at catchment scales. This process is very complex and is primarily driven by climatic, physiographic, topographic, soil characteristics, vegetation, geologic and land use practices. Water resources engineers usually rely on computational models to simulate the flow and

© Springer International Publishing AG 2017
G. Boracchi et al. (Eds.): EANN 2017, CCIS 744, pp. 429–440, 2017.
DOI: 10.1007/978-3-319-65172-9_36

conveyance of water for planning, designing and management of channels and in-stream structures, managing reservoirs, and conducting flood damage and risk analyses. Nevertheless, the simulation of fluid flow is a difficult and computationally intensive task, and simplifications are often necessary to reduce the assignment to one that is feasible regarding time management and data availability. Prone to significant high flooding, the Rio del Medio Basin (RDMB), also known as Rio Uvero sub-basin, is in the foot print of the Mina de Cobre Panamá (Fig. 1), owned by Minera Panama S.A-First Quantum Minerals Ltd (MPSA-FQML). The proposed Project is planning to mine and process copper sulfide ore in the Petaquilla Concession, Panamá. This concession covers an area of approximately 130 square kilometers (km^2) and is in the District of the Donoso region, Colón Province, in north-central Panamá. The concession contains at least three spatially distinct copper ore bodies (Colina, Botija and Valle Grande) and three conventional open pit mines are currently planned to exploit these ore bodies [1].

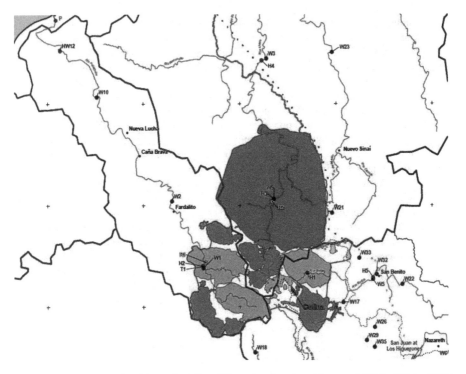

Fig. 1. Map showing the del Medio River H3 station location (Source: MPSA, ESIA, 2010)

To assess the impacts of flood prone areas, there are available several software packages for computing steady and unsteady flows; of which the most popular packages are listed in Table 1. These packages are usually operated through a graphical user interface and often have specific input and output data structuring requirements. These software packages are capable of simulating complex river networks of irregular

Table 1. Software packages for modeling steady and unsteady flow.

Package name	Propietary	License type	Open source	OS platform
FLOW3D [5]	Flow Science	Commercial	No	Windows
HEC-RAS [6]	USACE	Free	No	Windows
MIKE HYDRO RIVER [7]	DHI	Commercial	No	Windows
OPENFOAM [8]	The OpenFOAM Foundation	Free	Yes	Linux, Windows, Mac OSx
REEF3D [9]	REEF3D	Free	Yes	Linux, Windows, Mac OSx

channels and are therefore of great utility to water resources engineers and water practitioners working on large-scale projects, but they can render certain disadvantage in respect to accessibility in terms of cost, functionality and computer hardware requirements as well as additional data requirements (i.e. separate calibration data, bathymetry, wind, etc.).

Besides the software packages, another modeling approach which have being gaining popularity in water resources studies is Computational Intelligence (CI). For example, in [2] is illustrated the ability of artificial neural networks (ANN) to cope with missing data and to "learn" from the event currently being forecast in real time makes it a better choice than conventional lumped or semi distributed flood forecasting models. Researches in [3] did a comparative study of short-term rainfall prediction models for real time flood forecasting, established that the time series analysis technique based on ANN provides significant improvement in the flood forecasting accuracy in comparison to the use of simple rainfall prediction approaches. Another study in [4] in which they applied various artificial neural network (ANN) models for prediction and analysis of meteorological and oceanographic data, showed that ANN technique is very useful.

Due to relative short periods, missing data and inconsistencies of hydrometric record, to the knowledge of the authors, no studies have been reported in the literature that has investigated the accuracy of multilayer perceptron (MLP) neural networks model, enhanced with data imputation techniques (IM) and applied to missing hydrometric records of the del Medio river basin. Henceforth, in this experiment we present the use of data-driven modeling (DDM) techniques, mainly artificial neural networks (ANNs) with multilayer perceptron (MLP), cross-validation (CV) and multiple imputation (MI) to enhance the quality of the impaired available hydrometric records for the H3 station (del Medio/Uvero River) for the period 2012 to 2016 for flow forecasting. For the ease in perusing this manuscript it´s being organized in the following manner: Sect. 2 describes the background of the study area, data quality issues and handling and CI algorithms used to deal with the rest of the data problems; Sect. 3 shows the ANN setup task used for flow forecasting; Sect. 4 refers to conclusions and future work.

2 Background

2.1 Data Source and Study Area

The hydrometric data employed in this experiment was obtained from a real-time data acquisition system which monitors, logs and transmit 15 min and 1 h interval rainfall, river water level and discharge data during low and high flow periods from installed rain gauge, flow meter and surface water elevation radar. It should be noted that the 15 min interval data is more complete than the 1 h interval. The station also as capability for monitoring turbidity, temperature and conductivity data. The monitoring station (Station H3) as labelled in the environmental assessment study is located at the upper part of del Medio River in Panama (08° 52' 07.2"N and 80° 39' 57.1"W) and it is part of the Donoso region. Its actual hydrometric data record spans from March 30th, 2012 to December 3th, 2016. Del Medio river flow predominantly northward and combine, along with other tributaries to form the Caimito River about 7 km from the Caribbean coast. At present, there are major project components located at the upper sub-basin, including the mill, waste rock storage facility and the tailings management facility (TMF) and it is scheduled the construction of other future facilities. The drainage area of the H3 station is 15 km^2, accounting for approximately 30% of the total drainage area of the del Medio River basin. In Table 2 it is summarized the annual rainfall associated to each of the long-term regional climate stations in the Donoso region. The del Medio river station (H3), being relatively new is not mentioned in the table and therefore, it does not possess an extensive hydrometric data record. Consequently, based on this information, mean annual rainfall in the Donoso region varies from approximately 5,000 mm at the coast to 3,200 mm inland. Annual rainfall amounts associated with extreme wet and dry conditions are shown in Table 3.

Table 2. Mean annual rainfall in the Donoso Region, Panama.

Station	# Years of record	Approximate distance from coast [km]	Elevation [m]	Mean annual rainfall [mm]
Cocle del Norte	43	0	2	4,989
San Lucas	43	10	30	4,716
Boca de Toabré	43	20	30	4,413
Coclesito	33	30	60	3,171

Source: Minera Panamá S.A. Estudio de Impacto Ambiental (ESIA), 2010.

The del Medio river problem is centered on the short hydrometric record of the station, and the presence of too many missing instances in the record. The problems associated with the missing values and gaps are known to arise for many different reasons including but not limited to instrument damage that results in data gaps, data inconsistencies, incorrect logging of time stamp, duplicates and data lost. The field technician may also attribute the data gaps to several factors such as malfunctioning

Table 3. Extreme annual rainfall in the Donoso Region, Panama.

Return period	Cocle del Norte	Annual rainfall [mm]		
		San Lucas	Boca de Toabré	Coclesito
Number years of record	33	40	39	33
Highest recorded	8,836	6,715	6,239	5,195
Average	4,989	4,716	4,416	3,171
Lowest recorded	3,164	3,420	2,990	2,491

Source: Minera Panamá S.A. Estudio de Impacto Ambiental (ESIA), 2010.

and impairment of field equipment, or mishandling of observed records, natural phenomena (e.g., forest fires, earthquakes landslides, and severe floods), and to human induced factors such as vandalism, and theft. Therefore, these gaps and cutoffs in data availability, are triggers that lead to problems in planning and management of water resources schemes, design of water supply systems and design of hydraulic structures. Thereafter, we realize that the lack of consistent and complete hydrometric data can represent the loss of valuable and necessary information to carry out models of hydrological processes and phenomena in any stage of planning and construction of Hydraulic works and for the implementation of decision support systems for the prevention of floods risk assessment in highly vulnerable zones.

2.2 Neural Network Model

The neural networks algorithm employed in the study is the multilayer perceptron (MLP) and is part of a function within a library managed by the WEKA software package version 3.9.1 workbench and further descriptions are given in [10].

2.3 Models Performance Evaluation

To evaluate an intelligent system is, in fact, to assure that the system performs the intended task for which it was implemented. A formal evaluation of the implemented modeling system is normally accomplished with the test cases selected. The system's performance is compared against the performance criteria that were agreed upon at the end of the prototyping phase. The evaluation criteria habitually reveal the modeling system's limitations and weaknesses, so it is revised and optimized as necessary. In this sense, the performance of the neural networks models is evaluated using a diversity of standard statistical estimators. In our experiment, the models were evaluated using four estimators, correlation coefficient (R), mean absolute error (MAE), index of agreement (d) and root mean square error (RMSE). The R is a measure of the strength and direction of a linear relationship between two variables. MAE measures how close forecasts or predictions are to eventual outcomes, the d is a nondimentional and bounded measure with values closer to 1 indicating better agreement between the observed and simulated values. The RMSE has the same units as the measured and calculated data. Smaller values indicate better agreement between measured and calculated values.

2.4   Data Imputation

Missing data issues can be found in various disciplines like the social, behavioral, sciences, medical sciences and engineering. For years, researchers have relied on a variety of improvised techniques that attempt to amend the data by discarding incomplete instances or by filling in the missing gaps. Unfortunately, most of these techniques require a relatively strict assumption about the cause of missing data and are susceptible to significant bias. These methods have increasingly fallen out of favor in the methodological literature [11, 12] but they continue to enjoy widespread use in published research articles [13, 14]. Three types of missingness mechanisms are common [15]: missing completely at random (MCAR): when cases with missing values can be thought of as a random sample of all the cases; missing at random (MAR): when conditioned on all the data we have, any remaining missingness is completely random; that is, it does not depend on some missing variables. Missingness can be modelled using the observed data, missing not at random (MNAR): when the data seems not to be MCAR nor MAR. This is difficult to handle because it will require strong assumptions about the patterns of missingness. MI can handle all the three types of missing data situation. However, packages that do MI are usually not designed for MNAR case, as it is more complicated mechanism. In this study, we are assuming the data are not MAR, but are of missingness that depends on the unobserved predictors. Table 4 depicts the overall missing pattern of the data record prior to imputation, in which each row corresponds to a missing data pattern (1 = observed, 0 = missing). Rows and columns are sorted in increasing amounts of missing information. The last column and row contain row and column counts, respectively. The table shows that there are 147,086 out of 166,756 rows that are complete. There are 95 rows for which only RN15 is missing, 8,241 rows are eventually missing for the three variables, and there are 5,778 rows for which only RN15 is known. The variable with most missing values is for WSEL15.

Table 4. Missing data pattern for the H3 station hydrometric record.

No. of instances	RN15	Q15	WSEL15	Missing
147086	1	1	1	0
95	0	1	1	1
2977	1	1	0	1
2579	1	0	1	1
5778	1	0	0	2
8241	0	0	0	3
Total of missing	8336	16598	16996	41930

2.4.1   Imputations Results

Missing data was imputed with the Mice package from R statistical computing language [16]. The multiple imputations were done using three of the built-in univariate imputation methods; predictive mean matching (pmm), bayesian linear regression (norm) and linear regression, non-bayesian (norm.nob), a complete review of these functions can be found in [17]. The variables are labelled RN15, for 15-min rainfall, Q15, for the 15-min discharge and WSEL15 for the 15 min' water surface elevation. After selecting the Mice parameters and the three imputation methods for the imputing process, the original dataset with imputed instances resulting from the pmm imputation method (Fig. 2) is selected in accordance with methodology from the Mice package and is used along with the non-imputed dataset as complete inputs to the neural network for the forecasting purposes.

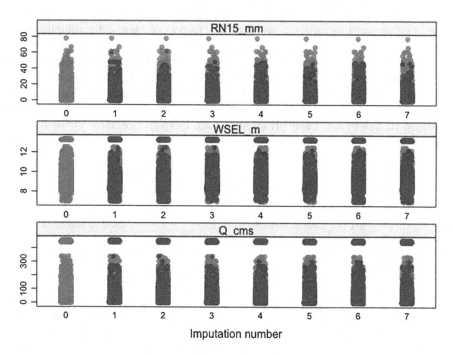

Fig. 2. A stripplot showing the distribution of the observed versus the imputed datasets with the pmm method for three variables. Observed data in blue, imputed data in red. (Color figure online)

By fitting a model to each of the imputed dataset and then pooling the results together, an adjusted R^2 of 0.873 was obtained. The summary of this model fit also showed the response variable discharge as influenced by the predictors rain and water surface elevation with $P(= 0.00)$. This explains a significant interaction between the two predictor variables exits; so, the relationship between the rain predictor and the response variable (discharge) depends on the level of the water surface elevation predictor. Therefore, our imputation model for predicting Q is $Q15t = -543.83 - 0.64RN15t + 65.89WSEL15t$.

3   Experimental Setup and Results

3.1   Setup for Non-imputed and Imputed Data with ANN

As mentioned before, the dataset used for the model setup and experimentation is composed of 166,756 instances of 15 min rainfall and discharge data. At this point after imputing all variables, we excluded the water surface elevation data, as our purpose was to clearly use the actual and past rain and discharge values in the forecasting of flows. So, we have created two datasets from the original; the first dataset is the complete data with missing values and the second dataset, containing imputed values; these were prepared in matrix form with their respective lags. After completing the process of data preparation and cleaning, the series was reduced to a total of 166,712 instances. From the correlation analysis done on the lagged rain series, it indicated that 6 to 7 instances back in time was enough and the autocorrelation function for the lagged discharge instances showed 4 to 5 time steps back was sufficient. No training or validation set were prepared, the two complete datasets are used in two different scenarios: (i) non-imputed and (ii) imputed data. From thence, three different inputs combinations were setup:

	Inputs	Out puts
(1)	R_t, R_{t-1}, R_{t-2}, R_{t-3}, R_{t-4}, R_{t-5}, R_{t-6}, R_{t-7}	Q_{t+1}, Q_{t+2}, and Q_{t+4}
	Q_t, Q_{t-1}, Q_{t-2}, Q_{t-3}, Q_{t-4}, Q_{t-5}	
(2)	R_t, R_{t-1}, R_{t-2}, R_{t-3}, R_{t-4}, R_{t-5}	Q_{t+1}, Q_{t+2}, and Q_{t+4}
	Q_t, Q_{t-1}, Q_{t-2}, Q_{t-3}	
(3)	R_t, R_{t-1}, R_{t-2}, R_{t-3}	Q_{t+1}, Q_{t+2}, and Q_{t+4}
	Q_t, Q_{t-1}	

These were feed as inputs including the previous month's rain and discharge to the neural network, MLP, with 10-fold cross validation, keeping the default parameters values in WEKA ANN suit unchanged. The output node was assigned for discharge at time steps 1, 2 and 4 h (t + 1, t + 2 and t + 4), ahead. An architecture of three models (as shown in Tables 5, and 6 respectively), were generated automatically by WEKA and the forecasting experiments for both data setups (non-imputed and imputed data) are run with these. The ANN models can be illustrated by the following expressions:

$$Q_{t+n} = ANN_{i,h,o}(Q_t, Q_{t-1}, \ldots, Q_{t-n}, R_t, R_{t-1}, \ldots, R_{t-n}) \tag{1}$$

where Q_{t+n} is the discharge $(m^3 \cdot s^{-1})$ to be forecasted for the respective hour ahead; R_t, R_{t-n} (n = 1, 2, 3 ... 7) are the current and pass rainfall; Q_t, Q_{t-n} (n = 1, 2 ... 5) are the current and pass discharge; $ANN_{i,h,o}$ represents the function of the *ANN* model with its respective input nodes, hidden node and output node.

3.2   Experimental Results

The purpose of this experiment was to analyze how good the present dataset was for forecasting the discharge at the del Medio river H3 station for 1, 2 and 4 h ahead, given the problems of missing data in the current record, analyze the effects of imputation of missing instances to enhance forecasting of the ANN. MLP, with 10-fold cross-validation was applied to the entire dataset under the two selected approaches (See Sect. 3.1). The results with each ANN, under both methodologies were compared with each other. In Fig. 3, there is an example of an output between the observed and 1 h ahead discharge forecasting with the ANN.

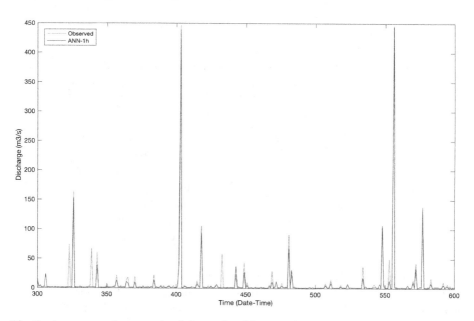

Fig. 3. A representative example of observed and forecasted 1 h ahead flow from time span 300 to 600 with the imputed dataset

3.3   Results for the Forecasted Non-imputed and Imputed Data with ANN

The ANN architectures generated by the WEKA software for each modeling scheme 1, 2 and 3 was for the discharge $(m^3 \cdot s^{-1})$ to predict the next hour ahead; in this sense, $ANN_{14,7,1}$ represents the ANN architecture with fourteen inputs, seven, nodes in one hidden layer and one output node; $ANN_{10,5,1}$ characterizes a network composed of ten inputs, five nodes in one hidden layer and one output node; and finally, $ANN_{6,3,1}$ is represented by six inputs, three nodes in one hidden layer and one output node respectively. Table 5 shows the performance of the ANN models for the non-imputed data approach and in Table 6 there are the results obtained with the imputed dataset. For the non-imputed data, all three models show acceptable results in terms of the performance metrics. The ranges are as fallows, for the R (= 0.77 to 0.96), MAE (1.68 to 5.61), d (0.86 to 0.98) and RMSE (8.85 $m^3 \cdot s^{-1}$ to 19.39 $m^3 \cdot s^{-1}$), in statistical

Table 5. Summary of ANN model's statistical performance for non-imputed data.

Scheme	ANN models	Prediction time steps	R	MAE	d	RMSE
1	$ANN_{14,7,1}$	Qt + 1	0.92	2.29	0.96	12.07
		Qt + 2	0.85	3.07	0.91	15.91
		Qt + 4	0.79	5.19	0.88	18.70
2	$ANN_{10,5,1}$	Qt + 1	0.96	1.68	0.98	8.85
		Qt + 2	0.84	3.41	0.90	16.53
		Qt + 4	0.79	5.11	0.88	18.57
3	$ANN_{6,3,1}$	Qt + 1	0.93	1.90	0.96	11.44
		Qt + 2	0.85	3.38	0.91	16.03
		Qt + 4	0.77	5.61	0.86	19.39

Table 6. Summary of ANN model's statistical performance for imputed data.

Scheme	ANN models	Prediction time steps	R	MAE	d	RMSE
1	$ANN_{14,7,1}$	Qt + 1	0.81	6.31	0.89	18.79
		Qt + 2	0.81	4.85	0.89	17.75
		Qt + 4	0.72	5.57	0.81	20.96
2	$ANN_{10,5,1}$	Qt + 1	0.84	5.26	0.91	16.60
		Qt + 2	0.77	6.53	0.87	19.69
		Qt + 4	0.70	6.13	0.80	21.49
3	$ANN_{6,3,1}$	Qt + 1	0.82	4.86	0.90	17.56
		Qt + 2	0.75	7.11	0.85	21.08
		Qt + 4	0.66	6.99	0.77	22.83

terms, it can be noted from Table 5 that the overall ANN models for the non-imputed data performs better than the imputed data approach, especially for forecasting the 1 and 2-hour lead times. The R although below 80% under the three schemes, for the 4 h lead time forecast, outperforms that which was reported for the imputed data approach. Besides, the MAE and RMSE measures disagree with the univariate statistics on the potentially better forecasting performance with non-imputed dataset having both a lower MAE and RMSE. The final index of agreement measure, d, also indicates that the non-imputed case is a bit better model approach to handle better the missing data instances. The $ANN_{10,5,1}$ model for lead time 1 h, showed the best forecasting results.

As shown in Table 6, the strength and direction of the linear relationship between the observed and ANN discharge forecast ranged 0.66 to 0.84. These values are acceptable except for the $ANN_{6,3,1}$ under scheme 3 whose R value was less than 70% at predicting the four-hour ahead discharge lead time. For the other statistical estimators, MAE, ranged from 4.85 to 7.11, the d from 0.77 to 91 and the RMSE values ranged from 16.60 to 22.83 $m^3 \cdot s^{-1}$. In contrast with the non-imputed scenario, both MAE and RMSE values for the imputed data scenario, presented a percentage difference of (65 to 21% and 46 to 15% respectively) larger.

The lower performance observed for the ANN with the imputed dataset with respect to the non-imputed data, can habitually be attributed to introduced uncertainties from the imputations in to the input data. However, in either cases the values of the performance measures employed to compare both modeling scenarios, suggest the non-imputed data set-up to rule out the imputed data scheme. The three ANN models under the non-imputed scheme shows to perform better than when applied to the imputed data for predicting the discharge values at 1, 2 and 4 h ahead. Table 6 also shows the $ANN_{6,3,1}$ model for the imputed data not to perform well at predicting a 4 h discharge lead time.

4 Conclusions and Future Work

The rainfall-runoff processes are complex systems, as it is characterized by various forcing function, as well as other phenomena and can vary significantly given the rainfall regime which is unique to a given region, specifically the tropics.

As water resources fluctuations are becoming severely influenced by climate change, the occurrence of hydrological disasters is continuing to cause sufferings and severe losses to economic, livestock's and the civilian population. For these reasons, it is necessary to propose effective forecasting methods to predict flows under different circumstances, especially when data is scarce, possesses impairments or is unavailable.

To predict water flows, it is important to have feasible and consistent hydrological data records, otherwise, the task to forecast might be hindered by the availability of hydrometric data that are impaired, as they can possess missing instances and large inconsistencies in respect to date-time of recording and events not responsive of accurate precipitations recordings and subsequent flood flows peaks. So as this was the case of del Medio river station H3, we concluded that accurate estimation of missing hydrometric data records would be an essential component to aid in flood forecasting, as this is vital to decision support system for efficient water resources management and future planning of water borne systems.

The objective of the paper was to investigate the performance of artificial neural networks (ANN) with MLP and 10-fold cross-validation in predicting the flow at 1, 2 and 4 h ahead, with a dataset with missing instances, in contrast to employing the same data record but with estimated missing values using multiple imputation techniques. The rain and flow data of the H3 station was used as inputs to the ANN to estimate flow under three distinct models as was pointed out in Sect. 3. The potential of the ANNs for estimating the flow has demonstrated in this study with the MLP applied to two distinct data scenarios (the dataset with missing instances and the same data with imputed values) showed the scenario with the non-imputed data to rule out that of the imputed data scenario. For the ANN using the non-imputed dataset, all three input combinations as was shown by the correlation coefficient (values of R = 0.77 to 0.96), outperformed the ANN models feed with the imputed dataset. For the imputed dataset, the worst fit was for model three for estimating the flow for 4 h ahead lead time, it didn't do well in terms of the value of the correlation coefficient (R = 0.66). In general, this work indicates that accurate estimations of missing hydrometric records under the complex hydrological condition of del Medio hydrometric station are necessary for obtaining good estimates of the missing observed flows versus the forecasted flow using MLP.

The study also suggests that in order to enhance hydrometric records with high data missingness, with multiple imputation techniques, it is encouraged, the technique to be optimized further to avoid the generation of unbiased data and uncertainties introduced and the use of robust imputation schemes to replace missing instances prior to performing any respective forecasting of flow events.

Some aspects still remain that require further studies and applications, such like the imputation methods applied, types of neural network schemes and topologies. There are other robust imputation techniques that could be used to estimate the missing records, and other data processing methodologies, like wavelet transform can be used to filter out noise and inconsistencies prior to data imputation.

Acknowledgement. The authors of this experiment will like to express their appreciation to Minera Panama S.A., Environmental Department for providing the necessary data. This work has been partially supported by the Spanish MICINN under projects: TRA2015–63708-R, and TRA2016-78886-C3-1-R.

References

1. Minera Panama, Environmental Impact Assessment Study (2010)
2. Christian, W.D., Robert, W.: An artificial neural network approach to rainfall runoff modeling. Hydrol. Sci. J. **43**(1), 47–66 (1998)
3. Toth, E., Brath, A., Montanari, A.: Comparison of short-term rainfall prediction models for real-time flood forecasting. J. Hydrol. **239**, 132–147 (2000). Elsevier
4. Hsieh, W.H., Tang, B.: Applying neural network models to prediction and analysis in meteorology and oceanography. Bull. Amer. Met. Soc. **79**, 855–1870 (1998)
5. FLOW3D Homepage. https://www.flow3d.com/. Accessed 18 Apr 2017
6. HEC-RAS Homepage. http://www.hec.usace.army.mil/software/hec-ras/downloads.aspx. Accessed 18 Apr 2017
7. MIKE HYDRO RIVER Homepage. https://www.mikepoweredbydhi.com/products/mike-hydro-river. Accessed 18 Apr 2017
8. OPENFOAM Homepage. http://www.openfoam.com/. Accessed 18 Apr 2017
9. REEF3D Homepage. https://reef3d.wordpress.com/. Accessed 18 Apr 2017
10. Frank, E., Hall, M.A., Witten, I.H.: The WEKA workbench. In: Data Mining: Practical Machine Learning Tools and Techniques, 4th edn. Morgan Kaufmann, Burlington (2016). (Online Appendix)
11. Little, R.J.A., Rubin, D.B.: Statistical Analysis with Missing Data, 2nd edn. Wiley, Hoboken (2002)
12. Wilkinson, L., Task Force on Statistical Inference. Statistical methods in psychology journals: guidelines and explanations. Am. Psychol. **54**, 594–604 (1999)
13. Bodner, T.E.: Missing data: prevalence and reporting practices. Psychol. Rep. **99**, 675–680 (2006)
14. Peugh, J.L., Enders, C.K.: Missing data in educational research: a review of reporting practices and suggestions for improvement. Rev. Educ. Res. **74**, 525–556 (2004)
15. Rubin, D.B.: Inference and missing data. Biometrika **63**(3), 581–592 (1976)
16. R Core Team. R: a language and environment for statistical computing. R Foundation for Statistical Computing, Vienna, Austria (2017). https://www.R-project.org/
17. van Buuren, S., Groothuis-Oudshoorn, K.: Mice: multivariate Imputation by chained equations in R. J. Stat. Softw. **45**(3) (2011). doi:10.18637/jss.v045.i03

Information Feature Selection: Using Local Attribute Selections to Represent Connected Distributions in Complex Datasets

Ioannis M. Stephanakis[1(✉)], Theodoros Iliou[2],
and George Anastassopoulos[2]

[1] Hellenic Telecommunication Organization S.A. (OTE),
99 Kifissias Avenue, 151 24 Athens, Greece
stephan@ote.gr
[2] Medical Informatics Lab, Democritus University of Thrace,
681 00 Alexandoupolis, Greece
{tiliou,anasta}@med.duth.gr

Abstract. Clustering algorithms like k-means, BIRCH, CLARANS and DBSCAN are designed to be scalable and they are developed to discover clusters in the full dimensional space of a database. Nevertheless their characteristics depend upon the size of the database. A DB/data warehouse may store terabytes of data. Complex data analysis (mining) may take a very long time to run on the complex dataset. One has to obtain a reduced representation of the dataset that is much smaller in volume - but yet produces the same or almost the same analytical results - in order to accelerate information processing. Reduced representations yield simplified models that are easier to interpret, avoid the curse of dimensionality and enhance generalization by reducing overfitting. Data reduction methods include data cube aggregation, attribute subset selection, fitting data into models, dimensionality reduction, hierarchies as well as other approaches. Feature selection is considered as a specific case of a more general paradigm which is called Structure Learning in cases of an outcome associated to a set of attributes. Feature selection aims at selecting a minimum set of features such that the probability distribution of different classes given the values of those features is as close as possible to the original distribution given the values of all features. A combined approach based upon representing complex datasets in DB as a minimal set of connected attribute sets of reduced dimensions is herein proposed. Value-Difference (VD) Metrics based upon binary, categorical and continuous values are used for subspace clustering. Each cluster can be represented by a different set of object features/attributes maximizing the information which is rendered by the cluster representation. Numerical data regarding a test-bed system for anomaly detection are provided in order to illustrate the aforementioned approach.

Keywords: Mutual information-based feature selection · Dimensionality reduction · Data mining

© Springer International Publishing AG 2017
G. Boracchi et al. (Eds.): EANN 2017, CCIS 744, pp. 441–450, 2017.
DOI: 10.1007/978-3-319-65172-9_37

1 Introduction

Data clustering concerns how to group a set of objects based on their similarity attributes and their proximity in the universal space of an information system. Main methods are classified as partitioning approaches (like k-means), hierarchical approaches (like BIRCH - Balanced Iterative Reducing and Clustering using Hierarchies [1], ROCK [2] and others) as well as Density-based approaches (like DBSCAN [3] and others). A good clustering method will produce high quality clusters with high intra-class similarity and low inter-class similarity. Feature selection is used to choose a suitable subset of relevant features for effective classification of data [4, 5]. The performance of a classifier often depends on the feature subset used for classification in the cases of high dimensional data classification. Binary, categorical as well as scalar continuous attributes of DB objects are considered as separate features in the context of an anomaly detection system. Thus classification space is a subspace of the entire universal space of the information system which is characterized by compound Value-Difference (VD) metrics (as defined in Subsect. 2.2). Feature selection methods use several variations of mutual information measures in order to find an optimal subset of features that minimize redundancy and - at the same time - maximize the relevance among them. The novelty of the proposed approach consists of its piecewise analysis of compact clusters (which are obtained by one of the aforementioned methods) in order to increase overall Shannon's mutual information-entropy.

An expert system is formally considered as an information system, which can be written as a quadruple $IS = (U, A, V, f)$, where: U is a non-empty finite set of objects, called a universe, A is a non-empty finite set of features, V is a union of feature domains such that $V = \cup_{a \in A} V_a$ where V_a denotes the value domain of feature $a, f : U \times A \to V$ is an information function such that $f(x, a) \in V_a$ for every $a \in A$ and $x \in U. f(x, a)$ can be defined upon scalar, vectorial or binary-word attributes. One may split set A of features into two subsets $C \subset A$ and $D = A - C$, i.e. conditional set of features and decision (or class) features respectively. The conditional features represent measured features of the objects, while the decision features are a posteriori outcome of classification. In the sequel of this paper, we try to determine a dataset of objects $\{x_1, x_2, x_3 \ldots \ldots\} \in U$ - along with corresponding subspaces $\{B(x_1), B(x_2), B(x_3) \ldots \ldots\} \subseteq C$ - that provide a reduced representation of a connected dataset. Attributes of $B(x_j)$ may differ in number and modality, i.e. $\dim(B(x_i)) \leq \dim(U)$ may be different from $\dim(B(x_j)) \leq \dim(U)$. It is assumed that the corresponding subspaces are equi-probable, i.e. $\Pr\{B(x_1)\} = \Pr\{B(x_2)\} = \{\Pr B(x_3)\} \ldots \ldots$

2 Information Based Feature Selection

2.1 Background and Definitions

The following algorithmic steps are common in several subspace clustering algorithms:

- Identification of subspaces that contain clusters.
- Identification of clusters

– Generation of minimal description for the clusters

Evaluation metrics distinguish between the following three main categories of feature selection algorithms for minimal description of clusters, i.e. wrappers, filters and embedded methods [4, 5]. Wrapper methods use a predictive model to score feature subsets; filter methods use a proxy measure instead of the error rate to score a feature subset (like for example mutual information and point-wise mutual information) whereas embedded methods are a catch-all group of techniques which perform feature selection as part of the model construction process. Shannon's mutual information is a symmetric measure of the dependence between two random variables X and Y according to the following formal definition,

$$I(X;Y) = H(X) + H(Y) - H(X,Y), \tag{1}$$

where $H(x)$ is the entropy of X. Feature selection methods that use mutual information, correlation, or distance/similarity scores to select features aim is to penalize a feature's relevancy by its redundancy in the presence of the other selected features. Data density is important as well. There are no dense 2-D units in Fig. 1. There are however 1-D dense units should the data be projected on the salary dimension. Two 1-D clusters are formed in such a case. There are no clusters in the age subspace.

Minimum-redundancy-maximum-relevance (mRMR) feature selection is based upon the general concept of mutual information [6, 7]. The relevance of a feature set S for the class c is defined by the average value of all mutual information values between the individual feature f_i and the class c as follows:

$$D(S,c) = \frac{1}{|S|} \sum_{f_i \in S} I(f_i;c), \tag{2}$$

The redundancy of all features in the set S is the average value of all mutual information values between the feature f_i and the feature f_j:

$$R(S) = \frac{1}{|S|^2} \sum_{f_i, f_j \in S} I(f_i;f_j), \tag{3}$$

The mRMR criterion is a combination of two measures given above and is defined as follows:

$$mRMR = \max_{S} \left[\frac{1}{|S|} \sum_{f_i \in S} I(f_i;Y) - \frac{1}{|S|^2} \sum_{f_i, f_j \in S} I(f_i;f_j) \right] \tag{4}$$

Suppose that there are n full-set features. Let x_i be the set membership indicator function for feature f_i, so that $x_i = 1$ indicates presence and $x_i = 0$ indicates absence of the feature f_i in the globally optimal feature set. Let $c_i = I(f_i;c)$ and $a_{ij} = I(f_i;f_j)$. The above relationship may then be written as an optimization problem:

$$mRMR = \max_{x \in \{0,1\}^n} \left[\frac{\sum_{i=1}^{n} c_i x_i}{\sum_{i=1}^{n} x_i} - \frac{\sum_{i,j=1}^{n} \alpha_{ij} x_i x_j}{\left(\sum_{i=1}^{n} x_i\right)^2} \right] \quad (5)$$

Computation of the mutual information between all candidate feature subsets by a greedy selection algorithm is intensive. Proposed variants attempt to overcome this obstacle:

– Mutual Information-Based Feature Selection (MIFS) which uses supervised neural networks [8]
– MIFS-U [9]
– a parameter-free method (referred to as Maximum-Relevance Minimum-Redundancy criterion - MRMR)
– improved version of MRMR, called NMIFS, which is dividing the normalized feature-feature mutual information to achieve a balance between the relevance and the redundancy term

A quite simple solution is to look iteratively for an attribute whose difference between its relevance and its redundancy with already selected attributes is maximal. If Y is the output to predict, A the set of indices of all features and S the set of indices of already selected features, a possible approach is to give each feature f_j, $j \in A \setminus S$ the score [10]

$$Score(f_j) = I(f_j; Y) - \frac{1}{|S|} \sum_{s \in S} I(f_j; f_s). \quad (6)$$

Whenever S is incremented by an additional feature, scores are reevaluated. Another approach is CLIQUE (CLustering In QUEst) [11] which is based upon data density and finds automatically subspaces with high density clusters in high dimensional DB. It produces identical results irrespective of the order in which the input records are presented and it does not presumed any canonical distribution for input data. It generates cluster descriptors in the form of DNF (Disjunctive Normal Form) expressions.

2.2 Subspace Clustering and Value-Difference (VD) Metrics

Contrary to hierarchical and density based approaches, clustering using neural networks and SOFMs requires value difference metrics (VDM). The notion was introduced by Stanfill and Waltz [12] in order to provide an appropriate distance function D (x, y) on nominal attributes. A simplified version (without the weighting schemes) of the VDM is defined as follows: $VDM = \sum_{f \in A} d_f(x_f, y_f)$ where A is the set of all features in the problem domain, and x and y are any two objects between which distance is calculated. For any feature $f \in A$, $d_f(x_f, y_f)$ is defined as the distance between the probability density of object x on feature f at x_f, $P(x_f)$, and the probability density of object y on feature f at y_f, $P(y_f)$. One dimensional, two dimensional as well as multi-dimensional histograms are used in the sequel of this paper defined upon the selected set of features. The number of samples (DB objects) within an 1-D interval, a 2-D area or multidimensional hyperbox defined upon the possible values of selected features are the outputs of such histograms. One may use entropy or information as defined in Eq. 1, the Canberra distance or the Jaccard distance in order to quantify similarity between different distributions. 1-D histograms over age and salary in Fig. 1 may be used as examples. 1-D histograms over age are similar for both clusters and may not be used to distinguish between them as the 1-D histograms over salary. On the other hand 2-D histograms are indicative of noise and yield no additional information as compared with 1-D histograms over salary. The Canberra distance between classes c' and c for a selected set of features (attributes) is defined as

$$d_{Canberra} = \frac{1}{\# \, non \, zero \, bins} \sum_{l=1...L} \frac{\left| h^{c'}_{set \, of \, features}(bin_l) - h^c_{set \, of \, features}(bin_l) \right|}{\left| h^{c'}_{set \, of \, features}(bin_l) + h^c_{set \, of \, features}(bin_l) \right|}. \qquad (7)$$

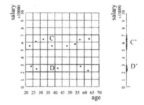

Fig. 1. Identification of clusters in subspaces projections of the original data space

The Jaccard distance measures the dissimilarity between sample sets. It is obtained by dividing the difference of the sizes of the union and the intersection of two sets by the size of the union,

$$d_J(S, S') = 1 - J(S, S') = \frac{|S \cup S'| - |S \cap S'|}{|S \cup S'|}. \qquad (8)$$

It can be used for categorical and non-scalar attributes.

3   The Proposed Algorithm

The proposed algorithm requires preprocessing in order to determine local sections of the universal space that are equi-probable, i.e. they feature roughly the same number of samples. One may use neural networks, SOFM or coarse hierarchical clustering (BIRCH) in order to accomplish such a task. Local dimensionalities are then estimated trying to maximize entropy/information or mRMR for the same number of average features/attributes. Let one assume the joint information (as defined in Eq. 1) for a set of optimal features $(a_1^*, a_2^* \cdots)$ for each cluster corresponding to the equi-probable sections:

I - Order features/attributes using data density or the Minimum-redundancy-maximum relevance criterion locally

- Find optimal $I(a_j;Y)$ $\forall a_j$ within the same section
- Find optimal
  $I(a_1^*,a_j;Y)$ $\forall a_j$ within the same section excluding a_1^*
- Find optimal
  $I(a_1^*,a_2^*,a_j;Y)$ $\forall a_j$ within the same section excluding a_1^*,a_2^* ......
  - Construct the entropy/information diagram over ordered features attributes for the same section

II - Find slopes for entropy/information over ordered features for all sections

- $I(a_1^*,a_2^*;Y)-I(a_1^*;Y)$ ......

III- Threshold slopes starting from higher to lower values and assign features/attributes locally while maintaining the same average dimensionality budget

4   Numerical Simulations

The proposed approach is evaluated against experimental data (network traces) obtained from a controlled testbed resembling a cloud environment. Network attacks are emulated during virtual machine (VM) migration [13]. The testbed allows the traces to be labelled with ground truth, about both the expected anomalies and the presence of a migration. It consists of two hosts, which serve as compute nodes running multiple VMs. Another host acts as a controller which initiates migrations and generates background traffic. A fourth host generates attack traffic. Each physical node runs KVM as virtualization infrastructure, and QEM provides hardware emulation. Migration is achieved with *libvirt*. Traces obtained at the virtual bridges are fed into detector to observe its reactions to normal/anomalous traffic. Network traces are split into 1-second bins, and, then, a set of statistical properties (features) of the traffic in each bin is computed (Table 1). Background traffic is created by running several HTTP servers and several clients repeatedly requesting dynamically created documents of varying size[1]. The training set consists of a total of 1,400 samples. 1-D histograms of feature values are illustrated in Fig. 2. Ordering of features using entropies of data densities as well as Minimum-redundancy-maximum-relevance (Eq. 4) for the global data space yields the diagrams of Fig. 3 (for network traces corresponding to normal operation). The same ranking of attributes is obtained for both ordering approaches (see Global mRMR of Table 2).

---

[1] One may see www.seccrit.eu//publications/presentatons for more details.

Table 1. Feature description

Feature #	Feature description
Feature 1	Number of packets
Feature 2	Number of bytes
Feature 3	Number of active flows in time stamp
Feature 4	Entropy of source IP address distribution
Feature 5	Entropy of destination IP address distribution
Feature 6	Entropy of source port distribution
Feature 7	Entropy of destination port distribution
Feature 8	Entropy of packet size distribution

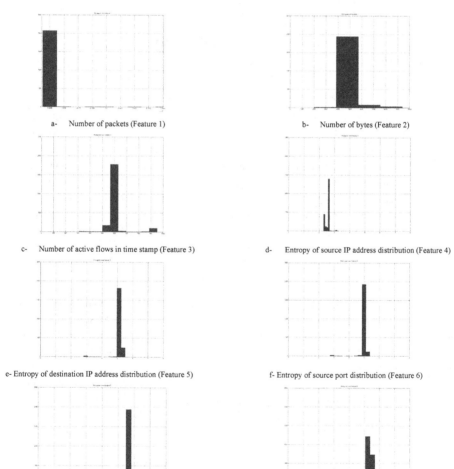

a-   Number of packets (Feature 1)

b-   Number of bytes (Feature 2)

c-   Number of active flows in time stamp (Feature 3)

d-   Entropy of source IP address distribution (Feature 4)

e- Entropy of destination IP address distribution (Feature 5)

f- Entropy of source port distribution (Feature 6)

g- Entropy of destination port distribution (Feature 7)

h- Entropy of packet size distribution (Feature 8)

Fig. 2. 1-D histograms over the entire space

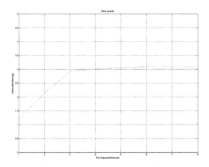 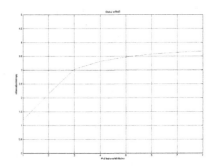

a - Entropies/information based on data densities as obtained from 1-D, 2-D, 3-D … histogrames

b - Minimum-redundancy-maximum-relevance

Fig. 3. Entropies and Minimum-redundancy-maximum-relevance (Eq. 4) over subspace dimensionality (global approach)

Table 2. Feature ordering according to mRMR

Order	Global mRMR	Local mRMR (Cluster 1)	Local mRMR (Cluster 2)	Local mRMR (Cluster 3)	Local mRMR (Cluster 4)
1	Feature 3	Feature 4	Feature 3	Feature 3	Feature 8
2	Feature 4	Feature 8	Feature 4	Feature 4	Feature 4
3	Feature 8	Feature 3	Feature 2	Feature 8	Feature 3
4	Feature 2	Feature 2	Feature 1	Feature 2	Feature 7
5	Feature 7	Feature 7	Feature 8	Feature 6	Feature 1
6	Feature 6	Feature 5	Feature 7	Feature 7	Feature 2
7	Feature 1	Feature 6	Feature 6	Feature 5	Feature 5
8	Feature 5	Feature 1	Feature 5	Feature 1	Feature 6

The proposed approach is applied by splitting the original distribution to four (4) equi-probable cluster partitions (which are estimated using a 2-D SOFM). One may use fewer or more cluster partitions without loss of generality depending upon such ad hoc criteria as data homogeneity. A family of curves (like those illustrated in Fig. 5) is obtained for different cluster partitions. Features are ordered locally according to the mRMR criterion (see Fig. 4). Different orders of features are obtained as illustrated in Table 2. Since data clusters are equi-probable one may prefer higher slopes for the same dimensionality of local subspaces. The proposed approach sets a threshold value for the entropy slopes and selects local features that correspond to entropy slopes above the selected threshold. Thus local dimensionalities may differ. The average dimensionality is calculated by weighting local dimensionalities with cluster probabilities. The threshold for the entropy slopes is gradually lowered in order to obtain the diagrams in Figs. 3, 4 and 5. Figure 5 is provided for illustrative purposes. It compares entropies obtained according to the mRMR criterion using global as well as local representations of the specific distribution. Describing the distribution of normal operation traces by subspaces of varying local dimensionality clearly yields more information with respect to data attributes.

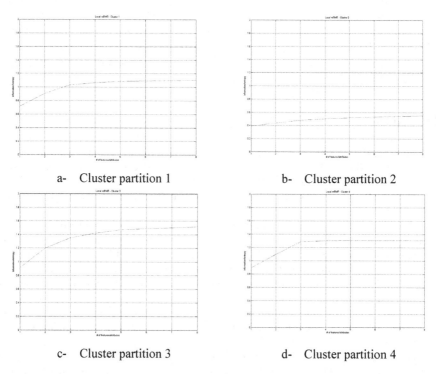

a-   Cluster partition 1   b-   Cluster partition 2

c-   Cluster partition 3   d-   Cluster partition 4

Fig. 4. Minimum-redundancy-maximum-relevance (Eq. 4) over subspace dimensionality (local approach)

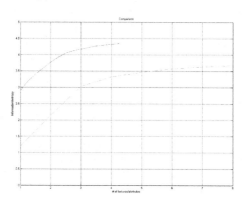

Fig. 5. Minimum-redundancy-maximum-relevance (Eq. 4) over subspace dimensionality comparison (local approach - solid line - and global approach dashed line)

5 Conclusion

Global as well local representations of data distributions are compared in the context of this paper in order to estimate efficient representations of data for fixed average dimensionalities. Using local dependent dimensionalities to describe distributions of DB objects is better since it provides more information. The aforementioned algorithm is directly extended to allow for more efficient classification and mining in the case of multiple distributions of data.

References

1. Zhang, T., Ramakrishnan, R., Livny, M.: BIRCH: an efficient data clustering method for very large databeses. In: Proceedings of the ACM SIGMOD Conference on Management of Data, Montreal, Canada, June 1996
2. Guha, S., Rastog, R., Shim, K.: ROCK: a robust clustering algorithm for categorical attributes. In: Proceedings ICDE 1999, pp. 512–521 (1999)
3. Ester, M., Kriegel, H.-P., Sander, J., Xu, X.: A density-based algorithm for discovering clusters in large spatial databases with noise. In: Proceedings of the 2nd International Conference on Knowledge Discovery in Databases and Data Mining, Portland, Oregon, August 1996
4. Yu, L., Liu, H.: Feature selection for high-dimensional data: a fast correlation based filter solution. In: Proceedings of the 20th International Conference on Machine Learning, pp. 56–63 (2003)
5. Dash, M., Choi, K., Scheuermann, P., Liu, H.: Feature selection for clustering - a filter solution. In: Proceedings of the Second International Conference on Data Mining, pp. 115–122 (2002)
6. Auffarth, B., López, M., Cerquides, J.: Comparison of redundancy and relevance measures for feature selection in tissue classification of CT images. In: Perner, P. (ed.) ICDM 2010. LNCS, vol. 6171, pp. 248–262. Springer, Heidelberg (2010). doi:10.1007/978-3-642-14400-4_20
7. Peng, H., Long, F., Ding, C.: Feature selection based on mutual information: criteria of max-relevance and min-redundancy. IEEE Trans. Pattern Anal. Mach. Intell. **27**, 1226–1238 (2005)
8. Battiti, R.: Using mutual information for selecting features in supervised neural net learning. IEEE Trans. Neural Netw. **5**, 537–550 (1994)
9. Kwak, N., Choi, C.H.: Input feature selection for classification problems. IEEE Trans. Neural Netw. **13**(1), 143–159 (2002)
10. Doquire, G., Verleysen, M.: Mutual information based feature selection for mixed data. In: ESANN 2011 Proceedings, European Symposium on Artificial Neural Networks, Computational Intelligence and Machine Learning, Bruges, Belgium, 27–29 April 2011. ISBN 978-2-87419-044-5
11. Agrawal, R., Gehrke, J., Gunopulos, D., Raghavan, P.: Automatic subspace clustering of high dimensional data for data mining applications. In: Proceedings of the 1998 ACM SIGMOD International Conference on Management of Data, SIGMOD 1998, Seattle, Washington, USA, 1–4 June 1998, pp. 94–105 (1998)
12. Stanfill, C., Waltz, B.: Towards memory based reasoning. Commun. ACM **29**, 1213–1228 (1986)
13. Shirazi, N., Simpson, S., Oechsner, S., Mauthe, A., Hutchison, D.: A framework for resilience management in the cloud. Electrotech. Informationstechnik **132**(2), 122–132 (2015). doi:10.1007/s005002-015-0290-9

Optimization of Freight Transportation Brokerage Using Agents and Constraints

Amelia Bădică[1], Costin Bădică[1(✉)], Florin Leon[2], and Daniela Dănciulescu[1]

[1] University of Craiova, Craiova, Romania
costin.badica@software.ucv.ro
[2] Technical University "Gheorghe Asachi" of Iaşi, Iaşi, Romania

Abstract. In this paper we address the problem of declarative modeling of freight transportation brokering using agents and constraints. Our model can be used for the optimization of vehicle assignments to customer orders that request the transportation of freight from source to destination points. The advantage is that a single vehicle can serve multiple customer orders on its multi-hop route that is part of a solution schedule. Our model is mapped to the ECLiPSe constraint logic programming system such that optimal schedules can be automatically computed using the available constraint solvers. We propose a method and protocol for integrating this constraint-based scheduler into a multi-agent system.

Keywords: Combinatorial optimization · Pickup and delivery problem · Constraint logic programming

1 Introduction

The results reported here were obtained in a project that aims to develop a multi-agent system that provides logistics brokerage services for the efficient allocation of transport resources (vehicles or trucks) to the transport applications. The system architecture contains several types of interacting agents [11]. It incorporates a freight broker agent to coordinate transportation arrangements of customers (usually shippers and consignees) with resource providers or carriers, following the freight broker business model.

The scheduling function of the freight broker agent can be formulated as a special vehicle routing with pickup and delivery problem. This research is based on our declarative model of the scheduling function using constraint logic programming – CLP [4]. This model allows the computation of the set of feasible transportation schedules.

The contribution of this paper is twofold: (i) we upgrade our model with a hybrid representation of feasible solutions that combines integer interval constraints with set constraints; the hybrid constraints are augmented with a declarative representation of optimal transport schedules to effectively compute optimal schedules; (ii) we propose a method for incorporating the scheduler into

© Springer International Publishing AG 2017
G. Boracchi et al. (Eds.): EANN 2017, CCIS 744, pp. 451–464, 2017.
DOI: 10.1007/978-3-319-65172-9_38

a freight broker agent integrated into a multi-agent system. Overall, this system combines ECLiPSe-CLP [14] and Jason agent-oriented programming [2] for providing intelligent freight brokering services.

2 Background and Related Works

The work reported in this paper is based on our previous results [3,4,10,11]. The idea of a multi-agent system architecture for freight transportation brokering was initially outlined in [11]. The proposal of enhancing the freight transportation brokering agent with a service for mathematical optimization, by combining software service and agent technologies was presented in [10]. The formulation of the scheduler function as a vehicle routing with pickup and delivery problem was introduced in [4], together with an ECLiPSe CLP implementation for computing feasible solutions. The declarative optimization model based on CLP was initially formulated in [3].

There are three major directions of developing declarative languages for the specification and solution of constraint satisfaction problems: CLP (or CLP) [6], answer set programming (ASP) [7] and satisfiability checking (SAT) [9]. An experimental comparison of CLP with finite domains and ASP for solving combinatorial NP-complete problems is presented in [5]. Our proposal belongs to the CLP approach. According to [6], CLP is one of the most successful branches of Logic Programming that attracted a lot of research and produced many practical applications during the last two decades.

There are also approaches of combining constraint satisfaction and evolutionary algorithms. One notable paper is [1] that proposes an optimization method by hybridizing genetic algorithms with CLP. This method uses constraint satisfaction for searching feasible solutions on a subspace of the search space, and then uses genetic algorithms for efficiently exploring the space of subspaces. The method is suitable for implementation in the CHIP – Constraint Handling in Prolog programming language [12].

3 Formal Model Using Constraints on Sets and Integers

3.1 Problem Definition

Definition 1. *A vehicle routing with pickup and delivery problem is a tuple* $\langle \mathcal{L}, \mathcal{O}, \mathcal{T}, \Delta \rangle$:

a. $\mathcal{L}$ *is the* set of locations of interest, *including the pickup points, the delivery points, and the truck home locations. We assume that* $\mathcal{L} = \{1, 2, \ldots, k+h\}$ *such that* $k > 0$ *and* $h \geq 0$. *The set* $\mathcal{P} = \{1, 2, \ldots, k\}$ *contains the pickup, as well as the delivery points, while the set* $\mathcal{H} = \{k+1, \ldots, k+h\}$ *contains the truck home locations not already included in the set of pickup and delivery points. The elements of* $\mathcal{L}$ *represent locations of a certain geographical region.*

b. $\mathcal{O}$ is the set of customer orders, $|\mathcal{O}| = n > 0$. Each order is a triple (OS_i, OD_i, C_i) such that $OS_i, OD_i \in \mathcal{P}$, $OS_i \neq OD_i$ are the pickup and the delivery points and $C_i > 0$ is the requested capacity of order i, for all $1 \leq i \leq n$. Note that $2 \leq k \leq 2n$.

c. $\mathcal{T}$ is the set of trucks, $|\mathcal{T}| = t > 0$. Each truck is a pair (H_i, Γ_i) such that $H_i \in \mathcal{L}$ and $\Gamma_i > 0$ are the home location or origin and respectively the maximum provided transportation capacity of truck $i = 1, \ldots, t$.

d. Δ is an $(k+h) \times (k+h)$ positive real matrix such that $\Delta_{ij} > 0$ is the distance between any two locations $1 \leq i \neq j \leq k + h$.

Let us consider a sample problem with $n = 3$ orders, $k = 3$ pickup and delivery points and $t = 2$ trucks. Let us assume that the set of orders is $\mathcal{O} = \{(1, 2, 5), (1, 3, 2), (2, 3, 6)\}$. Note that $O_2 = (1, 3, 2)$, i.e. it requires the transport of freight of capacity $C_2 = 2$ from pickup point $OS_2 = 1$ to delivery point $OD_2 = 3$.

Let us now assume that both trucks have the same home location with index 4. So there are only $h = 1$ home locations, while the total number of locations of interest is $k + h = 4$, i.e. $\mathcal{L} = \{1, 2, 3, 4\}$, $\mathcal{P} = \{1, 2, 3\}$, and $\mathcal{H} = \{4\}$. The set of trucks is $\mathcal{T} = \{(4, 7), (4, 5)\}$. Note that $T_2 = (4, 5)$, i.e. truck 2 has home location 4 and it can transport a maximum capacity $\Gamma_2 = 5$.

Definition 2. A schedule of the vehicle routing with pickup and delivery problem $\langle \mathcal{L}, \mathcal{O}, \mathcal{T}, \Delta \rangle$ can be represented as a tuple $\langle X, M, S, D \rangle$ such that:

a. $X \in \{1, 2, \ldots, k\}^m$ is a vector of size m that captures all the sequences of hops that determine the truck routes to serve all the customer orders. Each of the k locations must be visited, so $m \geq k$. Moreover, each order requires two hops, one for pickup, the other for delivery, so maximum $2n$ hops are needed. It follows that $k \leq m \leq 2n$.

b. $M \in \{0, 1, \ldots, m\}^t$ is a vector of size t such that M_l is the number of hops of each truck $1 \leq l \leq t$. The total number of hops is m so $\sum_{l=1}^{t} M_l = m$. If $M_l = 0$ then truck l is not part of the solution. So, setting $M_l \geq 0$ allows solutions using "at most" t trucks, while setting $M_l \geq 1$ constraints solutions to use "exactly" t trucks.

c. $S, D \subseteq \{0, 1, \ldots, n\}^m$ are two arrays of sets such that:

$$S_i = \{1 \leq j \leq n \mid X_i = \text{pickup point of order } j\}$$
$$D_i = \{1 \leq j \leq n \mid X_i = \text{delivery point of order } j\} \tag{1}$$

We partition the interval $[1, m]$ into t intervals $I_l = [A_l, B_l]$ defined as: $A_l = 1 + \sum_{\alpha=1}^{l-1} M_\alpha$ and $B_l = \sum_{\alpha=1}^{l} M_\alpha$, for $1 \leq l \leq t$. If $M_l = 0$ then $I_l = \emptyset$, so truck l does not contribute to the schedule. On the other hand, each interval $I_l \neq \emptyset$ defines the hops of truck l. The route of truck l (adding its home and departure/return segments from/to home point) is $H_l \to X_{A_l} \to X_{A_l+1} \to \cdots \to X_{B_l} \to H_l$. The total number of hops of truck l, excluding its departure from and return to its home point, is equal to $B_l - A_l + 1$.

For example, in our sample problem, the number m of hops satisfies $3 \leq m \leq 6$. Let us consider a schedule with $m = 5$, $m_1 = 3$, and $m_2 = 2$ hops. Interval $[1, 5]$ is partitioned into $[A_1, B_1] = [1, 3]$ and $[A_2, B_2] = [4, 5]$. Hops 1, 2, 3 define the route of truck 1, while hops 4 and 5 define the route of truck 2. So, the route of truck 2, including the departure and return to home point, is $4 \to X_4 \to X_5 \to 4$.

3.2 Constraints

Following [4], we can now formulate the set of constraints involving the pickup and delivery points and requested capacities of all the customer orders, as well as the available trucks and their capacities. Concerning the quantified variables used for the specification of constraints, we assume that $1 \leq i, k \leq m$ represent hops, $1 \leq j \leq n$ represents orders, and $1 \leq l \leq t$ represents trucks.

The first two constraints given by Eqs. (2) and (3) follow directly from the definition of sets S_i and D_i according to Eq. (1).

For each hop i and order j the pickup point of order j is OS_j so:

$$(\forall i, j)((j \in S_i) \Rightarrow (X_i = OS_j)) \tag{2}$$

Similarly, for each hop i and order j the delivery point of order j is OD_j so:

$$(\forall i, j)((j \in D_i) \Rightarrow (X_i = OD_j)) \tag{3}$$

There exists at least one load or unload operation in each hop i, so:

$$(\forall i)(S_i \cup D_i \neq \emptyset) \tag{4}$$

The freight is correctly delivered, i.e. the pickup point must always precede the delivery point. So for all hops i, k if there exists an order j such that i is the pickup point of order j and k is the delivery point of order j then hop i must precede hop k, i.e.:

$$(\forall i, k)(((\exists j)(j \in S_i \cap D_k)) \Rightarrow (i < k)) \tag{5}$$

Actually Eq. (5) can be simplified by removing the existential quantifier using the equivalence of $p \Rightarrow q$ with $\neg q \Rightarrow \neg p$ and then moving the universal quantifiers in front of the equation. We obtain the following simpler constraint equivalent with (5):

$$(\forall i, j, k)((i \geq k) \Rightarrow ((j \notin S_i) \vee (j \notin D_i))) \tag{6}$$

Each order must be handled exactly once. So for each order j there is a unique load point and a unique unload point, i.e.:

$$(\forall j)(\exists! i)(j \in S_i) \wedge (\forall j)(\exists! i)(j \in D_i) \tag{7}$$

Each order is completely served by a unique truck. So, for each truck l, the orders with load point assigned to l have also the unload point assigned to l, i.e.:

$$(\forall l) \cup_{i \in I_l} S_i = \cup_{i \in I_l} D_i \tag{8}$$

There are also the constraints stating that trucks capacities are not over-flowed along each route. Let T_i denote the transported capacity of truck l between hops $i - 1$ and i, for all $i = A_l, \ldots, B_l$. Initially $T_0 = 0$. For each hop i the value T_i is obtained from T_{i-1} by adding the capacities loaded and subtracting the capacities unloaded in hop i, i.e.:

$$T_0 = 0$$
$$T_i = T_{i-1} + \sum_{j \in S_i} C_j - \sum_{j \in D_i} C_j, 1 \leq l \leq t, A_l \leq i \leq B_l \qquad (9)$$
$$T_i \leq \Gamma_i, 1 \leq l \leq t$$

Note that for each truck $l = 1, \ldots, t$, the remaining capacity to be transported along the last segment from the last hop to the truck home location is 0, i.e. $T_{B_l} = 0$ holds.

If a schedule specifies that truck l can perform a load as well as an unload in hop j then, according to the constraint stating that there must be at least one pickup or delivery in each hop, the solver can generate distinct solutions representing the same schedule. For example, a solution could specify only one hop j containing both the pickup and the delivery, as well as two successive identical hops, the first defining the pickup and the second defining the delivery. The redundancy can be discarded by stating that for each truck l, each of its two consecutive hops i and $i + 1$ with $j \in [A_l, B_l - 1]$ are different:

$$(\forall i \in [A_l, B_l - 1])X_i \neq X_{i+1} \qquad (10)$$

Definition 3. *A feasible schedule of the vehicle routing with pickup and delivery problem is a schedule that satisfies the constraints (2), (3), (4), (6), (7), (8), (9), and (10).*

Let us now assume for our sample problem that $S_1 = \{1\}$, $S_2 = \{3\}$, $S_3 = \emptyset$, $S_4 = \{2\}$, and $S_5 = \emptyset$, while $D_1 = \emptyset$, $D_2 = \{1\}$, $D_3 = \{3\}$, $D_4 = \emptyset$, and $D_5 = \{2\}$. Now the values of the elements of X can be defined as follows: $X_1 = 1$, $X_2 = 2$, $X_3 = 3$, $X_4 = 1$, and $X_5 = 3$. They defines the trucks routs as follows: route of truck 1 is $4 \to 1 \to 2 \to 3 \to 4$, and route of truck 2 is $4 \to 1 \to 3 \to 4$. These routes clearly show that truck 1 handles orders 1 and 3, while truck 2 handles order 2.

The sample schedule is shown in Fig. 1. The figure shows the routes of the trucks, the operations in each hop, the transported capacities between hops, and the maximum capacities that can be transported by each truck. The figure

Truck t	1			2	
Γ	Maximum capacity 7			Maximum capacity 5	
T	Capacity 5	Capacity 6	Capacity 0	Capacity 2	Capacity 0
S, D	Load 1: 1,5	Unload 1: 2,5 Load 3: 2,6	Unload 3:3,6	Load 2: 1,2	Unload 2: 3,2
X	1	2	3	1	3
Hop i	1	2	3	4	5

Fig. 1. Example schedule

clearly shows that the problem constraints are satisfied, so this schedule is a feasible solution. The departure and return of trucks from, respectively to their homes are no shown on the figure.

3.3 Optimization Criterion

Many different optimization criteria can be set for the vehicle routing with pickup and delivery problem. Each optimization criterion defines a different problem. Example criteria can take into account: total travelled distance of the trucks, profit of the truck company, delivery time, a.o. Here we are interested to optimize the total distance travelled by trucks. The method can be generalized to other criteria, depending on the problem.

The total distance D_l travelled by truck $l = 1, 2, \ldots, t$, as well as the total distance $DIST$ travelled by all the trucks, can be determined as follows:

$$D_l = \begin{cases} 0 & \text{if } I_l = \emptyset \\ \Delta_{H_l X_{A_l}} + \Delta_{X_{B_l} H_l} + \sum_{i=A_l}^{B_l - 1} \Delta_{X_i X_{i+1}} & \text{if } I_l \neq \emptyset \end{cases} \qquad (11)$$
$$DIST = \sum_{l=1}^{t} D_l$$

Definition 4. *An* optimal schedule *of the vehicle routing with pickup and delivery problem is a feasible schedule that minimizes the cost function DIST given by Eq. (11).*

4   CLP Implementation

4.1   Model Development

According to our initial model proposed in [11], the function of assigning customer orders to vehicles is the responsibility of the *FBAgent*. In our system, the *FBAgent* incorporates a CLP engine represented by the state-of-the-art ECLiPSe CLP system [14].

In what follows we propose a CLP model of the reasoning capabilities of the of the *FBAgent*. This model is in fact an upgrade of our proposed model already introduced before in [4]. While initially we formulated our ECLiPSe implementation using only integer constraints, here we consider a hybrid implementation that mixes integer constraints with set constraints. For that purpose we formulate the model using two interoperable ECLiPSe constraint solving libraries:

- ic[1] library that provides a solver for integer interval arithmetic.
- ic_sets[2] library that provides a solver for sets of integers.

A CLP program is a set of logic statements – facts and rules, composed of predicates. CLP distinguishes between normal Prolog predicates encountered in

---

[1] http://www.eclipseclp.org/doc/bips/lib/ic/index.html.
[2] http://www.eclipseclp.org/doc/bips/lib/ic_sets/index.html.

standard Prolog programming and constraints that are specific to CLP. Constraints are handled by specialized constraint satisfaction algorithms that provide more efficient problem solving methods than the standard Prolog's backtracking search algorithm.

Following [13], an ECLiPSe-based model contains: (i) definition of variables and domains; (2) definition of constraints and cost variable; (4) search for optimal solution.

A solution is represented in ECLiPSe by a tuple of logic variables, as follows:

- M is the number of hops, N is the number of orders, and T is the number of trucks;
- MM is a T-sized ECLiPSe array [14] such that for each I from 1 to T, $MM[I]$ is the number of hops on the route of truck I; this number is between 1 and M;
- S and D are M-sized ECLiPSe arrays of sets of integers with values from 1 to N that capture the load / unload points associated to each hop;
- K is the number of points of interest (or cities);
- X is an M-sized ECLiPSe array of integers with values from 1 to K that captures the hops, and
- $Dist$ the cost of the optimal solution.

Following [4], we assume that the data of the scheduling problem handled by $FBAgent$ is represented as a set of Prolog facts built using the following predicates:

- number_of_orders(N) for representing the number N of customer orders.
- number_of_cities(K) for defining the number K of cities that represent pickup and delivery points or hops.
- number_of_homes(H) for representing the number H of truck home locations not already included in the set of pickup and delivery points.
- number_of_trucks(T) for representing the number T of trucks.
- order(OrderIndex, LoadIndex, UnloadIndex, Capacity) for representing the set of customer orders.
- truck(TruckIndex, HomeIndex, MaximumCapacity) for representing the set of available trucks.
- distance(LocationI, LocationJ, DistanceIJ) for representing the map of distances between any two cities.

```
configure_solution(M,MM,S,D,X,Dist) :-
    domains_and_variables(M,N,T,MM,S,D,X,K,Delta),
    constraints(M,N,T,MM,S,D,X),
    compute_distance_variable(M,T,MM,Delta,X,K,Dist).
```

Listing 1. Specification of the solution predicate

Following the methodology, a solution can be defined as shown in Listing 1. Here:

- domains_and_variables/9 initializes the problem parameters N, K, T, and *Delta*. Variable *Delta* represents the matrix of distances between the points of interest, including pickup and delivery points, as well as truck homes. This predicate generates all possible assignments of variables M and MM such that M is between K and $2N$, while array MM has T elements with values between 1 and M summing M. Variables M and MM define the top level part of our search tree for solutions.
- constraints/7 generates the problem constraints following their definitions according to Eqs. (2), (3), (4), (6), (7), (8), (9), and (10).
- compute_distance_variable/7 defines the distance variable that represents the optimization criterion using Eq. (11).

4.2 Defining the Constraints

Some constraints can be collectively enforced by a single predicate. We consider constraints (2) and (3) that share a similar structure. They are implemented by a single predicate constraint_1(M,N,S,D,X), as shown in Listing 2. Constraint (4) involves sets S_i and D_i that are immediately available following the evaluation of constraints (2) and (3), so its specification was also defined by predicate constraint_1/5.

The integer and set constraints are represented using special operators. The $\leq$ and $=$ integer constraints are represented using operators #=</2 and #=/2. The set membership constraint is represented using in/2 operator, while set union is represented using operator. In the definition of predicates check_source and check_destination we have used reified constraints, i.e. a constraint that is normally evaluated by the inference engine as true or false is reified into an integer value, either 1 or 0. An example of an implication constraint that is using reified constraints as arguments is (J in SI) => (X[I] #= OSJ) from the definition of check_source predicate.

In our ECLiPSe models we are heavily using logical loops [14]. They are needed for the implementation of the universal quantification that is present in our constraints. They are also useful for the definition of the optimization criterion that specifies constraints on summations of variables.

```
constraint_1(M,N,S,D,X) :-
  (for(I,1,M), param(X,N,S,D) do
    arg(I,S,SI), arg(I,D,DI),
    (for(J,1,N), param(I,X,SI,DI) do
      check_source(I,J,SI,X), check_destination(I,J,DI,X)
    ),
  #(SI \/ DI,C), 1 #=< C).
check_source(I,J,SI,X) :-
  order(J,OSJ,_,_), (J in SI) => (X[I] #= OSJ).
check_destination(I,J,DI,X) :-
  order(J,_,ODJ,_), (J in DI) => (X[I] #= ODJ).
```

Listing 2. Specification of constraints (2), (3) and (4)

In Listing 2, predicates check_source(I, J, SI, X) and check_destination (I, J, DI, X) specify the constraints (2) and (3) for given values of I and J. Moreover, note that unfortunately ECLiPSe does not allow to use the array indexing operator in the arguments that are passed to predicates. So, for example, instead of writing directly check_source(I, J, S[I], X) we must first extract the I-element SI of array S using the standard Prolog predicate arg/3, and then pass it to the predicate via a separate variable SI. Finally note that checking of constraint (4) is achieved by constraining the cardinality of $S_i \cup D_i$ such that $1 \leq |S_i \cup D_i|$.

4.3   Defining the Optimization Criterion

The model contains also a declarative representation of the optimization criterion that defines the optimal transportation schedules. We consider the total distance criterion defined by Eq. (11), as example. Nevertheless, the same approach can be used to define other suitable optimization criteria, for specific problems.

The cost of a schedule is computed using Eq. (11). Although apparently simple, using this equation has a tricky aspect: the variables X_i that are part of the solution occur as subscripts of matrix Δ, so $Dist$ cannot be directly expressed by an algebraic expression of X, as required by the general format of ECLiPSe constraints. Nevertheless, the problematic values $\Delta_{H_l X_{A_l}}$, $\Delta_{X_{B_l} H_l}$, and $\Delta_{X_j X_{j+1}}$ occurring in Eq. (11) can be rewritten algebraically using Eq. (12). Here the Boolean expression $x = y$ is evaluated to 1 if x equals y, otherwise it evaluates to 0.

$$\Delta_{H_l X_{A_l}} = \sum_{x=1}^{k} \Delta_{H_l x}(x = X_{A_l})$$

$$\Delta_{X_{B_l} H_l} = \sum_{x=1}^{k} \Delta_{H_l x}(x = X_{B_l}) \tag{12}$$

$$\Delta_{X_i X_{i+1}} = \sum_{x,y=1}^{k} \Delta_{xy}(x = X_i)(y = X_{i+1})$$

Note however that using Eq. (12) has the drawback of incurring an $O(k)$ computational overhead, as compared with the evaluation using Eq. (11). The detailed ECLiPSe-CLP model following Eq. (12) is presented in reference [3].

Table 1. Data set

#	$n = $ # orders	$k = $ # cities	$t = $ # trucks	$m \in k \ldots 2n$	Cost	# opt. solutions
1	5	6	2	$6 \ldots 10$	821	2

4.4   Defining the Queries

We define a special predicate query(M, MM, S, D, X, Dist) to invoke the constraint solver. We can define a separate query for each search strategy, independently of the declarative representation of the problem. For experimental purposes, the query can be defined with two additional parameters: $Config$ for defining the parameters of the search algorithm and T for determining the user runtime of the search process.

We use the built-in bb_min(Goal, Cost, Options) ECLiPSe predicate for performing a branch-and-bound search for optimal schedules. This means that whenever a new and better solution is found, it is remembered together with its (currently optimal) cost. Then the search continues using a supplementary constraint requiring that the future solutions must have a lower cost than the current solution.

The search process involves two stages. We start with `configure_solution` for instantiating M, MM and $Dist$, as well as for defining incompletely instantiated structure templates for S, D and X. Then we employ a search strategy implemented by predicate `solve` for exploring the possible values of S, D and X that satisfy the scheduling constraints. The general format of the query is shown in Listing 3. Parameter Tr returns the user cpu time. It is retrieved using the `statistics/2` ECLiPSe built-in predicate.

```
query(M,MM,S,D,X,Dist,Config,Tr) :-
  bb_min(
    (config_solution(M,MM,S,D,X,Dist),
      bb_min(solve(M,X,S,D,Config),Dist,_)),
    Dist,_)
```

Listing 3. General format of the search query

In fact our predicate `solve` is a hybrid solver that must combine an integer interval constraint solver for determining X, available in the `ic` ECLiPSe library, and a set constraint solver for determining S and D, based on the `ic_sets` ECLiPSe library.

We have used the `search/6` predicate that is available in the `ic` ECLiPSe library, for solving integer interval constraints. It takes the following parameters: $Select$, that denotes the variable selection method, $Choice$ that denotes the value-to-variable assignment method during the search process, and $Method$ that denotes the search method.

We have used the `insetdomain/4` predicate that is available in the `ic_sets` ECLiPSe library, for solving set constraints. It takes the following parameters: Set denotes the set variable that must be instantiated by the search algorithm; $CardSel$ determines how are sets enumerated depending on their cardinality. It can be `any`, `increasing` or `decreasing`; $ElemSel$ determines which potential set elements are firstly considered for inclusion or exclusion and it can takes the following values: `small_first`, `big_first`; $Order$ determines either if it is firstly tried to make the selected potential element a set member, or to firstly exclude it from the set.

The search predicate `labelsets` for solving set constraints is shown in Listing 4. It considers each element of the array of sets Set by constraining it to be instantiated to a value of its set domain. The values are explored using the configuration parameters of the ECLiPSe predicate `insetdomain`, as specified by the configuration parameters.

```
labelsets(Set,M,conf(CardSel,ElemSel,Order)) :-
  (for(I,1,M),param(Set,CardSel,ElemSel,Order) do
    arg(I,Set,X),insetdomain(X,CardSel,ElemSel,Order)).
```

Listing 4. Search predicate for set constraints

The `solve` predicate is presented in Listing 5. We can define 6 versions of `solve` by changing the ordering of the goals `search` and `labelsets` in the body of the definition of `solve`. Intuitively, the best option is to first constrain the elements of X.

```
solve(M,X,S,D,[Select,Choice,Method,CardSel,ElemSel,Order]) :-
  search(X,0,Select,Choice,Method),[]),
  labelsets(S,M,conf(CardSel,ElemSel,Order)),
  labelsets(D,M,conf(CardSel,ElemSel,Order)).
```

Listing 5. Hybrid search predicate for all the constraints

4.5   Experiments

We experimented using the 64-bit version of ECLiPSe 6.1_224 on an *x64*-based PC with Intel(R) Core(TM) i7-5500U CPU at 2.40 GHz running Windows 10. We considered the data set described in Table 1. Even these small parameter values determine a large search space. For $n = 5$ and $k = 6$ this size is $\sum_{m=k}^{2n} 2^{mn} \times 2^{mn} \times m^k > 2^{118}$!

Table 2. Experimental results

#	Ordering	Select	Choice	Method	Total runtime [ms]	Optimal solution runtime [ms]
1	S, X, D	input_order	indomain	complete	103409	5265
2	S, D, X	input_order	indomain	complete	103111	5593
3	X, S, D	input_order	indomain	complete	117891	4531
4	S, X, D	max_regret	indomain_median	lds(1)	123924	6328
5	X, S, D	max_regret	indomain_median	lds(2)	5500	781
6	X, S, D	max_regret	indomain_median	lds(1)	2047	250
7	X, S, D	most_constrained	indomain_median	lds(2)	4562	812
8	X, S, D	first_fail	indomain_median	lds(2)	5377	750

The experiment was focused on determining the optimal schedule that satisfies the constraints. We considered the solutions consisting of exactly t trucks. We performed a series of searches using different versions of the `query` predicate. We varied:

- The order of searching variables X, S, and D.
- The configuration options for searching the integer variables, i.e. elements of X, in the integer interval constraints part.
- The configuration options for searching the set variables, i.e. elements of S and D, in the set constraints part.

Our experiments revealed that the first two aspects in combination were the most important for the efficiency of our approach. So, our conclusion is that the shortest execution times for computing the optimal solution were obtained by firstly exploring the elements of X, in combination with using the incomplete search based on the limited discrepancy search algorithm [8] that is available in ECLiPSe, for exploring the elements of X. Using these options, the configuration parameters for exploring the set variables were less important, with minor differences observed in the execution times. Some results are summarized in Table 2.

5   Solution Outline of Scheduler Integration with an Agent System

Let us consider the sample scenario from Fig. 2. This scenario involves 2 $CAgent$ agents – $CA1$ and $CA2$ that issue 2 freight transportation orders – $order1$ and $order2$ to the $FBAgent$ named FBR. Moreover, two $FTPAgent$ transportation provider agents $FTP1$ and $FTP2$ declare the availability of 2 trucks, by providing their descriptions $truck1$ and $truck2$ to the FBR agent. The FBR agent formulates a scheduling problem based on the information available in the customer orders, as well as in the truck descriptions, adding also information about the distances between the locations of interest. This geographical information is available from a local database of the FBR agent.

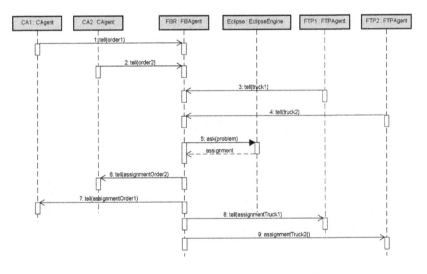

Fig. 2. Agents interaction protocol

The FBR agent incorporates an $EclipseEngine$ for solving constraint satisfaction problems, in particular to determine optimal schedules for the freight transportation problem. Firstly FBR has to map the $problem$ description into the schema accepted by our proposed scheduler, using a set of Prolog facts, as described in Sect. 4, and then load it into the $EclipseEngine$. Then FBR issues a query to the *EclipseEngine* for retrieving the set of solutions. Note that this query will retrieve all the possible solutions.

The *EclipseEngine* engine returns a list of solutions to the *FBR* agent. This agent selects the optimal solution and extracts from it the information that is relevant to each order addressed in this solution, as well as to each truck used in this solution. Referring to *order2*, *FBR* determines what truck should be responsible with handling this order and then send this information to the corresponding customer, *CA2* in this case. Moreover, referring to *truck1*, *FBR* determines all the orders that should be handled by *truck1* and then sends this information to the provider of *truck1* – *FTP1* agent.

Our proposed multi-agent system is based on Jason agent-oriented programming language [2]. Jason is logic-based, so this simplifies the interface between the agents and the *EclipseEngine* as concerning data formatting. The interface for querying *EclipseEngine*, retrieving solutions and converting them to the format recognized by *FBR* is programmed using the Java APIs of Jason and ECLiPSe CLP.

6   Conclusion and Future Works

We proposed a method for declarative modeling and optimization of freight transportation brokering using agents and constraint logic programming. Using this approach, the knowledge of the freight broker is explicitly captured as a set of constraints mapped to ECLiPSe CLP system. This enables the automated computation of optimal schedules using available constraint solvers. The logical representation provides increased flexibility of the solution by enabling the natural expansion of the broker knowledge to fit changing requirements of the transportation problem. Finally, using logic-based knowledge representation facilitates the integration of the constraint-based scheduler into a Jason-based multi-agent system. In the near future we plan to strengthen our scheduler, by adapting it to more realistic scenarios, where too high computation times are not acceptable. We will also focus on the technical aspects of the integration of Jason and ECLiPSe CLP, thus enabling Jason agents to use CLP reasoning.

References

1. Barnier, N., Brisset, P.: Optimization by hybridization of a genetic algorithm with constraint satisfaction techniques. In: Proceedings of IEEE International Conference on Evolutionary Computation 1998, IEEE World Congress on Computational Intelligence, pp. 645–649. IEEE (1998)
2. Bordini, R.H., Hübner, J.F., Wooldridge, M.: Programming Multi-Agent Systems in AgentSpeak Using Jason. Wiley Series in Agent Technology. Wiley, Hoboken (2007)
3. Bădică, A., Bădică, C., Leon, F.: Modeling and optimization of pickup and delivery problem using constraint logic programming. In: Proceedings of Tenth International Conference on Large Scale Scientific Computations - LSSC 2017 (2017)

4. Bădică, C., Bădică, A., Leon, F., Luncean, L.: Declarative representation and solution of vehicle routing with pickup and delivery problem. In: Proceedings of the International Conference of Computational Science 2017 - ICCS 2017. Procedia Computer Science, Elsevier (2017)

5. Dovier, A., Formisano, A., Pontelli, E.: A comparison of CLP(FD) and ASP solutions to NP-complete problems. In: Gabbrielli, M., Gupta, G. (eds.) ICLP 2005. LNCS, vol. 3668, pp. 67–82. Springer, Heidelberg (2005). doi:10.1007/11562931_8

6. Gavanelli, M., Rossi, F.: Constraint logic programming. In: Dovier, A., Pontelli, E. (eds.) A 25-Year Perspective on Logic Programming. LNCS, vol. 6125, pp. 64–86. Springer, Heidelberg (2010). doi:10.1007/978-3-642-14309-0_4

7. Gouidis, F., Patkos, T., Flouris, G., Plexousakis, D.: Declarative reasoning approaches for agent coordination. In: Likas, A., Blekas, K., Kalles, D. (eds.) SETN 2014. LNCS, vol. 8445, pp. 489–503. Springer, Cham (2014). doi:10.1007/978-3-319-07064-3_42

8. Harvey, W.D., Ginsberg, M.L.: Limited discrepancy search. In: Proceedings of the 14th International Joint Conference on Artificial Intelligence - IJCAI 1995, pp. 607–613 (1995)

9. Hölldobler, S., Manthey, N., Steinke, P.: A compact encoding of pseudo-Boolean constraints into SAT. In: Glimm, B., Krüger, A. (eds.) KI 2012. LNCS, vol. 7526, pp. 107–118. Springer, Heidelberg (2012). doi:10.1007/978-3-642-33347-7_10

10. Leon, F., Bădică, C.: A freight brokering system architecture based on web services and agents. In: Borangiu, T., Drăgoicea, M., Nóvoa, H. (eds.) IESS 2016. LNBIP, vol. 247, pp. 537–546. Springer, Cham (2016). doi:10.1007/978-3-319-32689-4_41

11. Luncean, L., Bădică, C., Bădică, A.: Agent-based system for brokering of logistics services – initial report. In: Nguyen, N.T., Attachoo, B., Trawiński, B., Somboonviwat, K. (eds.) ACIIDS 2014. LNCS, vol. 8398, pp. 485–494. Springer, Cham (2014). doi:10.1007/978-3-319-05458-2_50

12. Marriott, K., Stuckey, P.J., Wallace, M.: Constraint logic programming. In: Rossi, F., van Beek, P., Walsh, T. (eds.) Handbook of Constraint Programming, pp. 407–450. Elsevier, Amsterdam (2006)

13. Niederliński, A.: A Gentle Guide to Constraint Logic Programming via ECLiPSe, 3rd edn. Jacek Skalmierski Computer Studio, Gliwice (2014)

14. Schimpf, J., Shen, K.: ECLiPSe - from LP to CLP. Theory Pract. Logic Program. **12**(1–2), 127–156 (2012)

Driving Mental Fatigue Classification Based on Brain Functional Connectivity

Georgios N. Dimitrakopoulos[1,2] (ORCID), Ioannis Kakkos[1],
Aristidis G. Vrahatis[3,4] (ORCID), Kyriakos Sgarbas[2], Junhua Li[1] (ORCID),
Yu Sun[1] (ORCID), and Anastasios Bezerianos[1,3(✉)] (ORCID)

[1] Singapore Institute for Neurotechnology (SINAPSE), National University of
Singapore, 28 Medical Dr. #05-COR, Singapore 117456, Singapore
ioakakkos@gmail.com,
{juhalee,lsisu,tassos.bezerianos}@nus.edu.sg
[2] Department of Electrical and Computer Engineering, University of Patras,
25600 Patras, Greece
{geodimitrak,sgrabas}@upatras.gr
[3] School of Medicine, University of Patras, 25600 Patras, Greece
vrachatis@ceid.upatras.gr
[4] Computer Engineering and Informatics Department, University of Patras,
25600 Patras, Greece

Abstract. EEG techniques have been widely used for mental fatigue monitoring, which is an important factor for driving safety. In this work, we performed an experiment involving one hour driving simulation. Based on EEG recordings, we created brain functional networks in alpha power band with three different methods, partial directed coherence (PDC), direct transfer function (DTF) and phase lag index (PLI). Then, we performed feature selection and classification between alertness and fatigue states, using the functional connectivity as features. High accuracy (84.7%) was achieved, with 22 discriminative connections from PDC network. The selected features revealed alterations of the functional network due to mental fatigue and specifically reduction of information flow among areas. Finally, a feature ranking is provided, which can lead to electrode minimization for real-time fatigue monitoring applications.

Keywords: EEG · Functional connectivity · Classification · Mental fatigue · Driving

1 Introduction

Mental fatigue usually results in drowsiness, lapses of attention, longer reaction time and reduced mental performance [1, 2]. Since car driving requires considerable cognitive effort and attention from the driver, mental fatigue is a significant cause of fatal car accidents [3]. Therefore, interest in mental fatigue tracking technologies has increased, aiming to improve driving safety [4]. In order to study the effects of mental fatigue during driving, driving simulators have been employed, as studies have shown

© Springer International Publishing AG 2017
G. Boracchi et al. (Eds.): EANN 2017, CCIS 744, pp. 465–474, 2017.
DOI: 10.1007/978-3-319-65172-9_39

that fatigue can be equally evaluated in real and simulated driving environments [5]. Moreover, driving simulators provide safe, low cost, well-controlled conditions and data collection convenience.

A large number of studies based on EEG reported increases in lower alpha and theta band power with increasing levels of fatigue [6–8]. Progressive increases in the lower alpha band have specifically been implicated with decreasing arousal and alertness during periods when the attentional system is challenged [9]. In [10] it was observed increment of the alpha power in function of the time spent on the task as well as an increment in the theta band but in a lesser extent than in the alpha band. Finally, many fatigue related studies focus on alpha spindles detection, which are short time oscillations in alpha frequency band [11]. So far, studies based on EEG have shown that mental fatigue manifests mainly in the frontal areas [8], as well as frontal-central and parietal areas [12].

Network theory has effectively been used to study the brain functional organization in complex tasks involving many brain regions [13]. Based on EEG data from various problems, functional connectivity networks have been built representing the interactions among brain areas, revealing patterns during various brain states under high temporal solution. Regarding mental fatigue, several studies used different network methods to detect alterations between alert and fatigue states. For instance, Li et al. [14] using phase lag index to measure functional connectivity strengths in alpha band found that important nodes in fatigue state were localized in the frontal cortex. Liu et al. [15] performed direct transfer function to investigate the directionality and strength of cortical information flow with increasing fatigue. Furthermore, partial directed coherence has been used to describe the transmission of information in the effective connections in a vigilance task [16]. For mental fatigue detection, several studies performed classification, with the majority using power-based features [8, 17]. More recently, entropy-based features have also been successfully used for classification of driving mental fatigue [18, 19]. However, only few studies attempted classification using the network connectivity as features [19–21].

Here, we aim to detect functional connections able to distinguish alertness and fatigue states. We tried three different methods to create functional brain networks in alpha frequency band. Then, we performed feature selection and classification using the network connections as features. Thus, we isolated a small number of discriminative connectivity features that differentiate between the two states providing high accuracy. Hence, we obtained a comprehensive overview of the fatigue effects during driving in the functional network.

2 Methods

2.1 Subjects

The data were acquired from 20 male subjects (23.4 ± 5.4 years old) with the use of non-invasive EEG. Participants did not have any history of alcohol or drug addiction or fatigue-related disorders. Written informed consent was obtained from all subjects, prior to the start of the experiment, and they were reimbursed SGD$10/hour for their participation. The experiments were approved by the Institutional Review Board of the National University of Singapore.

2.2   Experimental Design

The aim of the experimental design was to induce mental fatigue after execution of a long-time task. Specifically, participants had to engage in 1 h driving simulation. In this study, City Car Driving 1.5 (http://citycardriving.com/) was used for the driving task with Logitech G27 Racing Wheel equipment, which included driving wheel, pedals and gear box. To make driving easier and to reduce movements of subject, automated clutch was used. Subjects were instructed to drive stably and with maximum speed up to 100 km/h. Finally, all sessions were carried out in the afternoon, to reduce variance due to time of the day.

2.3   Data Acquisition

High-density continuous EEG recordings were acquired from 64 scalp channels according to the international 10–20 system. (ASA Lab 4.9.2, ANT Software BV, the Netherlands). Additionally, vertical and horizontal electrooculogram (EOG) was recorded bipolarly. All electrophysiological recordings (EEG, EOG) were recorded at 512 Hz sampling rate. Electrode impedance was maintained below 10 kΩ throughout the experiment. During the data acquisition, anti-aliasing with a band-pass filter (0.5–70 Hz), and a 50 Hz notch filter was applied.

2.4   EEG Data Pre-processing

The EEG recorded signals were downsampled to 256 Hz and re-referenced to the average of mastoid electrodes, M1 and M2. Then, data were detrended and filtered using a band pass FIR filter with cut-off frequencies at 1 and 40 Hz. Artifact and noisy data rejection was applied by detecting and removing continuous parts of data with power over 6 db in high frequencies (20–40 Hz). Artifacts due to eye blinks were removed via independent component analysis (ICA), rejecting the components that presented high correlation with EOG signals [22]. Finally, driving continuous data were divided into 1 min segments. An overview of the methods applied on the EEG data is shown in Fig. 1. All EEG data were pre-processed using EEGLAB [23].

2.5   EEG Functional Connectivity Estimation

The connectivity values for every pair of sensors were calculated using Phase Locking Index (PLI) [24], Directed Transfer Function (DTF) [25] and Partial Directed Coherence (PDC) [26] metrics.

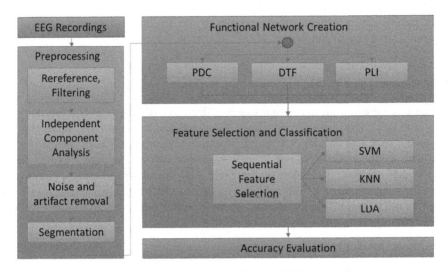

Fig. 1. Flowchart of our method.

PLI for two sensors A, B is defined as:

$$PLI_{A \to B} = \frac{1}{T} \sum_{t=1}^{T} e^{sign(\phi_A(\tau) - \phi_B(\tau))} \tag{1}$$

where φ is the phase obtained by Hilbert transformation. When there is a consistent nonzero phase difference between the two signals, PLI tends to 1, otherwise to 0.

For DTF and PDC metrics, first we define a multivariate autoregressive model (MVAR) of order p for multi-channel series $X(t) = [x_1(t) \ldots x_N(t)]$:

$$X(t) = - \sum_{\tau=1}^{p} A_\tau X(t - \tau) + E(t) \tag{2}$$

This model associates values of X with previous values from all channels. The transfer matrix $H(f)$ of the MVAR model in the frequency domain is defined as the inverse of $A(f)$.

DTF is defined as the ratio between the inflow from channel j to channel i to all the inflows to channel i:

$$DTF_{ij}^2(f) = \frac{|H_{ij}(f)|^2}{\sum_{m=1}^{N} |H_{im}(f)|^2} \tag{3}$$

where $H_{ij}(f)$ is an element of a transfer matrix of MVAR model. DTF ranges in the interval [0,1], with higher values indicating higher correlation of the signals.

In order to distinguish between direct and cascade flows, PDC metric was proposed, which emphasizes on direct flows between channels. PDC is defined in terms of MVAR coefficients transformed to the frequency domain:

$$PDC_{ij}(f) = \frac{A_{ij}(f)}{\sqrt{a_j^*(f)a_j(f)}} \tag{4}$$

where A(f) is Fourier transform of MVAR model coefficients A(t), aj(f) is j-th column of A(f) and asterisk denotes the transpose and complex conjugate operation. PDC ranges in the interval [0,1].

In this work, we focused our analysis on networks from alpha band (8–12 Hz), as they are more relevant for mental fatigue detection as aforementioned. One network was created for each time window of 1 min. Therefore, 62×62 weighted adjacency matrices were created, resulting in ($62 \times 62=$) 3844 features for directed metrics (PDC, DTF) and ($62 \times (62 - 1)/2=$) 1981 for undirected ones (PLI).

2.6 Mental Fatigue Classification Based on Network Connectivity

To predict fatigue level as high or low, a classification approach was employed, using the connectivity values as features. The connectivity values were considered to belong into alertness state in the start of the experiment and fatigue state towards the end, similarly with [11, 16]. Specifically, the dataset for classification was formed using the 5 first and 5 last networks (5 min data) for each subject.

Due to the large number of features, the Sequential Forward Selection (SFS) [27] method was employed to determine the subset of significant features that provided higher classification accuracy. Initially, classification is applied using each feature separately and the best feature is selected as initial subset. Then, all features are examined and a feature is added to the selected subset if this inclusion leads to maximal increment of classification accuracy. This way, features that carry little or no useful information as well as highly correlated with previously selected are not included.

For classification, we used (a) linear discriminant analysis (LDA), (b) k nearest neighbors (KNN) and Support Vector Machines (SVM) with (c) Sequential Minimal Optimization (SMO) and (d) Least Squares (LS) learning methods. To find maximum accuracy, we tried several values for the KNN parameter number of neighbors k = {11, 13, 15}. For SVM, radial basis function (RBF) kernel was used and the kernel width σ was varied from 1 to 4 with step of 0.5, while the soft margin regularization parameter C was varied from 10^{-2} to 10^2 with factor 10. The classification accuracy was computed and evaluated with a leave-one out cross-validation (LOOCV). All algorithms were implemented in MATLAB.

3 Results

The maximum classification accuracies for each method and network metric are presented in Fig. 2. The SVM-SMO algorithm displayed the highest accuracy across methods (0.81–0.847). The SVM-LS method displayed slightly lower accuracy (0.79–0.838), while LDA and KNN provided considerably lower accuracies (0.69–0.767 and 0.652–0.681 respectively). It is noted that PDC networks result in better classification accuracy than DTF and PDC ones in all cases.

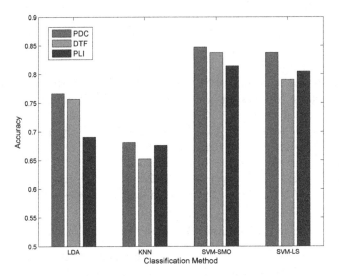

Fig. 2. Maximum accuracy achieved by different classification methods and network metrics. SVM-SMO provided maximum accuracy of 0.847 based on PDC metric.

In detail, focusing on SVM-SMO, the accuracy during feature selection varying the σ parameter (while $C = 1$) is shown in Fig. 3. The best performance was achieved by PDC metric for $\sigma = 3$, with accuracy of 0.847 (sensitivity 0.857, specificity 0.857) and using 22 features. Similarly, PDC networks provided the best or close to the best performance in terms of accuracy compared to DTF and PLI in all cases.

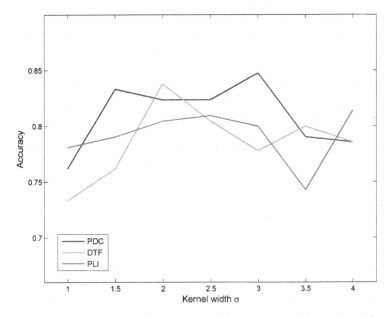

Fig. 3. Accuracy of best method, SVM-SMO, while varying parameter σ of the RBF kernel. Maximum accuracy of 0.847 was achieved for PDC network metric with $\sigma = 3$.

The 22 connectivity features resulting in maximum accuracy are provided in Table 1, ordered by SFS. The selected features are depicted in Fig. 4 in scalp maps, where color encodes PDC values in the two classes, alertness and fatigue. Moreover, the accuracy during feature selection by SFS is displayed in Fig. 5.

Table 1. PDC connectivity features ordered by SFS.

Order	Connection		PDC	
	Source	Target	Alert	Fatigue
1	F5	FT8	0.106	0.141
2	T7	PO3	0.204	0.150
3	T7	FC5	0.251	0.166
4	C6	Fp2	0.104	0.082
5	T7	CP1	0.223	0.168
6	CP3	PO4	0.072	0.057
7	P4	C4	0.083	0.107
8	FT8	CP5	0.163	0.124
9	Cz	AF8	0.087	0.068
10	Fp1	FC4	0.094	0.140
11	CP3	CP4	0.076	0.056
12	P1	Cz	0.072	0.068
13	Fp1	CPz	0.093	0.141
14	F8	P1	0.146	0.120
15	PO5	FT8	0.084	0.101
16	P4	P5	0.086	0.104
17	C6	F6	0.103	0.085
18	FC3	POz	0.088	0.099
19	Fp1	PO4	0.101	0.144
20	Fp1	FC6	0.113	0.155
21	CP3	AF8	0.073	0.064
22	CP5	C3	0.102	0.089

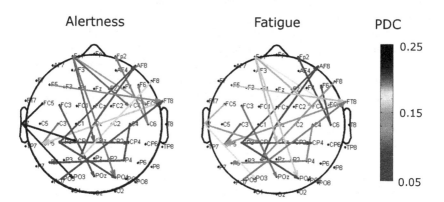

Fig. 4. Connectivity features resulting in maximum accuracy are depicted in scalp maps for alertness and fatigue states. The PDC value of each connection is color-encoded. (Color figure online)

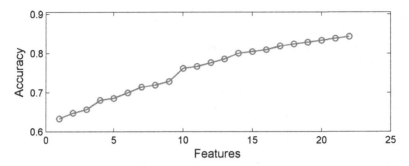

Fig. 5. Classification accuracy is presented according to the number of connectivity features selected by sequential feature selection algorithm. After 22 features, accuracy cannot be further increased.

4   Discussion

We performed classification on fatigue level, using the networks of the first 5 min of EEG recordings as alertness state and the 5 last ones as fatigue. Very high accuracy was achieved by SVM-SMO method as well as SVM-LS, in comparison to LDA and KNN, which are simpler methods with less parameters to fine-tune. However, the fact that different methods provide accuracies considerably over chance level, shows that accurate classification is feasible based on network features. Additionally, PDC metric had better performance over PLI and DTF. PLI is an undirected metric, thus it provides less information compared to directed network metrics. Also, PDC is more sophisticated metric compared to DTF, since it is designed to distinguish between direct and indirect connections between two nodes.

Next, we focused on the network topology of the most discriminative features, in the case of SVM-SMO and PDC classification. It is noted that most connections include frontal or parietal-occipital electrodes, in accordance with literature [8, 12]. Specifically, Fp1 electrode is involved in 4 connections, while three electrodes (FT8, T7 and CP3) in 3. Despite that some connections have been enhanced in fatigue condition in terms of PDC value, the majority (14 out of 22) has been weakened. Moreover, in most cases the selected features involve connections between anterior and posterior areas as well as between right and left hemisphere, namely they are "long" connections linking remote areas. Specifically, 10 connections connect left and right areas, 5 lie in the left hemisphere and 7 in the right. Additionally, information flow has a directionality from left to right in the majority of the connections. Therefore, information flow between areas has been reduced due to fatigue, especially between remote areas. This corroborates the findings of our previous study examining PDC networks in a Psychomotor Vigilance Task, where inter-hemispheric asymmetry was found, with fatigue-related connectivity located in right hemisphere and frontal-parietal areas [16].

We note that the small number of required electrodes is critical for effective real-time practical applications [28]. Therefore, depending on the desired accuracy level, it is possible to choose a small number of electrodes. For example, as shown in

Fig. 5, 70% accuracy can be achieved using only the top 6 features or 75% with the top 10 features from Table 1.

From this mental fatigue study, it can be concluded that fatigue induced by 1 h monotonous driving could alter network characteristics. Moreover, we performed classification and feature selection based on brain connectivity features, achieving high accuracy with a small number of features. We highlight that a feature ranking is provided by the sequential way of feature selection, thus considering the trade-off between accuracy and features number, the number of required electrodes can be minimized while retaining the desired accuracy level.

Acknowledgment. This work was supported by the National University of Singapore (NUS) and Defense Science Organization (DSO) for Cognitive Engineering Group at Singapore Institute for Neurotechnology (SINAPSE).

References

1. Dorrian, J., Roach, G.D., Fletcher, A., Dawson, D.: Simulated train driving: fatigue, self-awareness and cognitive disengagement. Appl. Ergon. **38**(2), 155–166 (2007)
2. Borghini, G., Astolfi, L., Vecchiato, G., Mattia, D., Babiloni, F.: Measuring neurophysiological signals in aircraft pilots and car drivers for the assessment of mental workload, fatigue and drowsiness. Neurosci. Biobehav. Rev. **44**, 58–75 (2014)
3. Kostyniuk, L.P., Streff, F.M., Zakrajsek, J.: Identifying Unsafe Driver Actions that Lead to Fatal Car-Truck Crashes, vol. 9, no. 10. AAA Foundation for Traffic Safety (2002)
4. Vanlaar, W., Simpson, H., Mayhew, D., Robertson, R.: Fatigued and drowsy driving: a survey of attitudes, opinions and behaviors. J. Safety Res. **39**(3), 303–309 (2008)
5. Philip, P., Sagaspe, P., Taillard, J., Valtat, C., Moore, N., Akerstedt, T., Charles, A., Bioulac, B.: Fatigue, sleepiness, and performance in simulated versus real driving conditions. Sleep (N. Y., Westchest.) **28**(12), 1511 (2005)
6. Liu, T.: Positive correlation between drowsiness and prefrontal activation during a simulated speed-control driving task. NeuroReport **25**(16), 1316–1319 (2014)
7. Boksem, M.A., Meijman, T.F., Lorist, M.M.: Effects of mental fatigue on attention: an ERP study. Cogn. Brain. Res. **25**(1), 107–116 (2005)
8. Craig, A., Tran, Y., Wijesuriya, N., Nguyen, H.: Regional brain wave activity changes associated with fatigue. Psychophysiology **49**(4), 574–582 (2012)
9. Klimesch, W.: EEG alpha and theta oscillations reflect cognitive and memory performance: a review and analysis. Brain Res. Rev. **29**(2), 169–195 (1999)
10. Charbonnier, S., Roy, R.N., Bonnet, S., Campagne, A.: EEG index for control operators' mental fatigue monitoring using interactions between brain regions. Expert Syst. Appl. **52**, 91–98 (2016)
11. Simon, M., Schmidt, E.A., Kincses, W.E., Fritzsche, M., Bruns, A., Aufmuth, C., Bogdan, M., Rosenstiel, W., Schrauf, M.: EEG alpha spindle measures as indicators of driver fatigue under real traffic conditions. Clin. Neurophysiol. **122**(6), 1168–1178 (2011)
12. Jap, B.T., Lal, S., Fischer, P., Bekiaris, E.: Using EEG spectral components to assess algorithms for detecting fatigue. Expert Syst. Appl. **36**(2 Part 1), 2352–2359 (2009)
13. Sporns, O.: Contributions and challenges for network models in cognitive neuroscience. Nat. Neurosci. **17**(5), 652–660 (2014)

474   G.N. Dimitrakopoulos et al.

14. Li, J., Lim, J., Chen, Y., Wong, K., Thakor, N., Bezerianos, A., Sun, Y.: Mid-task break improves global integration of functional connectivity in lower alpha band. Front. Hum. Neurosci. **10**, 304 (2016)
15. Liu, J.P., Zhang, C., Zheng, C.X.: Estimation of the cortical functional connectivity by directed transfer function during mental fatigue. Appl. Ergon. **42**(1), 114–121 (2010)
16. Sun, Y., Lim, J., Kwok, K., Bezerianos, A.: Functional cortical connectivity analysis of mental fatigue unmasks hemispheric asymmetry and changes in small-world networks. Brain Cogn. **85**, 220–230 (2014)
17. Trejo, L.J., Kubitz, K., Rosipal, R., Kochavi, R.L., Montgomery, L.D.: EEG-based estimation and classification of mental fatigue. Psychology **6**(5), 572 (2015)
18. Khushaba, R.N., Kodagoda, S., Lal, S., Dissanayake, G.: Driver drowsiness classification using fuzzy wavelet-packet-based feature-extraction algorithm. IEEE Trans. Biomed. Eng. **58**(1), 121–131 (2011)
19. Chai, R., Nguyen, T.N., Craig, A.: Driver fatigue classification with independent component by entropy rate bound minimization analysis in an EEG-based system. IEEE J. Biomed. Health Inform. **21**(3), 715–724 (2016)
20. Shen, K.Q., Li, X.P., Ong, C.J., Shao, S.Y., Wilder-Smith, E.P.: EEG-based mental fatigue measurement using multi-class support vector machines with confidence estimate. Clin. Neurophysiol. **119**(7), 1524–1533 (2008)
21. Sun, Y., Lim, J., Meng, J., Kwok, K., Thakor, N.: Discriminative analysis of brain functional connectivity patterns for mental fatigue classification. Ann. Biomed. Eng. **40**(10), 2084–2094 (2014)
22. Jung, T.P., Makeig, S., Westerfield, M., Townsend, J., Courchesne, E., Sejnowski, T.J.: Removal of eye activity artifacts from visual event-related potentials in normal and clinical subjects. Clin. Neurophysiol. **111**(10), 1745–1758 (2000)
23. Delorme, A., Makeig, S.: EEGLAB: an open source toolbox for analysis of single-trial EEG dynamics including independent component analysis. J. Neurosci. Methods **134**(1), 9–21 (2004)
24. Stam, C.J., Nolte, G., Daffertshofer, A.: Phase lag index: assessment of functional connectivity from multi channel EEG and MEG with diminished bias from common sources. Hum. Brain Mapp. **28**(11), 1178–1193 (2007)
25. Kaminski, M., Blinowska, K.J.: A new method of the description of the information flow in brain structures. Biol. Cybern. **65**(3), 203–210 (1991)
26. Baccalá, L.A., Sameshima, K.: Partial directed coherence: a new concept in neural structure determination. Biol. Cybern. **84**(6), 463–474 (2001)
27. Jain, A.K., Duin, R.P.W., Mao, J.: Statistical pattern recognition: a review. IEEE Trans. Pattern Anal. Mach. Intell. **22**(1), 4–37 (2000)
28. Ko, L.W., Lai, W.K., Liang, W.G., Chuang, C.H., Lu, S.W., Lu, Y.C., Hsiung, T.Y., Wu, H.H., Lin, C.T.: Single channel wireless EEG device for real-time fatigue level detection. In: 2015 IEEE International Joint Conference on Neural Networks (IJCNN), pp. 1–5 (2015)

Recommendation Systems

A Package Recommendation Framework Based on Collaborative Filtering and Preference Score Maximization

Panagiotis Kouris[1,2]([✉]), Iraklis Varlamis[2], and Georgios Alexandridis[1]

[1] Intelligent Systems Laboratory, School of Electrical and Computer Engineering,
National Technical University of Athens, Iroon Polytechniou 9,
15780 Zografou, Athens, Greece
{pkouris,gealexandri}@islab.ntua.gr
[2] Department of Informatics and Telematics, Harokopio University of Athens,
Omirou 9, 17778 Athens, Greece
varlamis@hua.gr

Abstract. The popularity of recommendation systems has made them a substantial component of many applications and projects. This work proposes a framework for package recommendations that try to meet users' preferences as much as possible through the satisfaction of several criteria. This is achieved by modeling the relation between the items and the categories these items belong to aiming to recommend to each user the top-k packages which cover their preferred categories and the restriction of a maximum package cost. Our contribution includes an optimal and a greedy solution. The novelty of the optimal solution is that it combines the collaborative filtering predictions with a graph based model to produce recommendations. The problem is expressed through a minimum cost flow network and is solved by integer linear programming. The greedy solution performs with a low computational complexity and provides recommendations which are close to the optimal solution. We have evaluated and compared our framework with a baseline method by using two popular recommendation datasets and we have obtained promising results on a set of widely accepted evaluation metrics.

Keywords: Recommendation system · Package recommendations · Top-k packages · Collaborative filtering

1 Introduction

Recommendation systems (RSs) have become popular since they can personalize user's experience by providing automated recommendations. RSs operate by analyzing user preference data, trying to identify correlations between them. User preference is expressed in various forms such as the history of purchases, usage logs and numerical ratings in a predefined scale (e.g. five star rating system). The RSs, in return, may propose a variety of items; for example, an on-line shop could

© Springer International Publishing AG 2017
G. Boracchi et al. (Eds.): EANN 2017, CCIS 744, pp. 477–489, 2017.
DOI: 10.1007/978-3-319-65172-9_40

suggest books or movies to users based on their profile, their previous purchases, the preferences of user's friends and other users with similar interests.

Package Recommender Systems extend the classical RSs by proposing to their users sets of items (packages) instead of single items. Package recommendation is extremely useful in a number of application domains (e.g. recommendation of packages of academic courses to students, packages of meals for a weekly diet, travel packages and sets of movies or books). This work focuses precisely on package recommender systems that are a specific category of RSs.

This work presents a recommendation framework which can be applied on systems that group items into categories and make package recommendations to users under various constraints (e.g. time, money). The proposed framework composes packages by selecting from a pool of items that may belong to multiple categories, using as input only the users' previous preference data (ratings of the items), the package size and a package cost threshold. The recommended packages comprise of items that match user preferences and satisfy the package cost restriction, as well as the package composition criteria that concern the categories of the selected items. The recommended packages are expected to achieve the highest preference score of the user to whom they are proposed to, while satisfying all user restrictions. Some examples are movie packages (a number of movies of different categories with a maximum total duration), weekly diets (a number of meals per day comprising of plates of a nutritional value and predefined calories) etc.

An optimal and a greedy solution, which are based on the user preferences and given restrictions in order to recommend the top-k packages, are proposed. The prediction of a user's preference score for an item is performed using the technique of item-based collaborative filtering [1]. The optimal solution matches items to categories so as to obtain the top-k packages of items with maximum preference score and at the same time it satisfies the constraint of maximum package cost. This optimal approach tries to solve the problem of recommendations by modeling it as a minimum cost flow problem that is solved by integer linear programming. On the other hand, the greedy solution performs in low running time due to low computation complexity and its recommendations are close to those of optimal solution. The proposed solutions were evaluated by estimating a set of measures and compared with a baseline approach, which is based on the popularity of items, without taking into account the current user preferences. For evaluation purposes, we have used two popular datasets and we have confirmed the robustness of the solution.

2   Related Work

Package recommendation has attracted the attention of the scientific community in recent years. In [2], the authors propose a system for recommending a team of experts, who have a set of predefined skills and a minimum communication cost. These experts are connected and communicate with each other within a social network. Their approach bears a similarity to our methodology as the skills may correspond to the categories of our work.

Other systems try to satisfy hard constraints among proposed packages of fixed size [3,4]. The packages can be the top-k tuples of entities that match user queries [3], or the result of rank join queries that formulate user aggregation constraints [4]. When a maximum cost (budget) restriction is added [5–7], the algorithms create packages that maximize user preference score satisfying the given constraints. In [8,9], authors introduce restrictions on prerequisite items of package recommendations. These restrictions are valid in certain application domains, e.g. in academic course recommendation, where an advanced course is offered only if other elementary courses have been completed successfully. Another course recommendation system that is based on the maximum flow algorithm is presented in [10]. Our proposed methodology differs from the aforementioned systems as it identifies the minimum cost flow, it includes the restriction of maximum package cost and it is more generic; it can be adapted and applied to a wide range of recommendation domains.

Flexible recommender systems, a special case of package recommender systems, do not obey hard constraints and can adapt to the application domain. In [11] authors introduced an intuitive user interface for travel package recommendation and in [12], they proposed a top-k package recommender that performs user preference elicitation. Our system is different since it is capable of directly producing recommendations based on past user evaluations, without requiring any additional user interaction. In [13], a versatile recommendation system for proposing packages of items has been developed. This system is flexible and does not refer to any particular application, but it is adapted to the requirements of each user. It is a *content-based* recommender system [14] that proposes packages adhering to user-provided constraints. A main difference of this system from our work is that it is based on content while we use collaborative filtering predictions.

Our model and system differentiates the other approaches in the field, since the compatibility constraints apply to the categories in which the items belong. The fact that an item may belong to more than one categories (i.e. item may fit to different packages with different roles), formulates a new package composition problem, which has not yet been discussed in the literature.

3   Package Recommendation

3.1   Problem Definition

The proposed recommender system assumes that a user u from a set of users U is interested in packages of fixed size, composed of items t_i from a set T. All items in T are considered to be of the same type (e.g. movies, plates or food portions, POIs) but may belong to one or more categories from a set C of l categories. Each user has provided ratings for several items of T and we can use these ratings to understand the user preferences, for both the specific items and the categories these items belong to. The rating information can be typically represented with a rating matrix R of size $n \times m$, where n is the total number of users and m is the total number of items. This matrix is usually sparse, since users typically rate only a few items, but by using a collaborative filtering algorithm, ratings for items that users have not rated yet are possible to be predicted [15]. Each item t_i has an associated cost i_{cost} (e.g. the movie duration, the distance to reach a POI or the money which is spent there or a combination of the two). Also each item has an item value i_{value}, which can be the item rating (actual or predicted). As a result, each package is a fixed-size set of s items that has a total cost p_{cost} (the sum of the costs of the items that each package contains) and a total value p_{value} (the sum of the selected item values).

The input parameters are the package size s, a maximum package cost max_{pcost}, optionally the number of alternative packages k and the user preferences which are represented by the number of categories per package or specifically by the number of items from each category (n_{cj}). The system recommends the package (or the *top-k* packages in non increasing p_{value} order) that covers user preferences, not surpassing max_{pcost} and maximizing the total package value p_{value}. Table 1 summarizes the notations used in the proposed model.

Table 1. Notation summary

Symbol	Description
U	The set of all users, $U = \{u_1, ..., u_n\}$, $|U| = n$
T	The set of items, $T = \{t_1, ..., t_m\}$, $|T| = m$
C	The set of categories $C = \{c_1, ..., c_l\}$, $|C| = l$
R	The *user-item* rating matrix (dimension $n \times m$)
i_{cost}	The cost of having item t_i in the package
i_{value}	The value that item t_i adds to the package
p	A recommended package with items $p = \{t_1, ..., t_s\}$
p_{cost}	The total package cost
p_{value}	The total package value
s	The package size (number of items in the package)
max_{pcost}	The maximum allowed package cost
k	The maximum number of recommended packages
n_{cj}	The number of items of category c_j in the package

A package p is a set of s items $\{t_1, ..., t_s\}$, where each item t_i belongs to one or more categories from C, but within the package each item represents a specific category c_j. The package can be recommended when it satisfies the compatibility and aggregation constraints of Eq. 1, where $|t_{cj}|$ is the number of items that represent the category c_j.

$$|t_{cj}| = n_{cj}, \forall c_j \in \{c_1, ..., c_l\}$$
$$p_{cost} \leq max_{pcost}, \text{ where } p_{cost} = \sum_{t_i \in p} i_{cost} \tag{1}$$
$$|p| = s$$

The total package value is given by Eq. 2 and the recommender system proposes the *top-k* packages in non increasing p_{value} order.

$$p_{value} = \sum_{t_i \in p} i_{value} \tag{2}$$

Package Compatibility Constraints: The main package composition constraint according to Eq. 1 refers to the number of items per category in the package $(n_{cj}: c_j \in \{c_1, ..., c_l\})$. These numbers can be given as input by the user or they may be derived implicitly from past user preferences. In this case, the popularity of each category is determined first, followed by the distribution of items among the q most popular categories. The popularity of a category c_j may depend on the number of items of this category that the user had rated in the past, the sum of the ratings the user has provided for the items in this category or any other method derived from the actually rated items. Once the category popularity for each user has been computed, then the top-q categories can be selected for composing the package, where q may be provided by the user explicitly. Based on the popularity scores for the top-q categories, we can derive the number of items for each category in the package, which may be proportional to the category popularity, or the same for all popular categories.

Package Aggregation Constraint: This constraint refers to the total package cost p_{cost}, which must not exceed a maximum threshold value max_{pcost}.

Item Value: The value of an item i_{value} can be the rating explicitly provided by the user for this item, or implicitly predicted by the system, e.g. using a collaborative filtering algorithm. An alternative method for defining the value of an item i, that combines the specific user value $r_{u,i}$ with the item popularity is given by the following equation:

$$i_{value} = r_{u,i} + PF \cdot \frac{n_i}{n} \tag{3}$$

where $PF \in \{0, 1, 2, ..., 10\}$ is a popularity factor, n_i is the number of users that have rated item t_i and n is the total number of users.

4   Solving the Package Composition Problem

4.1   Minimum Cost Flow Model

The problem of finding a package of items that satisfies the restrictions of Eq. 1 and at the same time maximizes the package value of Eq. 2 is an optimization problem which may add a big computational load to a recommendation system. A good algorithm that finds the optimum solution for each user is the key to the efficiency and effectiveness of the recommender system. The key concept is that each category may be optimally matched to a number of items according to the parameters of the problem. Essentially, this is a problem of optimal matching, which can be reduced to a minimum cost flow problem [16].

Since the proposed model aims at maximizing the package value for a predefined package size s, the total flow must be equal to the package size and the edge cost must be inverse to the item value in order for a minimum cost flow model to be used. The basis of the flow network is the bipartite graph $G(T, C, E)$, where T and C are the sets of item and category vertices respectively and E is the set of edges. An edge e_{ij} connects item t_i with category c_j and denotes that the item belongs to the category.

The bipartite graph is extended, for each user, to a flow graph (Fig. 1), by connecting the source vertex Src to all item vertices that the user has not yet valued as well as all the category vertices that are in the interest of the user to the termination vertex Trm. The edges carry the information $fc_{i,j}/max_{capij}/min_{capij}$, where $fc_{i,j}$ is the flow cost over the edge e_{ij}, max_{capij} and min_{capij} are the maximum and minimum edge capacity.

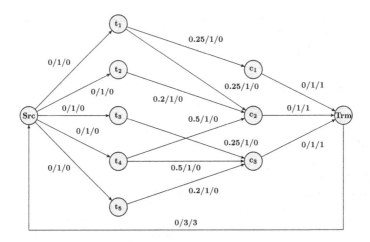

Fig. 1. An example of the minimum cost flow problem formulation

The maximum flow from Src to Trm may be reduced to the optimal matching between items and categories which, in turn, may lead to the creation of the optimal recommendation package. The derivation of this optimal matching between items and categories from the maximal flow problem is due to the validity of the integrality theorem [16] which claims that if all the capacities of the edges of a flow graph have integer values, then the maximum flow also assumes an integer value. Therefore, it is this theorem that allows the optimal matching between an integer number of items and categories. Adding a cost for passing the flow through a graph edge transforms the problem to a *minimum cost flow* one, and its solution results in the recommendation of the top package that maximizes the item value.

The cost for passing the flow through an edge that is connected to the source Src or terminal Trm vertices is zero ($fc_{Src,i} = fc_{j,Trm} = 0$, $\forall i \in T$, $\forall j \in C$). The flow cost $fc_{i,j}$ through the edge e_{ij} that connects item t_i with category c_j is the inverse of the (predicted) item value (i_{value}) and it is the same for all the edges that connect t_i with a category vertex (see Eq. 4).

$$fc_{i,j} = \frac{1}{i_{value}}, \forall i \in T, \forall j \in C \tag{4}$$

The maximum capacity of an edge connecting Src with an item vertex is equal to the maximum number of times the same item can be accessed (e.g. in the movie package this is always 1), while the minimum capacity is 0. The maximum capacity of edges connecting items to categories is always 1 since an item may belong to the same category only once (the minimum is 0). Finally, the maximum capacity of edges connecting category vertices and the Trm vertex is the number of items per category in the package (the minimum is 0). Figure 1 shows an example of a minimum-cost flow graph for finding the best package of size $s = 3$ with exactly 1 item from each of the categories c_1, c_2 and c_3. The predicted user ratings for items $\{t_1, t_2, t_3, t_4, t_5\}$ are $\{4, 5, 4, 2, 5\}$ respectively in the 0–5 scale. The optimal package for this problem includes items t_1, t_2, t_5, which minimize the cost of the flow with the minimum cost of 0.65 (i.e. $0.25 + 0.2 + 0.2$). The respective package value $p_{value} = 14$ is the maximum possible.

Integer Linear Programming Formulation: In the general case, the problem of discovering the minimum cost flow of the graph may be re-formulated in terms of linear programming [17]. In addition, the proposed model has an aggregation constraint; the maximum package cost threshold (max_{pcost}), which relates to budget or time restrictions that apply to the recommended package as a whole. The linear programming objective function and the constraints are depicted in Eq. 5, where $V = \{T, C, Src, Trc\}$ is the set of min-cost flow graph nodes, f_{ij} is the flow passing from an edge e_{ij} and i_{cost} is the cost of item $t_i \in T$.

$$\begin{aligned}
\text{minimize:} \quad & \sum_{i,j \in V, i \neq j} fc_{ij} \cdot f_{ij} \\
\text{subject to:} \quad & min_{cap_{ij}} \leq f_{ij} \leq max_{cap_{ij}}, \forall i, j \in V \,\&\, i \neq j \\
& \sum_{j \in V, i \neq j} f_{ij} = 0, \forall i \in V \\
& \sum_{i \in T, j \in C} f_{ij} \cdot i_{cost} \leq max_{pcost} \\
& f_{ij} \in \mathbb{Z}_{\geq 0}
\end{aligned} \tag{5}$$

4.2   Greedy Solution and Baseline

A greedy alternative begins by adding to the package items from the most wanted category (as defined by the user preferences) and continues with the other categories in non increasing i_{value} order, satisfying the composition restrictions and without violating the aggregation restriction (i.e. maximum package cost). If the package cost exceeds max_{pcost}, then the item with the minimum $i_{value}/cost$ value is replaced with the next item (in non increasing rating order) of the same category (in order to keep composition restrictions and improve towards the aggregation restriction). The algorithm terminates successfully when all the composition restrictions have been met (i.e. the requested number of items from each category has been added to the package) without exceeding max_{pcost}. It fails when all the items of a requested category have been examined but they cannot fit for the package.

Finally, a baseline method, which ignores user preferences and assumes that an item value (i_{value}) is proportional only to the overall popularity of the item, replace the predicted ratings for each user in the greedy algorithm, giving us a non-personalized baseline.

Data: u, T, C, R, s, max_{pcost}, k
Result: top-k packages
forall the $c_j \in C$ **do**
  | $T_{cj} \leftarrow$ sort items in decreasing value (rating) order for category c_j;
end
forall the Pi, $i = \{1, ..., k\}$ **do**
  | $Pi \leftarrow$ top n_{cj} items for each category c_j;
  | **while** $Pi_{cost} > max_{pcost}$ **do**
  | | replace t_x of minimum $\frac{i_{value}}{i_{cost}}$ with t_{x+1}: $category_{t_x} = category_{t_{x+1}}$
  | **end**
  | Update T_{cj} sets to produce the next package;
end

Algorithm 1. Greedy algorithm for the top-k package creation.

4.3   Composing Top-k Packages

In the minimum cost flow model, the problem of composing the *top-k* packages can be solved by repetitively using the minimum cost flow formulation for slightly modified flow networks. Since it produces the optimum solution each time, it is necessary to update the graph on each iteration by removing vertices and edges based on what was selected in the previous packages. The Greedy algorithm follows a similar strategy (e.g. items with the minimum i_{value}/i_{cost} scores that have already been recommended can be removed from the set of candidate items).

4.4   Computational Complexity

The problem of the Optimal solution which is formulated as integer linear programming is NP-Complete [18]. In our case, the pruning of the equations by using a certain number of items with higher values, the problem is solved in reasonable time as it is demonstrated in Sect. 5. The complexity of the Greedy algorithm for the first loop (sort ratings by category) is $O(|C| \cdot |T| \cdot log|T|)$ and for the second loop is $O(k \cdot |T|)$, in the worst case where all items belong to all categories and all items of the categories are examined for composing packages. Since $|C|$, $k \ll |T|$, the computational complexity of the Greedy algorithm is $O(|T| \cdot log|T|)$. This complexity allows the algorithm to perform in low running times as shown in the experimental evaluation.

5   Evaluation

In order to implement and evaluate the proposed model, we have developed a Java application[1] that allows us to test the proposed solution in a wide range of application scenarios, by modifying system parameters through a GUI. It also permits an incomplete rating dataset (not all users rate all items) to be imported and the missing ratings to be predicted using a Collaborative Filtering (CF) algorithm[2]. The application is connected to an external linear programming module[3], which solves the optimization problem of the optimal approach.

Datasets: The first dataset, Movielens 1M[4], includes almost 1 million ratings provided by 6,040 users for 3,900 movies. The durations of films are retrieved from the OMDB[5] web service and used as the cost of each movie item. The second dataset is a subset (1M ratings) of the Anime[6] dataset, including more than 7 million ratings provided by 73,516 users on 12,294 series. The cost for an Anime series is set to be the number of its episodes. It could also be a monetary cost, or any other quantity that best fits the application domain.

Experimental Parameters: Since the proposed solution is generic, it is not possible to examine its performance in all possible parameter setups (e.g. the number of packages, the package composition strategy etc.). Although we evaluated many scenarios, due to space limitation, we fix some parameters, which remain the same in all the experiments and evaluate the effect of the aggregation and composition parameters in system performance. More specifically, we evaluate the top-10 packages, comprising of one item from each of the top-3 most popular categories for each user and the choice is made from the top-500 highest rated items (prediction) for the user.

---

[1] The application jar file, source code, usage instructions and a sample dataset, which was also used for the evaluation, are available for downloading at https://goo.gl/IMbxq1.

[2] Item-based CF implementation of Apache Mahout (https://mahout.apache.org).

[3] IBM ILOG CPLEX solver.

[4] https://grouplens.org/datasets/movielens/.

[5] http://www.omdbapi.com/.

[6] https://www.kaggle.com/CooperUnion/anime-recommendations-database.

Evaluation Methodology: The item-item CF algorithm, an algorithm with proven performance [19], with the Tanimoto coefficient for measuring item similarity [20] being employed for predicting item ratings. For the evaluation of the package recommender algorithms, we choose randomly 50 users who have rated more than 500 items. For these users the 40% of their ratings is hidden (random stratified sampling per user) and all the remaining ratings in the dataset are used for training. We repeat this Monte Carlo cross-validation technique five times [21] and report the average performance of the package recommender algorithms. The metrics employed are precision and the value of each proposed package (typical for the performance evaluation of package recommender systems [10]). In order to evaluate the time performance, we also report the *total running time*. The *package value*, is given in Eq. 2. *Precision*, given in Eq. 6 (where T_p is the set of items in the package and T_h is the set of highly rated "hidden items"), counts the number of items in a package p that have been rated by the user, but the respective ratings have not been used for training ("hidden items"). Since we do not want to recommend items that have actually received low ratings by the user, we consider only those "hidden items" that have been rated highly (rating ≥ 4).

$$Precision_p = |T_p \cap T_h|/|T_p| = |T_p \cap T_h|/s \tag{6}$$

5.1   Results

As far as the maximum package cost is concerned, we have experimented with various thresholds and the results in Fig. 2 show that the precision increases by relaxing this aggregation constraint as expected. The Optimal algorithm performs better than the Greedy and both algorithms outperform the baseline approach.

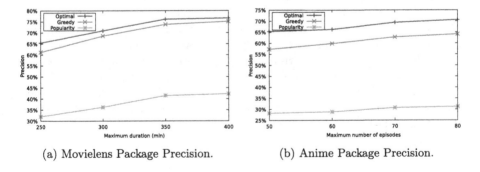

(a) Movielens Package Precision.   (b) Anime Package Precision.

Fig. 2. Precision with varying maximum package cost.

A comparison of the three approaches, in recommending the top-k packages of size 3 (1 item from the top-3 categories), as depicted in Fig. 3, reveals the superiority of the proposed approach, when compared against the Greedy and

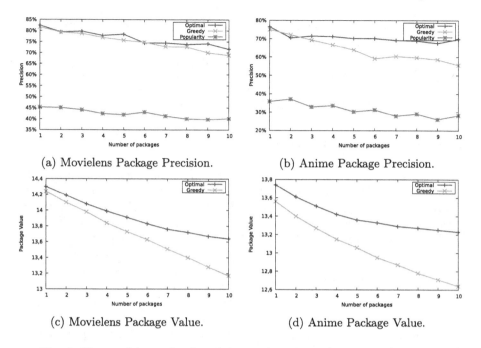

(a) Movielens Package Precision.

(b) Anime Package Precision.

(c) Movielens Package Value.

(d) Anime Package Value.

Fig. 3. The precision and value of the top-k packages (for various k values).

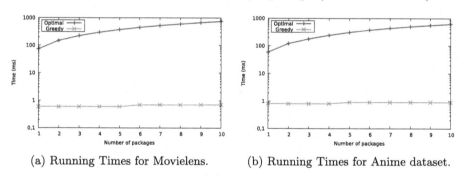

(a) Running Times for Movielens.

(b) Running Times for Anime dataset.

Fig. 4. Running time for creating the top-k packages (for various k values).

the Item-Popularity alternatives. Although the precision of the greedy algorithm is comparable to the optimum in the top-3 packages, the respective package value quickly deteriorates, which means that the greedy algorithm recommends an equal number of "hidden items" but with lower ratings.

As for the time complexity of each algorithm is concerned, Fig. 4 shows that the greedy solution is much faster due to low computational complexity and returns the top-10 packages in less than 1 ms, whereas the Optimal Solution needs 700 ms for the top-10 packages.

Based on the experimental results, we conclude that both the Optimal and the Greedy solution operate efficiently, achieve low running times and high precision rates. The Greedy solution can be applied in applications that require low running times and low computational complexity even though it may lose in package quality, whereas the Optimal solution can be applied in applications that require a maximum recommended package quality.

6 Conclusions and Future Work

This work proposes a package recommender framework that employs user preferences. Our model is not designed for a particular domain but can be applied widely as it is comprised of a number of parameters which can be configured accordingly, allowing it adapt to various recommendation problems. An integrated recommendation system, which incorporates all the required functionality, was developed in order to support the implementation and evaluation of our approach. The results of the evaluation experiments showed that the proposed model fulfills its purpose and works effectively and efficiently.

The first positive evaluation results of our approach lead us to think about further extensions. It is on our next plans this model to be applied and evaluated in the nutrition domain as it may be a solution for recommending personalized gastronomic diets. We are also working on automatically setting the system parameters which depend on the application, the dataset and the user preferences. This may allow us to achieve high performance.

References

1. Ekstrand, M.D., Riedl, J.T., Konstan, J.A., et al.: Collaborative filtering recommender systems. Found. Trends HCI 4(2), 81–173 (2011)
2. Lappas, T., Liu, K., Terzi, E.: Finding a team of experts in social networks. In: 15th ACM SIGKDD, pp. 467–476. ACM (2009)
3. Angel, A., Chaudhuri, S., Das, G., Koudas, N.: Ranking objects based on relationships and fixed associations. In: 12th EDBT, pp. 910–921. ACM (2009)
4. Xie, M., Lakshmanan, L.V., Wood, P.T.: Efficient rank join with aggregation constraints. VLDB Endow. 4(11), 1201–1212 (2011)
5. Xie, M., Lakshmanan, L.V., Wood, P.T.: Breaking out of the box of recommendations: from items to packages. In: 4th RecSys Conference, pp. 151–158. ACM (2010)
6. Xie, M., Lakshmanan, L.V., Wood, P.T.: CompRec-Trip: a composite recommendation system for travel planning. In: 27th ICDE Conference, pp. 1352–1355. IEEE (2011)
7. Benouaret, I., Lenne, D.: A package recommendation framework for trip planning activities. In: 10th RecSys Conference, pp. 203–206. ACM (2016)
8. Parameswaran, A.G., Garcia-Molina, H.: Recommendations with prerequisites. In: 3rd RecSys Conference, pp. 353–356. ACM (2009)
9. Parameswaran, A.G., Garcia-Molina, H., Ullman, J.D.: Evaluating, combining and generalizing recommendations with prerequisites. In: 19th ACM CIKM, pp. 919–928. ACM (2010)

10. Parameswaran, A., Venetis, P., Garcia-Molina, H.: Recommendation systems with complex constraints: a course recommendation perspective. ACM Trans. Inf. Syst. (TOIS) **29**(4), 20 (2011)
11. Xie, M., Lakshmanan, L.V., Wood, P.T.: IPS: an interactive package configuration system for trip planning. VLDB Endow. **6**(12), 1362–1365 (2013)
12. Xie, M., Lakshmanan, L.V., Wood, P.T.: Generating top-k packages via preference elicitation. VLDB Endow. **7**(14), 1941–1952 (2014)
13. Interdonato, R., Romeo, S., Tagarelli, A., Karypis, G.: A versatile graph-based approach to package recommendation. In: 2013 IEEE 25th International Conference on Tools with Artificial Intelligence (ICTAI), pp. 857–864. IEEE (2013)
14. Melville, P., Sindhwani, V.: Recommender systems. In: Sammut, C., Webb, G.I. (eds.) Encyclopedia of Machine Learning, pp. 829–838. Springer, New York (2011). doi:10.1007/978-0-387-30164-8_705
15. Schafer, J.B., Frankowski, D., Herlocker, J., Sen, S.: Collaborative filtering recommender systems. In: Brusilovsky, P., Kobsa, A., Nejdl, W. (eds.) The Adaptive Web. LNCS, vol. 4321, pp. 291–324. Springer, Heidelberg (2007). doi:10.1007/978-3-540-72079-9_9
16. Dasgupta, S., Papadimitriou, C.H., Vazirani, U.V.: Algorithms. McGraw-Hill, New York (2008)
17. Bertsekas, D.P.: Network Optimization: Continuous and Discrete Models. Athena Scientific, Belmont (1998)
18. Matousek, J., Gärtner, B.: Understanding and Using Linear Programming. Springer Science & Business Media, Heidelberg (2007). doi:10.1007/978-3-540-30717-4
19. Seminario, C.E., Wilson, D.C.: Case study evaluation of Mahout as a recommender platform. In: RUE@ RecSys, pp. 45–50 (2012)
20. Anil, R., Owen, S., Dunning, T., Friedman, E.: Mahout in Action. Manning Publications Co., Shelter Island (2012)
21. Dubitzky, W., Granzow, M., Berrar, D.P.: Fundamentals of Data Mining in Genomics and Proteomics. Springer Science & Business Media, New York (2007). doi:10.1007/978-0-387-47509-7

Deriving Business Recommendations for Franchises Using Competitive Learning Driven MLP-Based Clustering

Haidar Almohri and Ratna Babu Chinnam[✉]

Industrial and Systems Engineering Department, Wayne State University,
Detroit, MI 48201, USA
{haidar.almohri,ratna.chinnam}@wayne.edu

Abstract. Finite Mixture of Regression (FMR) models account for the target variable and cluster the data points based on the relationship between inputs and target variables. However, extant FMR models mostly rely on linear regression models. We propose a competitive learning algorithm to perform FMR modeling, which allows nonlinear models such a Multi-Layer Perceptrons (MLPs) to carry out the regression. We demonstrate the proposed method using a real-world case study that aims to derive tailored recommendations for franchises to improve their profitability and sales effectiveness.

Keywords: Competitive learning · Model-based clustering · Regression · Neural network · Grouped observations · Franchise · Recommendations · Performance management · Automotive dealerships

1   Introduction

The franchise industry contributes to about 3% of the total U.S. GDP, estimated at $521 billion in 2015. The roughly 782 thousand establishments in the top ten U.S. franchise sectors alone employ over 8.8 million workers. While the industry is expected to grow in the near term, a number of chains are struggling to perform and have seen high default rates in recent years on loans by their franchisees. Although some of these problems have been attributed to the recent recession, the industry is lacking comprehensive and effective performance management systems. While Retail Information Systems (RIS) are effective in helping to manage core store activities, many chains need advanced analytics platforms to analyze the growing and diverse data streams.

In the era of big data, the ability to collect and process data from all corners of the franchise has become relatively inexpensive and accessible for a lot of chains. While standard retail information systems can address the core needs for most chains, businesses that can analyze this wealth of structured (e.g., POS transactions, time-stamped process data) and unstructured (e.g., customer feedback) data to inform and guide real-time operations, tactical plans, and longer-term growth strategies can be game changers with the potential to disrupt industry.

© Springer International Publishing AG 2017
G. Boracchi et al. (Eds.): EANN 2017, CCIS 744, pp. 490–497, 2017.
DOI: 10.1007/978-3-319-65172-9_41

While the opportunities and solutions can be sector specific, the potential for fundamental transformation is unquestionably vast.

We report here modeling experiences from assisting a supplier of strategic and operational planning solutions for the automotive sector. Firm is a leader in providing dealership location and network analysis to many automotive original equipment manufacturers (OEMs). It wanted to develop a dealership performance management analytics platform to analyze monthly statements from thousands of dealers nationally to understand factors that jointly improve profitability for the dealership while also improving new vehicle sales to satisfy OEM requirements. The traditional approach to dealer performance assessment involves setting sales targets as a function of brand penetration within the immediate territory (e.g., by state). However, this simplistic approach does not account for demographic differences and other factors (e.g., proximity to other dealers) that drive dealership performance. In the absence of objective and effective data-driven analytics platforms, the current practice mainly relies on experienced dealership consultants and ad hoc guidance from field personnel.

Given the sparsity of data in our setting, i.e. monthly financial statements from individual dealers, it is not prudent to build dedicated predictive models for individual dealers. A typical monthly financial statement can report hundreds if not thousands of parameters from different departments of a store (e.g., the new product sales department can supply information regarding products carried, product mix, inventory levels, sales personnel, personnel experience etc.). A common approach in the retail/franchise sector is to segment stores by region, demographics, and/or products carried and build performance predictive models for each segment for providing guidance to individual stores. Regional segmentation does make sense in the automotive sector for products (including optional content) can vary by region depending on overall weather and climate factors (e.g., colder states prefer more heating equipment whereas customers in warmer climate might choose more convertibles). However, there are other factors that could lead to variation in performance across dealerships. Both internal factors (e.g., product carried, quality of workforce, operational processes) that are under more control of the management and external factors (e.g., demographics and proximity of competitive dealers) are at play. We posit that performance of analytics platforms for deriving tailored recommendations for individual dealers can be improved by further establishing sub-groups of franchisees within the local regions using model-based clustering techniques.

The core idea behind standard model-based clustering methods is that sample observations arise from a distribution that is a mixture of two or more components. Each component can be described by a density function and has an associated probability in the mixture. They are still 'unsupervised' techniques in that there is no target variable. In our application setting, the goal is to identify factors (or key performance indicators or KPIs) driving the performance of dealerships and derive effective recommendations. Typical performance measures of interest include profitability of the dealer and sales effectiveness (e.g., number of new vehicles sales). The task is to group similar dealers by considering the relation between the KPIs and performance metrics. Finite Mixture of Regression (FMR) models [1] were introduced to account for the target variable and cluster the data points based on the relationship between inputs and target variables. The core contribution of this article is the introduction of a competitive learning algorithm to perform FMR using the Multi-Layer Perceptron (MLP) as the underlying regression model. In addition, the proposed method also naturally supports any requirement that seeks to assign groups of observations to individual clusters rather than individual observations. For example, all the data points corresponding to a particular store or dealership should be assigned to the same cluster and not allocated to different clusters.

The rest of the manuscript is organized as follows: Sect. 2 further elaborates model-based clustering and finite mixture of regression techniques, Sect. 3 describes the completive learning algorithm, Sect. 4 describes results from a case study, and finally Sect. 5 offers some concluding remarks.

2 Model-Based Clustering and Finite Mixture of Regression

Clustering is a technique that is widely used in data mining applications and seeks to find subsets of a given data set, with the hope that the within-subset data points are somehow similar and the cross-subset data points are different. There is a large number of clustering techniques that exist in the literature and they mostly rely on heuristic or geometric procedures and typically focus on dissimilarity measures between pairs of observations. A popular dissimilarity measure is based on the distance between groups, introduced by Ward [13] for hierarchical clustering. k-means algorithm [8] is perhaps the most popular clustering algorithm among the geometric procedures. There is however little systematic guidance associated with these methods for solving important practical questions that arise in cluster analysis, such as how many clusters are there, which clustering method should be used, and how should outliers be handled [4]. Model-based clustering methods are developed to partially overcome these difficulties. Finite mixture models (FMMs) in particular have gained a lot of attention by the researchers [10,12]. FMMs are used to represent heterogeneity in a finite number of so called latent classes. It concerns modeling a statistical distribution by a weighted sum of other distributions. The general framework of FMMs is of the form:

$$f(x) = \sum_{i=1}^{I} \lambda_k f(x_i; \theta_k) \tag{1}$$

where $0 < \lambda_i < 1$ and $\sum_{i=1}^{I} \lambda_i = 1$. The goal is to estimate the parameters of each $f(x_i)$, and ultimately find the probability of data point y_j belonging to components (subsets) $1, \ldots, I$.

Most of the clustering algorithms including FMM are considered unsupervised techniques, meaning that the target variable is not present. Finite Mixture of Regression (FMR) models [1] were introduced to account for the target variable and cluster the data points based on the relationship between the input and target variable. While most parametric FMR models rely on linear statistical regression models, they might not adequately explain the variation in the target variable. Non-parametric regression techniques such as Multi-Layer Perceptrons (MLPs) might provide better performance, and hence, employed within the proposed framework.

In regular clustering, each data point is assigned to a cluster. However, sometimes it is desirable to link some of the data points so that they stay in the same cluster. For example, in some applications the background knowledge and domain expertise can be used to realize that some data points must end up being in the same cluster. This method can be used in various applications such as in document classification, Gene classification, clustering GPS trace data from automobiles, etc. (see [2]). In such cases, a constraint has to be imposed to the clustering algorithm so that it forces the "must-link" observations to stay in the same cluster. In the context of our dealership application, there are 60 monthly financial reports (observations) for each dealer corresponding to five years of monthly data. Since our goal is to cluster the dealers (not the individual monthly observations), we cannot use the standard mixture models for this purpose.

This problem is known as "clustering with constraints" in the literature (see [3,11]). There are two main constraint clustering problems that have been addressed by the researchers: "Cannot-link" and "Must-link" constraints. In "Cannot-link" constraint, the aim is to prevent the group members to be in the same cluster, where as in "Must-link" constraint, the group members have to stay in the same cluster. We refer to [7] for further reading. In this work, we developed an algorithm that performs model-based clustering and guarantees to keep the observations that belong to the same dealer in the same cluster. In other words, the outcome of the algorithm is the clustered dealers (not the individual monthly observations). This is a novel, non parametric, method to solve the "must-link" constraint clustering in the context of FMR.

3   Competitive Learning Algorithm for MLP-Based Clustering

Competitive learning is a form of unsupervised learning in artificial neural networks, in which nodes of the network compete for the right to respond to a subset of the input data [9]. A variant of Hebbian learning, standard competitive learning algorithms work by increasing the specialization of each node in the network (typically composed a single layer of neurons) and is well suited to finding clusters within data. There are three basic elements to the standard competitive learning rule [6]: (1) A set of neurons that are all the same except for some randomly distributed synaptic weights, and which therefore respond differently to a given set of input patterns; (2) A limit imposed on the 'strength' of each neuron, and (3) A mechanism that permits the neurons to compete for the right to respond to a given subset of inputs, such that only one output neuron, is active (i.e. 'on') at a time. Typically, the neuron that wins the competition is called a 'winner-take-all' neuron. Accordingly, during the training cycle, the individual neurons of the network learn to specialize on ensembles of similar patterns and in so doing become 'feature detectors' for different classes of input patterns.

The standard competitive learning algorithm can be readily extended to multi-layer networks to enhance the ability of the algorithm to improve the quality of the resulting clusters. Essentially, one can replace each neuron in a standard competitive learning network with a non-linear model such as a Multi-Layer Perceptron (MLP). We illustrated the approach using our dealership example. Let us suppose that we wish to group the dealers into k clusters. The algorithm starts by randomly selecting k initial dealers (each of which has $N = 60$ observations) and trains an MLP to each set to obtain k initial models and associated clusters. Next, a random dealer (that is not among the initial k dealers) is selected (along with all its observations). The target variables for the observations of the selected dealer are predicted by each of the k MLP models and the error for each case is recorded. The dealer is then assigned to the cluster that is associated with the model with the least prediction error. In essence, the winning MLP model 'takes' all the associated dealer observations. The rest of the dealers are assigned to one of k clusters in the same fashion to complete the first iteration. In the next iteration, the initial k models are trained using all the observations that are gathered in each of the k clusters. The dealers are again selected in random order one at a time and the target variable is predicted in the same way to assign the dealers to one of the k clusters. This process is repeated until the algorithm converges (i.e., there are no dealer reassignments across consecutive iterations). There are clearly parallels here to clustering algorithms such as the k-means clustering technique. Note that the proposed approach can readily employ any type of a supervised learning algorithm for regression (e.g., radial-basis-function network, support-vector-machine, or even a random forest).

To assess and compare the performance of the predictive models, several criterion such as minimum sum of squared errors (SSE) or minimum Akaike Information Criterion (AIC) can be employed by evaluating the models on a testing dataset. In our experiments, AIC proved to be quite effective. Founded on information theoretic principles, it offers a relative estimate of the information lost when a given model is used to represent the process that generates the data. In doing so, it deals with the trade-off between the goodness of fit of the model and the complexity of the model. After assigning each group of observations belonging to a dealer to the cluster with the least AIC, the 'Cluster AIC' is calculated for each cluster by averaging the AIC for all the assigned dealers within each of the K clusters. The 'Overall AIC' is the sum of the Cluster AICs: $Overall\ AIC = \sum_{j=1}^{K} Cluster.AIC_j, \forall j = 1, \ldots, K$. The stopping criteria is the relative change in overall AIC between iteration t and $t + 1$, i.e. detecting that there is no improvement in assigning the groups to new clusters, or if the maximum number of iterations has been reached.

To select the best initial index k, the algorithm is run several times for different values of k and the quality of predictions (on testing data set) are recorded. The index value that produced the best results is selected to initialize the algorithm.

4   Case Study

As mentioned in the introduction, the motivation for developing the described algorithms comes from a project that involves modeling the performance of individual automotive dealerships of a particular Original Equipment Manufacturer (OEM) across the United States. The idea comes from the fact that depending on certain characteristics such as demographics of the location, the competition in the market, etc., these dealerships are managed and perform in different manners. To better understand the relationship between the Key Performance Indicators (KPIs) and the outcome performance of the dealerships, the dealerships are to be assigned to different clusters. The goal is to create clusters that contain similar dealers (in terms of the relationships between KPIs and dealer performance).

The dataset consists of about $S = 3000$ dealerships, with 60 consecutive monthly observations for each. Subject-matter-experts deemed 281 KPIs (independent variables) within each financial statement or observation to be effective candidates for modeling. Therefore, the number of observations totaled $N = 3000 * 60 = 180,000$. There were several missing entries in the data set. The missing values are imputed using the matrix completion via soft thresholding SVD technique, using the 'softImpute' package in R [5].

To apply the competitive learning algorithm, k dealers are selected initially, and k neural networks are fit to each of them. Multi-layer Perceptron (MLP) with a single hidden layer is found to be sufficient in modeling the data. At each iteration, each of the remaining dealers are assigned to one of the k models as described in Sect. 3. To find the optimal number of clusters (k), the data is split into training (first four years of financial statements), and testing (statements from the last year). The parameters are then selected using cross validation, by evaluating the quality of the models on predicting the testing data. The parameters that produced the best result (highest R^2 value) on the testing dataset are selected.

Once the parameters are selected, the algorithm is applied to the data set. It is observed that in most cases, the algorithm converges in less than 15 iterations for this data set (with 180,000 rows and 281 columns). To evaluate the effectiveness of the algorithm, the dealers are further divided into three sub-groups based on the number of expected vehicle registrations in their territories (the client firm chose to not employ the demographic variables for dealer clustering and instead felt that forming dealer sub-groups using expected vehicle registrations will lead to more robust clustering). Dealers from each of the three sub-groups (by expected registrations) are then subject to the proposed competitive learning driven MLP-based clustering.

Note that there are two dependent variables (ROS measuring return on sales or profitability and SE measuring sales effectiveness). Tables 1 and 2 show the results of modeling ROS and SE on the testing dataset, respectively.

Table 1. Comparison of clustering vs. single model, dealership dataset focusing on ROS

Registration group	Algorithm	Number of clusters	R^2
1	Clustering	$k = 3$	0.59
	Single model	$k = 1$	0.56
2	Clustering	$k = 2$	0.64
	Single model	$k = 1$	0.63
3	Clustering	$k = 2$	0.50
	Single model	$k = 1$	0.50

Table 2. Comparison of clustering vs. single model, dealership dataset focusing on SE

Registration group	Algorithm	Number of clusters	R^2
1	Clustering	$k = 2$	0.18
	Single model	$k = 1$	0.24
2	Clustering	$k = 2$	0.12
	Single model	$k = 1$	0.05
3	Clustering	$k = 2$	0.15
	Single model	$k = 1$	0.02

Clustering the dealers has generated equal or better result in all the cases except for Group 1 under SE. While the results from Table 1 suggest no advantage to modeling ROS using a mixture of regression models, there is marked improvement in R^2 for two of the three registration groups for SE modeling using the proposed competitive learning driven MLP-based clustering approach (see Table 2).

For registration group 3, the competitive learning algorithm has increased the R^2 from 0.02 using a single model to 0.15, an increase of 750%, which can an impact on the performance of the dealership.

Overall, it appears that the dealers are not so different in terms of the factors at play for generating profit (ROS). However, selling new cars seems to be more sensitive to the type of dealer and its management strategies.

Using the output of the algorithm, the clustered dealers can be studied individually to identify the key drivers for the performance of each group. Furthermore, investigating the differences between the clusters can provide valuable information for deriving successful management strategies.

Given the models, deriving effective recommendations for individual franchises (dealers in our case) would involve studying the input-output relationships and carrying out multi-objective optimization where one seeks to jointly maximize ROS and SE while constraining the input variables (e.g., operating capital) to feasible domains.

5 Conclusion

We propose a competitive learning algorithm to perform finite mixture of regression modeling. In addition, the algorithm can support nonlinear models such as the Multi-Layer Perceptron (MLP) as the underlying regression model. In addition, the proposed method also naturally supports any requirement that seeks to assign groups of observations to individual clusters rather than individual observations. The method performs a non-parametric Finite Mixture of Regressions (FMR) with "Must-link" constraint clustering.

The proposed method is also illustrated using a real-world case study that aims to develop an analytics platform for providing tailored business recommendations for retail franchises. Results show a significant improvement in prediction quality in modeling the Sales Effectiveness for the dealerships.

Future research can extend the proposed algorithm to embrace a maximum likelihood approach and allow joint determination of the optimal number of clusters.

References

1. Bar-Shalom, Y.: Tracking methods in a multitarget environment. IEEE Trans. Autom. Control **23**(4), 618–626 (1978)
2. Berkhin, P.: A survey of clustering data mining techniques. In: Kogan, J., Nicholas, C., Teboulle, M. (eds.) Grouping Multidimensional Data, pp. 25–71. Springer, Heidelberg (2006). doi:10.1007/3-540-28349-8_2
3. Dhillon, I.S., Fan, J., Guan, Y.: Efficient clustering of very large document collections. Data Min. Sci. Eng. Appl. **2**, 357–381 (2001)
4. Fraley, C., Raftery, A.E.: Model-based clustering, discriminant analysis, and density estimation. J. Am. Stat. Assoc. **97**(458), 611–631 (2002)
5. Hastie, T., Mazumder, R.: softImpute: matrix completion via iterative soft-thresholded SVD (2015). R package version 1.4
6. Haykin, S.S.: Neural Networks and Learning Machines, vol. 3. Pearson, Upper Saddle River (2009)
7. Leisch, F., Grün, B.: Extending standard cluster algorithms to allow for group constraints (2006, n.a.)
8. MacQueen, J., et al.: Some methods for classification and analysis of multivariate observations. In: Proceedings of the Fifth Berkeley Symposium on Mathematical Statistics and Probability, Oakland, CA, USA, vol. 1, pp. 281–297 (1967)
9. Rumelhart, D.E., McClelland, J.L., PDP Research Group, et al.: Parallel Distributed Processing, vol. 1. IEEE (1988)
10. Sarstedt, M.: Market segmentation with mixture regression models: understanding measures that guide model selection. J. Target. Meas. Anal. Mark. **16**(3), 228–246 (2008)
11. Segal, E., Wang, H., Koller, D.: Discovering molecular pathways from protein interaction and gene expression data. Bioinformatics **19**(Suppl. 1), i264–i272 (2003)
12. Tuma, M., Decker, R.: Finite mixture models in market segmentation: a review and suggestions for best practices. Electron. J. Res. Methods **11**(1), 2–15 (2013)
13. Ward Jr., J.H.: Hierarchical grouping to optimize an objective function. J. Am. Stat. Assoc. **58**(301), 236–244 (1963)

The 50/50 Recommender: A Method Incorporating Personality into Movie Recommender Systems

Orestis Nalmpantis and Christos Tjortjis[(✉)]

School of Science and Technology, International Hellenic University,
14th km Thessaloniki - Moudania, 57001 Thermi, Greece
{o.nalmpantis, c.tjortjis}@ihu.edu.gr

Abstract. Recommendation systems offer valuable assistance with selecting products and services. This work checks the hypothesis that taking personality into account can improve recommendation quality. Our main goal is to examine the role of personality in Movie Recommender systems. We introduce the concept of combining collaborative techniques with a personality test to provide more personalized movie recommendations. Previous research attempted to incorporate personality in Recommender systems, but no actual implementation appears to have been achieved. We propose a method and developed the 50/50 recommender system, which combines the Big Five personality test with an existing movie recommender, and used it on a renowned movie dataset. Evaluation results showed that users preferred the 50/50 system 3.6% more than the state of the art method. Our findings show that personalization provides better recommendations, even though some extra user input is required upfront.

Keywords: Personalization · Collaborative filtering · Recommendation systems · Data mining

1 Introduction

The massive growth and impact of the World Wide Web had as a result the handling and distribution of huge amounts of data and information. Although this may seem as an improvement for computer technology, it certainly came with some drawbacks.

This is where a recommender system comes in place, an assistive device that directs and guides the user in their search for useful information. Personalization in recommender systems tends to be a new trend, and it is mostly based on the theory of human-computer interaction, which states that computers will and should always work with and for humans.

The main motivation behind this work is the lack of personalization in current recommender systems. We hypothesize that a recommender system should factor in the basic personality traits of the active user and that this will help in producing better recommendations.

Our aim is to use the Big Five Personality test [1], to find the user's personality traits and then connect these traits with his/her movie genre preferences. Then, based

© Springer International Publishing AG 2017
G. Boracchi et al. (Eds.): EANN 2017, CCIS 744, pp. 498–507, 2017.
DOI: 10.1007/978-3-319-65172-9_42

on the movie reviews the user gave to the system and the personality test results, we recommend a list of movies by applying our own formula. This list takes into account 50% his personality and 50% his movie. We also experimented with how users react to a list that is filled with movies that are 80% based on their personality and 20% on their movie reviews to see how far we can go with the addition of Personality in producing final recommendations, and compared these to just using standard k-NN based recommendations.

The remaining of the paper is organized as follows: in Sect. 2, we provide the background to our research, which includes information on CF Recommender Systems, k-NN and Personalization. In Sect. 3, we detail the proposed 50/50 Movie Recommender System. In Sect. 4, we present results and discuss the pros and cons of our method, along with threats to validity. Finally, Sect. 5 concludes with ideas for future improvements.

2 Background

A Recommender system is "any system that produces individualized recommendations as output or has the effect of guiding the user in a personalized way to interesting or useful objects in a large space of possible options" [2]. This is the main concept behind CF Recommender systems.

2.1 Collaborative Filtering Recommender Systems

The most common way of getting a recommendation in real life is by asking a friend for their opinion, preferably someone who usually likes the same things as you. This is exactly the idea behind CF [13]. The basic premise of CF is that if two users have the same opinion about a bunch of products, then they are likely to have similar opinions about other products too. The objective of the algorithm is to predict the active user's ratings for products they have not yet rated. With these predicted ratings, you can sort the products and recommend the top picks for the active user. The following sub-section shows the Basic Steps in CF as discussed in [9, 10], which are also the basic steps we follow initially for our recommender.

CF Basic Steps

1. The set of ratings for the active user is identified.
2. k-NN is used to select k users who are most similar to the active user, according to a similarity function called Pearson Correlation shown in formula (1).
3. Identify the products that these similar users liked.
4. A prediction is generated with a Prediction Rating Formula, meaning ratings that would hypothetically be given by the active user for each of these products.
5. A set of top N products is recommended based on the top highest predicted ratings of the products in the previous step.

In step 2, the k-NN Algorithm and Pearson Correlation are used to find similar users to the active user. Generally, the k-NN algorithm represents all the users as Data Points in an X- Y plot and Pearson Correlation is used as a similarity distance metric to find which of these Data Points are closer to the active user. The closer they are, the more similar they are to the active user.

$$Corr(x, y) = \frac{\sum_i (xi - \bar{x})(yi - \bar{y})}{\sqrt{\sum (xi - \bar{x})^2}\sqrt{\sum (yi - \bar{y})^2}} \tag{1}$$

Then in Step 4, once we find the k-nearest neighbors of the active user, we use those neighbors to find the rating that the active user will give to any product, using the following formula (2):

$$p_{a,i} = \bar{r}_a + \frac{\sum_{u \in U}(r_{u,i} - \bar{r}_u).w_{a,u}}{\sum_{u \in U}|w_{a,u}|} \tag{2}$$

This is the foundation of our Recommender, and later this will be enhanced by the addition of the personality feature of the active user.

2.2 Personalization

Computer scientists try to model the human psychological aspects and include them in recommender systems, so to make the recommender more personalized. As described in [3], Recommender systems are not always capable of generating good recommendations for the user based only on raw data. The authors provide their own framework, which is called Human-Recommender Interaction (HRI). It's their way to connect the user needs with the recommender algorithms.

Their two main assumptions are first, that recommenders cannot understand the reason a user asks for recommendations and second, that recommenders should be trained to have personalities and interact with the users in a conversational way. This paper appreciates the use of recommender systems and suggests that people should establish a relationship of trust with the recommendation engine, but also the engine must be able to adapt to the user needs. The second assumption made earlier, is the main reason we use the Big Five personality test so that the recommender "knows" and operates accordingly to the user's personality type.

2.3 The Big Five Personality Test

The Big Five personality test is a collection of 50 questions that follows the Big Five model theory, and when answered, it helps understand how you think and operate as also as how your personality is structured.

In [4], the authors present a study among Facebook users, where they find correlations between their Big Five Scoring and their preferences towards movies, TV shows, books and music. They come up with the table in Fig. 2, which is critical for our 50/50 Recommender System. Each row on the table is a vector, and the values for each cell are in the range of 1–5, and they represent the average score of the Big Five personality traits of the users who liked the corresponding genres.

Fig. 1. Main flow chart

3 The Proposed 50/50 Movie Recommender Methodology

In this section, we analyze the proposed methodology to incorporate personality into the Movie Recommender System. First, we present the data we used and an analysis on the basic operation of the 50/50 Recommender. Then we give an example with some results where we can notice the differences between the k-NN movie recommendations and our own 50/50 and 80/20 recommendations.

3.1 Dataset

The dataset we used is the MovieLens 100k database [5, 6]. It includes 100.000 ratings from 1000 users on 1700 movies. We also used the *Most rated movies table* which is a table that includes the most rated movies from the Movielens Dataset.

3.2 Main Flow Chart

Here, we present visually the operation of own formula.

Movie Rating Function

This function is used to solve the cold start problem [7] meaning that the system knows nothing about the active user and there needs to be a way for the user to provide information to the system. It also includes the basic CF procedure explained in Sect. 2.1. So, first a list of 10 movies is presented to the user. Here we use the *Most rated movies* table, and we present one movie from this table and one random from the whole dataset alternately. That way, we avoid bias, because each time the user enters the system, the system provides a list with unique movies for him to rate. This goes on until the user has rated 20 movies, and we use these 20 reviews and CF to create a table with predicted ratings for each movie. We call this table *Predicted Ratings* (Table 1). The flow chart is shown in Fig. 1.

Table 1. *Predicted ratings* table head

Movie id	Rating
1	7.29552
2	2.87429
3	9.87429
4	2.87429
5	2.87429
6	2.87429
7	−2.34085

Big Five Personality Test Function

This function presents to the user all the questions from the Personality Test. The active user answers all the questions, and then the system calculates his big five traits scores as, following the procedure described in [1].

Combining Personality with k-NN Flow Chart

This is our main contribution. First, we take the Big Five Scoring of the active user and we subtract each of his traits from the corresponding trait of each genre in Fig. 2. Then we take the absolute value of these results and add them. That way, we convert them into a distance metric, so to calculate which genres the user prefers. The lower the value, the more the active user prefers this genre.

Next step is to rearrange the *Predicted Ratings* table, based on these genre preferences. For the 50/50, we take the predicted rating of the movie and divide it by 2, and we also take the number which represents the genre preference of the user and divide it by 2. If we add these numbers, we get a new predicted rating for the movie, where the k-NN scoring of the movie and his genre preferences count 50% each. For the 80/20, we take the predicted rating of the movie and multiply it by 0.2 and the genre preference of the user and multiply it by 0.8. That way, personality is now 80% and k-NN is 20%.

3.3   Example

The user enters the system and reviews 20 movies. CF is applied and the *Predicted Ratings* table is created. This is the most important table in our system. The following table (Table 1) is an example of the first 7 entries of the *Predicted Ratings* table.

The higher the value of the rating the most likely he will like this movie. This table (Table 1) is then rearranged in a descending order and has only the first 10 movies saved in a new table called the k-NN *Movie List*. Then the user is presented with the Big Five Personality test, and after he completes it he gets the results presented in Table 2.

Table 2. User's personality traits

	ope	con	ext	agr	neu
User	3.87	3.56	3.53	4.7	2.89

Based on these results, we check the following table (Fig. 2) from [4] and see which genres the active user prefers based upon his personality.

We always pick the genre that has the closest value to the user's Personality Traits score. For example, the active's user score on Agreeableness is 4.7, which as we can see is high and there is not a corresponding value to the table. That simply means that we will pick the greater value for this trait, which is 3.68 and corresponds to the Adventure genre.

Now, in the following table (Table 3), you can see the scoring for the action and adventure genre. We will use that as an example to show the subtractions we need to do to find the final distance metric between the user and each genre. This will lead us to his final genre preferences.

MOVIE GENRE	OPE	CON	EXT	AGR	NEU
action	3.87	3.45	3.57	3.58	2.72
adventure	3.91	3.56	3.54	3.68	2.61
animation	4.04	3.22	3.26	3.35	3.02
cartoon	3.95	3.33	3.49	3.57	2.81
comedy	3.88	3.44	3.58	3.60	2.75
cult	4.27	3.10	3.45	3.40	3.16
drama	3.99	3.43	3.66	3.60	2.86
foreign	4.15	3.46	3.47	3.54	2.81
horror	3.90	3.38	3.52	3.47	2.91
independent	4.31	3.59	3.51	3.55	2.69
neo-noir	4.34	3.35	3.33	3.37	2.97
parody	4.13	3.36	3.35	3.28	2.73
romance	3.84	3.48	3.62	3.62	2.85
science fiction	3.99	3.55	3.33	3.57	2.73
tragedy	4.40	3.34	3.27	3.52	3.11
war	3.82	3.51	3.49	3.50	2.71

Above header: **All users**

Fig. 2. Correlation between Genre and Big Five Traits

In the next table (Table 4), we only have to subtract each trait from Table 2 from the corresponding trait Table 3. We then take that absolute value of each result and add it so that we get a final number which is a distance metric. The closer to 0, the better the user likes this specific genre.

Table 3. Example genre trait scores

	ope	con	ext	agr	neu
Action	3.87	3.45	3.57	3.58	2.72
Adventure	3.91	3.56	3.54	3.68	2.61

Table 4. Calculating 50/50 2

	ope	con	ext	agr	neu	Pers. score
Action	|3.87–3.87|	|3.56–3.45|	|3.53–3.57|	|4.7–3.58|	|2.89–2.72|	1.44
Adventure	|3.87–3.91|	|3.56–3.56|	|3.53–3.54|	|4.7–3.68|	|2.89–2.61|	1.35

We can see that the user has a slight preference towards the adventure genre. We do the same calculations for all the genres and then create a table (Table 5) that includes these results in an ascending order. These are the genre preferences of the active user based upon his personality.

Next step is to rearrange the k-NN *Movie List* based on these genre preferences. Let's assume that the k-NN algorithm predicted that the active user will provide a rating of 4 on Star Wars and a rating of a 4 on Psycho. For the 50/50, our recommender takes the predicted rating of the Star Wars movie from *Predicted Ratings* table and divides it by 2. Because Star Wars is a Sci-Fi movie, it also takes the value of the sci-fi genre from Table 5 and divides it with 2. We then add these two numbers and the new pseudo rating for Start wars is (4/2) + (1/2) = 2.5. By applying the same concept to the Psycho movie, we can see that the new pseudo rating for the psycho movie is 2.715. The higher the value, the less you will like this movie.

Table 5. 50/50 user genre preferences

Cartoon	0.8
Sci-fi	1
Drama	1.3
Horror	1.43
Action	1.44
War	1.5

If we only had the k-NN suggestions, Psycho and Star Wars would be equally presented to the user as they both have the same predicted rating. However, with the addition of the personality factor, Star Wars will be suggested first and then Psycho. We follow the same procedure for the 80/20 recommender, with the only difference that now we multiply the k-NN predicted rating with 0.2 and the Personality genre rating with 0.8. On the following table, we can see the final recommendations (Table 6).

The k-NN recommendations list is the one that we get by simply applying CF. On the 50/50 list, we see the Sci-fi movies rising in the rankings and even horror movies like Alien and Psycho appeared in the list, where previously on the k-NN list, they

Table 6. Final Recommandations

k-NN recommendations	50/50 recommendations	80/20 recommendations
Usual suspect, The	Blade runner	Lion King, The
Blade runner	Silence of the lambs, The	Blade runner
Silence of the lambs, The	Empire strikes back, The	Silence of the lambs, The
Empire strikes back, The	Alien	Wallace and Gromit: best animations
Amadeus	Psycho	Wrong trousers, The
Schindler's list	Amadeus	L.A. Confidential
One flew over the cuckoo's	Schindler's list	Close shave, A
Casablanca	One flew over the cuckoo's	Maltese Falcon, The
It's a wonderful life	Casablanca	Beauty and the beast
Rear window	It's a wonderful life	Faust

weren't included. This is because the user has a preference on Sci-fi and Horror movies based on Table 5. We can clearly see that the list is now rearranged based on the user's personality.

On the 80/20 list, we see that a cartoon movie is now suggested first, because cartoon is the favorite genre of the active user, as we see again on Table 5. We also see the appearance of more cartoon movies.

On the next and final Section, we discuss the future work on our movie recommender.

4 Evaluation

For the evaluation part, we sent a modified version of the recommender system to 50 different people. 32 responded, but 2 did not follow the instructions, thus their responses were discarded. Out of these 30 people, 10 were female and 20 were male, aged between 18 and 65. The modified version differs from the final version of the recommender is that it did not reveal which recommendation list was produced by which method. Users were presented with three different lists of movies, and they had to rank those in order of preference, according to how well these lists reflected their actual preferences.

The users awarded their first preference 3 points, the second preference 2 points and the last preference, 1 point. They could award equal points to tie cases, i.e. if lists A and B were equally preferable they were both awarded 3 points each, just as they could both be awarded 2 points each, if they were both second best. There were 15 such ties. We then added the scores for each list to find the total number of points for each method, and the respective percentage of the total number of points.

As it can be seen in Table 7, top preferred method was the 50/50 with 36.92% of the total points, second came k-NN with 34.36%, while the 80/20 method received 28.72% of the points. This means that the 50/50 method outperforms k-NN, whilst the 80/20 method, despite being heavily biased towards personality, did not perform poorly. This confirms our hypothesis that taking personality into account would improve recommendation quality. Table 8 shows results if we remove the 80/20 method, leaving 50/50 with 51.8% and k-NN lagging with 48.2%, 3.6% less than 50/50.

Table 7. 50/50 and 80/20 vs. k-NN Evaluation

	50/50	k-NN	80/20
# Points	72	67	56
%	36.92	34.36	28.72

4.1 Discussion

Observing the evaluation results we can conclude that personality plays a significant part in recommender systems, as the 50/50 was the favorite method for most users and

Table 8. 50/50 vs. k-NN Evaluation

	50/50	k-NN
# Points	72	67
%	51.8	48.2

even the 80/20 fared well. It is possible that a different mix could give even better results. However, it requires a significant overhead initially, when users need to take the personality test, even though this only need to take place once and the results could be used in many subsequent recommendations, not necessarily only movie related. In any case, there is a tradeoff between investing time to take the test and improving subsequent recommendation experience.

Threats to Validity

Our approach could be subject to a few threats to validity. First and foremost was the selection bias in the evaluation phase. No matter how many people you target to participate to an evaluation like the one described previously, there are few safeguards that the sample is random and representative of various demographics.

Last, we cannot neglect the Hawthorne Effect [11, 12], where users might complete the personality test with dishonest answers as they know they are being observed. In the next section, we discuss some important improvements that can be done in the future.

5 Future Work

As we discussed in Sect. 4.1, the time needed to answer the personality test was a reason why some users dropped off. An improvement for our recommender would be to reduce the number of questions in the personality test to the ones that are most important. This can be achieved by using decision trees for classification to find questions that have the strongest impact on classification, (possibly the ones with the widest range of answers, which provided the highest information gain).

We also plan to explore different mixes of x% personality and $(100 - x)$% k-NN recommender, and potentially plot a performance graph for various $x/100 - x$ spreads in order to better understand how personality affects the outcome and find the optimal balance. So far, we experimented with 0/100, 50/50 and 80/20 and noticed that the users' preference raises on the 50/50 and on the 80/20 there is a decrease. It would be interesting to see at which percentage exactly we have the peak of preference. An improvement would also be to include more metadata, which can be gathered from IMDB [8].

The MovieLens database includes also gender, age and occupation information for each user. In the future, these data could be used so to cluster the users and make more accurate predictions.

Acknowledgments. The authors would like to thank the Hellenic Artificial Intelligence Society (EETN) for covering part of their expenses to participate in EANN 2017.

References

1. Goldberg, L.R.: The development of markers for the Big-Five factor structure. Psychol. Assess. **4**(1), 26 (1992)
2. Burke, R.: Hybrid recommender systems: survey and experiments. User Model. User-Adap. Interact. **12**(4), 331–370 (2002)
3. McNee, S.M., Riedl, J., Konstan, J.A.: Making recommendations better: an analytic model for human-recommender interaction. In: CHI 2006 Extended Abstracts on Human Factors in Computing Systems (CHI EA 2006), pp. 1103–1108. ACM, New York (2006)
4. Cantador, I., Fernández-Tobías, I., Bellogín, A.: Relating personality types with user preferences in multiple entertainment domains. In: UMAP Workshops (2013)
5. MovieLens 100K Dataset: Grouplens. https://grouplens.org/datasets/movielens/100k/. Accessed 20 Apr 2017
6. MovieLens: Non-commercial, personalized movie recommendations. https://movielens.org/. Accessed 20 Apr 2017
7. Nadimi-Shahraki, M.-H., Bahadorpour, M.: Cold-start problem in collaborative recommender systems: efficient methods based on ask-to-rate technique. J. Comput. Inf. Technol. CIT **22**(2), 105–113 (2014)
8. IMDB. http://www.imdb.com/. Accessed 20 Apr 2017
9. Su, X., Khoshgoftaar, T.M.: A survey of collaborative filtering techniques. Adv. Artif. Intell. **2009**, Article ID 421425 (2009). 19 p.
10. Melville, P., Sindhwani, V.: Recommender systems. In: Sammut, C., Webb, G.I. (eds.) Encyclopedia of Machine Learning, pp. 829–838. Springer US, New York (2011). doi:10.1007/978-0-387-30164-8_705
11. Monahan, T., Fisher, J.A.: Benefits of "observer effects": lessons from the field. Qual. Res. (QR) **10**(3), 357–376 (2010)
12. Gerogiannis, V.C., Karageorgos, A., Liu, L., Tjortjis, C.: Personalised fuzzy recommendation for high involvement products. In: IEEE International Conference Systems, Man, and Cybernetics (SMC 2013), pp. 4884–4890 (2013)
13. Hill, W., Stead, L., Rosenstein, M., Furnas, G.: Recommending and evaluating choices in a virtual community of use. In: Proceedings of the SIGCHI Conference on Human Factors in Computing Systems (CHI 1995), pp. 194–201. ACM Press/Addison-Wesley Publishing Co. (1995)

Recommender Systems Meeting Security: From Product Recommendation to Cyber-Attack Prediction

Nikolaos Polatidis[1(⊠)], Elias Pimenidis[2], Michalis Pavlidis[1],
and Haralambos Mouratidis[1]

[1] School of Computing Engineering and Mathematics, University of Brighton,
Brighton BN2 4GJ, UK
{N.Polatidis,M.Pavlidis,H.Mouratidis}@Brighton.ac.uk
[2] Department of Computer Science and Creative Technologies,
University of the West of England, Bristol BS16 1QY, UK
Elias.Pimenidis@uwe.ac.uk

Abstract. Modern information society depends on reliable functionality of information systems infrastructure, while at the same time the number of cyber-attacks has been increasing over the years and damages have been caused. Furthermore, graphs can be used to show paths than can be exploited by attackers to intrude into systems and gain unauthorized access through vulnerability exploitation. This paper presents a method that builds attack graphs using data supplied from the maritime supply chain infrastructure. The method delivers all possible paths that can be exploited to gain access. Then, a recommendation system is utilized to make predictions about future attack steps within the network. We show that recommender systems can be used in cyber defense by predicting attacks. The goal of this paper is to identify attack paths and show how a recommendation method can be used to classify future cyber-attacks. The proposed method has been experimentally evaluated and it is shown that it is both practical and effective.

Keywords: Recommender systems · Cyber security · Attack graph · Exploit · Vulnerability · Attack prediction · Classification

1 Introduction

Recommender systems are decision support systems available on the web to assist users in the selection of item or service selection in online domains. In doing so recommender systems assist users in overcoming the information overload problem [1, 2]. Collaborative filtering (CF) is the most widely used method for providing personalized recommendations. In CF systems, a database of user submitted ratings is used and the generated recommendations are generated on how much a user will like an unrated item based on previous common rated items. Thus, the recommendation process is based on assumptions about previous rating agreements and if these agreements will be maintained in the future. In addition, the ratings are used to create an n × m matrix with user ids, item ids and ratings, with an example of such a matrix shown in

© Springer International Publishing AG 2017
G. Boracchi et al. (Eds.): EANN 2017, CCIS 744, pp. 508–519, 2017.
DOI: 10.1007/978-3-319-65172-9_43

Table 1. An example of a ratings matrix

	Item 1	Item 2	Item 3	Item 4
User 1	1	2	5	–
User 2	4	5	4	1
User 3	–	–	3	2
User 4	1	1	2	5

Table 1. This database has four users and four items with values from 1 to 5. The matrix is used as input when a user is requesting recommendations and for a recommendation to be generated the degree of similarity between the user who makes the request and the other users' needs to be predicted using a similarity function such as the Pearson Correlation Similarity (PCC) [3]. At the next step a user neighborhood which consists of users having the highest degree of similarity is created with the requester. Finally, a prediction is generated after computing the average values of the nearest neighborhood ratings about an item, resulting in a recommendation list of items with the highest predicted rating values.

Even though, recommender systems have been used for product or service recommendation, in the current era where cyber-attacks have been increasing we show how they can assist in the prediction of future attacks.

1.1 Problem Definition and Contributions

Cyber-attack prevention methods are based on graph analysis to identify attack paths or use previous attacker knowledge in combination with intrusion alerts to provide defense actions in real time. A gap is identified in attack prediction and mitigation which can be solved with the use of suitable recommendation technologies. We have made the following contributions:

1. We identify all attack paths in a graph according to constraints.
2. We use the attack paths in combination with common vulnerability data to build a recommender system that predicts future attacks.

1.2 Paper Structure

In Sect. 2 relevant background work is analyzed. In Sect. 3 the proposed method is explained. Section 4 presents the experimental evaluation and Sect. 5 contains the conclusions and future work parts.

2 Background

2.1 Collaborative Filtering

As explained above a database of ratings and a similarity function such as PCC are the two essential parts of the CF recommendation process. Except for the classical recommendation method, PCC, another similar method found in the literature is weighted

PCC (WPCC) which extends PCC by setting a statically defined threshold of common rated items. However, since the definitions of PCC and WPCC numerous approaches have been proposed with the aim of improving the recommendations. TasteMiner is a method that efficiently mines rating for learning partial users tastes to restrict the neighborhood size, thus reducing complexity and improving the accuracy of the recommendations [4]. Another CF approach that aims to improve the accuracy of the recommendations is entropy based can be found in the literature. In this approach an entropy driven similarity used to calculate the difference between ratings and a Manhattan distance model is then used to address the fat tail problem [5]. One more similarity measure for improving the accuracy of CF has been proposed with the name PIP. This measurement is based on Proximity, Impact and Popularity (PIP). Initially the proximity factor is applied to calculate the absolute difference between two ratings, then the impact factor is applied to show how strongly an item is preferred and finally the popularity factor is applied to how common the user ratings are. These three factors are then combined to calculate a final value [6]. HU-FCF is a hybrid fuzzy CF method for improved recommendations [7]. In this method, CF is extended with a fuzzy similarity that is calculated on user demographic data. A CF recommendation method based on singularities has been proposed [8]. In this method, the traditional similarities can be improved if contextual information from the entire user body are used to calculate singularities. Thus, the larger the singularity between users then the impact of it in the similarity is larger. Additionally, the use or power law augments to similarity values can be found in the literature with the name PLUS [9]. PLUS, is a method applied to user similarities to adjust their value using a power function and achieves a tradeoff between accuracy and diversity of the recommendations. Yet another approach for improved recommendations is the use of Pareto dominance [10]. Pareto dominance is used initially as a pre-filtering service were the less promising users are eliminated from the user neighborhood. Then, the rest are used in a typical CF recommendation process. An additional recommendation approach includes the breakup of the user neighborhood in multiples levels [11]. This can be done either using a static approach or a dynamic one [11, 12]. In both approaches the user similarities are adjusted either in a positive or a negative way based on the number of co-rated items and the PCC values and are assigned to one of multiple levels based on the final computed value. Thus, the predictions are made using the new user neighborhood and the recommendations are improved. An additional method that can be used to improve the quality of the recommendations is natural noise removal [13]. Items and users are characterized based on their profiles and a defined strategy is used to eliminate natural noise, thus receiving more accurate recommendations. Also, other traditional approaches exist that can be used to improve CF and include the use of content-boosted CF or the utilization of sparsity measures [14, 15]. COUSIN is a recommendation model that improves both the accuracy and the diversity of the recommendations by using a regression model that affectively removes weak user relationships [16]. There is also an approach in the literature called Trinity that uses historical data and tags to provide personalized recommendations based on a three-layered object-user tag network [17]. In addition to the methods mentioned already the use of user-item subgroups has been proposed as a way of providing improved recommendation systems [18].

2.2   Attack Graph Generation and Analysis

Cyber-attack prevention technologies typically use attack graph generation and analysis methods to identify all possible paths that attackers can exploit to gain unauthorized access to a system [19]. There are numerous methods available for attack graph generation and analysis. In [20] the authors use a general graph model, which is based on the JIGSAW specification language. Sample attack scenarios are created using different methods such as substitution, distribution and looping. In [21] the authors developed an intrusion correlator for intrusion alerts, which produces correlation graphs as output. Then, they use these graphs to create attack strategy graphs. The authors in [22] utilize modeling based approach that is used to perform an analysis of the security of the network. This is done using model checking tools and a model is presented that describes the vulnerability to attack of the network. In [23] the authors developed a tool called NuSMV. This is a model checking tool that implements an algorithm for automatic generation of attack graphs. A logic-based approach is proposed in [24]. In this approach, the authors use logic rules to compute the attack graph and use logic deduction to reach the final facts from the initial facts. Although, this approach suffers from performance issues as the state grows. In [25] a Breadth-first search solution is used by the authors to build the attack graph.

A layered solution is proposed where the bottom layer contains attacker privileges and the upper layer contains the privileges computer after each step of the algorithm. Once again, as the size of the graph grows there are performance issues. In [26] the authors propose an algorithm that only creates a graph containing the worst case scenarios. This approach performs better in terms of performance, but it cannot guarantee that all relevant paths will be returned. In [27] the authors try to reduce complexity by introducing the concept of group reachability. This method uses a breadth first method and uses prerequisite graphs that express reachability conditions among network hosts. The authors in [28] develop further the prerequisite graphs by adding information about client-side attacks, firewalls and intrusion detection. In [29] the authors use a distributed attack graph generation algorithm based on a multi-agent system, a virtual shared memory abstraction and hyper-graph partitioning to improve the overall performance of the system. The method is based on depth first search and it is shown that the performance is improved with the use of agents after a specific graph size. In [30] the authors use a bidirectional search method to generate the attack graph. They also apply a restriction about the depth of the search, which limits the algorithm from identifying less possible attacks. In [31] an approach that is based on artificial intelligence with the name Planner is applied to generate the attack graph. Customized algorithms are used to generate attack paths in polynomial time.

In [32] the authors propose a graph-based approach to analyze vulnerabilities, that can analyze risk to a specific asset and examine possible consequence of an attack. In [33] the use of a probabilistic model is proposed. This model measures risk security, computes risk probability and considers dynamic network features. A somewhat different approach is proposed by the authors in [34]. The use of dynamic generation algorithm is proposed, that returns the top K paths. Furthermore, it is not required to calculate the full attack graph to return the top attack paths. NetSPA is a network security planning architecture that very efficiently generates the worst case attack graphs [35]. To do this the system uses information from software types and versions, intrusion detection systems, network connectivity and firewalls. In [36] the use Bayesian attack graph generation for dynamic security risk management.

In [37] the authors developed a MulVAL, a logic-based network security analyzer. This is a vulnerability analysis tool that models the interaction of software bugs along with network configurations. The data about the software bugs are provided by a bug-reporting community, while all the other relevant information is enclosed within the system. In addition to MulVAL, TVA is another tool for generating attack graphs [19, 38]. TVA is based on topological analysis of network attack vulnerability and the idea is to exploit dependency graph to represent preconditions and postconditions and then exploit. At the next step, a search algorithm finds attack paths that exploit multiple vulnerabilities.

3   Proposed Method

Our proposed method takes elements from both collaborative filtering recommender systems and attack path discovery methods to identify attacks paths and predict attacks. Initially, we use an attack path discovery method that has unique characteristics, such as the attacker location, the attacker capability and which the entry and target points are. The, attack path discovery method returns all non-circular attack paths that exist between assets that belong to the specified characteristics. At the next step, we use the attack paths along with a recommender system to predict future attacks and to classify them.

3.1   Attack Path Discovery

Attackers can use a set of basic privileges that can satisfy some initial input require-ments to gain unauthorized access to a system. Attack graphs show every possible path that an attacker can use to gain further privileges [19, 39]. In general, various vul-nerabilities, such as software vulnerabilities or inappropriate configuration settings, exist in information systems and can be exploited by attackers to gain access. An infrastructure it typically comprised of numerous nodes that can be exploited to intrude into the network. In addition, the number of vulnerabilities that exist on the network and the reachability conditions that occur are the factors that determine the size of the attack graph. In, addition as the graph becomes larger, the possibility of more exploitation options for an attacker increases. To build the attack graph we use direct conditions and utilize information from open sources. Initially, the weaknesses defined in the Common Weakness Enumeration (CWE) [40] are used and at the second step, Information from the Common Vulnerabilities and Exposures (CVE) [41] database are used. A model is introduced where an attacker can gain access to information system sources and move in a directed path. Moreover, a set of preconditions are specified, which include the length of the path, the location and capability of the attacker. The pseudocode of the attack path discovery is shown in Algorithm 1.

Algorithm 1: Attack path discovery

Input: Asset graph (G), attacker location, attacker capability
Output: Graph, affected assets, attack paths

#We create two empty lists to hold attack paths and assets
attackpaths = [] affectedassets = []
#We return all paths from source to target
for e in parameters entry points
    If attacker location < required level of attacker location OR attacker capability <
    required attacker capability
    return empty graph
        else
        get single source shortest path length
        set propagation length for entry point e
            for target point t
#Create a list with all non-circular paths from entry e to target t
get all paths in the graph G from entry e to target t that are up to the pre-specified
path length
for the size of paths found
    add paths to attackpaths [] list, **add** affected assets to affectedassets [] list
#Return the graph, the affected assets and the attack paths found as a direct input to
#the attack visualization algorithm
return Graph, affected assets, attack paths

3.2   Attack Prediction

To recommend attack predictions we use a parameterized version of multi-level collaborative filtering method described in [11], although other methods could be applied
according the scenario and the available data. This method applies collaborative filtering and then rearranges the order of the k nearest neighbors according to the similarity value and the number of co-rated items. We use characteristics from the
above-mentioned method to classify attacks. To do that we initially apply classical
collaborative filtering using PCC defined in Eq. 1. In PCC Sim (a, b) is the similarity of
users a and b, $r_{a,p}$ is the rating of user a for product p, $r_{b,p}$ is the rating of user b for
product p and $\bar{r}a$, $\bar{r}b$ represent user's average ratings. P is the set of all products. At the
next step, we check the similarity values returned by Eq. 1 and the number of co-rated
vulnerabilities. Depending on the similarity value returned and the common vulnerabilities, we classify these attacks from very high to very low. Finally, we check if there
are any attack paths between the assets before the classification process is finished.
A detailed explanation of the steps can be found in Algorithm 2 which provides the
pseudocode of the attack prediction recommender system.

$$Sim\frac{PCC}{a,b} = \frac{\sum p \in P(ra, p - \bar{r}a)(rb, p - \bar{r}b)}{\sqrt{\sum p \in P(ra, p - \bar{r}a)^2}\sqrt{\sum p \in P(rb, p - \bar{r}b)^2}} \tag{1}$$

Algorithm 2: Attack prediction

Input: attack paths, affected assets, vulnerabilities
Output: predicted attacks

#Vulnerabilities refers to common vulnerabilities between assets
load vulnerabilities
apply equation 1 using vulnerabilities as input
    get similarity values
    #If there are common vulnerabilities, then typically these receive the same score
    #between assets, thus, resulting in absolute similarities
    #Then we rearrange the order of the similarity by adding the number of co-rated
    #items as a constraint
    #classification refers to predicted attack classification, which is from very high to
    #very low
  then #n is the number of co-rated items and x1, x2, x3 and x4 are fixed integers
    if n>=x1 **then** classification == very high
      else if n<x1 && n>=x2 **then** classification == high
        else if n<x2 && n>=x3 **then** classification == Medium
          else if n<x3 && n>=x4 **then** classification == Low
    else classification == very low
then
  get attack paths
    if attack path exists
      set classification == very high
    else if attack path does not exist && classification == very high then classification == high
    else classification == classification
Return predicted attacks

---

4   Experimental Evaluation

The experiments took place in a simulated environment using a Pentium i7 2.8 GHz with 12 gigabytes of RAM, running windows 10. The data used were supplied by the maritime supply chain IT infrastructure and more particular from the port of Valencia and the experiments were conducted within a cyber-security maritime supply chain risk management system. The dataset contains 26 hardware and software assets, numerous vulnerabilities, with some of them being common within the assets. Initially, we evaluate the attack path discovery method in terms of performance, the results of which are shown in Table 2. Then, we present a case study that shows how to predict attacks utilizing the data from the maritime supply chain IT infrastructure.

4.1   Case Study: The Maritime Supply Chain IT Infrastructure

The maritime supply chain infrastructure it typically comprised of numerous assets that can be exploited to gain access and reach specific assets by popping from one to another.

Table 2. Performance evaluation results

No. of test	Attacker capability	Propagation length	Running time (sec)
1	Low	3	<1
2	Low	4	<1
3	Low	5	<1
4	Medium	3	<1
5	Medium	4	<1
6	Medium	5	1
7	High	3	<1
8	High	4	1
9	Hugh	5	1.2

For the case study, we have used a snippet of data derived from the Valencia port IT infrastructure. In Table 3 the data used show the common vulnerabilities between assets and their respective score. Assets 1, 2 and 3 are hardware assets, while the description column represents the vulnerable software asset that is installed on the respective hardware asset. Furthermore, the assets and attacks paths between them are a vital part of risk assessment. The following non-circular attack paths are present in the system:

Table 3. Common vulnerabilities

Assets	Description	CVE 2015-1769	CVE 2015-2423	CVE 2015-2433	CVE 2015-2485
Asset 1 (Desktop PC)	Windows 10 Installed on Desktop PC	10	2.9	2.9	10
Asset 2 (Laptop 1)	Windows 10 Installed on Laptop 1	10	2.9	2.9	10
Asset 3 (Laptop 2)	Windows 10 Installed on Laptop 2	10	2.9	2.9	–

1. Asset1 → Asset2
2. Asset2 → Asset3
3. Asset2 → Asset1

However, it should be noted that attack paths might vary according to the specific settings used, such as the propagation length, attacker location, capability, entry and target points.

Then the administrator executed Algorithm 2 to predict very high and high classification attacks. Moreover, for the case study we have assigned the minimum number of co-rated items to be 3 for very high classification and 2 for high classification. Thus, Algorithm 2 classified:

1. Asset1 → Asset2 as very high
2. Asset2 → Asset1 as very high
3. Asset1 → Asset3 as high
4. Asset3 → Asset1 as high

5. Asset2 $\rightarrow$ Asset3 as high
6. Asset3 $\rightarrow$ Asset2 as high

At the next step, the method checked for attack path relations between the assets and rearranged the classifications. Thus, the administrator received the following final predictions:

1. Asset1 $\rightarrow$ Asset2 as very high
2. Asset2 $\rightarrow$ Asset1 as very high
3. Asset2 $\rightarrow$ Asset3 as very high
4. Asset1 $\rightarrow$ Asset3 as high
5. Assct3 $\rightarrow$ Asset1 as high
6. Asset3 $\rightarrow$ Asset2 as high.

4.2 Discussion

Cyber-attack prediction systems are important in risk management to provide mitigation solutions. To do that the identification of possible attack scenarios and providing defensive solutions for assets protection are the two most important parts. Furthermore, it is important for this to take place within a reasonable amount of time. It is shown that within a small amount of time the attack path discovery method delivers the non-circular attack paths between assets. Furthermore, at the next stage a classification list is created that provides a prediction list of attack movement between assets. For example, the likelihood that an attacker who gained access to asset 1 to explore the possibility of gaining access to asset 4 is higher when compared to gaining access to either asset 2 or asset 3. However, the possibility of common vulnerabilities receiving different scores in different assets should be further exploited since this will result in different classification scales.

5   Conclusions and Future Work

Various online services use recommender systems for product or service recommendation. However, the use of such systems in the cyber-defense domain has not been explored. In this paper, we proposed a collaborative filtering based recommender system that uses common vulnerabilities between assets, identifies attack paths and combines the information to recommend future attacks. Although, the method is practical, it could become more effective if certain aspects are extended. Thus, in the future, we aim to investigate the following directions:

Path length recommendation. We aim to apply recommendation techniques to dynamically identify the length of the path that should be searched, thus making the attack path discovery process faster.

Attack recommendation. A part of our research will concentrate on more intelligent approaches for cyber-attack predictions based on advanced methods.

Defense recommendation. Another research direction will focus on defense strategy recommendation.

Prediction Validation. We aim to validate the attack predictions using real data from real world scenarios along with expert consultation.

Acknowledgement. This work has received funding from The European Union's Horizon 2020 research and innovation program under grant agreement No. 653212.

References

1. Lu, J., Wu, D., Mao, M., Wang, W., Zhang, G.: Recommender system application developments: a survey. Decis. Support Syst. **74**, 12–32 (2015)
2. Polatidis, N., Georgiadis, C.K.: Recommender systems: the importance of personalization on e-business environments. Int. J. E-entrepreneursh. Innov. **4**, 32–46 (2013)
3. Su, X., Khoshgoftaar, T.M.: A survey of collaborative filtering techniques. Adv. Artif. Intell. **2009**, 1–19 (2009)
4. Shams, B., Haratizadeh, S.: TasteMiner: mining partial tastes for neighbor-based collaborative filtering. J. Intell. Inf. Syst. **48**, 165–189 (2017)
5. Wang, W., Zhang, G., Lu, J.: Collaborative filtering with entropy-driven user similarity in recommender systems. Int. J. Intell. Syst. **30**, 854–870 (2015)
6. Liu, H., Hu, Z., Mian, A., Tian, H., Zhu, X.: A new user similarity model to improve the accuracy of collaborative filtering. Knowl.-Based Syst. **56**, 156–166 (2014)
7. Son, L.H.: HU-FCF: a hybrid user-based fuzzy collaborative filtering method in recommender systems. Expert Syst. Appl. **41**, 6861–6870 (2014)
8. Bobadilla, J., Ortega, F., Hernando, A.: A collaborative filtering similarity measure based on singularities. Inf. Process. Manag. **48**, 204–217 (2012)
9. Gan, M., Jiang, R.: Improving accuracy and diversity of personalized recommendation through power law adjustments of user similarities. Decis. Support Syst. **55**, 811–821 (2013)
10. Ortega, F., Sánchez, J.L., Bobadilla, J., Gutiérrez, A.: Improving collaborative filtering-based recommender systems results using Pareto dominance. Inf. Sci. (N.Y.) **239**, 50–61 (2013)
11. Polatidis, N., Georgiadis, C.K.: A multi-level collaborative filtering method that improves recommendations. Expert Syst. Appl. **48**, 100–110 (2016)
12. Polatidis, N., Georgiadis, C.K.: A dynamic multi-level collaborative filtering method for improved recommendations. Comput. Stand. Interfaces **51**, 14–21 (2017)
13. Toledo, R.Y., Mota, Y.C., Martínez, L.: Correcting noisy ratings in collaborative recommender systems. Knowl.-Based Syst. **76**, 96–108 (2015)
14. Melville, P., Mooney, R.J., Nagarajan, R.: Content-boosted collaborative filtering for improved recommendations. In: Proceedings 18th National Conference on Artificial Intelligence (AAAI), pp. 187–192 (2002)
15. Anand, D., Bharadwaj, K.K.: Utilizing various sparsity measures for enhancing accuracy of collaborative recommender systems based on local and global similarities. Expert Syst. Appl. **38**, 5101–5109 (2011)
16. Gan, M.: COUSIN: a network-based regression model for personalized recommendations. Decis. Support Syst. **82**, 58–68 (2016)
17. Gan, M.-X., Sun, L., Jiang, R.: Trinity: walking on a user-object-tag heterogeneous network for personalised recommendations. J. Comput. Sci. Technol. **31**, 577–594 (2016)

18. Xu, B., Bu, J., Chen, C., Cai, D.: An exploration of improving collaborative recommender systems via user-item subgroups. In: Proceedings of 21st International Conference on World Wide Web - WWW 2012, p. 21 (2012)
19. Ou, X., Singhal, A.: Attack graph techniques. In: Ou, X., Singhal, A. (eds.) Quantitative Security Risk Assessment of Enterprise Networks. SpringerBriefs in Computer Science, pp. 13–23. Springer, New York (2011). doi:10.1007/978-1-4614-1860-3_2
20. Templeton, S.J., Levitt, K.: A requires/provides model for computer attacks. In: Proceedings of 2000 Workshop on New Security Paradigms - NSPW 2000, pp. 31–38 (2000)
21. Ning, P., Xu, D.: Learning attack strategies from intrusion alerts. In: Proceedings of the 10th ACM Conference on Computer and Communication Security - CCS 2003, p. 200 (2003)
22. Ritchey, R.W., Ammann, P.: Using model checking to analyze network vulnerabilities. In: Proceedings 2000 IEEE Symposium on Security and Privacy, S&P 2000, pp. 156–165 (2000)
23. Sheyner, O., Haines, J., Jha, S., Lippmann, R., Wing, J.M.: Automated generation and analysis of attack graphs. In: Proceedings of IEEE Symposium on Security and Privacy, pp. 273–284 (2002)
24. Ou, X., Boyer, W.F., McQueen, M.A.: A scalable approach to attack graph generation. In: 13th ACM Conference on Computer and Communications Security, pp. 336–345 (2006)
25. Ammann, P., Wijesekera, D., Kaushik, S.: Scalable, graph-based network vulnerability analysis. In: Proceedings of 9th ACM Conference on Computer and Communication Security - CCS 2002, p. 217 (2002)
26. Ammann, P., Pamula, J., Ritchey, R., Street, J.: A host-based approach to network attack chaining analysis. In: Proceedings of Annual Computer Security Applications Conference, ACSAC, pp. 72–81 (2005)
27. Ingols, K., Lippmann, R., Piwowarski, K.: Practical attack graph generation for network defense. In: Proceedings of Annual Computer Security Applications Conference, ACSAC, pp. 121–130 (2006)
28. Ingols, K., Chu, M., Lippmann, R., Webster, S., Boyer, S.: Modeling modern network attacks and countermeasures using attack graphs. In: Proceedings of Annual Computer Security Applications Conference, ACSAC, pp. 117–126 (2009)
29. Kaynar, K., Sivrikaya, F.: Distributed attack graph generation. IEEE Trans. Dependable Secur. Comput. **13**, 519–532 (2016)
30. Xie, A., Zhang, L., Hu, J., Chen, Z.: A probability-based approach to attack graphs generation. In: 2nd International Symposium on Electronic Commerce and Security, ISECS 2009, pp. 343–347 (2009)
31. Ghosh, N., Ghosh, S.K.: A planner-based approach to generate and analyze minimal attack graph. Appl. Intell. **36**, 369–390 (2012)
32. Phillips, C., Swiler, L.P.: A graph-based system for network-vulnerability analysis. In: Proceedings of 1998 Workshop on New Security Paradigms, pp. 71–79 (1998)
33. Almohri, H.M.J., Watson, L.T., Yao, D., Ou, X.: Security optimization of dynamic networks with probabilistic graph modeling and linear programming. IEEE Trans. Dependable Secur. Comput. **13**, 474–487 (2016)
34. Bi, K., Han, D., Wang, J.: K maximum probability attack paths dynamic generation algorithm. Comput. Sci. Inf. Syst. **13**, 677–689 (2016)
35. Artz, M.L.: NetSPA : a network security planning architecture, pp. 1–97 (2002)
36. Poolsappasit, N., Dewri, R., Ray, I.: Dynamic security risk management using Bayesian attack graphs. IEEE Trans. Dependable Secur. Comput. **9**, 61–74 (2012)
37. Ou, X., Govindavajhala, S., Appel, A.W.: MulVAL: a logic-based network security analyzer. In: Proceedings of the 14th Conference on USENIX Security Symposium, vol. 14, p. 8 (2005)

38. Jajodia, S., Noel, S., O'Berry, B.: Topological analysis of network attack vulnerability. In: Kumar, V., Srivastava, J., Lazarevic, A. (eds.) Managing Cyber Threats, pp. 247–266. Springer, Heidelberg (2005). doi:10.1007/0-387-24230-9_9
39. Barik, M.S., Mazumdar, C.: A graph data model for attack graph generation and analysis. In: Martínez Pérez, G., Thampi, S.M., Ko, R., Shu, L. (eds.) SNDS 2014. CCIS, vol. 420, pp. 239–250. Springer, Heidelberg (2014). doi:10.1007/978-3-642-54525-2_22
40. Common Weakness Enumeration, CWE. http://cwe.mitre.org/. Accessed 20 Apr 2017
41. Common Vulnerabilities and Exposures, CVE. https://cve.mitre.org/. Accessed 20 Apr 2017

Robotics and Machine Vision

Machine Vision for Coin Recognition with ANNs: Effect of Training and Testing Parameters

Vedang Chauhan[⊠], Keyur D. Joshi, and Brian Surgenor

Queen's University, Kingston, ON K7L 3N6, Canada
{vedang.chauhan, surgenor}@queensu.ca,
joshikeyurd@gmail.com

Abstract. Pattern recognition is a branch of machine learning that focuses on the recognition of patterns and regularities in data for object recognition, classification and computer vision segmentation. Features are extracted from input data and used for object classification purposes. Artificial Neural Networks (ANNs) and Deep Neural Networks (DNNs) are popular tools for pattern recognition applications. The performance of the networks is usually defined in terms of the classification accuracy. However, there are no real design guidelines for training and testing protocols. This research set out to evaluate the effect on accuracy of the design parameters, including: size of the database, number of classes, quality of images, type of network, nature of training and testing strategy. A coin recognition task was used for the evaluation. A set of guidelines for part recognition tasks is presented based on experience with this task.

Keywords: Pattern recognition · Machine learning · Object detection and classification · ANNs · DNNs · Testing protocols

1 Introduction and Background

Pattern recognition and classification is a very active area of research. A number of mathematical approaches are available to solve the classification problem, such as decision tree [1], template matching [2], clustering techniques [3], Artificial Neural Networks (ANNs) [4], Support Vector Machine (SVM) [5], Bag of Visual Words (BOVW) [6], and Deep Neural Networks (DNNs) [7]. For pattern classification problems, the results are usually presented in terms of how accurately a method classifies the pattern. However, there are some underlying questions related to the testing protocols for pattern recognition and classification applications. The questions are: (a) what is the effect of size (number of images) of the database on the accuracy? (b) what is the effect of the number of classes on the accuracy? (c) what is the effect of the quality of the images on the results? (d) does DNN give better results than ANN for the same database? (e) does the image distribution strategy between training and testing images effect the accuracy? (f) what are the guidelines for the testing protocols? This research project set out to obtain answers to the above questions. It is not a study of transferability, where the focus is on the effect of the network architecture [8]. Such a

© Springer International Publishing AG 2017
G. Boracchi et al. (Eds.): EANN 2017, CCIS 744, pp. 523–534, 2017.
DOI: 10.1007/978-3-319-65172-9_44

study is of the interest to computer vision community, while the research work of this paper focuses on the interest of machine vision community. Hence, it examines the effects of inputs to the network on the classification accuracy for an application.

Pattern recognition has been applied to a variety of applications such as classification of food products [9], resistors [10], gears [11], solder joins [12], welding chips [13] and bearings [14]. All these applications made use of one form or another of Machine Vision Inspection (MVI) techniques. The research work presented in this paper is based on the task of Indian coin recognition using MVI techniques. Indian coins currently circulating in the market consist of five denominations ₹0.5, ₹1, ₹2, ₹5, and ₹10. ₹ (Rupee) is the symbol for Indian currency [15]. The total number of classes, including various styles, for Indian coins is more than 30.

Literature surveys confirm that the most popular approaches for classification are ANNs-based. Bremananth et al. [16] reported 92.43% accuracy for Indian coin classification with ANN, based on numerical information extracted from the coins. In their study, they observed that increasing the number of features increased the classification accuracy. Cai-ming et al. [17] achieved 83% accuracy using an ANNs-based method. Velu et al. [18] used a multi-level counter propagation ANN and obtained 99.47% accuracy for the classification of Indian coins. Modi [19] obtained 81% average recognition rate by using intensity values of 100 pixels as a feature vector input to an ANN for Indian coins. To generate the feature vector, the coin images were shrunk in size to 10×10 pixels. Modi and Bawa [20] were able to improve on this result and achieved 98% accuracy when an image size of 20×20 pixels was used. Both obverse and reverse sides of Indian coins were used for the classification. However, the test conditions used in their research work were not considered very practical: (a) images were generated by scanner (i.e. coins were stationary), (b) images were digitally rotated (not physically rotated, to introduce the effect of shadowing) and (c) there was overlap in the training and testing databases.

DNN is a relatively new approach for application to the problem of pattern recognition. DNN identifies discriminative features of an image directly from the training image dataset (instead of being identified manually by the user). By eliminating the step of manual feature extraction, DNN-based methods are claimed to be more accurate than conventional classification methods. A DNN is an ANN with multiple hidden layers between the input and output layers. Like ANNs, DNNs can model complex non-linear relationships. They are popular in the field of image classification [21], speech recognition [22] and sentiment classification [23]. However, concerns have been raised within the computer vision community over the generality of DNNs and their tendency to misclassify under certain conditions. Nguyen et al. [24] demonstrated that discriminative DNN models are easily fooled, that is, they classify many unrecognizable images with near certainty as member of a recognizable class. Szegedy et al. [25] suggested that the explanation for this was that the set of negatives (or fooling images) was of extremely low probability. In other words, the likelihood of the negatives appearing in the training set was of low probability, and thus was rarely or never observed in the training dataset.

One objective of this research was to compare the performance of ANNs and DNNs as applied to the task of Indian coin recognition. A preliminary analysis of this problem was presented in [26]. This paper is a continuation of that work.

2   Experimental Setup

The experimental setup for the coin recognition task is shown in Fig. 1. The coins were fed to the conveyor using a gravity feeder. A Dorner 2200 series conveyor with a flush mounting package was selected with two key features: (1) there were no sidewalls which enabled flexible lighting and camera arrangements, and (2) conveyor speed was adjustable in the range from 0.5 to 50 m/min. A proximity sensor with a range of 4 mm was mounted on the frame. The sensor was read using an Arduino Uno R3 micro-controller. The purpose of sensor was to detect the presence of a coin and signal the camera via the Arduino to acquire an image. The sensor also served to reject non-metallic objects, which were introduced to mimic counterfeit coins.

Fig. 1. Experimental setup for coin recognition task

A Basler puA1280-54um camera with the maximum resolution of 1280 × 960 at 54 frames/sec (fps) was used for image acquisition. The camera resolution was set to 400 × 400 to minimize the background detail. An industrial grade red dark field light DFI RL1600 from Advanced illumination was used for illumination. The light was mounted 20 mm above the surface of the coins and the working distance between the camera and the coins was set to 160 mm. OpenCV with C++ was used for image acquisition. Figure 2 shows a set of sample coin images for 14 classes. Images were acquired at a rate of 200 coins/min. More detailed information on the design of the experimental setup is given in [27].

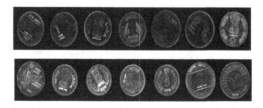

Fig. 2. Indian coins

3   Methodology

3.1   Networks: ANN and DNN

Pattern recognition using ANN and DNN will be discussed in this section. ANNs are widely used for classification problems, functional approximation, clustering and pattern association. Typical ANNs have at least three layers of nodes: input layer, hidden layer and output layer. Each layer has several processing nodes and these layers are connected to each other through nodes using connection weights. Figure 3, on left, shows a typical ANN structure.

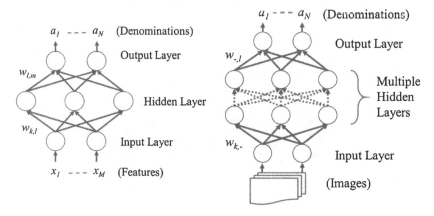

Fig. 3. ANN and DNN structures

For the Indian coin recognition task, a multilayer backpropagation type ANN was used. The input layer was fed with features extracted from the coin images. A feature is an important property of an image (e.g. size or shape of the part). An image can be described by many features. The selection of the features is one of the critical step in pattern recognition and it varies from application to application. If the input database is known to contain redundant data, then it can be transformed into a reduced representation, known as feature vectors. The output layer in the ANN provides the predicted denominations. The number of neurons in the input layer is the number of extracted features while the number of neurons in the output layer is equal to the number of output classes.

DNN has the potential to learn patterns and classify objects with a high degree of accuracy. It takes images as inputs and it has a multiple hidden layer structure. Figure 3, on right, shows a typical DNN structure. One of main advantages of DNN is that is can learn some useful features from the data automatically. To compare the results with ANNs, AlexNet DNN was used for coin classification application [7]. It contains 5 convolutional layers and 3 fully connected layers. For the coin classification application, AlexNet was implemented and trained using NVIDIA DIGITS 2.0. The training images were input to the AlexNet and test images were used to estimate classification accuracy. For the same database, DNN is known to provide higher classification accuracy than ANN [7].

3.2   Feature Extraction

The images of coins were acquired while the coins were fed on the conveyor at a rate of 200 coins/min. The FOV of the camera was set to 400 × 400 pixel resolution. The coin can be offset within the FOV. Hence, the next step was to crop the image in such a way that the coin appears at the center of the cropped image. The cropped image was of size 100 × 100 pixel resolution. A feature extraction step was implemented to generate the feature vector from the cropped images. To generate the feature vector, the image was first reduced to size 20 × 20 using a pattern averaging method. In this method, every pixel in the 20 × 20 was generated by averaging the corresponding 5 × 5 neighborhood pixels from the original image to ensure that significant details are not lost. The 20 × 20 image was then reshaped as 400 × 1 feature vector, with no loss of information. The feature vectors generated from all images of the database were input to the ANN. Hence, the ANN had 400 neurons in the input layer. The number of neurons in the hidden layer was set to 30 and the number of neurons in the output layer was set to the number of distinct classes in the database. MATLAB's ANN toolbox was used for creating, training and testing the ANN. The Scaled Conjugate Gradient method was selected as the training algorithm.

3.3   Training and Testing Strategies

There are two training-testing strategies considered for the database distributions for the ANN and DNN: (a) Mixed Train-Test (MTT) strategy and (b) Different Train-Test (DTT) strategy. Many published papers do not explicitly specify the strategy, but it is generally assumed that if not otherwise stated, a MTT strategy was used for training and testing the networks.

Figure 4 serves to illustrate the MTT and DDT strategies in the context of the coin application. The upper left corner of the figure shows three different physical coins of the same class. The upper right corner shows 9 images generated from 3 physical coins

Train (6 images) and test (3 images in box) distribution strategies

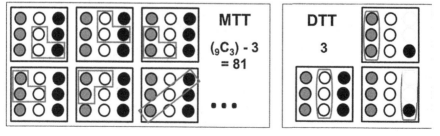

Fig. 4.   Training and testing strategies: Mixed Train Test (MTT) vs Different Train Test (DTT)

by rotating each coin to 3 different angles. Out of these 9 images, 6 images are used for training and 3 images are used for testing the network accuracy. The way training and test images are selected determines the type of strategy (MTT vs DTT). There are 84 possibilities ($_9C_3$) for image distribution between the training and testing databases. Out of these 84 possibilities, in 81 possibilities, one or more test images are the part of the training database. The testing images are not the same as the training images but they are the rotated versions of the training images, as shown in the lower left corner of Fig. 4. This type of image distribution strategy is called the MTT strategy. In the remaining 3 possibilities, testing images are not the part of training database. i.e. testing images are generated from different coins from those used to generate training images, as shown in the lower right corner of Fig. 4. This type of image distribution strategy is called the DTT strategy. MTT results in a high accuracy of classification as there is similarity between training and testing images. The more realistic strategy is the DTT strategy, wherein testing images are generated from physically different coins with the same class.

3.4   Database Description

The database description for the different databases is given in the Table 1. The first row in the table is the database used by Modi and Bawa [20]. They have considered 14 classes of "coins" and for each class acquired 5 distinct images. To increase the size of the database, each image was digitally rotated in the steps of 5° in the range of 0 to 360°. This resulted in 72 rotated images for each unique image. Thus, the database had a total 5040 images (i.e. 70 unique images × 72 rotations). The level of difficulty in the database was low as the MTT strategy was adopted for the image distribution within the database.

Database D1 to D6 were generated for the coin detection and classification task using the experimental setup shown in Fig. 1. Database D1 was generated by following the same guidelines used in [20]. i.e. acquiring 5 unique images per class for a total of 14 classes and digitally rotating each image in 5° steps. The total size of database D1 was 5040 images.

Table 1. Databases

DB#	Classes	Unique images	Angle	Total images
Ref.	14	5 × 14 = 70	5	70 × 72 = 5,040
D1	14	5 × 14 = 70	5	70 × 72 = 5,040
D2	14	(5 + 3) × 14 = 112	5 (train) 60 (test)	4788 (train) + 252 (test) = 5,040
D3	14	(21 + 4) × 14 = 350	18	5880 (train) + 1120 (test) = 7,000
D4	5	(21 + 4) × 5 = 125	18	2100 (train) + 400 (test) = 2,500
D5	36	(70 + 30) × 36 = 3600	10	90,720 (train) + 38,800 (test) = 129,000

Note: Ref. and D1 is of MTT strategy while D2 to D5 are of DTT strategy

Database D2 was generated for 14 classes by acquiring 8 unique images per class. The DTT strategy was used for the image distributions within the database. Out of 8 unique images, 5 images were considered for training only and 3 images were selected for testing purpose. Each training image was rotated in 5 degree steps and each test image was rotated in 60 degree steps between 0 to 360°. Thus, 360 training images per class were generated (i.e. 5×72). To keep the overall size of the database the same as the previous database, 342 images out of 360 images were selected per class. Hence, the overall size of database D2 was 5,040 images (4,788 training images + 252 testing images).

Database D3 was generated by considering more images per class and using the DTT strategy. For this database, a total of 25 unique images per class were acquired for 14 classes. Out of the 25, 21 images were considered for training and 4 images were considered for testing. Each image in the training database was digitally rotated in steps of 18 degree between 0 to 360°. The overall size of database D3 was 7,000 images (5,880 training images + 1,120 test images).

To study the effect of fewer classes on the accuracy, database D4 was generated with the DTT strategy. Five output classes were considered and 25 unique images per class were acquired. The same database distribution approach as mentioned in for the database D3 was used. The overall size of database D4 was 2,500 images (i.e. 2,100 training images + 400 test images).

To study the effect of increasing the number of classes, database D5 was generated based on the DTT strategy. For this database, 36 classes were considered. For each class 100 unique image were acquired. One hundred images were divided into 70 training images and 30 testing images per class. Each image was rotated in steps of 10°. The overall size of database D5 was 129,600 images including training and testing images.

4   Results and Discussion

The experimental results for the coin detection and classification task are given in Table 2. For comparative purposes, the results of Modi and Bawa [20] are also presented. The goal of the experiments was to study the effect of the following on the classification test accuracy: (a) different training and testing strategies (MTT vs. DTT); (b) varying database distributions; (c) varying images per class; (d) reducing the

Table 2. Experiments and results

Exp. #	DB#	Distribution	Epochs	Accuracy	Images per class
Ref.	Ref.	90-5-5	148	98	360
E1	D1	90-5-5	122	100	360
E2	D2	90-5-5	158	60 ± 4	360
E3	D3	90-5-5	141	74 ± 2	500
E4	D4	68-16-16	50	100	500
E5	D5	63-07-30	507	91 ± 2	3600
E6	D5	63-07-30	50	99 ± 0.8	3600

number of classes; (e) increasing the number of classes and (f) training with different networks (ANNs vs. DNNs). A total of six experiments were conducted and they are numbered E1 to E6. For each experiment, the results given are the average of 5 different runs.

Modi and Bawa [20] adopted the MTT strategy and consequently their experiments were considered to be at a lower level of difficulty than for those conducted with the DTT strategy. The database distribution was set to 90%-5%-5% (training-validation-testing). The images from the training database were used for learning the relationship between the inputs and outputs. The training set was used for adjusting the weights of the network. The validation database was used to determine how well the network trained and when to stop the training, as based upon the network performance (i.e. mean square error). A validation step helps to prevent overtraining of the network. The testing database was used to assess the performance of the fully trained network. The classification accuracy of the network reported is based on the performance of the testing database. The training of the ANN was conducted up to 148 epochs. A testing accuracy of 98% was reported.

In an attempt to replicate the results given in [20] for the coin recognition task, experiment E1 was conducted with 14 output classes and 360 images per class. The ANN was setup, trained and tested following the methodology given in [20]. The level of difficulty was considered low as the MTT strategy was used for training and testing. Database distribution was set to 90%-5%-5% (train-validation-testing). The network was trained up to 122 epochs and a classification accuracy of 100% was achieved for the testing database (i.e. more than the 98% reported by Modi and Bawa [20]).

The more realistic training and testing strategy is DTT. Hence, experiment E2 was conducted with 14 classes and 360 images per class with the DTT strategy. The level of difficulty for classification was high for the DTT strategy. The network was trained up to 158 epochs and a classification accuracy of 60% was reported. The accuracy dropped significantly (from 100% to 60%) due to the DTT strategy.

To study the effect of increasing the number of images per class on the network accuracy for the same number of output classes, experiment E3 was conducted. The same DTT strategy was used for training and testing and the 14 output classes with 500 images per class were implemented. The percentage of database distribution was set to 90%-5%-5% (training-validation-testing). The network was trained up to 141 epochs and a classification accuracy of 74% was achieved. In other words, accuracy was improved by 23% with the increase in the number of images.

It is understood that accuracy also depends on the number of output classes. Therefore, to evaluate the effect of reducing the number of classes and keeping the same number of images per class, experiment E4 was conducted. For the experiment, the number of classes are reduced to 5 with 500 images per class. The DTT strategy was adopted and the network was trained up to 50 epochs. A classification accuracy of 100% was achieved with experiment E4. This result confirmed the expectation that accuracy improves with a reduction in the number of output classes for the same size of database.

Based on the observations from experiment E4, the next step was to study the effect of an increase in the number of classes and images per class. Therefore, in experiment E5, 36 output classes are considered (that is the maximum possible with the coin application) and 3600 images per class are collected for the database. The percentage of database distribution was set to 68%-16%-16% (training-validation-testing) and the network was trained up to 507 epochs with the DTT strategy. A classification accuracy of 91% was achieved with this experiment. The accuracy did not improve much from this point after training the network several times and for more epochs. The possible reason for stagnation in the accuracy is the simple structure of the ANN and its limited learning capability.

Experiment E6 was conducted with the same database as that of experiment E5 and AlexNet DNN was used for training and testing. The number of output classes was 36 and for each class 3600 images were generated. AlexNet was trained for 50 epochs. The result was a 99% classification accuracy. The results for all the experiments are shown in Fig. 5.

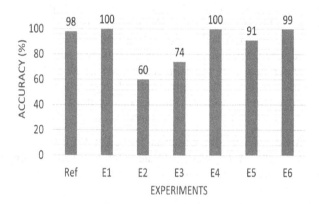

Fig. 5. Classification accuracy summary results

5   Summary and Conclusions

MVI for pattern recognition has been the subject of this paper. ANNs and DNNs are popular tools for pattern recognition using MVI techniques. Several researchers have applied these tools to various applications and reported high classification accuracies. This research has been carried out to answer some of the underlying questions related to pattern recognition applications and provides some guidelines for testing protocols. An Indian coin detection and classification task was used to provide the basis for these guidelines. A number of experiments were conducted and the accuracy results were obtained. It was observed that the classification accuracy result depended upon the size of the database, the quality of the images in the database, the number of classes, the number of images per class, intra-class variations, the type of training-testing strategies (MTT or DTT) and the complexity of the networks (ANNs or DNNs).

In the context of the Indian coin task, 100% classification accuracy was obtained with 360 images per class for 14 classes with ANN trained and tested using the MTT strategy. Most researchers report the results with the MTT strategy for training and testing of the ANNs. The more realistic approach is to use the DTT strategy for training and testing. Using the DTT strategy, the accuracy dropped to 60% for 14 classes. One solution to improve the accuracy with DTT was to increase the number of images per class. Increasing images per class to 500 from 360 for 14 classes and using the DTT strategy increased the accuracy to 74% from 60%. The 23% improvement in the accuracy was due to the increase in the number of images per class.

The number of classes also affected the network's ability to accurately classify images. A higher accuracy is achieved with a lower number of output classes. For example, a setup with 5 output classes resulted in 100% classification accuracy. When the number of classes was increased to 36 and with 3,600 images per class, 91% classification accuracy was achieved with DTT. Improving accuracy beyond this point was difficult due to the high degree of similarity between classes, that is to say there weren't many intra-class variations. Moreover, it is limited due to the simple structure of the ANN and its learning capabilities.

For pattern recognition applications with less intra-class variations, a better classification accuracy can be obtained if the DNN is selected for training and testing. Using the same database as the ANN and implementing DNN for 36 classes, an accuracy of 99% was obtained. The DNN multi-layer complex structure, automatic feature selection and many images per class were the main contributors to the high accuracy.

Based upon the results to date, the recommended guidelines for a testing protocol for pattern recognition and classification in the context of part recognition applications are as follows:

- MTT will provide higher accuracy than DTT, but DTT provides more realistic results, because testing images are generated with physically different coins.
- Classification accuracy can be improved if the number of images per class is increased.
- For the same number of images per class, a higher classification accuracy can be achieved with a lower number of output classes.
- Images per class should be increased along with an increase in the number of classes. As a rule of thumb, the number of images per class should be 100 times the number of classes.
- For less intra-class variation applications, it is difficult to achieve an accuracy on the order of 99% with the traditional ANNs, even though many images per class are used.
- DNN provides a higher accuracy than ANN for the same database. However, DNNs have their limitations and DNN should be tested with unknown classes to provide for realistic results.

6 Future Work

In order to test the validity of the guidelines, further tests should be conducted with different part recognition applications. It also is acknowledged that the quality of images in the database can significantly affect classification accuracy. Thus, further tests should be conducted to investigate the effect of image quality. In the context of the coin recognition task, this can be accomplished by collecting low, medium and high quality images, by dividing the coins into mint quality, medium wear and high wear classes. An image quality index should be used to quantify quality in order to provide quantitative guidelines. There are image quality indices available in the literature (e.g. [28]).

Acknowledgement. The research was partially funded by Mitacs and partially funded by Queen's University.

References

1. Davidsson, P.: Coin classification using a novel technique for learning characteristic decision trees by controlling the degree of generalization. In: 9th International Conference on Industrial and Engineering Applications of Artificial Intelligence and Expert Systems, pp. 403–412, Fukuoka, Japan (1997)
2. Van Der Maaten, L., Boon, P.: Coin-o-matic: a fast system for reliable coin classification. In: Proceedings of the MUSCLE CIS Coin Competition Workshop, pp. 7–17, Berlin, Germany (2006)
3. Tang, P., Steinbach, M., Kumar, V.: Introduction to Data Mining, pp. 487–568. Pearson Addison Wesely, London (2006). Chap. 8
4. Bianchini, M., Scarselli, F.: On the complexity of shallow and deep neural network classifiers. In: European Symposium on Artificial Neural Networks, Computational Intelligence and Machine Learning (ESANN), pp. 371–376, Bruges, Belgium (2014)
5. Csurka, G., Dance, C., Fan, L., Willamowski, J., Bray, C.: Visual categorization with bags of keypoints. In: Workshop on Statistical Learning in Computer Vision, pp. 1–22, Prague, Czech Republic (2004)
6. Nilsback, M., Zisserman, A.: A visual vocabulary for flower classification. In: IEEE Conference on Computer Vision and Pattern Recognition (CVPR), vol. 2, pp. 1447–1454, New York, USA (2006)
7. Krizhevsky, A., Sutskever, I., Hinton, G.: Imagenet classification with deep convolutional neural networks. In: Advances in Neural Information Processing Systems, pp. 1097–1105 (2012)
8. Azizpour, H., Razavian, A., Sullivan, J., Maki, A., Carlsson, A.: Factors of transferability for a generic convnet representation. IEEE Trans. Pattern Anal. Mach. Intell. **38**(9), 1790–1802 (2016)
9. Nashat, S., Abdullah, A., Aramvith, S., Abdullah, M.: Support vector machine approach to real-time inspection of biscuits on moving conveyor belt. Comput. Electron. Agric. **75**(1), 147–158 (2011)
10. Niklaus, P., Ulli, G.: Automated resistor classification. Group thesis, Swiss Federal Institute of Technology, Computer Engineering and Networks Laboratory, Zurich, Switzerland (2015)
11. Wu, W., Wang, X., Huang, G., Xu, D.: Automatic gear sorting system based on monocular vision. Digit. Commun. Netw. **1**(4), 284–291 (2015)
12. Kim, T.-H., Cho, T.-H., Moon, Y., Park, S.: Visual inspection system for the classification of solder joints. Pattern Recogn. **86**(11), 2278–2324 (1998)
13. Iyshwerya, K., Janani, B., Krithika, S., Manikanandan, T.: Defect detection algorithm for high speed inspection in machine vision. In: International Conference on Smart Structures and Systems (ICSSS), pp. 103–107, Chennai, India (2013)

14. Shen, H., Li, S., Gu, D., Chang, H.: Bearing defect inspection based on machine vision. Measurement **45**(4), 719–733 (2012)
15. Competition for design: retrieved from Indian government website. http://finmin.nic.in/the_ministry/dept_eco_affairs/currency_coinage/Comp_Design.pdf. Accessed 01 Aug 2016
16. Bremananth, R., Balaji, B., Sarkari, M., Chitra, A.: A new approach to coin recognition using neural pattern analysis. In: IEEE Indicon Conference, pp. 366–370, Chennai, India (2005)
17. Cai-ming, C., Shi-qing, Z., Yue-fan, C.: A coin recognition system with rotation invarience In: International Conference on Machine Vision and Human Interface, pp. 755–757, Kaifeng, China (2010)
18. Velu, C., Vivekanandan, P., Keshwan, K.: Indian coin recognition and sum counting system of image data mining using artificial neural networks. Int. J. Adv. Sci. Technol. **31**, 67–80 (2011)
19. Modi, S.: Automated coin recognition system using ANN. Master's thesis, Department of Computer Science Engineering, Thapar University, Patiala, India (2011)
20. Modi, S., Bawa, S.: Automated coin recognition system using ANN. Int. J. Comput. Appl. **26**(4), 13–28 (2011)
21. Shah, S., Bennamoun, M., Boussaid, F.: Iterative deep learning for image set based face and object recognition. Neurocomputing **174**, 866–874 (2015)
22. Noda, K., Yamaguchi, Y., Nakadai, K., Okuno, H., Ogata, T.: Audio-visual speech recognition using deep learning. Appl. Intell. **42**, 722–737 (2015)
23. Zhou, S., Chen, Q., Wang, X.: Active deep learning method for semi-supervised sentiment classification. Neurocomputing **120**, 536–546 (2013)
24. Nguyen, A., Yosinski, J., Clune, J.: Deep neural networks are easily fooled: high confidence predictions for unrecognizable images. In: IEEE Conference on Computer Vision and Pattern Recognition (CVPR), pp. 427–436, Boston, USA (2015)
25. Szegedy, C., Zaremba, W., Sutskever, I., Bruna, J., Erhan, D., Goodfellow, I., Fergus, R.: Intriguing properties of neural networks. arXiv preprint arXiv:1312.6199, (2014)
26. Joshi, K., Surgenor, B., Chauhan, V.: Analysis of methods for the recognition of Indian coins: a challenging application of machine vision to automated inspection. In: 23rd International IEEE Conference on Mechatronics and Machine Vision in Practice (M2VIP), pp. 1–6, Nanjing, China (2016)
27. Joshi K., Chauhan V., Surgenor, B.: Real-time recognition and counting of Indian currency coins using machine vision: a preliminary analysis. In: Proceedings of the Canadian Society for Mechanical Engineering (CSME) International Congress, Kelowna, Canada (2016)
28. Mittal, A., Soundararajan, R., Bovik, A.: Making a completely blind image quality analyzer. IEEE Sig. Process. Lett. **20**(3), 209–212 (2013)

Particle Swarm Optimization Algorithms for Autonomous Robots with Leaders Using Hilbert Curves

Doina Logofatu[1(✉)], Gil Sobol[1], and Daniel Stamate[2]

[1] Computer Science Department of Frankfurt University of Applied Sciences,
1 Nibelungenplatz, 60318 Frankfurt, Germany
[2] Department of Computing, Goldsmiths College, University of London,
London SE146NW, UK

Abstract. The approaches in this work combine the swarm behavior principles of Craig W. Reynolds with space filling curves movements. We intend to evaluate how the entire swarm moves by including a deterministic *leader* behavior for some agents. Therefore, we examine different combinations of Hilbert Curves with the classical swarm algorithms. We introduce a practical problem, the collection of manganese nodules on the sea ground by using autonomous agents. Some relevant experiments, combining different parameters for the leaders were run and the results are evaluated and described. Finally, we propose further developments and ideas to continue this research.

Keywords: Particle swarm optimization · Changing environments · Autonomous agents · Hilbert Curves · Leaders · Application

1 Introduction

At the same time with the research of renewable resources, it would be useful to find new ways in order to open up fossil ones. For example, manganese can be found on the sea bottom in form of nodules. Actually, this is a chemical element with the periodic table symbol M_n and atomic number 25. The biggest application area is for rust and corrosion prevention on steel [9]. Degradation can be prevented by collecting these manganese nodules from the sea ground using specialized robots. It is necessary to find appropriate ways how to handle the action of these agents. The results can be generalized and cover other collecting tasks as well. Consequently, this work focuses on optimizing the swarm behavior of autonomous agents. The solution can be found by either improving the achievements or reducing the effort by keeping the same size of achievements. The background for our approach is a framework for simulation and improvement of swarm behavior in changing environments [1]. It simulates the swarm behavior after the principles of Reynolds [2] later pointed out in Sect. 2.2. The main issue the associated application to the framework does, is to deploy agents with a specific strategy and then to gather them. While gathering, the agents are collecting the manganese which is distributed on every position in the coordinate system. Once they are gathered together, there is no more movement and the simulation ends. The intention of our work is to redesign and extend this framework. Manganese occurs in

© Springer International Publishing AG 2017
G. Boracchi et al. (Eds.): EANN 2017, CCIS 744, pp. 535–543, 2017.
DOI: 10.1007/978-3-319-65172-9_45

form of nodules, so it is not really realistic that they are distributed uniformly. Therefore, it is reasonable to implement a *Manganese-Nodule-Model*, where each nodule is represented stand-alone. It should also be considered that not all nodules have the same size (value). For this new design of the manganese distributions, benchmarks have to be created for having the opportunity to compare the results. The next issue is to improve the collecting procedure itself. The more meters the agents pass, the higher is the chance to find manganese. Consequently, we try to find a way to pass through a bigger area. The less complex solution would be to give each agent his own route. This would probably scatter the swarm because of the bad orientation and the uneven surface. Most of the research activities about swarm behavior are inspired by nature like genetic algorithms or particle swarm optimization. These researches focus on bird flocks or fish schools. An alternative discussion could be focusing on a pack of wolves, for example. A pack of wolves consists of autonomous individuals with a specific hierarchy. Not every wolf has the same power of decision for the pack. Normally there is one wolf who leads the group and the others are followers [11]. This work aims to study this concept more closely. We want to have one or more leaders who will move after a specific route, but still being part of the swarm and the rest calculates its new position, that means every iteration in consideration of all agents. We imagine an area where we know that there should be a big amount of manganese nodules. We need to find proper methods to explore systematically and carefully through a given area. One of the first things that comes to one's mind is the specialist mathematical field of space filling curves. Summarized, a space filling curve is a curve that covers recursively an entire 2-dimensional square. We are focusing on the one of the most famous space filling curves developed by David Hilbert (Sect. 2.4).

2 Background

This section describes the previous work the application is based on. It includes three main topics: Moving Algorithms, Particle Swarm Optimization, Hilbert Curves.

2.1 Framework for Adaptive Swarms Simulation and Optimization

The starting application is based on [1]. The framework is an application that runs a simulation of agents using moving algorithms Random, Square, Circle, Gauss, and Bad Centers [1]. It has different deployment strategies implemented from where the moving algorithms start. The frontend is based on the open source framework of *processing.org* [4]. The whole visualization part is done in the *Visualization* class with support of its derived class *VisualRobot*, which helps to represent the robots in the visualization. The whole simulation part is managed by a class with the same name. It creates the chosen deployment strategies and calculates the movement of the robots, as well as the collection of manganese. Manganese is located on every position in the coordinate system. In addition, it also counts the distance in walked meters of all agents together. It is also possible to set at the beginning the number of agents. This number must be between 2 and 100.

2.2   Moving Algorithms

Artificial systems are, for example, needed when it is wanted to solve problems which are beyond the capabilities of a single individual. In our case it is actually required to build a swarm of agents, where each agent individually moves forward with consideration of the other agents of the swarm. There are several efficient algorithms for swarm behavior and movement of agents that could be implemented in the application [4]. The previous work [1] uses a simplification of the bird flock movement described by Reynolds [2]. The idea was to develop algorithms that simulate swarm behavior inspired by flocks of birds or schools of fish. Therefore, three criteria every robot follows at each iteration were settled up. The contribution implemented three different algorithms that run simultaneously: cohesion, separation, alignment.

2.3   Particle Swarm Optimization

Particle Swarm Optimization (PSO) was first proposed in 1995 by Kennedy and Eberhart [6]. The idea was to build swarm behavioral algorithms for solving problems by iteratively improving a candidate's solution until termination criteria is satisfied [7]. It is similar to a genetic algorithm regarding that both algorithms are initialized with a random population, in PSO called particles. The difference is that in PSO algorithms, each particle is assigned to a randomized velocity and the particles move through hyperspace. Each particle consists of its position, its velocity, its current objective value and its personal best value of all time. PSO also keeps track of the global best value that is the best objective value of all particles and also the corresponding position.

$$x^{(i)}(n+1) = x^{(i)}(n) + v^{(i)}(n+1), \quad n = 0,1,2,\ldots,N-1 \tag{2.1}$$

The formula above describes a classical iteration for particle movement. The next position $x^{(i)}(n+1)$ is made from the current position $x^{(i)}(n)$ and the velocity vector $v^{(i)}(n+1)$ of a specific particle i. The velocity vector gets created by the following iteration:

$$v^{(i)}(n+1) = \underbrace{v^{(i)}(n)}_{inertia} + \underbrace{r_1^{(i)}(n)[x_p^{(i)}(n) - x^{(i)}(n)]}_{\substack{personal\ influence \\ n=0,1,2,\ldots,N-1,}} + \underbrace{r_2^{(i)}(n)[x_g^{(i)}(n) - x^{(i)}(n)]}_{global\ influence} \tag{2.2}$$

where x_p represents the individual and x_g the global best position. $[x_p^{(i)}(n) - x^{(i)}(n)]$ calculates a vector towards the personal best which is influenced by the random vector $r_1^{(i)}(n)$, that contains values uniformly distributed between 0 and 1.

$[x_g^{(i)}(n) - x^{(i)}(n)]$ calculates a vector towards the global best which is also influenced by some randomness $r_2^{(i)}(n)$. PSO has two options to focus on every iteration. The first option is *diversity*, that means particles are scattered, searching a large area but imprecise. The second option is *convergence* that means particles are close together, searching a small area very precise. The best result can be achieved through a combination of both.

2.4 Hilbert Curves

A Space Filling Curve is a special line of the mathematical calculus that fully covers a two or three dimensional area. Giuseppe Peano (1858–1932) was the first to discover them in 1890. He wanted to create a continuous mapping construction from the unit interval onto the unit square [7]. Space Filling Curves have a wide field of purpose in computer science. They are used specially to linearize multidimensional data, e.g. matrices, images and tables. With their help it is possible to simplify data operations like load-store operations, matrix multiplications and updating and partitioning of data sets by finding an efficient way to go through the data.

Definition 2.2. *Hilbert Curve* [10]. The unit square is divided into congruent sub squares $Q_n^{(k)}$ with side length 2^n. The only condition is, that neighboring sub intervals are mapped onto neighboring sub squares, whereby the square that is next to the zero position is always the first and the one that is next to the point (1, 0) is always the last. If we are now connecting the center of these squares in the right order, we get unequivocal curves C_n (Fig. 1).

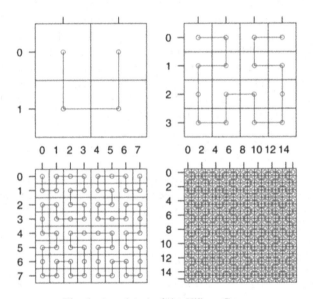

Fig. 1. Level 1–4 of the Hilbert Curve

3 Implementation Details

This section shows the practical changes and extensions that were necessary to implement for the experimental procedure. At first, some new classes had to be implemented to lay the basis for the new *Manganese-Nodule-Model*. These new classes help us to represent the nodules on the map as well as for the calculations in the back. In Sect. 3.3 is described how these new classes get connected to the existing simulation and visualization.

3.1   New Classes

The class *DeployRing* deploys the robots in a ring shaped way. It is part of the *DeploymentStrategy* interface. The deployment algorithm is similar to the *DeployCircle*, but has a specific radius right from the beginning. Objects of type *ManganeseNodule* represent the manganese nodules in the backend simulation. Each *ManganeseNodule* object also contains a *Coordiantes* Object, which specifies the exact position of the nodule in the coordinate system. The class *VisualManganModule* helps to visualize the manganese nodule in the simulation. For each available *ManganeseNodule* object in the corresponding list, a new *VisualManganNodule* object gets created every iteration (Fig. 2).

Fig. 2. DeployRing; top left initial deploy.

Objects of type *ManganeseNodule* represent the manganese nodules in the backend simulation. They have two attributes. One is the size of the nodule that ranges from 1–7 as an integer. The other one is the status if the nodule is available or already collected. Each *ManganeseNodule* Object also contains a *Coordiantes* Object, which specifies the exact position of the nodule in the coordinate system (Fig. 3).

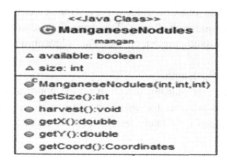

Fig. 3. Visual class diagram *ManganeseNodule*.

The class *VisualManganModule* helps to visualize the manganese nodule in the simulation. For each available *ManganeseNodule* object in the corresponding list, a new *VisualManganNodule* object gets created every iteration. This class includes the conversion from the size as an integer to the corresponding grey tone for placing it into the map in the application.

3.2   General Evolution

The *getMangan()* method was advised due to the new structuring of the benchmarks. As there is not manganese on every position anymore, this function needs to check if there is a manganese nodule on this position. If so, the size of it is also returned. The *step()* method helps creating the route to an specific space filling algorithm described later.

3.3   Benchmarking

The benchmarks are provided as independent files. It was necessary to create new classes *ManganeseNodule* and *VisualManganNodule*. These two classes help us to simulate the collection of manganese by our autonomous agents. The class *ManganesNodule* is thereby necessary for all backend happenings and the class *VisualManganNodule* is necessary for visualization in the graphical user interface. The visualization part is done by the *VisualManganNodule* class. The files are deposited in the project archive. Each line represents a *y*-value and each char represents a *x*-value in the coordinate system of the graphical user interface. The lines are filled with numbers from 0 to 7 in accordance with the size of the nodule, where zero means that no nodule can be find on this position. The user can choose between three options: MAP 1, MAP 2 or MAP 3 and load them. Then the associated file gets scanned and the nodules created (Fig. 4).

3.4   Hilbert Algorithm

The Hilbert algorithm is implemented with a recursive function which follows the description of Sect. 2.4. The function is called every time when the agent moves into the next unit square. The function calls varying from clockwise rotation to negative rotation which means counterclockwise. The ground structure of how going through the 9 sub-squares is fixed implemented. The algorithm function receives four parameters:

double len: Initial step length.
int direction: Specifies the starting direction on the coordinate system in degree.
int rot: Indicates whether the curve should run clockwise or not.
int deep: Determines how many levels deep the algorithm should go (Fig. 5).

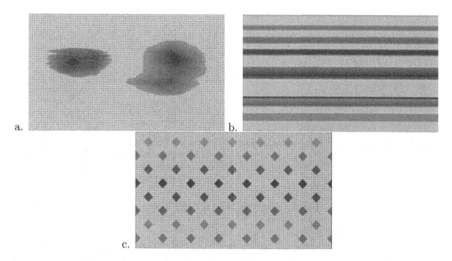

Fig. 4. Benchmarks: Fields (a), Lines (b), Diamond (c). All three graphics represent benchmark maps. The benchmarks include manganese nodules from size 1–7.

```
hilbertAlgorithm(length, direction, rotation, deep){
if under lowest level then
|        return;
end
        hilbertAlgorithm(length, direction, clockwise rotation, deep-1);
        step forward with given length and direction;

        direction turn clockwise with given rotation degree;
        hilbertAlgorithm(length, direction, counterclockwise rotation, deep-1);
        step forward with given length and direction;

        direction turn clockwise with given rotation degree;
        hilbertAlgorithm(length, direction, clockwise rotation, deep-1);
        step forward with given length and direction;

        hilbertAlgorithm(length, direction, counterclockwise rotation, deep-1);
```

Fig. 5. Hilbert algorithm

4 Experimental Results

The measured variables are the distance and the collected amount of manganese of all robots in one pass. The difference of robots between *Rob Total* and *Rob Hilbert* are robots behaving after the principles of Moving Algorithms in Sect. 2.2.

We increase successively the number of Hilbert Robots and ran 1000 iterations with every increase. It runs with the benchmark Diamonds and the deployment strategy Square. This experiment runs with the benchmark Diamonds and the deployment strategy Square. With every increase of the number of Hilbert Robots, the covered distance of all robots increases by 40,000–60,000 m with an average increase of 56,569.85 m. The collected manganese does not increase constantly as well. The global maximum of 5335 kg is reached with a constellation of 46 Hilbert Robots (see Fig. 6). The biggest jump is between the first and the second measurement, with an increase of

a. Collected amount of manganese of Diamond, Square, Hilbert 0-50.

b. Relation between the total amount of collected manganese and the distance all robots have covered, for the experiment Diamond, Square, Hilbert 0-50.

Fig. 6. Analysis Hilbert 0–50 increase.

562%. If we have a look at the efficiency in Fig. 6 (left diagram) we see a raising graph with some flat parts, all amounts of one flat part have the same efficiency, that is the case for the amount of 3–6 Hilbert Robots (average absolute deviation 3.2 m), for the amount of 15–21 Hilbert Robots (average absolute deviation 5.2 m) or 34–40 Hilbert Robots (average absolute deviation 2.9 m) (Fig. 7).

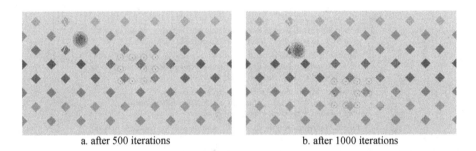

a. after 500 iterations   b. after 1000 iterations

Fig. 7. Diamond, square, Hilbert 0–50: screenshots experimental procedure.

The correlating local minimum occurs with the amount of 42 Hilbert Robots and 464 m per kg manganese. In this case, the total amount of manganese breaks down roughly 21% compared to the simulation run with 49 Hilbert Robots.

5 Conclusion and Future Work

As this work was focused only on combining swarm behavior with specific space filling curves, there are further things to develop in order to get deeper into this topic. It would be conceivable to think of different leaders with different weightings. For example, the leader who collected the most manganese in the last 10 iterations could get the highest weight in calculating the next position of each agent of the swarm. In addition to that, a distributed system could be implemented and thereby the communication between the agents would be extended. To really get a maximum amount of manganese, the swarm could divide and follow different leaders. The number of agents who join a specific leader could vary. If the leader loses strength, more and more agents join another swarm. The leader stays on his route; this keeps the chance high to find new manganese nodule fields. If a leader does not collect any manganese for a longer period, he may become a follower and joins a swarm. This could also be possible the other way around. If there is a big swarm, new leaders could be chosen to search in a specific direction. Another interesting direction would be to combine space filling curves with genetic algorithms. It would be imaginable to build populations with different amounts of deterministic and swarm agents, like this work already did, but keep on developing the next generation after the principles of genetic algorithms. As extension, we can consider a changing environment with hundreds or thousands of agents.

References

1. Canyameres, S., Logofătu, D.: Platform for simulation and improvement of swarm behavior in changing environments. In: Iliadis, L., Maglogiannis, I., Papadopoulos, H. (eds.) AIAI 2014. IAICT, vol. 436, pp. 121–129. Springer, Heidelberg (2014). doi:10.1007/978-3-662-44654-6_12
2. Reynolds, W.: Boids (simulated flocking). http://www.red3d.com/cwr/boids. Accessed 10 June 2017
3. Shyr, W.-J.: Parameters Determination for Optimum Design by Evolutionary Algorithm. doi:10.5772/9638. Accessed 10 June 2017
4. Fry, B., Reas, C.: Processing. https://processing.org/. Accessed 10 June 2017
5. Rodriguez, F.J., García-Martínez, C., Blum, C., Lozano, M.: An artificial bee colony algorithm for the unrelated parallel machines scheduling problem. In: Coello, C.A.C., Cutello, V., Deb, K., Forrest, S., Nicosia, G., Pavone, M. (eds.) PPSN 2012. LNCS, vol. 7492, pp. 143–152. Springer, Heidelberg (2012). doi:10.1007/978-3-642-32964-7_15
6. Kennedy, J., Eberhart, R.: Particle swarm optimization. In: IEEE Conference on Neural Networks, vol. 4, pp. 1942–1948
7. Barnsley, M.F.: Fractals Everywhere, Dover Books on Mathematics, New Edition. Dover Publications Inc., Mineola (2012). ISBN 978-0486488707
8. Detailed requirements for the first prototype. http://informaticup.gi.de/fileadmin/redaktion/Informatiktage/studwett/Aufgabe_Manganernte_.pdf. Accessed 10 June 2017
9. Rossum, J.R.: Fundamentals of Metallic Corrosion in Fresh Water. http://www.roscoemoss.com/wp-content/uploads/publications/fmcf.pdf. Accessed 10 June 2017
10. Kim, M.J., Kim, J.G.: Effect of manganese on the corrosion behavior of low carbon steel in 10 wt.% sulfuric acid. Int. J. Electrochem. Sci. **10**, 6872–6885 (2015)
11. Muro, C., Escobedo, L., Spector, L., Coppinger, R.P.: Wolf-pack (canis lupus) hunting strategies emerge from simple rules in computational simulations. Behav. Process. **88**(3), 192–197 (2011)

A Neural Circuit for Acoustic Navigation Combining Heterosynaptic and Non-synaptic Plasticity That Learns Stable Trajectories

Danish Shaikh[✉] and Poramate Manoonpong

Embodied AI and Neurorobotics Laboratory, Centre for BioRobotics,
Maersk Mc-Kinney Moeller Institute, University of Southern Denmark,
5230 Odense M, Denmark
{danish,poma}@mmmi.sdu.dk
http://ens-lab.sdu.dk/

Abstract. Reactive spatial robot navigation in goal-directed tasks such as phonotaxis requires generating consistent and stable trajectories towards an acoustic target while avoiding obstacles. High-level goal-directed steering behaviour can steer a robot towards the target by mapping sound direction information to appropriate wheel velocities. However, low-level obstacle avoidance behaviour based on distance sensors may significantly alter wheel velocities and temporarily direct the robot away from the sound source, creating conflict between the two behaviours. How can such a conflict in reactive controllers be resolved in a manner that generates consistent and stable robot trajectories? We propose a neural circuit that minimises this conflict by learning sensorimotor mappings as neuronal transfer functions between the perceived sound direction and wheel velocities of a simulated non-holonomic mobile robot. These mappings constitute the high-level goal-directed steering behaviour. Sound direction information is obtained from a model of the lizard peripheral auditory system. The parameters of the transfer functions are learned via an online unsupervised correlation learning algorithm through interaction with obstacles in the form of low-level obstacle avoidance behaviour in the environment. The simulated robot is able to navigate towards a virtual sound source placed 3 m away that continuously emits a tone of frequency 2.2 kHz, while avoiding randomly placed obstacles in the environment. We demonstrate through two independent trials in simulation that in both cases the neural circuit learns consistent and stable trajectories as compared to navigation without learning.

Keywords: Behaviour-based robotics · Reactive navigation · Phonotaxis · Lizard peripheral auditory system · Synaptic plasticity · Correlation-based learning

1 Introduction

Navigating towards a sound source is relevant is several real-world applications. For example, mobile robots in the home that autonomously navigate to a human

© Springer International Publishing AG 2017
G. Boracchi et al. (Eds.): EANN 2017, CCIS 744, pp. 544–555, 2017.
DOI: 10.1007/978-3-319-65172-9_46

speaker in response to voice commands can be useful in the human-robot interaction context. Such robots could also navigate towards people living alone that are immobilised due to injury sustained by falling down but are able to vocalise—a common problem among the elderly population. Furthermore, search-and-rescue robots looking for visually undetectable survivors amongst rubble and large debris in natural disaster scenarios such as earthquakes can use audition as a sensor modality for navigation in such cluttered environments.

Navigation is one of the earliest problems that has been investigated in the mobile robotics community. Control architectures (see [17] for a comparative review) for mobile robot navigation can be broadly classified into three types— reactive, deliberative and hybrid (a combination of deliberative and reactive). These architectures differ in the level of sensing, planning and acting performed [1]. Deliberative and hybrid control architectures can generate optimal robot trajectories but require precise a priori knowledge of the environment respectively at the global or local level. These approaches also consume significant computational power to achieve this optimality. Purely reactive controllers on the other hand are advantageous in that they do not rely on formal path planning algorithms that necessitate such precise and local/global knowledge of the environment a priori. Such controllers also tend to employ behaviour-based architectures [3] such as the widely known subsumption architecture [6]. The neural circuit presented here differs from the subsumption architecture in that the low-level obstacle avoidance behaviour modulates the parameters of the high-level goal-directed steering behaviour via unsupervised learning. Therefore, no explicit hand-tuning of the high-level behaviour is necessary. There has also been significant research in neurobiologically-inspired map-based navigation for mobile robots (see [24] for a review on the research in the last decade) where the focus has been on specialised neurons for spatial awareness and navigation in rodents.

Path-planning and obstacle avoidance are core components in mobile robot navigation and a number of techniques have been developed. Reviews of various approaches from different perspectives can be found in [13,22]. Path planning can be global or local, where the environment is respectively fully or locally known. Global path planning is usually performed offline due to high computational requirements while local path planning is performed online due to relatively lower computational requirements. Here we focus on online reactive path planning using only local information about the distance of the closest approaching obstacle and the direction of the target.

There are several analytical approaches in the literature on reactive acoustic navigation for mobile robots with obstacle avoidance [2,4,11,12,26,27], differing in the number of microphones as well as number and type of distance sensor(s) used. Our approach focusses on developing a purely reactive control architecture with two sound sensors and one distance sensor. This architecture is in the form of a neural circuit that implements two behaviours—high-level goal-directed steering and low-level obstacle avoidance. The neural circuit combines two brain-inspired mechanisms involved in learning and memory—heterosynaptic plasticity and non-synaptic plasticity. These two mechanisms allows a simulated non-holonomic mobile robot to learn stable trajectories towards an acoustic target, placed 3 m away, emitting a continuous tone of 2.2 kHz. The neural circuit is validated in simulation and its performance with and without learning is compared. Sound direction information is extracted by a previously developed model of the lizard peripheral auditory system [23]. This model has been extensively studied via various robotic implementations [21]. Braitenberg sensorimotor mappings [5] between the extracted sound direction and the robot's motor velocities are used to generate the goal-directed steering behaviour (phonotaxis), while the parameters of these mappings are modulated via Input Correlation (ICO) learning [19] during the obstacle avoidance behaviour. Thus, interaction with obstacles is explicitly exploited to fine-tune parameters of the high-level behaviour and learn consistent and stable trajectories. ICO learning is unsupervised, closed-loop, correlation learning adapted from differential Hebbian learning [14,15].

This paper is organised in the following manner. Section 2 describes the lizard peripheral auditory system model and its response characteristics. Heterosynaptic and non-synaptic plasticity are also described here. Section 3 describes the neural circuit and the experimental setup. Simulation results are presented and discussed in Sect. 4. Section 5 summarises the research and outlines further work.

2   Background

2.1   Lizard Peripheral Auditory System

The lizard peripheral auditory system (Fig. 1A, left) has two eardrums (TM) connected via internal air-filled Eustachian tubes (ET) opening into a central cavity. Sound passes through the ET, linking the two eardrums acoustically. This acoustical coupling maps small phase differences (corresponding to inter-aural time differences in the μ-sec scale) between sound waves impressing at either ear into relatively larger differences of up to 20 dB in sound amplitudes sensed at either ear. The magnitude of the phase difference and therefore the sensed amplitude difference corresponds to the relative direction from which sound arrives.

A lumped-parameter electrical model [9] (Fig. 1A, right) of the lizard peripheral auditory system mimics its filtering effects. Voltages V_I and V_C respectively model sound pressures P_I and P_C at the ipsilateral (towards the location of the sound source) and contralateral (away from the location of the sound source) ears. P_I and P_C respectively trigger the flow of currents i_I and i_C, that model eardrum vibrations, through impedances Z_r and Z_v. Z_r model the total acoustic filtering due to the stiffness of the ET and the eardrum mass while Z_v models the acoustic filtering effects of the central cavity. The resultant sound pressure in this cavity as modelled by voltage V_{cc}, due to the superposition of internal sound pressures experienced from either end triggers the flow of current i_{cc}. This current models the sound wave propagation inside the central cavity due to variations in the sound pressure inside it. This model generates as outputs $|i_I|$ and $|i_C|$ (Fig. 1B), which respectively are the sensed ipsilateral and contralateral amplitudes at the corresponding ear. Sound direction information is encoded into these quantities as described earlier and formulated as

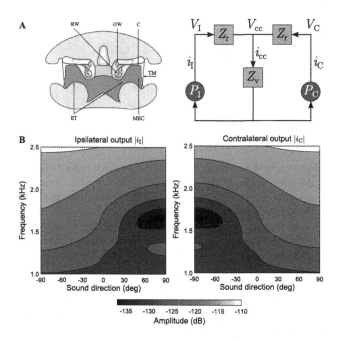

Fig. 1. Sound direction information driving phonotaxis behaviour. (A) Cross-section [7] (left) of a lizard (genus *Sceloporus*) peripheral auditory system and its electrical equivalent [9] (right). (B) The outputs $|i_I|$ (left) and $|i_C|$ (right) as given by (1).

$$|i_I| = |G_I \cdot V_I + G_C \cdot V_C| \equiv 20 \log |i_I| \text{ dB and}$$
$$|i_C| = |G_C \cdot V_I + G_I \cdot V_C| \equiv 20 \log |i_C| \text{ dB.} \tag{1}$$

G_I and G_C respectively are sound frequency-specific (1–2.2 kHz) ipsilateral and contralateral gains. These terms are experimentally derived via laser vibrometry [7] measurements of ear vibrations and are implemented as 4th-order digital infinite impulse response bandpass filters. The symmetry of the model implies that its outputs $|i_I|$ and $|i_C|$ are identical for a sound signal arriving directly from the front or the back in the azimuth plane, i.e. there is front-back ambiguity. Furthermore, when $|i_I| > |i_C|$ the sound signal coming from the ipsilateral side and when $|i_C| > |i_I|$ it is coming from the contralateral side. $|i_I|$ and $|i_C|$ vary symmetrically but non-linearly with the sound direction, respectively reaching maxima and minima towards the extremes of the relevant range of $[-90°, +90°]$.

2.2 Plasticity and Learning in the Biological Brain

Biological neurons are interconnected via synapses, that act as a bridge between two individual neurons (see Fig. 2A[1]). Electrical signals are propagated between

---

[1] modified from https://en.wikipedia.org/wiki/Nonsynaptic_plasticity#/media/File:Neurons_big1.jpg.

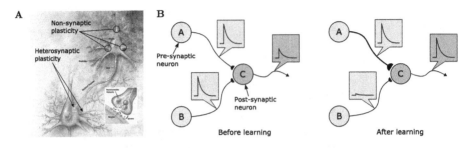

Fig. 2. Neuroplasticity and learning. (A) Illustration showing heterosynaptic and non-synaptic plasticity sites in the brain. (B) Hypothetical neural circuit before and after input correlation learning. During learning, temporally correlated activities of sensory neurons A and B leads to gradual increase in strength of synapse (depicted as a thick line) between A and motor neuron C. This leads to stronger activation of neuron C, causing correspondingly stronger behavioural response. After learning the behavioural response is strong enough to avoid activation of neuron B.

a pre-synaptic neuron and a post-synaptic neuron via biochemical processes occurring in these synapses. Changes in the properties of these synapses influence (but not exclusively) whether the post-synaptic neuron will fire or not in response to synaptic input from a pre-synaptic neuron. This phenomenon of activity-induced changes in the synapses is referred to as *synaptic plasticity* [20]. These changes can make the synapse relatively weaker or stronger, respectively decreasing or increasing the efficacy of the synaptic connection between the pre-synaptic and post-synaptic neuron. Synaptic plasticity is believed to be the core biochemical process underlying learning and memory.

Change in the synaptic strength between two neurons is typically dependent on the activity of the pre-synaptic neuron. However the activity of a third neuron can releases chemical neuromodulators that induce changes in synaptic strength between two other neurons as well, a phenomenon referred to as *heterosynaptic plasticity*. The intrinsic excitability, i.e. sensitivity to synaptic input, of neurons can also be altered. This is manifested as changes in the firing characteristics of the neuron itself. Such *non-synaptic plasticity* has been observed across many species and brain areas [25]. Non-synaptic plasticity is different from synaptic plasticity in that here the neuron's properties are altered, while in the latter the properties of synapses coupling two neurons together are altered. This form of plasticity is not well understood and has been suggested as a regulation mechanism to maintain a neuron's average firing rate at a target level.

Multimodal sensory stimuli while interaction with the environment modify the structure of the brain and change the stimulus-action relationship. These stimuli strengthen or weaken synapses between neurons and allow an organism to learn new associations between sensory inputs of different modalities in an unsupervised manner. Hebbian learning [10] is considered to be one form of such unsupervised associative learning, and is theorised to be the underlying mechanism for synaptic plasticity in the brain. One form of Hebbian-like learning is ICO learning [19], which is an online unsupervised differential Hebbian learning algorithm that implements heterosynaptic plasticity. Here, change in synaptic strength between a pre-synaptic neuron and a post-synaptic neuron is driven by the temporally correlated activity of all neurons projecting on to the post-synaptic neuron (see Fig. 2B). The post-synaptic neuron's output is a linear combination of all inputs, and indirectly influences the activity of both the input neurons via a defined behavioural response. ICO learning produces a behavioural response that nullifies the activation of the input neuron whose response temporally lags that of the other input neuron. ICO learning is fast, stable and can successfully generate adaptive behaviour in real robots [16,18].

3   Materials and Methods

The simulated mobile robot is modelled as a differential drive robot with two wheels (see Fig. 3), which imposes non-holonomic kinematic constraints. The distance l between the centres of the two wheels is defined to be 16 cm. The robot has two virtual acoustic sensors that functionally mimic microphones and receive auditory signals from the acoustic target towards which the robot must navigate. The separation d between these sensors is 13 mm because the parameters of the lizard peripheral auditory model have been derived from an animal with a 13 mm separation between its ear. The acoustic sensor separation must match that value, otherwise the actual ITD cue and the ITD cues to which the peripheral auditory model is tuned will be unmatched. These auditory signals are processed by the lizard peripheral auditory model and its outputs are fed as inputs to the neural circuit described next. Phonotaxis is performed exclusively via these auditory signals. The robot also has a distance sensor located at its centre that provides as outputs the distance of obstacles from the robot, within a 180° field-of-view in front of the robot, and its relative location. This sensor functionally mimics a laser range finder, a common distance sensor used in mobile robots for navigation purposes. White Gaussian noise at 20 dB and 3 dB is respectively added to the auditory inputs and to the distance sensor input to simulate noisy sensors. The forward kinematic model for differential drive mobile robots [8] as given by (2) is used to determine the pose $[x, y, \theta]$ of the robot, where (x, y) are the two-dimensional coordinates and θ is the heading.

$$\begin{bmatrix} x \\ y \\ \theta \end{bmatrix} = \begin{bmatrix} cos(\omega\delta t) & -sin(\omega\delta t) & 0 \\ sin(\omega\delta t) & cos(\omega\delta t) & 0 \\ 0 & 0 & 1 \end{bmatrix} \begin{bmatrix} Dsin(\theta) \\ -Dcos(\theta) \\ \theta \end{bmatrix} + \begin{bmatrix} x - Dsin(\theta) \\ y + Dcos(\theta) \\ \omega\delta \end{bmatrix}$$

where, angular velocity $\omega = \dfrac{(v_r - v_l)}{l}$ and,   (2)

distance D from instantaneous center of curvature $= \dfrac{l}{2}\dfrac{(v_r + v_l)}{(v_r - v_l)}$.

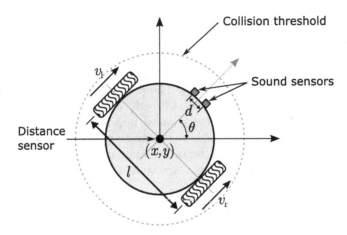

Fig. 3. Mobile robot kinematics.

Figure 4 depicts the neural circuit for reactive navigation with obstacle avoidance. The Braitenberg sensorimotor cross-couplings implement a mapping between the auditory inputs (sound direction information in decibels coming from the lizard peripheral auditory system) and the motor outputs (linear wheel velocities in cm/s). The sensorimotor couplings are defined such that the ipsilateral auditory input proportionally drives the contralateral motor output. The robot respectively turns either left or right, based on whether the left auditory signal is greater than the right auditory signal ($|i_\mathrm{I}| > |i_\mathrm{C}|$) or vice versa ($|i_\mathrm{C}| > |i_\mathrm{I}|$). The radius of the turn is proportional to the relative magnitudes of the auditory signals. When the robot is directly facing the target, the two acoustic sensors are equidistant from the target. In this condition there is no phase difference between the auditory signals arriving at the two acoustic sensors and thus the peripheral auditory model's outputs are identical. Therefore the wheel velocities are also identical, resulting in the robot moving in a straight line. These sensorimotor cross-couplings, implemented as micro-circuits with two sensorimotor neurons having non-linear sigmoid transfer functions, generate phonotaxis behaviour. The two transfer functions are respectively defined as

$$v_\mathrm{l} = \frac{4}{1 + \beta_\mathrm{r} e^{-|i_\mathrm{C}|}} \mathrm{and} v_\mathrm{r} = \frac{4}{1 + \beta_\mathrm{l} e^{-|i_\mathrm{I}|}} \tag{3}$$

where the parameters β_l and β_r determine the amount of respective horizontal shifts. The transfer functions have finite limits between 0 cm/s to 4 cm/s. The strengths of these couplings are determined by the intrinsic excitability of the individual sensorimotor neurons, which is modelled as the horizontal shift in the transfer functions. The two identical micro-circuits that implement ICO learning modify β_l and β_r and shift these transfer functions during learning.

The micro-circuits implementing obstacle avoidance behaviour respectively inhibit and excite the ipsilateral and contralateral outputs of the sensorimotor neurons. If the closest approaching obstacle is detected to be within the collision

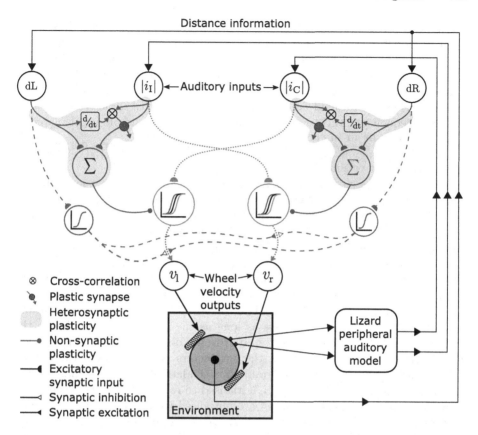

Fig. 4. The neural circuit for reactive navigation with obstacle avoidance embodied in the environment. The sensorimotor mappings between the auditory inputs $|i_I|$ and $|i_C|$ and the motor outputs v_l and v_r (dotted blue lines) implement phonotaxis behaviour. The micro-circuits implementing ICO learning (solid lines inside the shaded areas) modify the strengths, modelled by β_l and β_r, of these mappings. The obstacle avoidance behaviour is implemented by the sensorimotor mappings, with fixed strengths, between the distance sensor inputs and motor outputs (dashed red lines). (Color figure online)

threshold, a obstacle avoidance reflex is triggered that makes the robot turn sharply away from the obstacle. This reflex is implemented by enhancing the motor velocity output to the wheel on the same side as the obstacle to a relatively high but fixed value of 4 cm/s while the motor velocity output to the opposite wheel is suppressed to a relatively low but fixed value of 0.1 cm/s. This causes the robot to sharply turn left when it detects an obstacle to its right and vice versa. The collision threshold (see Fig. 3) is the perimeter of radius 20 cm, centred on the robot. The navigation goal is to move towards an acoustic target located 3 m away from the initial position of the robot, while avoiding any obstacles in the environment. Since the strength of the Braitenberg couplings determines the straightness of the robot's trajectories, the ICO learning algorithm must learn the best possible coupling that generates as straight trajectories as possible.

The initial pose of the robot is set to $[0, 0, 0°]$. A random number of circular obstacles are randomly distributed in the environment between the robot and the target as well as around the target. The size of the obstacles are randomly chosen but are below a threshold equal to the wavelength of the sinusoidal acoustic signal of frequency 2.2 kHz emitted by the target. This threshold is calculated as $\dfrac{\text{speed of sound in air in cm/s}}{\text{sound frequency in Hz}} = \dfrac{34000}{2200} = 15.45\,\text{cm}$. Wheel slip is not modelled to maintain relative simplicity.

The learning is performed for a maximum of 50 iterations to keep simulation time reasonable. The learning is stopped either when the robot reaches the acoustic target without triggering the obstacle avoidance behaviour or when all 50 iterations are complete. Each iteration runs for a maximum of 1000 time steps. At each simulation time step of 1 s, the sound direction information is extracted and wheel velocities are computed via the sigmoid transfer functions. Then the new pose of the robot based on current wheel velocities is determined by (2). If the robot encounters an obstacle within its collision threshold, the obstacle avoidance reflex overrides the calculated wheel velocities as described earlier. The obstacle avoidance reflex lasts as long as the obstacle remains inside the collision threshold. At each time step during the obstacle avoidance maneuver, ICO learning respectively updates the parameters β_l and β_r using the temporal correlation between the ipsilateral and contralateral auditory signal and the distance sensor signal. β_l and β_r are initialised to random values between 0 and 1, while synaptic weights w_l and w_r are initialised to random values between 0 and 0.1. The learning rate η is set to 0.01. β_l and β_r are updated as

$$
\begin{aligned}
\frac{\delta \beta_l}{\delta t} &= w_l\,|i_I| + \mathrm{dL} \quad,\text{where}\,\frac{\delta w_l}{\delta t} = \eta\,|i_I|\,\frac{\delta \mathrm{dL}}{\delta t} \quad,\text{and} \\
\frac{\delta \beta_r}{\delta t} &= w_r\,|i_C| + \mathrm{dR} \quad,\text{where}\,\frac{\delta w_r}{\delta t} = \eta\,|i_C|\,\frac{\delta \mathrm{dR}}{\delta t} \;.
\end{aligned}
\tag{4}
$$

Since (4) is simply positive feedback, allowing the ICO learning algorithm to run without restraint will cause β_l and β_r to increase uncontrollably. This will continuously push the transfer function curves to the right such that the both v_l and v_r, and thus the robot's linear velocity, will eventually become infinitesimally small. To counter this, β_l and β_r are exponentially decreased as function of time during learning. This is done by multiplying β_l and β_r with an time varying scaling factor defined as $e^{(-t/k)}$, where t is the current time step and k is chosen via trial-and-error to be 60000. This negative feedback prevents the uncontrollable shift in the transfer functions and maintains homeostasis.

4   Results and Discussion

Figure 5 depicts the performance of the neural circuit in two independent trials. When the learning is not used, the neural circuit produces more winding

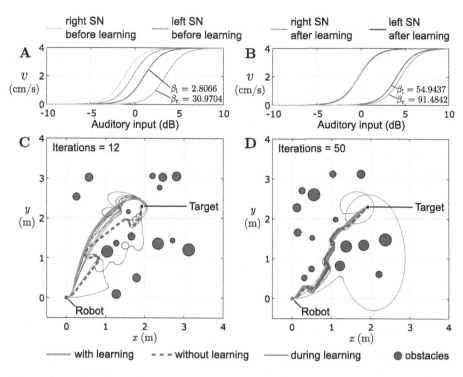

Fig. 5. Acoustic navigation with the proposed neural circuit. (A) and (B) The transfer functions of the left SN (red) and right SN (blue) before learning (dotted) and after learning (solid). (C) and (D) Robot trajectories without learning (dotted), during learning (solid blue) and after learning (solid green) (Color figure online).

trajectories as compared to when using learning. In this case the robot always attempts to move in a straight line towards the target due to the strong and constant Braitenberg sensorimotor couplings between the auditory inputs and the motor outputs. This straight trajectory is heavily modified by the obstacle avoidance behaviour such that robot makes sharp turns while avoiding obstacles and reorienting towards the target. When learning is enabled, the neural circuit learns relatively smoother trajectories. The robot is able to either completely suppress its obstacle avoidance behaviour (Fig. 5, A and C) or minimise it to a relatively large extent (Fig. 5, B and D). This is due to the modification of the strengths of the sensorimotor couplings by the ICO learning algorithm during interaction with the obstacles. The neural circuit tries to learn the best possible sensorimotor couplings that minimise sharp turns during movement. Early in the learning process the robot generates very winding trajectories, since the Braitenberg sensorimotor couplings are initially relatively weak. This enforces circular trajectories with the turning radius being dependent on the strength of the couplings. As the learning progresses, the sensorimotor couplings adapt to the environment and the robot moves in progressively straighter trajectories.

5   Conclusions and Future Directions

We have presented a neural circuit for reactive acoustic navigation with obstacle avoidance. The circuit uses simple Braitenberg acoustomotor couplings to realise goal-directed navigation behaviour towards an acoustic target located 3 m away, emitting a continuous 2.2 kHz tone. The strength of these couplings are learned through interaction with obstacles in the environment, realised as obstacle avoidance behaviour. The circuit learns stable and consistent trajectories using only noisy sensor information about sound direction, extracted via a model of the lizard peripheral auditory system, and distance from closest obstacle. In this manner the low-level obstacle avoidance behaviour modifies the high-level goal-directed steering behaviour to resolve the conflicting trajectories that these behaviours enforce. The circuit allowed a simulated mobile robot to follow relatively smooth trajectories during phonotaxis while avoiding obstacles. In comparison, trajectories generated without learning were relatively less smooth. The immediate next steps are to validate the neural circuit in real robot trials.

ewThe trajectories generated by the proposed neural circuit are sub-optimal and characteristic of reactive navigation. Strong Braitenberg couplings tend to enforce straight-line trajectories that are difficult to maintain in cluttered environments. Smoother trajectories may be obtained by correcting them pre-emptively to circumvent the obstacle avoidance behaviour. ICO learning can be used to modify the sensorimotor couplings for obstacle avoidance behaviour so that the robot turns progressively earlier in time in response to approaching obstacles.

Acknowledgements. This research was supported with a grant for the SMOOTH project (project number 6158-00009B) by Innovation Fund Denmark.

References

1. Alves, S., Rosario, J., Ferasoli Filho, H., Rincon, L., Yamasaki, R.: Conceptual bases of robot navigation modeling, control and applications. In: Barrera, A. (ed.) Advances in Robot Navigation. InTech (2011)
2. Andersson, S., Shah, V., Handzel, A., Krishnaprasad, P.: Robot phonotaxis with dynamic sound source localization. In: Proceedings 2004 IEEE International Conference on Robotics and Automation, ICRA 2004, vol. 5, pp. 4833–4838, April 2004
3. Arkin, R.: Behavior-Based Robotics. MIT Press, Cambridge (1998)
4. Bicho, E., Mallet, P., Schner, G.: Target representation on an autonomous vehicle with low-level sensors. Int. J. Robo. Res. **19**(5), 424–447 (2000)
5. Braitenberg, V.: Vehicles: Experiments in Synthetic Psychology. MIT Press, Cambridge (1984). Bradford Books
6. Brooks, R.: A robust layered control system for a mobile robot. IEEE J. Robot. Autom. **2**(1), 14–23 (1986)
7. Christensen-Dalsgaard, J., Manley, G.: Directionality of the lizard ear. J. Exp. Biol. **208**(6), 1209–1217 (2005)
8. Dudek, G., Jenkin, M.: Computational Principles of Mobile Robotics, 2nd edn. Cambridge University Press, New York (2010)

9. Fletcher, N., Thwaites, S.: Physical models for the analysis of acoustical systems in biology. Q. Rev. Biophys. **12**(1), 25–65 (1979)

10. Hebb, D.: The Organization of Behavior: A Neuropsychological Theory. Psychology Press, Abingdon (2005)

11. Huang, J., Supaongprapa, T., Terakura, I., Wang, F., Ohnishi, N., Sugie, N.: A model-based sound localization system and its application to robot navigation. Robot. Auton. Syst. **27**(4), 199–209 (1999)

12. Hwang, B.-Y., Park, S.-H., Han, J.-H., Kim, M.-G., Lee, J.-M.: Sound-source tracking and obstacle avoidance system for the mobile robot. In: Tutsch, R., Cho, Y.-J., Wang, W.-C., Cho, H. (eds.) Progress in Optomechatronic Technologies. LNEE, vol. 306, pp. 181–192. Springer, Cham (2014). doi:10.1007/978-3-319-05711-8_19

13. Janis, A., Bade, A.: Path planning algorithm in complex environment: a survey. Trans. Sci. Technol. **3**(1), 31–40 (2016)

14. Klopf, A.: A neuronal model of classical conditioning. Psychobiology **16**(2), 85–125 (1988)

15. Kosko, B.: Differential Hebbian learning. In: AIP Conference Proceedings, vol. 151, no. 1, pp. 277–282 (1986)

16. Manoonpong, P., Wörgötter, F.: Adaptive sensor-driven neural control for learning in walking machines. In: Leung, C.S., Lee, M., Chan, J.H. (eds.) ICONIP 2009. LNCS, vol. 5864, pp. 47–55. Springer, Heidelberg (2009). doi:10.1007/978-3-642-10684-2_6

17. Nakhaeinia, D., Tang, S., Noor, S., Motlagh, O.: A review of control architectures for autonomous navigation of mobile robots. Int. J. Phys. Sci. **6**(2), 169–174 (2011)

18. Porr, B., Wörgötter, F.: Fast heterosynaptic learning in a robot food retrieval task inspired by the limbic system. Biosystems **89**(1–3), 294–299 (2007). (In: Selected Papers Presented at the 6th International Workshop on Neural Coding)

19. Porr, B., Wörgötter, F.: Strongly improved stability and faster convergence of temporal sequence learning by utilising input correlations only. Neural Comput. **18**(6), 1380–1412 (2006)

20. Purves, D., Augustine, G., Fitzpatrick, D., Hall, W., LaMantia, A., White, L.: Synaptic plasticity. In: Neuroscience, 5th edn., pp. 163–182. Sinauer Associates, Sunderland (2012)

21. Shaikh, D., Hallam, J., Christensen-Dalsgaard, J.: From "ear" to there: a review of biorobotic models of auditory processing in lizards. Biol. Cybern. **110**(4), 303–317 (2016)

22. Tang, S., Kamil, F., Khaksar, W., Zulkifli, N., Ahmad, S.: Robotic motion planning in unknown dynamic environments: existing approaches and challenges. In: 2015 IEEE International Symposium on Robotics and Intelligent Sensors (IRIS), pp. 288–294, October 2015

23. Wever, E.: The Reptile Ear: Its Structure and Function. Princeton University Press, Princeton (1978)

24. Zeno, P., Patel, S., Sobh, T.: Review of neurobiologically based mobile robot navigation system research performed since 2000. J. Robot. 2016 (2016)

25. Zhang, W., Linden, D.: The other side of the engram: experience-driven changes in neuronal intrinsic excitability. Nat. Rev. Neurosci. **4**(11), 885–900 (2003)

26. Zu, L., Yang, P., Zhang, Y., Chen, L., Sun, H.: Study on navigation system of mobile robot based on auditory localization. In: 2009 IEEE International Conference on Robotics and Biomimetics (ROBIO), pp. 321–326, December 2009

27. Zuojun, L., Guangyao, L., Peng, Y., Feng, L., Chu, C.: Behavior based rescue robot audio navigation and obstacle avoidance. In: Proceedings of the 31st Chinese Control Conference, pp. 4847–4851, July 2012

MHDW2017

An Implementation of Disease Spreading over Biological Networks

Nickie Lefevr[1], Spiridoula Margariti[2], Andreas Kanavos[1(✉)],
and Athanasios Tsakalidis[1]

[1] Computer Engineering and Informatics Department,
University of Patras, Patras, Greece
`nick.lefevr@gmail.com`, {`kanavos,tsak`}`@ceid.upatras.gr`
[2] Department of Computer Engineering,
Technological Educational Institute of Epirus, Arta, Greece
`smargar@teiep.gr`

Abstract. Complex networks can be considered as a new field of scientific research inspired by the empirical study of real-world networks such as computer, social as well as biological ones. More to this point, the study of complex networks has expanded in many disciplines including mathematics, physics, biology, telecommunications, computer science, sociology, epidemiology and others. An important type of complex networks are called biological dealing with the mathematical analysis of connections - interfaces that are ecological, evolutionary and physiological studies, such as neural networks or network epidemic models. The analysis of biological networks in connection with human diseases has led to expand science and examine medical supplies networks for their deeper understanding. In this paper, an implementation of epidemic/networks models is introduced concerning the HIV spreading in a sample of people who are needle drug users.

Keywords: Complex networks · Biological networks · Neural networks · AIDS · HIV · Epidemic models

1 Introduction

The theory of graphs in which complex networks are based, is the main theoretical part of complex network analysis as it facilitates the visualization of the structures and also solves many other practical problems. As a typical example of a graph, one can consider the Internet or even the human brain.

In this paper, we analyze the structure and evolution of complex networks through a comprehensive report of graphs theory introducing alongside their main types. It is common knowledge that network structure can be seen in all aspects of life, e.g. the brain is a network of connected nerve cells, where even those cells alone make up a network whose job is to make biochemical reactions. Complex networks, with their extremely high popularity as well as their rapid

© Springer International Publishing AG 2017
G. Boracchi et al. (Eds.): EANN 2017, CCIS 744, pp. 559–569, 2017.
DOI: 10.1007/978-3-319-65172-9_47

development, give scientists a unique opportunity to study their properties and their capabilities. Their in-depth understanding of structure and development may lead to useful information for further evaluating and improving platforms, tools and the results of investigations formed. These aforementioned reasons drive us to the implementation which graphically shows the spreading of HIV in a population of individuals who are drug users.

The rest of the paper is organized as follows: in Sect. 2 we discuss the Models of Complex Networks as well as the Epidemic Models in Biological Networks. In Sect. 3, Epidemics is introduced while in Sect. 4, the datasets for validating our framework are presented. Moreover, Sect. 5 presents the evaluation experiments conducted and the results gathered. Ultimately, Sect. 6 presents conclusions and draws directions for future work.

1.1 Preliminaries

A graph G is a way to set the relationship between a data collection consisting of N nodes and E edges. Specifically, a graph consists of a set of objects, called nodes. Several pairs of these objects are associated with links, called edges [5], [16]. More to this point, if the edges are oriented in one direction, then the graph is considered as a directed one [4]; on the other hand, if the edges are not oriented, then the graph is considered as an undirected one. In many applications, each end of a graph has an associated numerical value, called a weight. In this specific case, the edges are non-negative integers and the graph can be considered as a weighted one. This graph can be either directed or undirected as well as the weight of an edge is frequently referred to as the edge's "cost"[1].

Furthermore, a path is a sequence of nodes that have the status of each successive pair $(V_i, V_i + 1)$, i.e. an edge of a graph. This path doesn't only include the nodes, but also the sequence of edges connecting these nodes. In some cases, there are paths where some specific nodes are repeated; but in most paths, we prefer nodes to be non repeated ones. Concretely, if we want to emphasize the fact that a non-simple path overpasses a specific node, then we can refer to it by presenting it as a non-simple path [5].

The adjacency matrix of a simple labeled graph is a table with rows and columns where the information we want to analyze is in the vertices of the graph. The analysis of the graph in the table is represented by the use of 0 or 1 at positions (i, j) according to whether i and j are contagious or not.

If the type of the graph edges is known, then we can calculate a variety of useful quantities that describe the characteristics of the network. More specifically, the concept of Centrality is used in order to figure out which nodes of the graph of a specific network are important [11]. In addition, the Degree Distribution is one of the most basic properties of the graph [12]. The fraction of the network nodes that have a degree equal to k is denoted by p_k. The total of these quantities p_k gives the degree distribution in order to plot the degree

---

[1] http://www.cs.vt.edu/undergraduate/courses.

distribution of a large network as a function of k. So, according to this, degree distribution does not give us insight on the full network structure [11].

2   Related Work

2.1   Models of Complex Networks

The modeling of complex networks gives us a lot of information and can help us thoroughly understand the complex networks. More to the point, it enables the simulation, the analysis as well as the prediction of the behavior of processes taking place in them, such as diffusion or information retrieval.

2.1.1   The Erdös - Rényi Model

The model of Erdös and Rényi is one of the first network models and represents the random graph [6]. Specifically, the random graph model is characterized by the number of nodes as well as by the probability of connecting two randomly selected nodes. Each one of the pairs of nodes is associated with an equal probability to each other, independently of the other pairs [4,6]. The Erdös - Rényi model has been accepted because of its properties, which facilitate the network modeling. Random graphs do not reflect the structure of real networks, because the degrees of random graph nodes follow the Poisson distribution [8], instead of the power-law distribution [1][2]. So, the Erdös - Rényi model does not reflect the effect of clustering, instead the random graph can be considered as a perfect model choice for studying complex networks [6].

2.1.2   The Small - World Model

The Small - World model is a random graph that generates other graphs in terms of the status of the small world. That means that the graphs use the shortest path length and the high rate of clustering [4,17]. In [18], authors created a class of graphs obtained by the interpolation between a regular lattice having a high clustering rate and a random graph in the capacity of the small world. The graphs are starting from a ring lattice in which each edge has degree k with probability p for each edge to be refit. In this way, the small values of probability p seems sufficient so as to substantially reduce the lengths of the shortest path length between the edges, while the clustering rate remains high [18]. The small world phenomenon is also known as six degrees of separation [19]. It explains that if someone chooses any two individuals anywhere on Earth at a random place, it will find a path of at most six intimates (nodes) between them [7,15].

2.2   Epidemic Models in Biological Networks

Another kind of complex networks is the biological neural networks of living organisms, as well as the networks of epidemics that occur in them. There are

---

[2] http://www.necsi.edu/guide/concepts/powerlaw.html.

two main types of biological networks that are presented in the next paragraphs. Also, an even reference to networks based on the inheritance of people through the genetic material (DNA) and general characteristics people carry from their ancestors is introduced. Moreover, the epidemic activity of complex networks through the sensitivity that their nodes have is discussed. Additionally, examples are considered, such as transferable filial diseases (airborne or sexually transmitted) between humans and filial constructs transmitted through internet complex networks, social networks, or from computer to computer. As biological networks, one can characterize that they make networking between communities in living organisms or between the functions of these bodies.

2.2.1 Biological Neural Networks

As it is well known, the human brain is one of the most complex networks that exist with its functions being so complex that the approach of this network or its running processes is very difficult to be "copied" by a computer. Neurons are particular types of cells, which constitute the main information processing system the human nervous system possesses. A neuron consists of three main parts: the body, the dendrite and the axon [2]. Each dendrite has a very small gap, called synapse and it is through these small gaps that the transmission of electrical signals from the axon recipient takes place. The rate of the electrical activity transmitted to the dendrite, is called synaptic weight. These are the hallmark that gives neural network the capacity for environmental development and adaptation.

2.2.2 Artificial Neural Networks

The Artificial Neural Networks are originally proposed as a mathematical simulation model for the complex function of the human brain. The structure of the artificial neural network emulates that of biological neural network, thus showing similar properties. The degree of interaction is different for each pair of neurons and in following is determined by synaptic weights. In Artificial Neural Networks, two ways exist for their training. In the first way, the education is supervised and the network is supplied with a set of conditions. So the network is able to acquire knowledge from these examples through a learning algorithm. In the second way, the education is considered without supervision and the network is invited to identify similarities, differences and patterns in data that are more meaningful [10]. Thus, it must be adapted so as to separate these similarities, differences and patterns in groups until the classification of the data is observed.

The main advantage of neural networks is that they can store knowledge and experience from the environment, which can then be withdrawn. Moreover, neural networks have the ability to extract the essential features of a system, even when they are hidden among noisy-corrupted data [10].

3   Epidemics

When we are dealing with epidemic diseases, we should be aware of infectious diseases caused by biological pathogens, which are passed on from person to person. Epidemics are powerful enough to wipe out a population, or can dormantly exist without manifest for long periods. In extreme cases, a single outbreak of disease can have a significant effect on an entire culture, such as epidemics triggered by the arrival of Europeans in America [3].

3.1   Diseases Transmitted by Networks

The way of how a disease can be spread from person to person, inside the population of a complex network, can be determined as presented in [13]. The opportunity that a disease has in order to spread out, comes from the contact network [1]. This indicates that the modeling accuracy of the underlying network is critical to understanding the spread of the epidemic. Similar is the study mode of malware spread transmitted between computers [2,9].

3.2   Branching Processes

The Branching Process is the simplest model of transmission of a disease, and particularly regarding an airborne disease. Such a network may be referred to as a tree. This model operates by transmission waves having the following details: when a group of disease carriers enters a healthy population, then there is a possibility for transmitting the disease to a susceptible portion of this population, in accordance with the random transmission of disease [5] (Fig. 1).

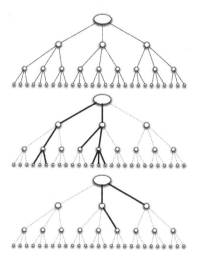

Fig. 1. The spreading of an epidemic with a tree network [5] (Fig. 21.1, pg. 570)

3.3   The SIR Model

An independent node in the branching model goes through three specific steps during the epidemic. The first step, called sensitivity, concerns the nodes - individuals that have not yet developed the disease, but are prone to contamination from their neighbors. Thereafter, the node, which is a person afflicted with the disease, is converted to the disease entity and has a chance to infect any susceptible (S) neighbor; that is the second step, called infection (I). Finally in the third step, the disease has passed its full infectious period where the node-person is removed (R) from consideration since there is now a threat of future infections. Thus, this step is called removal [5]. By using these three phases of the "life cycle" of a disease at each node-person, a model for epidemics in networks can be defined. Based on the above, the spreading of the epidemic is controlled by the structure of a network of contacts with two additional quantities: the possibility of transmitting and the length of the infection.

3.4   The SIS Model

A simple variant of epidemic models subsection cited, allows the assumption that the nodes affected by an epidemic, may be sick more than once. Such epidemics are considered when the nodes alternate between sensitivity (S) and infection phase (I). So this is the reason why this model is named SIS. The procedure followed by this model is that the network exhibits some nodes in the infectious phase (I) and all the others are in sensitivity mode (S). Each node-person who enters the infectious phase remains contaminated for a constant period. During this time, the infected nodes-users have the possibility of catching the disease by any susceptible neighbor. After this period, infectious nodes, i.e. those that are no longer contaminated, can return to their original phase as sensitive ones. Like the SIR model, the SIS model can be expanded in order to handle more complex types of cases: either different transmission probabilities between different node pairs-individuals, or probability of recovery from illness [5].

3.5   The Transient Contacts Model

The diseases spreading through a population over longer time scales, such as sexually transmitted diseases, e.g. the HIV/AIDS, taking several years to evolve the network, resulting in their course, depend largely upon the properties of sexual intercourse per pair. Most people have few contacts at any time, while the identities of those contacts may change during disease progression that new relationships are formed and others are broken up [5]. For the modeling of these networks, it is important to consider the fact that the contacts are transient and this does not necessarily last through the whole course of the epidemic, but only for specified periods.

4 Implementation - Simulation of HIV Transmission in Biological Network

In this section, the implementation for the transmission of the Acquired Immunodeficiency Syndrome (AIDS) virus, caused by the virus entitled Human Immunodeficiency Virus (HIV), in a sample with injecting illicit drug users, is presented. Three examples of virus transmission with different parameters and studies of the syndrome during transmission are thoroughly described. We have chosen to conduct a study of virus transmission through syringes used among injecting drug users. Studies have shown that 18 million people in the European Union (EU) are users of illegal drugs, whereas the proportion attributable to injecting drug users by 18 million is 5%[3]. The 5% is the 0.63% attributable to users who have been infected from AIDS, by sharing needles [14].

The proposed model is based on the model of Erdös - Rényi random graph, where each human pair shares an equal probability of contact, thus independently of the other pairs. While the transmission of infection in the sample was made according to the epidemic transiently contact model, its spread takes a fair amount of time to evolve. The sample used in the implementation is 300 users, who are injecting drug users, and amongst them, there is 0.63% disease possibility by the AIDS virus, according to the study aforementioned. It is assumed that there is a random choice of a common syringe between people through the setup-people process; but in the case of use through common infected syringe in "healthy" people, they contract the virus with probability equal to 100%.

The implementation uses Boolean variables in order to verify the following questions:

– Whether the person is infected without being aware of it.
– If the infection is known (or not) to the person.
– If the person has shared syringe (or not) with someone else.

These are also basic variables that the implementation is based on. Moreover, the tendency of population infection manually, i.e. what part of the population will become ill, using a variable in order to further present it, is checked. Variations of input data give different results according to the following queries and problem probabilities.

– How long the person will use the syringe.
– How likely the person is to share their syringe with others at the same time.
– How many will use clean (new) needles.
– How often they are tested by a physician.

According to these variables input into the model, the user can change the values of the auxiliary variables and in following alternate the model data thus gather other information from the original set in the model. During the simulation, the opportunity to monitor the state changes between the samples through

---

[3] http://www.emcdda.europa.eu/.

the chromatic process is given. There, people set off as "healthy" and if infected by the virus, they turn into a new color. When it is cross checked through a test that they are infected, the color alters once more. The proposed system presents the contamination percentage as well as the number of contaminants on a weekly basis. The percentage varies until the time another change occurs.

5   Results

As mentioned above in Sect. 4, three different examples have been implemented whereas the only stable and unchanged data are the following:

1. If a person uses an infected needle, then they have 100% possibility of becoming infected.
2. 0.63% of the population is infected without knowing it.
3. All the individuals are to be tested when they are asked for.
4. The sample is 300 people.
5. The symptoms will appear after 100 weeks.

Moreover, the unchanged data are the percentage of people who shared the same syringe with others and in following, the percentage of the population which use a clean (new) syringe (Table 1).

Table 1. Data of the three examples

Features	1st	2nd	3rd
Sample	300	300	300
Shared syringe	9	0	7
Weeks	100	100	100
New syringe	0	9	2
Tests (times/year)	1	1	1

First Example: In this extreme case as shown in Table 2, the users who share the same syringe are 9 and no one has ever used a clean (new) syringe. The rate of infection begins with 0.67% and not 0.63% because the infection probability lies between 1.89 and 300 (under the $300 * 0.0063 = 1.89$). The evolution of the phenomenon reaches approximately value equals to 55.67% and the joint use is randomly made among the people so that there is one standard deviation of the percentage each time the current program runs for the same example. As a result, it can be depicted from the first example what was actually expected; healthy people become ill very quickly.

Second Example: In this second extreme case (Table 3), the users who share the same syringe are 0 and the number of those who use a clean (new) syringe is 9. The rate of infection starts with 0.67% (as in the first example) and remains

Table 2. People regarding the first example for different weeks

Option of illness	100	390	1390	2390	3930
AIDS−	296	285	223	155	140
AIDS+	0	2	49	134	160
AIDS?	4	13	28	11	0

Table 3. People regarding the second example for different weeks

Option of illness	100	390	1390	2390	3930
AIDS−	298	298	298	298	298
AIDS+	0	2	2	2	2
AIDS?	2	0	0	0	0

Table 4. People regarding the third example for different weeks

Option of illness	100	390	1390	2390	3930
AIDS−	296	287	217	148	128
AIDS+	0	9	55	141	172
AIDS?	4	6	28	11	0

therein. The results in this example are likewise. We expected that none of the individuals are to be infected and we found ourselves correct.

Third Example: In the third case (Table 4), the data are closer to the reality, as the users who share the same syringe are 7 and the number of those who use a clean (new) syringe is 2. The rate of infection starts with 0.67% (as the above examples) and reaches approximately a percentage equal to 62.33%. The results in the third example are likewise too. We anticipated that the rate of the infected people would be raised extremely fast and maybe a dangerous result for the virus spreading among our sample would probably be shown.

6   Conclusions

Complex networks are not only popular and attractive but also very important. Their study can be useful in many areas of science as well as in problems of the real world. Complex networks also provide study possibilities for each science networks, according to the inserted data and depending on the requirements set by each sector of the real world. Essentially, they give insight for arbitrarily formed networks, i.e. the entities that make up entering the network and create connections without predetermined manner. Additionally, these entities may at any time be withdrawn from the network.

Furthermore, in molecular biology, networks consist of counter interacting parts, such as proteins, enzymes, genes, etc. By understanding the processes that complex networks use, a challenge for researchers regarding the study of problem solving (mutations, disease transmission) within these networks, arises. Epidemiology, as part of biology, is therefore not only studied with the help of complex networks, but it also utilizes the corresponding networks as tools for analysis, modeling and simulation.

Concluding, we have used and tested the transmission of a disease (AIDS) in a group of people. The initial population is chosen as a sample out of the total, with the assumption that the relationships between them follow the Erdös - Rényi model random graph and the transmission of the disease among the sampled people follow the epidemic contacts transient model. As discussed above, measurements were made for different conditions and examples between the sample, thus obtaining data through simulation, which cannot be studied and realized in the real world. The simulation results show the progression of the infection of young people (not diseased) in time. Such information is particularly useful for anticipating and launching preventive measures in many kinds of diseases.

Future directions include the extension of this work to other epidemic models as well as the study of multiple viruses on a single network. Moreover, analysis from dynamic systems theory or even different algorithmic analytical tools can be incorporated. A more extensive experimental evaluation can additionally be considered as direction for future work. Finally, the adoption of efficient heuristics on time-varying graphs is very promising and thus can be introduced in our proposed work.

References

1. Adamic, L.A., Lukose, R.M., Puniyani, A.R., Huberman, B.A.: Search in power-law networks. Phys. Rev. E **64**(4), 046135 (2001)
2. Anastasio, T.J.: Tutorial on Neural Systems Modeling. Sinauer Associates, Incorporated (2009)
3. Diamond, J., Guns, G.: Steel: The Fates of Human Societies. W. W. Norton, New York (1997)
4. Dorogovtsev, S.N.: Lectures on Complex Networks, vol. 24. Oxford University Press, New York (2010)
5. Easley, D.A., Kleinberg, J.M.: Networks, Crowds, and Markets - Reasoning About a Highly Connected World. Cambridge University Press, Cambridge (2010)
6. Erdös, P., Rényi, A.: On random graphs i. Publ. Math. Debrecen **6**, 290–297 (1959)
7. Gurevitch, M.: The social structure of acquaintanceship networks. Ph.D. thesis, Massachusetts Institute of Technology (1961)
8. Haight, F.A.: Handbook of the poisson distribution (1967). https://books.google.gr/books/about/Handbook_of_the_Poisson_distribution.html?id=l8Y-AAAAIAAJ&redir_esc=y
9. Kephart, J., Sorkin, G., Chess, D., White, S.: Fighting Computer Viruses, pp. 88–93. Scientific American, New York (1997)
10. Kriesel, D.: A brief introduction on neural networks (2007). http://www.dkriesel.com/_media/science/neuronalenetze-en-zeta2-2col-dkrieselcom.pdf
11. Newman, M.E.J.: Networks: An Introduction. Oxford University Press Inc., Oxford (2010)

12. Newman, M.E.J.: The structure and function of complex networks. SIAM Rev. **45**(2), 167–256 (2003)
13. Pastor-Satorras, R., Vespignani, A.: Evolution and Structure of the Internet: A Statistical Physics Approach. Cambridge University Press, Cambridge (2007)
14. Patel, P., Borkowf, C.B., Brooks, J.T., Lasry, A., Lansky, A., Mermin, J.: Estimating per-act hiv transmission risk: a systematic review. Aids **28**(10), 1509–1519 (2014)
15. Travers, J., Milgram, S.: An experimental study of the small world problem. Sociometry **32**, 425–443 (1969)
16. Wasserman, S., Faust, K.: Social Network Analysis: Methods and Applications, vol. 8. Cambridge University Press, Cambridge (1994)
17. Watts, D.J.: Small Worlds: The Dynamics of Networks between Order and Randomness. Princeton University Press, Princeton (1999)
18. Watts, D.J., Strogatz, S.H.: Collective dynamics of "small-world" networks. Nature **393**(6684), 440–442 (1998)
19. Zhang, L., Tu, W.: Six degrees of separation in online society (2009). http://journal.webscience.org/147/2/websci09_submission_49.pdf

Combining LSTM and Feed Forward Neural Networks for Conditional Rhythm Composition

Dimos Makris[1], Maximos Kaliakatsos-Papakostas[2], Ioannis Karydis[1(✉)], and Katia Lida Kermanidis[1]

[1] Department of Informatics, Ionian University, Corfu, Greece
{c12makr,karydis,kerman}@ionio.gr
[2] Institute for Language and Speech Processing, R.C. "Athena", Athens, Greece
maximos@ilsp.gr

Abstract. Algorithmic music composition has long been in the spotlight of music information research and Long Short-Term Memory (LSTM) neural networks have been extensively used for this task. However, despite LSTM networks having proven useful in learning sequences, no methodology has been proposed for learning sequences conditional to constraints, such as given metrical structure or a given bass line. In this paper we examine the task of conditional rhythm generation of drum sequences with Neural Networks. The proposed network architecture is a combination of LSTM and feed forward (conditional) layers capable of learning long drum sequences, under constraints imposed by metrical rhythm information and a given bass sequence. The results indicate that the role of the conditional layer in the proposed architecture is crucial for creating diverse drum sequences under conditions concerning given metrical information and bass lines.

Keywords: LSTM · Neural networks · Deep learning · Rhythm composition · Music information research

1 Introduction

Attempting to imitate human creativity has long been an interesting direction for computer scientists. Music has received significant attention in relation to other arts such as graphical arts, painting, dance, or architecture due to its rigorous formalisation, available from the early stages of its evolution [19]. The work of Hiller and Isaacson [7], on the composition of a musical piece using a computer program, was published as early as shortly after the introduction of the very first computer.

Among numerous definitions of Algorithmic (or Automatic) Music Composition (AMC) [10,23], D. Cope[1] provided an interesting alternative, "a sequence (set) of rules (instructions, operations) for solving (accomplishing) a [particular]

---

[1] Panel Discussion in the ICMC'93.

© Springer International Publishing AG 2017
G. Boracchi et al. (Eds.): EANN 2017, CCIS 744, pp. 570–582, 2017.
DOI: 10.1007/978-3-319-65172-9_48

problem (task) [in a finite number of steps] of combining musical parts (things, elements) into a whole (composition)" [20].

AMC is a complex problem with tasks ranging from melody and chords' composition to rhythm and lyrics, among others [2]. Its applications include varying degrees of combination of computational creation (e.g. fully automated background/harmonisation music or assisted co-composition) with humans' creativity for the production of musical works. AMC has been approached with a plethora of methods from various points of view.

Artificial Neural Networks (ANNs) acted as an extra powerful computation tool to extend the available methods which are usually probabilistic models. A number of research works have been published using ANNs and, especially lately, Deep Learning architectures for composing music (e.g. [6,13]) and most of these are using LSTM layers.

Recurrent neural networks, especially LSTM networks, have been utilized for music generation (see [22] for further references), since they are capable of modeling sequences of events. However, hitherto proposed architectures focus on the generation of sequences *per se*, without considering constraints. For instance, the performance of human drummers is potentially influenced by what the bass player plays. Additionally, human drummers could play drums on a time signature they have never played before, e.g. 15/8, by utilising the knowledge they have obtained only by practicing (learning) 4/4 beats; this happens because human drummers have an *a priori* understanding of metric information which helps them perform drum rhythms on meters they have never seen before. In this work, we propose the combination of different neural network layers, i.e. feedforward and recurrent, for composing drum sequences based on external information, i.e. metric and bass information.

The remainder of this paper is organised as follows: Sect. 2 presents background information and existing research on Automated Musical Composition and related notions. Next, Sect. 3 presents the proposed combination of LSTM and Feed Forward Neural Networks for conditional rhythm composition. Section 4 details the experimental evaluation of the proposed method, while the work is concluded in Sect. 5.

2   Related Work

Algorithmic Musical Composition refers to the type of creativity that does not focus on "flash out of the blue", as would be the result of inspiration or genius, but a process of incremental and iterative revision, more similar to hard work [10] in the post-digital computer age. Artificial neural networks (ANNs) have long been utilised for the purposes of automated composition[2]. Their extended use in AMC is due on the capability of ANNs to resemble human creative activities despite the expensive training required in order to do so.

---

[2] Papadopoulos and Wiggins [20] compiled an extensive such list, dating back to 1992.

A very common version of ANNs is the feedforward ANN, which include neurons/units that are commonly divided in three types of layers: input, hidden, and output. Units between different layers are interconnected by weights that are multiplied with the values of their respective input unit. The resulting output is then propagated to a transfer function through units to the output unit. The "learning" procedure of feedforward ANNs refers to their ability to modify the connecting weights for producing output that optimally matches known target values (supervised learning), by minimising the error within a threshold. Recurrent Neural Networks (RNNs) on the other hand, have a similar architecture but the hidden layers are more sophisticated, including recurrent connections for remembering past events. RNNs have been extensively utilised for AMC purposes such as generation of chord sequences [15] and melodies [18].

Long Short-Term Memory (LSTM) is a form of RNN initially proposed by Hochreiter and Schmidhuber [8]. Following the principles of RNNs, LSTM networks are generic in the sense that given adequate network units, LSTMs allow for any conventional computation. In contrast to RNNs, LSTMs are more suited to learning from experience to classify, as well as to processing and predicting time-series, when there are time lags of unknown size and bound between important events. LSTMs feature relative insensitivity to gap length, thus being advantageous to alternative forms of RNNs or hidden Markov models. Accordingly, LSTMs have been utilised for AMC purposes such as learning chord progressions [5], drum progressions [2] and learning generic percussion tracks [14] as well as creating musically-appealing four-part chorales in the style of Bach [6] and folk tunes [22].

One of the key advantages of applying ANNs for ACM is the lack of requirement for a priori knowledge, such as rules, constraints, of the domain since ANNs learn through the examination of input examples. Nevertheless, ANNs' application in AMC is not without drawbacks, namely the diminished capability of mapping input sequences to output higher-level features of music and the difficulty of generalisation related to the input examples' size [20].

3   Conditional Rhythm Composition with LSTM and Feed Forward Neural Networks

In this paper we introduce an innovative architecture for creating drum sequences by taking into account the drum generation of previous time steps along with the current metrical information and the bass voice leading. Section 3.1 describes the collection and preprocessing methodology for the training data, while Sect. 3.2 presents the model's architecture.

3.1   Data Representation

The utilised corpus consists of 45 drum and bass patterns with 16 bars each, in 4/4 time signature from three different rock bands. All the pieces were collected manually from web tablature learning sources[3] and then converted to MIDI

---

[3] http://www.911tabs.com/.

files by keeping the drum and bass tracks only. For each track, we selected a characteristic 16 bar snippet which is usually detected on the chorus part of a rock song. The selection was done with the help of a student of the Music Department of the Ionian University, Greece.

We used two different input spaces to represent the training data and feed them to two different two different types of ANNs, that are afterwards merged into a common hidden layer. The first one, an LSTM network, corresponds to the drum representation while the second one, the feedforward network, represents information of the bass movement, and the metrical structure information.

As far as the drums were concerned, their representation was based on text words, as proposed by [2] with the following alterations. To encode simultaneous events in a track into texts, we used the binary representation of pitches, i.e., standard components of drums - kick, snare, hi-hats, cymbals and toms. We also limited the number of events in a bar to 16 by quantising every track to the closest 16th-note. This process lead to 256 word music events for each learning input. Due to the limited training data, and for efficient representation and learning, only five components were retained; kick, snare, any tom event, open or closed hi-hats, and crash or ride cymbals. For example, 10010 and 01010 represents a time step with simultaneous playing of kick and hi-hat followed by simultaneous playing of snare and hi-hat. There can be theoretically $2^5 = 128$ words, but there are indeed much fewer, since the combinations of drum components that are actually played are limited.

Moving on to the bass, we use information regarding the voice leading (VL) of bass (which has proven a valuable aspect in harmonisation systems [9,17]). Specifically, VL was defined by calculating the pitch difference of the bass between two successive time steps, representing this information in a 1×4 binary vector. The first digit of this vector declares the existence of a bass or rest event, while the three remaining digits show the calculation of the bass voice leading in the following 3 different cases: [000] steady VL, [010] upward VL and [001] downward VL. For example, if the bass pitch from 42 changes successively to 35, then to 40 and finally to a rest, the Bass VL vectors occurring are [1001], [1010] and [0000].

In addition to the bass information we included a 1×3 binary vector representing metrical information for each time step. This information ensures that the network is aware of the beat structure at any given point. Considering the simple beat structure (resolution of 16ths and 4/4 time signature) of the examined examples, the first digit declares the start of a bar, the 2nd of half-bar, and finally the 3rd of quarter bar. So for example the first 9 time steps of a sequence could be translated to metrical information as: [111], [000], [000], [000], [001], [000], [000], [000], [011]. This representation can be arbitrarily expanded or modified to reflect additional or other information. For instance, a 4th digit can be included that indicates, e.g., the beginning of the chorus. Additionally, the second and third digits can be used for representing beat accents rather than beat positions. For instants, different binary representations can be used for denoting two different accentuation patterns in 7/8 time signature, e.g., 4-3 and 3-2-2. The examination of such possibilities is left for work.

3.2 Proposed Architecture

The proposed architecture consists of 2 separate modules for predicting the next drum event. An LSTM module learns sequences of consecutive drum events, while a feedforward layer takes information on the metrical structure and bass movement. The output of the network is the prediction of the next drum event. We used the Theano [1] deep learning framework and Lasagne [3] library. Figure 1 shows a diagram of our proposed architecture.

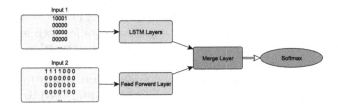

Fig. 1. Proposed deep neural network architecture for conditional drum generation.

For the drum input space we used 2 stacked LSTM layers with 128 or 512 Hidden Units and a dropout of 0.2, similarly to [25]. The LSTM layers have 16 time steps memory, which correspond to a full bar. Accordingly, the LSTM attempts to predict the next step in a stochastic manner. In each prediction for time index n, the network outputs the probabilities of every state. The bass and metrical input space is then fed into the Hidden module on a dense fully connected layer. Subsequently, a merge layer is used in order to take the output of each module that is then passed through the *softmax* nonlinearity to generate the probability distribution of the prediction. During the optimisation phase, the AdaGrad loss function [4] is calculated as the mean of the (categorical) cross-entropy between the prediction and the target.

4   Results

The LSTM methodology has been used for generating sequences that reflect a learned style, specifically using similar representations of rhythms as the one followed herein. Since the aim of the conditional rhythm composition methodology is to employ conditional information in a neural network fashion, our experimentation focuses on the following questions:

1. Are the training and generation processes negatively affected by the introduction of the feed forward (conditional) layer?

2. Does the feed forward (conditional) layer play any role in the rhythms generated by the system and if so, in what aspects?
3. How are the capabilities of the system affected when trained on datasets with different characteristics?

To answer the aforementioned questions, simple examples of generated rhythms were analysed, allowing the exploration of conditional layer's impact. Then, two larger-scale experiments are conducted where the features of composed and "ground-truth" rhythms were compared, focusing on the role of the conditional layer and the characteristics of different training scenarios, respectively. Those experiments allowed a deeper view into what the proposed methodology achieves and revealed the main weakness of related methodologies (including the proposed one): their incapability to capture high-level structure; suggestions for future enhancements are proposed in the concluding section of the paper (Sect. 5).

In the following examples and experiments, 5 different networks were trained:

1. PTNN: trained with 15 excerpts from the Porcupine Tree band.
2. FloydNN: trained with 15 excerpts from the Pink Floyd band.
3. QueenNN: trained with 15 excerpts from the Queen band.
4. allNN15: trained with 15 excerpts in total – 5 randomly selected from each aforementioned band.
5. allNN: trained with all 45 excerpts of the aforementioned bands.

Especially for the experiments in Sect. 4.2, where the role of the conditional layer was examined, a custom version of those networks is tested that does not include the conditional layer, i.e. those networks were trained and generated as typical LSTM networks without the conditional parts. Those networks are marked with a "no" suffix, e.g. the "PTNN" without the conditional layer is named as "PTNNno".

4.1 Examples of Generated Rhythms

Figure 2 shows three different examples of drum generation of the *allNN* model of two bars from a song of Queen which was not included on the training dataset. Starting from bottom to top in the piano-roll drum depiction, we identify in the following order kick, snare, toms, hi-hat and crash/ride events. According to the Ground Truth of the example, the rhythm is very simple with a standard pattern on kick snare and hi-hat events and a crash in the end of the first bar. In addition the bass harmony is also simple with few onsets and large durations.

As shown in Fig. 2's epoch progressions, the network is adjusted from the early stages of learning. On *epoch*10 it is able to understand the metric structure of the song and create simple and repeating drum patterns. Moving forward on the training the network gives more complex generations. On *epoch*30 the generation is almost the same as the original with minor difference in some additional cymbals (crash-ride) on the start and at end of the excerpt. At this stage, the Condition Layer seems to contribute a lot on the generation, as evident in the last example.

Then, we edited the bass track and replaced it with a complex playing with several more onsets in total. The Conditional Layer acted accordingly, forcing the network to create drum events which can follow this particular bass harmony. This is evident in Fig. 2(d) by the additional kick events created in both bars.

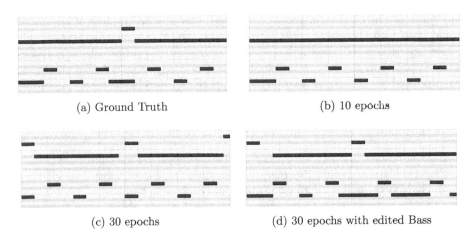

(a) Ground Truth   (b) 10 epochs

(c) 30 epochs   (d) 30 epochs with edited Bass

Fig. 2. Piano roll drum generation samples of the proposed network.

4.2   Training with vs. Without the Conditional Layer

The characteristics of drums' rhythms can be examined by extracting qualitative features, similarly to the ones presented in [11]. However, in order to extract such features the representation of rhythms requires transformation in a compatible representation, where only three basic drum elements are included: hi-hat, snare and kick drum (H, S and K). Since the current representation incorporates 5 elements, the snare and tom elements (rows) are merged into a single snare row, the cymbals rows (hi-hat and crash) are mapped into a single hi-hat row while the kick row remains as is.

Following the methodology found in [11], 32 features were extracted (described in Table 1), forming a vector representing each rhythm in the database, $r_i = \{f_1^{(i)}, f_2^{(i)}, \ldots, f_{32}^{(i)}\}$, where i is the index of the rhythm; those features incorporate the concepts of "density", "syncopation" [21], "symmetry" [12] and the "weak-to-strong" [11] ratios (measuring the ratios of the total intensities of weak over the total intensities of distinctive events), in the distinctive beats of the rhythm and in each separate percussive element (features 1–16). The distinctive beats of a rhythm are considered to be the beats in the snare or the kick drum that exceed the 70% of the maximum intensity in this rhythm. Since rhythms in this work are assumed binary, the distinctive beat simply describes the locations of snare and/or kick, expressed as a 1-row binary array. Features 17–19 and features 24–26 capture simultaneous pairs and isolated onsets respectively, while features 20–23 describe transition probabilities between the S and K elements. The mean values and standard deviations of the intensities of each percussive element are described in features 27–32. Since the examined rhythms are excerpts of 16 bars, each rhythm is represented by an array of $32 \cdot 16 = 512$ values.

Table 1. The employed drums' features.

Feature indexes	Feature description
1–4	Density, syncopation, symmetry and weak-to-strong ratio of the distinctive beat
5–16	Density, syncopation, symmetry and weak-to-strong ratio of each drum element
17–19	Percentages of simultaneous pairs of drums onsets (H–K, H–S and S–K)
20–23	Percentages of transitions between all combinations of K and S
24–26	Percentages of isolated H, S or K onsets
27–32	Intensity mean value and standard deviation for each drum element

In order to examine the efficiency of the introduced conditional layer, 6 "ground-truth" pieces were used – two from each band (Porcupine Tree, Pink Floyd and Queen) – none of which was included in the training sets. Both the simple LSTM and the conditional versions of all networks generated drum rhythms with initial seeds given by the ground-truth pieces, while the conditional networks were also given the metrical and bass voice information of the ground-truth. The compositions of each networks after every 5 epochs of training were extracted until epoch 30. The features of every composition were then extracted and mapped, along with the features of the ground truth rhythms, in two dimensions using the t-SNE [24] technique.

Figure 3 illustrates the 2D mapping of all features (ground-truth and compositions in all epochs by all network versions) with colors according to the 5 clusters assigned by k-means clustering [16]; this number of clusters was used partially because of the fact that 5 different networks were trained and partially because of the clarity of the provided results after experimentation with different cluster numbers. The assessment of the characteristics of each cluster is performed by counting the occurrences of different training "keywords" in each cluster, i.e., the trained network name and the training epochs that composed a rhythm.

Fig. 3. Two-dimensional mapping of the features of all network compositions with and without (circumscribed by triangles) the conditional layer. The ground-truth features mapping is illustrated with "×"s. Coloring is based on k-means clustering with 5 clusters. (Color figure online)

The aggregated keywords in each cluster are shown in Table 2. The cluster of main interest is C_3 since this is the cluster where the ground truth rhythms are. It is evident from the third column in Table 2 that the *allNN* network was mostly able to capture the characteristics of the ground truth rhythms. The pure LSTM rhythms (triangles), composed without using the conditional layer, are mainly placed far away from the ground-truth cluster. This fact indicates that the conditional layer does indeed contribute in the production of rhythms that more accurately resemble the characteristics of the trained style; at the same time, those rhythms have relatively diverse characteristics, which is evident by the dispersion in C_3. Therefore, the use of the conditional layer is shown to improve the efficiency of the network in capturing the characteristics of the training rhythms, while the diversity of the produced rhythms is preserved.

4.3   Examining the Role of the Conditional Layer

To examine the role of the conditional layer, a new version of the 3 out of the 6 ground truth rhythms was produced by the first author of this work, with the bass being modified. The modification in all 3 cases was mild and compatible with the style/original version of each piece. The modifications in the bass were expected to affect the system when composing, since the conditional layer changes. The influence of the changes in the bass are illustrated in Fig. 4, where the features of all 16 bars of the rhythms were reduced to 2 dimensions. The coloring is again the result of clustering with the k-means with 5 clusters,

Table 2. The six most common labels in each cluster in Fig. 2 along with the numbers of their occurrences.

C_1		C_2		C_3		C_4		C_5	
QueenNN:	29	epoch5:	13	allNN:	24	epoch30:	21	allNNno:	18
epoch5:	19	epoch10:	13	epoch20:	17	epoch20:	19	FloydNNno:	18
allNN15:	18	FloydNNno:	13	epoch25:	16	epoch15:	17	epoch25	17
epoch10:	17	allNN15no:	11	epoch10:	15	PTNN:	16	PTNNno:	15
FloydNN:	16	QueenNNno:	9	epoch5:	14	FloydNN:	15	epoch30:	13
epoch25:	15	epoch30:	4	epoch15:	13	PTNNno:	14	epoch20:	11

while the rhythms in circles are the ones composed with the use of the modified bass in the conditional layer.

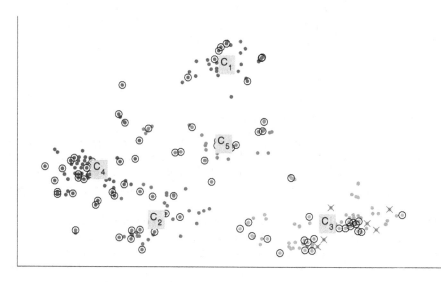

Fig. 4. Two-dimensional mapping of the features of all network compositions with original and modified bass (circumscribed by circles). The ground-truth features mapping is illustrated with "×"s. Coloring is based on k-means clustering with 5 clustes. (Color figure online)

The most common keywords in each cluster (composing network and epoch of training) are show in Table 3. The cluster of interest is C_3 where the ground truth rhythms are included. Again the *allNN* has the strongest presence in this cluster, a fact that, combined with the findings in Table 2, indicates that more training data (*allNN* is trained with 3 times as much data as all other networks) lead to the generation of rhythms that are closer to "real rhythms" of the learned style. Additionally, the presence of circled rhythms (composed with modified bass in the conditional layer) within this cluster indicates that the proposed system does not get "lost" when presented alternative scenarios and continues to compose rhythms that are close to "real rhythms" (ground truth). It should be noted that the evaluation of the rhythms composed with the modified bass are different from the ones composed with the original bass – i.e. the circled rhythms are not on top of non-circled rhythms. Therefore, the modifications in the bass potentially affect the network, but in a way that appears to be meaningful.

Table 3. The six most common labels in each cluster in Fig. 3 along with the numbers of their occurrences.

C_1		C_2		C_3		C_4		C_5	
FloydNN:	19	allNN15:	19	allNN:	34	QueenNN:	30	allNN15:	16
PTNN:	16	QueenNN:	12	PTNN:	20	FloydNN:	19	epoch10:	11
epoch30	13	epoch20:	9	epoch25:	16	epoch5:	15	PTNN:	11
epoch20:	12	epoch10:	9	epoch5:	14	epoch30:	11	epoch15:	7
epoch15:	12	FloydNN:	9	epoch10:	14	epoch25:	11	epoch5:	6
allNN:	8	epoch5:	8	epoch20:	12	epoch10:	10	epoch30:	6

5   Conclusions

In this work, we introduced an innovative architecture approach for creating drum sequences. The proposed Neural Network consists of a Recurrent module with LSTM layers which learns sequences of consecutive drum events and a Feed Forward Layer which takes information on the metrical structure and the bass movement.

Despite the few training data, the experiments showed promising results highlighting the importance of the Feed Forward Layer. LSTM systems without the conditional layer learn and compose music regardless of given conditions. Therefore, a drumming system that is based on a simple LSTM architecture does not get affected by, e.g., potential changes in the bass, which is not the case with human drummers.

Additionally, the preservation of a metrical structure in simple LSTM systems is only dependent on their ability to learn the metric structure these are trained on. The conditional layer enables the LSTM networks to simulate humans in both tasks: respond to changes in other instruments (e.g. bass) and "tune-in" to certain metrical structures.

Our future work will include the extension of the available training data, along with the expansion of network architecture with additional Feed Forward Layers featuring other rhythm instruments (e.g. guitar) or rhythm information. Finally more experiments will be conducted to test the system's behavior in metrical structures not trained on (e.g. 15/8) as well as how the accentuation patterns can be represented and learned.

References

1. Bergstra, J., Bastien, F., Breuleux, O., Lamblin, P., Pascanu, R., Delalleau, O., Desjardins, G., Warde-Farley, D., Goodfellow, I., Bergeron, A., et al.: Theano: deep learning on GPUs with python. In: NIPS 2011, BigLearning Workshop, Granada, Spain, vol. 3. Citeseer (2011)
2. Choi, K., Fazekas, G., Sandler, M.: Text-based LSTM networks for automatic music composition. arXiv preprint arXiv:1604.05358 (2016)
3. Dieleman, S., Schlüter, J., Raffel, C., Olson, E., Sønderby, S.K., Nouri, D., Maturana, D., Thoma, M., Battenberg, E., Kelly, J., et al.: Lasagne: First Release. Zenodo, Geneva (2015)
4. Duchi, J., Hazan, E., Singer, Y.: Adaptive subgradient methods for online learning and stochastic optimization. J. Mach. Learn. Res. **12**(Jul), 2121–2159 (2011)
5. Eck, D., Schmidhuber, J.: A first look at music composition using LSTM recurrent neural networks. Istituto Dalle Molle Di Studi Sull Intelligenza Artificiale, p. 103 (2002)
6. Hadjeres, G., Pachet, F.: DeepBach: a steerable model for Bach chorales generation. arXiv preprint arXiv:1612.01010 (2016)
7. Hiller, L.A., Isaacson, L.M.: Experimental Music; Composition with an Electronic Computer. Greenwood Publishing Group Inc., Westport (1979)
8. Hochreiter, S., Schmidhuber, J.: Long short-term memory. Neural Comput. **9**(8), 1735–1780 (1997)
9. Hörnel, D.: Chordnet: learning and producing voice leading with neural networks and dynamic programming. J. New Music Res. **33**(4), 387–397 (2004)
10. Jacob, B.L.: Algorithmic composition as a model of creativity. Organised Sound **1**(03), 157–165 (1996)
11. Kaliakatsos–Papakostas, M.A., Floros, A., Vrahatis, M.N.: evoDrummer: deriving rhythmic patterns through interactive genetic algorithms. In: Machado, P., McDermott, J., Carballal, A. (eds.) EvoMUSART 2013. LNCS, vol. 7834, pp. 25–36. Springer, Heidelberg (2013). doi:10.1007/978-3-642-36955-1_3
12. Kaliakatsos-Papakostas, M.A., Floros, A., Vrahatis, M.N., Kanellopoulos, N.: Genetic evolution of L and FL-systems for the production of rhythmic sequences. In: Proceedings of the 2nd Workshop in Evolutionary Music (GECCO 2012), Philadelphia, USA, pp. 461–468, 7–11 July 2012
13. Kalingeri, V., Grandhe, S.: Music generation with deep learning. arXiv preprint arXiv:1612.04928 (2016)
14. Lambert, A., Weyde, T., Armstrong, N.: Perceiving and predicting expressive rhythm with recurrent neural networks. In: Proceedings of the 12th International Conference in Sound and Music Computing, SMC 2015. SMC15 (2015)
15. Lewis, J.P.: Algorithms for Music Composition by Neural Nets: Improved CBR Paradigms. Michigan Publishing, University of Michigan Library, Ann Arbor (1989)
16. MacQueen, J., et al.: Some methods for classification and analysis of multivariate observations. In: Proceedings of the Fifth Berkeley Symposium on Mathematical Statistics and Probability, Oakland, CA, USA, vol. 1, pp. 281–297 (1967)
17. Makris, D., Kaliakatsos-Papakostas, M.A., Cambouropoulos, E.: Probabilistic modular bass voice leading in melodic harmonisation. In: ISMIR, pp. 323–329 (2015)
18. Mozer, M.C.: Neural network music composition by prediction: exploring the benefits of psychoacoustic constraints and multiscale processing. In: Musical Networks: Parallel Distributed Perception and Performance, p. 227 (1999)

19. Pachet, F., Roy, P.: Musical harmonization with constraints: a survey. Constraints **6**(1), 7–19 (2001)
20. Papadopoulos, G., Wiggins, G.: Ai methods for algorithmic composition: a survey, a critical view and future prospects. In: AISB Symposium on Musical Creativity, Edinburgh, UK, pp. 110–117 (1999)
21. Sioros, G., Guedes, C.: Complexity driven recombination of MIDI loops. In: Proceedings of the 12th International Society for Music Information Retrieval Conference (ISMIR), pp. 381–386. University of Miami, Miami, October 2011
22. Sturm, B., Santos, J.F., Korshunova, I.: Folk music style modelling by recurrent neural networks with long short term memory units. In: 16th International Society for Music Information Retrieval Conference (ISMIR) (2015)
23. Supper, M.: A few remarks on algorithmic composition. Comput. Music J. **25**(1), 48–53 (2001)
24. van der Maaten, L.: Learning a parametric embedding by preserving local structure. RBM **500**(500), 26 (2009)
25. Zaremba, W., Sutskever, I., Vinyals, O.: Recurrent neural network regularization. arXiv preprint arXiv:1409.2329 (2014)

Efficient Identification of k-Closed Strings

Hayam Alamro[1], Mai Alzamel[1], Costas S. Iliopoulos[1], Solon P. Pissis[1],
Steven Watts[1(✉)], and Wing-Kin Sung[2]

[1] Department of Informatics, King's College London, London, UK
{hayam.alamro,mai.alzamel,costas.iliopoulos,solon.pissis,
steven.watts}@kcl.ac.uk
[2] Department of Computer Science, National University of Singapore,
Singapore, Singapore
ksung@comp.nus.edu.sg

Abstract. A closed string contains a proper factor occurring as both a prefix and a suffix but not elsewhere in the string. Closed strings were introduced by Fici (WORDS 2011) as objects of combinatorial interest. In this paper, we extend this definition to k-closed strings, for which a level of approximation is permitted up to a number of Hamming distance errors, set by the parameter k. We then address the problem of identifying whether or not a given string of length n over an integer alphabet is k-closed and additionally specifying the border resulting in the string being k-closed. Specifically, we present an $\mathcal{O}(kn)$-time and $\mathcal{O}(n)$-space algorithm to achieve this along with the pseudocode of an implementation.

1 Introduction

Closed strings (or closed words) are bordered strings that satisfy an additional property that their border does not occur elsewhere in the string. A bordered string x is such that there exists a prefix of x which is also a suffix of x. There are a number of earlier studies dealing with closed strings. Fici in [8] introduced the notion of closed strings in addition to characterisations of this class.

The more practical relevance of closed strings was established via their relationship with palindromic strings. The number of closed factors in a string is minimised if these factors are also palindromic. Additionally it was shown that the upper bound on the number of palindromic factors of a string coincides with the lower bound on the number of closed factors (see [3] and references therein). Thus the study of closed strings shows potential applications in connection with applications of palindromes [1]. On the algorithmic side, Badkobeh et al. in [2] presented (among others) an algorithm for the factorisation of a given string of length n into a sequence of longest closed factors in time and space $\mathcal{O}(n)$ and another algorithm for computing the longest closed factor starting at every position in the string in $\mathcal{O}(n\frac{\log n}{\log\log n})$ time and $\mathcal{O}(n)$ space.

Here we extend the definition of closed strings to k-closed strings, for which a level of approximation is permitted up to a number of Hamming distance errors,

© Springer International Publishing AG 2017
G. Boracchi et al. (Eds.): EANN 2017, CCIS 744, pp. 583–595, 2017.
DOI: 10.1007/978-3-319-65172-9_49

set by the parameter k. The main contribution is an $\mathcal{O}(kn)$-time and $\mathcal{O}(n)$-space algorithm for identifying whether or not a given string of length n over an integer alphabet is k-closed. We also specify the border that results in the string being k-closed.

The rest of this paper is organised as follows. In Sect. 2, we present basic definitions and notation on strings as well as an overview of the Kangaroo method [9,11] utilised in our algorithm. Section 3 presents the extension from closed to k-closed strings and the problem statement. In Sect. 4, a detailed implementation of our algorithm is presented. Our final remarks are noted in Sect. 5.

2   Preliminaries

2.1   Basic Terminology

We begin with basic definitions and notation from [6]. Let $x = x[0]x[1]\ldots x[n-1]$ be a *string* of length $|x| = n$ over a finite ordered alphabet. We consider the case of strings over an *integer alphabet*: each letter is replaced by its lexicographical rank in such a way that the resulting string consists of integers in the range $\{1, \ldots, n\}$.

For two positions i and j on x, we denote by $x[i \mathinner{.\,.} j] = x[i]\ldots x[j]$ the *factor* (sometimes called *substring*) of x that starts at position i and ends at position j (it is of length 0 if $j < i$), and by ε the *empty string* of length 0. We recall that a *prefix* of x is a factor that starts at position 0 ($x[0 \mathinner{.\,.} j]$) and a *suffix* of x is a factor that ends at position $n - 1$ ($x[i \mathinner{.\,.} n - 1]$). A prefix (resp. suffix) is said to be *proper* if its length is strictly less than n. We denote the *reverse* string of x by x^R, i.e. $x^R = x[n - 1]x[n - 2]\ldots x[1]x[0]$.

Let y be a string of length m with $0 < m \le n$. We say that there exists an *occurrence* of y in x, or, more simply, that y *occurs in* x, when y is a factor of x. Every occurrence of y can be characterised by a starting position in x. Thus we say that y occurs at the *starting position* i in x when $y = x[i \mathinner{.\,.} i + m - 1]$.

The *Hamming distance* between two strings x and y of the same length is defined as the number of corresponding positions in x and y with different letters, denoted by $\delta_H(x, y) = |\{i : x[i] \ne y[i], i = 0, 1, \ldots, |x| - 1\}|$. For the sake of completeness, if $|x| \ne |y|$, we set $\delta_H(x, y) = \infty$. If two strings x and y are at Hamming distance k or less, we call this a *k-match*, written as $x \approx_k y$.

The *longest common extension* (LCE) between two suffixes of a string x starting at positions i and j is defined as the length of the longest prefix common to both suffixes. Formally, for a given string x:

$$\mathrm{LCE}(i, j) = \max\{l : x[i \mathinner{.\,.} i + l - 1] = x[j \mathinner{.\,.} j + l - 1]\}.$$

We may generalise the concept of LCE to that of an LCE with k errors. In this case, the LCE is similarly defined but considers a common prefix between two suffixes to be valid if they match within k or less errors in terms of their Hamming distance. Formally, for a given string x:

$$\mathrm{LCE}_k(i, j) = \max\{l : x[i \mathinner{.\,.} i + l - 1] \approx_k x[j \mathinner{.\,.} j + l - 1]\}.$$

2.2 Tools

The algorithm described in this paper makes substantial use of the *kangaroo method*, a well-established method used to perform multiple LCE_k queries on a given string x [9,11]. This is done by initially preprocessing the string x to build a *suffix tree* data structure in $\mathcal{O}(n)$ time and space [7] (Fig. 1).

The suffix tree provides a compact representation of the set of suffixes of the string x, in which each leaf node of the tree uniquely corresponds to one of the n suffixes of x, which may be reconstructed by following the unique path from the root to the leaf and concatenating the edge labels as they are encountered. To ensure every suffix corresponds uniquely to a leaf node, the string x may be appended with a unique $ letter. Full technical details of the suffix tree data structure can be found in [7].

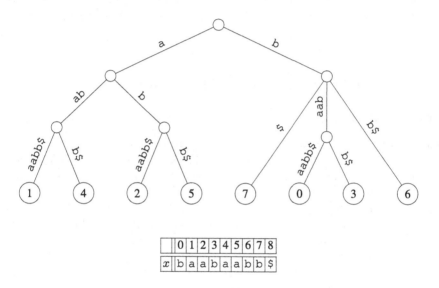

Fig. 1. Suffix tree for the string $x = $ baabaabb$.

The suffix tree of x allows us to perform LCE queries in $\mathcal{O}(1)$ time. For a query $LCE(i, j)$, we first identify two distinct leaf nodes corresponding to the suffixes i and j; then, the lowest common ancestor node in the tree for these two leaf nodes has a string depth equal to $LCE(i, j)$. The calculation of the lowest common ancestor and its depth may be performed in $\mathcal{O}(1)$ time after $\mathcal{O}(n)$ preprocessing time [5], hence the LCE query can be computed in $\mathcal{O}(1)$ time (Fig. 2).

The kangaroo method extends this methodology, allowing the calculation of LCE_k queries. Precisely, we have the following lemma:

Lemma 1. *Given the suffix tree of x, $LCE_k(i, j)$ can be computed in $\mathcal{O}(k)$ time.*

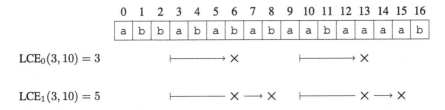

Fig. 2. Longest common extensions for $k = 0$ and $k = 1$.

Proof. $\mathrm{LCE}_k(i, j)$ can be defined recursively as follows:

We denote $l_r = \mathrm{LCE}_r(i, j)$ for any $r \geq 0$. Then, we have $l_0 = \mathrm{LCE}(i, j)$. Also, we have $l_r = l_{r-1} + 1 + \mathrm{LCE}(i + l_{r-1} + 1, j + l_{r-1} + 1)$.

By the above recursive formula, $\mathrm{LCE}_k(i, j)$ can be computed by performing LCE queries k times. Since each LCE query requires $\mathcal{O}(1)$ time, the lemma follows. □

3 From Closed Strings to k-Closed Strings

3.1 Closed Strings

If a string b is both a proper prefix and a proper suffix of a non-empty string x, then b is called a *border* of x. A string x is said to be *closed* if and only if it is empty or if there exists a border b of x that occurs exactly twice in x (i.e. only as a prefix and suffix). In other words, b satisfies (1) $b = x[0 \mathinner{\ldotp\ldotp} |b| - 1] = x[|x| - |b| \mathinner{\ldotp\ldotp} |x| - 1]$ and (2) $b \neq x[i \mathinner{\ldotp\ldotp} i + |b| - 1]$, for all $1 \leq i \leq |x| - |b| - 1$. If x is closed, we call such a b the *closed border* of x. We additionally define the special case of a single letter $a \in \Sigma$ to be closed, with the empty string ε as the border of a.

For instance, string ACA is closed, since the factor A occurs only as a prefix and as a suffix. The word ACAA, on the contrary, is not closed: A has an internal occurrence.

3.2 k-Closed Strings

The definition of closed strings can be generalised to *k-closed strings*, where k expresses a Hamming distance error bound. This is useful for dealing with strings where errors or approximations in the data may occur.

Definition 1. *A string x of length n is called k-closed if and only if $n \leq 1$ or the following properties are satisfied for some k' where $0 \leq k' \leq k$:*

1. *There exists some proper prefix u of x and some proper suffix v of x of length $|u| = |v|$, such that $\delta_H(u, v) \leq k'$.*
2. *Except for u and v, there exists no factor w of x of length $|w| = |u| = |v|$ such that $\delta_H(u, w) \leq k'$ or $\delta_H(v, w) \leq k'$.*

For the above definition, the pair u and v for the smallest k' is called the k-closed border of x. In the case where $n \leq 1$ we assign ε as the k-closed border.

This works as a generalisation of closed strings, also now known as 0-closed strings. It is clear from the definition, that a smaller value of k corresponds to k-closed being a stronger statement on the nature of x. Therefore Lemma 2 follows trivially from Definition 1.

Lemma 2. *A string x that is k-closed is also r-closed for all $r > k$.*

It additionally follows from Definition 1 that the k-closed border of a string x is unique by Lemma 3.

Lemma 3. *A length-n k-closed string x, $n > 1$, has exactly one k-closed border, i.e. there exists exactly one prefix u and one suffix v satisfying the conditions in Definition 1 for the smallest $k' \leq k$.*

Proof. Since x is k-closed, it has at least one k-closed border and an associated smallest $k' \leq k$ for which the conditions are satisfied. Let us consider the longest of these k-closed borders, and call u and v the prefix and suffix respectively, comprising the longest k-closed border with length $|u| = |v|$. Let us assume a second k-closed border exists, comprised of the prefix and suffix, u' and v' respectively. We know that $|u'| = |v'| < |u| = |v|$ and $u' = u[0..|u'| - 1]$. Since $u \approx_{k'} v$ it is trivially true that $u[0..|u'| - 1] \approx_{k'} v[0..|u'| - 1]$ and therefore $u' \approx_{k'} v[0..|u'| - 1]$. Thus we see that u' k'-matches the prefix of v of the same length, and this corresponds to an occurrence of u' within x, i.e. $u' \approx_{k'} x[n - |v|..n - |v| + |u'| - 1]$, where n is the length of x, which is an internal occurrence of u' in x. We arrive at a contradiction due to Condition 2 of Definition 1 being violated, therefore no second k-closed border can exist. □

Remark 1. Let x be a non-empty k-closed string of length n. The following characterisations follow easily from Definition 1 and Lemma 3:

1. x has exactly one k-closed border.
2. If $n > 1$, there exists a string w with $|w| < n$ and a natural number k', with $0 \leq k' \leq k$, such that $w \approx_{k'} x[i..i + |w| - 1]$ for exactly two values of i and no others; specifically $i = 0$ and $i = n - |w|$.
3. There exists a natural number k', with $0 \leq k' \leq k$, such that the longest repeated prefix (resp. suffix) of x within k' errors, is equal to u (resp. v), where u and v are the prefix and suffix, respectively, comprising the k-closed border.
4. There exists a natural number k', with $0 \leq k' \leq k$, such that any repeated prefix (resp. suffix) of x within k' errors is necessarily a prefix (resp. suffix) of u (resp. v), where u and v are the prefix and suffix, respectively, comprising the k-closed border.

We display an example in Fig. 3. Note that for the string GTGAGTGGTA we illustrate only that a border length of 3 with error 1 is not a possible 1-closed

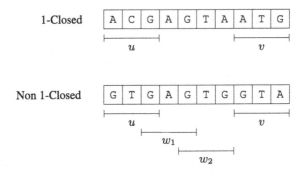

Fig. 3. Closed and non-closed strings for $k = 1$.

border. To fully verify it is non 1-closed, all combinations of border lengths and error levels $0 \le k' \le 1$ must be considered. It is in fact possible to show that no borders of any length exist that satisfy the 1-closed criteria, therefore the string is indeed non 1-closed (and by Lemma 2 also non 0-closed).

We are now in a position to formally define the problem solved in this paper. For computation purposes, we focus only on the case when $k > 0$.

---

k-CLOSED BORDER
Input: A string x of length n and a natural number k, $0 < k < n$
Output: The k-closed border or -1 if x is not k-closed

---

4   Algorithm

In addition to our definition of k-closed strings, we further define some variants of Definition 1 which proved useful in obtaining the main result.

Definition 2. *A string x of length n is called k-weakly-closed if and only if $n \le 1$ or the following properties are satisfied:*

1. *There exists some proper prefix u of x and some proper suffix v of x of length $|u| = |v|$, such that $\delta_H(u, v) \le k$.*
2. *Both factors u and v occur only as a prefix and suffix respectively within x, i.e. no internal occurrences of u or v exist in x.*

We call such a pair u and v a k-weakly-closed border of x. In the case where $n \le 1$, we assign ε as the k-weakly-closed border.

Definition 2 describes a scenario where the border may have errors, but internal occurrences are considered to not have errors. Note that under this definition there may be multiple k-weakly-closed borders.

Definition 3. *A string x of length n is called k-strongly-closed if and only if $n \le 1$ or the following properties are satisfied:*

1. *There exists some border b of x.*
2. *There exists no factor w of x of length $|w| = |b|$ such that $\delta_H(b, w) \leq k$, except the prefix and suffix of x.*

We call b the k-strongly-closed border of x. In the case where $n \leq 1$, we assign ε as the k-strongly-closed border.

Definition 3 describes a scenario where the border does not have errors, but internal occurrences may have errors. Note that under this definition there is only one k-strongly-closed border.

Definition 4. *A string x of length n is called k-pseudo-closed if and only if $n \leq 1$ or the following properties are satisfied:*

1. *There exists some proper prefix u of x and some proper suffix v of x of length $|u| = |v|$, such that $\delta_H(u, v) \leq k$.*
2. *Except for u and v, there exists no factor w of x of length $|w| = |u| = |v|$ such that $\delta_H(u, w) \leq k$ or $\delta_H(v, w) \leq k$.*

We call such a pair u and v the k-pseudo-closed border of x. In the case where $n \leq 1$, we assign ε as the k-pseudo-closed border.

Definition 4 may be regarded as a merge of Definitions 2 and 3. Both the border and internal occurrences may have errors. Note that Condition 1 is *less selective* and Condition 2 is *more selective*. Therefore the requirement to satisfy both conditions implies that a 0-closed string is not necessarily k-pseudo-closed and a k-pseudo-closed string is not necessarily 0-closed (hence the *pseudo* term). For instance, ABAC is 1-pseudo-closed with a border comprised of the prefix AB and suffix AC, but not 0-closed. In a contrary example, ABBA is 0-closed but not 1-pseudo-closed.

Note that there is a similarity in the construction of Definitions 1 and 4 that permits us to trivially conclude the following crucial lemma:

Lemma 4. *x is k-closed $\iff$ $\exists k'$ where $0 \leq k' \leq k$, such that x is k'-pseudo-closed.*

We initially begin by constructing the suffix tree of x. As has been discussed, this is constructible in $\mathcal{O}(n)$ time and space. Recall that once the suffix tree is constructed it can be pre-processed within the same complexity to answer any $\mathrm{LCE}_k(i, j)$ query by applying the Kangaroo method in $\mathcal{O}(k)$ time (see Lemma 1).

For the purpose of this algorithm, we draw attention to a specific subset of the possible LCE_k queries and store their values in two related data structures. These structures are the *longest prefix k-match array* and *longest suffix k-match array* of string x, denoted by $\mathrm{LPM}_k(x)$ and $\mathrm{LSM}_k(x)$, respectively.

$\mathrm{LPM}_k(x)[j]$ (respectively $\mathrm{LSM}_k(x)[j]$) is defined as the length of the longest factor of x starting (ending) at index j, which matches the prefix (suffix) of x of the same length within k errors, with the exception of the index j corresponding to the prefix (suffix) itself, for which we set a value of -1. Note that within the literature, the LPM array is similar to the *k-prefix table* [4] with the exception of using the -1 flag.

$$\text{LPM}_k(x)[j] = \begin{cases} \max\{l : \delta_H(x[0..l-1], x[j..j+l-1]) \le k\} & j \in [1, n-1] \\ -1 & j = 0 \end{cases}$$

$$\text{LSM}_k(x)[j] = \begin{cases} \max\{l : \delta_H(x[n-l..n-1], x[j-l+1..j]) \le k\} & j \in [0, n-2] \\ -1 & j = n-1 \end{cases}$$

Note that it follows from the definition that the LSM array for a string x is equal to the reverse of the LPM array for the reverse of x, with the opposite also being true:

$$\text{LSM}_k(x)[j] = \text{LPM}_k(x^R)[n-1-j]$$
$$\text{LPM}_k(x)[j] = \text{LSM}_k(x^R)[n-1-j].$$

Using these identities, we may express the LPM and LSM in terms of the familiar LCE queries, making it possible to apply the Kangaroo method to construct them:

$$\text{LPM}_k(x)[j] = \text{LCE}_k(0, j) \text{ of } x \qquad j \in [1, n-1]$$
$$\text{LSM}_k(x)[j] = \text{LCE}_k(0, n-1-j) \text{ of } x^R \quad j \in [0, n-2].$$

Using the method for answering LCE_k queries, we can calculate a single value of LPM or LSM in $\mathcal{O}(k)$ time, implying a total time of $\mathcal{O}(kn)$ required to fully calculate both arrays. In fact the complexity of the full algorithm is bounded by this procedure.

A further set of identities allows us to compute the LPM_{k+1} and LSM_{k+1} arrays from the LPM_k and LSM_k arrays in $\mathcal{O}(1)$ time per entry, such that the arrays are progressively constructed, with each intermediate step yielding valuable information:

$$\text{LPM}_{k+1}(x)[j] = p + 1 + \text{LCE}(p+1, j+p+1) \text{ of } x$$
$$\text{LSM}_{k+1}(x)[j] = s + 1 + \text{LCE}(s+1, n-j+s) \text{ of } x^R$$

where $p = \text{LPM}_k(x)[j]$ and $s = \text{LSM}_k(x)[n-1-j]$.

After computing $\text{LPM}_{k'}$ and $\text{LSM}_{k'}$, for $0 \le k' \le k$, we may determine if a given string x of length $n \ge 2$ is a k-closed string by checking against three conditions for each k', as shown by Lemma 5. Recall that in the case when $n = 0$ or $n = 1$, x is trivially k-closed by definition.

Lemma 5. *Given a string x of length $n \ge 2$ and a natural number k, $0 \le k < n$, x is k-closed if and only if there exists some $j \in \{1, \ldots, n-1\}$ and some $k' \in \{0, \ldots, k\}$ such that all the following conditions hold:*

(1) $j + \text{LPM}_{k'}(x)[j] = n$
(2) $\forall i < j, \ \text{LPM}_{k'}(x)[i] < \text{LPM}_{k'}(x)[j]$
(3) $\forall i > n-1-j, \ \text{LSM}_{k'}(x)[i] < \text{LSM}_{k'}(x)[j].$

Proof. Recall that $n \geq 2$. The three conditions can be seen to be necessary and sufficient for a string to be k-closed by considering the cases individually.

($\implies$) Suppose Conditions 1–3 hold. We need to show that x is k-closed. We first prove that the conditions imply x is k'-pseudo-closed. In other words, we need to find a prefix u of x and a suffix v of x such that:

(I) $u \approx_{k'} v$

(II) Except for u and v, there exists no length-$|u|$ factor w of x such that $w \approx_{k'} u$ or $w \approx_{k'} v$.

First, Condition 1 implies that the longest prefix match within k' errors starting at j terminates at position $n - 1$ in x. This implies that $u = x[0 .. n - j - 1] \approx_{k'} x[j .. n - 1] = v$. Hence, (I) is true.

By contrary of (II), we have either (1) a factor w starting at position $i < j$ such that $w \approx_{k'} u$ or (2) a factor w ending at position $i < n - 1 - j$ such that $w \approx_{k'} v$.

For (1), this means that $\mathrm{LPM}_{k'}(x)[i] \geq \mathrm{LPM}_{k'}(x)[j]$. However, this contradicts Condition 2.

For (2), this means that $\mathrm{LSM}_{k'}(x)[i] \geq \mathrm{LSM}_{k'}(x)[j]$. However, this contradicts Condition 3.

Hence, both (I) and (II) are true. This implies that x is k'-pseudo-closed. Since $0 \leq k' \leq k$, we may further imply by Lemma 4 that x is k-closed.

($\impliedby$) If x is k-closed, there must exist some k', where $0 \leq k' \leq k$, such that x is k'-pseudo-closed, by Lemma 4. For such a k', there is an associated k'-pseudo-closed-border consisting of some proper prefix u and some proper suffix v with equal length, such that $\delta_H(u, v) \leq k'$. We denote j where $v = x[j .. n - 1]$ and consequently $u = x[0 .. n - j - 1]$. The longest prefix match $\mathrm{LPM}_{k'}(x)[j]$ starting at j must be greater than or equal to $|u|$ as $u \approx_{k'} v$, yet it may not exceed the bounds of x and is therefore less than or equal to $|v|$. Therefore $\mathrm{LPM}_{k'}(x)[j] = |u| = |v| = n - j \implies \mathrm{LPM}_{k'}(x)[j] + j = n$ which implies Condition 1. From the definition of k-closed strings we also conclude that there exists no factor w of x with length $|w| = |u| = |v|$ such that $\delta_H(u, w) \leq k'$ or $\delta_H(v, w) \leq k'$. Therefore if we choose $i < j$ it must be the case that $\mathrm{LPM}_{k'}(x)[i] < \mathrm{LPM}_{k'}(x)[j]$, since otherwise we would have a $w \approx_{k'} v$ starting at i which cannot be the case, and therefore we conclude Condition 2. Similarly if we choose $i > n - 1 - j$ it must be the case that $\mathrm{LSM}_{k'}(x)[i] < \mathrm{LSM}_{k'}(x)[j]$, since otherwise we would have a $w \approx_{k'} v$ ending at i which cannot be the case, and therefore we conclude Condition 3. Thus all three conditions are satisfied. $\qquad\square$

j	0	1	2	3	4	5	6	7	8	9	10	11	12	13	14	
$x[j]$	a	b	b	a	b	a	a	b	a	b	a	a	b	a	b	
$\mathrm{LPM}_2[j]$	-1	3	4	7	2	10	4	4	7	2	5	4	3	2	1	
$\mathrm{LSM}_2[j]$	1	2	3	4	5	2	7	6	2	10	2	5	7	2	-1	
$j + \mathrm{LPM}_2[j] = n$	F	F	F	F	F	T	F	F	T	F	T	T	T	T	T	Cond.1
$\mathrm{LPM}_2\_\mathrm{peaks}[j]$	T	T	T	T	F	T	F	F	F	F	F	F	F	F	F	Cond.2
$\mathrm{LSM}_2\_\mathrm{peaks}[n-1-j]$	T	T	T	F	F	T	F	F	F	F	F	F	F	F	F	Cond.3
2-Closed Borders	F	F	F	F	F	T	F	F	F	F	F	F	F	F	F	

Fig. 4. 2-closed border of length $n - j - 10$ found at $j = 5$ for string x of length $n = 15$. This corresponds to strings **abbabaabab** and **aababaabab** which are at Hamming distance 2.

4.1   Main Result

Theorem 1. *Given a string x of length n over an integer alphabet and a natural number k, $0 < k < n$, the k-closed border of x, if it exists, can be determined in $\mathcal{O}(kn)$ time and $\mathcal{O}(n)$ space.*

Proof. By Lemma 5, the time taken to determine whether a string x of length n is k-closed (and determine the k-closed border itself) is bounded by the computation of the $\mathrm{LPM}_{k'}(x)$ and $\mathrm{LSM}_{k'}(x)$ arrays, for all $0 \leq k' \leq k$. For a single k', Condition 1 trivially requires $\mathcal{O}(n)$ time to check across all possible j. Conditions 2 and 3 can be answered for each j in $\mathcal{O}(1)$ time by first preprocessing the $\mathrm{LPM}_{k'}(x)$ and $\mathrm{LSM}_{k'}(x)$ arrays in $\mathcal{O}(n)$ time to determine where the appropriate peaks lie (inspect Fig. 4 for an example). Therefore a total of $\mathcal{O}(kn)$ time is required to check across all possible k' and j as shown in Lemma 5. The $\mathrm{LPM}_{k'}(x)$ and $\mathrm{LSM}_{k'}(x)$ arrays can be updated for one k' value to the next one and so the space required is only $\mathcal{O}(n)$.   □

4.2   Implementation

A full implementation of our algorithm was produced and the resulting pseudocode is presented here. The main function of the pseudocode is GETBOR-DER. This accepts a string x of length n in addition to a parameter k specifying the maximum number of errors. The length that determines the k-closed border is returned. Note that if the string x is not k-closed, the function returns -1. We make use of some functions for which the pseudocode is not given and detail those functions here:

REVERSE(x). Standard library function. Accepts a string or array x of length n and returns the reversed string or array, respectively.

LCE(x, i, j). Longest common extension function. Given a string x returns the length of the longest common prefix between the ith and jth suffixes of x (details in Sect. 2). Open-source implementations of LCE are available [10].

5   Final Remarks

We have presented an algorithm for finding the k-closed border of a given string x of length n within Hamming distance k. The proposed algorithm was dependent on building two simple data structures, namely, $\text{LPM}_k(x)$ and $\text{LSM}_k(x)$. Given these data structures, it takes a further $\mathcal{O}(n)$ time to determine the k-closed border.

The main improvement could therefore be in the construction of these two tables, currently requiring $\mathcal{O}(kn)$ time. Decreasing this time complexity appears to be a reasonable, however non-trivial, goal for any future work on this problem, as any faster computation of $\text{LPM}_k(x)$ and $\text{LSM}_k(x)$ would imply a major breakthrough in approximate string matching.

Algorithm. k-Closed Border

```
 1: function GETBORDER(x, n, k)
 2:     if n = 0  or  n = 1 then                              ▷ trivial cases
 3:         return  0
 4:     end if
 5:
 6:     lpm = integer array of length n filled with -1 at every position
 7:     lsm = integer array of length n filled with -1 at every position
 8:
 9:     for i = 0 to k do
10:         lpm = GETNEXTLPM(lpm, x, n)
11:         lsm = REVERSE(GETNEXTLPM(REVERSE(lsm), REVERSE(x), n))
12:
13:         lpm_peaks = GETPEAKS(lpm)
14:         lsm_peaks = GETPEAKS(lsm)
15:
16:         for j = 1 to n − 1 do                   ▷ check 3 conditions for every j
17:             if j + lpm[j] == n and lpm_peaks[j] and lsm_peaks[n − 1 − j] then
18:                 return n − j
19:             end if
20:         end for
21:     end for
22:
23:     return −1
24: end function
```

Algorithm. k-Closed Border

1: **function** GETNEXTLPM(lpm, x, n)
2:     **for** $i = 1$ to $n - 1$ **do**
3:         **if** $lpm[i] == n - i$ **then**
4:             **continue**
5:         **end if**
6:
7:         $lpm[i] = lpm[i] + \text{LCE}(x, lpm[i] + 1, i + lpm[i] + 1) + 1$ ▷ update LPM
8:     **end for**
9:
10:     **return** lpm
11: **end function**

Algorithm. k-Closed Border

1: **function** GETPEAKS($values$)   ▷ $values$ is array of integers
2:     $peaks$ = boolean array with same length as $values$
3:     $max\_val = -1$
4:
5:     **for** $i = 0$ to $n - 1$ **do**
6:         **if** $values[i] > max\_val$ **then**
7:             $peaks[i] = \text{True}$
8:             $max\_val = values[i]$
9:         **else**
10:             $peaks[i] = \text{False}$
11:         **end if**
12:     **end for**
13:
14:     **return** $peaks$
15: **end function**

References

1. Almirantis, Y., Charalampopoulos, P., Gao, J., Iliopoulos, C.S., Mohamed, M., Pissis, S.P., Polychronopoulos, D.: On avoided words, absent words, and their application to biological sequence analysis. Algorithms Mol. Biol. **12**(1), 5 (2017)
2. Badkobeh, G., Bannai, H., Goto, K., Tomohiro, I., Iliopoulos, C.S., Inenaga, S., Puglisi, S.J., Sugimoto, S.: Closed factorization. Discret. Appl. Math. **212**, 23–29 (2016). Stringology Algorithms
3. Badkobeh, G., Fici, G., Lipták, Z.: A note on words with the smallest number of closed factors. CoRR, abs/1305.6395 (2013)
4. Barton, C., Iliopoulos, C.S., Pissis, S.P., Smyth, W.F.: Fast and simple computations using prefix tables under hamming and edit distance. In: Kratochvíl, J., Miller, M., Froncek, D. (eds.) IWOCA 2014. LNCS, vol. 8986, pp. 49–61. Springer, Cham (2015). doi:10.1007/978-3-319-19315-1_5
5. Bender, M.A., Farach-Colton, M.: The LCA problem revisited. In: Gonnet, G.H., Viola, A. (eds.) LATIN 2000. LNCS, vol. 1776, pp. 88–94. Springer, Heidelberg (2000). doi:10.1007/10719839_9

6. Crochemore, M., Hancart, C., Lecroq, T.: Algorithms on Strings. Cambridge University Press, Cambridge (2007)
7. Farach, M.: Optimal suffix tree construction with large alphabets. In: 38th Annual Symposium on Foundations of Computer Science, Proceedings, pp. 137–143. IEEE (1997)
8. Fici, G.: A classification of trapezoidal words. In: Proceedings 8th International Conference Words 2011, Prague, Electronic Proceedings in Theoretical Computer Science, vol. 63, pp. 129–137 (2011)
9. Galil, Z., Giancarlo, R.: Improved string matching with *k* mismatches. SIGACT News **17**(4), 52–54 (1986)
10. Gog, S., Beller, T., Moffat, A., Petri, M.: From theory to practice: plug and play with succinct data structures. In: Gudmundsson, J., Katajainen, J. (eds.) SEA 2014. LNCS, vol. 8504, pp. 326–337. Springer, Cham (2014). doi:10.1007/978-3-319-07959-2_28
11. Landau, G.M., Vishkin, U.: Efficient string matching with *k* mismatches. Theoret. Comput. Sci. **43**, 239–249 (1986)

Bloom Filters for Efficient Coupling Between Tables of a Database

Eirini Chioti[1], Elias Dritsas[1], Andreas Kanavos[1(✉)], Xenophon Liapakis[3],
Spyros Sioutas[2], and Athanasios Tsakalidis[1]

[1] Computer Engineering and Informatics Department,
University of Patras, Patras, Greece
eldritsas@gmail.com, {chiotie,kanavos,tsak}@ceid.upatras.gr
[2] Department of Informatics, Ionian University, Corfu, Greece
sioutas@ionio.gr
[3] Interamerican, Athina, Greece
liapakisx@interamerican.gr

Abstract. Nowadays, digital data are the most valuable asset of almost every organization. Database management systems are considered as storing systems for efficient retrieval and processing of digital data. However, effective operation, in terms of data access speed and relational database is limited, as its size increases significantly [6]. Bloom filter is a special data structure with finite storage requirements and rapid control of an object membership to a dataset. It is worth mentioning that the Bloom filter structure has been proposed with a view to constructively increase data access in relational databases. Since the characteristics of a Bloom filter are consistent with the requirements of a fast data access structure, we examine the possibility of using it in order to increase the SQL query execution speed in a database. In the context of this research, a database in a RDBMS SQL Server that includes big data tables is implemented and in following the performance enhancement, using Bloom filters, in terms of execution time on different categories of SQL queries, is examined. We experimentally proved the time effectiveness of Bloom filter structure in relational databases when dealing with large scale data.

Keywords: Databases · Bloom filters · RDBMS · SQL queries optimization

1 Introduction

The business data, associated with all of the business activities, are typically stored in relational databases in order to manage them using the SQL language, and more specifically perform SQL queries to the database. The relational databases are particularly effective in their operation. However, their efficiency is limited if they store "big data" with complex correlations [14]. An SQL query can be very expensive in execution cost, and concretely in time and access to

© Springer International Publishing AG 2017
G. Boracchi et al. (Eds.): EANN 2017, CCIS 744, pp. 596–608, 2017.
DOI: 10.1007/978-3-319-65172-9_50

resources, if the execution plan is not optimized. Possible delays in the accomplishment of SQL queries may have impact on application performance using relational databases, thus reducing business performance.

The main way to improve the performance of an SQL query is to reduce the number of required operations/calculations that should be performed during the execution of the corresponding query. However, further reduction of the required commands in an SQL query is not always possible and also requires additional techniques for SQL query performance optimization in a database [5]. In [11], authors investigate this specific problem and recommend the use of IN, EXITS, EQUAL and OPERATOR-TOP along with indexes. Moreover, the bloom filter structure is used in databases such as Google Big Data or Apache HBase in order to decrease searching (in disk) for non-existent records, optimizing in this way the performance of executed SQL queries [3].

The traditional database systems store data in the form of a table with records. Each record corresponds to a different entity object that holds information in a relational table. The relative organization of the databases is effective when there are performing queries on tables with a small number of records. However, as the number of records increases, e.g. hundreds of thousands or millions of records, SQL queries usually search in a much larger number of records in order to locate and access a small number of records or fields [9].

The best way to improve the execution speed of SQL queries in a database is the definition of indexes in fields, which are part of the search criteria of an SQL query. When indexes are not set in a database, then the database management system operates as a reader trying to find a word in a book by reading the entire book. By integrating an index term at the back of a book, the reader can complete the procedure much more quickly. The benefit of using indexes when searching records in a table becomes greater as the number of table entries increases[1]. The role of indexes, in a database is to direct access records according to the search criteria of the SQL query. However, when a table in a database contains millions of records, despite the use of indexes, then the identification of records that meet the search criteria, requires to access thousands of records of the relational table[2]. Therefore, in order to improve the efficiency of execution speed of relational SQL queries, the, in advance, exclusion of a significant number of records that do not meet the search criteria, would be particularly useful. To this purpose, the implementation of Bloom filter structure is suggested; this structure is based on records of the tables and it is further used for the exclusion of records that do not meet the criteria of relevant SQL queries.

The purpose of this research is to examine to what extent the structure of Bloom filter tables in relational databases can affect the performance of data access queries for data tables with millions of records. To achieve the aim of this survey, our contributions lie in the following bullets: (i) implementation of Bloom filter to a relational database, (ii) experimental evaluation of queries with

---

[1] http://odetocode.com/articles/237.aspx.

[2] http://dataidol.com/tonyrogerson/2013/05/09/reducing-sql-server-io-and-access-times-using-Bloom-filters-part-2-basics-of-the-method-in-sql-server.

or without the support of Bloom filter and table recording of execution time of queries and (iii) graphic visualization of results to show Bloom filter effectiveness (in terms of integration time) in executing SQL queries on tables with millions of records.

The rest of the paper is organized as follows: in Sect. 2 the properties and basic components of Bloom filters are introduced. In Sect. 3, Relational Databases and SQL framework is presented. Moreover, Sect. 4 presents the evaluation experiments conducted and the results gathered. Ultimately, Sect. 5 presents conclusions, constraints and draws directions for future work.

2   Bloom Filters Background

2.1   Bloom Filter Elements

The Bloom filter structure, devised by Burton Howard Bloom in 1970, is used for rapid check whether an element is present in a data set or not [1]. It also permits checking if an item certainly does not belong to it. Although the Bloom filters allow false positive responses, the space savings they offer outweigh any downside [8]. A Bloom filter is composed of two parts: a set of k hash functions and a bit vector. The number of hash functions and the length of bit vector are chosen according to the expected number of keys to be added to the Bloom filter and the level of acceptable error rate per case[3].

A number of important components need to be properly defined in order for a bloom filter to operate correctly. These parameters are briefly and comprehensively described in the following paragraphs.

2.1.1   Hash Functions
A hash function takes as input data of any length and returns as output an ID smaller in length and fixed in size, which can be employed with the aim to identify elements[4].

The main features that a hash function should have, are the following:

- Return the same value at each iteration with the same data input.
- Quick execution.
- Generate output with uniform distribution in the potential range it produces.

Some of the most popular algorithms for implementing hash functions are: $SHA1$ and $MD5$. These functions differ in safety level and hash value calculation speed. Also, some algorithms homogeneously distribute the values generated by the hash function, but they are impractical. In each case, the selected hash function should satisfy the application requirements.

As for the hash functions number, the bigger this number is, then the hash values are generated in a slower way and the binary vector fills in a faster way.

---

[3] https://www.perl.com/pub/2004/04/08/bloom_filters.html.
[4] https://blog.medium.com/what-are-bloom-filters-1ec2a50c68ff.

However, this decision increases the incorrect predictions on the existence of an object in a dataset[5]. The optimal number of hash functions derives from the following formula in [8]:

$$k = \frac{m}{n} \ln(2) \tag{1}$$

where m is the binary vector length and n the number of inserted keys in bloom filter. When selecting the number of hash functions to be used, we also calculate the probability of false positive predictions. The previous step is repeated until we get an accepted value for the probability index of false positive responses [4].

2.1.2   Binary Vectors Length

The length of the binary values of a Bloom filter vector affects the pointer value of false positive responses of the filter. The greater the length of the binary vector values, the lower the probability of false positive responses. Conversely, as the length of the vector is shrinked, the relative probability is increased. Generally, a Bloom filter is considered complete when 50% values of bits in the array are equal to 1. At this point, further addition of objects will result in the increase of false positive responses rate [10].

2.1.3   Key Insertion

We initialize a Bloom filter by setting the values of binary vector equal to 0. To insert a key into a Bloom filter, the relevant k hash functions are originally performed and positions of the binary vector, which corresponds to hash values, change from 0 to 1. If the relevant bit is already set to 1, then the value of the relevant bit does not further alter(See footnote 3). Each bit of the vector can simultaneously encode multiple keys, which makes the Bloom filter compact as shown in Fig. 1 [2].

The overlapping values do not permit a key removal from the filter, since it is not known whether the relevant bits are not activated by other key values. The only way to remove a key from a Bloom filter is to rebuild the filter from scratch, thus not incorporating the key to be removed from the Bloom filter. For checking the possibility for a key in the Bloom filter to be present, the following

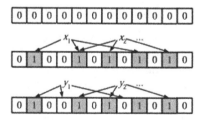

Fig. 1. Bloom filter overview

---

[5] https://llimllib.github.io/bloomfilter-tutorial.

procedure is applied. Initially, the hash functions are applied to the search key, and then we check the relevant bits generated by the hash functions to be all activated. Concretely, if at least one of the bits is disabled, it is certain that the corresponding key is not included in the filter. If all bits are turned on, then we know that with high probability, the key has been introduced.

2.2   Space-Time Advantages and Constraints

The implementation of a Bloom filter is relatively simple in comparison with other relevant search structures. In addition, the use of a Bloom filter ensures the fast membership checking of a value and in following absolute reliability of the non-existence of an object in it (no false negatives) [13]. Concerning the time required for adding a new item or to control whether a point belongs to a set of data, it is independent of the number of elements in the filter[6]. More to the point, a strong advantage of Bloom filters is the storage space saving in comparison with other data structures such as sets, hash tables, or binary search trees.

The insertion of an element into a Bloom filter is an irreversible process[7]. The size of data in a Bloom filter must be known in advance for determining the vector length and the number of hash functions. However, the number of objects that will be imported into a Bloom filter are not always known in advance. It is theoretically possible to define an arbitrarily large size, but it would be wasteful in terms of space and would overturn the main advantage on the Bloom filter, which is storage economy. Alternatively, a Dynamic Bloom filter structure could be adopted, which, however, is not always possible. There is a variant of the Bloom filter, called Scalable Bloom filter, which dynamically adjusts its size for different number of objects. The use of a relative Bloom filter could alleviate some of its shortcomings. A Bloom filter cannot produce the list of items imported, but it can only check whether an item has been introduced in a dataset. Finally, the Bloom filter cannot be used for answering questions about the properties of the objects.

3   Bloom Filters and RDBMS

3.1   Relational Database Management Systems

The relational database management systems have been a common choice for storing information in databases used for a wide range of data such as financial, logistic information, personal data, and other forms of information, since 1980. The relational databases have replaced other forms such as hierarchical or network databases, as they are easier in understanding and their use is convenient. The main advantage of relational data model is that it allows the user to make query-in data access command, without the need to define access paths to

---

[6] https://prakhar.me/articles/bloom-filters-for-dummies.

[7] http://bugra.github.io/work/notes/2016-06-05/a-gentle-introduction-to-bloom-filter.

stored data or other additional details [7]. Furthermore, the relational databases keep their data in form of tables. Each table consists of records, called tuples, and each record is uniquely identified by a field, i.e. primary key, which has a unique value. Each panel is usually connected to at least another database table in relation to the form: (i) one-by-one, (ii) one-to-many, or (iii) many-to-many.

These relationships grant users unlimited ways of data access and dynamic combination amongst them from different tables. Nowadays, the market provides more than one hundred RDBMS systems and the most popular of them are the following: (i) Oracle, (ii) MySQL, (iii) Microsoft SQL Server, (iv) PostgreSQL, (v) DB2 and (vi) Microsoft Access (DB-Engines 2016), etc.[8].

The SQL language is used for user communication with a relational database [12]. An SQL query demands no knowledge of the internal operation of database or the relevant data storage system [15]. According to ANSI (American National Institute Standards) standards, the SQL is a standard language for relational database management systems. Moreover, the SQL language is used in order to query a database for the management of such data and also for the data update or retrieval from a database. Some examples of relational databases that use SQL are: Oracle, Sybase, Microsoft SQL Server, Access and Ingres.

The most important commands of SQL query language are[9]: SELECT, UPDATE, DELETE, INSERT INTO, CREATE DATABASE, ALTER DATA-BASE, CREATE TABLE, ALTER TABLE, DROP TABLE, CREATE INDEX, DROP INDEX.

The SQL commands are classified into the following basic types:

- **Query Language with key command:** where the Select command for accessing information from the database tables is used.
- **Data Manipulation Language with key commands:** (i) Insert-introduction of new records, (ii) Update-modify records, and (iii) Delete-delete records.
- **Data Objects Definition with key commands:** (i) Create Table, and (ii) Alter Table.
- **Safety Control of Database with key commands:** (i) Grand, Revoke for user rights management to database objects, and (ii) Commit, Rollback for transactions management.

3.2 Queries Language-SQL

3.2.1 Membership Queries

The command SQL IN controls whether an expression matches any value from a list of values. Furthermore, it is used in order to prevent multiple use of the OR command in SELECT, INSERT, UPDATE or DELETE queries[10]. Besides checking should an expression belong to a set of values registered directly to a

---

[8] http://db-engines.com/en/ranking/relational+dbms.

[9] http://www.w3schools.com/sql/sql_syntax.asp.

[10] https://www.techonthenet.com/sql/in.php.

relevant query SQL, it may also check if an expression is part of a set of values from other tables.

3.2.2   Join Queries

The union queries, which combine values from two or more data tables based on a JOIN criterion, usually concern relationships between relevant tables. More to the point, JOIN queries are distinguished in four categories:

1. **Inner Join:** returns the values from Table A and Table B that satisfy the joining criteria.
2. **Left Join:** returns all the values from Table A and the values of the Table B meeting the joining criteria.
3. **Right Join:** returns all the values from Table B and the values of the Table A that meet the joining criteria.
4. **Outer Join:** returns all the values from Table A and Table B regardless if they satisfy the relevant criteria combination.

3.2.3   Exist Queries

The existence control queries are used in conjunction with a secondary query. It is considered that the control condition is satisfied when the secondary query returns at least one relevant registration. The verification can be used in terms of the following queries: SELECT, INSERT, UPDATE or DELETE[11].

3.2.4   Top Queries

The command TOP limits the number of records that a query will return, that is to a specified number of rows or a specified percentage of records from the 2016 version of SQL Server[12]. When the command TOP is used in combination with the ORDER BY command, then the first N records are returned according to the sorting arrangement provided by the ORDER BY command. Otherwise, N unsorted records are returned.

In addition, the TOP command specifies the number of records returned by a SELECT statement or affected by a plethora of command statements, such as INSERT, UPDATE, JOIN, or DELETE. The TOP SELECT command can be particularly useful in large tables with thousands of records. The access and choice of a large number of records can adversely affect the performance execution of a query.

3.3   Indexes Table

Indexes are auxiliary structures in a relational database management system with the aim of increasing data access performance to the database. Relevant helping structures are created in one or more fields (columns) of a table or a

---

[11] https://www.techonthenet.com/sql/exists.php.

[12] https://docs.microsoft.com/en-us/sql/t-sql/queries/top-transact-sql.

database. Moreover, an index provides a quick way to search data based on the values in the specific fields that are part of the index.

For example, if an index on the primary key of a table is created, and then a series of data based on the values of the corresponding fields is found, then the SQL Server finds the value of the index field first and in following it uses the relevant index so as to quickly locate the whole relevant table entries. In this way, without the index marker field, it would require a scan of the entire table line by line, directly influencing the performance of the relevant query execution[13].

Furthermore, an index consists of a set of pages that are organized into B-tree data structure. The relevant structure is hierarchical, comprising a root node at the top of the tree and the leaf nodes at the lower level, as illustrated in the above Fig. 2. When a query, including a search criterion, is executed, then the query starts delving into relevant records from the root node and navigates through intermediate nodes, which are the leaf nodes of the B-tree structure. After locating the relevant leaf node, the query will access the interrelative record either directly (in the case of clustered index), or through a pointer to the relevant data record (if it is a non clustered index).

A table in an SQL Server database can have at most one clustered index and more than one non clustered index, depending on the version of SQL Server that is used.

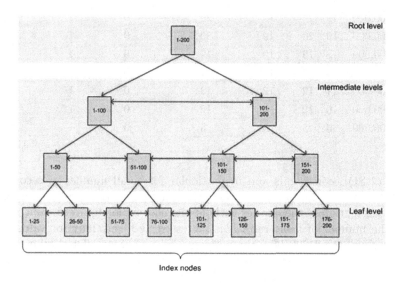

Fig. 2. B-tree overview

---

[13] https://www.simple-talk.com/sql/learn-sql-server/sql-server-index-basics.

4   Experimental Evaluation in SQL Server

In this section, the results of the experiments conducted in the context of this research in order to evaluate the use of Bloom filters, are presented. We perform a series of common SQL database queries with and without the support of the Bloom filter and graphically present the resulted time performance of executed SQL queries. The SQL queries utilized are the following: *In*, *Inner Join*, *Left Join*, *Right Join*, *Exists* and *Top*.

The following Tables 1 and 2 as well as Fig. 3 show the execution times of the questions described previously. In particular in the corresponding Tables, the results are shown with and without the use of Bloom filter by introducing the label *BF* as the relative number of records is changed.

Table 1. SQL queries execution time results vs data size

Execution time in seconds						
Data	In	In BF	Inner Join	Inner Join BF	Left Join	Left Join BF
10.000.000	44	24	44	24	1	1
9.000.000	38	24	41	26	1	1
8.000.000	26	21	26	24	1	1
7.000.000	19	21	19	20	0	1
6.000.000	19	20	19	20	0	1
5.000.000	18	19	19	19	0	1
4.000.000	13	14	13	12	0	1
3.000.000	11	12	12	13	0	1
2.000.000	10	12	11	12	0	1
1.000.000	3	3	3	3	0	1

For all SQL commands and in particular for small number of records, we observed that the adoption of Bloom filter structure overloaded the system and thus, the execution of the queries without the use of Bloom filter is much faster.

As the number of table records is increased and especially for values more than (or equal to) 8,000,000, the performance advantage offered by employing the Bloom filter structure increases significantly and the difference in speed execution of queries is obvious as it also rises exponentially.

It is important to consider the fact that during the repetitive execution of the same queries, we observed the same runtime, but sometimes there was a gap of about two seconds between results. In these cases, we decided to take the average values in the relevant cases.

Table 2. SQL queries execution time results vs data size

Execution time in seconds						
Data size	Right join	Right join BF	Exists	Exists BF	Top	Top BF
10.000.000	44	42	43	23	26	6
9.000.000	37	27	38	24	18	6
8.000.000	26	26	25	25	7	6
7.000.000	19	20	18	20	7	5
6.000.000	19	20	19	20	7	5
5.000.000	19	19	19	19	5	5
4.000.000	13	12	12	12	5	5
3.000.000	12	13	11	12	5	5
2.000.000	11	12	12	12	3	3
1.000.000	3	3	2	3	1	2

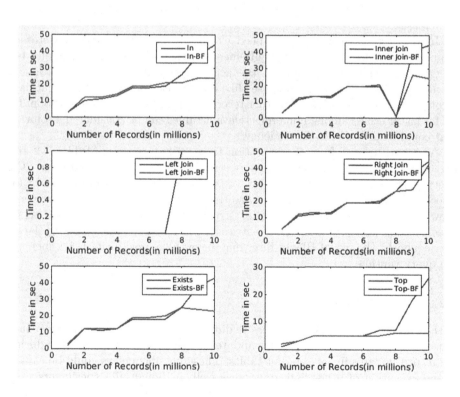

Fig. 3. Queries execution time vs records size

5   Conclusions

5.1   Research Conclusions

The large response times of SQL queries in relational databases affects not only the users, but also other applications that may run on the same computer or the network itself hosting the relevant database. The Bloom filter, capacity wise, is an effective solution and it has been used in numerous applications in the past, especially when immediate control of an object membership was required.

The relevant experiments suggest that the inclusion of Bloom filter structure in an SQL Server database (with large number of records - 10, 000, 000 records) that may increase its data access performance. The optimization of query execution time to a database, using Bloom structure, allows users to quickly extract the needed information and increase the efficiency of relevant database. The Bloom structure in a relational database acts as a filter that removes from the join, membership or existence control queries the need to access-process records that do not meet the criteria of relevant questions. The potential profit from this restriction involved in accessing-searching records through an SQL query highly depends on the false positive records that the control of the Bloom filter returns. This relative number is limited as the length of the binary values of Bloom filter is increased.

In this topic, an acceptable speed execution as well as balanced storage requirements, according to the requirements of each database instance and the user requirements, which concern the access speed of the relevant database, should be chosen. Especially, in cases like a database containing historical data records with no probability of further record updates, the adoption of Bloom filter for faster access among numerous relevant tables can be considered a solution that could lead to increased efficiency.

As it can be seen from the execution times of SQL queries (Tables 1, 2 and Fig. 3), the benefit of the, in advance, restriction of records involved in an SQL query is greater than including all records of data tables and indexes used for direct access to them. It should be noted that as the experimental measurements show, the application of Bloom filter structure in a database deserves to be selected only when the number of entries in the relevant tables are very large. Consequently, the use of Bloom filter may have the opposite effect, i.e. increase the query runtime.

5.2   Research Constraints

In the evaluation of the Bloom filter, we did not take into account possible delays caused by maintenance and regular updates of the Bloom filter structure during the record updates in the relevant tables. These possible delays could be caused in the execution of other SQL queries as well. Although all experiments were performed on the same machine, the ones with Bloom filter that were performed at different times, may have been affected (in performance) by possible processes running in the background. These relevant deviations do not directly affect the performance comparison among the same queries with or without the use of Bloom filter, but mostly between different SQL commands used.

5.3   Future Extensions

A promising and useful step would be to investigate the applicability of Bloom filters in other relational database management systems (like Oracle, SysBase, MySQL) with the aim of generalizing previous conclusions drawn from experimentation on the SQL Server relational database management system. Also, a possible review of the actual performance of database operations with millions of records used to store application data, will allow more reliable conclusions about the use of Bloom filter structure in relational databases. Thus possible delays of the system during the application operation can be taken into account.

References

1. Blustein, J., El-Maazawi, A.: Bloom Filters: A Tutorial, Analysis and Survey (2002). https://www.cs.dal.ca/research/techreports/cs-2002-10
2. Broder, A.Z., Mitzenmacher, M.: Survey: network applications of bloom filters: a survey. Internet Math. 1(4), 485–509 (2003)
3. Chang, F., Dean, J., Ghemawat, S., Hsieh, W.C., Wallach, D.A., Burrows, M., Chandra, T., Fikes, A., Gruber, R.E.: Bigtable: a distributed storage system for structured data. ACM Trans. Comput. Syst. (TOCS) 26(2), 4:1–4:26 (2008)
4. Christensen, K.J., Roginsky, A., Jimeno, M.: A new analysis of the false-positive rate of a bloom filter. Inf. Process. Lett. 110(21), 944–949 (2010)
5. Gupta, M.K., Chandra, P.: An empirical evaluation of like operator in Oracle. BVICAM's Int. J. Inf. Technol. 3(2) (2011). https://scholar.google.co.in/citations?view_op=view_citation&citation_for_view=jab7XG0AAAAJ:MXK_kJrjxJIC
6. Khan, M., Khan, M.N.A.: Exploring query optimization techniques in relational databases. Int. J. Database Theory Appl. 6(3), 11–20 (2013)
7. Kim, W.: On optimizing an SQL-like nested query. ACM Trans. Database Syst. (TODS) 7(3), 443–469 (1982)
8. Kirsch, A., Mitzenmacher, M.: Less hashing, same performance: building a better bloom filter. In: Azar, Y., Erlebach, T. (eds.) ESA 2006. LNCS, vol. 4168, pp. 456–467. Springer, Heidelberg (2006). doi:10.1007/11841036_42
9. Larson, P., Clinciu, C., Hanson, E.N., Oks, A., Price, S.L., Rangarajan, S., Surna, A., Zhou, Q.: SQL server column store indexes. In: ACM SIGMOD International Conference on Management of Data, pp. 1177–1184 (2011)
10. Lyons, M.J., Brooks, D.M.: The design of a bloom filter hardware accelerator for ultra low power systems. In: International Symposium on Low Power Electronics and Design, pp. 371–376 (2009)
11. Oktavia, T., Sujarwo, S.: Evaluation of sub query performance in SQL server. In: EPJ Web of Conferences, vol. 68 (2014)
12. Ramakrishnan, R., Donjerkovic, D., Ranganathan, A., Beyer, K.S., Krishnaprasad, M.: SRQL: sorted relational query language. In: International Conference on Scientific and Statistical Database Management (SSDBM), pp. 84–95 (1998)
13. Roozenburg, J.: A literature survey on bloom filters. Res. Assign. Comput. Sci. (2005). https://scholar.google.gr/scholar?cluster=8721372506493746175

14. Vicknair, C., Macias, M., Zhao, Z., Nan, X., Chen, Y., Wilkins, D.: A comparison of a graph database and a relational database: a data provenance perspective. In: ACM Southeast Regional Conference, p. 42 (2010)
15. Winand, M.: SQL Performance Explained: Everything Developers Need to Know about SQL Performance (2012). http://sql-performance-explained.com/img/9783950307825_preview.pdf

A Random Forest Method to Detect Parkinson's Disease via Gait Analysis

Koray Açıcı[1], Çağatay Berke Erdaş[1(✉)], Tunç Aşuroğlu[1],
Münire Kılınç Toprak[2], Hamit Erdem[3], and Hasan Oğul[1]

[1] Department of Computer Engineering, Başkent University, Ankara, Turkey
{korayacici,berdas,tuncasuroglu,hogul}@baskent.edu.tr
[2] Department of Neurology, Başkent University, Ankara, Turkey
mkilinc@baskent.edu.tr
[3] Department of Electrical and Electronics Engineering, Başkent University,
Ankara, Turkey
herdem@baskent.edu.tr

Abstract. Remote care and telemonitoring have become essential component of current geriatric medicine. Intelligent use of wireless sensors is a major issue in relevant computational studies to realize these concepts in practice. While there has been a growing interest in recognizing daily activities of patients through wearable sensors, the efforts towards utilizing the streaming data from these sensors for clinical practices are limited. Here, we present a practical application of clinical data mining from wearable sensors with a particular objective of diagnosing Parkinson's Disease from gait analysis through a sets of ground reaction force (GRF) sensors worn under the foots. We introduce a supervised learning method based on Random Forests that analyze the multi-sensor data to classify the person wearing these sensors. We offer to extract a set of time-domain and frequency-domain features that would be effective in distinguishing normal and diseased people from their gait signals. The experimental results on a benchmark dataset have shown that proposed method can significantly outperform the previous methods reported in the literature.

Keywords: Parkinson's Disease · Gait analysis · Remote care · Wireless sensor

1 Introduction

Parkinson's disease occurs due to the deterioration of cells that generate dopamine in the brain. After the Alzheimer disease, it is the second most common geriatric disease. Since enough dopamine could not be produced in the brain, some disorders in motor activities such as tremor, rigidity, slowness of movement, postural disturbance, walking abnormalities and balance problems arise. While the exact causes of the disease are unknown, a method has not been known yet for cure. Therefore, the medical treatment to Parkinson's Disease (PD) patients usually involves the elimination or reduction of disease symptoms. While constant monitoring of the severity of symptoms is essential to this end, a typical symptom evaluation requires a weighty effort and time. PD

© Springer International Publishing AG 2017
G. Boracchi et al. (Eds.): EANN 2017, CCIS 744, pp. 609–619, 2017.
DOI: 10.1007/978-3-319-65172-9_51

patients are usually hospitalized during a one-day visit to the hospital, which should be held every 3 months. However, these evaluations are not routinely practiced regularly because of the functional deficiencies of PD patients and the lack of accessibility to specialists everywhere. If the regular assessments are not made, this can lead to incorrect drug amounts and timing. Another problem is that the grading of specialist doctors for the disease is not sufficiently objective. Besides assessment problems, these PD patients are known to have problems such as falling down, walking lock, and when they are alone, this situation creates vital problems. Therefore, it is important to remotely detect the physical activity of PD patients to treat the patients more adequately and to increase their quality of life.

In recent years, there have been various studies on remote monitoring and evaluation of PD patients more objectively. The frequent use of wearable sensors on activity recognition has led to the consideration of these interventions in health practices including Parkinson's Disease [1–4]. These practices have been used both in assessing disease symptoms and in informing PD patients about the current state of health [5–9]. Although the idea of the use of variable sensors, specific for Parkinson's disease, arises in the 1990s, it has not been widespread until recently due to the size, speed and similar constraints of the respective hardware units [10, 11]. Miniaturized sensor technologies with improvements in wireless communication, signal processing and pattern recognition have come into play in recent years [12–15]. An accelerometer mounted on the knee has been experimentally demonstrated in several studies where the values recorded from the sensor are significantly different when compared to healthy and PD patients marches and therefore can be used for long-term gait analysis of accelerometer sensors [16, 17]. In conducted studies, motion measurements made through accelerometer sensors have been shown to be associated with Parkinson's symptoms and the severity of these symptoms. LeMoyne et al. observed that PD patients and healthy people had significant differences in the frequency domain according to the accelerometer signals they were using with their iPhone devices, which they were mounting on the wrists [18].

In this study, we address a specific problem that seeks solution to classify a person as Parkinson's or not through the signals acquired from a set of ground reaction force (GRF) sensors worn under the foot. We introduce an ensemble classification framework that uses a number of time-domain and frequency-domain features extracted from pre-processed sensor signals.

2   Previous Works

The previous works on Parkinson's disease classification are presented in this section. In all of the related works explained here, the same benchmark dataset was used for performance evaluation. The dataset contains gait signals which measure GRF in Newtons instead of accelerometer signals.

Lee and Lim have presented a classification method by using neural network with weighted fuzzy membership functions [19]. The authors have used vertical ground reaction force for all the sixteen sensors. Then they applied wavelet transform to extract features. Finally, a neural network classifier is proposed to classify inputs. They divided their model into three stages. In the first stage, they obtained values by sub-tracting two signals each of which represent eight sensors' outputs from one foot. In the second stage, they used only the left foot. Using eight sensors, they subtracted maximum and minimum records to obtain a value. In the final stage, firstly they applied stage two to both feet and then they obtained two signals. After that, stage one was applied to those two signals. They achieved the best classification accuracy as 77.33% by applying stage three, outperforming other stages.

Daliri has presented a new technique for identification of Parkinson's disease, using Gait force measurements [20]. He used Short Time Fourier Transform (STFT) to extract features. Histogram of the extracted features was evaluated. For classification he proposed a SVM kernel which computes chi-square distance between feature histograms. The results of classification accuracy were compared according to kernel type. Chi-Square Distance kernel has the best classification accuracy as 91.20%, outperforming linear, quadratic, polynomial, and radial basis function.

Jane et al. have presented a Q-backpropagated time delay neural network for classification of gait signals [21]. They evaluated performance of this proposed system for Hoehn Yahr scale values of PD patients. They achieved 91.49% classification accuracy. They outperformed other training algorithms like Levenberg-marquardt, Gradient Descent, Gradient Descent with Momentum and Scaled Conjugate Gradient in a neural network.

Ertugrul et al. transformed data into Local binary patterns (LBP) and then performed feature extraction on LBP histograms [22]. Statistical features extracted from each LBP histogram to classify Parkinson's disease. After feature extraction, they performed correlation-based feature selection to reduce the dimensionality of data set. They conducted several experiments by using different LBP neighborhood values and assessed classification accuracy with several learning methods. Using all features as inputs and Multilayer Perceptron (MLP) as classifier they obtained the best accuracy result, 88.88%. On the other hand, using selected features as inputs and Random Forest as classifier they obtained the best accuracy result, 87.58%.

3   Methods

3.1   Feature Extraction

Feature extraction is widely used in the recognition of a signal which is collected from human movement, and moreover it is capable of classifying people who have Parkinson's disease or not. To extract features, we exploited time domain features owing to the fact that they are basis of Signal characteristics and productive. Furthermore, we decided to use frequency domain features due to the fact that they are in alignment with time domain features.

Time Domain Features. In order to use for classification, 16 features were extracted from each sensor; Mathematical/Statistical attributes such as mean, median, minimum value and its index, maximum value and its index, range, RMS, IQR, MAD, skewness, kurtosis, entropy, energy, power and harm mean.

Frequency Domain Features. In addition to the time domain features, widely used frequency domain features in the literature have been extracted [23]. In doing so, the signal is transformed from the time domain to the frequency domain by using Fast Fourier Transform (FFT). After this step, the following 7 attributes were extracted from each signal: mean, maximum value, minimum value, normalized value, energy, phase, band power.

3.2 Feature Selection

In order to create more fertile feature subset and get rid of relatively unnecessary attributes in term of classification, we use feature selection. According to Attal et al., feature selection is one of the most effective ways in classification process to diminish cost and enhance the performance of classification [24]. While we're doing the process, we preferred the classifier based feature selection. The classifier based feature selection takes all the features which are extracted from time domain and frequency domain, and manufacture the most noteworthy features as output. For this dataset, the feature selection algorithm that is mentioned above found only one attribute called 'median' from sensor number one. Median feature is one of the eminent envelope metrics that are extracted from time domain.

When we sort the samples in data set in ascending order, Median is the value that separates the data from the center. If a data diversity distribution is not symmetric and shows skewness, then the median is a more appropriate centric position measure than the arithmetic average. Non-symmetry occurs for the sorted data values either with the smallest values or with the largest values farther away than the others. These unexpected small and big values are called as outliers. If the data distribution includes asymmetric outliers, Median becomes a robust centroid position measure relative to arithmetic mean. In order to calculate median, once the data numbers are sorted from small to large, the sequence number of the median value in data set, with k observations, is found as (1).

$$Median\ Position = \frac{k+1}{2} \tag{1}$$

If the number of observations is odd, the sequence number of the median will be an integer, and the median is directly located. Otherwise the sequence number of median will be a float. In this case, the arithmetic mean of the two values around this number is median.

3.3 Classification

We use a bagging-based ensemble learning method to classify the input feature vectors. Bagging is meta-learning technique that includes training several classifiers on different partitions of the training data and employing a majority voting on the results of all those classifiers to report the final decision for a given query. The sub classifier is usually a decision tree that is built from current training partition.

Random Forest is an extended version of bagging algorithm with randomness property injected [25]. Random Forest splits every node to branches using the best one from randomly selected variables on every node instead of using the best branch through all variables. Each dataset is generated from the original one by replacement. Random feature selection is used to extend trees. Extended trees are not pruned [25, 26]. This strategy makes Random Forest's accuracy. Random Forest is also very fast and robust to overfitting that takes its source from decision trees' nature. It can consist of as many trees as you want.

Random Forest algorithm requires two parameters to set before starting. The first one is the number of variables (m) used in every node for determining the best split. The second one is the number of trees (N) to be extended.

Random Forest uses CART (Classification and Regression Tree) algorithm to produce trees [25]. On every node, branches are generated according to GINI index of CART algorithm criteria. GINI index measures class homogeneity. If GINI index increases, class heterogeneity will increase also. If GINI index decreases, class homogeneity will increase. A branch is said to be victorious, if GINI index of a sub node is smaller than GINI index of an upper node. When GINI index reaches zero, in other words when there will be one class on every leaf node, the branching process stops [27]. When all trees (N trees) are generated, class of the input is determined according to all tree predictions as shown in Fig. 1.

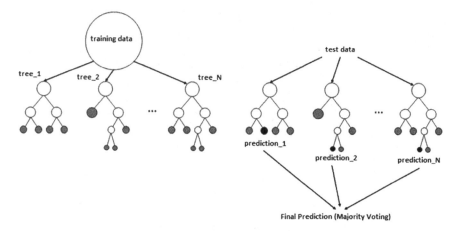

Fig. 1. Random forest classification algorithm [28].

4 Results

4.1 Dataset

The dataset [29–32] we used consists of measurements that collected by Physionet [33] and the Laboratory for Gait & Neurodynamics, Movement Disorders Unit of the TelAviv Sourasky Medical Center. It includes gait measurements of PD patients and control groups. Average age of PD patients is 66.3 and 63% of PD group population consists of men. For the control group these numbers are 66.3 as age and 55% for men population distribution. There are 93 subjects who have PD and 73 subjects who are in control group. The dataset consists of vertical ground reaction force records. These records were taken during the walks of the people on a flat ground for 2 min. There are 8 sensors under each foot, and they measure force in Newtons as a function of time. Output of all 16 sensors have been digitalized and recorded, sampling rate is 100 samples per second. Also there are 2 signal values which are total measurements of 8 sensors in each foot. By using this information, the force value can be examined as a function of time and the timing criteria for each foot can be subtracted from the center of pressure value (for example: Swing Time). Swing time is a period of time when a foot is removed from the floor and then touched again. Figure 2, shows an example swing time graph for PD and Control group subject.

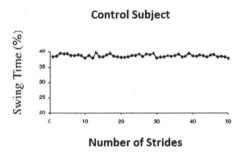

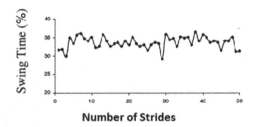

Fig. 2. Swing time graph for PD and control subjects [33].

The position of two sensors is as follows; assuming person standing up with two legs parallel to each other, the point of origin is exactly in the middle of the legs and the person is facing towards the positive side of the Y axis. X and Y coordinates of each sensor is given in Table 1. L letter stands for left foot and R stands for right foot.

The X and Y values are artificial coordinates that can be used to determine the location of the sensors. This artificial coordinate system has been developed to facilitate the calculation of the COP (center of pressure) position of each foot. The dataset that is used in this study is normalized before feature extraction phase. Normalization process

Table 1. Arrangement of sensors under feet.

Sensor name	X coordinate	Y coordinate
L1	−500	−800
L2	−700	−400
L3	−300	−400
L4	−700	0
L5	−300	0
L6	−700	400
L7	−300	400
L8	−500	−800
R1	500	−800
R2	700	−400
R3	300	−400
R4	700	0
R5	300	0
R6	700	400
R7	300	400
R8	500	800

includes making mean value zero and standard deviation one. All classification processes that are used, include 10 fold cross validation.

4.2   Evaluation

To evaluate the performance of classification we used five different metrics. They are accuracy, Kappa statistic, sensitivity, specificity and F-Meaure.

Kappa Statistic: The Kappa statistic is used to measure the agreement between predicted and observed categorizations of a dataset (2).

$$K = \frac{P(A) - P(E)}{1 - P(E)} \tag{2}$$

where P(A) is the observed agreement and P(E) is the expected agreement by chance.

TP, FP, TN and FN stand for number of true positives, false positives, true negatives and false negatives respectively. TP refers to correctly classified person with PD. FP refers to person classified as PD but actually belongs to control group. TN refers to correctly classified person without PD. FN refers to person classified as non-PD but actually belongs to PD class.

Accuracy: It is the success rate of the classification process (3).

$$Accuracy = \frac{TN + TP}{TN + TP + FN + FP} \tag{3}$$

Sensitivity: Refers to the ratio of correctly classified people with Parkinson's disease to all people with PD (4).

$$Sensitivity = \frac{TP}{TP + FN} \tag{4}$$

Specificity: Refers to the ratio of correctly classified people without Parkinson's disease to all people belong to control group (5).

$$Specificity = \frac{TN}{TN + FP} \tag{5}$$

F-Measure: It can be accepted as a weighted average of the specificity and Sensitivity, where an F-measure ranges between 0 (which is the worst value) and 1 (which is best value) (6).

$$F\text{-}Measure = \frac{2 * TP}{2 * TP + FP + FN} \tag{6}$$

4.3   Empirical Results

As shown at Table 2, classification performance results which obtained from three different feature subsets are indicated. The subsets mentioned above, are all features, all median features and selected median feature respectively. All features subset includes 368 features which were extracted from both of the time and frequency domains for all sixteen sensors. All median features subset includes only sixteen features each of which is median of each sensor. Selected median feature subset includes one and only median feature which belongs to sensor number one. When using the all features' subset, classification accuracy rate has reached the peak point with 98.04%. When feature selection called Classifiersubseteval was applied to all features subset, it was come up that median feature from sensor number one was the most effective feature so that the classification accuracy while using mentioned selected median feature became 74.51%. The use of all the median features from all sensors, is predicted to increase classification performance rate rather than using only selected median feature from sensor number one. As a result, all median features subset has a classification performance rate 94.12% outperforming selected median feature.

Table 2. Results of feature subsets by using 10 fold CV with random forest classifier

10 fold CV-RF	Accuracy	Kappa statistic	Sensitivity	Specificity	F-measure
All features	98.04%	0.953	0.991	0.957	0.980
All median features	94.12%	0.858	0.972	0.870	0.941
Selected median feature	74.51%	0.390	0.822	0.565	0.744

The results of previous methods on the same dataset are shown in Table 3. As seen in this table, our study outperforms the previous studies in terms of accuracy, sensitivity, and specificity. Our approach can lead to a 7% increase in prediction accuracy compared with the best previous work.

Table 3. Classification performance according to different methods on the same dataset.

	Accuracy (%)	Sensitivity	Specificity
Lee and Lim [19]	74.32	0.816	0.738
Daliri [20]	89.92	0.917	0.912
Jane et al. [21]	91.53	–	–
Ertugrul et al. [22]	88.89	0.889	0.822
This study (all features)	**98.04**	0.991	0.957

5 Conclusion

Since the PD patients usually suffer from losing basic motor abilities, their remote monitoring is a recent challenge to serve a satisfactory home care and clinical support. Use of wearable sensors is a promising idea in general for telemedicine applications. In this study, we present an application of this idea which aims at making directly clinical inference from streaming sensor data to support medical decision making for PD. While current computational research on the topic has been focused on developing systems and algorithms for recognizing the activities of daily living using motion sensors, we here present a solution that directly diagnosis the PD via GRF sensors. Our solution is based on the use of an ensemble classifier, namely the Random Forest, on a set of time-domain and frequency-domain features to improve the prediction accuracy. Main contribution of the study is twofold. First, the accuracy of proposed method has improved upon the state-of-the-art for PD classification via GRF sensors. Second, we have shown that conventional time-domain and frequency-domain features extracted from GRF signals can present a satisfactorily descriptive space for diagnosis of PD.

Acknowledgement. This study was supported by the Scientific and Technological Research Council of Turkey (TUBITAK) under the Project 115E451.

References

1. Bonato, P.: Advances in wearable technology and its medical applications. In: 32nd Conference of the IEEE EMBS, Buenos Aires, Argentina, pp. 2021–2024 (2010)
2. Aminian, K., Najafi, B.: Capturing human motion using body-fixed sensors: outdoor measurement and clinical applications. Comput. Animation Virtual Worlds **15**(2), 79–94 (2004)
3. Baga, D., Fotiadis, D.I., Konitsiotis, S., Maziewski, P., Greenlaw, R., Chaloglou, D., Arrendondo, M.T., Robledo, M.G., Pastor, M.A.: PERFORM: personalised disease management for chronic neurodegenerative diseases: the Parkinson's disease and amyotrophic lateral sclerosis cases. In: eChallenges e-2009 Conference, Istanbul (2009)

4. Godfrey, A., Conway, R., Meagher, D., Olaighin, G.: Direct measurement of human movement by accelerometry. Med. Eng. Phys. **30**(10), 1364–1386 (2008)

5. Zhao, W., Adolph, A.L., Puyau, M.R., Vohra, F.A., Butte, N.F., Zakeri, I.F.: Support vector machines classifiers of physical activities in preschoolers. Physiol. Rep. **1**(1), e00006 (2013)

6. Patel, S., Lorincz, K., Hughes, R., Huggins, N., Growdon, J., Standaert, D., Akay, M., Dy, J., Welsh, M., Bonato, P.: Monitoring motor fluctuations in patients with Parkinson's disease using wearable sensors. IEEE Trans. Inf. Technol. Biomed. **13**(6), 864–873 (2009)

7. Shima, K., Tsuji, T., Kan, E., Kandori, A., Yokoe, M., Sakoda, S.: Measurement and evaluation of finger tapping movements using magnetic sensors. In: IEEE EMBS Conference, pp. 5628–5631. IEEE, Vancouver (2008)

8. Lee, S.W., Mase, K.: Activity and location recognitions using wearable sensors. IEEE Pervasive Comput. **1**(3), 24–32 (2002)

9. Greenlaw, R., et al.: PERFORM: building and mining electronic records of neurological patients being monitored in the home. In: Dössel, O., Schlegel, W.C. (eds.) World Congress on Medical Physics and Biomedical Engineering 2009, IFMBE Proceedings, vol. 25/9, pp. 533–535. Springer, Heidelberg (2009). doi:10.1007/978-3-642-03889-1_143

10. Ghika, J., Wiegner, A.W., Fang, J.J., Davies, L., Young, R., Growdown, J.H.: Portable system for quantifying motor abnormalities in Parkinson's disease. IEEE Trans. Biomed. Eng. **40**(3), 276–283 (1993)

11. Spieker, S., Jentgens, C., Boose, A., Dichgans, J.: Reliability, specificity and sensitivity of long-term tremor recordings. Electroencephalogr. Clin. Neurophysiol. **97**(6), 326–331 (1995)

12. Bonato, P., Sherrill, D.M., Standaert, D.G., Salles, S.S., Akay, M.: Data mining techniques to detect motor fluctuations in Parkinson's disease. In: Engineering in Medicine and Biology Society, pp. 4766–4769. IEEE, San Francisco (2004)

13. Patel, S., Sherrill, D., Hughes, R., Hester, T., Huggins, N., Lie-Nemeth, T., Standaert, D., Bonato, P.: Analysis of the severity of dyskinesia in patients with Parkinson's disease via wearable sensors. In: International Workshop on Wearable and Implantable Body Sensor Networks. IEEE, Cambridge (2006)

14. Patel, S., Chen, B.R., Buckley, T., Rednic, R., McClure, D., Tarsy, D., Shih, L., Dy, J., Welsh, M., Bonato, P.: Home monitoring of patients with Parkinson's disease via wearable technology and a web-based application. In: Engineering in Medicine and Biology Society (EMBC), pp. 4411–4414. IEEE, Buenos Aires (2010)

15. Keijsers, N.L., Horstink, M.W., van Hilten, J.J., Hoff, J.I., Gielen, C.C.: Detection and assessment of the severity of levodopa-induced dyskinesia in patients with Parkinson's disease by neural networks. Mov. Disord. **15**(6), 1104–1111 (2000)

16. Moore, S.T., MacDougall, H.G., Gracies, J.M., Cohen, H.S., Ondo, W.G.: Long-term monitoring of gait in Parkinson's disease. Gait Posture **6**(2), 200–207 (2007)

17. Cancela, J., Pastorino, M., Arredondo, M.T., Nikita, K.S., Villagra, F., Pastor, M.A.: Feasibility study of a wearable system based on a wireless body area network for gait assessment in Parkinson's disease patients. Sensors **14**(3), 4618–4633 (2014)

18. LeMoyne, R., Mastroianni, T., Cozza, M., Coroian, C., Grundfest, W.: Implementation of an iPhone for characterizing Parkinson's disease tremor through a wireless accelerometer application. In: Engineering in Medicine and Biology Society (EMBC), pp. 4954–4958. IEEE, Buenos Aires (2010)

19. Lee, S.H., Lim, J.S.: Parkinson's disease classification using gait characteristics and wavelet-based feature extraction. Expert Syst. Appl. **39**(8), 7338–7344 (2012)

20. Daliri, M.R.: Chi-square distance kernel of the gaits for the diagnosis of Parkinson's disease. Biomed. Sig. Process. Control **8**(1), 66–70 (2013)

21. Jane, Y.N., Nehemiah, H.K., Arputharaj, K.: A Q-backpropagated time delay neural network for diagnosing severity of gait disturbances in Parkinson's disease. J. Biomed. Inform. **60**, 169–176 (2016)

22. Ertugrul, O.F., Kaya, Y., Tekin, R., Almali, M.N.: Detection of Parkinson's disease by shifted one dimensional local binary patterns from gait. Expert Syst. Appl. **56**, 156–163 (2016)

23. Figo, D., Diniz, P.C., Ferreira, D.R., Cardoso, J.M.P.: Preprocessing techniques for context recognition from accelerometer data. Pers. Ubiquit. Comput. **14**(7), 645–662 (2010)

24. Attal, F., Mohammed, S., Dedabrishvili, M., Chamroukhi, F., Oukhellou, L., Amirat, Y.: Physical human activity recognition using wearable sensors. Sensors **15**, 31314–31338 (2015)

25. Breiman, L.: Random forests. Mach. Learn. **55**(1), 5–32 (2001)

26. Archer, K.J., Kimes, R.V.: Empirical characterization of random forest variable importance measures. Comput. Stat. Data Anal. **52**(4), 2249–2260 (2008)

27. Watts, J.D., Powell, S.L., Lawrence, R.L., Hilker, T.: Improved classification of conservation tillage adoption using high temporal and synthetic satellite imagery. Remote Sens. Environ. **115**, 66–75 (2011)

28. Mennitt, D., Sherrill, K., Fristrup, K.: A geospatial model of ambient sound pressure levels in the contiguous United States. J. Acoust. Soc. Am. **135**(5), 2746–2764 (2014)

29. Frenkel-Toledo, S., Giladi, N., Peretz, C., Herman, T., Gruendlinger, L., Hausdorff, J.M.: Effect of gait speed on gait rhythmicity in Parkinson's disease: variability of stride time and swing time respond differently. J. Neuroeng. Rehabil. **2**(23) (2005). doi:10.1186/1743-0003-2-23. https://www.ncbi.nlm.nih.gov/pmc/articles/PMC1188069/

30. Frenkel-Toledo, S., Giladi, N., Peretz, C., Herman, T., Gruendlinger, L., Hausdorff, J.M.: Treadmill walking as a pacemaker to improve gait rhythm and stability in Parkinson's disease. Mov. Disord. **20**(9), 1109–1114 (2005)

31. Hausdorff, J.M., Lowenthal, J., Herman, T., Gruendlinger, L., Peretz, C., Giladi, N.: Rhythmic auditory stimulation modulates gait variability in Parkinson's disease. Eur. J. Neurosci. **26**(8), 2369–2375 (2007)

32. Yogev, G., Giladi, N., Peretz, C., Springer, S., Simon, E.S., Hausdorff, J.M.: Dual tasking, gait rhythmicity and Parkinson's disease: which aspects of gait are attention demanding? Eur. J. Neurosci. **22**(5), 1248–1256 (2005)

33. Physionet Gait in Parkinson's Disease. https://physionet.org/pn3/gaitpdb/. Accessed 01 May 2017

Efficient Computation of Palindromes
in Sequences with Uncertainties

Mai Alzamel[(✉)], Jia Gao, Costas S. Iliopoulos, Chang Liu, and Solon P. Pissis

Department of Informatics, King's College London, London, UK
{mai.alzamel,jia.gao,costas.iliopoulos,chang.2.liu,
solon.pissis}@kcl.ac.uk

Abstract. In this work, we consider a special type of uncertain sequence called weighted string. In a *weighted string* every position contains a subset of the alphabet and every letter of the alphabet is associated with a probability of occurrence such that the sum of probabilities at each position equals 1. Usually a *cumulative weight threshold* $1/z$ is specified, and one considers only strings that match the weighted string with probability at least $1/z$. We provide an $\mathcal{O}(nz)$-time and $\mathcal{O}(nz)$-space off-line algorithm, where n is the length of the weighted string and $1/z$ is the given threshold, to compute a smallest maximal palindromic factorization of a weighted string. This factorization has applications in hairpin structure prediction in a set of closely-related DNA or RNA sequences. Along the way, we provide an $\mathcal{O}(nz)$-time and $\mathcal{O}(nz)$-space off-line algorithm to compute maximal palindromes in weighted strings.

1 Introduction

A palindrome is a sequence that reads the same from left to right and from right to left. Detection of palindromic factors in texts is a classical and well-studied problem in algorithms on strings and combinatorics on words with a lot of variants arising out of different practical scenarios. In molecular biology, for instance, palindromic sequences are extensively studied: they are often distributed around promoters, introns, and untranslated regions, playing important roles in gene regulation and other cell processes (see e.g. [2]). In particular these are strings of the form $s\bar{s}^R$, also known as complemented palindromes, occurring in single-stranded DNA or, more commonly, in RNA, where s is a string and $\bar{s}^R$ is the reverse complement of s. In DNA, C-G are complements and A-T are complements; in RNA, C-G are complements and A-U are complements.

A string $x = x[0]x[1] \ldots x[n-1]$ is said to have an initial palindrome of length k if its prefix of length k is a palindrome. Manacher first discovered an on-line algorithm that finds all initial palindromes in a string [19]. Later Apostolico et al. observed that the algorithm given by Manacher is able to find all maximal

---

M. Alzamel and C.S. Iliopoulos—Partially supported by the Onassis Foundation.

G. Boracchi et al. (Eds.): EANN 2017, CCIS 744, pp. 620–629, 2017.
DOI: 10.1007/978-3-319-65172-9_52

palindromic factors in the string in $\mathcal{O}(n)$ time [4]. Gusfield gave an off-line linear-time algorithm to find all maximal palindromes in a string and also discussed the relation between biological sequences and gapped palindromes (i.e. strings of the form $sv\bar{s}^R$ where the complemented palindromes are separated by v) [15].

The problem that gained significant attention recently is the factorization of a string x of length n into a sequence of palindromes. We say that $x_1, x_2, \ldots, x_\ell$ is a (maximal) palindromic factorization of string x, if every x_i is a (maximal) palindrome, $x = x_1 x_2 \ldots x_\ell$, and ℓ is minimal. In biological applications we need to factorize a sequence into palindromes in order to identify *hairpins*, patterns that occur in single-stranded DNA or, more commonly, in RNA. Alatabbi et al. gave an off-line $\mathcal{O}(n)$-time algorithm for finding a maximal palindromic factorization of x [1]. Fici et al. presented an on-line $\mathcal{O}(n \log n)$-time algorithm for computing a palindromic factorization of x [14]; a similar algorithm was presented by Tomohiro et al. [16]. In addition, Rubinchik and Shur [21] devised an $\mathcal{O}(n)$-sized data structure that helps locating palindromes in x; they also showed how it can be used to compute a palindromic factorization of x in $\mathcal{O}(n \log n)$ time.

In this work, we consider a special type of uncertain sequence called weighted string (also known as position weight matrix or PWM). In a *weighted string* X every position contains a subset of the alphabet and every letter of the alphabet is associated with a probability of occurrence such that the sum of probabilities at each position equals 1. For example, we write $X = $ a[(a,0.5),(b,0.5)] $\ldots$ to denote that the probability of occurrence of a at the first position is 1 while at the second one is $1/2$, and so on. X thus represents many different strings, each with probability of occurrence equal to the *product* of probabilities of its letters at subsequent positions of X. A great deal of research has been conducted on weighted strings for indexing [7,17], for alignments [3,12], for pattern matching [8,9,18], and for finding regularities [5,10].

Our Problem. Muhire et al. [20] showed how a set of virus species can be clustered using multiple sequence alignment (MSA) to obtain subsets of viruses that have common hairpin structure (see Fig. 1(a)). A more compact representation of an MSA can be trivially obtained using weighted strings (see Fig. 1(b)). The non-trivial computational problem thus arising is how to factorize a weighted string in a sequence of palindromes.

Our Contribution. Usually a *cumulative weight threshold* $1/z$ is specified, and one considers only strings that match the weighted string with probability at least $1/z$. In this paper, we generalize Alatabbi et al's solution for standard strings [1] to compute a maximal palindromic factorization of a weighted string. In particular, we provide an $\mathcal{O}(nz)$-time and $\mathcal{O}(nz)$-space off-line algorithm, where n is the length of the weighted string and $1/z$ is the given threshold. Along the way, we provide an $\mathcal{O}(nz)$-time and $\mathcal{O}(nz)$-space off-line algorithm for computing maximal palindromes in weighted strings.

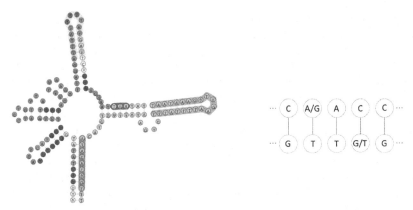

(a) Hairpins common to *Malvastrum yellow vein virus*, *Cotton leaf curl Multan* virus isolate, and *Bhendi yellow vein India* virus; figure taken from [20].

(b) Hairpin represented as a weighted string: C[(A,0.5),(G,0.5)]ACC (top) and GTT[(G,0.5),(T,0.5)] G (bottom).

Fig. 1. Hairpins that are common to a set of closely-related sequences can be represented compactly as weighted strings.

2   Preliminaries

Let $x = x[0]x[1]\ldots x[n-1]$ be a *string* of length $|x| = n$ over a finite ordered alphabet Σ of size $|\Sigma| = \sigma = \mathcal{O}(1)$. For two positions i and j on x, we denote by $x[i\mathinner{\ldotp\ldotp}j] = x[i]\ldots x[j]$ the *factor* (sometimes called *substring*) of x that starts at position i and ends at position j. We recall that a *prefix* of x is a factor that starts at position 0 ($x[0\mathinner{\ldotp\ldotp}j]$) and a *suffix* is a factor that ends at position $n-1$ ($x[i\mathinner{\ldotp\ldotp}n-1]$). We denote the *reversal* of x by string x^R, i.e. $x^R = x[n-1]x[n-2]\ldots x[0]$. The *empty string* (denoted by ε) is the unique string over Σ of length 0. The *concatenation* of two strings x and y is the string of the letters of x followed by the letters of y. It is denoted by $x.y$ or, more simply, by xy.

Let y be a string of length m with $0 < m \leq n$. We say that there exists an *occurrence* of y in x, or, more simply, that y *occurs in* x, when y is a factor of x. Every occurrence of y can be characterised by a starting position in x. Thus we say that y occurs at the *starting position* i in x when $y = x[i\mathinner{\ldotp\ldotp}i+m-1]$.

A string w is said to be a *palindrome* if and only if $w = w^R$. If factor $x[i\mathinner{\ldotp\ldotp}j]$, $0 \leq i \leq j \leq n-1$, of string $x[0\mathinner{\ldotp\ldotp}n-1]$ is a palindrome, then $\frac{i+j}{2}$ is the *center* of $x[i\mathinner{\ldotp\ldotp}j]$ in x and $\frac{j-i+1}{2}$ is the *radius* of $x[i\mathinner{\ldotp\ldotp}j]$. Moreover, $x[i\mathinner{\ldotp\ldotp}j]$ is called a *palindromic factor*. It is said to be a *maximal palindrome* if there is no other palindrome in x with center $\frac{i+j}{2}$ and larger radius. Hence x has exactly $2n-1$ maximal palindromes. A maximal palindrome w can be encoded as a pair (c,r), where c is the center of w and r is the radius of w. By $\mathcal{MP}(x)$, we denote the set of center-distinct maximal palindromes of string x. The sequence $x_1, x_2, \ldots, x_\ell$ of ℓ non-empty strings is a *(maximal) palindromic factorization* of a string x if all strings x_i are (maximal) palindromes, $x = x_1 x_2 \ldots x_\ell$, and ℓ is minimal.

Definition 1. *A weighted string X on an alphabet Σ is a finite sequence of n sets. Every $X[i]$, for all $0 \le i < n$, is a set of ordered pairs $(s_j, \pi_i(s_j))$, where $s_j \in \Sigma$ and $\pi_i(s_j)$ is the probability of having letter s_j at position i. Formally, $X[i] = \{(s_j, \pi_i(s_j)) \mid s_j \ne s_l \text{ for } j \ne l, \text{ and } \Sigma\pi_i(s_j) = 1\}$. A letter s_j occurs at position i of X if and only if the occurrence probability of letter s_j at position $i, \pi_i(s_j)$, is greater than 0.*

Note that for clarity we use upper case letters for weighted strings, e.g. X, and lower case letters, e.g. x, for standard strings.

Definition 2. *A string u of length m is a* factor *of a weighted string X if and only if it occurs at starting position i with cumulative probability $\prod_{j=0}^{m-1} \pi_{i+j}(u[j]) > 0$. Given a cumulative weight threshold $1/z \in (0, 1]$, we say factor u is z-valid, if it occurs at position i with cumulative probability $\prod_{j=0}^{m-1} \pi_{i+j}(u[j]) \ge 1/z$.*

Example 1. Let $X = \texttt{ab[(a,0.5),(b,0.5)][(a,0.5),(b,0.5)]bab}$ and $1/z = 1/8$. String $u = \texttt{baaba}$ is a z-valid factor of X since u occurs at position 1 with cumulative probability $1/4 \ge 1/z = 1/8$.

Definition 3. *Given a cumulative weight threshold $1/z \in (0, 1]$, a weighted string X of length m is a z-palindrome if and only if there exists at least one z-valid factor u of X of length m which is a palindrome.*

Example 2. Let $X = \texttt{a[(a,0.5),(b,0.5)]bab[(a,0.4),(b,0.6)]a}$ of length $m = 7$ and $1/z = 1/8$. $u = \texttt{abbabba}$ is a z-valid factor of X of length 7 and u is a palindrome. Hence we say X is a z-palindrome.

If the weighted string $X[i \mathinner{.\,.} j]$ is a z-palindrome, we analogously define the number $\frac{i+j}{2}$ as the center of $X[i \mathinner{.\,.} j]$ in X and $\frac{j-i+1}{2}$ as the radius of $X[i \mathinner{.\,.} j]$.

Definition 4. *Let X be a weighted string of length n, $1/z \in (0, 1]$ a cumulative weight threshold, and $X[i \mathinner{.\,.} j]$, where $0 \le i \le j \le n - 1$, a z-palindrome. Then $X[i \mathinner{.\,.} j]$ is a maximal z-palindrome if there is no other z-palindrome in X with center $\frac{i+j}{2}$ and larger radius.*

A maximal z-palindrome can thus also be encoded as a pair (c, r). We study the following computational problem.

SMALLEST MAXIMAL z-PALINDROMIC FACTORIZATION
Input: A weighted string X of length n and a cumulative weight threshold $1/z \in (0, 1]$
Output: $X_1, X_2, \ldots, X_\ell$, if any, such that $X = X_1 X_2 \ldots X_\ell$, X_i, for all $1 \le i \le \ell$, is a maximal z-palindrome, and ℓ is minimal.

We call this output sequence $X_1, X_2, \ldots, X_\ell$, i.e. when ℓ is minimal, a *smallest maximal z-palindromic factorization* of X.

The *suffix tree* $\mathcal{T}(x)$ of a non-empty string x of length n is a compact trie representing all suffixes of x. The nodes of the trie which become nodes of the suffix tree are called *explicit* nodes, while the other nodes are called *implicit*. Each edge of the suffix tree can be viewed as an upward maximal path of implicit nodes starting with an explicit node. Moreover, each node belongs to a unique path of that kind. Thus, each node of the trie can be represented in the suffix tree by the edge it belongs to and an index within the corresponding path. We let $\mathcal{L}(v)$ denote the *path-label* of a node v, i.e., the concatenation of the edge labels along the path from the root to v. We say that v is path-labelled $\mathcal{L}(v)$. Additionally, $\mathcal{D}(v) = |\mathcal{L}(v)|$ is used to denote the *string-depth* of node v. Node v is a *terminal* node if its path-label is a suffix of x, that is, $\mathcal{L}(v) = x[i \mathinner{.\,.} n - 1]$ for some $0 \leq i < n$; here v is also labelled with index i. It should be clear that each factor of x is uniquely represented by either an explicit or an implicit node of $\mathcal{T}(x)$. The *suffix-link* of a node v with path-label $\mathcal{L}(v) = \alpha w$ is a pointer to the node with path-label w, where $\alpha \in \Sigma$ is a single letter and w is a string. The suffix-link of v is defined if v is an explicit node of $\mathcal{T}(x)$, different from the root. In standard suffix tree implementations, we assume that each node of the suffix tree is able to access its parent. Once $\mathcal{T}(x)$ is constructed, it can be traversed in a depth-first manner to compute $\mathcal{D}(v)$ for each node v. The suffix tree of a string of length n can be computed in time and space $\mathcal{O}(n)$ [13]. It can also be preprocessed in time and space $\mathcal{O}(n)$ so that *lowest common ancestor* (LCA) queries for any pair of explicit nodes can be answered in $\mathcal{O}(1)$ time per query [11].

Fact 1 [15]. *Given a string x, $\mathcal{MP}(x)$ can be computed in time $\mathcal{O}(|x|)$.*

3 $\mathcal{O}(nz)$-Time and $\mathcal{O}(nz)$-Space Algorithm

In this section, we present an algorithm to compute a smallest maximal z-palindromic factorization of a given weighted string X of length n for a given cumulative threshold $1/z \in (0, 1]$. Our algorithm follows the one of Alatabbi et al. for computing a smallest maximal palindromic factorization of standard strings [1] with some crucial modifications.

Why the Algorithm of Alatabbi et al. Cannot be Applied for Weighted Strings. Odd-length maximal palindromes centered at position i of a standard string x can be computed by finding the longest common prefix of suffixes $x[i \mathinner{.\,.} n - 1]$ and $x^R[n - i - 1 \mathinner{.\,.} n - 1]$. The longest common prefix of two suffixes can be found in $\mathcal{O}(1)$ time after $\mathcal{O}(n)$-time pre-processing of the suffix tree of $x\#x^R\$$, where $\#, \$ \notin \Sigma$, using LCA queries; using a similar computation, we can find all even-length maximal palindromes (see [15] for the details).

The length of the longest common z-valid prefix of any two suffixes of our weighted string X can be computed in time $\mathcal{O}(z)$ after $\mathcal{O}(nz)$-time pre-processing using the suffix-tree-based Weighted Index (WI) of [7] (inspect also Fig. 2). However, this does not guarantee that the two corresponding common z-valid prefixes shall form a maximal z-palindrome: the two prefixes are z-valid by definition of the WI but their concatenation that forms a palindrome *may not* be z-valid because its occurrence probability drops below $1/z$.

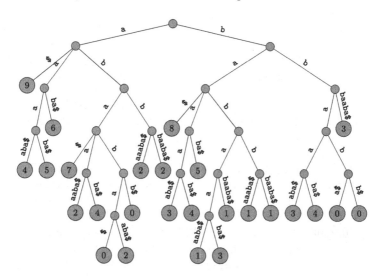

Fig. 2. The WI for X and $1/z$ shown in Example 3 (labels of edges to terminal nodes are appended with a letter $\$ \notin \Sigma$ for convenience).

We hence proceed as follows. By $\mathcal{MP}(X, z)$, we denote the set of center-distinct maximal z-palindromes of our weighted string X. Recall that we can represent a z-palindrome with center c and radius r by (c, r). For each position of X we define the *heaviest letter* as the letter with the maximum probability (breaking ties arbitrarily). We consider the string obtained from X by choosing at each position the heaviest letter. We call this the *heavy string* of X.

We define a collection $\mathcal{Z}_X$ of $\lfloor z \rfloor$ *special-weighted strings* of X, denoted by $\mathcal{Z}_k$, $0 \le k < \lfloor z \rfloor$. Each $\mathcal{Z}_k$ is of length n and it has the following properties. Each position j in $\mathcal{Z}_k$ contains at most one letter with positive probability and it corresponds to position j in X. If f is a z-valid factor occurring at position j of X, then f occurs at position j in some of the $\mathcal{Z}_k$'s. The combinatorial observation telling us that this is possible is due to Barton et al. [6]. For clarity of presentation we write $\mathcal{Z}_k$'s as standard strings.

Example 3. Given the weighted string

$$X = [(\texttt{a},0.5),(\texttt{b},0.5)]\texttt{bab}[(\texttt{a},0.5),(\texttt{b},0.5)][(\texttt{a},0.5),(\texttt{b},0.5)]\texttt{aaba}$$

and a cumulative weight threshold $1/z = 1/4$, we have:

$$\mathcal{Z}_X = \{\mathcal{Z}_0, \mathcal{Z}_1, \mathcal{Z}_2, \mathcal{Z}_3\} = \{\texttt{ababaaaaba}, \texttt{ababbaaaba}, \texttt{bbababaaba}, \texttt{bbabbbaaba}\}.$$

Lemma 2 ([6]). *Given a weighted string X of length n and a cumulative weight threshold $1/z \in (0, 1]$, the $\lfloor z \rfloor$ special-weighted strings of X can be constructed in time and space $\mathcal{O}(nz)$.*

Fact 3. *Given a weighted string X of length n and a cumulative weight threshold $1/z \in (0, 1]$, we have that $\mathcal{MP}(X, z) \subseteq \mathcal{MP}(\mathcal{Z}_0, z) \cup \mathcal{MP}(\mathcal{Z}_1, z) \cup \ldots \cup \mathcal{MP}(\mathcal{Z}_{\lfloor z \rfloor - 1}, z)$.*

Proof. Suppose $U = X[i..j]$ is a maximal z-palindrome of center $c = \frac{i+j}{2}$ and radius $r = \frac{j-i+1}{2}$. By definition of U there must exist a z-valid palindromic factor u of U of radius r. Therefore, by definition of the special-weighted strings of X, u must be a z-valid factor of some $\mathcal{Z}_k$ and thus $(c, r) \in \mathcal{MP}(\mathcal{Z}_k, z)$. □

There are two steps for the correct computation of $\mathcal{MP}(X, z)$. First, we compute the set $\mathcal{A}_k$ of all maximal palindromes of the heavy string of $\mathcal{Z}_k$, for all $0 \le k < \lfloor z \rfloor$, using Fact 1. We then need to adjust the radius of each reported palindrome for $\mathcal{Z}_k$ to ensure that it is z-valid in X (the center should not change). To achieve this, we compute an array $\mathcal{R}_k$, for each $\mathcal{Z}_k$, such that $\mathcal{R}_k[2c]$ stores the radius of the longest factor at center c in $\mathcal{Z}_k$ which is a z-valid factor of X at center c, e.g. $\mathcal{R}_k[2c] = \frac{j-i+1}{2}$, $c = (i + j)/2$, if $\mathcal{Z}_k[i..j]$ is a z-valid factor of X centered at c, and $\mathcal{Z}_k[i - 1..j + 1]$ is not a z-valid factor of X. By Fact 3, we cannot guarantee that all (c, r) in $\mathcal{MP}(\mathcal{Z}_k, z)$ are necessarily in $\mathcal{MP}(X, z)$. Hence, the second step is to compute $\mathcal{MP}(X, z)$ from $\mathcal{MP}(\mathcal{Z}_k, z)$ by taking the maximum radius per center and filtering out everything else.

Lemma 4. *Given a weighted string X of length n, a cumulative weight threshold $1/z \in (0, 1]$, and the special-weighted strings $\mathcal{Z}_X$ of X, each $\mathcal{R}_k$, $0 \le k < \lfloor z \rfloor$, can be computed in time $\mathcal{O}(n)$.*

Proof. By $< i, c, j >$, where $0 \le i \le c \le j \le n - 1$, we denote a factor of $\mathcal{Z}_k$ that has starting position i, ending position j and center $c = (i + j)/2$. We further denote the occurrence probability of $< i, c, j >$ in $\mathcal{Z}_k$ by $\Pi_{<i,c,j>} = \prod_{q=i}^{j} \pi_q(\mathcal{Z}_k[q])$. A factor $< i, c, j >$ of $\mathcal{Z}_k$ is called a special maximal z-valid factor of $\mathcal{Z}_k$ if $\Pi_{<i,c,j>} \ge 1/z$ and $\Pi_{<i-1,c,j+1>} < 1/z$.

For each $\mathcal{Z}_k$, we compute $\mathcal{R}_k$ from left to right. If we have $\Pi_{<0,0,0>} \ge 1/z$, we set $\mathcal{R}_k[0] = \frac{1}{2}$. If not, we go to the next position until we find a valid letter, say at position ℓ; then we have $\mathcal{R}_k[0] = \cdots = \mathcal{R}_k[2\ell - 1] = 0$ and $\mathcal{R}_k[2\ell] = \frac{1}{2}$. Note that this corresponds to the first special maximal z-valid factor. Suppose we have a special maximal z-valid factor $< i, c, j >$ and $\mathcal{R}_k[2c] = \frac{j-i+1}{2}$, we show how to compute $\mathcal{R}_k[2c + 1]$, which is the length of the special maximal z-valid factor at center $c' = \frac{2c+1}{2}$. We add the letter after $< i, c, j >$, so we have $< i, c', j+1 >$. We compute $\Pi_{<i,c',j+1>}$, which is simply $\Pi_{<i,c,j>} \times \pi_{j+1}(\mathcal{Z}_k[j+1])$. If $\Pi_{<i,c',j+1>} \ge 1/z$, the special maximal z-valid factor at center c' should be $< i, c', j + 1 >$ and $\mathcal{R}_k[2c + 1] = \mathcal{R}_k[2c] + \frac{1}{2} = \frac{j-i+2}{2}$. Factor $< i - 1, c', j + 2 >$ cannot be z-valid, since if $\Pi_{<i-1,c',j+2>} \ge 1/z$, we must have $\Pi_{<i-1,c,j+1>} \ge \Pi_{<i-1,c',j+2>} \ge 1/z$, which gives a longest special maximal z-valid at center c, namely $< i-1, c, j+1 >$, a contradiction. For $\Pi_{<i,c',j+1>} < 1/z$, the special maximal z-valid factor at center c' is $< i+1, c', j >$ since it always holds that $\Pi_{<i+1,c',j>} \ge \Pi_{<i,c,j>} \ge 1/z$. Therefore $\mathcal{R}_k[2c + 1] = \mathcal{R}_k[2c] - \frac{1}{2} = \frac{j-i}{2}$.

Each center needs only to be considered once and there exist $2n - 1$ distinct centers in each $\mathcal{Z}_k$. Therefore each $\mathcal{R}_k$ can be computed in $\mathcal{O}(n)$ time. □

Fact 5 (Trivial). *Let $x[i..j]$ be a palindrome of string x with center c and let u, $|u| < j - i + 1$, be a factor of x with center c. Then u is also a palindrome.*

After computing $\mathcal{A}_k$ and $\mathcal{R}_k$, we perform the following check for each palindrome $(c, r) \in \mathcal{A}_k$. If $r > \mathcal{R}_k[2c]$, the palindrome with radius r is not z-valid but the factor with radius $\mathcal{R}_k[2c]$ is z-valid and maximal (by definition) and palindromic (by Fact 5); if $r \leq \mathcal{R}_k[2c]$, the palindrome with radius r_i must be z-valid and it is maximal. Therefore we set $(c, r) \in \mathcal{MP}(\mathcal{Z}_k, z)$, such that $r = \min\{r, \mathcal{R}_k[2c]\}$, $0 \leq 2c \leq 2n - 2$, and $r \geq 1/2$.

To go from $\mathcal{MP}(\mathcal{Z}_k, z)$ to $\mathcal{MP}(X, z)$ we need to take the maximum radius for each center. Therefore for each center $c/2$, $0 \leq c \leq 2n - 2$, we set $(c/2, r) \in \mathcal{MP}(X, z)$, such that $r = \max\{r_k | (c/2, r_k) \in \mathcal{MP}(\mathcal{Z}_k, z), 0 \leq k < \lfloor z \rfloor\}$. We thus arrive at the first result of this article.

Theorem 6. *Given a weighted string X of length n and a cumulative weight threshold $1/z \in (0, 1]$, all maximal z-palindromes in X can be computed in time and space $\mathcal{O}(nz)$.*

After the computation of $\mathcal{MP}(X, z)$, we are in a position to apply the algorithm by Alatabbi et al. [1] to find a smallest maximal z-palindromic factorization. We define a list $\mathcal{F}$ such that $\mathcal{F}[i]$, $0 \leq i \leq n - 1$, stores the set of the lengths of all maximal z-palindromes ending at position i in X. We also define a list $\mathcal{U}$ such that $\mathcal{U}[i]$, $0 \leq i \leq n - 1$, stores the set of positions j, such that $j + 1$ is the starting position of a maximal z-palindrome in X and i is the ending position of this z-palindrome. Thus for a given $\mathcal{F}[i] = \{\ell_0, \ell_1, \ldots, \ell_q\}$, we have that $\mathcal{U}[i] = \{i - \ell_0, i - \ell_1, \ldots, i - \ell_q\}$. Note that $\mathcal{U}[i]$ can contain a "-1" element if there exists a maximal z-palindrome starting at position 0 and ending at position i. Note that the number of elements in $\mathcal{MP}(X, z)$ is at most $2n - 1$, and, hence, $\mathcal{F}$ and $\mathcal{U}$ can contain at most $2n - 1$ elements. The lists $\mathcal{F}$ and $\mathcal{U}$ can be computed trivially from $\mathcal{MP}(X, z)$.

Finally, we define a directed graph $\mathcal{G}_X = (\mathcal{V}, \mathcal{E})$, where $\mathcal{V} = \{i \mid -1 \leq i \leq n - 1\}$ and $\mathcal{E} = \{(i, j) \mid j \in \mathcal{U}[i]\}$. Note that (i, j) is a directed edge from i to j. We do a breath first search on $\mathcal{G}_X$ assuming the vertex $n - 1$ as the source and identify the shortest path from $n - 1$ to -1, which gives a factorization.

We formally present the above as Algorithm SMPF for computing a smallest maximal z-palindromic factorization and obtain the following result.

Theorem 7. *Given a weighted string X of length n and a cumulative weight threshold $1/z \in (0, 1]$, Algorithm SMPF correctly solves the problem* SMALLEST MAXIMAL z-PALINDROMIC FACTORIZATION *in time and space $\mathcal{O}(nz)$.*

Proof. The correctness follows from Theorem 6 for computing $\mathcal{MP}(X, z)$ and from the correctness of the algorithm in [1] for computing a smallest maximal palindromic factorization.

By Lemma 2, the construction of the special-weighted strings can be done in time and space $\mathcal{O}(nz)$. Computing $\mathcal{A}_k$ and $\mathcal{R}_k$, for all $0 \leq k < \lfloor z \rfloor$, can be done in total time $\mathcal{O}(nz)$ by Fact 1 and Lemma 4, respectively. From there on, computing $\mathcal{MP}(X, z)$ can be done in time $\mathcal{O}(nz)$. The lists $\mathcal{F}$ and $\mathcal{U}$ can be computed in time $\mathcal{O}(n)$ since the size of $\mathcal{MP}(X, z)$ is no more than $2n - 1$. There exist in total $n + 1$ vertices in $\mathcal{G}_X$. The number of edges $|\mathcal{E}|$ depends on $\mathcal{U}$, which contains no more than $2n$ elements; we have $|\mathcal{E}| = \mathcal{O}(n)$. Therefore, the construction of $\mathcal{G}_X$ and the breadth first search can be done in time $\mathcal{O}(|\mathcal{V}|+|\mathcal{E}|) = \mathcal{O}(n)$. The identification of the desired path can also be done easily if we do some simple bookkeeping during the breadth first search. The total running time of Algorithm SMPF is thus $\mathcal{O}(nz)$ and the space required is $\mathcal{O}(nz)$.   □

---

1   **Algorithm** $SMPF(X, n, 1/z)$

2   Construct the set $\mathcal{Z}_X$ of special-weighted strings of X;

3   **foreach** $\mathcal{Z}_k \in \mathcal{Z}_X$ **do**

4       $\mathcal{A}_k \leftarrow$ maximal palindromes of the heavy string of $\mathcal{Z}_k$;

5   Compute $\mathcal{R}_k$ for $\mathcal{Z}_k$;

6       $\mathcal{MP}(\mathcal{Z}_k, z) \leftarrow$ EMPTYLIST();

7   **foreach** $(c, r) \in \mathcal{A}_k$ **do**

8           $r \leftarrow \min\{r, \mathcal{R}_k[2c]\}$;

9   **if** $r \geq \frac{1}{2}$ **then**

10   Insert (c, r) in $\mathcal{MP}(\mathcal{Z}_k, z)$;

11 $\mathcal{MP}(X, z) \leftarrow$ EMPTYLIST();

12 **foreach** $c \in [0, 2n - 2]$ **do**

13      $r \leftarrow \max\{r_k | (c/2, r_k) \in \mathcal{MP}(\mathcal{Z}_k, z), 0 \leq k < \lfloor z \rfloor\}$;

14   Insert $(c/2, r)$ in $\mathcal{MP}(X, z)$;

15 $\mathcal{F} \leftarrow$ EMPTYLIST();

16 $\mathcal{U} \leftarrow$ EMPTYLIST();

17 **foreach** $(c, r) \in \mathcal{MP}(X, z)$ **do**

18      $j \leftarrow \lfloor c + r \rfloor$;

19   Insert $2r$ in $\mathcal{F}[j]$;

20   Insert $j - 2r$ in $\mathcal{U}[j]$;

21 Construct directed graph $\mathcal{G}_X = (\mathcal{V}, \mathcal{E})$, where $\mathcal{V} = \{i \mid -1 \leq i \leq n - 1\}$, $\mathcal{E} = \{(i, j) \mid j \in \mathcal{U}[i]\}$ and (i, j) is a directed edge from i to j;

22 Breadth first search on $\mathcal{G}_X$ assuming the vertex $n - 1$ as the source;

23 Identify the shortest path $P \equiv \langle n - 1 = p_\ell, p_{\ell-1}, \ldots, p_2, p_1, p_0 = -1 \rangle$;

24 **Return** $X[0 .. p_1], X[p_1 + 1 .. p_2], \ldots, X[p_{\ell-1} + 1 .. p_\ell]$;

References

1. Alatabbi, A., Iliopoulos, C.S., Rahman, M.S.: Maximal palindromic factorization. In: PSC, pp. 70–77 (2013)
2. Almirantis, Y., Charalampopoulos, P., Gao, J., Iliopoulos, C.S., Mohamed, M., Pissis, S.P., Polychronopoulos, D.: On avoided words, absent words, and their application to biological sequence analysis. Algorithms Mol. Biol. **12**(1), 5 (2017)
3. Amir, A., Gotthilf, Z., Shalom, B.R.: Weighted LCS. J. Discrete Algorithms **8**(3), 273–281 (2010)

4. Apostolico, A., Breslauer, D., Galil, Z.: Parallel detection of all palindromes in a string. Theoret. Comput. Sci. **141**(1), 163–173 (1995)
5. Barton, C., Iliopoulos, C.S., Pissis, S.P.: Optimal computation of all tandem repeats in a weighted sequence. Algorithms Mol. Biol. **9**(21), 21 (2014)
6. Barton, C., Kociumaka, T., Liu, C., Pissis, S.P., Radoszewski, J.: Indexing Weighted Sequences: Neat and Efficient. CoRR, abs/1704.07625 (2017)
7. Barton, C., Kociumaka, T., Pissis, S.P., Radoszewski, J.: Efficient index for weighted sequences. In: CPM. LIPIcs, vol. 54, pp. 4:1–4:13. Schloss Dagstuhl-Leibniz-Zentrum fuer Informatik (2016)
8. Barton, C., Liu, C., Pissis, S.P.: Linear-time computation of prefix table for weighted strings and applications. Theoret. Comput. Sci. **656**, 160–172 (2016)
9. Barton, C., Liu, C., Pissis, S.P.: On-line pattern matching on uncertain sequences and applications. In: Chan, T.-H.H., Li, M., Wang, L. (eds.) COCOA 2016. LNCS, vol. 10043, pp. 547–562. Springer, Cham (2016). doi:10.1007/978-3-319-48749-6_40
10. Barton, C., Pissis, S.P.: Crochemore's partitioning on weighted strings and applications. Algorithmica (2017). doi:10.1007/s00453-016-0266-0
11. Bender, M.A., Farach-Colton, M.: The LCA problem revisited. In: Gonnet, G.H., Viola, A. (eds.) LATIN 2000. LNCS, vol. 1776, pp. 88–94. Springer, Heidelberg (2000). doi:10.1007/10719839_9
12. Cygan, M., Kubica, M., Radoszewski, J., Rytter, W., Walen, T.: Polynomial-time approximation algorithms for weighted LCS problem. Discrete Appl. Math. **204**, 38–48 (2016)
13. Farach, M.: Optimal suffix tree construction with large alphabets. In: FOCS, pp. 137–143. IEEE Computer Society (1997)
14. Fici, G., Gagie, T., Kärkkäinen, J., Kempa, D.: A subquadratic algorithm for minimum palindromic factorization. J. Discrete Algorithms **28**, 41–48 (2014)
15. Gusfield, D.: Algorithms on Strings, Trees, and Sequences: Computer Science and Computational Biology. Cambridge University Press, New York (1997)
16. Tomohiro, I., Sugimoto, S., Inenaga, S., Bannai, H., Takeda, M.: Computing palindromic factorizations and palindromic covers on-line. In: Kulikov, A.S., Kuznetsov, S.O., Pevzner, P. (eds.) CPM 2014. LNCS, vol. 8486, pp. 150–161. Springer, Cham (2014). doi:10.1007/978-3-319-07566-2_16
17. Iliopoulos, C.S., Makris, C., Panagis, Y., Perdikuri, K., Theodoridis, E., Tsakalidis, A.: The weighted suffix tree: an efficient data structure for handling molecular weighted sequences and its applications. Fundamenta Informaticae **71**(2, 3), 259–277 (2006)
18. Kociumaka, T., Pissis, S.P., Radoszewski, J.: Pattern matching and consensus problems on weighted sequences and profiles. In: ISAAC. LIPIcs, vol. 64, pp. 46:1–46:12. Schloss Dagstuhl-Leibniz-Zentrum fuer Informatik (2016)
19. Manacher, G.: A new linear-time "on-line" algorithm for finding the smallest initial palindrome of a string. J. ACM **22**(3), 346–351 (1975)
20. Muhire, B.M., Golden, M., Murrell, B., Lefeuvre, P., Lett, J.-M., Gray, A., Poon, A.Y.F., Ngandu, N.K., Semegni, Y., Tanov, E.P., et al.: Evidence of pervasive biologically functional secondary structures within the genomes of eukaryotic single-stranded DNA viruses. J. Virol. **88**(4), 1972–1989 (2014)
21. Rubinchik, M., Shur, A.M.: EERTREE: an efficient data structure for processing palindromes in strings. In: Lipták, Z., Smyth, W.F. (eds.) IWOCA 2015. LNCS, vol. 9538, pp. 321–333. Springer, Cham (2016). doi:10.1007/978-3-319-29516-9_27

A Genetic Algorithm for Discovering Linguistic Communities in Spatiosocial Tensors with an Application to Trilingual Luxemburg

Georgios Drakopoulos[3(✉)], Fotini Stathopoulou[4], Giannis Tzimas[2],
Michael Paraskevas[1,2], Phivos Mylonas[3], and Spyros Sioutas[3]

[1] Computer Technology Institute & Press "Diofantus", Patras, Greece
mparask@cti.gr, mparask@teiwest.gr
[2] Technological and Educational Institution of Western Greece,
Antirio Campus, Patras, Greece
tzimas@teimes.gr
[3] Department of Informatics, Ionion University,
Tsirigoti Square 7, 49100 Corfu, Greece
{drakop,fmylonas,sioutas}@ionio.gr
[4] Faculty of English Language and Literature,
University of Athens, Zografou Campus, 15784 Athens, Greece
fstathop@uoa.gr

Abstract. Multimodal social networks are omnipresent in Web 2.0 with virtually every human communication action taking place there. Nonetheless, language remains by far the main premise such communicative acts unfold upon. Thus, it is statutory to discover language communities especially in social data stemming from historically multilingual countries such as Luxemburg. An adjacency tensor is especially suitable for representing such spatiosocial data. However, because of its potentially large size, heuristics should be developed for locating community structure efficiently. Linguistic structure discovery has a plethora of applications including digital marketing and online political campaigns, especially in case of prolonged and intense cross-linguistic contact. This conference paper presents TENSOR-G, a flexible genetic algorithm for approximate tensor clustering along with two alternative fitness functions derived from language variation or diffusion properties. The Kruskal tensor decomposition serves as a benchmark and the results obtained from a set of trilingual Luxemburgian tweets are analyzed with linguistic criteria.

Keywords: Language variation · Multilingual social networks · Cross cultural communication · Geolocation edges · Tensor algebra · Multilayer graphs · Functional analytics · Genetic algorithms · Heuristics · Spatiosocial data

G. Boracchi et al. (Eds.): EANN 2017, CCIS 744, pp. 630–644, 2017.
DOI: 10.1007/978-3-319-65172-9_53

1   Introduction

Diachronically language remains the primary communication vehicle. Thus, not only is by definition a complex social phenomenon, but also a major generator of massive and structured, or often semistructured, humanistic data. The latter is evident in the case of multimodal social media like Twitter, Facebook, and LinkedIn where language tends to be short and informal with non-linguistic elements such as memes and hashtags replacing a sentence or a part thereof. Although text is largely displaced by images in Instagram or by video in Snapchat, aided by the deployment of 4G mobile networks [22], language is far from expunged from the digital sphere.

Multilingualism is present for various social or historical reasons in various countries such as Canada, Switzerland, Belgium, and Luxemburg to name a few. This is strongly reflected in Twitter netizen interaction, as there is no single *lingua franca* in terms of data volume. Instead, in this sense Japanese and Spanish are roughly equivalent with English and Indonesian follows closely [13,27]. This operational frame leads to two major effects. First, language undergoes a ceaseless alteration driven by external factors [7]. This flux of linguistic changes includes syntax, forms, emoticons, abbreviations, phonetic spellings, and neologisms [20]. Second, digital linguistic communities are formed whose compactness depends heavily on social, spatial, and linguistic factors such as overall status, region, and dialect respectively [48]. This form of online activity which consists of social, namely linguistic, and spatial components is termed *spatiosocial*.

Traditionally, Twitter interaction is represented as a *follow* or a *reply* graph or a combination thereof. In the latter case edge weight or valence is determined by the intensity of these two network functions. However, for complex network functionality, such as that between multilingual and geographically dispersed netizens, a sophisticated representation is required. One solution is multilayer graphs, namely graphs whose edges have one and only one label and each vertex pair can be connected with more than one edges as long as these edges have pairwise distinct labels. The algebraic counterpart of a multilayer graph, which is of combinatorial nature, is an *adjacency tensor*, which is the analogous of adjacency matrices for ordinary social graphs.

The primary contribution of this conference paper is TENSOR-G, a genetic algorithm tailored for locating linguistic communities in multilayer graphs containing spatiosocial data and represented as compressed adjacency tensors. Two alternative fitness functions based on sociolinguistic notions, specifically of how resistant to change a language is with social networks acting as an explanatory framework, are outlined as part of TENSOR-G, though other appropriate ones can be selected. The proposed genetic algorithm has been applied to Twitter data from Luxemburg, a trilingual country with rich online activity.

The structure of this conference paper follows. The scientific literature is reviewed in Sect. 2. Fundamental sociolinguistic concepts necessary to develop and evaluate the performance of TENSOR-G are introduced in Sect. 3, whereas the heuristic algorithm itself is outlined in Sect. 4. The conference paper concludes with Sect. 6 where the groundwork for future work is laid. Finally, paper notation is summarized in Table 1. Tensors are printed in capital italics and vectors in small boldface.

Table 1. Paper notation.

Symbol	Meaning
$\triangleq$	Definition or equality by definition
$\{s_1, \ldots, s_n\}$	Set consisting of elements $s_1, \ldots, s_n$
$|S|$ or $|\{s_1, \ldots, s_n\}|$	Set cardinality
$S_1 \setminus S_2$	Asymmetric set difference S_1 minus S_2
τ_{S_1, S_2}	Tanimoto similarity coefficient between sets S_1 and S_2
$\langle x_k \rangle$	Sequence of elements x_k
$(s_1, \ldots, s_n)$	Tuple of elements $s_1, \ldots, s_n$
$\|\mathcal{T}_F\|$	Tensor Frobenius norm
$\circ_n$	Vector outer product along dimension n
$\mathcal{H}(x_1, \ldots, x_n)$	Harmonic mean of $x_1, \ldots, x_n$
$\mathrm{E}[X]$	Mean value of random variable X
$\mathrm{Var}[X]$	Variance of random variable X

2 Previous Work

A mainstay of sociolinguistics is the language evolution process [35,40]. The latter is treated as crucial to understanding language itself [28]. As [43] states some linguistic patterns may only make sense with knowledge from outside the discipline. Change diffusion among communities by correlating linguistic variation with social factors is examined in [39]. The mechanisms of language maintenance and change in the multilingual community of Palau are studied in [36]. The universality of language change as a social phenomenon is treated in [1,12]. Since speech communities and their digital reflections differ widely, linguistic change is expected to be universally modeled [33,34]. Moreover, the latter can be cast in quantitative terms [25]. The propagation and diffusion of this change through a speech community is influenced by the structural and social properties of that community [47]. In other words, the processes of change might be the same, but the social conditions may be different enough to render variation-change-diffusion models non-transferable across languages [38]. Finally, in certain historical cases linguistics are the focus of sociopolitical dispute as suggested for instance in [30] which attributes to [45] considerable changes in the scientific administration of the former USSR.

Multilingual digital interaction is a related yet distinct line of research [25]. The ways netizens arbitrate among language groups in their social networks, focusing on social network properties, language choice, and information diffusion are the focus of [21]. The study [26] analyzed topic-based cross-language linkings among blogs and concluded that designing for cross-cultural awareness has an impact on the underlying network structure. The connections in online social media tend to be geographically assortative [2,37]. A probabilistic characterization of macro-scale linguistic connections with respect to demographic and geographic predictors is given in [29]. The valence of Twitter connections in conjunction with linguistic changes is the focus of [23]. The diffusion of linguistic changes and their social correlations comparatively in Tudor and Stuart London is explored in [41].

Multilinear algebra or tensor analysis is the current evolution step of linear algebra [18,31]. Signal processing applications of tensor algebra include MIMO radars [42], blind source separation [6], and biomedical image analysis [46,49]. In data mining tensors have been applied to dimensionality reduction [11–44]. Within the context of social network analysis, multilayer graphs were the tool for sentiment analysis in Twitter [14]. In information retrieval third order tensors extended the term-document linear algebraic model to term-author-document [15] and to term-keyword-document [16] models. Finally, higher order statistics, which are closely associated with tensors, have been used to assess the performance of operating system level process scheduling policies [17].

Genetic algorithms are an offshoot of numerical optimization and machine learning [10]. This class of heuristics imitates Darwinian evolution [8,9]. Operations include *selection, crossover,* and *mutation* and are applied on candidate problem solutions termed *genes* [5]. Finally, the close ties between genetic algorithms and machine learning are overviewed in [24].

3   Linguistic Notions and Spatiosocial Data

In order to facilitate further discussion as well as the analysis of TENSOR-G, some preliminary notions should be outlined.

Definition 1. *Spatiosicial data have spatial and social components at minimum.*

Definition 2. *A multilayer graph G is the ordered quintuple*

$$G \triangleq (V, E, Q, \Sigma, f) \tag{1}$$

where V is the vertex set, $E \subseteq V \times V \times Q$ the edge set, Q the label set, and Σ the edge value set. The function $f : E \to \Sigma$ assigns to edges a value.

Thus, (u_1, u_2, q) denotes that u_1 and u_2 are connected by an edge whose label is q, whereas $(u_1, u_2, *)$ means that there exists at least one edge connecting u_1 and u_2 regardless of its label.

Let $L(u)$ denote the language set of vertex u, where $|L(u)| \geq 1$. Also, let $\ell(v) \in L(v)$ be the *predominant language*, namely the language more often used in online communication. Finally, if there are n vertices in total, then the total number of languages L_0 is

$$L_0 \triangleq |L(1) \cup \ldots \cup L(n)| = \left| \bigcup_{k=1}^{n} L(k) \right| \tag{2}$$

Assumption 1. *Each vertex u has a single predominant language.*

This does not imply that a netizen is obliged to post only in one language, but indicates instead which language is the most frequent.

Definition 3. *Two vertices u_1 and u_2 are coherent if and only if $\ell(u_1) = \ell(u_2)$.*

Definition 4. *The set of neighbors of u is denoted as $\Gamma(u)$, while that of coherent neighbors as $\tilde{\Gamma}(u)$. Then $0 \leq \left|\tilde{\Gamma}(u)\right| \leq |\Gamma(u)|$.*

Definition 5. *The coherency $c_\ell(S)$ of a set of vertices $S \subseteq V$, $S \neq \emptyset$ for a language $\ell \in \bigcup_{s \in S} L(s)$ is the average ratio of the coherent neighbors to the total number of neighbors.*

$$c_\ell(S) \triangleq \frac{1}{|s \in S \wedge \ell = \ell_0(s)|} \sum_{s \in S \wedge \ell = \ell_0(s)} \frac{\left|\tilde{\Gamma}(s)\right|}{|\Gamma(s)|}, \qquad 0 \leq c_\ell(S) \leq 1 \qquad (3)$$

Definition 6. *The density $d_\ell(S)$ of a set of vertices $S \subseteq V$, $S \neq \emptyset$ for a language $\ell \in \bigcup_{s \in S} L(s)$ is ratio of the vertices whose predominant language is ℓ to $|S|$.*

$$d_\ell(S) \triangleq \frac{|s \in S \wedge \ell = \ell_0(s)|}{|S|}, \qquad 0 \leq d_\ell(S) \leq 1 \qquad (4)$$

Next a social, in particular linguistic, and a spatial factor are introduced, upon which the fitness function of TENSOR-G will be built.

Definition 7. *Factor $\phi_\ell(u)$ expresses how easy is for a linguistic change to spread from u for language ℓ. The contagion depends on the ratio of the number of coherent neighbors to that of noncoherent ones. Thus*

$$\phi_\ell(u) \triangleq \begin{cases} \frac{1}{2} + \frac{1}{2\pi} \arctan\left(\frac{\left|\tilde{\Gamma}(u)\right|}{\left|\Gamma(u) \setminus \tilde{\Gamma}(u)\right|}\right), & \tilde{\Gamma}(u) \subset \Gamma(u) \\ 1, & \Gamma(u) = \tilde{\Gamma}(u) \end{cases} \qquad (5)$$

The continuous function of $\arctan(\cdot)$ has been selected since it monotonously maps $\mathbb{R}$ to $\left[-\frac{\pi}{2}, \frac{\pi}{2}\right]$. Alternatively, a related metric $\phi'_\ell(u)$ is defined as

$$\phi'_\ell(u) \triangleq \tau_{\Gamma(u), \tilde{\Gamma}(u)} \triangleq \frac{\left|\Gamma(u) \cap \tilde{\Gamma}(u)\right|}{\left|\Gamma(u) \cup \tilde{\Gamma}(u)\right|} = \frac{\left|\Gamma(u) \cap \tilde{\Gamma}(u)\right|}{|\Gamma(u)| + \left|\tilde{\Gamma}(u)\right| - \left|\Gamma(u) \cap \tilde{\Gamma}(u)\right|} \qquad (6)$$

Although both $\phi_\ell(u)$ and $\phi'_\ell(u)$ return a value in $[0, 1]$, the latter is preferrable due to its superior numerical stability, especially when $\tilde{\Gamma}(u) \ll \Gamma(u)$.

Definition 8. *Factor $\psi(u_1, u_2)$ is a function of the geographic distance $d(u_1, u_2)$ between two netizens u_1 and u_2 as follows*

$$\psi(u_1, u_2) \triangleq \begin{cases} 1, & (u_1, u_2, *) \in E \wedge 0 \leq d(u_1, u_2) \leq \delta_0 \\ \frac{\delta_0}{d(u_1, u_2)}, & (u_1, u_2, *) \in E \wedge d(u_1, u_2) > \delta_0 \\ 0, & (u_1, u_2, *) \notin E \end{cases} \qquad (7)$$

Finally, the following definition will be valuable in assessing the performance of the proposed genetic algorithm.

Definition 9. *When $\phi_\ell(u)$ or $\phi'_\ell(u)$ exceed a threshold η_0, then u is uncontested.*

4   Genetic Algorithm Tensor Clustering

4.1   Tensor and Multilayer Graph Representations

The spatiosocial multilayer graph was constructed as follows from Twitter data by a language sampling approach as described in the next section. First, the label set was decided to contain five elements, and thus $|Q| = 5$, namely

$$Q \triangleq \{:\text{location}, :\text{german}, :\text{french}, :\text{english}, :\text{social}\} \tag{8}$$

Notice that the spatial component is expressed by the geolocation distance, whereas there are four spatial components, namely the three languages most commonly spoken in Luxemburg, and one dimension for online interaction as indicated by the *follow* and *reply* functions. The edge labels follow the Neo4j notation [32]. The following two criteria determined whether there is digital interaction between any two netizens:

- If netizen u follows v or vice versa, then interaction is considered to exist.
- Alternatively, if either of netizens u and v mention the other, then interaction is also considered to take place.

Then, each netizen was mapped to one vertex of the graph so $|V| = n$. The connectivity conditions for any netizen pair u and v were the following:

- If u and v are interacting, then edge $(u, v, :\text{social})$ is added.
- If $d(u, v) < 8\delta_0$, then edge $(u, v, :\text{location})$ is added.
- If $\ell_0(u) = \ell_0(v)$, then the appropriately labeled edge is added.

The geolocation of each vertex was determined by the location Twitter metadata field and it was compared to the bounds of a rectangle circumscribing the borders of Luxembourg. Furthermore, the latitude and longitude pair of each tweet was checked against both the national and regional borders of Luxembourg as encoded by GIS files publicly available through the Global Administrative Areas database (GADM)[1].

Concerning the language set of each vertex, two methods were used. First, the corresponding tweets were analyzed as in [19]. The starting points were words unique to a specific language. Then, by dividing the findings into linguistic classes such as phonetic spellings and lexical words a profile for each candidate language was built and it was compared to that of the three languages under investigation. Second, the frequencies of digrams and trigrams, namely byte sequences encoding Unicode characters, were compared to these from standard corpora.

---

[1] www.gadm.org.

Finally, $\Sigma \triangleq \{0, 1\}$, as TENSOR-G focuses on edge labels, and Σ contains token values indicating whether an edge exists or not.

As is the case with ordinary social graphs, multilayer graphs can be also represented algebraically through adjacency tensors. A *tensor* is a multidimensional vector meaning that each entry is indexed by a tuple of p non permutable integers $(i_1 \ldots i_p)$ where $1 \leq i_k \leq I_k$. Formally

Definition 10. *A p-th tensor T, $p \in \mathbb{Z}^+$, is a linear mapping simultaneously connecting p not necessarily distinct linear spaces $\mathbb{S}_k$, $1 \leq k \leq p$.*

In this specific case the linear spaces combined to create the adjacency tensor are the netizen space, in fact twice, and the label space. Thus, the corresponding adjacency tensor T is of third order, specifically

$$T \in \Sigma^{|V| \times |V| \times |Q|} = \{0, 1\}^{n \times n \times (L_0 + 2)}. \tag{9}$$

4.2   Algorithmic Aspects

The proposed genetic algorithm TENSOR-G is outlined in Algorithm 1. As with any such scheme, its development is more an art then science. For instance, being a heuristic, TENSOR-G relies heavily on random numbers which largely decide the outcome of selection and crossover operators. Finally, observe that Algorithm 1 has a low conditional Kolmogorov complexity, since once the data, namely the tensor T and the random number sequence, are factored out, then the algorithm proper is of constant size and can be thus efficiently coded in the universal Turing machine. The use of $\langle \rho_k \rangle$ is implied throughout the algorithm.

Since the initial number of communities J_0 is unknown, it is selected semi-randomly based on knowledge from linguistics and the Gaussian distribution

$$f_{J_0}(j_0) = \frac{1}{4L_0 \sqrt{2\pi}} \exp\left(-\left(\frac{j_0 - 3L_0}{4L_0}\right)^2\right) \tag{10}$$

Namely, $\mathrm{E}[J_0] = 3L_0$ and $\mathrm{Var}[J_0] = 16L_0^2$. These parameters were selected based on statistical observations from [29], while the Gaussian distribution itself was chosen because it has the maximum differential entropy among all distributions with the same finite variance giving, thus, an upper limit to the variation of J_0.

Each of the J_0 communities can have an arbitrary number of vertices as long as it is not empty. Note that the n netizens can be distributed to J_0 with a very large number of ways, specifically

$$\binom{n}{L_0} = \frac{\partial^{L_0}}{\partial x^{L_0}}(1+x)^n \bigg|_{x=0} \approx n^{L_0}, \qquad L_0 \leq n \tag{11}$$

To avoid this, the vertices are randomly assigned to communities. Although this may lead to less than satisfactory fitness, it is computationally efficient and its effects are eventually nullified over some iterations.

Algorithm 1. Proposed Genetic Algorithm (TENSOR-G)

Require: Termination criterion τ_0, random sequence $\langle \rho_k \rangle$, tensor $\mathcal{T}$
Ensure: Linguistic communities are approximately discovered
  1: pick number of communities J_0 semirandomly
  2: partition $\mathcal{T}$ by assigning vertices to each community C_k, $1 \leq k \leq J_0$
  3: **repeat**
  4:    evaluate fitness of each community C_k
  5:    retain the $\lceil \alpha_0 J_0 \rceil$ fittest communities with probability p_α
  6:    retain the $\lceil \beta_0 J_0 \rceil$ least fit communities with probability p_β
  7:    crossover the remaining m communities to create each possible pair
  8:    select the m fittest of the $\Theta\left(m^2\right)$ new pairs
  9:    choose a community pair with probability p_γ **and** mutate the pair
 10:    **with** probability p_ζ:
 11:    **for all** community pairs **do**
 12:       **if** any two communities are spatiosocially close **then**
 13:          merge these communities **and** update J_0
 14:       **end if**
 15:    **end for**
 16: **until** τ_0 is **true**
 17: **return** $\{C_k\}$

There are two fitness functions for evaluating spatiosocial communities. The first is the harmonic mean of coherency and $\bar{\psi}$, the mean value of factor ψ

$$g_1(C_k) \triangleq \mathcal{H}\left(c_\ell(C_k), \bar{\psi}\right) = 2\left(\frac{1}{c_\ell(C_k)} + \frac{|C_k|}{\sum_{u_1, u_2 \in C_k} \psi(u_1, u_2)}\right)^{-1} \quad (12)$$

while the second is the harmonic mean of density and $\bar{\psi}$

$$g_2(C_k) \triangleq \mathcal{H}\left(d_\ell(C_k), \bar{\psi}\right) = 2\left(\frac{1}{d_\ell(C_k)} + \frac{|C_k|}{\sum_{u_1, u_2 \in C_k} \psi(u_1, u_2)}\right)^{-1} \quad (13)$$

In each iteration a portion $\lceil \alpha_0 J_0 \rceil$ of the fittest communities is kept unchanged with probability p_α. This is done in order to preserve a possibly very good solution. On the other hand, it entails the potential danger that TENSOR-G is trapped to a local maximum. For this reason, with probability p_β a segment of the $\lceil \beta_0 J_0 \rceil$ least fit communities is also retained in order to provide a (counterintuitive) escape from such a trap.

Since TENSOR-G is designed for tensor clustering, it is imperative that each netizem is assigned to one and only one language community and that no communities become void. To this end, the crossover, selection, and mutation operations were designed to uphold these constraints in spite of their inherent randomness. Specifically, the crossover operation creates selects each possible pair of the m communities. Inside each of the $\Theta\left(m^2\right)$ pairs a number of netizens, which may be random or deterministic, is selected to be swapped. Note that m is a random variable whose values and their associated probabilities are

$$m = \begin{cases} J_0 - \lceil \alpha_0 J_0 \rceil, & p_\alpha \\ J_0 - \lceil \beta_0 J_0 \rceil, & p_\beta \\ J_0 - \lceil \alpha_0 J_0 \rceil - \lceil \beta_0 J_0 \rceil, & p_\alpha p_\beta \\ J_0, & 1 - p_\alpha - p_\beta - p_\alpha p_\beta \end{cases} \tag{14}$$

This potentially large number of community pairs is then evaluated by either g_1 or g_2 and the m fittest are selected. Finally, with probability p_γ a random pair of the m new ones is selected and only one vertex is swapped between them.

The last and optional operation of community merge may come as a surprise to the reader familiar with genetic algorithms, the reason being that merge is a typical clustering operation and not part of the standard functions a genetic algorithm performs. However, since TENSOR-G is essentially a clustering algorithm, it is worth incorporating a clustering element with probability p_ζ. Thus, TENSOR-G can partially work as an agglomerative algorithm. The condition which determines the spatiosocial proximity of two communities is that they have the same predominant language and also that a random sample of b netizen pairs has low overall value for the ψ factor. In this case

$$b \triangleq \max \left\{ \log |C_i|, \log |C_j| \right\} \tag{15}$$

The primary termination criterion τ_0 was a combination of a hardcoded maximum number of M_0 iterations, with a minimum of μ_0 iterations, and of a condition that the average total fitness sum should exceed a threshold γ_0 during the past five iterations. Also, there was a secondary termination criterion τ_1 that stopped TENSOR-G when a partition achieved a fitness of γ_1.

4.3 Kruskal Decomposition

At this point the reference method, Kruskal tensor decomposition, is introduced. Given a p-th order tensor $\mathcal{T} \in \mathbb{R}^{I_1 \times \cdots \times I_p}$ and an integer $r_0 \leq p$, the Kruskal numer of the tensor rank, r_0 rank one tensors $\mathcal{G}_k$ and positive normalization scalars λ_k are computed such that

$$\mathcal{T} = \sum_{k=1}^{r_0} \lambda_k \mathcal{G}_k, \qquad \lambda_k > 0 \tag{16}$$

A rank one tensor $\mathcal{G}_p \in \mathbb{R}^{I_1 \times \cdots \times I_p}$ is one that can be written as a series of $p - 1$ vector outer products [31], namely

$$\mathcal{G}_p \triangleq \mathbf{v}_1 \circ_1 \mathbf{v}_2 \circ_2 \mathbf{v}_3 \ldots \mathbf{v}_{p-1} \circ_{p-1} \mathbf{v}_p, \qquad \mathbf{v}_k \in \mathbb{R}^{I_k}, \|\mathbf{v}_k\|_2 = 1 \tag{17}$$

Note that this is a direct generalization of a rank one matrix, essentially a second order tensor $\mathcal{G}_2$, which is written as

$$\mathcal{G}_2 \triangleq \mathbf{v}_1 \circ_1 \mathbf{v}_2 = \begin{bmatrix} \mathbf{v}_1[1]\mathbf{v}_2[1] & \cdots & \mathbf{v}_1[1]\mathbf{v}_2[I_2] \\ \vdots & \ddots & \vdots \\ \mathbf{v}_1[I_1]\mathbf{v}_2[1] & \cdots & \mathbf{v}_1[I_1]\mathbf{v}_2[I_2] \end{bmatrix}, \qquad \|\mathbf{v}_1\|_2 = \|\mathbf{v}_2\|_2 = 1 \tag{18}$$

Along similar lines, a third order tensor $\mathcal{G}_3$ is defined as

$$\mathcal{G}_3 \triangleq \mathbf{v}_1 \circ_1 \mathbf{v}_2 \circ_2 \mathbf{v}_3, \mathcal{G}_3[i_1, i_2, i_3] = \mathbf{v}_1[i_1]\mathbf{v}_2[i_2]\mathbf{v}_3[i_3], \qquad \mathbf{v}_k \in \mathbb{R}^{I_k}, \|\mathbf{v}_k\|_2 = 1 \tag{19}$$

However, both computing r_0 and the exact Kruskal decomposition are NP-hard problems. Therefore, for various estimates $\hat{r}_0$ the approximate decomposition is computed instead

$$\min_{\hat{r}_0, \lambda_k, \mathcal{G}_k} \left\| \mathcal{T} - \sum_{k=1}^{\hat{r}_0} \lambda_k \mathcal{G}_k \right\|_F = \min_{\hat{r}_0, \lambda_k, \mathbf{v}_{k,j}} \left\| \mathcal{T} - \sum_{k=1}^{\hat{r}_0} \lambda_k \mathbf{v}_{k,1} \circ_1 \mathbf{v}_{k,2} \dots \mathbf{v}_{k,p} \right\|_F \tag{20}$$

where the Frobenius norm $\|\mathcal{T}\|_F$ of a real valued tensor $\mathcal{T}$ is defined as

$$\|\mathcal{T}\|_F \triangleq \left(\sum_{i_1=1}^{I_1} \dots \sum_{i_p=1}^{I_p} \mathcal{T}^2[i_1, \dots, i_p] \right)^{\frac{1}{2}} = \left(\sum_{(i_1,\dots,i_p)} \mathcal{T}^2[i_1, \dots, i_p] \right)^{\frac{1}{2}} \tag{21}$$

4.4 Implementation Aspects

Regarding implementation, TENSOR-G was implemented in MATLAB. To conserve memory, the tensor was stored in quadruples of the form (i_1, i_2, i_3, ℓ) and the netizen properties such as their geolocation and predominant language were separately stored. In other words, the tensor was compressed with the coordinate scheme which requires four integers for every non-zero entry. Although more efficient tensor compression schemes exist [3], the coordinate method is balanced between memory conservation and simplicity. Each gene of TENSOR-G was encoded as a list quadruples for efficient manipulation. The Kruskal decomposition was already implemented in the MATLAB tensor toolbox [4].

Finally, the data was obtained by a Twitter crawler implemented in Python using the *tweepy*[2] library. The latter uses the OAuth authentication protocol in conjunction with the quadruple of Twitter generated tokens. Also, it is subject to the constraints placed by Twitter for batch data harvest.

5 Results

5.1 Data Synopsis

The tensor contains information about $n = 579$ Luxemburgian netizens, 217 of whom were identified as predominantly tweeting in English, 199 in German, and 163 in French. In overall this is a fairly balanced sample in terms of language representation. Table 2 contains more information about these netizens.

As it can be deduced, the above network is sparse since the average degree is 17. That justifies the coordinate compression scheme for the spatiosocial tensor.

---

[2] www.tweepy.org.

Table 2. Netizen statistics.

Property	Value
Follows and replies	7571
Spatial connections	1933
Min, max, avg degree	1, 31, 17
Monolinguals	29
Bilinguals	196
Trilinguals	354

5.2   Performance

The values and the characteristics of the thresholds and parameters used in
TENSOR-G are summarized in Table 3.

The number of communities which achieved the better overall fitness in
TENSOR-G for both fitness functions was 5. Using a range of values of ± 3
around this number as $\hat{r}_0$, the lowest Frobenius difference norm was achieved
for 7 communities. These number of uncontested vertices was computed using
ϕ'_ℓ. For the English, German, and French respectively the clustering obtained
by TENSOR-G returned 201, 170, and 126, whereas the Kruskal decomposition
yielded 192, 163, and 109. This can be attributed to the dispersion of predominantly French speaking netizens among the more adamant German speakers

Table 3. TENSOR-G parameters.

Parameter	Meaning	Value
α_0	Percentage of best fit clusterings kept in each iteration	0.1
β_0	Percentage of worst fit clusterings kept in each iteration	0.1
γ_0	Threshold that must be exceeded in τ_0 to continue	0.15
γ_1	Terminating threshold in criterion τ_1	0.85
δ_0	Geolocation distance for maximum assortativity	25 Km
η_0	Threshold for declaring a vertex uncontested	0.65
M_0	Maximum number of iterations in criterion τ_0	1024
μ_0	Minimum number of iterations in criterion τ_0	32
N_0	Number of instances of TENSOR-G executed	2048
b	Random sample size for merging communities C_i and C_j	Eq. (15)
L_0	Total number of languages in the tweets	3
p_α	Probability distribution for retaining best clusterings	Binomial
p_β	Probability distribution for retaining worst clusterings	Binomial
p_γ	Probability distribution for mutation	Poisson
p_ζ	Probability distribution for agglomeration check	Poisson

and the omnipresent English ones. Also, notice that Kruskal decomposition was designed with another minimization property in mind. Specifically, the constraint for rank 1 tensors led to more communities which are more compact but left many vertices in a contested state.

6 Conclusions

This conference paper presents TENSOR-G, a genetic algorithm for spatiosocial sparse tensor clustering. The latter contains trilingual Twitter data in English, French, and German from Luxemburg, a country with thriving language communities and strong digital presence. The communities obtained by TENSOR-G using two different fitness functions based on languistic criteria were compared to those obtained by Kruskal tensor decomposition. Although the proposed methodology is slower and more memory intensive than the benchmark, the communities of TENSOR-G were more compact from a linguistic viewpoint and also make more sense in geolocation terms.

This work can be improved in many aspects. A more detailed description of digital interaction would include separate labels for *follow* and *mention* options and possibly additional layers for other Twitter functions. Also, bidirectional connections between netizens would reveal more communication patterns, for instance how differs the communication between netizens and between a netizen and an institution or a company and whether Dunbar's number is a loose bound or not in the digital sphere. Moreover, better fitness functions can be designed utilizing observable and measurable language variations such as change in individual words and their spelling compared to template corpora. However, it is not necessarily true that language change proceeds horizontally in the different domains. Thus, any research of language change needs to incorporate both the similarities and the differences in mechanisms across different domains [23].

Regarding future research directions, language change results from the differential propagation of linguistic variants distributed among the linguistic repertoires of communicatively interacting netizens. From this it follows that language change is socially-mediated in two important ways. First, language is a social epidemiological process that takes place by propagating some aspect of communicative practice across a network and the organization of the social group in question can affect how a variant propagates. Second, sociocultural factors such as language ideologies, can encourage the propagation of particular variants at the expense of others in particular context. Variant selection leads to language change when it forms part of larger scale processes of differential variant propagation within the speech community. Since tensors are particularly suited to diffusion phenomena, their application to spatiosocial data in general and to the propagation of language changes should be thoroughly examined.

Acknowledgements. This conference paper has been developed within the framework of the project "Strengthening the Research Activities of the Directorate of the Greek School Network and Network Technologies", financed by the own resources of the Computer Technology Institute and Press "Diophantos" (project code 0822/001).

References

1. Androutsopoulos, J.: Language change and digital media: a review of conceptions and evidence. In: Standard Languages and Language Standards in a Changing Europe (2011)
2. Backstrom, L., Sun, E., Marlow, C.: Find me if you can: improving geographical prediction with social and spatial proximity. In: Proceedings of the 19th International Conference on World Wide Web, pp. 61–70. ACM (2010)
3. Bader, B.W., Kolda, T.G.: Efficient MATLAB computations with sparse and factored tensors. SIAM J. Sci. Comput. **30**(1), 205–231 (2007)
4. Bader, B.W., Kolda, T.G., et al.: MATLAB tensor toolbox version 2.5 (2012)
5. Booker, L.B., Goldberg, D.E., Holland, J.H.: Classifier systems and genetic algorithms. Artif. Intell. **40**(1–3), 235–282 (1989)
6. Cardoso, J.F.: Eigen-structure of the fourth-order cumulant tensor with application to the blind source separation problem. In: ICASSP-90, pp. 2655–2658. IEEE (1990)
7. Croft, W.: Mixed languages and acts of identity: an evolutionary approach. Mixed Lang. Debate: Theoret. Empirical Adv. **145**, 41 (2003)
8. Darwin, C.: On the origin of species by means of natural selection. John Murray, November 1859
9. Dawkins, R.: The Selfish Gene, 30th edn. Oxford University Press, Oxford (2006)
10. De Jong, K.: Learning with genetic algorithms: an overview. Mach. Learn. **3**(2), 121–138 (1988)
11. De Lathauwer, L., Vandewalle, J.: Dimensionality reduction in higher-order signal processing and rank-$(r_1, r_2, \ldots, r_n)$ reduction in multilinear algebra. LAA **391**, 31–55 (2004)
12. Dixon, R.M.: The Rise and Fall of Languages. Cambridge University Press, Cambridge (1997)
13. Donoso, G., Sánchez, D.: Dialectometric analysis of language variation in Twitter. arXiv preprint arXiv:1702.06777 (2017)
14. Drakopoulos, G.: Tensor fusion of social structural and functional analytics over Neo4j. In: Proceedings of the 6th International Conference of Information, Intelligence, Systems, and Applications, IISA 2016. IEEE, July 2016
15. Drakopoulos, G., Kanavos, A.: Tensor-based document retrieval over Neo4j with an application to PubMed mining. In: Proceedings of the 6th International Conference of Information, Intelligence, Systems, and Applications, IISA 2016. IEEE, July 2016
16. Drakopoulos, G., Kanavos, A., Karydis, I., Sioutas, S., Vrahatis, A.G.: Tensor-based semantically-enhanced PubMed retrieval. Computation, May 2017. Accepted
17. Drakopoulos, G., Megalooikonomou, V.: An adaptive higher order scheduling policy with an application to biosignal processing. In: SSCI 2016. IEEE, December 2016
18. Dunlavy, D.M., Kolda, T.G., Acar, E.: Temporal link prediction using matrix and tensor factorizations. TKDD **5**(2), 10 (2011)
19. Eisenstein, J.: Sociolinguistic variation in online social media. In: 2015 AAAS Annual Meeting (2015)
20. Eisenstein, J., O'Connor, B., Smith, N.A., Xing, E.P.: Diffusion of lexical change in social media. PLoS One **9**(11) (2014)
21. Eleta, I., Golbeck, J.: Bridging languages in social networks: how multilingual users of Twitter connect language communities? Proc. Am. Soc. Inf. Sci. Technol. **49**(1), 1–4 (2012)

22. Ge, X., Cheng, H., Guizani, M., Han, T.: 5G wireless backhaul networks: challenges and research advances. IEEE Netw. **28**(6), 6–11 (2014)
23. Goel, R., Soni, S., Goyal, N., Paparrizos, J., Wallach, H., Diaz, F., Eisenstein, J.: The social dynamics of language change in online networks. In: Spiro, E., Ahn, Y.-Y. (eds.) SocInfo 2016. LNCS, vol. 10046, pp. 41–57. Springer, Cham (2016). doi:10.1007/978-3-319-47880-7_3
24. Goldberg, D.E., Holland, J.H.: Genetic algorithms and machine learning. Mach. Learn. **3**(2), 95–99 (1988)
25. Hale, M.: Historical Linguistics: Theory and Method. Wiley-Blackwell, Hoboken (2007)
26. Hale, S.A.: Global connectivity and multilinguals in the Twitter network. In: Proceedings of the SIGCHI Conference on Human Factors in Computing Systems, pp. 833–842. ACM (2014)
27. Hong, L., Convertino, G., Chi, E.H.: Language matters in Twitter: a large scale study. In: ICWSM (2011)
28. Kershaw, D., Rowe, M., Noulas, A., Stacey, P.: Birds of a feather talk together: user influence on language adoption. In: Proceedings of the 50th Hawaii International Conference on System Sciences (2017)
29. Kershaw, D., Rowe, M., Stacey, P.: Language innovation and change in on-line social networks. In: Proceedings of the 26th ACM Conference on Hypertext and Social Media, pp. 311–314. ACM (2015)
30. Kirk, N.A., Mees, B.: Stalin, Marr and the struggle for a Soviet linguistics. Verbatim **31**(3) (2006)
31. Kolda, T.G., Bader, B.W.: Tensor decompositions and applications. SIAM Rev. **51**(3), 455–500 (2009)
32. Kontopoulos, S., Drakopoulos, G.: A space efficient scheme for graph representation. In: Proceedings of the 26th International Conference on Tools with Artificial Intelligence, ICTAI 2014, pp. 299–303. IEEE, November 2014
33. Labov, W.: Principles of linguistic change vol. 2: social factors. Lang. Soc. **29** (2001)
34. Labov, W.: Transmission and diffusion. Language **83**(2), 344–387 (2007)
35. Matras, Y.: Languages in contact in a world marked by change and mobility. Revue française de linguistique appliquée **18**(2), 7–13 (2013)
36. Matsumoto, K.: The role of social networks in the post-colonial multilingual island of Palau: mechanisms of language maintenance and shift. Multilingua-J. Cross-Cultural Interlang. Commun. **29**(2), 133–165 (2010)
37. Maybaum, R.: Language change as a social process: diffusion patterns of lexical innovations in Twitter. In: Annual Meeting of the Berkeley Linguistics Society, pp. 152–166 (2013)
38. Michael, L., Bowern, C., Evans, B.: Social dimensions of language change. In: Bowern, C., Evans, B. (eds.) Routledge Handbook of Historical Linguistics, pp. 484–502. Routledge (2014)
39. Milroy, J., Milroy, L.: Linguistic change, social network and speaker innovation. J. Linguist. **21**(02), 339–384 (1985)
40. Milroy, L.: Language and Social Networks, 2nd edn. Blackwell, Oxford (1980)
41. Nevalainen, T.: Social networks and language change in Tudor and Stuart London-only connect? English Lang. Linguist. **19**(2), 269–292 (2015)
42. Nion, D., Sidiropoulos, N.D.: Tensor algebra and multidimensional harmonic retrieval in signal processing for MIMO radar. IEEE Trans. Sig. Process. **58**(11), 5693–5705 (2010)

43. Pakendorf, B.: Historical linguistics and molecular anthropology. In: Bowern, C., Evans, B. (eds.) Routledge Handbook of Historical Linguistics. Routledge (2014)
44. Papalexakis, E., Doğruöz, A.S.: Understanding multilingual social networks in online immigrant communities. In: 24th WWW, pp. 865–870. ACM (2015)
45. Stalin, J.V.: Marxism and problems of linguistics. In: Pravda, May 1950
46. Tagkalakis, F., Papagiannaki, A., Drakopoulos, G., Megalooikonomou, V.: Augmenting fMRI-generated brain connectivity with temporal information. In: Proceedings of the 6th International Conference of Information, Intelligence, Systems, and Applications, IISA 2016. IEEE, July 2016
47. Trudgill, P.: Social structure, language contact and language change. In: The SAGE Handbook of Sociolinguistics, pp. 236–249 (2011)
48. Weinreich, U., Labov, W., Herzog, M.I.: Empirical foundations for a theory of language change. University of Texas Press, Austin (1968)
49. Westin, C.F., Maier, S.E., Mamata, H., Nabavi, A., Jolesz, F.A., Kikinis, R.: Processing and visualization for diffusion tensor MRI. Med. Image Anal. 6(2), 93–108 (2002)

Analyzing the Mobile Learning System Behavior: The Case of the Russian Verbs of Motion

Oxana Kalita[1]([✉]), Vladimir Denisenko[2], Anatoly Tryapelnikov[3],
Fotis Nanopoulos[4], and Georgios Pavlidis[4]

[1] School of Engineering, Patras University, Patras, Greece
kalitaxenia@gmail.com
[2] Russian Peoples' Friendship University, Moscow, Russia
denissenko@mail.ru
[3] Department of Russian Language and Intercultural Communication, Faculty of
Humanities and Social Sciences, Russian Peoples' Friendship University,
Moscow, Russia
tryapelnikov@yandex.ru
[4] Computer Engineering and Informatics Department, School of Engineering,
Patras University, Patras, Greece
nphotis@gmail.com, pvlds0l@upatras.gr

Abstract. The evolution of mobile technologies gives the opportunity for innovative approaches to the content, the process and the evaluation of the educational activity. In this paper we study the particular situation where learner (student) is in displacement, in various conditions (walking in the rain, driving a car, riding a bicycle etc.), and receives the educational content through his or her smart mobile devices. Such a dynamic educational process, that we call mobile learning, presents many new aspects to study. We focus on the study of the interaction of the student with the educational process, running under various influencing factors and restrictions. In particular, the research carried out on Greek students who are learning Russian language.

Keywords: Mobile learning · Student's paths · Learning conditions · Psychological influences · Statistical analysis

1 Introduction

Trainees, as all people do, go down different routes (trips, paths) under varying conditions and constrains. For the requirements of a mobile learning management system (m-learning), it is important to consider if these paths are new or old, the conditions in which they arise and when and how they occur. All these queries are based on:

A. The weather conditions prevailing in a geographical area, the speed at which the learner (student) moves, the interlocutor (escort, companionship), the transportation means used, the new destination he (or she) has set or the old route that he follows, the sights he meets and any other similar factors,

© Springer International Publishing AG 2017
G. Boracchi et al. (Eds.): EANN 2017, CCIS 744, pp. 645–654, 2017.
DOI: 10.1007/978-3-319-65172-9_54

B. The extent, the complexity and other endogenous characteristics of the particular educational material, the pedagogical situation and this includes not only the rules of behavior that the professor has defined with the team for the system development and support, namely the logics of response, the educational methodology, the available predefined exercises, tests, and other teaching factors,

C. The personal data, the cognitive background and the student's interests with the data regarding his current performance as reflected in his profile, in other words the degree of coverage of the educational material, how, where and when the student learns, which is his current performance, what problems, like failures and obstacles, are presented, and any other relevant data issues.

The knowledge owner (e.g. the professor), as a part of the development and support team of a specific m-learning system, is needed to combine and exploit the new technological possibilities and the new information that this system offers along with the innovative ideas, suggestions and initiatives of the student.

It is necessary for the educational content, to be more effectively transmitted to the students in order to ensure higher performance regarding the required exercises, tests and examinations.

In other words, in this paper, m-learning is considered as a process of "discovery" of opportunities for better and more efficient completion of the educational process.

2 Current Situation and Main Problems

A large number of the diverse paths followed by a student are recurring.

As a consequence, situations are created in which the development and system support team must intervene, otherwise the behavior of a m-learning begins to repeat, for example, one often makes the same question at a specific point on a path.

The research carried out on Greek students who are learning Russian language and, in particular, on the meaning and use of Russian verbs relating to movement, has found that these iterations adversely affect both their performance and psychology (Fig. 1).

As a result, when teaching Greek students, the development and system support team is compelled to prepare a wide range of exercises in order (a) to predict potential cross-referenced incompatibilities, (b) to identify mistakes when using verbs of motion in oral and written Russian, and (c) to determine the extent of their frequency and consistency in order to achieve better learning outcomes (performance).

In other words, one can see a significant reduction in students' participation in the educational process and, over time, loss of interest in their lessons.

This work investigates the recurrence and other similar phenomena and factors taking place in order to propose avoidance solutions in conjunction with principles of action that have to be taken by the system development and support team.

When the m-learning finds that a student follows an old path or when the student itself declares that it is at the beginning of a new trip, then a way to avoid unwanted, undesired situations is to examine this path as a "black box".

Of course, there is the case when some part of a new path is an "old" one, as well as the opposite situation, where some part of an old path is a "new" one, leading the m-learning system to unexpected behavior.

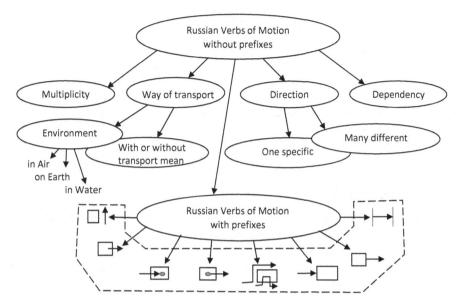

Fig. 1. Semantic features of Russian verbs of motion [1].

Therefore, it is necessary to look at what events have occurred and what events may follow so that they are all synchronized and coordinated with the student's performance and other information defined in his profile.

Analysis of the latter has resulted in the classification of the Greek students in accordance to their performance level, which – in our case – is excellent, very good, medium or low.

The conclusion of many psychological studies indicates that if an exercise, practically impossible to be solved, is given to a student, he gives up in general.

For example, if it is necessary to solve a "match" exercise for 15″ and after 10″ no satisfactory progress is achieved, it would be fair to assume that the final result will be negative.

On the other hand, if the duration of the exercise is 30″ instead of 15″ then after 10″ the progress that the student has made will be much better.

In other words, m-learning is called upon to transform the education process, as a set of predefined and fully controlled steps, into a dynamic and creative process.

A process that runs in an environment that provides a lot of new data, information and knowledge, does not have a psychological impact on the student and increases his overall performance.

Of course, the m-learning cannot guarantee that all students will achieve their final goals, but can help them significantly to better understand the current situation and not to abandon the educational process prematurely.

A smart m-learning system never gets the attention of one student while driving a car, or never asks time-consuming questions when someone travels by bus from which he will get off in a short period of time.

Therefore, the m-learning is a process that offers new capabilities and, consequently, new requirements for the management, development and support team [2].

If these capabilities are not exploited carefully and properly the m-learning system can, instead of contributing to the educational process, bring things to an immeasurable disaster.

Thus many questions arise such as: "Which points and which dimensions of the m-learning system require special attention?", "When does a professor in charge will be in a position to say with a high degree of certainty that the system has satisfactory and constructive behavior towards the students?"

2.1   The Important Role of the End User

The m-learning, as a study area, can be analyzed from different points of view.

Its success largely depends on the end user, i.e. the student.

When he has a proper judgment, memory, and strong scientific and technological background, when he is methodical and interested in learning, when he is actively involved in the educational process, then any wrong behavior of the m-learning system could not lead to failure.

In addition, student's comments and suggestions are of great importance to the m-learning system development and support team.

However, the design of the development team is required to intervene proactively to prevent any unwanted situations.

Of course, one could argue that the m-learning system, like any other artifact, might contain bugs and errors.

In this paper we are not concerned with identifying these problems (mistakes).

We are interested in ensuring that when the incoming data are known (student profile, current path, environmental conditions etc.) then the m-learning system behavior towards the student will be correct.

2.2   Approaching the Problem

The fact that someone know only part, not the whole set of incoming data, automatically leads to a "black box" system behavior analysis approach.

That is, if the m-learning system ignores the characteristics of the current path and/or the potential environmental conditions, then it will present the content of the teaching material in accordance with the professor's pre-defined instructions (Fig. 2).

In other words, every next step (part) in the educational process will depend, basically, from all the previous ones.

Considering the characteristics of the current path plus the current environmental conditions, then the m-learning development and support team will necessarily follow a "white-box" approach.

In this case, for example, the system knows that one of the students goes to university every working day in the morning on foot and returns by bus in the afternoon.

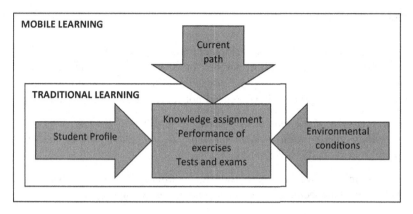

Fig. 2. The traditional as a part of a modern m-learning system.

In particular, the m-learning system knows that during walking someone takes 1 from a total of 5 possible paths (see Fig. 3).

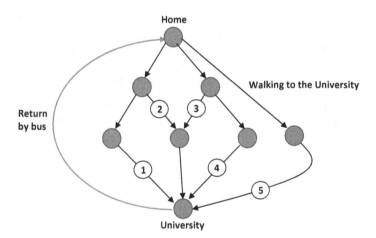

Fig. 3. Representing the paths followed from a student each working day.

Each path has its own duration, its own attractions, and the weather forecast for the next day indicates that it would rain, so the question then arises: What should be the correct behavior of the m-learning system?

Since paths, as well as environmental conditions, are constantly changing, the implementation of the "white-box" approach is practically impossible, so it is imperative to apply some kind of a mixed or combined – partially "black" and partially "white" – approach.

3 Basic Principles

During the m-learning system development our approach was combined (mixed), rational rather than optimal and, in parallel, based on a set of principles.

These principles also have a psychological and technological dimensions.

They are intuitively clear, but often do not receive the necessary attention.

1. *The description of the matches between the inputs and the characteristics of the expected behavior of the student must be an important part of the m-learning system.*

This principle is often breached, as designers often "see what they want to see" and not what really is happening to the environment and its end users.

2. *The student should be actively involved in the educational process, as the system development and support team is unable to resolve all possible, extraordinary and non-incidental events.*

An m-learning system is a product of collective effort, in which the development team is the one that composes and builds the platform and it is very normal to overlook or insists on something that is ultimately not correct.

The student, as the end user of the system, is the one who determines, decomposes and deconstructs the educational process.

Therefore, the student is the one who will immediately see the existence of a problem – gap, error – or a mistake in the system, and should report it to the development team in order to resolve it in the best possible way.

3. *A special purpose calendar should be supported with suggestions for improvement to the m-learning system and the corresponding interventions of the development and support team.*

Based on the logs of the interventions and after careful analysis, the whole system should be upgraded at regular time intervals.

4. *Both the positive and the negative comments of students about the behavior of the m-learning should be taken into account seriously by the development team.*

Only in this way can the development and support team determine under what conditions and in which category of students the m-learning system behaves properly, correctly etc.

Are there unnecessary paths, conditions and, accordingly, descriptions of potential behavior of the student that are unduly burdened by the system?

5. *The value of probability for discovering new, "uncovered" until now conditions is equal to the number of "covered" by the system conditions.*

For example, if for path 1 of one student the system covers 5 conditions, while for path 3 covers 1 condition, it is very likely that the latter will need to cover other new conditions (See Fig. 3).

In other words, this principle plays the role of retrospectively linking (feedback) to the development of the m-learning, as the team's efforts need to be satisfied.

6. *The development and support of a m-learning system are two creative and innovative processes.*

Compared to the initial design, it is highly likely that the adaptation of the educational process to path characteristics and current conditions will require much more creative effort from the development team.

4 Statistical Analysis Model

1. Data Model

The data model is provided by the learning process taking in account conditions, rules and interactions of the students with the projected system. We distinguish 5 components of data:

(A) Student's personal data. This data will be provided by electronic questionnaire as a mobile application.
(B) Description of the location and environmental conditions (LEC). This data will be provided by GPS-GIS facility incorporated in the mobile device of the student, and will be updated adequately. Other data like meteorological, cultural, historical, and any other relevant information, will be collected in real time and materialized via special classification schemes.
(C) A classification of "educational material" such as lessons, tests, FAQ, exams, and any other educational information, constructed on a dynamic way, will be used to select, on real time, the educational sequences proposed to the particular student. The aim will be to adapt the educational sequence to student's profile and LEC.
(D) Using knowledge from A-B-C the educational sequence will be applied and results will be registered.
(E) The final data creation work will consist of the evaluation of the student and the scoring of his/her performance within a peer population.

2. Analysis objectives

The analysis will aim at building *à priori* knowledge (machine learning) for optimally adapting "educational material" and "educational sequences" to student profile in order to maximize students learning performance (Y).

3. Statistical model

The statistical model is a vector $V = (Y, X1, ..., Xk)$ where the Xi's play the role of independent (explicative) variables and Y the dependent (explained) variables. The statistical unit used in this work will be the couple (s,t), where s is the student and t is the path in the learning graph.

It is worth noticing that the model admits missing data, as a student may quit at any stage of the process and restart. This will be handled via a special variable Z incorporated in the education sequence. So, to each student trial (s,t), corresponds a history h (s,t) which is a sequence of vectors:

X1(s,t) = Information as described in (A) above
X2(s,t) = Information as described in (B) above
X3(s,t) = Information as described in (C) above
X4(s,t) = Information as described in (D) above
X5(s,t) = Information as described in (E) above

Among these variables are those to be selected, which will be explained by the others. For clarity we will use the letter Y to indicate those variables.

4. Data collection - Describe How data will be collected.

The data will be split in two data sets:

(1) Those that have completed the process and performance may be computed.
(2) Those that have not completed the educational sequence, so that the independent variable explains the level of abandonment.
(3) The whole set is also of interest.

5. Treatment

A1: One-dimensional analysis

Each variable will be treated to compute histograms and parameters whenever possible.
Histograms will be used to reduce multiplicity of values and avoid over-fitting in later analysis.
Especially for the variables expressing performance, a time series of parameters will be recorded allowing for future analysis of educational methodology.

A2: Codification of variables.

This will provide codes for values in order to reduce high data variability and apply international classifications norms and standards. Losses off information will be measured using entropy indicators.

A3: Analysis of all couples of variables.

At first two way histograms will be computed and "functional correlation coefficients" will be evaluated using the formula:

$$\varphi^2(Y/X) = \frac{E\|Y^* - E(Y)\|^2}{E\|Y - E(Y)\|^2}, \text{ where } Y^* = E(Y/X) \text{ the conditional expectation of } Y$$

given X.

This is the well-known R2 coefficient in regression. The choice of this coefficient is that from a mathematical point of view it expresses the degree of approximation of a real vector Y by a function of X. This holds for any type of X, as the conditional expectation E(Y/X) depends only on the σ-field of X.

If Y is a function of X the $\varphi(Y/X) = 1$, and if they are independent $\varphi(Y/X) = 0$. Note the fcc is not symmetric.

Throughout this paper correlation of a variable Y with respect to another variable X, will be denoted as fcc(Y/X).

A4: Variables selection.

A table F(All Xs/All Xs) of fcc's of all couples of variables will be created and analyzed.

Using the table F we will chose which variables are the most correlated (explicative) for Y and among them we will eliminated those that are mostly explained by another in the list. Thus we will end up with a vector (Xi1, ..., Xik) that will be used as explicative variables of Y.

Finally $\varphi 0 = \varphi(Y/(Xi1, ..., Xik))$ will be computed.

The significance of the coefficient $\varphi 0$ will depend of the size N = number of observations, with respect to number M of probable cells in the hypercube of the values of the explicative variables, used at each stage.

A5: The Explicative Markov Tree (EMT)

To prevent the model from over fitting due to small N in comparison to M, we will apply a binary segmentation algorithm creating a "conditional tree" of two branches at each node, providing best variables for splitting and conditional distribution of Y, after each split. Special stopping rules will be used, so that at each level of iteration, leaves will be split, providing leaves to the next level, until no leaves can be split (Fig. 4).

The results will be manifold [3]:

(1) A hierarchy of variables and splits will provide information on what best explains students' performance.
(2) At each level l, a local fcc will be computed between Y and the so far obtained partition of the population.
(3) A pruning operation will take place to mix groups with similar distributions of Y which compared to $\varphi 0$, provides a stopping rule.

A final Markov Tree will be provided for use in the applications.

Note1: The construction of the Markov tree as a machine learning application may be timely updated providing a dynamic process with feedback.

Note2: The above described process can be applied also when the explained variable is not real, so that direct computation of conditional expectation cannot be applied. In this case we can associate a vectorial representation of the values of Y and thus FCC can be computed.

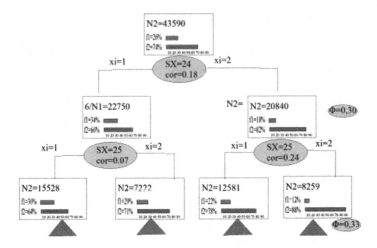

Fig. 4. Conditional decision tree.

5 Conclusion

This paper is based on the results and experiences of implementing a m-learning system for Greek students learning Russian language, in particular, Russian verbs of movements. It is stressed that the complexity of the situations is due to the existence of a large volume of automatically generated and incoming data. The active role of the student, as end user, in the whole process of adapting and optimizing the system productivity is confirmed. The question: What approach should be implemented and what principles should the development and maintenance team follow? is answered.

Finally, a model for statistical analysis is presented which increase the overall system productivity index in parallel with the performance of various categories of students. In the near future concrete results and measurements from the implementation of the statistical model will be presented.

References

1. Kalita, O.: Contributing in an ethno-oriented methodology: the case of Greek students who learn Russian verbs of motion. Int. J. Comput. Intell. Stud. **5**(1), 94–105 (2016)
2. Kalita, O., Balykxina, T., Pavlidis, G.: Mobile navigator for learning foreign languages. In: 6th IEEE International Conference on Information, Intelligence, Systems and Applications, Corfu, Greece (2015)
3. Nanopoulos, Ph: Méthodes et Programmes de Segmentation. Département de Mathématique. Centre de Calcul. Université Louis Pasteur, Strasbourg (1973)

5GPINE2017

Implications of Multi-tenancy upon RRM/Self-x Functions Supporting Mobility Control

Ioannis Chochliouros[1(✉)], Oriol Sallent[2], Jordi Pérez-Romero[2],
Anastasia S. Spiliopoulou[1], and Athanassios Dardamanis[3]

[1] Hellenic Telecommunications Organization (OTE) S.A.,
99, Kifissias Avenue, 151 24 Athens, Greece
ichochliouros@oteresearch.gr, aspiliopoul@ote.gr
[2] Universitat Politècnica de Catalunya (UPC),
c/Jordi Girona, 1-3, 08034 Barcelona, Spain
{sallent,jorperez}@tsc.upc.edu
[3] SmartNet S.A., 2, Lakonias Street, 17342 Agios Dimitrios, Attica, Greece
ADardamanis@smartnet.gr

Abstract. Based on the context of the original SESAME project research effort, in the present work we examined the implications of multi-tenancy upon the Radio Resources Management (RRM) and Self-x functions that support mobility control, as the latter is a fundamental functionality to ensure a seamless experience to the user equipments of the different operators when moving across the cells of a shared RAN (Radio Access Network) and when entering and leaving the shared infrastructure.

Keywords: Mobility control · Mobility Load Balancing · Multi-tenancy · Radio Resource Management (RRM) · Self-Organising Networks (SON) · "Self-x" functions · Small Cell (SC) · Virtual Network Function (VNF)

1 Introduction

Self-Organising Networks (SON) refers to a set of features and capabilities for automating the operation of a network so that operating costs can be reduced and human errors can be minimised [1]. With the introduction of SON features, classical manual planning, deployment, optimization and maintenance activities of the network can be replaced - and/or supported - by more autonomous and automated processes, thus making network operations simpler and faster. SON functions, denoted in the context of SESAME as "Self-x" functions, are organized around the following three main categories, which are based on previous references [1, 2]: (i) *Self-planning:* Automatization of the process of deciding the need to roll out new network nodes in specific areas, identifying the adequate configurations and settings of these nodes, as well as proposing capacity extensions for already deployed nodes (e.g. by increasing channel bandwidths and/or adding new component carriers). Specific functions belonging to this category include the planning of a new cell and the spectrum

© Springer International Publishing AG 2017
G. Boracchi et al. (Eds.): EANN 2017, CCIS 744, pp. 657–668, 2017.
DOI: 10.1007/978-3-319-65172-9_55

planning. (ii) *Self-optimization:* Once the network is in operational state, the self-optimization includes the set of processes intended to improve - or maintain - the network performance in terms of coverage, capacity and service quality by tuning the different network settings. Examples of functions include Mobility Load Balancing (MLB), Mobility Robustness Optimisation (MRO), Automated Neighbour Relation (ANR[1]), Coverage and Capacity Optimization, optimization of admission control, optimization of packet scheduling, inter-cell interference coordination and energy saving. (iii) *Self-healing:* Automation of the processes related to fault management (i.e., fault detection, diagnosis, compensation and correction), usually associated to hardware and/or software problems, in order to keep the network operational, while awaiting a more permanent solution to fix it and/or prevent disruptive problems from arising. Examples of self-healing functions include Cell Outage Detection and Cell Outage Compensation.

Self-x functions might automatically tune global operational settings of a small cell (e.g., maximum transmit power, channel bandwidth, electrical antenna tilt) as well as specific parameters corresponding to Radio Resource Management (RRM) functions (e.g., admission control threshold, handover offsets, etc.). Regarding the architectural models for implementing the Self-x functions, the following possibilities are distinguished [3, 4]: (i) Centralized SON (cSON): It is a solution where the Self-x algorithms are executed at the NMS (Network Management System) or at the EMS (Element Management System); (ii) Distributed SON (dSON): It is a solution where the Self-x algorithms are executed at the Network Element (NE) level (i.e. autonomously within a single Small Cell (SC) or in a distributed manner among several SCs), and; (iii) Hybrid SON: This combines cSON and dSON, in a way that part of the Self-x functionalities are distributed and reside at the SC, while others are centralized and reside at the EMS and/or the NMS. 1. In this case, the cSON functions can be used to provide guidelines and parameters to the distributed SON functions based on information retrieved from them in terms of, e.g., performance measurements.

2 The SESAME Architectural Approach

Based on the above model, Fig. 1 depicts a simplified view of the SESAME architecture [5] focusing on the relationship with Self-x functionalities. The SESAME project [6] focuses on the management of multi-tenant small cells. The Cloud Enabled Small Cell (CESC) is a complete Small Cell with necessary modifications to the data model to allow Multi-Operator Core Network (MOCN) radio resource sharing. The CESC is composed by a Physical SC unit and a micro-server. The physical aggregation of a set of CESCs (CESCs cluster) provides a virtualized execution infrastructure, denoted as "Light Data Centre" ("Light DC"), enhancing the virtualization capabilities and process power at the network edge. The functionalities of the CESC are split between SC Physical Network Functions (PNFs) and SC Virtual

---

[1] ANR is also referred to as Network Listen (NWL). It is a process by which a cell scans its radio environment to discover neighbouring cells.

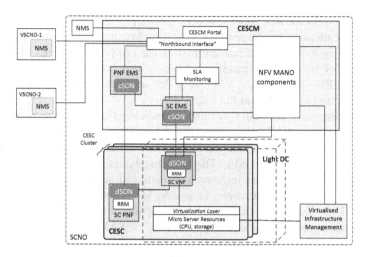

Fig. 1. SESAME architecture (simplified view) in relation to Self-x.

Network Functions (VNFs). SC VNFs are hosted in the environment provided by the Light DC. The CESC Manager (CESCM) is the central service management component in the architecture that integrates the traditional 3GPP network management elements and the novel functional blocks of the NFV-MANO (Network Function Virtualization - Management and Orchestration) framework. Configuration, Fault and Performance Management of the SC PNFs is performed through the PNF Element Management System while the management of the SC VNFs is carried out through the SC EMS. The EMSs provide performance measurements to the Service Level Agreement (SLA) Monitoring module that assesses the conformance with the agreed SLAs. EMSs are connected through the northbound interface with the Network Management Systems (NMS) of the Small Cell Network Operator (SCNO) and the different tenants, denoted in SESAME as Virtual Small Cell Network Operators (VSCNOs), providing each VSCNO with a consolidated view of the portion of the network that they are able to manage. Finally, the CESCM includes a portal that constitutes the main graphical frontend to access the SESAME platform for both SCNO and VSCNOs.

Automated operation of CESCs is made possible by different SON functions that will tune global operational settings of the SC (e.g., transmit power, channel bandwidth, electrical antenna tilt) as well as specific parameters corresponding to Radio Resource Management (RRM) functions (e.g., admission control threshold, handover offsets, packet scheduling weights, etc.). As shown in Fig. 1, the PNF EMS and SC EMS include the centralized Self-x functions (cSON) and the centralized components of the hybrid SON functions. In turn, the decentralized (dSON) functions - or the decentralized components of the hybrid functions - reside at the CESCs. The dSON functions can be implemented as PNFs or, if proper open control interfaces with the element (e.g. the RRM function) controlled by the SON function are established, they can also be implemented as VNFs running at the Light DC. The mapping of the specific RRM and SON functions in the different components of the architecture depends, in general, on the selected functional split between the physical and virtualized functions. It is worth mentioning that, depending on the implementation, and assuming that the SLA monitoring function is effectively part of the EMS, it could also be possible to associate cSON to the SLA monitoring, which also has visibility of Performance Management (PM) [7] counters, KPIs (Key Performance Indicators) and SLAs. This could relieve the SC EMS from the cSON decisions.

Furthermore, whatever Self-x function is considered of interest to be deployed, it can be implemented as a PNF or, if proper open control interfaces with the element controlled by the Self-x function are established, it can also be implemented as a Virtual Network Function (VNF). The implementation as VNFs provides: (i) An inherent flexibility through easy instantiation, modification and termination procedures; (ii) An inherent efficiency in hardware utilisation, since VNFs are executed on a pool of shared NFVI resources, and; (iii) An inherent capability to "add" new functionalities and/or extend/upgrade/evolve existing VNFs. In SESAME, this can be applied to distributed Self-x functions that would run as SC-VNFs in the Light DC.

Under the above framework, we focus upon certain initial approaches regarding the use of Self-x functions in the context of SESAME. Particular emphasis is put on the implications of multi-tenancy, because in multi-tenant scenarios - like those considered in SESAME - it should be distinguished between those Self-x functions that are tenant-*specific* (i.e., the configuration of parameters can differ from tenant to tenant) and those that are common to all tenants. In this respect, Sect. 3 analyses the implications of multi-tenancy over the Self-x functions related with mobility control which is assessed as a critical challenge among several ones in SESAME.

3 Analysis of Multi-tenancy Support in Self-x/RRM Functions for Mobility Control

This Section intends to analyse the implications of multi-tenancy on the RRM and Self-x functions that are related with mobility control. The main motivation behind this analysis is that, although mobility control is a fundamental functionality to ensure a seamless experience to the User Equipments (UEs) of the different operators when moving across the cells of a shared RAN (Radio Access Network) and when entering and leaving the shared infrastructure, no previous works in the literature have addressed this issue yet.

Let us consider the scenario depicted in Fig. 2. The RAN of the SCNO (Small Cell Network operator) is composed by different CESCs denoted in the following as small cells (e.g. small cells SA, SB, SC) and provides service to a Tenant (i.e. the VSCNO) e.g., within a stadium. The VSCNO is an MNO, with its own RAN around the stadium (e.g. cells TA, TB, TC). Figure 2 includes the relevant elements of the SESAME architecture for the analysis considered here. In particular, the interconnection of the SCs of the SCNO to the Evolved Packet Core (EPC) of the VSCNO is done through the S1 interface, delivering both data (e.g., transfer of end-users traffic) and control (e.g., activation of radio bearers) plane functions. Using current 3GPP principles [2], the support of MOCN at the SC cell is provided by using the S1-flex mechanism that allows connecting one SC to multiple EPC nodes (e.g. belonging to different

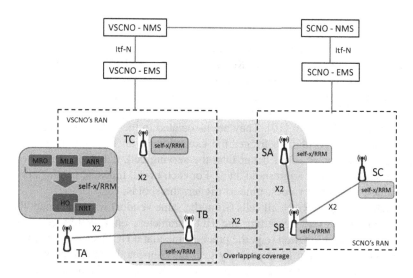

Fig. 2. Scenario for the analysis of multi-tenancy.

VSCNOs). The SCNO's SCs are connected with the RAN of the VSCNO through the X2 interface. X2 connectivity can be provided through the X2 GW (Gateway) [2] in case it is used. The X2 interface allows neighbour cells to exchange different types of information (e.g. load, interference, handover information, trace information, information to support self-optimisation, etc.) for coordination purposes and supports procedures/messages for parameter negotiation (e.g. to request handover parameter changes, etc.) [8]. As shown in Fig. 2, there is partial overlapping between the coverage of the SCNO's RAN and the VSCNO's RAN. Mobility of UEs between cells of both RANs is supported in both ways (i.e. referred to as incoming traffic when going from the VSCNO's RAN to the SCNO's RAN and outgoing traffic in the opposite direction).

Mobility control of connected terminals is realized through the handover (HO) function, which is one of the central RRM functionalities. The RRM-HO function, executed at each small cell, is used to determine the cell to which a given UE is connected to. Decisions made by the RRM-HO function are based on measurements that are compared through a set of parameters (e.g., thresholds, offsets). The RRM-HO function at a specific SC only considers as candidate cells those that are listed in the so-called Neighbour Relation Table (NRT) associated to that SC. Some parameters could be statically configured from the NMS/EMS (where EMS encompasses here both the PNF EMS and the SC EMS) or dynamically adjusted at runtime by Self-x functions such as ANR, MRO and MLB [9]. In a general case, the RRM-HO and Self-x functions implemented in the VSCNO's RAN and in the SCNO's RAN are likely to differ (e.g., different vendor's equipment in each RAN with vendor-specific implementations of RRM/Self-x functions).

3.1 RRM-HO Function

The HO function commonly considers the measurement reports provided by the UE including e.g., Reference Signal Received Power (RSRP) and Reference Signal Received Quality (RSRQ) values for the serving and neighbour cells. Furthermore, the HO algorithm can also consider as an input the load level at neighbour cells. In this case, load information is provided by neighbour cells via X2 interface. A detailed list of HO parameters is found in [10]. They are associated to the detection of certain events used by the HO algorithm to trigger the execution of an HO (e.g. detecting that a neighbour cell becomes offset better than the serving cell, detecting that a neighbour cell becomes better than a threshold, etc.). For each event, tuneable parameters include offset values, hysteresis values, time to trigger, thresholds, etc.

Typically, within a RAN, all cells (from the same vendor) will implement the same HO function. However, by setting the HO parameters on a cell-by-cell basis, the behaviour of the HO function (e.g. the precise time that a HO decision is made) can be different in each cell. In order to properly steer the connected UEs across the VSCNO's RAN and the SCNO's RAN, neighbourhood relationships shall be properly captured in the corresponding cells. For the example of Fig. 2, the NRT at TB should include SA and SB to enable an incoming handover to the SCNO's RAN. Similarly, the NRT at SB should include TB and TC to enable handovers to the VSCNO's RAN.

As long as coverage overlapping and traffic steering strategies between the SCNO's RAN and each of the VSCNOs are likely to differ, the RRM-HO function shall be VSCNO-*aware* (i.e. the RRM-HO shall be able to associate E-RABs (E-UTRAN Radio Access Bearer [11]) with VSCNOs and enforce the VSCNO-*specific* policies). Furthermore, some of the parameters used by the RRM-HO function (e.g. offset values, hysteresis values) shall also be parametrised per VSCNO when pursuing an optimized operation of the HO function. This latter aspect is analysed in the following from the perspective of the Self-x functions that impact on the adjustment and optimization of the different parameters used by the RRM-HO function.

3.2 Self-x Automated Neighbour Relation

The configuration of the NRT in each of the cells can be realized through the Automated Neighbour Relation (ANR) function, avoiding the burden of human interactions between the VSCNO and SCNO to exchange information about the cells in close vicinity. From the SCNO perspective, each SC will maintain a single NRT list that includes neighbour SCs from the SCNO and, in overlapping coverage areas, the cells of the different VSCNOs.

The ANR function relies on different procedures to find new NRs (Neighbour Relations), such as UE-*assisted* neighbour discovery that uses UE measurements to identify new NRs, network listen measurements done by the eNodeB (eNB), and X2 assisted network discovery (e.g., when a neighbour eNB attempts an X2 connection setup with another cell, it is automatically added in the NRT of this cell) [12].

When a new neighbour is detected, certain procedures are to be used to setup the X2 with the new cell. From the MME (Mobile Management Entity) point of view, the eNB sends an eNB Configuration Transfer message to the MME. If the MME receives the SON Configuration Transfer IE (Information Element), it shall transparently transfer the SON Configuration TransferIE towards the eNB indicated in the Target eNB-ID IE which is included in the SON Configuration Transfer IE. On the way back, the MME will send an MME Configuration Transfer. The purpose of the MME Configuration Transfer procedure is to transfer RAN configuration information from the MME to the eNB in unacknowledged mode. For instance, the eNB retrieves the IP address from MME to setup the X2 interface. Thus, in this case, the serving eNB gets in touch with the MME (if an X2 connection does not exist) to assist itself for creating a X2 tunnel with the target eNB. Once this tunnel is established, the serving cell eNB forwards the CGI²-info to the target eNB. Thus, both the eNB's updates its own respective NRT.

Regarding UE measurements, a UE receives instructions about how to configure the measurement process. The UE will be indicated what frequencies to measure, possibly specifying as well the list of cells to measure in a given frequency (i.e., the cells that are defined in the NRT). This is indicated through dedicated RRC (Radio Resource Control) signalling for UEs in connected mode. Similarly, the UE will be instructed about the reporting criterion (i.e. the event that triggers the UE to send a report). Example events can be detecting that a neighbour cell becomes offset better than the serving cell, detecting that a neighbour cell becomes better than a threshold.

Given that the SCNO and a VSCNO will usually operate at different frequencies, the automated detection of neighbour cells from different RANs requires inter-frequency measurements, for which the UEs must be properly instructed while staying in the overlapping coverage area between the VSCNO and the SCNO's RAN. Then, on the one hand, UEs perform measurements at the frequency of the VSCNO while they are connected to the SCNO's RAN. On the other hand, UEs perform measurements at the frequency of the SCNO's RAN while they are connected to the VSCNO's RAN. For the example of Fig. 2, TB should configure the UEs to measure on the frequency that the SCNO is operating. In this way, UEs will be able to detect SA and SB, report measurements from these cells and the ANR at TB will update the NRT at TB, *accordingly* [13]. Once this is accomplished, HO from TB to SA or SB will be possible. Equivalently, SB should configure the UEs depending on the VSCNO that they belong to, so that each UE will measure on the frequency of its VSCNO. In this way, the UEs of the VSCNO illustrated in Fig. 2 will be able to detect TB and TC, so that the NRT at SB is updated accordingly. Therefore, measurement configuration of the UEs will be different depending on the VSCNO that they belong to.

3.3 Self-x Mobility Robustness Optimization

Mobility Robustness Optimisation (MRO) function will automatically set HO parameters, aiming at avoiding different HO problems, such as connection failures due to mobility (too late handover, too early handover, handover to wrong cell), unnecessary HOs and ping-pongs [14]. In general, optimized HO parameters will be different at

---

² For more information see, *inter-alia*: https://en.wikipedia.org/wiki/Common_Gateway_Interface.

each SC because each SC exhibits a particular situation with respect to its neighbours. This will be particularly relevant to be considered for SCs deployed in the overlapping coverage areas between the SCNO and VSCNO's RAN. In this case, different coverage footprints from different VSCNOs will lead to different types of HO problems experienced by the UEs of each VSCNO. Therefore, the MRO function at the SCNO's RAN shall exploit HO-*related* performance measurements per VSCNO which can be useful in detecting the likely different HO-*related* issues arisen between the SCNO's cells and the cells of different VSCNOs. Consequently, VSCNO-*specific* HO parameter settings shall be supported to achieve optimized HO operation for all VSCNOs.

For example, let us consider the situation illustrated in Fig. 3, for a UE heading from the SC to a VSCNO's cell. The overlapping area between the cell of VSCNO B and the SC is very small, while there exists a large overlapping between the cell of VSCNO A and the SC. In such situation, one could expect high call dropping rate (CDR) for UEs from VSCNO B due to too late handovers. The degradation of the CDR for VSCNO B in that specific cell should lead the MRO at the SC to conclude that the HO offsets should be reduced (so that, as soon as the VSCNO B's cell is detected, the HO is quickly executed). Regarding VSCNO A, if a low HO offset were defined for its UEs, these UEs would be handed-over to the VSCNO A's RAN at a very early stage. Then, depending on the HO algorithm and its parametrization at the VSCNO A's RAN, a ping-pong effect might arise and UEs could be handed-over again to the SCNO's RAN. In such case, the observation of a relevant ping-pong effect should lead the MRO to conclude that the HO offset should be increased, so that UEs from VSCNO A would not be handed-over until a much stronger signal from VSCNO A's cell was received. Regarding the HOs between two SCs of the SCNO's RAN, if the HO parameters were not sufficiently optimized and HO problems occurred, this would affect in the same way to all UEs, regardless of the VSCNO they are associated to. Thus, the MRO function would be in charge of tuning the setting of HO parameters in the SC, without making distinctions among VSCNOs.

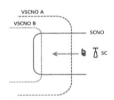

Fig. 3. Impact of different overlapping between the VSCNOs' cells and the small cells on the MRO function.

3.4   Self-x Mobility Load Balancing

The objective of Mobility Load Balancing (MLB) [15] is to distribute cell load evenly among cells or to transfer part of the traffic from congested cells to less loaded cells. MLB relies on exchanging cell specific load information between neighbour cells over the X2 interface (e.g. resource block usage separately for Guaranteed Bit Rate and non-Guaranteed Bit Rate E-RABs, available capacity that a cell can accept, etc.).

MLB function is a hybrid self-x function in which the MLB decisions are made at each SC according to policies controlled by the EMS and/or NMS [10]. MLB is supported by different procedures for transferring load between cells. For UEs in connected mode, MLB relies on the handover process, either by forcing HOs of specific UEs to a neighbour cell or by adjusting the HO parameters of a neighbour cell to facilitate that more UEs make HOs towards this cell. In this respect, the X2 interface includes a procedure to negotiate the changes in HO parameters between two neighbour cells, in order to facilitate a coordinated operation between them.

For UEs in idle mode, MLB relies on the modification of the cell reselection parameters (i.e. cell reselection offsets and cell reselection priorities) of each neighbour cell that are broadcast in the System Information Block (SIB) messages (e.g. SIB Type 4 for intra-frequency neighbours and SIB Type 5 for inter-frequency neighbours). When an idle mode UE detects a neighbour cell, it will use these parameters together with the received power to decide if it camps on this cell. Therefore, by adjusting these parameters, the MLB function can favour that more or less idle UEs camp in the different cells. However, in this case it is not possible to broadcast multiple parameters on a per VSCNO basis for a neighbour cell.

In a typical multi-tenant scenario, such as a stadium, high correlation among the traffic profiles (in time and space) associated to each VSCNO can be expected. Clearly, high load levels will be observed during e.g., a football match all over the stadium and for all the different VSCNOs simultaneously. However, some cases and situations (e.g., supporters from the visitor team are grouped in a certain area of the stadium, youth local supporters are usually grouped right behind the goalkeeper) as well as different market segments associated to the different competing operators acting as VSCNOs in the stadium (e.g., a low cost MNO will usually have youth customers, who in turn may stay in the areas of the stadium where attendees are standing) may lead to differences in the load levels associated to the different VSCNOs in the different cells. Therefore, the analysis of MLB strategies in multi-tenant scenarios requires further attention, since the load levels from the different VSCNOs in the different cells needs to be considered. At this point, it is worth remarking that X2 interfaces should be able to exchange load information on a per VSCNO basis.

To illustrate how the MLB actions can vary depending on the VSCNOs' load in different SCs, let us consider the example shown in Fig. 4a, with VSCNO A and B. Assuming the planned load level for each VSCNO, as shown in the left side of the figure, let us consider three different cases for the actual load distribution in small cell SC1, denoted as I, II, III. Case I corresponds to an overload situation in which the aggregate load of both VSCNOs in SC1 exceeds the maximum acceptable level in the cell (i.e., the overload situation is causing performance degradation). This overload situation is due to VSCNO A, whose load substantially exceeds its planned level. In order to handle this situation, RRM Congestion Control techniques can reduce the load by e.g. reducing the bit rate of best-effort traffic. Additionally, assuming that a neighbour small cell SC2 has some capacity available, the MLB can transfer part of the load of SC1 to SC2 through the adjustment of HO parameters. In this case, only the load of UEs located in the overlapping coverage area between the two cells can be transferred to SC2.

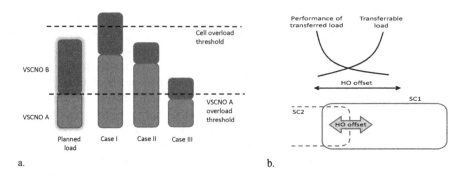

Fig. 4. (a) Illustration of MLB with different loads per VSCNO, and; (b) illustration of trade-off between transferable load and performance.

Given that the transfer of a UE from SC1 to SC2 may cause some performance degradation (e.g., the UE is transferred to SC2 even if the received signal from SC2 is worse than the signal from SC1), the adjusted parameters favouring the HO should initially be applied to UEs from VSCNO A, since it is the VSCNO originating congestion. The fraction of the load that can potentially be transferred to a neighbour cell depends on the UEs spatial distribution within the cell. In principle, UEs in SC1 that are in close vicinity to SC2 will be handed-over. By increasing the offset, UEs that are further away from SC2 can also be transferred, at the expense of a certain degradation in performance (e.g., lower peak bit rate), as Fig. 4a illustrates.

In Case II of Fig. 4a, the load of VSCNO A in SC1 exceeds its planned level, but this does not generate overload in the cell. In this case, MLB actions to transfer some of the excess load of VSCNO A to SC2 would be required if the excess of load of VSCNO A causes some degradation in the performance of VSCNO B. Finally, in Case III, neither the load of VSCNO A nor the load of VSCNO B exceeds the planned level in SC1. Therefore, MLB actions are not strictly needed in this situation. Still, in case that the load in SC2 were very low, transferring part of the load to this cell could be of interest if, by doing so, the performance observed by the UEs in SC1 could be improved.

In the case that load balancing involves two neighbour cells from different RANs (e.g., SB and TB in Fig. 2), the availability of X2 interface between the SCs of the SCNO and the cells of the VSCNO facilitates the coordination for MLB purposes. The X2 enables that an SC receives information about the available capacity that each neighbour cell can accept. Computation of the load per VSCNO should be supported and exchanged accordingly. Then, the MLB function at the SC can make decisions accordingly, thus minimising the risk that a HO is not accepted at a neighbour cell, or that a ping-pong occurs if the target cell of the VSCNO decides to make a HO back to the SC. Otherwise, in the absence of such load information, the SC can only initiate blind MLB actions to arbitrary neighbour cells, which would lead to ping-pong behaviour. Coordination of MLB at different SCs through the use of cSON will also avoid this undesired effect. Like in the previous example of Fig. 4b, MLB decisions will depend on how the UEs of each VSCNO are spatially distributed and on the overlapping coverage areas between the SC and cells of the different VSCNOs.

As discussed in the previous section, MLB is a self-optimization functionality that intelligently spread users across system resources to ensure QoS and improve edge users throughput. MLB is typically triggered in response to local instances of overload. This reactive approach enables overloaded cells to redirect a percentage of their load to neighbouring less loaded cells hence alleviating congestion problems. Traditionally, all users use the same set of handover parameters (e.g. hysteresis margin and time to trigger). Moreover, mobility and interference are normally treated separately. Ideally, a pro-active approach to MLB is needed for MLB offloading taking into account inter-related factors such as interference, load, speed and including an enhancement to small cell discovery and user association.

Standard MLB makes use of Cell Range Expansion (CRE), which is achieved by either cell coverage or mobility parameters adjustments. CRE increases the downlink coverage footprint of a low power cell by adding a positive bias value. Offloaded users may experience unfavourable channel from biased cells and strong interference from unbiased higher power cells. CRE forces alternate cell selection without considering loading or resource allocation in the corresponding cell. Re-association of a user to a cell other than the one offering the largest signal strength as is sometimes implemented by traditional MLB approaches described above, often leads to reduced desired signal level and an increase in interference level which results in an overall network performance degradation. Advanced MLB makes use of CRE together with the 3GPP release 10 Almost Blank Subframes (ABS) and reduce-power ABS (RP-ABS) features, designed for macrocell interference mitigation, in HetNets. ABS is a time domain interference avoidance technique, which improves the overall throughput of the off-loaded users by sacrificing the throughput of unbiased cells. Given an ABS ratio (i.e. a ratio of blank over total sub-frames), a user may select a cell with maximum ABS ratio. CRE together with ABS is classified as distributed cell association scheme. Advanced approaches such as multi cell load balancing in dense small cell deployments can help reduce blocking probability and improve network performance. Such action makes use of clustering of cells which in turn ensures that resources are appropriately allocated to groups of similar cells and the frequency of invocation of other SON algorithms is reduced, thereby minimising conflicts.

4 Conclusions

This work has analysed the implications of multi-tenancy on the RRM and Self-x functions that support mobility control. In particular, the key role of ANR function has been analysed and the requirement to configure the measurements of each UE depending on the tenant that they belong to has been addressed so that the neighbourhood relations are properly captured at both the VSCNO's RAN and SCNO's RAN. Regarding the MRO function, a distinction has to be made depending on the involved neighbour cells. For HOs involving the VSCNO's RAN and SCNO's RAN, the different coverage footprints of different VSCNOs will lead to different HO problems experienced by each VSCNO and, *consequently*, HO parameters have to be set differently for each VSCNO. For HOs between the small cells of the SCNO, the MRO function should not make distinctions among VSCNOs. Regarding the MLB function, it has been shown that MLB actions in a cell should be different for each VSCNO, taking into consideration the different load distribution of each VSCNO in each cell and the spatial distribution of the UEs within the cell.

Acknowledgments. The present work has been performed in the scope of the *SESAME* ("*Small cEllS CoordinAtion for Multi-tenancy and Edge services*") European Research Project and has been supported by the Commission of the European Communities (*5G-PPP/H2020, Grant Agreement No. 671596*).

References

1. Ramiro, J., Hamied, K.: Self-Organizing Networks. Self-Planning, Self-Optimization and Self-Healing for GSM. UMTS and LTE. Wiley, Hoboken (2012)
2. European Telecommunications Standards Institute (ETSI): TS 136 300: LTE; Evolved Universal Terrestrial Radio Access (E-UTRA) and Evolved Universal Terrestrial Radio Access Network (E-UTRAN); Overall description; Stage 2 (Release 13). ETSI (2016)
3. ETSI: TS 132 500: LTE; Self-Organizing Networks (SON); Concepts and Requirements (Release 12). ETSI (2015)
4. Small Cell Forum (SFC): SON API for Small Cells (Document 083.05.01). SFC (2015)
5. Chochliouros, Ioannis P., et al.: A model for an innovative 5G-oriented architecture, based on small cells coordination for multi-tenancy and edge services. In: Iliadis, L., Maglogiannis, I. (eds.) AIAI 2016. IAICT, vol. 475, pp. 666–675. Springer, Cham (2016). doi:10.1007/978-3-319-44944-9_59
6. SESAME Project (GA No. 671596). http://www.sesame-h2020-5g-ppp.eu/Home.aspx
7. ETSI: TS 132 425: Telecommunication Management; Performance Management (PM); Performance Measurements Evolved Universal Terrestrial Radio Access Network (E-UTRAN). ETSI (2012)
8. The Third Generation Partnership Project (3GPP): TS 36.420 v13.0.0: E-UTRAN; X2 General Aspects and Principles (Release 13). 3GPP (2015)
9. 3GPP: TS 32.522 v11.7.0: Self-Organizing Networks (SON) Policy Network Resource Model (NRM) Integration Reference Point (IRP); Information Service (IS) (Release 11). 3GPP (2013)
10. Sánchez-González, J., Pérez-Romero, J., Agustí, R., Sallent, O.: On learning mobility patterns in cellular networks. In: Iliadis, L., Maglogiannis, I. (eds.) AIAI 2016. IAICT, vol. 475, pp. 686–696. Springer, Cham (2016). doi:10.1007/978-3-319-44944-9_61
11. ETSI: TS 136 410: LTE; Evolved Universal Terrestrial Radio Access (E-UTRAN); s1 General Aspects and Principles. ETSI (2015)
12. Qualcomm Technologies Inc.: LTE Small Cell SON Test Cases. Functionality and Interworking. Qualcomm Technologies Inc., San Diego (2015)
13. 3GPP: TS 32.511 v11.0.0: Automatic Neighbour Relation (ANR) Management; Concepts and Requirements (Release 11). 3GPP (2011)
14. Zheng, W., Zhang, H., et al.: Mobility robustness optimization in self-organizing LTE femtocell networks. J. Wirel. Commun. Netw. **27**, 1–10 (2013)
15. Yamamoto, T., Komine, T., Konishi, S.: Mobility load balancing scheme based on cell reselection. In: Proceedings of ICWMC-2012, pp. 381–387. IARIA (2012)

Design of Virtual Infrastructure Manager with Novel VNF Placement Features for Edge Clouds in 5G

Ruben Solozabal[1], Bego Blanco[1(✉)], Jose Oscar Fajardo[1],
Ianire Taboada[1], Fidel Liberal[1], Elisa Jimeno[2], and Javier G. Lloreda[2]

[1] UPV/EHU, Bilbao, Spain
Begona.blanco@ehu.es
[2] Atos, Madrid, Spain

Abstract. This paper focuses on multi-tenant 5G networks with virtualization and mobile edge computing capabilities, in the scope of cloud-enabled small cell deployments. In this context, the work here presented deals with the service management and orchestration challenges that arise when handling service mapping on the multi-tenant distributed cloud-enabled radio access network architecture. For that aim, once analysed cloud edge services management and 5G network instantiation in the OpenStack platform, we modify the provided virtual infrastructure manager so as to incorporate virtual network function placement features of the SESAME environment. As main contributions, we adapt the OpenStack application instances to 5G Network Service instantiation, and we include an energy-aware and latency-constrained placement solution.

Keywords: Network Function Virtualization · Network Service instantiation · OpenStack · Cloud-enabled small cells

1 Introduction

In the framework of evolved 4G LTE architectures and future 5G mobile networks, different proposals are emerging aimed at overcoming the capacity limitations of current radio access networks (RANs) so as to improve user performance. The concept of Network Softwarization is being proposed to move some RAN-related functions back to shared hardware (HW) elements with high computational capacities, taking into account several trends such as Software Defined Networking (SDN) and Network Function Virtualization (NFV).

This kind of RAN systems concentrate different processing resources together to form a pool in micro servers. This aggregation of HW resources in shared locations not only reduces deployment costs, but also leverages low latency connections between different RAN processing units. Moreover, when these resources run over virtualized infrastructures, adding flexible and scalable HW resource management capabilities, the RAN becomes a Cloud RAN (C-RAN) [1].

The current trends towards cloudification of the RAN, enabling scalable and flexible RAN virtual network functions (VNFs), also allows reusing the available HW

© Springer International Publishing AG 2017
G. Boracchi et al. (Eds.): EANN 2017, CCIS 744, pp. 669–679, 2017.
DOI: 10.1007/978-3-319-65172-9_56

infrastructure for deploying service instances at the edge of the mobile network. In this way, networked applications could support low latency requirements in this mobile edge computing (MEC) paradigm. In this context, by making use of SDN, the data and control plane decoupling innovative feature of 5G is aligned with MEC principles. The application of MEC to 5G systems allows the physical separation of the planes, leaving the data plane close to the user in the network edge and uploading the centralized control plane to the cloud servers. Furthermore, when combining NFV with the aforementioned cloud computing concepts, a centralized NFV orchestrator (NFVO) and Virtual Infrastructure Manager (VIM) are the responsible for the on-boarding of new Network Services (NS), being the VIM in charge of the management of network service lifecycle, as stated in ETSI architecture standards [2].

The work presented here is part of the SESAME EU-funded H2020 project [3, 4], which contributes to the development of 5G networks that make use of NFV and MEC, focusing on multi-tenant cloud-enabled small cells (CESCs). In this way, the concept of CESC refers to a complete SC that contains a micro server denominated Light Data Center (Light DC), offering, thus, computing (Virtual Machines, VM), storage and radio resources. A number of CESCs form a cluster whose virtualized physical resources are shared and controlled by the VIM. In this manner, the VIM is responsible for controlling and managing the compute, storage, and network resources to allocate the VNFs within the CESC cluster.

As a cloud management system, the OpenStack cloud-computing software [5] integrates networking and inter-VM connectivity aspects, driving dynamically instantiated software switches within computing nodes (hosting VMs). It provides the instantiation of multi-tenant networks, and supports the creation of networks and their interconnection via virtual routers, which can route traffic between internal and external networks. Hence, even though OpenStack can fulfil the role of VIM, which provides application instances, it lacks on NS instantiation itself for the virtualized 5G framework under study.

Besides, a mechanism that suitably allocates the softwarized components of a NS onto the Light DC resources (for instance, mapping the VNFs that compose a NS into VMs) becomes essential, which is not even standardized. The OpenStack open source cloud computing platform includes the Nova scheduler that allows basic VNF placement solutions based on resources (CPU or/and RAM) usage. Nevertheless, designing intelligent VNF placement algorithms is fundamental to achieve a high level of quality of service.

Therefore, the contributions of the presented work are twofold. On the one hand, we provide the adaptation of OpenStack application instances to NS instantiation. On the other hand, we incorporate novel VNF placement features, including an energy-aware placement proposal subject to NS latency constraints.

This paper is organized in six sections. First, Sect. 2 deals with cloud architectures evolution, while Sect. 3 with edge computing in 5G network using OpenStack. Next, Sect. 4 analyses the VNF placement in SESAME environment, so as to then incorporate a placement solution in the OpenStack platform in Sect. 5. Finally, Sect. 6 summarizes the main conclusions.

2   Evolution of Mobile Network to Cloud Architectures

The high degree of centralization in C-RAN entails a network architecture where all the baseband processing is made by Base Band Units (BBU) at centralized data centres, and radio signals are exchanged with the Remote Radio Heads (RRH) over high speed low-latency connections that constitute the mobile fronthaul. This centralization of resources in cloud data centers imposes stringent requirements to the front-haul connections in terms of throughput and latency. Therefore, all those cells with limited network access would not be able to offer these types of services.

Alternatively to fully centralized RAN, the partially centralized RAN approach allows splitting the Evolved Node B (eNB) functionalities in a flexible way. Depending on the selected functional split, the associated requirements for the remote and central HW for the front-haul connection are variable.

Figure 1 below illustrates different alternatives for the centralization of RAN functions. In fully centralized RAN case, the RRH only performs the radiofrequency (RF)-related operations such as the transmission and reception of the radio signals. The upper protocol stack layers are performed in the centralized entity.

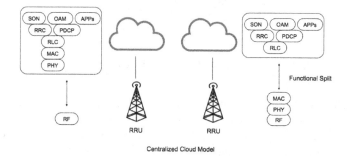

Fig. 1. Centralized C-RAN

Beyond the pure centralization of an eNB functions, one of the emerging technologies to cope with more personalized and user-centric service provisioning is the novel MEC. This may be exploited to deploy proximity-enabled services with close-to-zero latency characteristics, in order to optimize the management of future mobile networks. Regardless of the adopted architecture for C-RAN, MEC-driven service instances must be deployed over the cloud resources avail-able at the RAN side.

In this centralized solution, the upper RAN functions are located in powerful data centers that are ideally connected to the RRHs through high-speed and low-latency fronthauls. Yet, high fronthaul delays may degrade the performance of certain novel edge services that require close-to-zero latencies as prescribed by 5G objectives. Alternatively, nowadays CESCs architecture may become better suited for deploying mobile edge services. In that case, some processing and storage resources are placed close to the RRH, and thus, the fronthaul delay is significantly reduced. Deploying huge data centres implies a series of requirements in terms of space, energy, etc. Hence, this second option envisages the deployment of a series of HW resources with limited capacity and requirements and in a distributed configuration, which may ideally collaborate to provide some edge computing capabilities.

The proposed CESCs include a highly efficient microserver with a limited set of virtualized resources offered to the cluster of small cells. As a result, a Light DC is created and commonly used for deploying mobile edge computing functionalities (Fig. 2).

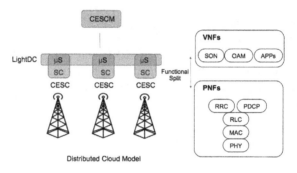

Fig. 2. Distributed C-RAN

This solution based on distributed cloud is especially relevant to enable flexible deployment of SCs, and particularly attractive for targeting currently deployed network architectures and special limited-access scenarios. In the former case, a SC operator may think about endowing its deployed network with novel mobile edge capabilities by gradually upgrading the Customer-Premises Equipment (CPE) without requiring changing wired connections. Distributed computing and storage capabilities associated to the CPEs arises as an affordable and scalable solution.

Hybrid approaches have been proposed to deal with the transitions from 4G specific hardware based architectures to software based 5G platforms. Both centralized and distributed clouds will coexist during this evolution to a completely softwarized central C-RAN. Multiple clouds can work together under orders from the same orchestrator to develop a hybrid cloud. In this model, VNFs can be spread between centralized and CESC distribute clouds. Depending on the functional split, central cloud can take part in a different layer of the softwarized upper layer protocols. Depending on the case, it can process just the VNFs form the upper protocol layers or the whole softwarized stack as initially proposed. A Hybrid NFV manager is proposed to orchestrate both cloud in a unified manner.

As the functional split encompass lower layers on the CESC Cluster architecture, centralized clusters will take on major relevance (Fig. 3).

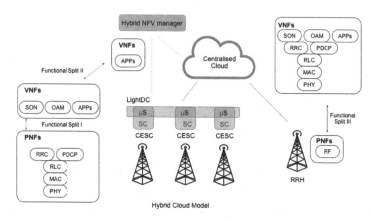

Fig. 3. Hybrid C-RAN

3 5G Edge Computing Using OpenStack

As stated before, one of the key elements in 5G is MEC, which uses NFV infrastructure to create a small cloud at the edge of the network. Experts are advocating for going from large-centralized cloud computing infrastructures to smaller ones massively distributed at the edge of the network. Referred to as 'fog/edge computing'. As seen in Sect. 2, the centralized and CESC distributed clouds will coexist on the edge of the network.

To favour the adoption of this decentralized model of the cloud computing paradigm, the development of a system in charge of turning a complex and diverse network of resources into a global cloud is critical.

This way multiple clouds residing in the edge could be seen as one by the CESC Manager (CESCM). VNFs could be deployed in a distributed manner between physically separated clouds. Having in mind Wide Area Network limitations (in terms of latency/bandwidth).

OpenStack Massively distributed work group (WG) [6] is currently working to extent current OpenStack mechanisms in order to include this feature.

As analysed in 2016 Austin OpenStack Summit [7] some alternatives could be taken. Orchestration of clouds are the first approaches that are considered when it comes to operate and use distinct clouds. Each micro DC hosts and supervises its own cloud and a brokering service is in charge of provisioning resources by picking them on each cloud (Fig. 4).

Another option chosen by the WG is delivering an OpenStack architecture that can natively cooperate with other instances, giving the illusion of a global cloud. The direction in which is working the project OpenStack is TriCircle, [8] to provide networking automation across Neutron in multi-region OpenStack deployments.

TriCircle provides a proxy for local OpenStack instances running on the edge. It provides a single API to the orchestrator, which now just need to talk with TriCircle component instead of doing with numerous endpoints.

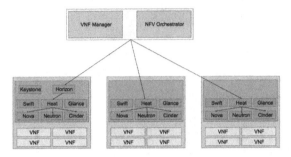

Fig. 4. Orchestration of multiple clouds. First approach

From the control plane view, Tricircle enables Neutron(s) in multi-region OpenStack clouds working as one cluster, and allows the creation of global network/router etc. abstract networking resources across multiple OpenStack clouds. From the data plane view (end user resources view), all VMs are provisioned in different cloud but can be inter-connected via the global abstract networking resources with tenant level isolation.

4 Placement Distinctive Features of the SESAME Environment and Implications in OpenStack

The mixed radio-cloud environment proposed in SESAME has some distinctive features due to the requirements of 5G communications and the distributed underlying virtualized architecture. This special characteristics impose some particularities in the VNF placement process that are not currently covered in OpenStack (Fig. 5).

As previously mentioned, SESAME environment offers a multi-tenant distributed C-RAN architecture to place intelligence at the network edge through the use of virtualization techniques [9]. This way, a SC Network Operator (SCNO) can open its network deployment to third party virtual operators in order to provide MEC services to

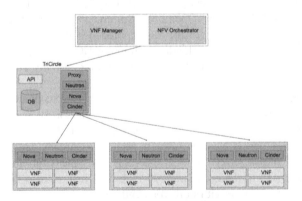

Fig. 5. TriCircle multiple cloud architecture

their end users and improve their 5G experience. Figure 6 shows an example scenario that hosts three Virtual SCNOs (VSCNO) requesting one or more NSs over the shared C-RAN.

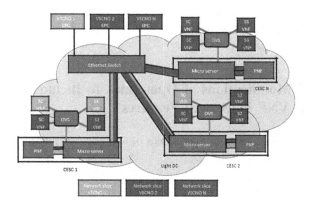

Fig. 6. Network slicing in SESAME

The role of the placement algorithm is to assign a network slice to each tenant according to the requirements of the NSs offered by the virtual operators. Each VSCNO offers a collection of NSs defined by a VNF chain, plus the correspondent Key Performance Indicators (KPIs) captured in the Service Level Agreement (SLA). The VNF chain includes both radio-related and service-related functionalities. In this sense, every CESC must allocate a Small Cell Common VNF (SC-C-VNF) to support the control plane communications. In addition, every NS must include a Small Cell VNF (SCVNF) at the beginning and the end of the service chain to manage the GTP tunnel of 5G user communications. Moreover, each VSCNO only instantiates a single SCVNF that is shared among all the NSs belonging this tenant. From the service perspective, the VNFs are characterized to use computing and storage resources according to the aggregated bit rate served by the correspondent NS. On the other hand, the service KPIs of the SLA between the VSCNO and the C-RAN operator for each NS de ne network parameters, such as aggregated user bit rate or the maximum accepted latency.

All these special characteristics of SESAME related to 5G impose a collection of constraints to the design of the placement algorithm. The remaining special features of the placement algorithm are related to the virtualized infrastructure.

The proposed placement mechanism must allocate all the VNFs that compose the requested NSs into the available resources in the CESC cluster, complying with the correspondent SLA. The virtualized resources of the cluster include a number of computing assets, RAM and storage. It is important to note that processor sharing for VNF instantiation is not considered in the scope of SESAME. Therefore, each CPU of the Light DC will execute a single VM serving one VNF. Nevertheless, if a single VM is not capable to support the traffic of a NS, the VNF instance can be distributed over more VMs. Finally, among the virtualized resources, the Light DC may also include eventual hardware appliances to help meeting the KPIs of the SLA. These hardware accelerators improve the performance of heavy VNFs in terms of latency, but at higher energy cost.

This complex context compels the design of a novel service function allocation strategy that adapts the default behaviour of OpenStack to fit the needs of SESAME environment. Next section presents the proposal of an algorithm that is integrated in OpenStack to perform the placement of multi-tenant edge computing services including the aforementioned features.

5 Proposed Modifications of OpenStack to Include SESAME VNF Placement Features

This section presents the operation principles of the proposed placement algorithm and its integration into OpenStack.

5.1 Algorithm Description

The general VNF placement problem is an optimal resource consumption problem. Since the SESAME Light DC is designed to reduce power consumption and cost, the optimization objective considered for the placement algorithm is energy consumption minimization with a maximum latency constraint per NS.

Figure 7 depicts the operation of the placement algorithm upon the reception of a NS request from the VSCNO. The NFVO gets the NS along with the KPIs from the SLA and launches a query to the NS catalogue, which returns the VNF chain of the correspondent NS.

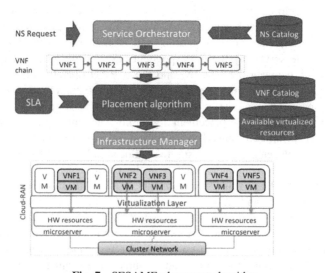

Fig. 7. SESAME placement algorithm

This ordered sequence of VNFs is passed to the placement algorithm to start the decision-making process about the allocation of the VNF instances. The placement algorithm checks the available resources in the NFVI catalogue and the instantiation requirements for the VNFs in the VNF catalogue. With this input, the algorithm applies the SESAME specific constraints and performs the optimization cycle to obtain the placement decision that minimizes energy consumption while observing the maximum latency constraint imposed by the SLA of the NS. Finally, the VIM maps the placement decision to the underlying physical resources that compose the NFV Infrastructure (NFVI).

SESAME uses OpenStack as the VIM of the CESC cluster. Therefore, the next section analyses the possibilities to insert the described placement algorithm into OpenStack operation.

5.2   Introduction of the Placement Algorithm into OpenStack

OpenStack software is used as the VIM to provide Infrastructure-as-a-Service (IaaS) capabilities. Coming from the cloud perspective, this implies that Open-Stack is designed to provide a transparent view of the set of available computing nodes, i.e. nodes running the OpenStack compute software to the tenant. In the event of a new VM or VNF deployment request, the tenant does not need to know, in fact typically does not care, in which node of the whole infrastructure the VM is going to be launched as long as it covers the hardware demands.

The perspective studied in SESAME however differs from the traditional cloud environment. Because of the virtual infrastructure is composed by a set of CESCs distributed over a specific area, the localization of end users connected to the platform, and therefore, the CESCs that have more active users, becomes relevant, e.g. to reduce delay by placing VNFs as closest as possible to end users, to redistribute the location of existing VNFs to reduce energy consumption.

Due to the low level requirements that are considered to deploy the VNFs in terms of information about the status of the infrastructure, there has been decided to enhance OpenStack's Nova scheduler algorithm [10] with the logic needed for the SESAME scenario.

Nova scheduler module takes two steps to determine the appropriate host, i.e. compute node, in which to deploy a VM. First, it applies a series of filters to the set of available compute nodes to eliminate those hosts that do not comply with the requirements of the VM, for example in terms of available memory or due to lack of specific hardware. The nodes that result as output of this process are then weighted based on configurable metrics, yielding an ordered list of candidate hosts depending on suitability (Fig. 8).

Following the design of OpenStack architecture, in order to extend the functionality of the stock filters, the implementation needs to inherit from the abstract class BaseHostFilter, where the host passes method has to be implemented; this method returns, in binary, the hosts that has passed the filter based on two parameters (host state and filter properties dictionary), the filter will reject the host not considered. To integrate the placement algorithm, two configuration steps are needed to import the custom new filter; first, the implemented algorithm needs to be place inside the

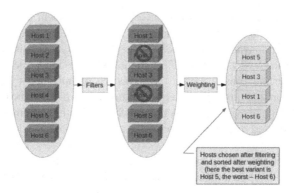

Fig. 8. Filtering workflow

nova/scheduler/filters catalogue, and etc./nova/nova.conf needs to be updated with the new configuration this INI le is copied in every compute node. This le is designed to be configurable and easy to use, and it will run the nova-* services:

- scheduler.driver = nova.scheduler.FilterScheduler
- filter_scheduler.available_filters = nova.scheduler.Filters.all_filters
- filter_scheduler.available_filters = my_lter.MyFilter
- filter_scheduler.enabled_filters = RamFilter, ComputeFilter, MyFilter.

This configuration allows Nova to use the FilterScheduler, default and customized filters, for the scheduler driver. The RamFilter, ComputeFilter, and MyFilter are used by default when no filters are specified in the request. The order the filters has been defined in the configuration parameters, defines the iteration over the filter scheduler to endorse the possible hosts; this procedure sift out based on criteria of reducing the non-valid hosts for placement. Filter Scheduler refers to the hosts weights in the selection process. Hosts are given a value that determines their priority in the list of available hosts. OpenStack offers the possibility of customizing the value of the default weights by simply changing configuration les as well as defining own weights. For that, similar to the filtering process, the class BaseHostWeigher should be inherited, in order to implement the weight object method. All the weights are normalized before-hand. As result of the implementation process, the algorithm will sort selected host by the largest weight as suitable placement solution.

6 Conclusions and Foreseen Future Research Lines

This paper has analysed the evolution of mobile networks to cloud architectures and the use of OpenStack to instantiate 5G network services in the scope of SESAME project. The proposed mixed radio-cloud architecture offers valuable possibilities of introducing intelligence in the proximity of the end user through the instantiation of MEC services, but also presents new challenges related to orchestration and management issues. In particular, the SESAME architecture introduces significant design requirements into the placement mechanism of the VNF chains that compose the NSs. These requirements are related to 5G communication characteristics, multi-tenant nature of the service providers and the particularities of the distributed NFVI. We propose a novel energy-aware and latency-constrained placement approach that addresses all the aforementioned design specifications and suggests a modification to OpenStack's Nova scheduler. Future work will address the refinement of the placement algorithm and its proper integration in the work ow of OpenStack.

Acknowledgements. The research leading to these results has been supported by the EU funded H2020 5G-PPP projects SESAME (grant agreement no. 671596) and ESSENCE (grant agreement no. 761592) and national Spanish project 5RANVIR (no. TEC2016-80090- C2-2-R).

References

1. Wu, J., Zhang, Z., Hong, Y., Wen, Y.: Cloud radio access network (C-RAN): a primer. IEEE Netw. **29**(1), 35–41 (2015)
2. ETSI: Network Functions Virtualization (NFV); Management and Orchestration (2014)
3. Giannoulakis, I., Fajardo, J.O., Lloreda, J.G., Khodashenas, P.S., Ruiz, C., Betzler, A., Kafetzakis, E., Perez-Romero, J., Albanese, A., Paolino, M., et al.: Enabling technologies and benefits of multi-tenant multi-service 5G small cells. In: 2016 European Conference on Networks and Communications (EuCNC), pp. 42–46. IEEE (2016)
4. Khodashenas, P., Ruiz, C., Riera, J.F., Fajardo, J., Taboada, I., Blanco, B., Liberal, F., Lloreda, J., Perez-Romero, J., Sallent, O., et al.: Service provisioning and pricing methods in a multi-tenant cloud enabled RAN. In: 2016 IEEE Conference on Standards for Communications and Networking (CSCN), pp. 1–6. IEEE (2016)
5. OpenStack. https://docs.openstack.org/
6. Massively Distributed Cloud OpenStack. https://wiki.openstack.org/wiki/MassivelyDistributedClouds
7. Khan, T., Hoban, A., Prithiv Mohan, P., Thulasi, A., Willis, P.: Distributed NFV and OpenStack: challenges and potential solutions. OpenStack Summit - Austin (2016). https://www.openstack.org/assets/presentation-media/OpenStack-2016-Austin-D-NFV-vM.pdf
8. OpenStack TriCircle. https://wiki.openstack.org/wiki/Tricircle
9. Blanco, B., Fajardo, J.O., Giannoulakis, I., Kafetzakis, E., Peng, S., Perez-Romero, J., Trajkovska, I., Khodashenas, P.S., Goratti, L., Paolino, M., Sfakianakis, E., Liberal, F., Xilouris, G.: Technology pillars in the architecture of future 5G mobile networks: NFV, MEC and SDN. Computer Standards & Interfaces (2017). http://www.sciencedirect.com/science/article/pii/S0920548916302446
10. Scheduling-OpenStack Configuration Reference-kilo. https://docs.openstack.org/kilo/cong-reference/content/sectioncompute-scheduler.html

On Introducing Knowledge Discovery Capabilities in Cloud-Enabled Small Cells

Jordi Pérez-Romero[1(✉)], Juan Sánchez-González[1], Oriol Sallent[1], and Alan Whitehead[2]

[1] Universitat Politècnica de Catalunya (UPC),
c/Jordi Girona, 1-3, 08034 Barcelona, Spain
jorperez@tsc.upc.edu
[2] ip.access, Cambridge, UK

Abstract. The application of Artificial Intelligence (AI)-based knowledge discovery mechanisms for supporting the automation of wireless network operations is envisaged to fertilize in future Fifth Generation (5G) systems due to the stringent requirements of these systems and to the advent of big data analytics. This paper intends to elaborate on the demonstration of knowledge discovery capabilities in the context of the architecture proposed by the Small cElls coordinAtion for Multi-tenancy and Edge services (SESAME) project that deals with multi-operator cloud-enabled small cells. Specifically, the paper presents the considered demonstration framework and particularizes it for supporting an energy saving functionality through the classification of cells depending on whether they can be switched off during certain times. The framework is illustrated with some results obtained from real small cell deployments.

Keywords: Knowledge discovery · Small cells · Classification · Energy saving

1 Introduction

As a next step in the evolution of cellular communication systems, industry and academia are focused on the development of the 5th Generation (5G) of mobile systems that targets a time horizon beyond 2020. 5G intends to provide solutions to the continuously increasing demand for mobile broadband services associated with the massive penetration of wireless equipment such as smartphones, tablets, the tremendous expected increase in the demand for wireless Machine To Machine communications and the proliferation of bandwidth-intensive applications including high definition video, 3D, virtual reality, etc. Requirements of future 5G system have been already identified and discussed at different fora [1, 2].

It is expected that 5G networks will also be fueled by the advent of big data and big data analytics [3]. The volume, variety and velocity of big data are simply overwhelming. Nowadays, there are already tools and platforms readily available to efficiently handle this big amount of data and turn it into value by gaining insight and understanding data structures and relationships, extracting exploitable knowledge and deriving successful decision-making. While applications of big data and big data

© Springer International Publishing AG 2017
G. Boracchi et al. (Eds.): EANN 2017, CCIS 744, pp. 680–692, 2017.
DOI: 10.1007/978-3-319-65172-9_57

analytics are already present in different sectors (e.g. entertainment, financial services industry, automotive industry, logistics, etc.), it is envisaged that they will play a key role in 5G to extract the most possible value of the huge amount of available data generated by mobile networks and for efficiently delivering mobile services.

In this context, this paper supports the idea that Artificial Intelligence (AI) mechanisms, which intend to develop intelligent systems able to perceive and analyse the environment and take the appropriate actions, will fully fertilize in the 5G ecosystem. In [4] the authors presented a general framework for the application of AI-based knowledge discovery mechanisms relying on machine learning as a means to extract models that reflect the user and network behaviours. The paper identified different candidate tools and discussed the applicability in the development of Self-Organizing Network (SON) functionalities, also known as Self-X functionalities, for automating the operation of a cellular network [5]. In turn, a particularization of this general framework was presented in [6] focusing on extracting knowledge from the time domain traffic patterns of the different cells in a network. Two applicability use cases were elaborated, dealing with energy saving and spectrum management. Similarly, [7] focused on the identification of user mobility patterns in cellular networks by means of clustering techniques and on its applicability in the context of SON.

Relying on these prior works, this paper intends to further elaborate on the demonstration of the knowledge discovery capabilities in the context of the architecture proposed by the Small cEllS coordinAtion for Multi-tenancy and Edge services (SESAME) project [8] that deals with multi-operator cloud-enabled small cells. The proposed framework is particularized for supporting an energy saving Self-X functionality through the classification of cells depending on whether they can be switched off during certain times. To illustrate the operation of the process, the paper presents some results obtained from real small cell deployments.

The rest of the paper is organized as follows. Section 2 summarizes the architecture of the SESAME project, while Sect. 3 presents the considered demonstration framework for introducing knowledge discovery capabilities in this architecture. Then, Sect. 4 particularizes the framework for the energy saving use case and discusses the implementation of the building blocks for classifying the different cells. This is followed by Sect. 5, which provides some illustrative results of the proposed framework. Finally, conclusions are summarized in Sect. 6.

2 SESAME Architecture

The SESAME project [8] focuses on the provision of Small Cell as a Service (SCaaS) under multi-tenancy, exploiting the benefits of Network Function Virtualisation (NFV) and Mobile Edge Computing (MEC). For that purpose, it proposes the Cloud-Enabled Small Cell (CESC) concept, a new multi-operator enabled Small Cell (SC) that integrates a virtualized execution platform for executing novel applications and services inside the access network infrastructure. In general terms, SESAME scenarios assume a certain venue (e.g. a mall, a stadium, an enterprise, etc.) where a Small Cell Network Operator (SCNO) is the SCaaS provider that has deployed a number of

CESCs that provide wireless access to end users of different Virtual Small Cell Network Operators (VSCNOs), according to specific Service Level Agreements (SLAs).

The SESAME architecture is presented in Fig. 1 [9]. The CESC consists of a Small Cell Physical Network Function (SC PNF) unit, where a subset of the SC functionality is implemented via tightly coupled software and hardware, and a micro server that supports the execution of Virtualised Network Functions (VNFs), which provide the rest of the SC functionality together with other added-value services. The CESCs support the Multi-Operator Core Network (MOCN) sharing model of 3GPP [10], which allows them to offer access over shared radio channels to multiple operators' core networks. Accordingly, each CESC is connected with the Evolved Packet Core (EPC) of each VSCNO through an S1 interface.

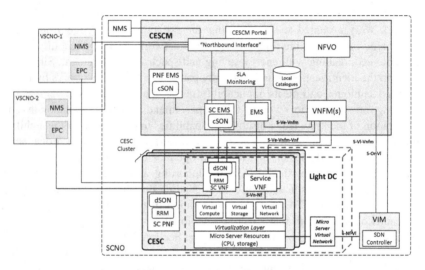

Fig. 1. SESAME architecture

The physical aggregation of a set of CESCs, denoted as a CESC cluster, gives the possibility to jointly operate the computational, storage and networking resources of the micro servers as a single virtualised execution infrastructure, denoted as Light Data Centre (Light DC).

The CESC Manager (CESCM) is the central service management component in the architecture that integrates the traditional 3GPP network management elements and the novel functional blocks of the NFV-MANO (Network Function Virtualization Management and Orchestration) framework. Configuration, Fault and Performance management of the SC PNFs and VNFs is performed through the Element Management System (EMS). In turn, the lifecycle management of the VNFs is carried out by the VNF Manager (VNFM), while the Network Functions Virtualization Orchestrator (NFVO) composes service chains constituted by one or more VNFs running in one or several CESCs and manages the deployment of VNFs over the Light DC with the support of the Virtualized Infrastructure Manager (VIM).

The CESCM is connected to the Network Management System (NMS) of the SCNO and the VSCNOs. Besides, it includes a portal that constitutes the main graphical frontend to access the SESAME platform for both SCNO and VSCNOs.

The SESAME architecture supports Self-X functions to tune global operational settings of the SC (e.g., transmit power, channel bandwidth, electrical antenna tilt) as well as specific parameters corresponding to Radio Resource Management (RRM) functions (e.g., admission control threshold, handover offsets, packet scheduling weights, etc.). Self-X functions can be centralised (cSON) at the EMS, distributed (dSON) at the CESCs or hybrid if they include both centralised and distributed components.

3 Implementing Knowledge Discovery Capabilities as Part of the SESAME Demonstration Framework

The introduction of knowledge discovery capabilities in a wireless network provides the ability to smartly process input data from the environment and come up with knowledge that can be formalized in terms of models and/or structured metrics that represent the network behaviour. This allows gaining in-depth and detailed knowledge about the network, understanding hidden patterns, data structures and relationships, and using them for a making smart network planning and optimisation decisions. In this way, the extracted knowledge models can be used to drive the decision-making of the actions associated to different Self-X functionalities.

Knowledge discovery is supported by machine learning tools to perform the mining of the data. Extracted knowledge models can be defined at different levels: cell level (contains the characterisation of the conditions on a per cell basis), cell cluster level (characterisation of groups of cells built according to their similarities) and user level (contains the characterisation of the conditions experienced by individual users).

Based on the above, Fig. 2 presents the considered framework for demonstrating the introduction of knowledge discovery capabilities in the context of SESAME. It is associated to the EMS, which encompasses both the PNF EMS and the SC EMS modules of the architecture shown in Fig. 1. The different elements of the considered framework are discussed in the following.

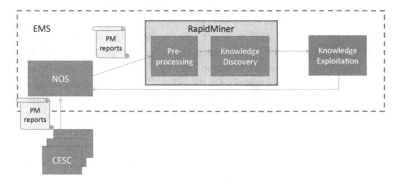

Fig. 2. Demonstration of knowledge discovery in SESAME

3.1   Network Orchestration System (NOS)

The SESAME EMS is based on the EMS of small cell vendor ip.access. The ip.access Network Orchestration System (NOS) provides configuration, fault and performance management features for small cells and related network elements. In SESAME, these elements include the SESAME CESC and the VNFs hosted by the CESC plus a collection of virtual network operators (VNOs), their individual virtual cells and associated SLA data.

The configuration and fault management aspects of the NOS are based on the ITU X.730 series of recommendations [11–14]. It represents these elements and their functions as managed objects [11]. Elements and optional functions are provisioned by creating managed objects and defining the values of their configurable attributes [12]. Once in service, the element or function represented by a managed object is able to report its state to the NOS [13]. Similarly, when a network element or function encounters a fault condition an alarm is raised on the managed object that represents it in a manner consistent with [14]. The procedural aspects of performance management reports are based on the concepts set out in [15] and the file format used conforms to [16].

Managed objects are organised in a tree structure with the relationship between objects being captured by containment. In the SESAME context, there are two sub-trees of managed objects that have special significance.

- The CESC managed object is a collection object beneath which CESC objects are provisioned. Each CESC has a similar sub-tree that comprises (see Fig. 3): (i) Exactly one SC-PNF object. The PNF represents the configuration of the physical cell that is associated in a one-to-one relationship with the CESC. In SESAME, the PNF function is provided by the ip.access E40 LTE AP. (ii) Exactly one Small Cell Common VNF (SC-C-VNF) object. The function represented by this managed object interfaces with the physical cell to split the control plane into separate, per virtual network, slices. (iii) From one to six SC-VNF objects. Each such managed represents the control plane processing associated with a single virtual network slice and maintains a dedicated S1 connection into the associated EPC. As each of the above are discrete managed entities, there is also an associated Connection object for each function that represents the management link between the PNF or VNF and the NOS.
- The VSCNO managed object is a collection object beneath which VSCNO objects are provisioned. Each VSCNO object represents the data corresponding to a specific virtual network operator and is a sub-tree comprising (see Fig. 3): (i) A single VSCNO object. Each such object captures the key properties of a specific virtual network operator such as their name, Public Land Mobile Network IDentifier (PLMN ID) and access credentials for the NOS. This object also acts as a container for the child objects described below that provide details of the operator's virtual cells and SLAs. (ii) A single Virtual Cells (vCells) object. This is a collection object beneath which Virtual Cell objects are created in order to provision a new virtual cell. (iii) Zero or more Virtual Cell objects. Each such object represents the parameters a single virtual cell and contains a link to the CESC that hosts it. (iv) A single Mobility Management Entity (MME) Pools. (v) A single SLA (vSlas) object

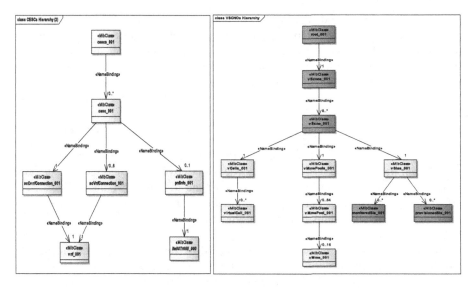

Fig. 3. Managed object sub-trees for the CESC (left) and the VSCNO (right).

beneath which SLA objects are created. (vi) At least one Provisioned SLA object. Each such object represents the details of a "network slice" that is applied when provisioning a new virtual cell. The network slice defines parameters such as the maximum number of UEs that are supported by the virtual cells and the maximum uplink and downlink bandwidth available to these UEs. (vii) Zero or more Monitored SLA objects. Each such object represents a set of criteria that are used to assess the performance of a set of virtual cells and the action to take when any of these criteria are not met.

3.2   Performance Management (PM) Reports

Performance Management reports are XML files conforming the format described in [16]. They are produced according to a configured Reporting Interval and each file contains one or more Granularity Periods. The Granularity Period defines the time frame across which measurements are collected and aggregated and is typically defined to be in the range of 5 min to 24 h. The granularity period of all the sample data in this study was set to one hour. The Reporting Interval is typically a multiple of the Granularity period. In the sample data used for this study, the Reporting Interval was set to either one hour or 24 h.

Within each file, performance measurements are organised into groups of related items known as packages. Each file may contain up to 128 performance counters organised in to 27 packages. Some example packages are: (i) Access Control and Admission Control Packages that record the number of attempted and failed attempt to access the cell and establish radio bearers. (ii) Hand-in and Hand-out packages that record the number of attempted and successful handover procedures. (iii) A GTP-U

Usage Package that records the number of uplink and downlink GTP-U packets sent and received, the number of packets lost plus the total number of octets sent and received. (iv) A User Plane Package that records the number of call attempts by Radio Access Bearer (RAB) type, the maximum and mean number of simultaneous calls, uplink and downlink bandwidth utilisation by RAB type.

3.3 Pre-processing, Knowledge Discovery and Knowledge Exploitation

The pre-processing stage takes as input the PM files generated by the NOS and extracts the relevant metrics to be used by the knowledge discovery depending on the use case in hand. For that purpose, this stage can combine multiple PM files associated to different cells and/or time periods. Then, the knowledge discovery stage includes the machine learning algorithms to carry out the mining of the input data and extract the knowledge models.

Both the pre-processing and the knowledge discovery stages are implemented by means of the RapidMiner Studio Basic tool [17]. It is a powerful visual design environment for rapidly building complete predictive analytic workflows and incorporates multiple pre-defined data preparation and machine learning algorithms.

Finally, the knowledge exploitation stage applies the obtained knowledge models to drive the decision-making associated to different Self-X functionalities. As shown in Fig. 2 this stage can interact with the NOS to configure specific SC parameters.

4 Use Case: Energy Saving

The considered use case to illustrate the operation of the proposed framework is the energy saving Self-X functionality, which intends to reduce the overall energy consumption associated to the small cells deployed by the SCNO. In this case, the energy reduction is achieved by switching off the cells that carry very little traffic at certain periods of the day (e.g. at night) and making the necessary adjustments in the neighbour cells so that the existing traffic can be served through another cell. In this context, the knowledge discovery framework applies a classification methodology for identifying candidate cells to be switched-off based on their time domain traffic patterns. The automation of this procedure based on expert criteria captured in a training set becomes particularly useful considering that networks in the envisaged ultra-dense scenarios for future 5G systems can comprise several tens of thousands of cells. Therefore, it is not practical that a human expert can make this classification manually.

Based on the above, the classification categorizes the cells in the following classes:

- Class A: Candidate cell to be switched off
- Class B: Cell that cannot be switched off.

It is worth mentioning that the final decision on whether or not to switch off a cell will make use of this classification as well as other possible inputs which are out of the scope of this paper (e.g. the neighbour cell lists to ensure that traffic generated in a cell that has been switched-off can be served through another cell).

The following sub-sections illustrate the different steps of the proposed classification approach, whose mathematical details are presented in [6].

4.1   Pre-processing Stage

The PM files generated by the NOS system include a number of XML files corresponding to different cells and periods of time. Each XML file includes metrics associated to different time instants. In turn, the considered classification process is based on the time domain traffic pattern of the different cells. For that purpose, the pre-processing stage is responsible for extracting the relevant metric to be used in the classification and presenting them in a format that is understandable by the classifier. Specifically, the selected metric considered in this work is the number of RAB admissions that have been accepted in a cell. Then, the pre-processing stage builds, for each cell, a time series $\mathbf{X_i} = (x_i(t), x_i(t-1), \ldots, x_i(t-(M-1)))$ composed of M samples of the number of RAB admissions in the cell i at different times t with a certain granularity.

Figure 4 illustrates the building blocks of the pre-processing stage implementation using RapidMiner. The first block (Loop XML Files) reads each of the input XML files, while the subsequent blocks perform different operations to merge multiple XML files, to select from each one the hourly samples of the number of RAB admissions, and to build the table with the pre-processed data.

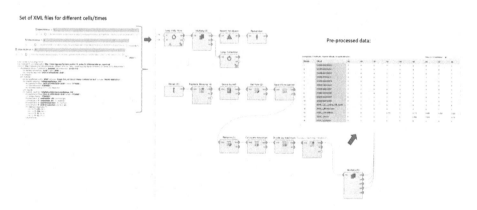

Fig. 4.   Pre-processing stage

4.2   Classification Stage

The classification stage performs the association between the input time series $\mathbf{X_i}$ of the i-th cell and the class $C(\mathbf{X_i}) \in \{A, B\}$ of the cell. The internal structure of the classifier is given by the specific classification tool being used and its settings are automatically configured through a supervised learning process executed during an initial training stage. This training uses as input a training set composed by S time series $\mathbf{X_j} \, j = 1, \ldots,$ S of some cells whose associated classes $C(\mathbf{X_j})$ are pre-defined by an expert. The supervised learning process will analyse this training set to determine the appropriate configuration of the classification tool. In this way, the resulting classifier after the training stage can be used for classifying other cells whose class is unknown.

Figure 5 illustrates the RapidMiner modules implemented for the performing the classification stage. The first module is the training stage that reads the cells from the training set and injects them to each classification in order to build the classification model (i.e. this is done in the first module shown inside a classification algorithm). Then, the last module of each classification algorithm takes as input the small cells to be classified from the pre-processing stage (output of Fig. 4) and applies the obtained classification model.

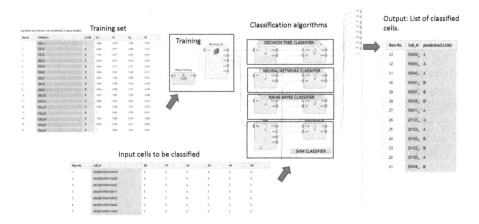

Fig. 5. Classification stage

As shown in Fig. 5, the following classifiers are implemented [18]:

- Decision tree induction: The classification is done by means of a decision tree, which is a flow-chart structure where each node denotes a test on a feature value, i.e. a component of vector X_i, each branch represents an outcome of the test, and tree leaves represent the classes. The tree structure is built during the supervised learning stage through a top-down recursive divide-and-conquer manner.
- Naive Bayes classifier: The classifier evaluates the probability $\text{Prob}(C(X_i) \mid X_i)$ that a given cell X_i belongs to a class $C(X_i)$ based on the values of the components of X_i. The resulting class is the one with the highest probability. The computation of this probability is done using Bayes' theorem under the assumption of class conditional independence. The different terms in the computation of the Bayes' theorem are obtained from the analysis of the training set.
- Support Vector Machine (SVM): A SVM is a classification algorithm based on obtaining, during the training stage, the optimal boundary that separates the vectors X_j of the training set in their corresponding classes $C(X_j)$. This boundary is used to perform the classification of any other input vector X_i. The optimal boundary is

found by means of a nonlinear mapping to transform the original training data into a higher dimension so that the optimal boundary becomes a hyperplane.

- Neural Network: The classification is done by means of a feed-forward neural network that consists of an input layer, one or more hidden layers and an output layer. Each layer is made up of processing units called neurons. The inputs to the classifier, i.e. each of the components of vector $\mathbf{X_i}$, are fed simultaneously into the neurons making up the input layer. These inputs pass through the input layer and are then weighted and fed simultaneously to a second layer. The process is repeated until reaching the output layer, whose neurons provide the selected class $C(\mathbf{X_i})$. The weights of the connections between neurons are learnt during the training phase using a back propagation algorithm.

5   Results

5.1   Scenario Description

The considered scenario considers three different small cell deployments. The PM files of the first deployment include 9 different small cells belonging to an operator providing service on an island in the Pacific Ocean. The cells were deployed mainly in office blocks, hotels and the residences of VIPs. Whilst hand-in and hand-out to the macro network was possible, the small cells did not perform handovers to other small cells. The second deployment includes one small cell belonging to a national operator in a central European country. It is deployed as stand-alone cell in a shop belonging to the operator and, typically, did not perform hand-overs to any other cells. The third deployment includes 23 small cells. They belong to an operator providing service on an island in Northern Europe and were used to provide service mainly to users in their homes, in public houses and restaurants. Whilst hand-in and hand-out to the macro network was possible, the small cells did not perform handovers to other small cells.

5.2   Classification Results

The available PM files for the considered small cells include the metrics for a total of one day. Then, the pre-processing stage shown in Fig. 4 builds, for each cell, a time series $\mathbf{X_i}$ composed of $M = 24$ samples with the hourly values of the traffic in the cell. In turn, the classification stage of Fig. 5 applies the four considered classifiers. As for the training set, it consists of a total of $S = 228$ cells from the deployment of [6].

As a first result, Fig. 6 shows the time domain pattern of two small cells classified as A and B by the decision-tree classifier. This appears as an adequate decision because the cell classified as A exhibits relatively long periods at night serving no traffic at all while the cell classified as B exhibits traffic during most of the time.

Table 1 summarizes the results obtained with the considered classifiers. In addition, the table also includes as a reference the "Expert classification", which indicates the result of the classification process if it was made by the expert. Results show that, in general the number of small cells classified as A or B is very similar for the

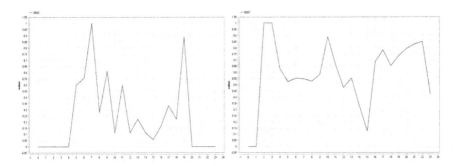

Fig. 6. Example of the time domain pattern of a cell classified as A (Left) and B (Right)

Table 1. Classification results.

Algorithm	Number of cells classified as A	Number of cells classified as B
Decision-tree	26	7
Bayes classifier	27	6
SVM	26	7
Neural network	23	10
Expert classification	27	6

decision-tree, Bayes and SVM classifiers, while there are some more discrepancies for the Neural Network classifier. To further analyse this result, Table 2 assesses the different classification tools by presenting the percentage of coincidences between every pair of tools. For example, the table shows that 90.91% of the cells (i.e. 30 out of 33 cells) have been classified equally by the SVM and the Neural Network. The table also presents the percentage of coincidences with respect to the classification made by the expert. It can be observed that the largest percentages of coincidences are obtained with SVM and the decision-tree.

Table 2. Percentage of total coincidences by every pair of classification tools.

Algorithm	SVM	Neural network	Bayes classifier	Decision-tree	Expert
SVM	–	90.91%	96.97%	93.94%	96.97%
Neural network	90.91%	–	87.88%	84.85%	87.88%
Bayes classifier	96.97%	87.88%	–	90.91%	93.94%
Decision-tree	93.94%	84.85%	90.91%	–	96.97%

6   Conclusions

This paper has presented a framework for demonstrating the use of knowledge discovery capabilities in the context of the architecture of the SESAME project. The proposed approach is based on pre-processing the PM files generated by a Network

Orchestration System to extract the relevant metrics that will be used by the knowledge discovery to obtain the adequate knowledge models making use of machine learning tools. The framework has been particularized for supporting an energy saving Self-X functionality through the classification of cells depending on whether they can be switched off during certain times. To illustrate the operation of the process, the paper has presented some results obtained from real small cell deployments, comparing the behaviour of different classification algorithms.

Acknowledgements. This work has been supported by the EU funded H2020 5G-PPP project SESAME under the grant agreement 671596 and by the Spanish Research Council and FEDER funds under RAMSES grant (ref. TEC2013-41698-R).

References

1. Fallgren, M., Timus, B. (eds.): Scenarios, requirements and KPIs for 5G mobile and wireless system. Deliverable D1.1. of the METIS project, May 2013
2. El Hattachi, R., Erfanian, J. (eds.): NGMN 5G White paper. NGMN Alliance, February 2015
3. Ericsson: Big data analytics. White paper, August 2013
4. Pérez-Romero, J., Sallent, O., Ferrús, R., Agustí, R.: Knowledge-based 5G radio access network planning and optimization. In: 13th International Symposium on Wireless Communication Systems (ISWCS-2016), Poznan, Poland, September 2016
5. Ramiro, J., Hamied, K.: Self-Organizing Networks. Self-Planning, Self-Optimization and Self-Healing for GSM, UMTS and LTE. Wiley, Hoboken (2012)
6. Pérez-Romero, J., Sánchez-González, J., Sallent, O., Agustí, R.: On learning and exploiting time domain traffic patterns in cellular radio access networks. In: Pérez-Romero, J., Sánchez-González, J., Sallent, O., Agustí, R. (eds.) Machine Learning and Data Mining in Pattern Recognition. LNCS, vol. 9729, pp. 501–515. Springer, Cham (2016). doi:10.1007/978-3-319-41920-6_40
7. Sánchez-González, J., Pérez-Romero, J., Agustí, R., Sallent, O.: On learning mobility patterns in cellular networks. In: 1st Workshop on 5G – Putting Intelligence to the Network Edge (5G-PINE 2016), Thessaloniki, Greece, September 2016
8. Small cEllS coordinAtion for Multi-tenancy and Edge services (SESAME). http://www.sesame-h2020-5g-ppp.eu/
9. Giannoulakis, I. (ed.): SESAME final architecture and PoC assessment KPIs. Deliverable D2.5 of SESAME, December 2016
10. 3GPP TS 23.251 v13.1.0: Network sharing; architecture and functional description (Release 13), March 2015
11. ITU-T X.730: Open Systems interconnection – systems management: object management function, January 1992
12. ITU-T X.732: Open systems interconnection – systems management: attributes for representing relationships, January 1992
13. ITU-T X.731: Open systems interconnection – systems management: state management function, January 1992
14. ITU-T X.733: Open systems interconnection – systems management: alarm reporting function, January 1992
15. 3GPP TS 32.401 v14.0.0: Performance management (PM). Concept and requirements (Release 14), March 2017

16. 3GPP TS 32.435 v14.0.0: Performance measurement; eXtensible Markup Language (XML) file format definition (Release 14), April 2017
17. RapidMiner Studio. http://www.rapidminer.com
18. Wilson, R.A., Keil, F.C.: The MIT Encyclopedia of the Cognitive Sciences. MIT Press, Cambridge (1999)

Are Small Cells and Network Intelligence at the Edge the Drivers for 5G Market Adoption? The SESAME Case

Ioannis Neokosmidis[1(✉)], Theodoros Rokkas[1],
Ioannis P. Chochliouros[2], Leonardo Goratti[3],
Haralambos Mouratidis[4], Karim M. Nasr[5], Seiamak Vahid[5],
Klaus Moessner[5], Antonino Albanese[6], Paolo Secondo Crosta[6],
and Pietro Paglierani[6]

[1] inCITES Consulting SARL, Strassen, Luxembourg
i.neokosmidis@incites.eu
[2] Hellenic Telecommunications Organization (OTE) S.A., Athens, Greece
[3] FBK Create-Net, Trento, Italy
[4] University of Brighton, Brighton BN2 4GJ, UK
[5] University of Surrey, Guildford GU2 7XH, UK
[6] ITALTEL, Castelletto, Milan, Italy

Abstract. Although 5G promises advanced features such as low latency, high data rates and reliability as well as high socio-economic value, the business opportunities of the proposed solutions have not yet been examined. In this paper, the SESAME approach along with spectrum sharing options and indicative use cases are initially described. The incentives for SESAME adoption along with the value proposition and creation are analyzed. Finally, a reference model describing the role and interactions of the involved players as well as the revenue streams is provided.

Keywords: 5G networks · Business model · Cloud Computing · NFV · SDN · Small Cells · Telecom market

1 Introduction

The "fifth generation" of telecommunications systems, or "5G", will be the most critical "building block" of our "digital society" in the next decade; it is strongly anticipated that 5G should radically modify the global e-communications framework, worldwide [1] and involve a paradigm shift, that is to establish a next generation network framework achieving reliable, omnipresent, ultra-low latency, broadband connectivity, capable of providing and managing critical and highly demanding applications and services. 5G will be the first illustration of a truly converged network environment where wired and wireless communications will make use of the same infrastructure, "driving further" the future networked society. It will so offer virtually ubiquitous, ultra-high bandwidth, "connectivity" not only to separate users but also to (Internet-) connected objects [2]. Therefore, it is assumed that the future 5G

© Springer International Publishing AG 2017
G. Boracchi et al. (Eds.): EANN 2017, CCIS 744, pp. 693–703, 2017.
DOI: 10.1007/978-3-319-65172-9_58

infrastructure will "serve" a multiplicity of services-applications and domains-sectors also including professional uses (e.g. assisted driving, eHealth, energy management, possibly safety applications, etc.). Nevertheless, it is anticipated that 5G, inter-alia, will bring innovative and exceptional network and service capabilities, in common with modern network management applications and related services/facilities. Mostly, it will safeguard user experience continuity in challenging situations such as high mobility, very "dense" or "sparsely populated" areas, and journeys covered by heterogeneous technologies [3]. In addition, 5G will be a "key enabler" for the Internet of Things (IoT) by offering a suitable and modern "platform" to connect an enormous number of sensors, rendering devices and actuators-or other sort of equipment- with strict energy and transmission constraints. Moreover, mission critical services requiring very high reliability, global coverage and/or very low latency, which are up-to-now handled by specific networks, typically public safety, will become natively supported by the 5G infrastructure. In addition, 5G will "integrate" networking, computing and storage resources into "one programmable and unified infrastructure". This sort of much promising "unification" will allow for an optimized and more enhanced and fully dynamic usage of all distributed resources, as well as for the anticipated "convergence" of all "underlying" fixed, mobile and broadcast services. Within this scope, 5G will also support multi-tenancy models [4], thus enabling operators and other market players to collaborate in new ways, enhancing opportunities for growth and development.

The fresh, ground-breaking advances in the field are expected to enforce revolutionary changes in network infrastructure and management, offering the power to align with a demanding set of diverse use cases and scenarios. Specifically, one of the envisaged key elements of the 5G technological framework is the capability to deliver intelligence directly to network's edge, in the form of virtual network appliances, jointly exploiting the emerging paradigms of Network Functions Virtualisation (NFV) [5] and Edge Cloud Computing. 5G network infrastructures need to offer rich virtualisation and multi-tenant capabilities, not only in term of partitioning network capacity among multiple tenants, but also offering dynamic processing capabilities on-demand, optimally deployed close to the user. The potential benefits from such an approach trigger the interest of Communications Service Providers (CSPs) such as Mobile Network Operators (MNO), Mobile Virtual Network Operators (MVNO) as well as content and service providers, allowing them to "gain" an extra share in the telecom market by pursuing emerging business models. Following this direction, novel business cases will produce added value from any kind of infrastructure or application that has the potential to be offered "as-a-Service". While the virtualisation of the communications infrastructure (core/edge segments as well as access points/macrocells) has been extensively studied by several industry and research initiatives up to now, the applicability of this paradigm to Small Cell (SC) infrastructures has received so far very limited attention [6].

The rest of the paper is organized as follows: In Sect. 2, the SESAME approach along with spectrum sharing issues for SCs and indicative use cases is presented. SESAME value proposition and a reference model illustrating the involved players, their relations and revenue streams are described in Sect. 3. Concluding remarks are provided in Sect. 4.

2   SESAME Approach

SESAME 5GPPP project (GA No. 671596) proposes a novel 5G platform based on Small Cells, featuring multi-tenancy and edge cloud capabilities, offered to both network operators and mobile users. In the SESAME system shown in Fig. 1, one key design principle is to support the innovative concept of Virtual Small Cell Network Operators (VSCNO). The SESAME approach enriches the SCs with computing and storage resources, located in micro-servers connected whereby a dedicated internal network to form the Light DC execution environment.

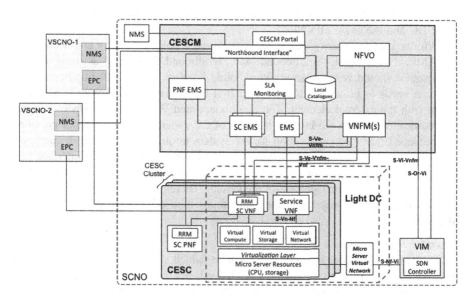

Fig. 1. SESAME system architecture [7].

2.1   Concept/Architecture

The system in Fig. 1 bases its architecture on the concept of the Cloud-Enabled Small Cell (CESC), a new multi-operator enabled environment that integrates a virtualised execution platform (i.e. the Light DC) for deploying VNFs, supporting automated network management and executing novel applications and services inside the access network infrastructure. The Light DC is composed of 64-bit non-x86 architecture ARMv8 technology micro-servers, that support hardware acceleration (e.g. GPU, DSPs and FPGAs) for time critical operations and provides a highly manageable clustered edge computing infrastructure. Besides, a suitable backhaul connection provides connectivity for the SESAME system to external packet data networks.

The CESC Manager (CESCM) provides optimized management of the CESC deployment and is one pillar of the architecture shown in Fig. 1. It implements orchestration, NFV management, virtualisation of management views per tenant, self-x features and radio access management techniques. The CESCM Portal is used by externals to request resources or apply (re)configuration of parameters. The North-Bound Interface (NBI) is the connecting point between orchestration, Service Level Agreements (SLAs) monitoring and VSCNOs' management systems. The SLA Monitoring module provides inputs to both VSCNOs and infrastructure providers about correct execution of the environment, and it enables the orchestration subsystem to react accordingly to possible changes in the network. Small Cells connect to each operator's Evolved Packet Core (EPC) and the Network Management System (NMS). Typically, the NMS is responsible for communicating with the Element Management System (EMS) in the Small Cells domain, in charge of Fault, Configuration, Accounting, Performance and Security (FCAPS) for physical and virtual functions. As shown in Fig. 1, SDN is used to configure the forwarding behaviour of traffic inside the Light DC and the chain of VNFs.

With respect to security and in order to define a holistic security architecture for SESAME, Northbound, Southbound and Westbound interfaces as well as the Orchestrator should be considered. Northbound interface is the most important interface that needs to be thoroughly secured. It is the interface that provides communication with the external element. The Southbound and Westbound interfaces provide communications within the internal modules of the system. The Orchestrator provides self-propelled management solution for the system through creating, monitoring, and deploying resources in the virtual environment. Securing the Northbound interface will provide secure communication with the external elements, and securing the orchestrator provides secure management of the system. Therefore, the Northbound Interface and the Orchestrator security is the main focus area of our security work in the SESAME project.

Essential Features
SESAME fully embraces network virtualization, which allows Small Cell functions to be decoupled from the physical hardware. Indeed, SC functions can be split between Physical Network Function (SC PNF) and virtual (SC VNF), with one SC VNF which can be connected to multiple SC PNF through the front-haul [8]. The fact that one single SC VNF can connect to different SC PNF offers several advantages: (i) improved coordination of radio functions (coordinated scheduling, inter-cell interference coordination, etc.); (ii) enhanced scalability of Small Cell deployments with simplified management, (iii) accelerated life-cycle upgrade enabling new features, and; (iv) flexibility make the workload placement. As shown in Fig. 1, the CESCM includes the ETSI Management and Orchestration (MANO) NFV Orchestrator (NFVO) and the VNF Manager (VNFM) [9], while the Virtualised Infrastructure Manager (VIM) is left outside. SESAME NFVO can deploy new VNF or NS upon receiving the request through the portal or in reaction to the information supplied by SLA Monitoring module through the NBI.

Spectrum Sharing Issues for Small Cells
The main target of spectrum assignment in heterogeneous networks (HetNets) comprising macro cells and Small Cells is to split the total bandwidth among tiers such that: (i) the capacity of the overall system is maximised; (ii) a level of global fairness is ensured between users in different tiers; and (iii) the QoS requirements of users in different tiers in terms of relative data rates are satisfied [10–13].

Frequency spectrum assignment/resource partitioning in HetNets can be accomplished in three main ways:

(1) Shared spectrum/Co-channel allocation: Spectrum resources are fully reused by all tiers in the cost of increased interference.
(2) Split spectrum with a specific Split Spectrum Ratio: It solves cross tier interference at the expense of reducing the amount of bandwidth available to each tier.
(3) Hybrid methods uses a mixture of co-channel and dedicated channel and aim to reuse the spectrum resources whenever feasible.

Shared spectrum solutions are receiving increased interest recently due to the extremely high cost and scarcity of dedicated licensed spectrum bands. It is expected that efficient and economic use of spectrum in 5G networks will rely on sharing rather than exclusive licenses to ease congestion in licensed bands and to increase capacity. Methods for mutually acceptable sharing strategies include looking up a central database with current location to find the permitted frequencies, RF power levels etc. To ensure maximum density with least interference, a more complex approach involves measurement of building penetration loss plus terrain path-loss models that are necessary to accurately estimate interference.

Co-primary Spectrum Sharing (CoPSS) is one approach where any operator is allowed to use shared spectrum. In CoPSS, primary license holders agree on the joint use of (or parts of) their licensed spectrum. This can be suitable for Small Cells especially when base stations have a limited coverage similar to that of WiFi access points and the frequency is dedicated to Small Cell use. Several techniques are reported in the literature for CoPSS [e.g. [15, 17] and references therein].

Two target bands for SC deployments are the 3.5 GHz and the 5 GHz bands. An example of LTE systems commercially deployed at 3.5 GHz includes those offered by UK Broadband in central London in dense urban areas, which is compatible with the short transmission range at those frequencies.

The idea of extending LTE-A specifications to operate in license-exempt bands has recently received considerable attention [14–17]. In this approach, Small Cells are capable of operating in both licensed and the 5 GHz license-exempt spectrum, with a primary use case known as License Assisted Access (LAA) and LTE-Unlicensed (LTE-U). License-exempt bands alongside the licensed bands are aggregated employing the same Carrier Aggregation (CA) techniques that are currently applied in licensed bands in the LTE-A.

LTE technology alternatives in unlicensed spectrum includes: LTE WiFi aggregation (LWA), LWA using IPSEc Tunnel (LWIP), LTE Licensed Assisted Access (LAA) and LTE in the unlicensed spectrum (LTE-U) [16]. These techniques aim to provide seamless aggregation of LTE and WLAN radio links, using WiFi and/or LTE in unlicensed spectrum. LAA is the 3GPP standard solution while LTE-U is a solution developed by Qualcomm.

LAA and LTE-U work in the unlicensed 5 GHz WiFi band and this is aggregated and controlled by LTE anchor carrier running in licensed bands at lower frequencies (UHF to 3.8 GHz). The main control channel and basic voice or data calls would remain on much lower frequency licensed LTE spectrum, but during peak traffic periods, supplemental data channels would be added using unlicensed spectrum. LAA is the 3rd Generation Partnership Project's (3GPP) effort to standardize operation of LTE in the Wi-Fi bands. It uses a contention protocol known as listen-before-talk (LBT), mandated in some European countries, to coexist with other Wi-Fi devices on the same band and to ensure fairness.

A controversial aspect of LTE-U compared to LAA is that it does not incorporate an LBT mechanism for coexistence and does not meet the regulatory requirements for using unlicensed spectrum in a significant part of the world. It uses instead proprietary techniques based on channel selection and Carrier Sensing Adaptive Transmission (CSAT).

Another emerging technique, called MulteFire, is planned to be built on LAA and 3GPP release 13 to operate solely in unlicensed spectrum without requiring an LTE anchor in licensed spectrum.

2.2   Use Cases

The core use cases that have been identified as relevant for the system in Fig. 1, which places intelligence at the network edge, are presented here. The first use case (UC1) reflects the raise of Sporadic Crowd Events. It is presupposed that the CESC infrastructure supporting multitenancy exists already in an area (e.g. stadium), and two or more VSCNOs wish to exploit the SESAME infrastructure to provide their services to the end users. Specific network services that suit the SLA of each individual virtual operator are automatically deployed through the CESC manager.

UC2 is called Radio and Light DC slicing. This use case is functional to demonstrate the capability of the edge infrastructure to deploy multiple tenant operators over the same physical infrastructure, ensuring traffic isolation, traffic classification and distribution of multimedia services directly at the edge with consequent reduction of possible unreliability factors.

UC3 focuses on Orchestration and Automatic VNFs Deployment into the CESC infrastructure providing on the top an SDN-enabled Service Function Chaining between VNFs. The key idea is to orchestrate and compose end-to-end services that optimise the delivery of video services to end-users through their mobile devices.

UC4 materialises the idea of mechanisms enabling Optimized Radio Network Capacity Planning and Operation of the Small Cell Network Operator. The objective of this use case is to illustrate the behavior of an Artificial Intelligence (AI)-based framework for knowledge discovery that can be used to support the optimization decisions made by self-x features.

3   Business Model

The technological changes of the last years leading to the idea and development of the 5G concept have greatly influenced the business environment of telecommunications industry. Virtualization and softwarization of networks urged the introduction of new players and change the roles and relationships of the existing ones. These advances along with the use of general-purpose computers instead of specialized devices have also significantly impact the cost centers. This results in a transformation of Capital Expenditures (CAPEX) to Operating Expenses (OPEX), which in turn reduces the cost of deployment. The derived X-as-a-Service paradigm adopted by several fields allowed the provision of services from different locations of the value chain terminating the bilateral relationship between cellular operators and their customers. It is thus evident that an investigation of new business model is of high importance for the viability of operators and other players of the telecom ecosystem.

3.1   Incentives and Value Proposition – SESAME Enablers

Even though the value generation will be specific to each player in the business model, the adoption of the SESAME solution, offers the following incentives for Operators, providers as well as for vendors and manufacturers:

- Increase of customer lifetime (avoid service termination or churn) and ARPU by improving QoS and QoE.
- Reduce of CAPEX and OPEX by using NFV.
- Increase of sales on slices, (e-)services delivery (health, security, entertainment, etc.) and equipment/related products (smart phones, sensors, cars, etc.).
- Create new players and Accelerate time to market.
- Deliver agility and flexibility.

Regarding the Value Proposition, SESAME is aiming to:

- Improve QoS and QoE mainly due to increased capacity, low latency and caching.
- Reduce energy consumption by moving computation and data at the edge.
- Improve network performance and availability of services due to the use of Small Cells.
- Improve health, saving lives and entertainment through: Fast deployment in case of natural disaster, Support of crowd events (festival, meeting, outdoors concert) through Small Cells increased capacity and provision of High Definition streams through caching.
- Reduce rural/urban divide: SESAME can provide services to locations which lack network coverage such as rural areas through the use of CESC clusters.
- Create new jobs: New players will enter the market e.g. NF developers and facilities managers.

In order to achieve the above value proposition, SESAME creates value by creating tools and providing solutions (SESAME enablers), such as: CESC, CESC cluster (it provides access to a geographical area with one or more operators), CESC Manager, Self-X features (leading to the automated operation of CESCs), Light DC, Edge computing capabilities and Multi-tenancy.

3.2 Reference Model

A reference model to coordinate the generation of revenues to provide a profit margin over its cost is necessary to guarantee the survival of a firm. In this framework, revenue streams are stated as sources to monetize the products and services a company offers, indicating the "value capture". In such a reference model:

- The direction of the arrows in the model represents the direction of service flow.
- Revenue flow is considered to be in the opposite direction. In some cases, revenue sharing exists between two players, resulting in a bidirectional flow.
- The ellipse represents a group of players/roles. Each player/role is depicted in rectangular boxes.

Taking into account SESAME architecture and use cases, Small Cell Network Operator (SCNO) is assumed to be the main player in SESAME ecosystem since it will mainly adopt the proposed solutions and tools.

The reference model of Fig. 2 describes a basic SESAME scenario for the delivery of services from one operator to another as well as from the operators/providers to end-users. In the latter case end-users subscribe for these services. Hence, SCNOs is assumed to be the main SESAME player while in general operators/providers act as the main responsible players towards the Subscriber providing telecom services like voice and video telephony, broadband access, etc.

In this model, the following type of service level agreement exists between subscribers, operators or/and service providers: A subscriber/operator enters into a contract or service level agreement (SLA) with another operator or the SP for the usage of

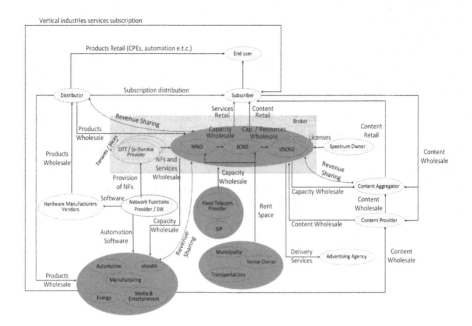

Fig. 2. SESAME reference model.

services/resources or/and infrastructure. Regarding the relation between subscribers and operators/SPs, the contract can be of two types: post-paid or pre-paid, based on the mode of subscription payment. In the case of post-paid contract, the subscriber is billed by the service operator for basic connectivity and communication services on a monthly basis. A service provider may adopt different pricing models to charge for the services provided. The billing service is taken care of by the service provider, thus incurring OPEX.

In fact, there is a wide range of choices for the service provisioning between operators and pricing scheme (flat rate, pay-as-you-go, hybrid) in a joint radio-cloud scenario. The similarities with the existing solutions of the individual radio and cloud systems can facilitate the establishment of mixed environments. Depending on the use case, VSCNO can decide to select one or another option. In general, the CESCIaaS model allows a better understanding of the use of NFVI resources to the VSCNO, but requires specialized VSCNO staff and systems. On the contrary, the MECaaS model hides the low level details of the NFVI to the VSCNO, which only need to care about the outcomes of the process at service level.

In multi-tenant environments with more than one SCNOs or/and VSCNOs more complex solutions (especially pricing schemes) should be adopted. In such a scenario, negotiations (through brokers) should be performed between the interested parties. In a 5G multi-tenant environment customers (VSCNO) will have a set of resource requirements, the number of customers and the duration of usage are unknown while both the pricing and the willingness to pay strategy of competitors should be taken into account. The main objective of all parties is to maximize their own profit. Unit price of resources of SCNOs can vary depending on the utilization.

Apart from brokers who will be a key player of the 5G ecosystem, new players are expected to enter into the market. Functions developers offering VNFs to operators and/or service providers as well as facilities managers (e.g. transportation players) providing Infrastructure-as-a-Service are indicative examples of such players. Other players can also use the operators network in order to provide a series of Value-added services such as location-based content, presence services, banking info/alerts, sports/match alerts, travel/transportation alerts, weather alerts, missed call alerts, contests and voting, online payment services, etc.

Content providers/aggregators are anticipated to greatly contribute to SESAME success since SESAME solution is closely related to content-oriented use cases such as an event in a stadium as described in section IIB. A revenue sharing model can also exist between these players for content usage. These revenue models may differ depending on the type of content provided. Alternatively, users can have a contract with both the operator and the service provider. Moreover, it is interesting to note that in the new 5G era, end-users can have a twofold (dual) role acting as both content consumers and producers, a feature that totally changes the business landscape.

The business prospects of the operator can be further enhanced if there is collaboration with an advertising agency by selling paid placement inside data feeds. The operator can opt to have contextually targeted overlay ads that will run on e.g., videos. In this case, advertising campaigns produced for the clients of the agency can be delivered via the operator's content, for which advertising delivery the advertising agency pays the operator. Similar collaborations also exist between advertising agencies and content providers or OTT.

Similarly, collaboration with vertical industries will allow revenue sharing. For example, the health care industry companies provide services to subscribers using operator's capacity. In particular, the operator resells a part of its capacity to the health care provider. The health care provider in turn has to pay for the capacity usage to the operator. A revenue sharing model can also exist between these two players.

4   Conclusions

In this paper, the business prospects of SESAME solution were analyzed. The incentives for SESAME adoption along with value proposition and creation were described. A reference model illustrating the involved players, their relations and revenue streams was developed and described. Several choices of service provisioning and pricing schemes are presented. The importance of brokers, verticals and content providers for the success of SESAME vision was highlighted. The results of this paper can be a valuable tool for decision and policy makers as well as for a number of stakeholders.

The SESAME EU-funded project promotes several important innovative features within the context of its activities, with significant impact upon the involved market actors, as already discussed in the previous sections. These are mainly evolved "around" the fundamental pillars of the core research effort that focus, inter-alia, on the inclusion - or placement - of network intelligence and applications in the network edge through Network Functions Virtualisation (NFV) and Edge Cloud Computing, within the broader 5G context.

This important fact also correlated to the enhanced and/or the extended use of small cells and together with the consolidation of multi-tenancy in modern communications infrastructures towards the forthcoming 5G, provides a remarkable variety of promising opportunities and important challenges to be easily adopted by the market sector, and so to become applicable and beneficial, in a short-term framework.

Moreover, the innovative features coming from the proper use of NFV offers a wide set of benefits for automated network management as well as for the inclusion of cognitive features within the various edge network elements, which is a prerequisite for the inclusion of intelligence within the modern 5G-oriented infrastructures. When this is strongly correlated to the market-related expectations of the involved "actors"/players, as in the case of SESAME, then it is evident that the modern network revolution can be become not just a forthcoming vision but an actual and explicit reality.

Acknowledgments. The research leading to these results has been supported by the EU funded H2020 5G-PPP project SESAME (grant agreement No. 671596).

References

1. The 5G Infrastructure Public Private Partnership (5G-PPP): 5G Vision: The next generation of communication networks and services. https://5g-ppp.eu/wp-content/uploads/2015/02/5G-Vision-Brochure-v1.pdf
2. Andrews, J.G., Buzzi, S., Chopi, W., Hanly, S.V., Lozano, A., Soong, A.C.K., Zhang, J.C.: What will 5G be? IEEE J. Sel. Areas Commun. **32**(6), 1065–1082 (2014)
3. El Hattachi, R., Erfanian, J.: NGMN 5G White paper. Next Generation Mobile Network (NGMN) Ltd. (2015). https://www.ngmn.org/uploads/media/NGMN_5G_White_Paper_V1_0.pdf
4. Capdevielle, V., Feki, A., Temer, E.: Enhanced resource sharing strategies for LTE picocells with heterogeneous traffic loads. In: Proceedings of IEEE 73rd Vehicular Technology Conference (VTC2011-Spring), pp. 1–5. IEEE (2011)
5. Mosharaf, N.M., Chowdhury, K., Boutaba, R.: A survey of network virtualization. Comput. Netw. **54**(5), 862–876 (2010)
6. Small Cell Forum (SCF): Release 7.0: Document 055.07.01 – Small Cells and 5G Evolution: A Topic Brief. http://scf.io/en/documents/055__Small_cells_and_5G_evolution_a_topic_brief.php
7. SESAME H2020 5G-PPP Project. http://www.sesame-h2020-5g-ppp.eu/Home.aspx
8. Small Cell Forum (SCF): Small Cell Virtualization Functional Splits and Use Cases, SCF 159.07.02 (2016)
9. European Telecommunications Standards Institute: ETSI GS NFV-MAN 001 (V1.1.1): Network Function Virtualisation (NFV); Management and Orchestration. ETSI, Sophia-Antipolis (2012)
10. Quek, T.Q.S., de la Roche, G., Güvenç, İ., Kountouris, M.: Small Cell Networks Deployment, PHY Techniques, and Resource Management. Cambridge University Press, Cambridge (2013)
11. Mahmoud, H.A., Guvenc, I.: A Comparative study of different deployment modes for femtocell networks. In: Proceedings of IEEE Indoor Outdoor Femtocells (IOFC) Workshop (co-located with PIMRC 2009), pp. 1–5. IEEE (2009)
12. Lopez-Perez, D., Valcarce, A., Ladányi, A., de la Roche, G., Zhang, J.: Intracell handover for interference and handover mitigation in OFDMA two-tier macrocell-femtocell networks. EURASIP J. Wirel. Commun. Netw. **4**, 1–15 (2010)
13. Lopez-Perez, D., Guvenc, I., de la Roche, G., Kountouris, M., Quek, T.Q.S., Zhang, J.: Enhanced intercell interference coordination challenges in heterogeneous networks. IEEE Wirel. Commun. **18**(3), 22–30 (2011)
14. Maladi, D.: Making the Best Use of Unlicensed Spectrum. Qualcomm Technologies, Inc. (2016)
15. Luoto, P., Pirinen, P., Bennis, M., Samarakoon, S., Scott, S., Latva-aho, M.: Co-primary multi-operator resource sharing for small cell networks. IEEE Trans. Wirel. Commun. **14**(6), 3120–3130 (2015)
16. Intel Corporation: Intel White paper - Alternative LTE Solutions in Unlicensed Spectrum: Overview of LWA, LTE-LAA and Beyond. Intel Corporation (2016)
17. Tehrani, R.H., Vahid, S., Triantafyllopoulou, D., Lee, H., Moessner, K.: Licensed spectrum sharing schemes for mobile operators: a survey and outlook. IEEE Commun. Surv. Tutor. **l18**(4), 2591–2623 (2016)

Putting Intelligence in the Network Edge Through NFV and Cloud Computing: The SESAME Approach

Ioannis P. Chochliouros[1(✉)], Anastasia S. Spiliopoulou[1],
Alexandros Kostopoulos[1], Maria Belesioti[1], Evangelos Sfakianakis[1],
Philippos Georgantas[1], Eirini Vasilaki[1], Ioannis Neokosmidis[2],
Theodoros Rokkas[2], and Athanassios Dardamanis[3]

[1] Hellenic Telecommunications Organization (OTE) S.A., 99, Kifissias Avenue,
15124 Maroussi, Athens, GR, Greece
{ichochliouros,alexkosto,mbelesioti,esfak,
fgeorgantas,evasilaki}@oteresearch.gr,
aspiliopoul@ote.gr
[2] INCITES Consulting S.A.R.L., 130, Route d'Arlon, 8008 Strassen,
Luxembourg
{i.neokosmidis,trokkas}@incites.eu
[3] SmartNet S.A., 2, Lakonias Street, 17342 Agios Dimitrios, Attica, GR, Greece
ADardamanis@smartnet.gr

Abstract. The core challenges in the actual SESAME EU-*funded* project is to develop an ecosystem to sustain network infrastructure openness, built on the pillars of network functions virtualization (NFV), mobile-edge computing (MEC) capabilities and cognitive network management that will provide multi-tenancy and flexible cloud-network interaction with highly-predictable and flexible end-to-end performance characteristics. Based on this aspect, we discuss the potential benefits of including NFV and MEC in a modern mobile communications infrastructure, through Small Cells coordination and virtualization, also focused upon realistic 5G-*oriented* considerations. Within the proposed SESAME architecture, we also assess the various advantages coming from a more enhanced network operation and management of resources, as it appears with the incorporation of cognitive capabilities embracing knowledge and intelligence.

Keywords: 5G · Edge cloud computing · Mobile edge computing (MEC) · Network functions virtualization (NFV) · Small cell (SC) · Self-X functions · Virtual network function (VNF)

1 Towards a Modern 5G-*Based* Automated World

In our modern societies, electronic communication networks are more than "*simply fundamental*" for the adequate offering of an extended "set of services and/or related facilities", for the benefit of all involved market "actors" (i.e., corporate users, residential users, the State and local authorities, etc.). Such kind of infrastructures are able

G. Boracchi et al. (Eds.): EANN 2017, CCIS 744, pp. 704–715, 2017.
DOI: 10.1007/978-3-319-65172-9_59

to support the provision of modern Internet, not only by covering necessary network-*related* aspects but also by involving the disposal of services/facilities, the use of numerous equipment/devices, the provision of content, etc., under a broader scope of "convergence". Moreover, apart from the well-known electronic communication networks there is also a variety of critical infrastructures (such as energy, transportation, health, water and many more) that are gradually becoming reliant upon Internet connectivity, adequacy and automation. Thus, Internet which is so correlated to an immense multiplicity of underlying networks, services and equipments can be perceived as a "key factor" for making real the progress and the evolution of the "digital economy", bringing huge socio-economic value. This also implicates for an enhanced - and occasionally for an automated- use of all related available resources, of any probable origin/nature. In order to face this significant challenge, the "fifth generation" of telecommunications systems, or "5G", has been assessed as the most critical building block of our "digital society" in the next decade [1] and it has been promoted as a core strategic perspective for an effective global growth. 5G should "bring together" wired and wireless communications by providing virtually ubiquitous, ultra-high bandwidth "connectivity", including not only distinct users but also numerous (Internet-) connected objects [2]. The future 5G infrastructure is expected to support an extended variety of converged networks/infrastructures able to support a diversity of services/applications and related equipment/devices, also including professional uses (e.g., eHealth, energy management, possibly safety applications, etc.). 5G is aiming to be reasonably different compared to any prior technology. It is about more than just *"raising the bar"* on previous generations, or extending them to a certain context [3]. Thus 5G will not only be a single "progression" -or a "simple evolutionary step"- of mobile broadband networks; nevertheless, it is expected that it should "bring" innovative and exceptional network/service capabilities, in common with modern applications and related facilities and should be an "enabler" of the Internet of Things (IoT).

In addition, 5G will "integrate" networking, computing and storage resources into *"one programmable and unified infrastructure"*. This sort of "unification" of functions will allow for an optimized, more enhanced, and fully dynamic usage of all distributed resources, as well as for the anticipated "convergence" of all "underlying" fixed, mobile and broadcast services. Within this scope, 5G will also support multi-tenancy models [4], thus enabling operators and other market players to collaborate in new ways, enhancing opportunities for growth and development within an intelligent network environment. Furthermore, leveraging upon the features of existing cloud computing, 5G will support further progress of the single digital market, e.g. by "paving the way" for virtual pan-European operators relying on nation-wide infrastructures. 5G should be designed in a way to be a sustainable and fully scalable technology. In view of this aim, cost reduction through human task automation and hardware optimization will allow for supportable business models for all ICT stakeholders to be involved [5].

In this paper, we discuss the challenge of incorporating intelligence at the network edge mainly via the consideration of Network Functions Virtualization (NFV) and Edge Cloud Computing. The introductory part discusses broader challenges for network development and growth, coming from the 5G perspective. Section 2 becomes more specific and correlates actual network operations and management challenges to the NFV and edge cloud computing features, also promoting the context of cognitive features and network autonomicity. The potential benefits coming from a combined approach of network virtualization and edge computing within modern mobile (5G-*based*) infrastructures -that are now assessed via the consideration of Small Cells (SCs)- and towards fulfilling the fundamental aim of improving network management are discussed in Sect. 3; in particular, the presented context is based upon the actual SESAME 5G-PPP EU-*funded* project research effort. Section 4 realizes a step further and discusses exact ways of implementing prior considerations within the SESAME's fundamental architectural framework.

2   NFV and Cloud Computing as "Enablers" of Network Intelligence in 5G Infrastructures

Future networks will need to be deployed much more densely than today's networks and, due to the economic constraints and the availability of sites, will need to become significantly "more heterogeneous" and use multi Radio Access Technologies (RATs). A network needs to be able to scale its operation, even for short time-periods, depending on the widely varying traffic capacity requirements, while it should also remain energy-efficient. Furthermore, in modern (5G) networks, devices are no longer connected to just one single access node; on the contrary, the full picture consists of a combination of multiple physical interfaces based on the same -or different- radio technologies. Fast selection and combination of most -if not of all- of the available interfaces can support an adaptive set of virtual interfaces and functions, subsequently depending on respective applications. SCs can contribute to the effort of making "best use" of novel applications offered by "denser" and more heterogeneous RATs while, *in parallel*, being able to efficiently support several widely varying traffic needs. Furthermore, SCs also support scalability issues ([6, 7]).

In addition, future network deployments have to allow for network/infrastructure/ resource sharing and potential re-utilization on all levels, so that to "fulfil" the fast growing demands on network resources management and operation. This is to take place simultaneously with the proper inclusion of cognitive capabilities in the network design on all layers, able to support a flexible network adaptation at low operational costs, towards providing exactly the performance required for the determined user context. The Operation and Management (OAM) of the wireless mobile network infrastructure plays an important role in suitably "addressing" network management and automation, in terms of constant performance optimisation, fast failure recovery, and fast adaptations to changes in network loads, architecture, infrastructure and technology. Self-Organising Networks (SON) are the first step towards the automation of networks' OAM tasks, for example via the introduction of closed control loop functions dedicated to self-configuration, self-optimization and self-healing. In brief, SON is a collection of procedures -or functions- for automatic configuration, optimization, diagnostication and healing of cellular networks [8]. SON is conceived as a major necessity in future mobile networks and operations, mainly due to possible savings in capital expenditure (CAPEX) and operational expenditure (OPEX). The tendency introduced with SON is to enable system's OAM at local level as much as possible.

SON functionalities are also referred to as "Self-x" functionalities and correspond to a set of features and capabilities for automating the operation of a network, so that operating costs can be reduced and human errors minimized [9]. With the introduction of "Self-x" features, classical manual planning, deployment, optimization and maintenance activities of the network can be replaced and/or supported by more autonomous and automated processes, thus making network operations simpler and faster. "Self-x" functions can automatically "tune" global operational SC settings (e.g., maximum transmit power, channel bandwidth, electrical antenna tilt) as well as specific parameters corresponding to Radio Resource Management (RRM) functions (e.g., admission control threshold, handover offsets, packet scheduling weights, etc.).

On the other hand, today's networks are populated with a great and growing diversity of proprietary hardware appliances. Launching a new network service often requires yet another variety of appliance, thus increasing the overall complexity of the network and causing a number of issues to be addressed. The shortage of skills necessary to design, integrate and operate increasingly complex hardware-*based* appliances, poses supplementary challenges. Moreover, hardware-*based* appliances rapidly reach their "end-of-life", requiring much of the procure-, design-, integrate- and deploy-cycle to be repeated with little or no revenue benefit. Worse, hardware lifecycles are becoming shorter as technology and services innovation accelerates, inhibiting the roll-out of new revenue earning network services and constraining innovation in an increasingly network-centric connected world [10].

Network Functions Virtualization aims to "address" these critical problems by leveraging standard IT virtualization technology to consolidate many network equipment types onto industry standard high volume servers, switches and storage, which could be located in data centres, network nodes and in a variety of end-user premises [11]. Actually, NFV is applicable to any data plane packet processing and control plane function both in fixed and mobile network infrastructures [12]. There are various challenges to implement NFV that need to be examined by the community interested in accelerating technological progress. Based on the original context of the SESAME 5G-PPP project [13] - which is also discussed in more details, in Sects. 3 and 4- we can, *among others*, identify the following meaningful cases: (i) Management and orchestration; (ii) the perspective of automation, *and*; (iii) the options of security and resilience. As of the former case, for a consistent management and orchestration architecture, NFV presents an opportunity, through the flexibility afforded by software network appliances operating in an open and standardized infrastructure, to rapidly align management and orchestration northbound interfaces to well-defined standards and abstract specifications. This can significantly reduce the cost and time to integrate new virtual appliances into a network operator's operating environment. Besides, Software Defined Networking (SDN) further extends this to streamlining the integration of packet and optical switches into the system [14]; e.g. a virtual appliance or NFV orchestration system may control the forwarding behaviors of physical switches by using SDN. Traditionally, SDN and NFV although not dependent on each other, are seen as "*closely related*" and/or as "complementary" concepts [15]. The orchestration and federation of network resources as "network functions" is an important aspect of the future network ecosystem. As such, research is relevant on the way in which resources and functions are described, protecting the "know-how" of the network and service providers and, *at the same time*, opening the right interfaces to enable new business models to appear. Service Level Agreements (SLAs) automated definition and monitoring/control of network functions is also a relevant topic, under the management and orchestration domain. Regarding the option of automation that has been identified in the second case, above, NFV will only scale if all of functions can be automated, while automation of processes is paramount towards improving OAM. As of the latter case (i.e., the third case) which is about security and resilience, it is assumed that network operators need to be assured that the security, resilience and availability of their networks are not reduced -or not harmed- when virtualized network functions are to be introduced. NFV can improve network resilience and availability by allowing network functions to be recreated "on demand", after a possible failure. In fact, a virtual appliance should be "as secure as a physical appliance" if the infrastructure, especially the hypervisor and its configuration, is secure. Network operators are thus seeking for tools to control and verify hypervisor configurations [16].

In addition to the above, ensuring stability of the network is not impacted when managing and orchestrating a large number of virtual appliances between different hardware vendors and hypervisors. This is particularly important when, *for example*, virtual functions are relocated, or during re-configuration events (e.g. due to hardware and software failures) or due to cyber-attacks. NFV is also beneficial for operators as it supports simplicity and integration: For the first case, a significant and topical focus for network operators is simplification of the plethora of complex network platforms and support systems, which have evolved over decades of network technology evolution, while maintaining continuity to support important revenue generating services. Regarding the second one, seamless integration of multiple virtual appliances onto existing industry standard high volume servers and hypervisors is a "key challenge" for NFV. Network operators need to be able to "mix & match" equipment and virtual appliances from different vendors without incurring significant integration costs and avoiding undesired lock-in. Therefore, the ecosystem needs to offer integration services and maintenance, third-parties support and will require mechanisms to validate new NFV products.

Edge cloud computing refers to data processing power at the edge of a network, instead of holding that processing power in a cloud or a central data warehouse. Edge computing places data acquisition and control functions, storage of high bandwidth content and applications closer to the end-user [17]. A fundamental future challenge is to "guarantee and constantly improve" customer experience offered by edge cloud-*based* services. Such experience relies on the End-to-End (E2E) QoS, and more generally on respective SLAs in place for a given service. This includes well-known characteristics, such as latency, throughput, availability and security, but by adopting the principles of Clouds, also elasticity, on-demand availability, lead- and disposal-times, multi-tenancy, resilience, recovery, and similar characteristics important especially in case of cloud-*based* services [18]. However, in order to guarantee this kind of service level, network-*based* service qualities may not be enough, but need to be aligned with platform-level and Cloud specific tenets, like dynamic discovery, replication, and on-demand sizing of Virtual Machines (VMs), since previous over-provisioning best-practices inherent to hosted and managed execution environments are no longer applicable [19].

Edge networks are expected to create distributed environments made of clouds of virtual resources (even operated by diverse players) interconnected by a simpler and less hierarchical core network [17]. Some business models necessitate federation and/or orchestration capabilities. In a federation context, the stakeholders agree on jointly providing a service. In an orchestration context, each entity keeps its service models, interfaces and SLAs and a specific component (called as the "broker"), will compose services from each stakeholder to be able to provision a requested service. Both approaches can be used to extend coverage, increase capacity or enhance quality (for example deploying functionality or locating content near by the customers). The broker functionality can be implemented by one of the players or by a third party. It, *therefore*, represents by itself a business opportunity. In future ecosystems, the operator will need to efficiently orchestrate its own resources not only for cost reduction purposes, but also for being able to open the network capabilities to enable third party services. The single domain orchestration has many challenges such as *how to describe the resources and define the interfaces* in such a way that the network capabilities are exposed to third parties or partners without exposing the level of detail that constitutes the operator's know-how and hence its market differentiation. Interface definition, resource/price discovery, publishing and negotiation and service level monitoring and assurance are also main components of the single domain orchestration. Key elements for the orchestration are the network and service modelling and key optimization algorithms used for resource embedding. The orchestration needs of the future network will involve not only connectivity (and its associated functions) but also computing resources enabling complex network functions ranging from platform to applications.

3   The Innovative Vision of the SESAME-*Based* Research

It is now widely accepted that both mobile data traffic and services have reached to a "critical" level of penetration to our daily activities [1], mainly due to the extreme adoption and use of a great variety of (personalized) applications, serving not only communication and information purposes but also those related to work, leisure, etc. For most -if not for all of these- the requirement is always to "preserve" an acceptable level of the quality of services (QoS) offered, together with the user experience and satisfaction. The evolution of previous technological frameworks (such as 3G and/or 4G) has mainly focused upon the support of network aspects, such as via the promotion of network coverage and capacity, in parallel with improved resource usage [10]. The strategic challenge for 5G, *however*, is not just to create a new paradigm shift via the establishment -and the validation- of a next generation network framework attaining consistent, omnipresent, ultra-low latency, broadband connectivity, able of offering and/or managing critical and highly demanding applications and services [20]. Apart from these explicit and useful aims, 5G intends to enact fundamental changes in network infrastructure management, in particular by supporting the ability to deal with a wide set of miscellaneous use cases and related scenarios. For these purposes, the 5G scenery needs to couple fast connectivity and optimized spectrum usage with cloud networking and high processing power, optimally combined in a converged environment. Precisely, a critical challenge among the essential ones in the 5G technological context is the ability to "bring intelligence" directly to underlying network's edge, via the inclusion of virtual network appliances suitably exploiting the evolving examples of NFV [12] and Edge Cloud Computing [17]. The future 5G network infrastructures should have the capability to provide enhanced virtualization and support multi-tenancy, not only in the scope of dividing/partitioning network capacity among multiple possible tenants, but also via the offering of (dynamic) processing capabilities on-demand, optimally deployed within the vicinity of the involved end-users. The corresponding advantages may be of prime importance for existing Communications Service Providers (CSPs), such as Mobile Network Operators (MNO), Mobile Virtual Network Operators (MVNO), Over-The-Top (OTT) content and service providers, as these actors -via a respective implementation- can have the ability to extend their business activities and acquire extra shares in the network market. Within this scope, the use/deployment of modern businesses should produce new beneficial revenues from any sort of network infrastructure and/or facility, able to be offered "*as-a-Service*".

Although the virtualization of the network infrastructure (mainly involving the core/edge segments as well as the access points/macrocells) has been broadly examined in the framework of related market or research initiatives, the applicability of this conceptual view to SC infrastructures has only been the case of some independent works with limited attention. However, the SC concept has become fundamental in today's existing 4G infrastructures [21]; in fact, SCs can bring better cellular coverage, capacity and applications for residential and corporate uses, along with rural public spaces and dense metropolitan areas. SCs are essential for offering services in domains/spaces like shopping malls, performance venues, stadiums and, generally speaking, places with (tactic or sporadic) high end-user density. For the above use cases it is expected that each separate telecom operator should deploy his dedicated infras-tructure, as a "complement" to the macro-cell network. The usual SC provisioning implicates for certain time- and money-consuming procedures (such as the provi-sioning of installation site, power supply, etc.). Involved operators also have to take care of costs of launching committed, high-capacity backhaul connections, as well as those relevant to radio resource management and interference mitigation, thus increasing their operational expenses. This sort of approach is based on the possession of the physical SC infrastructure and becomes "impractical", due to a variety of rea-sons; it usually results to increase of network CAPEX and sets obstacles to business agility, while it cannot support active scenarios of use. With the intention of responding to such sort of challenges, network operators can alternatively install, and for a certain period of time, a SC network to serve a related event/case, without necessarily owing the related infrastructure. Underlying facilities could be provided by a third party (such as, for example, the owner/operator of the venue). Such shared uses are expected to "play an important role" in 5G networks [21], following to related policy challenges.

Towards covering this high-demanding request and via the conceptual considera-tion of several fundamental features such as network functions virtualization, mobile-edge computing (MEC) and cognitive management [22], the actual SESAME's EU-funded research effort is the development and demonstration of modern architec-ture, able to provide SC coverage to multiple operators "*as-a-Service*". SESAME considers the logical partitioning of the localized SC network to several isolated slices as well as their delivery to some tenants. Apart from virtualizing and partitioning SC capacity, the SESAME-*based* effort supports enriched multi-tenant edge cloud services via the by upgrading of SCs with micro-servers [23]. SESAME develops a framework to perform multi-tenant cloud-*enabled* Radio Access Network(s) - RAN(s), via a major conceptual modification of the architecture of commercially offered SCs; it is realized by evolving SCs to the "Cloud-*Enabled* Small Cell" ("CESC"). This modification implicates for "placing" enhanced network intelligence and/or applications in the network edge, with the fundamental support of virtualization techniques and Network Function Virtualization (NFV). Therefore, an innovative architectural framework has been proposed with the aim of attracting network operators/service providers and "engaging" these in a modern multi-tenant ecosystem, able to fulfill 5G visions. The CESC concept is a new multi-operator enabled SC, able to integrate a virtualized execution platform (i.e., the so-called Light Data Centre (Light DC) for deploying Virtual Network Functions (VNFs), supporting strong "Self-x" management and performing innovative applications and services inside the access network infrastructure. The Light DC should feature low-power processors and hardware accelerators for time-critical operations; furthermore, it should structure a highly manageable clustered edge computing infrastructure, with may advantages. This concept allows new market actors to "get stakes" of the value chain as they can operate as "neutral host providers" in high traffic domains where densification of multiple networks is not technically or economically practical. The optimal management of a CESC deployment becomes a core issue for SESAME, implying for further evolution and development of new orchestration, NFV management, virtualization of management views per tenant, "Self-x" features and radio access management techniques.

In addition, the SESAME context with its anticipated distinct innovations manages to extend the "*Small Cell-as-a-Service*" ("*SCaaS*") model [24]; this, *in turn*, enables a the provisioning of shared radio access capacity by a third-party -or entity- to MNOs in certain localized areas, and this is conceived in parallel with the delivery of Mobile Edge Computing (MEC) services. MEC, also known as "*Fog computing*", is an innovative concept that extends the services, typically provided by the Cloud, to the network edge [25]. (In case of 5G wireless networks, the term "edge" usually means the RAN, while some part of the Cloud services is provided by cognitive base stations). Potential facilities to be offered may include storage, computing, data, and application services. Available MEC infrastructure permits running of applications closer to the end-user so that to reduce the E2E network latency as well as the backhaul capacity requirements. MEC also allows for more enhanced quality of experience (QoE) of fast moving end-users and enables highly-interactive real-time applications. Our architectural assumptions are based upon the *SESAME architecture*, as discussed in Sect. 4. Our analysis can be easily extended to "alternative" network architectures and even in the cases of macro-cells or combinations of macro- and small-cells. The SESAME architectural framework can lead to a variety of substantial features that can be beneficial for the involved industry and end-users; these could be relevant, *inter-alia*, to: a more efficient management of involved (network) resources; the fast inclusion of modern network function(s) and/or service(s); the ease and simplicity of network upgrades and maintenance; CAPEX/OPEX reduction, and; inclusion of openness within the corresponding ecosystem. Following to the design and related upgrades of the relevant architecture and of all the involved CESC modules, the SESAME framework will conclude with a prototype with all corresponding functionalities.

4 The Fundamental SESAME Architectural Context

The architecture provided so far by the SESAME project (Fig. 1), acts as a "*solid reference point*" for 5G multi-tenant small cell infrastructures with mobile edge computing capabilities [26]. It combines the current 3GPP framework for network management in RAN sharing scenarios and the ETSI NFV framework for managing virtualized network functions [27]. The CESC offers virtualized computing, storage and radio resources and the CESC cluster is considered as a cloud from the upper layers. This cloud can also be "sliced" to enable multi-tenancy [28]. The execution platform is used to support VNFs that implement the different features of the SCs, as well as to support the mobile edge applications of the end-users.

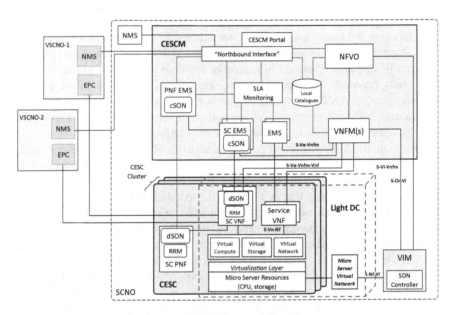

Fig. 1. The SESAME essential architecture.

The overall SESAME system architecture is shown in Fig. 1. The SESAME architecture foresees the split of the small cell physical and virtual network functions [29], respectively Physical Network Function (PNF) and VNF, based on the Multi-Operator Core Network (MOCN) requirements and associated RRM and OAM features, which need to be supported.

The Cloud Enabled Small Cell (CESC) is a complete Small Cell with necessary modifications to the data model to allow MOCN radio resource sharing. The CESC is composed by a Physical SC unit and a micro-server. The physical aggregation of a set of CESCs (i.e.: CESCs cluster) provides a virtualized execution infrastructure, denoted as the "Light DC" [30], enhancing the virtualization capabilities and process power at the network edge. The functionalities of the CESC are split between SC PNFs and SC VNFs; the latter are hosted in the environment provided by the Light DC.

The Light DC encompasses the micro-servers of the different CESCs in a cluster and provides a high manageable architecture optimized to reduce power consumption, cabling, space and cost. To achieve these requirements, it relies upon an infrastructure that aggregates and enables sharing of computing, networking and storage resources available in each micro-server belonging to the CESC cluster. The Light DC infrastructure provides also the backhaul and fronthaul resources for guaranteeing the requirements for connectivity in case of multi-operator (multi-tenancy) scenarios. The hypervisor computing virtualization extensions enable access of virtual machines to the hardware accelerators for providing an execution platform that can support the deployment of VNFs. Different types of VNFs can be deployed through the Virtual Infrastructure Manager (VIM), for carrying out the virtualization of the SC, for running the cognitive/"Self-x" ([9, 31]) management operations and for supporting computing needs for the mobile edge applications of the end-users. The combination of the proposed architecture allows achieving an adequate level of flexibility and scalability in the edge cloud infrastructure [32].

The CESC Manager (CESCM) is the central service management component in the architecture that integrates the traditional 3GPP network management elements and the novel functional blocks of the NFV-MANO (Network Functions Virtualization - Management and Orchestration) framework. Configuration, Fault and Performance management of the SC PNFs is performed through the PNF Element Management System (EMS), while the management of the SC VNFs is carried out through the SC EMS. The EMSs provide performance measurements to the SLA Monitoring module that assesses the conformance with the agreed SLAs. EMSs are connected through the northbound interface with the Network Management Systems (NMS) of the Small Cell Network Operator (SCNO) and the different tenants, denoted as Virtual Small Cell Network Operators (VSCNOs), providing each VSCNO with a consolidated view of the portion of the network that they are able to manage. Finally, the CESCM includes a portal that constitutes the main graphical frontend to access the SESAME platform for both SCNO and VSCNOs.

Automated operation of CESCs is made possible by different SON functions that will tune global operational settings of the SC (e.g., transmit power, channel bandwidth, electrical antenna tilt) as well as specific parameters corresponding to RRM functions (e.g., admission control threshold, handover offsets, packet scheduling weights, etc.). As shown in Fig. 1, the PNF EMS and SC EMS include the centralized "Self-x" functions (cSON) and the centralized components of the hybrid SON functions. In turn, the decentralized (dSON) functions - or the decentralized components of the hybrid functions - reside at the CESCs. The dSON functions can be implemented as PNFs or, if proper open control interfaces with the element (e.g. the RRM function) controlled by the SON function are established, they can also be implemented as VNFs running at the Light DC. The mapping of the specific RRM and SON functions in the different components of the architecture depends in general on the selected functional split between the physical and virtualized functions.

By summarizing, the SESAME project has proposed a detailed architectural framework to implement NFV and Edge Cloud Computing with the pure aim of providing intelligence at the network edge for modern 5G mobile networks that have currently been examined under the SC context [33]. In particular and within certain priorities, 5G also targets at offering rich virtualization and multi-tenant capabilities, not only in term of partitioning network capacity among multiple tenants, but also by offering dynamic processing capabilities on-demand, optimally deployed close to the end-user or the end-device. Furthermore, the SC concept, is expected to be enriched within 5G via the incorporation of virtualization and edge computing capabilities, aiming to provide better cellular coverage, capacity and applications for homes and enterprises, at various public spaces (both rural and urban). The potential benefits from such a combined approach of NFV, Edge Computing and SCs and with the aim of improving network management and operations are critical for most involved market players, as the latter may access new revenue streams. Such a conceptual framework as actually proposed in SESAME may be the "key" for promoting relevant 5G aspects in the market.

Acknowledgments. This work has been performed in the scope of the *SESAME* European Research Project and has been supported by the Commission of the European Communities (*5G-PPP/H2020, Grant Agreement No. 671596*).

References

1. European Commission and 5G-PPP: 5G Vision: The 5G-PPP Infrastructure Private Public Partnership: The Next Generation of Communication Network and Services (2015). https://ec.europa.eu/digital-single-market/en/towards-5g
2. Demestichas, P., Georgakopoulos, A., et al.: 5G on the horizon: key challenges for the radio-access network. IEEE Veh. Technol. Mag. **8**(3), 47–53 (2013)
3. Andrews, J.G., et al.: What Will 5G Be? IEEE JSAC Spec. Issue 5G Wirel. Commun. Syst. **32**(6), 1065–1082 (2014)
4. Capdevielle, V., Feki, A., Temer, E.: Enhanced resource sharing strategies for LTE picocells with heterogeneous traffic loads. In: Proceedings of IEEE 73rd Vehicular Technology Conference (VTC2011-Spring), pp. 1–5. IEEE (2011)
5. Commission of the European Communities: Communication on "A Digital Market Strategy for Europe" [COM(2015) 192 final, 06.05.2015]. http://eur-lex.europa.eu/legal-content/EN/TXT/?qid=1447773803386&uri=CELEX%3A52015DC0192
6. Quek, T.Q.S., de la Roche, G., Güvenç, İ., Kountouris, M.: Small Cell Networks Deployment, PHY Techniques, and Resource Management. Cambridge University Press, Cambridge (2013)
7. Small Cell Forum (SFC): Small Cells. What's the Big Idea? (SFC document 030.07.03). http://scf.io/en/documents/030_-_Small_cells_big_ideas.php
8. Østerbø, O., Grøndalen, O.: Benefits of self-organizing networks (SON) for mobile operators. J. Comput. Netw. Commun. **2012**(862527), 1–16 (2012)
9. Ramiro, J., Hamied, K.: Self-Organizing Networks. Self-planning, self-optimization and self-healing for GSM. UMTS and LTE. Wiley, Hoboken (2012)
10. 5G-Public Private Partnership (5G-PPP): Advanced 5G Network Infrastructure for the Future Internet – "Creating a Smart Ubiquitous Network for the Future Internet" (2013)

11. European Telecommunications Standards Institute: Network Functions Virtualisation - Introductory White Paper (2012). http://portal.etsi.org/NFV/NFV_White_Paper.pdf
12. Mosharaf, N.M., Chowdhury, K., Boutaba, R.: A survey of network virtualization. Comput. Netw. **54**(5), 862–876 (2010)
13. SESAME H2020 5G-PPP Project. http://www.sesame-h2020-5g-ppp.eu/Home.aspx
14. Nadeau, T.D., Gray, K.: Software Defined Networks, 1st edn. O'Reilly, Sebastopol (2013)
15. Haleplidis, E., Salim, J.H., Denazis, S., et al.: Towards a network abstraction model for SDN. J. Netw. Syst. Manag. **23**(2), 309–327 (2015)
16. https://www.vmware.com/pdf/hypervisor_performance.pdf
17. Manzalini, A., Minerva, R., Callegati, F., Cerroni, W., Campi, A.: Clouds of virtual machines in edge networks. IEEE Commun. Mag. **51**(7), 63–70 (2013)
18. Schubert, L., Jeffery, K.: Advances in Clouds – Research in Future Cloud Computing. European Commission (2012)
19. Loeffler, B: Cloud Computing: What is Infrastructure as a Service? https://technet.microsoft.com/en-us/library/hh509051.aspx
20. Chochliouros, I.P., et al.: Challenges for defining opportunities for growth in the 5G era: the SESAME conceptual model. In: Proceedings of the EuCNC-2016, pp. 1–5 (2016)
21. Small Cell Forum: Small Cells and 5G Evolution: A Topic brief. (Document 055.07.01). http://scf.io/en/documents/055__Small_cells_and_5G_evolution_a_topic_brief.php
22. Fajardo, J.O., Liberal, F., et al.: Introducing mobile edge computing capabilities through distributed 5G cloud enabled small cells. Mob. Netw. Appl. **21**(4), 564–574 (2016)
23. Costa, C.E., Goratti, L.: SESAME essential architecture features. In: Proceedings of the EuCNC-2016, pp. 1–5 (2016)
24. Erikksson, M.: Small cells as a service: rethinking the mobile operator business. Arctos Lab (2014). http://timelab-wp-media.s3-eu-west-1.amazonaws.com/arctoslabs/Small_Cells_as_a_Service.pdf
25. Vaquero, L.M., Rodero-Merino, L.: Finding your way in the fog: towards a comprehensive definition of fog computing. ACM SIGCOMM Comput. Commun. Rev. **44**(5), 27–32 (2014)
26. Giannoulakis, I., Khodashenas, P.S., et al.: Enabling technologies and benefits of multi-tenant multi-service 5G small cells. In Proceedings of the EuCNC-2016, pp. 1–5 (2016)
27. European Telecommunications Standards Institute: Network Functions Virtualization (NFV); Architectural Framework. (GS NFV 002 V1.1.1) (2013)
28. Krebs, R., Momm, C., Kounev, S.: Architectural concerns in multi-tenant SaaS applications. In: Proceedings of CLOSER 2012, pp. 1–6. SciTePress (2012)
29. Hoffmann, M., Staufer, M.: Network virtualization for future mobile networks: general architecture and applications. In: Proceedings of the ICC-2011, pp. 1–5. IEEE (2011)
30. Tso, F.P., et al.: Network and server resource management strategies for data centre infrastructures: a survey. Comput. Netw. **106**, 209–225 (2016)
31. Blanco, B., Fajardo, J.O., Liberal, F.: Design of cognitive cycles in 5G networks. In: Iliadis, L., Maglogiannis, I. (eds.) AIAI 2016. IAICT, vol. 475, pp. 697–708. Springer, Cham (2016). doi:10.1007/978-3-319-44944-9_62
32. Son, I., Lee, D., Lee, J.-N., Chang, Y.B.: Market perception on cloud computing initiatives in organizations: an extended resource-based view. Inf. Manag. **51**(6), 653–669 (2014)
33. Chochliouros, Ioannis P., et al.: A model for an innovative 5G-*Oriented* architecture, based on small cells coordination for multi-tenancy and edge services. In: Iliadis, L., Maglogiannis, I. (eds.) AIAI 2016. IAICT, vol. 475, pp. 666–675. Springer, Cham (2016). doi:10.1007/978-3-319-44944-9_59

Inclusion of "Self-x" Properties in the SESAME-Based Wireless Backhaul for Support of Higher Performance

Ioannis P. Chochliouros[1(✉)], Alan Whitehead[2], Oriol Sallent[3], Jordi Pérez-Romero[3], Anastasia S. Spiliopoulou[1], and Athanassios Dardamanis[4]

[1] Hellenic Telecommunications Organization (OTE) S.A., 99, Kifissias Avenue, 151 24 Athens, Greece
ichochliouros@oteresearch.gr, aspiliopoul@ote.gr
[2] IP.Access Ltd., Building 2020, Cambourne Business Park, Cambourne, Cambridge CB23 6DW, UK
alan.whitehead@ipaccess.com
[3] Universitat Politècnica de Catalunya (UPC), c/Jordi Girona 1-3, 08034 Barcelona, Spain
{sallent,jorperez}@tsc.upc.edu
[4] SmartNet S.A., 2, Lakonias Street, 17342 Agios Dimitrios, Attica, Greece
ADardamanis@smartnet.gr

Abstract. Based on the actual framework of the SESAME 5G-PPP EU-*funded* project, we identify the importance of the related wireless backhauling within the broader 5G innovative framework, with the pure aim of using small cells together with suitable network virtualization techniques for serving multiple tenants in a modern architectural approach. The virtualization of the network nodes and the wireless links allow for the development of a suitable SDN controller intending to perform network slicing, where the wireless backhaul resources are shared and assigned on a per-tenant basis. In order to apply SON features as they are also applied at the access radio level, the SDN controller is responsible for collecting and evaluating status information of the network (link qualities, status of wireless interfaces, ongoing traffic), thus resulting to self-planning, self-optimization and self-healing attributes.

Keywords: 5G · Cloud-Enabled Small Cell (CESC) · Network Functions Virtualization (NFV) · Multi-tenancy · Self-x properties · Small Cell (SC) · Software-Defined Networking (SDN) · Wireless backhauling

1 Introduction

A crucial component of the modern SESAME architecture [1, 2] is the backhauling. The backhauling infrastructure enables communications between the CESCs (Cloud-Enabled Small Cells) and the core network, but also interconnects CESCs with each other and with managing system components, i.e. the CESC Manager (CESCM). Both signalling traffic, to control and monitor the Physical Network Functions (PNFs)

© Springer International Publishing AG 2017
G. Boracchi et al. (Eds.): EANN 2017, CCIS 744, pp. 716–727, 2017.
DOI: 10.1007/978-3-319-65172-9_60

and Virtual Network Functions (VNFs) of CESCs, as well as the access traffic exchanged between User Equipments (UEs) and the core network, are carried over the backhaul. Further, the micro-servers forming the Light Data Centre (DC) are interconnected via the backhaul. In SESAME, this is necessary to offer features like Service Function Chaining (SFC), which enables chaining of different edge services running on separate machines.

An important candidate for 5G backhaul technologies are wireless radio communications [3]. In contrast to traditional wired backhauls relying on technologies like Ethernet over copper or fibre optics, wireless backhauls do not require the deployment of additional infrastructure. Instead, each CESC is equipped with at least one wireless radio transceiver dedicated to backhauling. In the following sections, we present the main concept of how wireless backhauling is applied in SESAME in the context of backhaul network virtualization and Self-x backhauling features. Before that, we give a short introduction into which are the wireless radio backhaul technologies that are considered for SESAME deployments.

2   Wireless Backhauling in the Forthcoming 5G-Era

In 5G deployments, the use of a large variety of wireless technologies is envisaged [4–6]. One of the main paradigms of 5G is to include above-6 GHz technologies in the wireless access spectrum, which also can be used for backhauling.

Among the wireless technologies, radio communications in the 60 GHz band are predestined for the use in 5G deployments [7]: they offer a high bandwidth and short- to medium-communication range, which makes them well suited for Small Cell (SC) deployments [8, 9] as they are also targeted in SESAME [2]. As an alternative to 60 GHz communications, sub 6 GHz technologies can be integrated in 5G deployments. More specific, wireless devices supporting the well-known IEEE 802.11 standard[1] can be used to implement the backhaul.

In 5G networks [10], the access traffic is expected to show significant changes over time in terms of quantity and quality, depending on the number of users and the type of services that they request, requiring a high degree of flexibility from the backhaul. In the following part, the investigation framework used to perform evaluations of wireless backhauling solutions, which satisfy these requirements and take into account SESAME-*specific* concepts like multi-tenancy, is presented.

2.1   Backhauling Framework

Classic wired backhauling solutions often make use of star-topologies, where a central switch, common to all network devices, is used to interconnect the network devices (in SESAME that would be the CESCs [11]) with each other and to connect to the external networks, such as the Internet. This network topology is often chosen due to its simplicity and because it is capable of satisfying the needs of simple point-to-point

---

[1] For more details, see: http://standards.ieee.org/about/get/802/802.11.html.

(P2P) connectivity. In wireless backhauls, *however*, we are able to implement different sorts of topologies without any further costs: A device equipped with wireless transceivers may be able to communicate not only with one, but with several other devices over a wireless link. By enabling each CESC in the network to act as a relay (router) node, a mesh network can be formed.

In addition to the CESCs acting as relay devices in the mesh network, one or several CESCs need to act as egress points (gateway nodes) for the wireless backhaul [12]. As depicted in Fig. 1, they are responsible for enabling communications between the cluster of CESCs and the core network or any SESAME managing entities of the CESCM, like the EMS (Element Management System) or the VNFM (Virtual Network Functions Manager). Gateways (GWs) are an essential part of the backhaul, because they are the only devices of the mesh that have access to the management/Internet backbone.

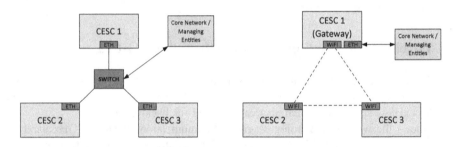

Fig. 1. Wired star topology and wireless mesh topology comparison

An implication of mesh topologies is that routing becomes necessary to interconnect CESCs with each other or with a gateway. Further, to compose routes with high stability and link qualities, it is necessary to know about the state of the wireless backhaul resources. This includes link qualities and availability of links, since they can change over time. In SESAME, we use SDN (Software-Defined Networking) technologies [13], which provide solutions for both of these requirements.

As such, an SDN controller is dedicated to manage the backhauling network, it is responsible for keeping track of the network topology and for performing centralized routing decisions in the data and control plane. Further, in order to apply SON features [14, 15] as they are applied at the access radio level, the SDN controller is responsible for collecting and evaluating status information of the network (link qualities, status of wireless interfaces, ongoing traffic). The gathered information is used to enhance the routing algorithms of the SDN controller, by aggregating link state information and information about existent traffic flows in order to assign high performance routes for access and control traffic.

There are two options for the placement of the SDN controller: That is, it runs either internally in the Light DC or on an external machine that manages the virtualized backhauls of all tenants. It is even possible that one particular SDN controller is assigned to each tenant, a decision on the design of the backhaul network that will be taken as its investigations progress over the course of the project.

As in wired networks that rely on SDN technology, applying SDN requires a certain degree of abstraction of the network [16]: The wireless interfaces used by every CESC are represented as virtualized ports of a wireless virtual switch, the wireless radio connections between two network nodes are abstracted to be treated like traditional, wired links. The virtualization of the network nodes and the wireless links allow the SDN controller to perform network slicing, where the wireless backhaul resources are shared and assigned on a per-tenant basis. This aspect of the wireless backhaul is discussed in the next subsection.

2.2   Per-tenant Virtualization of the Backhaul

Each tenant (i.e.: Virtual Small Cell Network Operator - VSCNO) signs an SLA (Service Level Agreement) that determines the type and quality of service they will obtain in the SESAME deployment. In an SLA, there is an agreement about which types and qualities of service each tenant will obtain from a SESAME deployment. Mainly, this "translates" into the assignment of radio (access) resources and a minimum service performance to each tenant, according to the signed SLAs.

The SLAs have also an impact on the wireless backhaul, which needs to be taken into account. Since wireless backhauling resources are limited in terms of bandwidth and minimum per-hop delays, it is necessary to assign backhaul resources efficiently to each tenant, so that the KPIs from the SLAs can be met. The SDN controller uses an API (Application Programming Interface) to exchange information about the active SLAs and other relevant data with the CESCM. Based on the overall view of the network and the knowledge about SLAs, the SDN controller performs network slicing [17–19].

In this slicing, the available backhaul resources are distributed among the tenants to assure that each tenant's performance requirements are met. On one hand, the SDN controller virtualizes the available (physical) topology, which includes the wireless links, the relays and the gateways of the backhaul network, as shown in the example topology (Fig. 2).

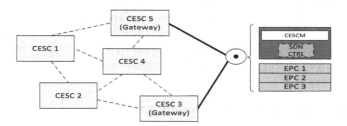

Fig. 2. Example topology for a SESAME deployment with several CESCs that act as relay nodes or gateways, respectively.

As a result, a subset of all the available virtualized networks elements is assigned to every tenant [20–22]. Thus, every tenant may have a different share of the backhaul

network, yet being agnostic about it. An example of such differing per-tenant virtu-
alizations applied to the example topology is shown in Fig. 3.

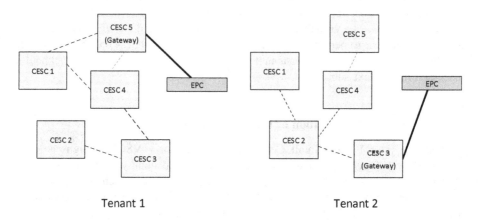

Tenant 1   Tenant 2

Fig. 3. Possible per-tenant virtualization of the example topology.

Each of the tenants is assigned a different set of wireless links (except the link
marked in red) and a different gateway to reach the tenant's EPC (Evolved Packet
Core) [23, 24]. In spite of the obvious differences in the topology, all CESCs are
included in each topology and each topology includes a gateway, required for the
connection between the CESC cluster and the tenant's correspondent EPCs.

On the other hand, the wireless radio resources are virtualized at the physical
(radio) level. This includes the virtualization of the bandwidth of wireless links in terms
of available data rates and the virtualization of frequency spectrum of the links.
The SDN controller monitors the state of these physical resources and performs slicing
to assign them to different tenants. For example, a wireless link may offer a transmission ratio of up to 1 Gbit/s. If the link is used by multiple tenants, the available data
rate needs to be shared among them in such a way that the requirements of ongoing
transmissions (e.g., video streams) and also the SLAs signed with each tenant can be
met. To achieve this, the SDN controller uses the knowledge about the state of the
wireless backhaul resources. This knowledge is not only necessary for the on-the-fly
assignment of wireless resources, in a more general view, but it is also required to apply
"Self-x" features [25, 26]) as discussed in the following section.

3   Inclusion of "Self-x" Backhauling Features

Self-Organising Networks (SON) denotes a set of features and capabilities for
automating the operation of a network so that operating costs can be reduced and
human errors can be minimised [14]. This purely implicates for an enhanced network
management. With the inclusion of SON features [27], standard manual planning,
deployment, optimization and maintenance activities of the network can be effectively

replaced -and/or supported- by more autonomous and automated processes, thus making network operations simpler and faster [28]. SON functions, denoted in the context of SESAME as "Self-x" functions, are organised around the following main categories, which are based on previous references (such as [14, 28]).

The automation of the management processes in wireless networks by means of "Self-x" functions is seen as a key element to deal with the tremendous complexity and the stringent requirements associated with future 5G networks [29].

In order to achieve a high and stable performance in the wireless backhaul, a series of "Self-x" features can be applied [30, 31]). Whereas at the access radio SON features are applied by the NMS (Network Management System) or at the EMS (Element Management System), in the wireless backhaul the SDN controller is the responsible for applying self-planning, self-optimization and self-healing. In the backhaul, centralized and decentralized Self-x approaches as introduced by cSON (centralized SON) and dSON (decentralized SON) solutions[2], as well as hybrid solutions are possible. The centralized part of the Self-x functions is executed in the SDN controller. The decentralized aspects of SON are executed in the CESCs. In the following, a set of basic "Self-x" features applying to the wireless backhaul are discussed. Self-x functions are able to automatically tune global operational settings of a small cell (e.g., maximum transmit power, channel bandwidth, electrical antenna tilt) as well as specific parameters corresponding to Radio Resource Management (RRM) functions (e.g., admission control threshold, handover offsets, etc.).

3.1 Self-planning

This usually implicates for automatization of the process of deciding the need to roll out new network nodes in specific areas, identifying the adequate configurations and settings of these nodes, as well as proposing capacity extensions for already deployed nodes (e.g., by increasing channel bandwidths and/or adding new component carriers). Specific functions belonging to this category include the planning of a new cell and the spectrum planning. Some more generalized examples of self-planning may include:

- *Planning a new cell:* This process intends to automatically make the decision that a new small cell has to be deployed in a certain area, specifying as well as its geographical position.
- *RF planning of a new cell:* This process intends to automatically select RF aspects such as transmit powers and antenna parameters of the new small cell.
- *Spectrum planning:* Spectrum planning should decide the type of spectrum that is needed (i.e. licensed, non-licensed, etc.), the required bandwidth, and the assignment of this spectrum to the different small cells.

Aware of the available wireless interfaces and the links of the mesh network formed by the CESCSs, the SDN controller can assign wireless channels to the interfaces to assure an efficient spectrum sharing and reuse of the backhaul during the planning

---

[2] For more details see, *for example*: https://en.wikipedia.org/wiki/Self-organizing_network.

phase [32]. In the same process, each active tenant is assigned a share (slice) of the backhaul network. However, due to the high dynamicity of the traffic occasioned by fluctuations in the number of users and varying traffic patterns, it is likely that the slice for a tenant is modified during network operation, which is part of the self-optimization discussed below. The self-planning applied in the wireless backhaul is mainly performed by the SDN controller that has a general network overview. Thus, we can put declare the self-planning to be a cSON feature.

3.2   Self-optimization

Once the network is in operational state, the self-optimization includes the set of processes intended to improve -or maintain- the network performance in terms of coverage, capacity and service quality by tuning the different network settings. Examples of functions include Mobility Load Balancing (MLB), Mobility Robustness Optimisation (MRO), Automated Neighbour Relation (ANR), Coverage and Capacity Optimization (CCO), optimization of admission control, optimization of packet scheduling, inter-cell interference coordination (ICIC) and energy saving.

- *Mobility Load Balancing (MLB):* This case aims at coping with dynamic load variations by optimizing the network parameters so that load from highly loaded cells can be shifted to low loaded cells. This is usually done by tuning handover and cell reselection thresholds.
- *Mobility Robustness Optimisation (MRO):* This case intends to optimize the handover parameters in order to minimize the handover failures and the inefficient utilization of network resources due to, e.g., ping-pong effects. In the context of SESAME both handovers between small cells and handovers from/to small cells and the rest of cells of the tenant can be considered.
- *Automatic Neighbour Relations (ANR):* This function is responsible for automatically building the Neighbour Relation Table of each cell. The proper identification of the neighbour relations of a cell is fundamental for mobility purposes because handovers can only be executed between neighbour cells.
- *Coverage and Capacity Optimization (CCO):* This functionality intends to provide optimal coverage and capacity in the radio network (RN). This is done by adjusting RF parameters of the small cells (i.e. transmit power, antenna tilt, antenna azimuth, etc.) in accordance with specific optimization targets and trying to avoid coverage holes, weak coverage areas, pilot pollution, overshoot coverage and download (DL)/upload (UL) channel coverage mismatch.
- *Optimization of admission control/congestion control/packet scheduling parameters:* This intends to optimize the setting of different RRM parameters such as the admission/congestion thresholds, the priorities of the quality of service (QoS) classes, etc.
- *Inter-Cell Interference Coordination (ICIC):* This functionality intends to configure the power, time and frequency resources in a coordinated way among different cells so that inter-cell interference can be minimized. The adjustment is done at a relatively slow rate.

- *Energy saving:* This use case aims at reducing the energy consumption in the deployed network. This is usually done by switching off the cells that carry very little traffic at certain periods of the day (e.g. at night) and making the necessary adjustments in the neighbour cells so that the existing traffic can be served through some other cell.

During the runtime of the actual SESAME deployment, the SDN controller gathers detailed information about the state of the backhaul links and the ongoing control and data transmissions. As a result, the controller is capable of sharing the available wireless backhaul resources among the tenants so that the requirements (SLAs) can be satisfied. The actions of the SDN controller not only limit to assigning different shares of the network topology and the physical bandwidth (in terms of wireless link data rates) to the tenants, but also on rerouting traffic throughout the network in such a way that other network policies, like congestion avoidance or energy saving, can be satisfied. This cSON approach is similar to the one followed in MLB algorithms for access traffic [33], yet applied to the wireless backhaul resources.

3.3   Self-healing

This is relevant to the automation of the processes related to fault management (i.e., fault detection, diagnosis, compensation and correction), usually associated to hardware and/or software problems, in order to keep the network operational, while awaiting a more permanent solution to fix it and/or prevent disruptive problems from arising. Examples of self-healing functions include Cell Outage Detection (COD) and Cell Outage Compensation (COC) as discussed below:

- *Cell Outage Detection (COD):* This consists in detecting poor performing cells, usually due to hardware or software faults (e.g. faults in the connectivity, in the radio boards, in the power supply, etc.).
- *Cell Outage Compensation (COC):* This refers to the actions to solve or alleviate the outage detected in one cell, usually by acting on the neighbouring cells in order that they serve the traffic of the cell in outage.

During network operation, it may be necessary to apply self-healing features in the wireless backhaul. While wireless links are stable in general, under certain meteorological conditions or upon obstruction of the line of sight between wireless transceivers, one or several links may disappear (temporarily). Another incident could be the physical failure of a wireless interface, which would cause the loss of all wireless links associated to the interface. In any of these cases, the controller needs to be able to react by redirecting traffic over alternative routes, taking into account how the new assignment of wireless backhaul resources will affect the overall network performance and whether the new state would be acceptable in terms of meeting SLA requirements. For this purpose, algorithms are designed to assure robustness of the network by, for example, calculating backup paths that are as disjoint as possible from main data paths in order to be able to provide alternative routes in case of failures. Self-Healing can be applied both as cSON and dSON features. In a cSON approach, failures detection and correction are responsibility of the SDN controller, whereas in a dSON approach CESCs can react to failures on their own. Such features can include fast-rerouting without waiting for new instructions from the SDN controller.

4   Updating of the SESAME Architecture

Regarding the architectural models for implementing the Self-x functions, the following possibilities are distinguished (as in [27, 34]):

- Centralized SON (cSON): This is a sort of solution where the Self-x algorithms are executed at the NMS or at the EMS.
- Distributed SON (dSON): This is a solution where the Self-x algorithms are executed at the Network Element level (i.e. autonomously within a single SC or in a distributed manner among several SCs).
- Hybrid SON: It combines cSON and dSON, in such a way that part of the Self-x functionalities are distributed and reside at the SC while others are centralized and reside at the EMS and/or the NMS. This is illustrated in Fig. 4. In this case, the cSON functions can be used to provide guidelines and parameters to the distributed SON functions based on information retrieved from them in terms of, e.g., performance measurements.

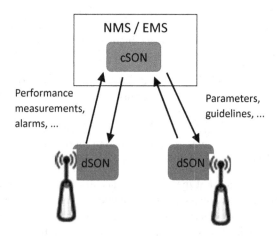

Fig. 4. Depiction of the hybrid SON model for the SESAME architecture.

Based on the above model, Fig. 4 depicts a simplified view of the SESAME architecture focusing on the relationship with Self-x functionalities [35].

As shown in the figure, the PNF EMS and SC EMS include the cSON functions and the centralized components of the hybrid functions. In turn, the dSON functions -or the decentralized components of the hybrid functions- reside at the CESCs. It is worth mentioning that, depending on the implementation, and assuming that the SLA monitoring function is effectively part of the EMS, it could also be possible to associate cSON to the SLA monitoring, which also has visibility of PM (Performance Management) counters, KPIs (Key Performance Indicators) and SLAs. This could relieve the SC EMS from the cSON decisions.

Furthermore, whatever Self-x function is considered of interest to be deployed in our SESAME framework, it can be implemented as a PNF or, if proper open control interfaces with the element controlled by the Self-x function are established, it can also be implemented as a VNF [36]. The implementation as VNFs provides certain benefits as identified below: (i) An inherent flexibility through easy instantiation, modification and termination procedures; (ii) An inherent efficiency in hardware utilisation, since VNFs are executed on a pool of shared NFV (Network Functions Virtualization) infrastructure resources, *and*; (iii) An inherent capability to "add" new functionalities and/or extend/upgrade/evolve existing VNFs. In SESAME, this can be applied to distributed Self-x functions that would run as SC-VNFs in the Light DC.

Acknowledgments. The present work has been performed in the scope of the *SESAME* (*"Small cElls CoordinAtion for Multi-tenancy and Edge services"*) European Research Project and has been supported by the Commission of the European Communities (*5G-PPP/H2020, Grant Agreement No. 671596*).

References

1. Chochliouros, I.P., et al.: A model for an innovative 5G-*oriented* architecture, based on small cells coordination for multi-tenancy and edge services. In: Iliadis, L., Maglogiannis, I. (eds.) AIAI 2016. IAICT, vol. 475, pp. 666–675. Springer, Cham (2016). doi:10.1007/978-3-319-44944-9_59
2. SESAME Project (GA No. 671596). http://www.sesame-h2020-5g-ppp.eu/Home.aspx
3. Naylon, G.: Why Wireless Backhaul Holds the Key to 5G (2016). https://www.wirelessweek.com/article/2016/03/why-wireless-backhaul-holds-key-5g
4. Jaber, M., Imran, M.A., Tafazolli, R., Tukmanov, A.: 5G backhaul challenges and emerging research directions: a survey. IEEE Access **4**, 1743–1766 (2016)
5. Gupta, A., Jha, R.K.: 5 survey of 5G network: architecture and emerging technologies. IEEE Access **3**, 1206–1232 (2015)
6. Chochliouros, I.P., Sfakianakis, E., et al.: Challenges for defining opportunities for growth in the 5G era: the SESAME conceptual model. In: Proceedings of the EuCNC-2016, pp. 1–5 (2016)
7. European Commission: Radio Spectrum Policy Group (RSPG) – Report on Spectrum Issues on Wireless Backhaul (RSPG15-607). European Commission (2015)
8. Next Generation Mobile Network Alliance (NGMN-A): Small Cell Backhaul Requirements. White Paper. NGMN-Alliance, Frankfurt, Germany (2012)
9. Jungnickel, V., Manolakis, K., Zirwas, W., Panzner, B., et al.: The role of small cells, coordinated multipoint, and massive MIMO in 5G. IEEE Commun. Mag. **52**(5), 44–51 (2014)
10. Soldani, D., Manzalini, A.: Horizon 2020 and beyond: on the 5G operating system for a true digital society. IEEE Veh. Technol. Mag. **10**(1), 32–42 (2015)
11. SESAME 5G-PPP Project: Deliverable 2.3: Specification of the CESC Components – First Iteration (2016)

12. Bernardos, C.J., De Domenico, A., Ortin, J., Rost, R., Wubben, D.: Challenges of designing jointly the backhaul and radio access network in a cloud-based mobile network. In: Proceedings of Future Network Summit 2013, pp. 1–10. IEEE (2013)
13. Dräxler, M., Karl, H.: Dynamic Backhaul Network Configuration in SDN-Based Cloud RANs. https://arxiv.org/pdf/1503.03309.pdf
14. Ramiro, J., Hamied, K.: Self-Organizing Networks. Self-planning, self-optimization and Self-healing for GSM, UMTS and LTE. Wiley, Hoboken (2012)
15. Sánchez-González, J., Pérez-Romero, J., Agustí, R., Sallent, O.: On learning mobility patterns in cellular networks. In: Iliadis, L., Maglogiannis, I. (eds.) AIAI 2016. IAICT, vol. 475, pp. 686–696. Springer, Cham (2016). doi:10.1007/978-3-319-44944-9_61
16. Medved, J., Tkacik, A., Varga, R., Gray, K.: Opendaylight: towards a model-driven SDN controller architecture. In: Proceedings of the WoWMoM-2014, pp. 1–6. IEEE (2014)
17. Bojic, D., Sasaki, E., Svijetic, N., et al.: Advanced wireless and optical technologies for small-cell mobile backhaul with dynamic software-defined management. IEEE Commun. Mag. **51**(9), 86–93 (2013)
18. Huawei Technologies Co., Ltd.: 5G Network Architecture - A High Level Perspective. Shenzen, China (2016)
19. Fajardo, J.O., Liberal, F., Giannoulakis, I., Kafetzakis, M., Pii, V., Trajkovska, I., Bohnert, T.M., Goratti, L., et al.: Introducing mobile edge computing capabilities through distributed 5G cloud enabled small cells. Mob. Netw. Appl. **21**(2), 564–574 (2016). Springer
20. Small Cell Forum (SFC): Virtualization for Small Cells: Overview (Document 106.05.1.01) (2015). http://scf.io/doc/106
21. Small Cell Forum (SFC): Small Cell Virtualization Functional Splits and Use Cases (Document 159.07.02) (2016). http://scf.io/doc/159
22. European Telecommunications Standards Institute (ETSI): Network Functions Virtualisation - Introductory White Paper. ETSI, Sophia-Antipolis (2012). http://portal.etsi.org/NFV/NFV_White_Paper.pdf
23. Basta, A., Kellerer, W., Hoffmann, M., Hoffmann, K., Schmidt, E.D.: A virtual SDN-enabled LTE EPC architecture: a case study for S−/P-gateways functions. In: Proceedings of SDN4FNS-2013, pp. 1–7. IEEE (2013)
24. Chourasia, S., Sivalingam, K.M.: SDN based evolved packet core architecture for efficient user mobility support. In: Proceedings of the 1st IEEE Conference on Network Softwarization (NetSoft-2015), pp. 1–5. IEEE (2015)
25. Pérez-Romero, J., Sallent, O., Ruiz, C., Betzler, A., et al.: Self X in SESAME. In: Proceedings of the EuCNC-2016, pp. 1–5 (2016)
26. Belschner, J., Arnold, P., Eckhardt, H., Kühn, E., Patouni, E., et al.: Optimization of radio access network operation introducing self-x functions. In: Proceedings of the 69th IEEE VTC, pp. 1–5. IEEE (2016)
27. European Telecommunications Standards Institute (ETSI): TS 132 500: LTE; Self-Organizing Networks (SON); Concepts and requirements (Release 12). ETSI, Sophia-Antipolis (2015)
28. The Third Generation Partnership Project (3GPP): TS 32.522 v11.7.0: Self-Organizing Networks (SON) Policy Network Resource Model (NRM) Integration Reference Point (IRP); Information Service (IS) (Release 11). 3GPP (2013)
29. 3GPP: TS 32.522 v11.7.0: Self-Organizing Networks (SON) Policy Network Resource Model (NRM) Integration Reference Point (IRP); Information Service (IS) (Release 11). 3GPP (2013)
30. Wilson, R.A., Keil, F.C.: The MIT Encyclopedia of the Cognitive Sciences. MIT Press, Cambridge (1999)

31. Biglieri, E., Goldsmith, A.J., Greenstein, L.J., Mandayam, N.B., Poor, H.V.: Principles of Cognitive Radio. Cambridge University Press, New York (2012)

32. Kumar, N., Nidhi, K.N., Acharya, S.: A survey on SDN: an unprecedented approach in networking. Int. J. Eng. Comput. Sci. **5**(2), 15668–15672 (2016)

33. Yamamoto, T., Komine, T., Konishi, S.: Mobility load balancing scheme based on cell reselection. In: Proceedings of ICWMC-2012, pp. 381–387. IARIA (2012)

34. Small Cell Forum (SFC): SON API for Small Cells (Document 083.05.01). SFC (2015). http://scf.io/doc/083

35. Blanco, B., Fajardo, J.O., Liberal, F.: Design of cognitive cycles in 5G networks. In: Iliadis, L., Maglogiannis, I. (eds.) AIAI 2016. IAICT, vol. 475, pp. 697–708. Springer, Cham (2016). doi:10.1007/978-3-319-44944-9_62

36. Drutskoy, D., Keller, E., Rexford, J.: Scalable network virtualization in software-defined networks. IEEE Internet Comput. **17**(2), 20–27 (2013)

The Role of Virtualization in the Small Cell Enabled Mobile Edge Computing Ecosystem

Leonardo Goratti[1], C.E. Costa[1(✉)], Jordi Perez-Romero[2],
P.S. Khodashenas[3], Alan Whitehead[4], and Ioannis Chochliouros[5]

[1] CREATE-NET, Via alla Cascata 56/D, Trento, Italy
ccosta@fbk.eu
[2] Universitat Politecnica de Catalunya (UPC), Barcelona, Spain
[3] i2CAT, Barcelona, Spain
[4] IP.Access, Cambourne, UK
[5] Hellenic Telecommunications Organization (OTE), Marousi, Greece

Abstract. Virtualisation is playing a fundamental role in the evolution of telecommunication services and infrastructures, bringing to rethink some of the traditional design paradigms of the mobile network and enabling those functionalities necessary for supporting new complex ecosystems where multiple actors can participate in a dynamic and secure environment. In Small Cell enabled Mobile Edge Computing deployments, the impact of virtualization technologies is significant in two main aspects: the design and deployment of the telecommunication infrastructure, and the delivery of edge services. Besides, the adoption of virtualization technologies has implications also in the implementation of Self Organizing Network (SON) services and in the enforcement of Service Level Agreement (SLA) policies, both critical in the automation of the delivery of multi-tenant oriented services in such complex infrastructure. From the work performed by the H2020 SESAME project, the beneficial use of virtualization techniques emerges in adding network intelligence and services in the network edge. SESAME relays on virtualization for providing Small Cell as a Service (SCaaS) and per operator Edge Computing services, consolidating the emerging multi-tenancy driven design paradigms in communication infrastructures.

Keywords: NFV · Small Cell · Virtualization · 5G

1 Introduction

Virtualization technologies decouple software from physical infrastructure, and create virtual resources dedicated to distinct services. This approach fosters an efficient, flexible and dynamically re-configurable usage of the physical resources, and a rapid and efficient service creation process.

While virtualization is a fundamental element of Cloud Computing, it is also making its way in the Telecommunication domain. Through virtualization, network elements, functions and components are released from the limitations of dedicated hardware and physical constraints, allowing the dynamic set up and reconfiguration of services and service providers, and the support the functionalities necessary for implementing multi-tenancy and autonomic management.

© Springer International Publishing AG 2017
G. Boracchi et al. (Eds.): EANN 2017, CCIS 744, pp. 728–733, 2017.
DOI: 10.1007/978-3-319-65172-9_61

Operators can leverage on virtualization for implementing new business models. An important novelty enabled by multi-tenancy is the possibility of decoupling the Mobile Network Operators (MNO) roles and functions between multiple operators: such as service providers, infrastructure provider and network provider. Besides the MNO, new types of Virtual Network Operators (MVNO) and Over-The-Top (OTT) service providers are enabled to enter in the value chain. For example, Small Cell (SC) operators (SCOs) can leverage on this framework for offering to existing MNOs on-demand access to network resources, acting as a neutral host provider. Venue owners, real estate companies, municipalities, etc., can become SCOs by deploying an appropriate Small Cell access infrastructure and by providing new services, especially in high traffic and with high end-users density delimited places such as office areas, dense urban areas, stadiums, shopping malls, and concert venues [1].

1.1   SESAME Project

With the introduction of new paradigms, such as edge computing and radio access (RAN) functional split, telecommunication services and infrastructures are evolving into a new hybrid Cloud-Telco breed of services, where Cloud Computing and telecom infrastructure virtualization are put together for supporting new multi-tenant business models and vertical markets, forcing to rethink some of the traditional design paradigms of the mobile network.

The framework envisaged by the SESAME 5G-PPP EU-funded project [5] supports a hybrid cloud-telecommunication infrastructure, with enhanced multi-tenant cloud services by combining SCs with micro-server facilities at the edge. Its target is to design and develop a novel 5G platform based on Small Cells and Edge Cloud Computing. Featuring a virtualised execution environment and multi-tenancy between network operators, it aims at offering to both network operators and mobile users services that can potentially take advantage of the location of computational resources at the edge of the network.

Various high-impact use cases for Small Cell enabled Edge Cloud Computing [2] and the key stakeholders have been identified. For satisfying their requirements, SESAME innovations focus around three central elements in the wider 5G context:

(i) The placement of network intelligence and applications in the network edge through Network Functions Virtualization (NFV) and Edge Cloud Computing; (ii) the substantial evolution of the Small Cell concept, already mainstream in 4G but expected to deliver its full potential in the challenging high dense 5G scenarios, *and*; (iii) the consolidation of multi-tenancy in communications infrastructures, allowing several operators/service providers to engage in new sharing models of both access capacity and edge computing capabilities.

The SESAME Small Cell infrastructure integrates, inside the access network infrastructure, a virtualized execution platform (i.e., the Light Data Centre (DC)): a highly manageable clustered edge computing infrastructure that features low-power processors and hardware accelerators for time critical operations. In the Light DC, Virtual Network Functions (VNFs) are deployed, and applications and services executed.

The SESAME approach allows new stakeholders to dynamically enter the value chain by acting as neutral host providers in high traffic areas where densification of multiple networks is not practical. The optimal management of a CESC deployment is a key challenge of SESAME, for which new orchestration, NFV management, virtualization of management views per tenant, Self-X features and radio access management techniques are developed.

1.2   Technology Trends in Small Cells

Some of technology trends that can be identified for the Small Cells are enabled by virtualization capabilities of the infrastructure:

Functional Split in Small Cells. The functional split defines which of the functions typically run in the eNodeB can be executed in the cloud. The Small Cell Forum (SCF) has recognized multiple functional splits of the Small Cell networking layers for different use cases [3, 4]. At which point of the protocol stack perform the functional split, together with the decision of where to execute virtual the network functions with respect to physical small cell functions, are two crucial aspects in Small Cell virtualization that influences the flexibility and the management of RAN optimization.

Multitenancy Capabilities. Traditional mobile operators deploy their own network infrastructure in competition with others. In multi-tenancy approaches, instead, infrastructure and resources are shared between multiple operators, encouraging a more dynamic and scalable market: operators can focus on their core business and differentiate their offer based rather delegating network connectivity maintenance to a third party. SESAME approach allow several operators/service providers to engage in new sharing models, obtaining higher capacity on the access side and exploiting edge computing capabilities.

Radio Access Network (RAN) Sharing. Multi-tenancy poses unprecedented challenges to the owner of the shared RAN in relation to network planning. The management of tenants may need to handle drastic variations in the network's traffic demand, e.g. due to the aggregation of new tenants. Moreover, the new aggregated traffic is tenant-specific, meaning that some characteristics (e.g. busy hour, type of services, etc.) may substantially differ from tenant to tenant, particularly if these tenants correspond to different vertical sectors. For these reasons, flexibility is crucial for an efficient and cost-effective multi-tenancy implementation. There are several approaches for RAN slicing in relation to different aspects such as the degree of isolation that can be achieved between tenants or the degree of customization, which impacts on the capability to customize RRM/Self-X functions on a per tenant basis.

Cloud-Enabled Small Cell. Making computing capabilities available at the network edge allows providers to locate services nearer to the mobile subscribers, thus enabling accelerated services, content and application. SESAME approach proposes to deploy a these capabilities between the mobile core and the Radio Access Network (RAN). Typical services which can benefit from mobile-edge computing include Internet-of-Things, augmented reality and data caching.

Network Slicing. The SC network is partitioned in multiple isolated slices: Virtualised resources and small cells capacity is partitioned into slices configured on the requirements of multiple vertical markets.

Self-Organising Network (or Self-X Functions). The complexity of the highly dense environments where Small Cells are envisaged to be deployed require the introduction mechanisms for reducing or even removing the need for manual network optimization tasks, thus enabling autonomic capabilities to the network. Through the implementation of SON features the network framework is are able to dynamically tune global operational settings of the SC (e.g., transmit power, channel bandwidth, electrical antenna tilt) as well as specific parameters corresponding to Radio Resource Management (RRM) functions (e.g., admission control threshold, handover offsets, packet scheduling weights, etc.). The development of Self-X functions benefits from the availability of a virtualized execution platform provided at the RAN and by the use of NFV and MEC technologies. The introduction of SON architectures have a positive impact in reducing the network operating costs by minimizing human errors [6].

2   The Role of Virtualization

Virtualization is fundamental for an effective and exploitable deployment of computation capabilities at the mobile network edge, allowing managing and orchestrating network services from different services providers in a dense small cell scenarios and different use cases.

Cloud computing approaches, such as Infrastructure-as-a-Service (IaaS), can be adopted in order to facilitate the decoupling between providers and their roles: i.e. the service provider, the infrastructure provider and the network provider.

To achieve a mapping of the conventional telecommunication services to the new Cloud-enabled infrastructures, Service Function Chaining techniques can be applied. Software Defined Networking (SDN) mechanisms for Service Function Chaining (SFC) are the most prominent candidates to enforce traffic steering through a logical network graph and to achieve certain service functionality among the virtualized components. Extending the concept of SFC in the context of the 5G ecosystem requires deeper understanding on the NFV concepts in such a scenario. Carefully identifying the requirements of the specific setup is important in order to choose the most suitable mechanisms and protocols to establish the desired functionality.

From the standardization point of view, the ETSI NFV Industry Specification Group (ISG) has developed the Management and Orchestration (MANO) framework, a reference for managing virtualised environments [7].

2.1   Virtualisation of the Communication Infrastructure

Virtualisation of the communication infrastructure, such as core/edge network elements and access points/macrocells, finds its application in Small Cells and Small Cell-as-a-Service (SCaaS) scenarios. If we consider mobile networks, there exist several virtualized deployments of the common LTE functional blocks, such as EPC, BSS, HSS, RAN, etc.

The management of dense SC networks includes mitigating interference and assigning resources dynamically. SC installation is an effective way to achieve greater performance and capacity to both indoor and outdoor places but this brings complexity. In order to share SC networks between multiple operators, it is necessary to logical partitioning of the SC network into multiple isolated slices. Virtualizing part of RAN functionalities allow partitioning small cells capacity between multiple tenants. Small Cell functions virtualization gives improved ability and agility to manage dense SC networks, e.g. mitigating interference and assign resources dynamically.

SESAME relays on virtualization to develop a SC operating model under which multi-operators (i.e., multi-tenancy) can coexist.

2.2   Virtualization of Edge Services

Virtualization of edge services offers computing capabilities at the network edge and brings different services near to the mobile subscribers such as augmented reality (AR), caching, security services, multimedia services, big data. In such a scenario, operators can differentiate their offer based on services rather than on network connectivity.

Network Functions (NFs) - such as caching proxies, firewalls, load balancers or intrusion detection systems (IDSs) - traditionally deployed as middleboxes (running in dedicated hardware), can be virtualized and linked together e.g. using SDN based approaches.

SESAME provides enhanced multi-tenant edge cloud services combining SCs with micro-server facilities, supporting virtualised execution environment. SESAME leverages on the capability to deliver intelligence directly to the network's edge, in the form of virtual network appliances.

Using Service Function Chaining (SFC) capabilities, it is possible to create a new complex service defining an ordered list of a network services (e.g. firewalls, load balancers) that are connected together in the network.

3   Conclusions

Virtualization plays a fundamental role in the new hybrid Cloud-Telco infrastructure, such as the Edge Clout Computing Small Cell architecture proposed by SESAME [8], and forces to rethink some of the traditional design paradigms of the mobile network. In particular, virtualization has a fundamental role in:

- Supporting an infrastructure that is sustainable and reconfigurable, both from the SC RAN and Services point of view;
- Making the SC infrastructure and resources shared between operators, with a transparent and neutral approach;
- Accelerating the creation of innovative services with superior quality of experience through mobile edge computing;
- Enabling Self-X functionalities, capable of optimise the usage of radio, network, storage and computing resources.

Acknowledgments. This work has been supported by the EU funded H2020 5G-PPP project SESAME under the grant agreement no. 671596

References

1. Small Cell Forum (SCF): Small Cell Virtualization: Functional Splits and Use Cases - SFC 159.06.02 (2016)
2. Belesioti, M., Chochliouros, I. (eds.): System Use Cases and Requirements, Deliverable D2.1 of the SESAME project, December 2015
3. Small Cell Forum (SCF): Virtualization in Small Cell Networks - SCF 154.07.02 (2015)
4. Small Cell Forum (SCF): Business case elements for small cell virtualisation - SCF 158.05.1.04 (2015)
5. Small cEllS coordinAtion for Multi-tenancy and Edge services (SESAME) 5G-PPP project. http://www.sesame-h2020-5g-ppp.eu/
6. Ramiro, J., Hamied, K.: Self-Organizing Networks: Self-Planning, Self-Optimization and Self-Healing for GSM. UMTS and LTE. Wiley, Hoboken (2012)
7. European Telecommunications Standards Institute (ETSI): ETSI GS NFV-MAN 001 (V1.1.1): Network Function Virtualisation (NFV); Management and Orchestration. ETSI, Sophia-Antipolis (2014)
8. Giannoulakis, I. (ed.): SESAME Final Architecture and PoC Assessment KPIs, Deliverable D2.5 of SESAME project, December 2016

Author Index

Printed in the United States
By Bookmasters